Chemistry of Natural Products

Volume III

Chemistry of Natural Products

Volume III

ASHUTOSH KAR

Formerly

Professor, School of Pharmacy, Addis Ababa University,
Addis Ababa **(Ethiopia)**

Dean, Chairman & Professor, Faculty of Pharmaceutical Sciences,
Guru Jambheshwar University, Hisar **(India)**

Professor, School of Pharmacy, Al Arab Medical University,
Benghazi **(Libya)**

Professor, College of Pharmacy (University of Delhi),
Delhi **(India)**

Professor & Head, Department of Pharmaceutical Chemistry,
Faculty of Pharmaceutical Sciences, University of Nigeria,
Nsukka **(Nigeria)**

CBS

CBS Publishers & Distributors Pvt Ltd

New Delhi • Bengaluru • Chennai • Kochi • Kolkata • Mumbai

Bhopal • Bhubaneswar • Hyderabad • Jharkhand • Nagpur • Patna • Pune • Uttarakhand • Dhaka (Bangladesh)

Disclaimer

Science and technology are constantly changing fields. New research and experience broaden the scope of information and knowledge. The author and contributors have tried their best in giving information available to them while preparing the material for this book. Although, all efforts have been made to ensure optimum accuracy of the material, yet it is quite possible some errors might have been left uncorrected. The author and the publisher will not be held responsible for any inadvertent errors, omissions or inaccuracies.

Chemistry of Natural Products
Volume III

ISBN: 978-81-239-2599-8

First Edition: 2015

Reprint: 2020

Published by Satish Kumar Jain and produced by Varun Jain for
CBS Publishers & Distributors Pvt Ltd
4819/XI Prahlad Street, 24 Ansari Road, Daryaganj, New Delhi 110 002, India.
Ph: 23289259, 23266861, 23266867 Website: www.cbspd.com
Fax: 011-23243014 e-mail: delhi@cbspd.com; cbspubs@airtelmail.in.
Corporate Office: 204 FIE, Industrial Area, Patparganj, Delhi 110 092
Ph: 4934 4934 Fax: 4934 4935 e-mail: publishing@cbspd.com; publicity@cbspd.com

Branches

- **Bengaluru:** Seema House 2975, 17th Cross, K.R. Road, Banasankari 2nd Stage, Bengaluru 560 070, Karnataka
  Ph: +91-80-26771678/79 Fax: +91-80-26771680 e-mail: bangalore@cbspd.com
- **Chennai:** 7, Subbaraya Street, Shenoy Nagar, Chennai 600 030, Tamil Nadu
  Ph: +91-44-26680620, 26681266 Fax: +91-44-42032115 e-mail: chennai@cbspd.com
- **Kochi:** 42/1325, 1326, Power House Road, Opp KSEB, Power House, Ernakulam 682 018, Kochi, Kerala
  Ph: +91-484-4059061-65 Fax: +91-484-4059065 e-mail: kochi@cbspd.com
- **Kolkata:** 6/B, Ground Floor, Rameswar Shaw Road, Kolkata-700 014, West Bengal
  Ph: +91-33-22891126, 22891127, 22891128 e-mail: kolkata@cbspd.com
- **Mumbai:** 83-C, Dr E Moses Road, Worli, Mumbai-400018, Maharashtra
  Ph: +91-22-24902340/41 Fax: +91-22-24902342 e-mail: mumbai@cbspd.com

Representatives

• Bhopal	0-8319310552	• Bhubaneswar	0-9911037372	• Hyderabad	0-9885175004
• Jharkhand	0-9811541605	• Nagpur	0-9421945513	• Patna	0-9334159340
• Pune	0-9623451994	• Uttarakhand	0-9716462459	• Dhaka (Bangladesh)	01912-003485

Printed at JS Offset, Patparganj Industrial Areal, Delhi, India

Dedication

The textbook is earnestly and humbly dedicated to the millions of men and women from each and every nation and varying culture who dedicated selflessly innumerable hours of most precious time and valuable resources to the ever-expanding advancement of the **Natural Products**-an inherited **Indispensable Gift of the Mother Nature.**

To them belongs the proud appellation derived from amator (the **Latin** terminology for *Lover:* "**amateuer**").

"YOU can't have a **Better Tomorrow**
If YOU are **Thinking About Yesterday All the Time**".

— **Charles F. Kettering**

"The **Prince of Wands**
Will **Take YOU Forward**
Progressively in
Professional Aspects".

— *Anonymous*

"**Always be Sensitive, Reflective and Receptive**"
-*A.Scoltish Adage*

Preface

The '**Chemistry of Natural Products: Volume III**' is a sequel to **Volumes 1 and II** and it comprises the following *seven* important group of **Natural Products,** namely:

- **Steroids and Hormones,**
- **Antibiotics,**
- **Lipids,**
- **Polynuclear Aromatic Hydrocarbons,**
- **Enzymes, Coenzymes, and Fermentation,**
- **Prostaglandins [PGE$_2$ and PGF$_{2\alpha}$],** and
- **Biosynthesis of Flavonoids.**

The present compilation constitutes the latest and up-to-date integral curriculum for both the *Postgraduate and B.Sc (Hons) Students of Chemistry* in practically all the **Indian Universities**. In addition, this well-elaborated, documented, and explicitly expatiated textual matter is ardently beneficial to the **M.Sc. students** grossly involved in the pursuation of their respective studies in: **Chemistry, Biochemistry, Pharamacy, Biotechnology, Molecular Biology**, as well as **other allied disciplines**. The *Research Scholars* belonging to a host of *integrated disciplines*, in both Ph.D., and Masters programmes may indeed find the subject matter equally gainful and of immense interest.

The author has duly taken a particular care and attention to include the contents as far as the **New Syllabi of University Grants Commission (UGC)** in order to make it rather more user-friendly amongst the *M.Sc. and B.Sc (Hons)* aspirants across India and abroad as well.

In order to apprise the *august* **end-users** special care has been undertaken to include most of the *Chemical Structures* right from the latest edition of *Merck Index* for strengthening their in-depth knowledge related to the **Chemistry of Natural Products.**

An exceptional coverage and proper attention has been duly endorsed towards the following critical aspects intelligently:

- development of a *Chapter* in a methodical manner,
- scientific and logical expansion of each chapter,

- constitution of a *Chemical Entity* and its *Total Synthesis* has been elaborated in a sequential manner, and
- explanation of *Total Synthesis* has been provided.

The **Chemistry of Natural Products: Vol. III** includes an well-documented with references given as a *Footnote;* while, each chapter concludes with *two* most vital and cardinal aspect such as:

- **Further Reading References** and
- **Review Questions**.

to enable the students for a proper guidance in their respective examinations.

The author most humbly acknowledges the excellent support by Shri Satish Kumar Jain MD, CBS-Publishers & Distributors, Pvt.Ltd., New Delhi, for bringing out this textbook in a record time-frame.

GURGAON

December, 2014

Ashutosh Kar

Expression of Novel Thoughts and Ideas in the Discovery of 'Natural Products'.....

Dr. John Bastyr (1910-1993) Professor of Medicine, at the National College of Naturopathic Medicine (Seattle Campus, USA) where he taught and practiced *Naturopathic Medicine.* His effective and result-oriented approach were broadly confined to:

- encyclopedic coverage of herbs,
- sticklers for botany and botanical description,
- scientists solely concerned with chemical structures and evidence-based medicines, and
- duly authenticated by a host of tried, tested, and double-blind placebo-controlled investigative studies.

Several concrete evidences from the survey of literatures do suggest vehemently a glimpse at the specific **Botanical Medicines**-as being practiced predominantly in the **Eclectic Euro-American Tradition** domain. However, certain modern usage of the *medicinal plants* reflect explicitly the so-called **eclectic phenomenon**-wherein the *Newer Therapies* have been integrated meticulously with the *Older Therapies-* based on clear-cut positive therapentic advantages.

The 'Natural Products'-Phenomenon- The beginning of the 21st century has witnessed a quantum-leap forward in the thoughtful and judicious usage of the enormous storehouse of the God-gifted **Natural Products** spread across the globe abundantly. An extensive and intensive search for a good number of excellent *drug products,* in the form of *modern-day pharamaceuticals*, such as:

- **Tablets** • **Capsules** • **Syrups** • **Creams** • **Ointments**
- **Gels** • **Granules** • **Powders** and • **Plant Extracts,**

are being developed and used in the **Herbal Medicines** –that has eventually gained a tremendous confidence and faith amongst the actual end-users for the therapeutic advantage of scores of dreadful human ailments.

Over the years the **Phytochemists, Pharmacognosists**, and **Researchers/Scientists** engaged deeply in the production of *life-saving* **dosage forms** as the **therapentic agents,** namely:

- **Neutraceuticals** • **Botanical Medicines** • **Herbal Drugs** • **Plant Medicines**
- **Plant-Based Pharmaceuticals (PMPs)** • **Ayurvedic Systems of Medicine** • **Unani Systems of Medicine** • **Chinese Systems of Medicine** • **Tibetan Herbal Drugs**
- **Arab Herbal Medicines** • **Cuban Drug Systems** • **African Herbal Drugs** and
- **Brazilian and Mexican Herbal Medicines.**

The Natural Products have set firmly their tentacles so deep and broad extent that a plethora of **Herbal Pharmacopoeas** *viz.,* **United States Herbal Pharmacopoeae, Herbal Pharmacopeas of India** etc., so as to regulate the authenticity of the **Plant Drugs** and their **usefulness.**

As-to-date, the candid and expounded knowledge pertaining to the **Herbal Drugs** are duly available throughout the entire world to provide a substantial and plausible relief from the various *human diseases,* such as:

- **Artharitis** • **Anti-agening Process** • **Anti-rheumaties** • **Anti-virals** • **Antispasmodics**
- **Anti-coagulants** • **Anti-neoplasms** • **Asthma** • **Bronchitis** • **Bile irregulraties**

- Chronic Constipation • Chronic Heart Diseases (CHDs) • Digestants (Digestive Agents)
- Emmenogogues • Expectotrants • Galactogogues • Genitourinary Agents • Glandulars
- Hemostatic Agents • Hydrogogue Cathartics • Hypoglycaemics • Laxatives • Liver Stimulants • Male Reproductive System Agents • Muscular Pains • Mood Elevators
- Mental Alertness • Nervines • Nutritive Botanicals • Ovarian Neuralgia Remedies
- Oxytocics • Osteroporesis • Pulmonary Disorders • Respiratory Stimulants
- Rubefacients and Vesicants • Revitalizing Agents • Skin Remedies • Spleen Remedies
- Stomach Remedies • Ulterine Tonics • Uterine Contraceptives and • Whooping Cough

The gallantly and progressive break-throughs acclaimed all over the globe in the field of *'Allopathic Drugs'* i.e., such **drug molecules** that are *designed* and *synthesized* based upon clue(s) derived from a galaxy of resources, namely:

❏ **'biologically active prototypes'** *viz., Mosphine, Cocaine, Oestradiols* etc.,

❏ **'adaptogens'** *viz.*, materials that critically enhance the human body's ability to cope-up with the physiological stress that we often encounter while deling with the **'toxins'** in the environment and/or with stress from life itself, and

❏ **'tailor-made synthetic drugs'** (*viz.*, **aza-steroids**) based on *Computer Aided Drug Designs (CADD)-* a latest globally recognized and adapted means *via* excellent and useful softwares, such as:

  ❏ **research-based drug-receptor** *docking methods*, or
  ❏ **drug discovery** *via* **enzyme inhibition** (*in-vivo*), or
  ❏ **pro-drug approach** in *drug-design modalities*, or
  ❏ **narcotic analgesiss** *viz., methadones*, or
  ❏ **molecular modeling and 3D drug-design**

     using meticulous amalgamation of computer-based softwares.

The following *three* textbooks entitled- *'Chemistry of Natural Products'*:

- **Chemistry of Natural Products: Vol: I** by Ashutosh Kar, **(2010)**
- * contains **Three Chapters;**
- **Chemistry of Natural Products: Vol: II** by Ashutosh Kar, **(2012)**
- * contains **Five Chapters; and**
- **Chemistry of Natural Products: Vol: II** by Ashutosh Kar, **(2015)**
- * contains **Seven Chapters,**

That may provide a substantial subject matter to enable the august *readers, teachers, researchers,* and *scientists* to enrich their knowledge in the domain of '**Natural Products".**

Author

Contents

2. Antibiotics

3. Lipids 269-321

4. Polynuclear Aromatic Hydrocarbons 323–401

5. Enzymes, Coenzymes, and Fermentation 403–440

Chemistry

of

Natural Products
Volume–III

Contents at a Glance

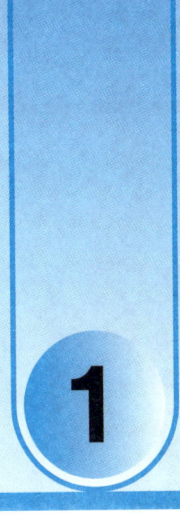

1 Steroids and Hormones

1. INTRODUCTION

The **steroids** designate *cyclic organic compounds* that comprises essentially a 'cyclopentanoperhydrophenanthrene' nucleus and constitutes an integral part of the structure of the *adrenal cortico-hormones, cardiac glycosides, cholesterol,* and *sex-hormones.*

In other words, the **steroids** do constitute a class of structurally related chemical entites (compounds) which are distributed extensively both in :

- **plant kingdom,** and
- **animal kingdom.**

Importantly, the so-called '**basic nucleus**' of most of these *physiologically potent* and *biochemically dynamic* therapeutic agents essentially exhibit a more or less identical **stereochemical relationship**. In true sense, the **steroids** have been found to consist of either partly or completely **hydrogenated 17 H-cyclopenta-phenanthrene nucleus.**

In a broader perspective, both **steroidal** hormones and related congeners designate and constitute one of the most glaring and abundantly used group of pharmacologically active potent agents that are employed in the control and management of :

- **Birth control,**
- **Inflammatory conditions,**
- **Hormone-replacement therapy (HRT),**
- **Neoplastic diseases (cancer),**
- **Contraceptives,**
- **Cardio-tonic agents,**
- **Sex-hormones,**
- **Oestrogens,**
- **Gestogens,** and
- **Adrenocortical hormones.**

How do the *steroids* **made in living cells**. The **natural steroids** are duly made in vivo by the help of **enzymes** present in the living cells and are very much confined in '**number**'.

Following are a few typical chemical structures pertaining to

- **1,2-Cyclopentanoperhydrophenanthrene** (I).
- **Diels' Hydrocarbon (II)**
- **Chrysene (III)** and
- **Picene (IV).**

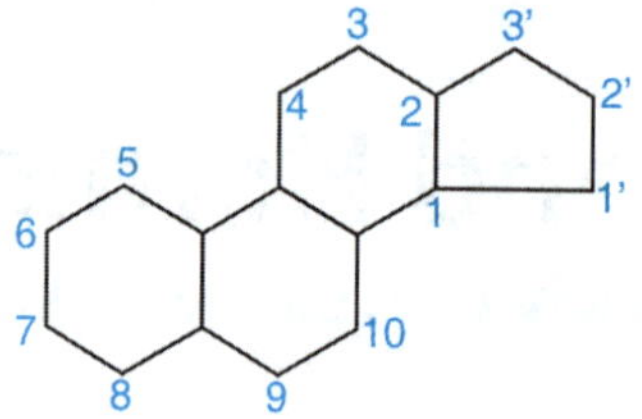

1,2-Cyclopentanoperhydrophenanthrene
(I)

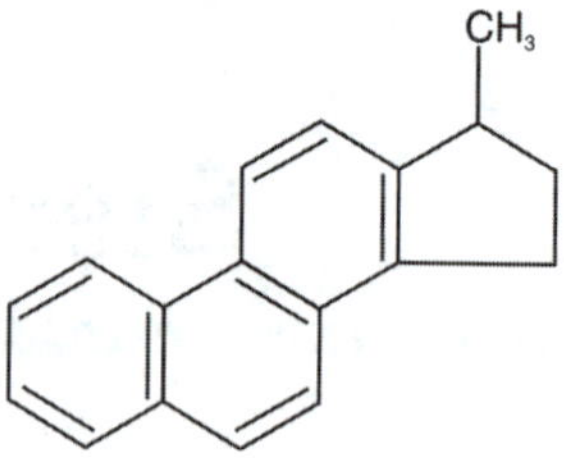

Diel's Hydrocarbon
(II)

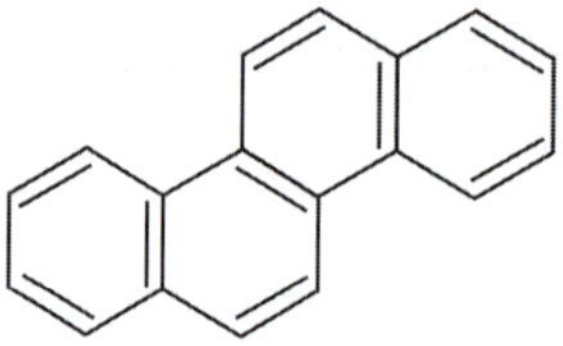

Chrysene
(III)

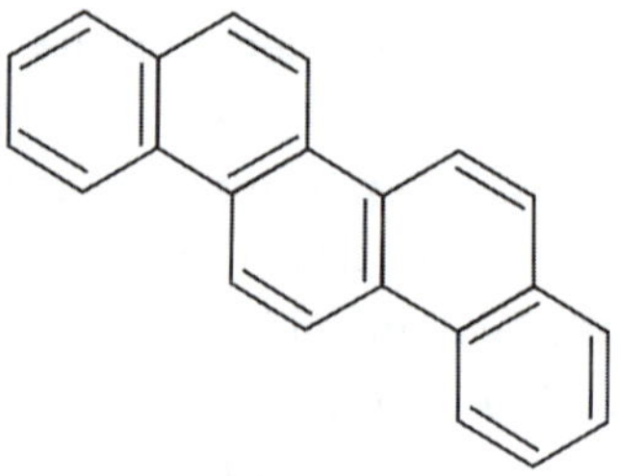

Picene
(IV)

It is, however, pertinent to state here that the plethora of these agents are solely based upon a specific common **structural nucleus** invariably known as the **'steroidal backbone';** nevertheless, the different **steroidal variants** attribute critically to the most:

- **specific,** and
- **unique,**

molecular targets.

2. STEROID : NOMENCLATURE – NUMBERING-DOUBLE BONDS

Following is the generalized chemical formula for the **basic structure** of the aforesaid *medicinal compounds* that may be represented duly as stated under:

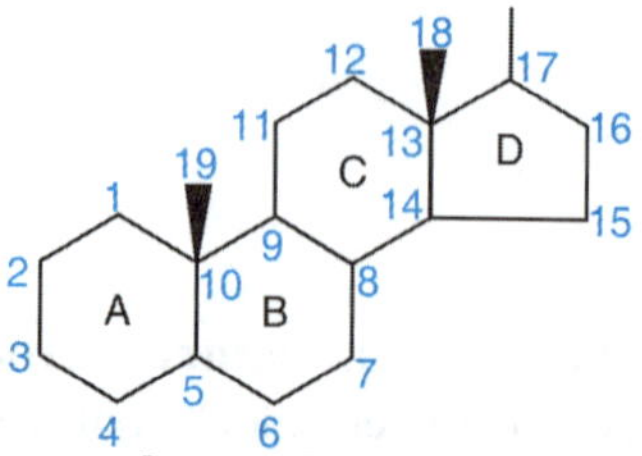

General Steroidal Formula

The various Trings present in the above **general steroidal nucleus** are conventionally *lettered* and *numbered* as shown above. In actual practice, however, the **basic structure of steroid** is *not planar*, but a *3D-conformational structure*.

Salient Features – These essentially include:

1. In the naturally occurring **steroidal compounds** the *substitutions* in the various rings invariably occur at **C-3, C-7, and C-11 positions.**

2. As per the usual '**standard convention**' the actual direction of the projection form the *plane of the ring system of the ensuing substituting moieties* located strategically at the **centres of symmetry** is normally designated by the **Greek Letter**s : α-and β.

3. Thus, we may have *two* **critical situations** :

   (a) **α-Substituting Groups** – These are viewed as *projecting* **beneath the ring plane**; and is conventionally designated by a **dotted line** (or **broken line**).

   (b) **β-Substituting Groups** – These are viewed as *projecting* **above the ring plane**; and is invariably represented by a **solid line.**

4. It has been duly established that all the **steroids** on being subjected to *dehydrogenation* with **Sepowder at 360°C** normally yield's **Diel's Hydrocarbon** (*i.e*, **3'-methyl-1,2-cyclopentanophen-anthrene**; whereas, at **420°C** the **steroids** mostly give rise to the formation of **chrysene** and **picene** (in small quantum, see structures under section 1).

In the following sections (a) through (d) critically include a few typical examples of some important 'steroidal drugs' together with their *nomenclature* and *numbering* are given as under :

   (a) **Common and Systematic Nomenclature**,

   (b) **Nomenclature and Numbering**,

   (c) **Nomenclature and Double Bonds**, and

   (d) **Nomenclature and Stereochemistry.**

(a) **Common and Systematic Nomenclature :**

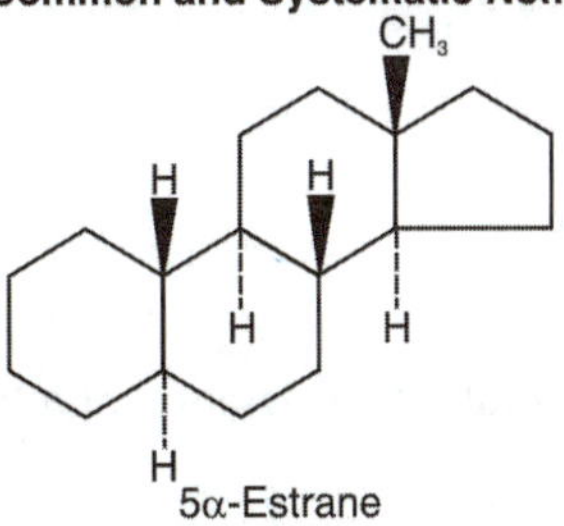

5α-Estrane 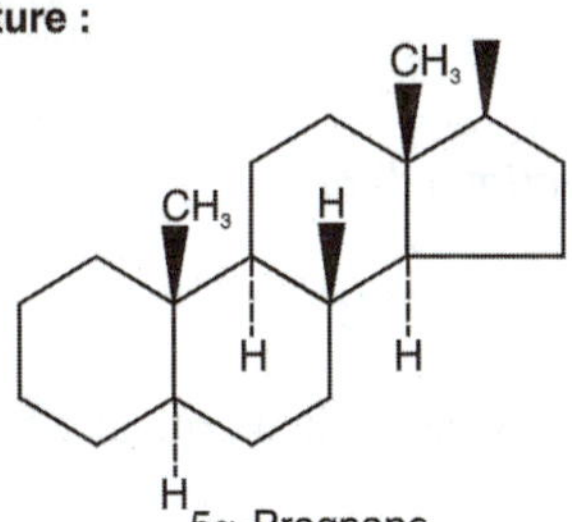   5α-Pregnane 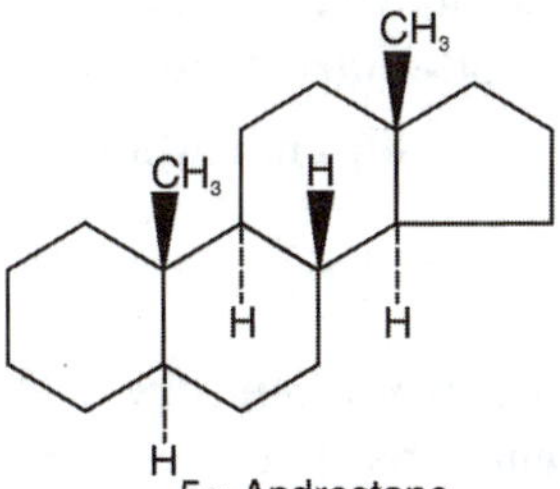   5α-Androstane

(b) **Nomenclature and Numbering :**

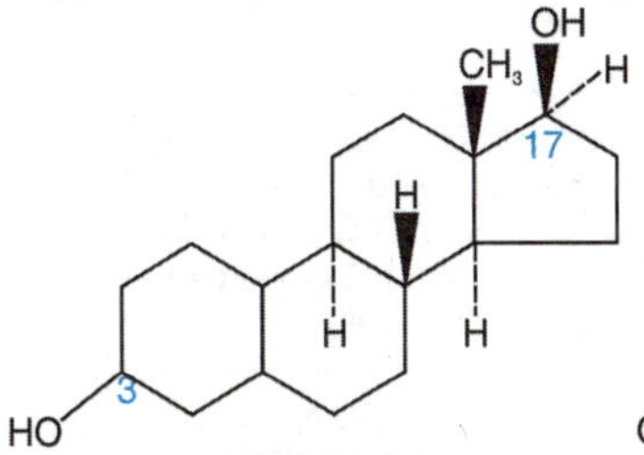

17β-Estradiol
[Estra-1,3,5(10)-triene-3, 17β-diol]

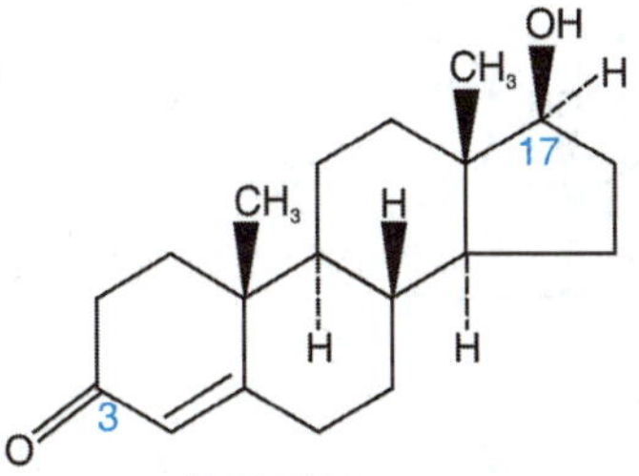

Testosterone
[17β-Hydroxyandrost-4-en-3-one]

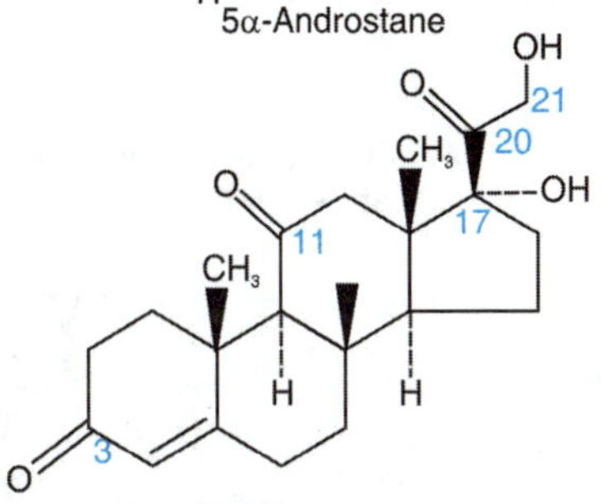

Cortisone
[17,21-Dihydroxypregn-4-ene-3,11,20-trione]

(c) Nomenclature and Double Bonds :

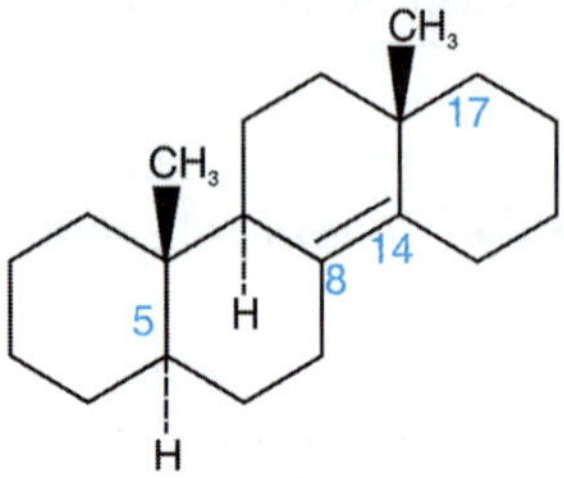

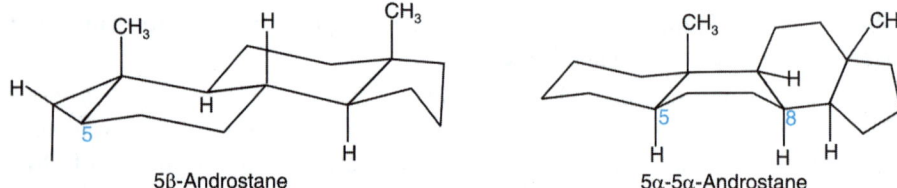

5α-Androst-8 (14)-ene
or
5α-$\Delta^{8(14)}$-Androstene

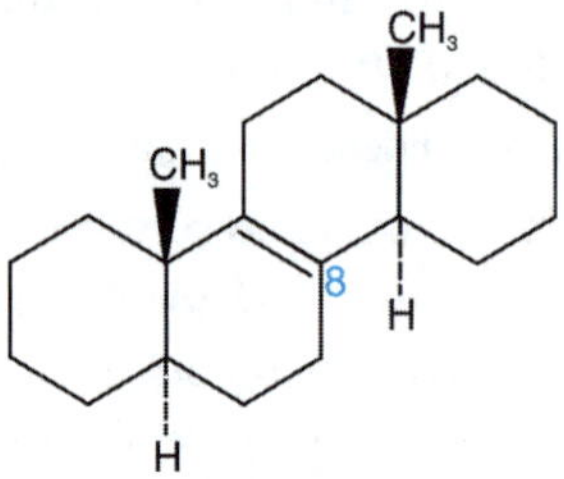

5α-Androst-8-ene
or
5α-Δ^{8}-Androstene

(d) Nomenclature and Stereochemistry :

5β-Androstane

5α-5α-Androstane

Points to Ponder – Now, with respect to the **IUPAC-guided nomenclature** and **stereochemisty** the following points need to be looked into carefully:

1. **Stereochemistry** of **H-atom at C-5** is usually included in the *nomenclature* itself.
2. **Stereochemistry** of the remaining **H-atoms is not** normally indicated until it is required essentially to differ from **5α-cholestane**, and
3. Changing the **stereochemistry** status at any of the '**ring-junction**' with a *heavy dark-line* (see Fig. in *section.2*) alters largely the prevailing–**shape of the steroid molecule** (as could be seen in the above cited examples:)
   - **5β-androstane,** and
   - **5α, 8b-androstane.**   *see section (d) above*

3. STEREOCHEMISTRTY OF STEROIDS

The **stereochemistry of steroids** may be studied critically by taking into consideration of a '**sterol**' (say, **cholesterol**) as given under :

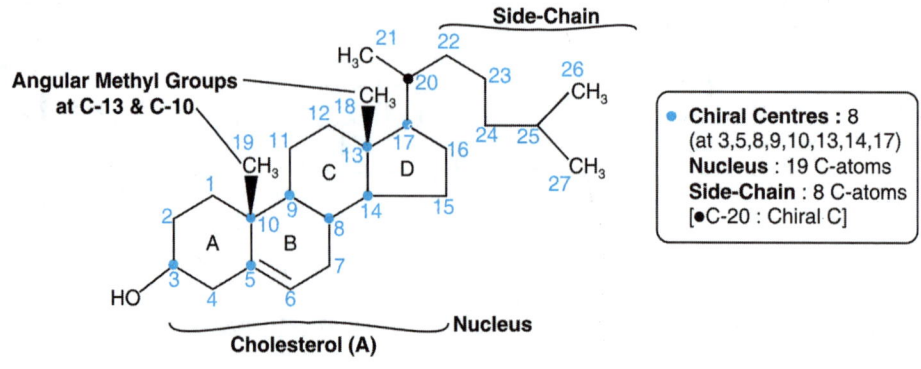

- **Chiral Centres :** 8
  (at 3,5,8,9,10,13,14,17)
  Nucleus : 19 C-atoms
  Side-Chain : 8 C-atoms
  [●C-20 : Chiral C]

Cholesterol (A)

Since, there exists explicitly *eight* **altogether dissimilar chiral centres** (·) present in the *steroidal nucleus* (with **19 C-atoms**) of the so-called **completely saturated sterol** located strategically at **C-atoms : 3,5,8,9,10,13,14, and 17** respectively; thereby suggesting the *possible optical isomers* to be **$2^8 = 256$**.

It is however, pertinent to state here that if one also considers the **side-chain of cholesterol** (containing **8C- atoms**), it has *one more chiral centre* at **C-20**, which concludes that the possible optical isomers shall now be **$2^9 = 512$**. Therefore, in order to study exhaustively so many **(512) optical isomers**, it is a lot easier and convenient to divide the ensuing **stereoisomerism of steroids** into *two* distinct categories, namely :

- ❏ **Category –I** : **which specifically deals in the manner whereby the rings are fused together;** and
- ❏ **Category –II** : **which essentially deals with the configuration of the substituent moieties, specifically those at C-3 and C-17.**

3.1. Configuration of the Nucleus

In case, one examines critically the nucleus of the *fully saturated nucleus* of **cholesterol (A)** there exists prominently *six chiral centres*; and , therefore, there will be **$2^6 (=64)$** optically active forms that are possible **theoretically**. Nevertheless, in actual practice one may lay hands on only a few of them; whereas, several others simply do not exist by virtue of the **steric hindrances** (or *steric limitations*). **Stereochemistry of the Steroid Nucleus** – Importantly, the so-called **oxidative degradation technique** provides ample valuable evidences with respect to the **stereochemistry of the steroid nucleus.**

Examples : Windaus (1926)* synthesized *four* **isomeric acids**, namely :

- **Lithobilianic Acid (B)**   • **Allolithobilianic Acid (C)**
- **Isolithobilianic Acid (D)** and   • **Alloisolithobilianic Acid (E).**

| NOTE | In all these **isomeric acids** R= CH.CH$_3$ (CH$_2$)$_2$ COOH. |

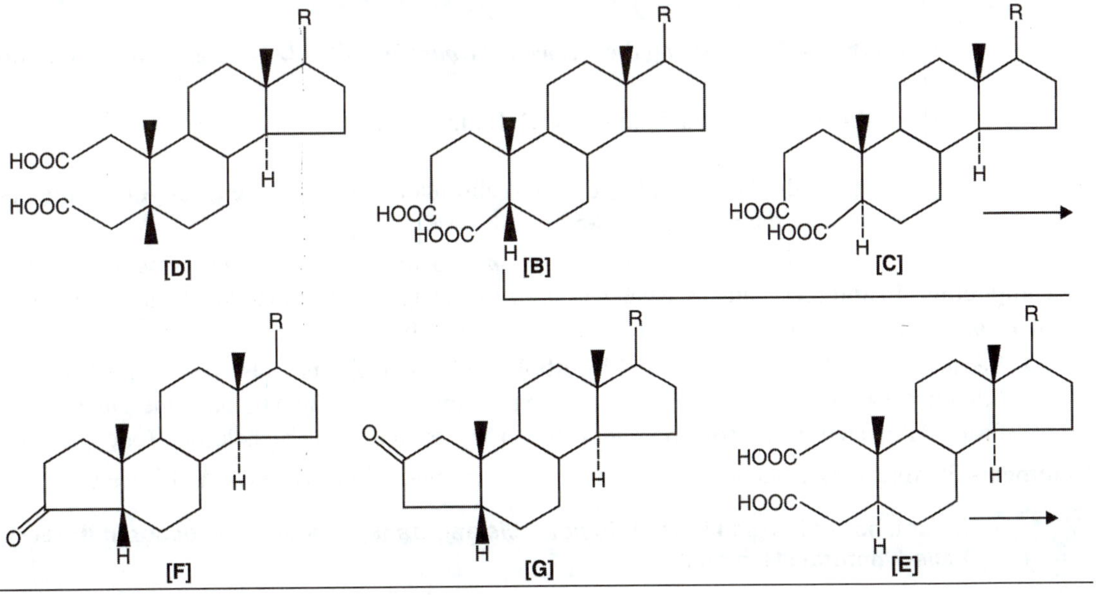

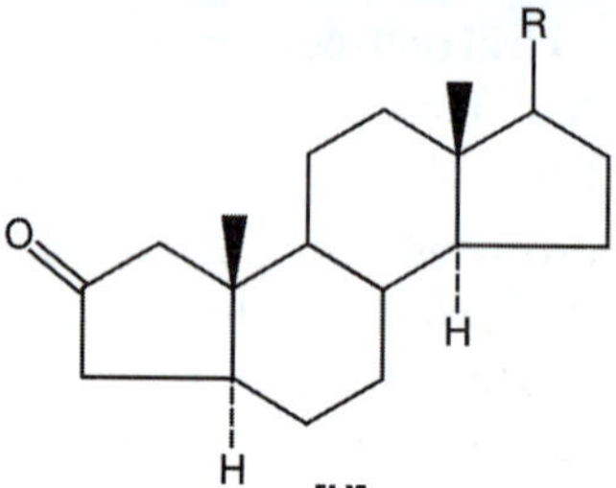

[H] 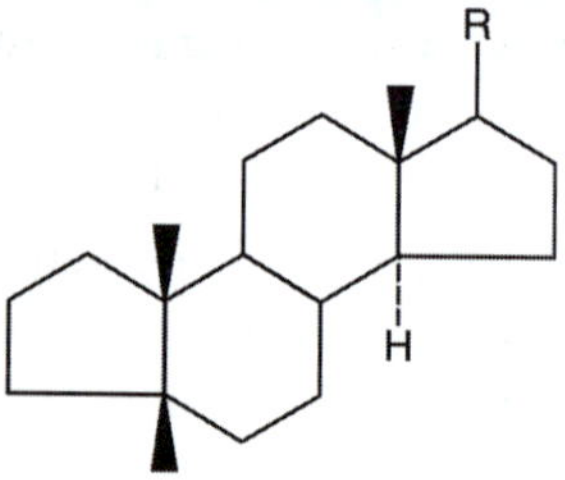   [I]

Explanation

Interestingly, all the *four* isomeric acids *viz.*, (**B**), (**C**),(**D**), and (**E**) on being subjected to heat one would obtain the following results, namely :

(i)   **Acid (B)** *i.e.*, **lithobilianic acid** gives *pyrolithobilianic acid* (**F**);
(ii) **Acid (C)** *i.e.*, **allolithobilianic acid** yields *pyrolithobilianic acid* (**F**) ;
(iii) **Acid (D)** *i.e.*, **isolithobilianic acid** gives *pyroisolithobilianic acid* (**G**); and
(iv) **Acid (E)** *i.e.*, **alloisolithobilianic acid** yields *pyroalloisolithobilianic acid* (**H**) .

Inferences: Based upon the foretold results one may conclude the following important inferences, such as :

❑ *Clemensen Reduction* of **products (F) and (G)** gave the identical product **deoxypyroacid (I)**, which suggests strongly that **products (F)** essentially possess a *cis*-**configuration**.

> **NOTE** It is further substantiated by the underlying fact that if 2 carboxyl (-COOH) moieties are duly positioned in the *cis*-position, one would expect cyclization to occur almost promptly.

❑ Since., the **compound (C)** (*i.e*, **allolithobilianic acid**) yielded **compound (F)** (*i.e*, **pyrolithobilianic acid**) – it means vividly that the said inversion must have occurred at **C-5** either :

- *via* the *enol*-form using the adjacent *carbonyl moiety* ($-\overset{\overset{\text{O}}{\|}}{\text{C}}-$) presently in compound (**F**), or
- by the help of the **carboxyl (-COOH)** moiety prior to *cyclization*. However, in *two* compounds **(G) and (H)** there exist absolutely no adjacent **carbonyl** ($-\overset{\overset{\text{O}}{\|}}{\text{C}}-$) **moiety** attached to the **H-atom at C-5.**

Comment – Therefore, the possible inversion *via* the *enol*-**form** will not come into play. Besides, both the **compounds (F) and (G)** –actually gave the *same product* upon **reduction,** thereby ascertaining the fact that **compound (G)** happens to be the *cis*-isomer positively.

❑ Because **lithobilianic acid (B)** and **isolithobilianic acid (D)** may also be prepared from the **lithocholic acid** (a *natural bile acid*)-it suggests that both of them do have the same *cis*-A/B fusion. Ultimately, these isomeric acids can also be prepared form **5β**-cholestane (or coprostane).

Comment-The foregoing evidences suggest that the compound also possess *cis*-A/B fusion

> **NOTE** In fact, both of these kinds of 'fusion' (*cis*-and *trans*) invariably do occur in the so-called 'natural steroids'.

Important Features-These essentially comprise:

1. **Fusion of Rings 'B' and 'C' has *trans*-Configuration** : Bernal *et al.* (1932) performed the X-ray **analysis** of the aforesaid compounds and reported the fusion of the **rings B and C** to have a distinct Steroids and Hormones 9 *trans*-**configuration**. In addition, as such the so-called **steroidal nucleus** (or **molecule**) to exist as a ***thin*** and **long** inherent structure *i.e*, it is more or less **flat**. In fact, the said observation could be further expatiated only when the actual fusion of **rings B and C** has taken up a ***trans*-configuration**.

> **NOTE** The *trans*-B/C fusion occurs only in the naturally occurring plant steroids.

2. ***trans*–Fusion of C/D Rings in Sterols and Bile Acids**-Wieland *et al.* (1993) provided solid chemical evidences to ascertain that the ***trans*-fusion C/D rings occur in plant sterols and bile acids**.

Examples: Following are a few typical examples:

(i) **Deoxycholic Acid (J)**-when subjected to *degradation* gave rise to the formation of **compound (K)** – that failed to form the 'anhydride' rapidly.

However, when the **compound (X)** was subjected to heating under vacuo produced the corresponding anhydride **(L)**. The resulting **compound (L)** upon *hydrolysis* yielded the respective **dicarboxylic acid (M)**, which was indeed found to be totally a different form of compound **(K)**.

Comment – It reveals that the respective **anhydride (L)** and the **dicarboxylic acid (M)** should by all means possess the ***cis*-configuration**; and, therefore, the **rings C and D** must have the ***trans*-configuration** in the **deoxycholic acid (J)**.

Thus, we may have the following *sequential reactions* :

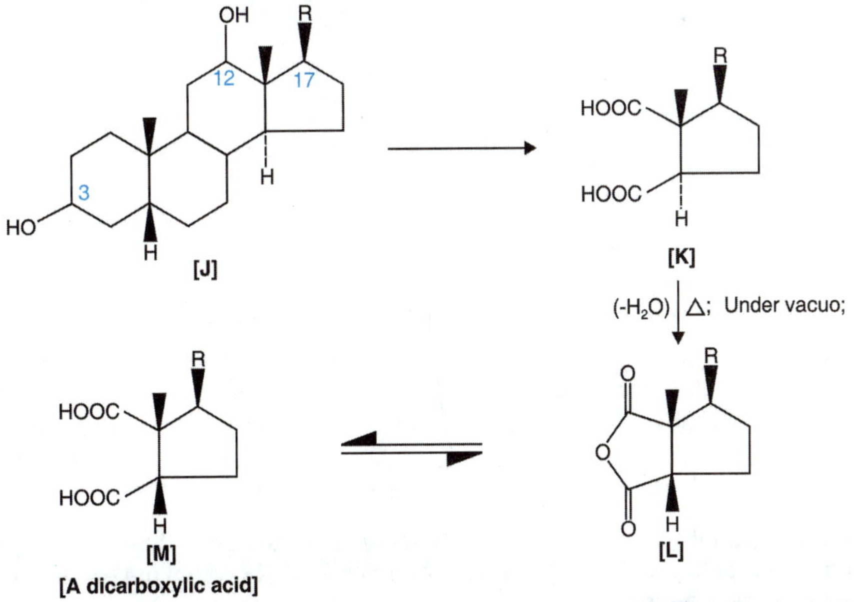

(ii) It has been duly ascertained that in a plethora of the 'steroids' (*i.e*, including the '*bile acids*') one may critically observe the *trans*- **C/D fusion**. Nevertheless, there are quite a few *glaring exceptions*, wherein the *cis*-**C/D fusion** takes place, such as :

• Cardiac Glycosides *viz* **Digoxin**; **Digitoxin**

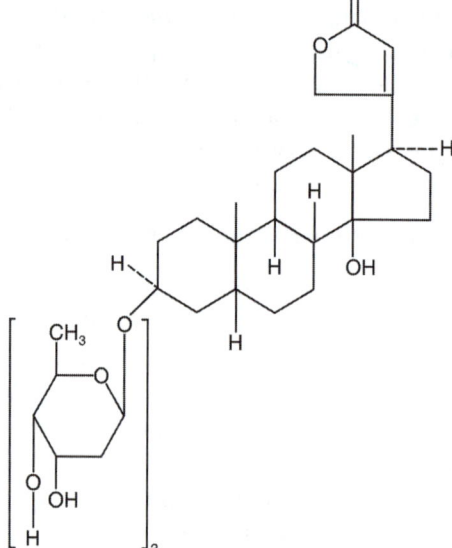

Digoxin [Lanoxin[(R)];SK-Digoxin[(R)]]   **Digitoxin [Crystodigin[(R)];Crystallin[(R)]]**

- Toad Poisons *viz*, Bufalin

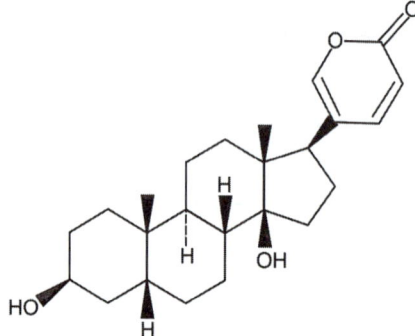

Bufalin
[A 'bufadienolide structure' found in the skin of toads (*Bufo.* spp.)]

(iii) **X-Ray Analysis of Steroids** – Bernal *et al*. (1940) meticulously studied the **X-ray analysis of several plant steroids** and demonstrated that the **H-atom** positioned at **C-9** has the **trans orientation** with respect to *methyl moiety* at **C-10**. Incidentally, it is further substantiated by the chemical evidence, which also ascertains that the methyl (CH_3) moieties located both at **C-10 and C-13** do have the *cis*-orientation.

NOTE   Based upon the *X-ray analysis* and *chemical evidences* of the *cholesteryl iodide* it has been duly proved that the *two* –angular methyl groups at C-10 and C-13 have the *cis*-orientation.

(iv) **NMR-Spectroscopy**-Amazingly, **nuclear magnetic resonance (NMR)-spectroscopy** has turned out to be extremely useful in the **steroidal chemistry**. Williamson *et al*. (1966) Steroids and Hormones 11 carried out a copious volume of meaningful research and amply demonstrated that the **angular methyl (CH$_3$) moieties positions at C-10 and C-13** essentially undergo coupling with certain segments on the *steroid nucleus*.

Important Observations – These are *two* important observations:

1. Assuming that the **C-10 methyl moiety** is capable of rotating freely, it could easily couple with an *axial H-atom* positioned duly at **C-1, C-5, and C-9**, in a situation when the actual fusions of **rings A and B** takes A/B is in the *trans*-**configuration** specifically. However, if rings **A/B is in the *cis*-configuration,** only *one axial H-atom located at C-9* may couple in this fashion.

2. Besides, it may also be found that the peak width (pertaining to the *two* angular methyl moieties) at 50% height for the typical *trans* **fused isomer** is definitely **higher** *vis-a-vis* the corresponding *cis*-**fused isomer**. In this manner, it becomes a lot easier to assign the respective *stereochemistry* of the **A/B junction present duly in both types of compounds.**

3.2. Configuration of Substituent Moiety

In the present context the critical discussion will prevail about the actual configuration of the **hydroxyl (OH) moiety** located at **C-3**. However, conventionally the **hydroxyl (OH) moiety** at **C-3** in **cholestanol** and **cholesterol** is regarded to be present in the *cis*-**configuration** with respect to the **methyl (CH$_3$) moiety** at **C-10**. Obviously, this kind of a *configuration* usually yields the β-**series** and is invariably found to occur in most '**natural sterols**;'. Importantly, in such particular instances the **prefix 'β'** normally indicates that the *substituent hydroxyl moiety* lies **above the plane of the molecule** (shown by a *solid line*) ; whereas, the *hydroxyl moiety* lies below the plane of the molecule (shown by a *dotted line*).

Further Confirmation of Actual Position of Hydroxyl (OH) Moiety

The **X-Ray analysis** of all the foretold compounds duly revealed that the **hydroxyl (OH) moiety** in *cholesterol* lies very much *above the plane of the molecule**, which has been further substantiated chemically by Shoppe (1948) by the following sequence of reactions.:

Cholesterol acetate → HNO$_3$; (Nitration) → 6-Nitro-Cholesterol acetate → Zn/CH$_3$COOH; (Reduction) →

Intermediate → H$_2$O; → 6-*keto*-derivative → (I) Br$_2$; (ii) AgNO$_3$ in Pyridine

(Contd....)

---

* That is, it is essentially present in the cis-**configuration** with respect to the methyl (CH$_3$) moiety located at **C-10** (3-**configuration**).

| γ-Lactone [M] | ← (CH₃CO)₂O Acetic anhydride ← | A Dicarboxylic Acid (Ring 'B'-ruptured) | ← NaOH; H₂O₂; ← | 5,7-Diketo deriv. |

Explanation – The various steps involved in the above sequential reactions may be explained as under :

1. **Cholesterol acetate** on nitration gives **6-nitro cholesterol**, which upon reduction with Zn /CH₃ COOH first gets converted into a **6-amine derivative** that rearranges to yield the corresponding **6- imine derivative**-as an **intermediate**.

2. The resulting *imine* intermediate upon hydrolysis with water (**H₂O**) gives rise to a **6-*keto* derivative**.

3. The **6-*keto* derivatie** first on *bromination* followed by treatment with **AgNO₃** in pyridine yields the **5,7-diketo derivative**, which on further reaction with an alkali (NaOH) and subsequent *oxidation* with **hydrogen peroxide (H₂O₂)** gives a **dicarboxylic acid** (with the *ring 'B'* ruptured).

4. The resulting product finally upon treatment with **acetic anhydride** gives **γ-lactone (M)**.

4. DIEL'S HYDROCARBON

The **Diel's Hydrocarbon** represents a solid substance having the melting point **126-127°C** and has a molecular formula **C₁₈H₁₆**. Based upon the results of *oxidation reactions, X-rays crystal analysis* coupled with the *absorption spectrum pattern* it was revealed duly that the **hydrocarbon** in question could be **3'-methyl-1,2-cyclopentanophenanthrene**. The next essential step was related to establish the structure of this compound by *synthesis e.g,* as suggested by Harper *et al.* (1934) who made use of the **Bogert-Cook Method** commencing from :

- **2-(1-naphthyl)- ethyl-magnesium bromide (A)**, and
- **2,5- dimethylcyclopentanone (B)**.

The interaction of **compounds (A) and (B)** usually react to give a condensed product which upon oxidation with P₂O₅ (phosphorus pentoxide) at **140°C** and subsequent distillation under reduced pressure gives an '**intermediate**'. The resulting intermediate undergoes '**cyclization**' first and later on when distilled with **Se powder yields** the desired **Diel's Hydrocarbon**.

The various sequential reactions are as stated under :

[A]   +   [B]   →

(I) P₂O₅;
(ii) △140°C;
(iii) Distilled under vacuo;

(Contd....)

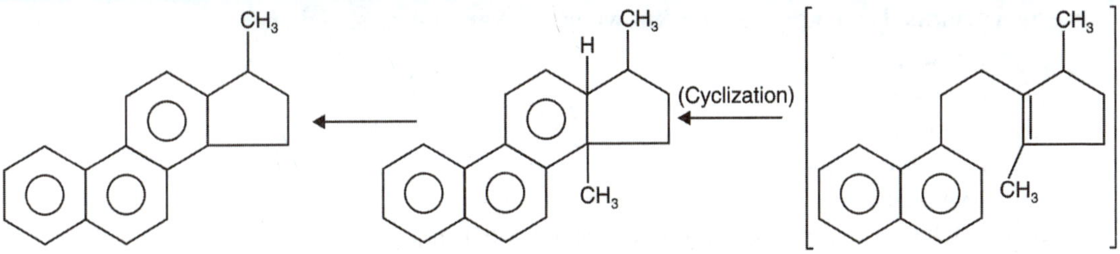

| Diel's Hydrocarbon | Cyclized nucleus | Intermediate |

5. STEROLS

Sterols represent the *solid unsaturated steroid alcohols* with a **hydroxyl (-OH) moiety** positioned at **C-3** which occur invariably as :

- **free state,**
- **esterified form,** or
- **glycosides,**

and are generally classified according to the specific organism in which they occur as : *mycosterols, phytosterols* etc.

The term **sterols** is derived from the *Greek-Stereos* meaning **solid** and *ols*-meaning **hydroxyl compounds** (or **sterols = steroidal alcohols**).

In a broader perspective, the **sterols** are duly present as components virtually in all **plant cells, animal cells;** and having the noticeable exception of the **microorganisms** wherein these are either:

- **absolutely absent,** or
- **exist in small quantum only.**

Besides, the sterols mostly occur as: **waxy, colourless solids, fairly soluble in practically all organic** solvents, and virtually insoluble in an aqueous medium.

Importantly, the **sterols** predominatly comprise:

➤ **one alcoholic functional moiety present duly as –**
- **saturated forms** *viz, plant sterols,* and
- **unsaturated forms** *viz, animal sterols (cholesterol), and several phytosterols (stigmasterol, ergosterol).*

| NOTE | Interestingly, the naturally occurring sterols are often found to be esterified with the *higher fatty acids.* |

Classification of Sterols-The **sterols**, in general, may be classified into *three* **categories**, namely :

(a) **Phytosterols**-The **sterol** that are solely obtained from the *plant sources viz,* **Ergosterol, Stigmasterol, β-Sitosterol.**

(b) **Zoosterols**-The **sterols** that are derived exclusively from the *animal sources viz,* **Cholesterol, Coprostanol, Cholestanol.**

(c) **Mycosterols**-The sterols which are duly obtained from the *yeast and fungi viz,* **Ergosterol.**

Sterol Variants-There are *five* **well-known sterol variants** as stated under :

    (i)   **Cholesterol,**

    (ii)   **Lanosterol,**

    (iii)   **Stigmasterol,**

    (iv)   **Ergosterol**, and

    (v)   **Bile Acids,**

which shall now be treated individually in the sections that follows :

5.1. Cholesterol ($C_{27}H_{46}O$)

Preamble – **Cholesterol** is invariably present in most **mammalian tissues** both in the '*free state*' and '*esterified state*' (with *higher fatty acids*).

The major sources of cholesterol are as follows:

-   **Human brain**   : ~ 17% ;
-   **Gallstone***   : ~ 90-100%;
-   **Spinal cord**   : significant quantum; and
-   **Fish Liver oils**   : good amount.

> **NOTE :**   **(1)** Almost all mammalian *steroidal hormones* and *bile acids* are critically obtained from the *cholesterol* enzymatically.
>
>   **(2)** Besides, it is duly established that the *steroidal sapogenins* and steroidal alkaloids – are based structurally upon the *cholesterol frame work*.
>
>   **(3)** In addition, being used largely for the commercial preparation of *vitamin* D$_3$; *cholesterol serve* as the pioneer source of the *hormonal steroids*.

Sources of Cholesterol- **Cholesterol** was initially isolated from **gall-stones** deposited in the *bile duct* or *gallbladder*. In fact, **cholesterol** derives its name from the **Greek-'Chole'** means **bile**, and *sterol*-refers to *steroidal* alcohol. However, the various other sources of **cholesterol** are :

-   **spinal cord of mammalians**
-   **brain** and
-   **fish-liver oils.**

Besides, *wool-fat* '**Lanoline**' comprises a mixture of **cholesteryl palmitate (C_{16}), stearate (C_{18}),** and **oleate (C_{18}-unsaturated acid).**

Characteristic Features – It essentially includes:

1.   **Cholesterol** is usually obtained as monohydrate pearly leaflets or plates from dilute ethanol.
2.   It becomes absolutely anhydrous at **70-80°C** to get a product having mp **148.5°C**.
3.   It gets sublimed as the *orthorhombic needles* (**bp$_{0.5}$233°C; bp$_{760}$360°C** (with s*ome decomposition*).
4.   It has d$_{19}^{19}$ 1.052 (anhydrous product).
5.   It shows $[\alpha]_D^{20}$ -31.5° (C=2 in either); and $[\alpha]_D^{20}$ -39.5 (C=2 in chloroform).
6.   *Solubility Profile* – Almost *insoluble in water* (~0.2mg/100mL water); *slightly soluble in ethanol* (1.29 % w/w at 20°C); and *more soluble in hot ethanol* (100g of saturated 96% ethanol solution contains 28g at 80°C).

---

*   **Gallstone**: A calculus formed in the **gallbladder or bile duct.**

However, 1g of **cholesterol** dissolves in:

- **1.5 mL pyridine**
- **2.8 mL solvent Ether**
- **4.5 mL chloroform.**

Colour Reactions of Cholesterol – Following are the *three* **typical colour reactions** of cholesterol:

❑ **Salkowski Reaction–**The careful addition of a few drops of concentrated H_2SO_4 to a *chloroform solution of cholesterol* gives an instant **red colouration.**

❑ **Liebermann–Bruchard's Reaction** – A *chloroform solution of cholesterol* on being treated with concentrated H_2SO_4 and a few drops of **acetic anhydride** produces a distinct **greenish** colouration.

❑ **Digitonide Formation Test** – An *ethanolic solution of cholesterol* when treated with an *ethanolic solution of digitonin* (a **saponin**) it gives rise to the formation of a distinct **white precipitate** (*cholesterol digitonide*). The resulting **white precipitate** when dissolved in **dimethylsulpho xide (DMSO)** – heated the solution on a steam – bath – cooled the contents to obtain **cholesterol as a precipitate** (Issidorides *et al.*, 1962).

Isolation of Cholesterol – First of all, the starting material (*i.e*, the source of **cholesterol**, say 'Fishliver') is cut into small pieces treated with ethanol (95% *v/v*), and centrifuged to obtain an **ethanolic solution of cholesterol**. Prepare separately an *ethanolic solution of digitonin* and add this gradually into the *ethanolic solution of cholesterol* (as obtained above) till the precipitation of **cholesterol digitonide** is absolutely complete.

> **NOTE** The 'cholesterol digitonide' is a unique molecular complex that predominantly comprises one mole each of :
> - **cholesterol, and**
> - **digitonin.**

Following are the *two* widely adopted means to recover cholesterol from the **cholesterol digitonide**, namely:

(a) **Breaking of cholesterol digitonide complex-** It is accomplished by incorporating *solvent ether* (*i.e*, **diethyl ether**) when **cholesterol** remains in the *ethereal solution* and the *digitonin* gets precipitated almost completely (being insoluble in nature). Ultimately the resulting ethereal solution is filtered-evaporated (*under vacuo*)-to obtain **cholesterol** as a solid residue.

(b) **DMSO-Method-**The *complex* **cholesterol digitonide** is first made to dissolve in **dimethyl sulphoxide (DMSO)** – heated on a steam-bath to allow the complex to undergo **dissociation**. The resulting solution when cooled allows perceptively the **cholesterol** to get deposited, which may be separated by filtration and purified.

Comments: It is, however, pertinent to state here that in both the above cited methods for the recovery of **cholesterol** from the **complex cholesterol digitonide** remains quite satisfactory; however, its further purification through recrystallization from an appropriate solvent is an absolute must.

CONSTITUTION OF CHOLESTEROL

The **constitution of cholesterol** was duly elaborated and ascertained by various researches: Wieland; and Windaus *et al.* almost stretched over a span of *three* **decades** in the twentieth century (1903-1932), whose meticulous findings indeed provided a meaningful input.

Therefore, in a most logical and systematic approach adopted for the critical determination of the *structure of cholesterol* (A) (see *section :3*), we may have to look at the following *four* **major aspects** both intensively and extensively :

   (i)   **Structure of the Steroidal Nucleus,**
   (ii) **Exact Position of Hydroxyl (OH) Moiety and Double Bond,**
   (iii) **Nature and Point of Attachement of Side-Chain to the Steroidal Nucleus,** and
   (iv) **Precise Location of Two Angular Methyl Moieties.**

[A] Structure of the Steroidal Nucleus

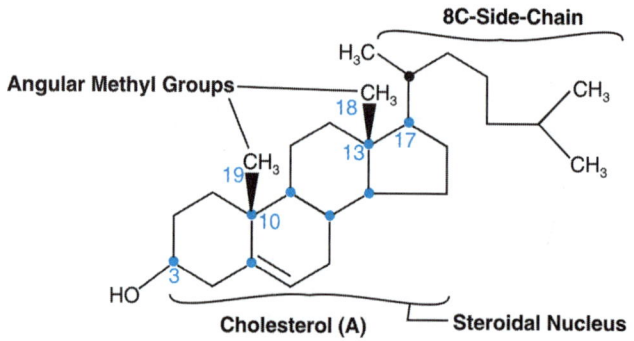

Cholesterol (A)

| NOTE | Keeping in view the *strength of simplicity* and *devoid of any sort of ambiquity*, it is always better to consider the established known structure of *cholesterol* at the very outset of the foregoing discussion. |

The structure of the *steroidal nucleus* is elucidated duly on the basis of the underlying facts, namely:

   (i)   **Molecular Formula** – Based on the elemental analysis data, the molecular formula of cholesterol is found to be $C_{27}H_{46}O$.
   (ii) **Presence of 1- Hydroxyl (OH) Moiety** – Cholesterol on being subjected to *acetylation* forms the respective '**monoacetate**' thereby suggesting the presence of one **hydroxyl (-OH) moiety** in it.
   (iii) **Presence of 1-Double Bond** – **Cholesterol** takes up *1 mole of Br_2 (or 2 Br-atoms)* so as to form **cholesterol dibromide**, which ascertains that it essentially comprises **1-double bond.** However, the presence of **1-double bond** may be further confirmed by the fact that **cholesterol [$C_{27}H_{46}O$]** on *hydrogenation catalytically* gives the corresponding cholestanol (B) [$C_{27}H_{48}O$].
   (iv) **Cholesterol (A)** – when *reduced* with H_2/Pt gives rise to the formation of **cholestanol (B)**; and the latter upon *oxidation* with CrO_3 (chromic acid) yields **cholestanone (C)** [$C_{27}H_{46}O$], which on *reduction with Zn-Hg/HCl* produces **cholestane (D) [$C_{27}H_{48}$].**

Thus, we may have the following *sequential conversions*:

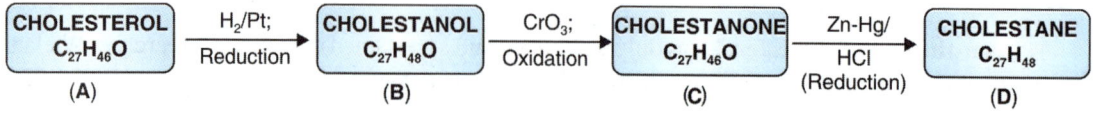

Inferences-Basesd on the aforesaid *sequential conversions* one may logically derive the following three **inferencs,** such as :

- ❏ The actual conversion of **(A)** *Cholesterol* to **(B)** *Cholestanol* reveals explicitly that (A) critically consists of **one double bond**;

- ❏ The oxidation of **(B)** *cholestanol* to **(C)** *cholestanone* (a **ketone**) reveals prominently that **(A)** *cholesterol* is a *secondary* **alcohol**, since a *sec*-**alcoholic moiety** upon oxidation gives a **ketone**.

- ❏ The saturated hydrocarbon **[D]** *cholestane* corresponds to the *generalized molecular formula* C_nH_{2n-6}, which obviously means that it must be *tetracyclic* in nature; and, therefore, **cholesterol [A]** is certainly proved to be a **tetracyclic alcohol.**

NOTE

The DBE of *cholestane* is found to be : 27+1-48/2=4

***i.e, cholestane* has a tetracyclic structure.**

(v) **Formation of Diel's Hydrocarbon and Chrysene** – Interestingly, when **cholesterol** is distilled with **Se at 360°C**, it gives the following *two* products:

- • **Diel's hydrocarbon**, and
- • **Chrysene**.

Thus, we may have the following expression:

Cholesterol $\xrightarrow[\text{(Dehydrogenation)}]{\text{Se; 360°C;}}$
[A]
$(C_{27}H_{46}O)$

Diel's Hydrocarbon ($C_{18}H_{16}$)

Comments : The following of the **Diel's hydrocarbon** may be explained by assuming that its nucleus is critically present in the **chemical structure of cholesterol.**

Chrysene [BP: 448°C; MP: 255°C] – Based on the crucial formation of **chrysene** (*a 4-membered cyclic hydrocarbon*), it was duly concluded by **Rosenheim and King (1932)** that the chemical structure of **cholesterol** essentially comprises the *chrysene nucleus.* However, after a short spell of time the said researchers opined strongly that **cholesterol** contains the so-called **cyclopentenophenanthrene nucleus**, which was also proposed by **Wieland** *et al.*

Chrysene

Comment – It is, however, pertinent to state here that the **cyclopentenophenanthrene nucleus** present in *cholesterol* adequately supports and fits practically most of the available evidences that is derived meticulously from detailed investigative studies pertaining to the so-called **oxidation products** of the *two* naturally occurring phytoconstituents :

- **sterols,** and
- **bile acids**.

The aforesaid facts are also confirmed by the *total synthesis of* **Cholesterol.**

Ascertaining Various Size of Rings (A,B,C,and D) in Cholesterol

The various size of rings (*viz,* A,B,C, and D) in **cholesterol** may be ascertained individually by definite and well-defined procedures :

➢ **Size of Ring 'A'**-Both **cholesterol** and **cholic acids** are duly converted into a respective **dicarboxylic acid (I),** which eventually gets transformed into a **cyclopentanone (II)** derivative (with *4-cyclic rings*) as given below :

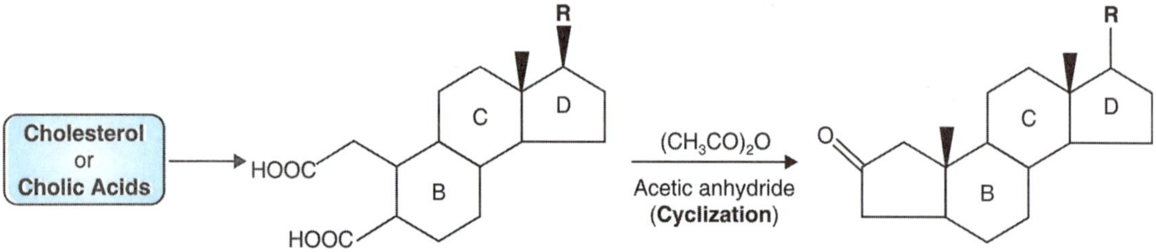

(I)   (II)

Comment:– The **cyclopentanone (II)** formation explicitly suggests that the ring 'A' happens to be a **six membered** ring; while, **R**-being the *suitable side-chain*.

➢ **Size of Ring 'B'**-**Cholesterol** may be converted duly into a corresponding **tricarboxylic acid (III),** that eventually gives rise to the formation of a corresponding **cyclopentanone** derivative **(iv)** (with *3-cyclic rings*) as shown under:

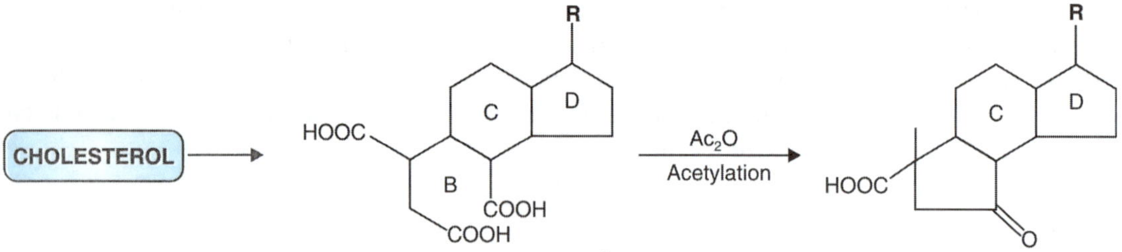

A Tricarboxylic Acid*   A Cyclopentanone
(III)   Derivative (IV)

Comment : Compound (III) on acetylation with *aceitic anhydride* (Ac$_2$O) aids in the closure of the **ruptured ring B** into a **closed 5-membered ring system** to yield a **cyclopentanone derivative (IV).**

---

* A **tricarboxylic acid (III)** is formed by the removal of **ring (A)** and rupture of **ring (B)** in cholesterol.

➤ **Size of Rings 'C'**-Since **deoxycholic acid (V)** may be duly converted into a **dicarboxylic acid (VI)**; it gives rise to the formation of a typical **cyclic anhydride (VII)** as shown below :

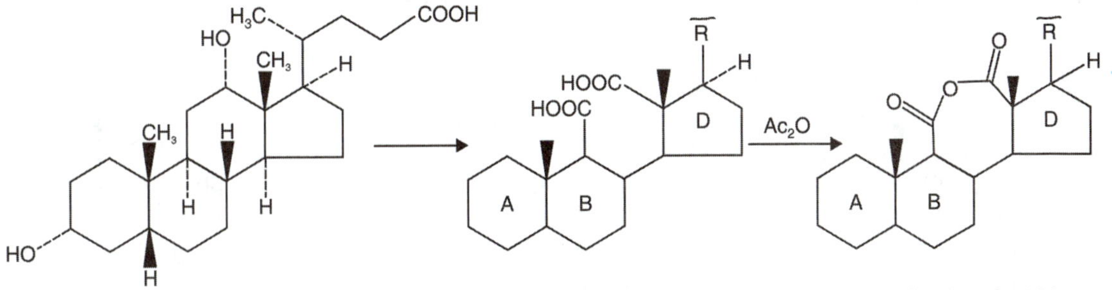

| Deoxycholic acid | A Dicarboxylic acid | A Cyclic anhydride |
| (V) | (VI) | (VII) |

Comment-Treatment of **dicarboxylic acid (VI)** with *acetic anhydride* yields a **cyclic anhydride (VII)** by the *elimination* of a mole of water and *closure* of **ring (C)** ; whereas, the side-chain **R** remains intact.

Based on the aforesaid reactions, Windaus and Wieland (1928)* put forward the following probable structure of cholesterol:

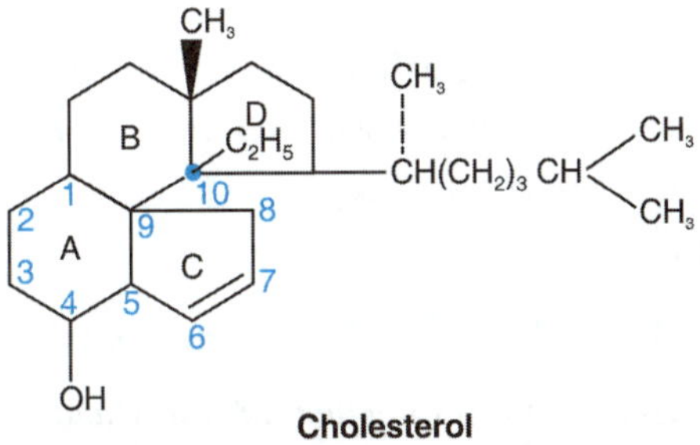

Cholesterol
[Structure proposed by Windaus and Wieland (1928)]

Points to Ponder – The above proposed structure for 'cholesterol' as suggested by **Windaus and Wieland (1928)*** obviously depicts a great extent of *uncertainty*, since at that material time no definite precise explanation could be assigned to the *two* **extra C- atoms.** Nevertheless, they were presumed to be present in the form of an *ethyl (-C$_2$H$_5$) moiety* located strategically at **C-10**. Later on, Wieland *et al.* (1930)** duly ascertained the total absence of the aforesaid **ethyl (C$_2$H$_5$) moiety at C-10**. Interestingly, the exact position of these *two* **C-atoms** could not be decided unless and until **Rosenheim and King,*** first, and foremost revealed that the *steroids* did possess the '**chrysene nucleus**' (see *Section1.*) ; and subsequently, proposed the **cyclopentanophenanthrene nucleus**. [see **structure (I)** *Section 1*].

Ultimately, Bernal (1932) performed the **X-ray analysis of cholesterol and ergosterol** etc., and suggested that the molecule of **cholesterol** plus **other steroids** was found to be *thin*. Amazingly, the aforesaid structure of the *steroidal nucleus* is certainly '*thick*' and not '*thin*'.

---

* Windaus and Wieland : *Physiol. Chem*, **167**:69,1928.

** Wieland *et al.* : *Physiol Chem*, **191** :69,1930.

*** Rosenheim and King : *Nat Inst Med Res*; King Harnold : *Nat. Inst Med Res.,*

➤ **Size of Ring 'D'**-Since, **coprostane** (or **5β-cholesterol**) **(VIII)** may be easily converted into the respective **etiobilianic acid (IX)** that eventually yields a **cyclic anhydride (X)** as shown under :

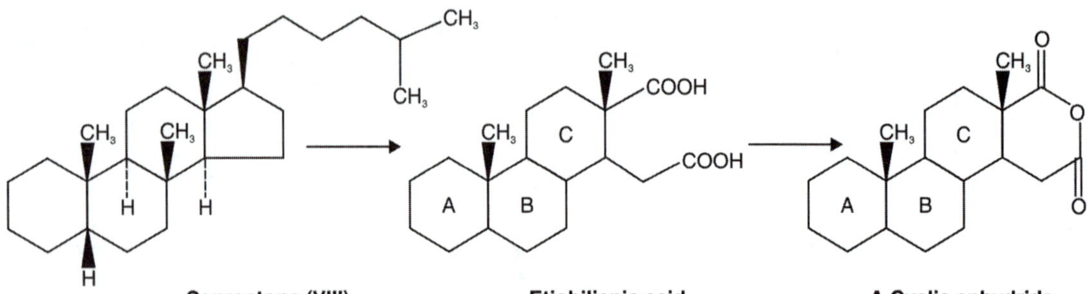

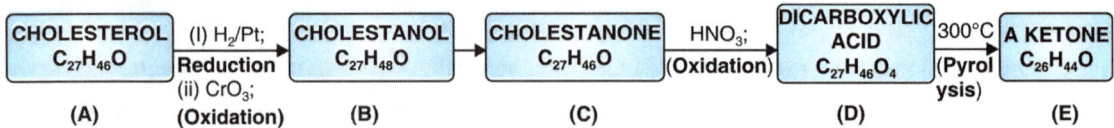

| **Coprostone (VIII)** | **Etiobilianic acid** | **A Cyclic anhydride** |
| [or 5β-Cholestane] | (IX) | (X) |

Comment – The above sequence of reactions starting from **(VIII)** to **(IX)** to **(X)** evidently suggests that **ring D** happens to be a **5- membered ring.**

[B] Positions of the Hydroxyl (-OH) Moiety and Double Bond

In the present context, the following *three* **cardinal aspects** need to be examined at length, namely :

❑ **Whether the hydroxyl (-OH) moiety is positioned at C-2 or C-3 ?**
❑ **Position of Hydroxyl (-OH) Moiety at C-3.**
❑ **Position of Double Bond.**

1. Whether the Hydroxyl (-OH) Moiety is Positioned at C-2 or C-3 ?

Obviously, when **cholesterol (A)** is subject to *reduction catalytically* and followed immediately by *oxidation* with *chromic acid (CrO₃)*, it gives **cholestanone (C)** *via* formation of **cholestanol (B)**. However, the *ketone i.e,* **cholestanone (C)** upon further oxidation with *nitric acid (HNO₃)* produces the respective **dicarboxylic acid (D)**, that essentially bears the **identical number of C-atoms** as present in the starting **original compound (A)** (*i.e,* **cholesterol**). Besides, the resulting **dicarboxylic acid (D)** on being subjected to *pyrolysis* at 300°C yields the **ketone (E)**.

Thus, we may have the following *sequence of reactions*:

CHOLESTEROL $C_{27}H_{46}O$	(I) H_2/Pt; Reduction (ii) CrO_3;	CHOLESTANOL $C_{27}H_{48}O$	CHOLESTANONE $C_{27}H_{46}O$	HNO_3; (Oxidation)	DICARBOXYLIC ACID $C_{27}H_{46}O_4$	300°C (Pyrolysis)	A KETONE $C_{26}H_{44}O$
(A)	**(Oxidation)**	**(B)**	**(C)**		**(D)**		**(E)**

Inferences – Based upon the aforesaid *sequence of reactions* on may critically draw the following **inferences**, namely:

1. **Cholestanone (C)** on oxidation to **dicarboxylic acid (D)** do essentially posses the **same number of C-atoms** as the **original ketone (C)** from which it has been derived duly; and this eventually suggests that the *ketonic functional moiety present in* **(C)** should be positioned in the **ring itself** Obviously, had the **ketonic moiety** been located in the *side-chain*, it would have resulted into an 'acid' having **lesser number of C-atoms.**

2. *Pyrolysis* of **dicarboxylic acid (D)** into a **ketone (E)** having a loss of *one C-atom* reveals prominently that **(E)** could be either a :

- **1,6 – dicarboxylic acid,** or
- **1,7 – dicarboxylic acid,**

according to the **Blanc's Rule***. However, the actual opening of the ring to produce the respective **dicarboxylic acid (D)** takes place from the respective *hydroxyl (OH) moiety*. Hence the **OH-moiety** by any stretch of imagination cannot be present in the ensuing **ring D**, since it may result into the formation of *1,5-dicarboxylic* acid instead of **1,6- or 1,7- dicarboxylic acid** by the aforesaid treatment.

Inference – Thus, it may be inferred vehemently that the said **hydroxyl (-OH) moiety** in **cholesterol (A)** may be present only in the **ring A, B, or C.**

3. **Cholesterol (C)** upon oxidation actually yields *two* **possible isomeric dicarboxylic** acids. Interestingly, the critical formation of the said **isomeric dicarboxylic** acids strongly suggests that the **keto (>C=O) function** moiety duly present in **cholestanone (C)** is obviously flanked on

either side by a **methylene (=CH$_2$)** moiety [*i.e*, the presence of **grouping (-CH$_2$-$\overset{\overset{\text{O}}{\|}}{\text{C}}$-CH$_2$-) in (C)**].

At this point in time, if one critically examines the **proposed structure (A) for cholesterol,** one may safely conclude that such an arrangement is only possible if the said **hydroxyl (-OH) moiety** is duly present in **ring A** either at **C-2 or C-3 position.**

Thus, we may have the following *two* **probabilities** for the **cholestanone (C)** such as :

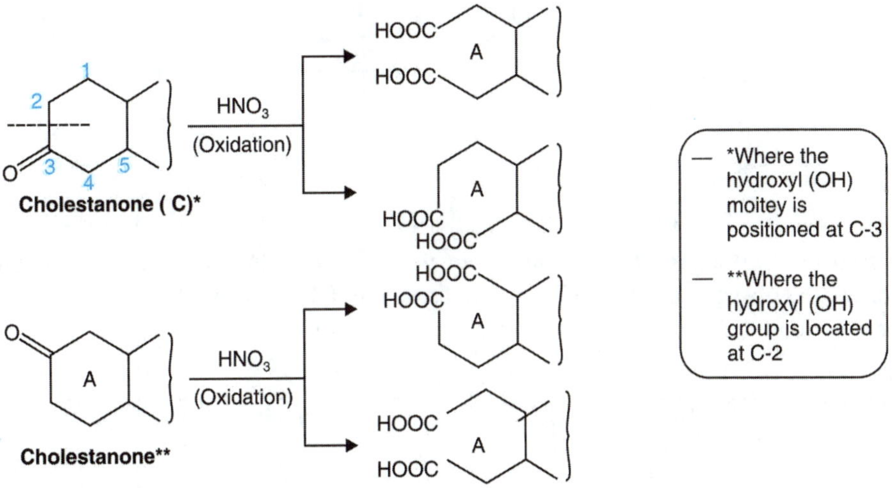

Cholestanone (C)*

Cholestanone**

— *Where the hydroxyl (OH) moitey is positioned at C-3

— **Where the hydroxyl (OH) group is located at C-2

---

* **Blanc's Rule : Blanc** observed that acids of the *glutaric type,* when submitted to the same process of distiallation with *acetic ahydride,* are converted into *anhydrides* and not into *cyclobutanone derivatives.* He proposed these reactions as a **diagnostic test** for distinguishing between the **dibasic acids** having a chain of **5C-atoms** and those having 6 or **more C-atoms** in the chains.

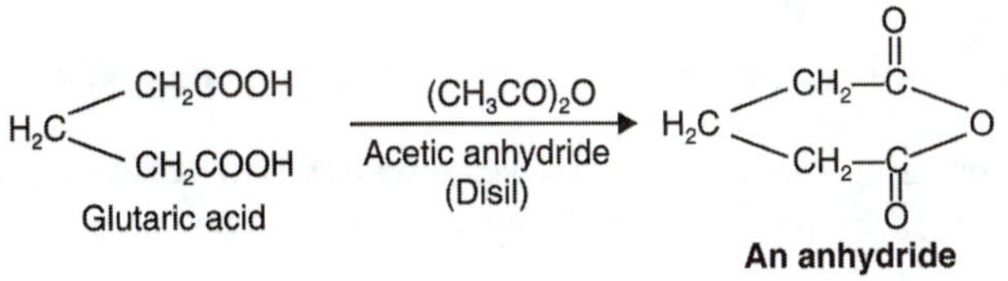

Glutaric acid

Acetic anhydride (Disil)

An anhydride

2. **Position of Hydroxyl (OH) Moiety at C-3 :** Kon *et al* (1937,1939) first and foremost confirmed the exact position of the hydroxyl (OH) moiety at **C-3 position** by certain well-defined experimented evidences, namely :

   ❑ **Cholesterol (A)** on being subject to reduction gives rise to the formation of **cholesterol (B).**

   ❑ **Cholestanol (B)** upon *oxidation* produces **cholestanone (C)** which upon treatment with **methyl magnesium bromide [CH$_3$-Mg-I],** a *Grignard Reagent*, followed immediately by *dehydrogenation* with **Se** yields **3,7-dimethylcyclopentenophenanthrene (D).** However, the chemical structure of **compound (D)** has been duly established by its synthesis.

   In fact, all the above reactions may be formulated in a sequential manner as stated under provided the position of the **hydroxyl (OH) moiety** is positioned at **C-3** which eventually corresponds with the **C-7 position** in **compound (D).**

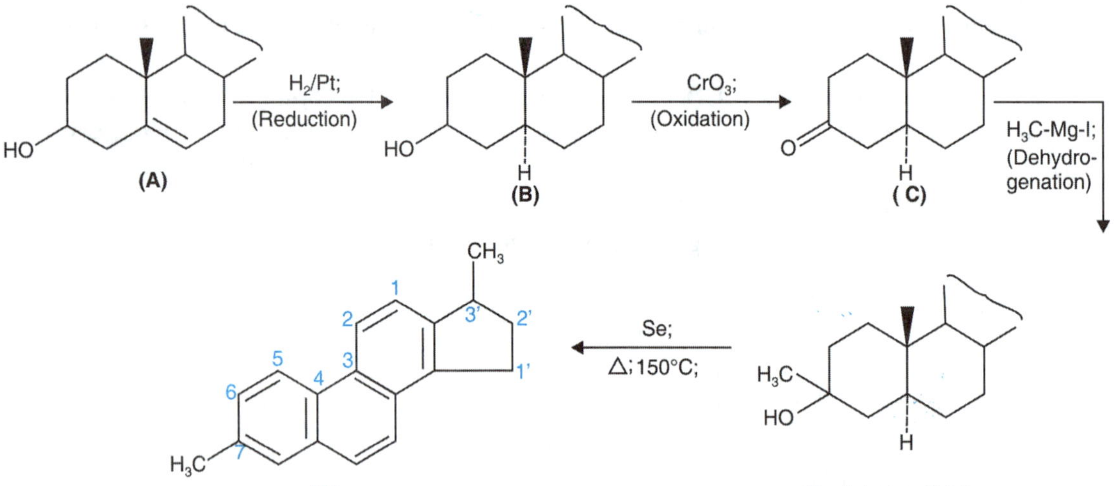

3. **Position of Double Bond :** It is, however, pertinent to state here that in an attempt to locate precisely the **position of the double bond,** present in **Cholesterol (A)** one may have to take into consideration the following set of the *sequential reactions* :

Inferences – The above reactions lead to the following important inferences:

1. **Cholesterol (A)** on being converted to **cholestanetriol (E)** *via oxidation* with H$_2$O$_2$ and CH$_3$COOH, whereby the so-called **hydroxylation of double bond in (A)** takes place.

Comment – In **cholesterol (A)**, there exists *only one hydroyxl moiety;* and, therefore, in **Cholestanetriol (E),** the *two* **additional hydroxyl (-OH) moieties** might have been generated by the hydroxylation of the double bond between C-5 and C-6 in (A).

2. **Cholestanetriol (E)** upon *oxidation* with *chromic acid (CrO₃)* gives rise to the formation of a *diketone i.e,* **hydroxyl-cholestanedione (F).**

Comment – The aforesaid observation reveals critically that in **(F)** the *two* **hydroxyl moieties** are essentially the *secondary alcoholic in nature i.e,* two of these are present as the *secondary alcoholic group* in **cholesterol (A)** ; whereas, the *third one*, which being resistant to oxidation, could be present as the *tertiary* **alcoholic group.**

3. **Compound (F)** on being subjected to *dehydration* followed by *reduction* with Zn/CH₃COOH yields **cholestanedione (G),** which upon oxidation with **chromic acid (CrO₃)** gives a **tetracarboxylic acid (H)** amazingly without the *loss of any C-atom.*

Comments – These essentially comprise :

❑ Obviously, the *two* **ketonic (=C=O)** functions present in **(G)** (*i.e,* **cholestanedione**) should be present in different rings.

❑ In case, these are present in the *same ring*, the actual *loss of C-atoms* would never have taken place upon **oxidation**; and hence, the **tetracarboxylic acid (H)** could never be produced.

❑ Therefore, one may duly infer that the prevalent *double bond* and *hydroxyl (OH) moiety* should be present in **cholesterol (A).**

❑ Since, it has already been established beyond any reasonable doubt that the **hydroxyl (OH) moiety** of **cholesterol** is perceptively present in '**ring A**'. Hence, by all means the *double bond* should be present critically either in **ring B, C, or D.**

4. **Formation of Pyridazine Derivative** – Since, **cholestanedione (G)** gives rise to the formation of *pyridazine derivative* by treatment with **hydrazine (H₂N.NH₂),** it evidently shows that **compound (G)** (*i.e,* **cholestanedione**) happens to be a '**diketone**' *i.e,* the *two* **keto moieties in (G)** are present strategically in **γ-positions** with respect to each other. Obviously, this could only be possible provided the *double bond* is located in between **C-5 and C-6** ; and, therefore, all the above mentioned reactions may be expressed explicitly as given under :

• **Hydroxyl (OH) moiety is located at C-3 position,** and

• **Double bond present between C-5 and C-6 of cholesterol (A).**

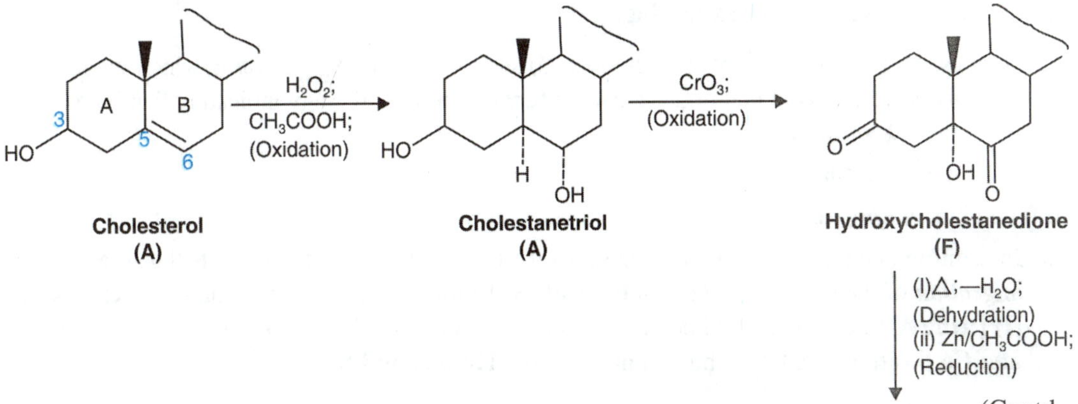

Cholesterol
(A)

H_2O_2;
CH_3COOH;
(Oxidation)

Cholestanetriol
(A)

CrO_3;
(Oxidation)

Hydroxycholestanedione
(F)

(I)△;—H_2O;
(Dehydration)
(ii) Zn/CH_3COOH;
(Reduction)

(Contd....)

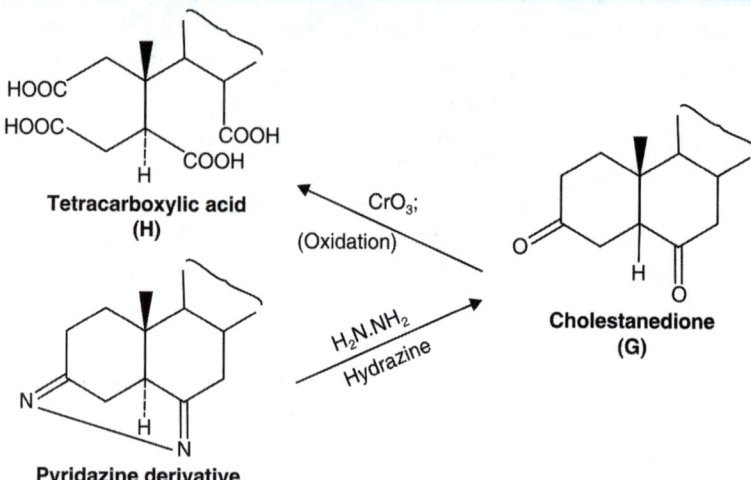

Tetracarboxylic acid
(H)

CrO₃;
(Oxidation)

H₂N.NH₂
Hydrazine

Cholestanedione
(G)

Pyridazine derivative

5. **Pyridazine Deirivative–A 'Polymer'**–Noller (1939) established critically, that the so-called **pyridazine derivative** happens to be a '**polymer**'. Hence it leads to the interpretation that **cholestanedione (G)** being a '*diketone*' – appears to be **incorrect**.

6. **Cholestenone (I) bears the *keto*–group and double-bond in the same ring :** Further critical evidence with respect to the aforesaid interpretation has been duly put forward by the underlying fact that **cholesterol (A)**, on being heated with *copper oxide (CuO)* at 290°C, gives **cholestenone (I)**, which upon subsequent oxidation with **potassium permanganate (KMnO₄)** yields a **keto acid (J)** due to the loss of **one C-atom**. The above reactions may be expressed as below :

$$\text{Cholesterol} \xrightarrow[\substack{290°C;}]{\substack{\text{Copper Oxide} \\ \text{(CuO)}}} \text{Cholestenone} \xrightarrow[\text{(Oxidation)}]{\text{KMnO}_4; \text{[O]}} \text{Keto acid} + \text{CO}_2$$
$$\quad\quad\text{[A]} \quad\quad\quad\quad\quad\quad\quad\quad\quad\quad \text{[I]} \quad\quad\quad\quad\quad\quad\quad\quad\quad \text{[J]}$$

Comments – These essentially comprise :

1. Since, the **keto acid (J)** is duly produced by the *oxidation* of **cholestenone (I)** with the critical loss of **one C-atom** thereby indicating that :

    • *keto*-**moiety, and**
    • **double bond,**

present in **(I)** are located in the **same ring.**

2. Besides, the **UV-spectrum** of **cholestenone (I)** shows explicitly λ$_{max}$ 240nm, which suggest that **(I)** virtually corresponds to that of an α, β-**unsaturated ketone**, theregy indicating that both :

    • *keto*-**moiety, and**
    • **double bond,**

are found in a *conjugated status*.

3. In addition one may also expatiate these results by assuming that there exists the so-called actual *migration* of the *double bond* in **cholesterol (A)** ['ring B'] during the formation of **cholestenone (I)** ['ring A'] ; and this could be only possible provided the **double bond** is present between **C5 and C6** position (*i.e*, C5 being common to both **rings A and B).**

Thus, we may have the following expressions :

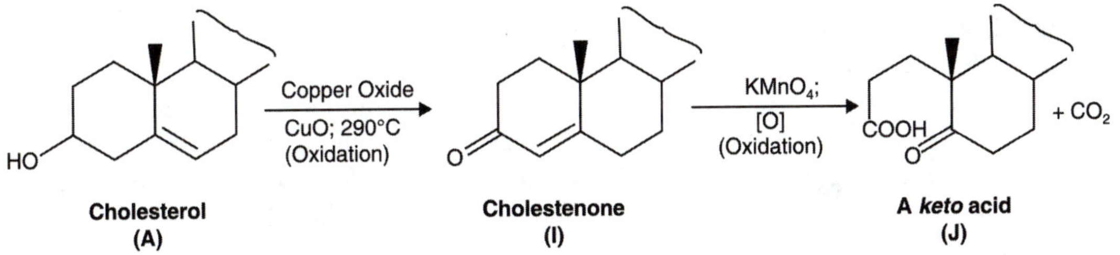

Cholesterol	**Cholestenone**	**A *keto* acid**
(A)	**(I)**	**(J)**

[C] Nature and Position of Side – Chain

The precise **nature and position of side-chain** in *cholesterol* [A] may be duly established by the aid of following sequential steps, such as :

❑ **Acetylation :** Cholesterol [A] on being subjected to acetylation yields **cholesterol acetate**. The resulting *acetylated product* upon oxidation with *chromium trioxide (CrO₃)* gives *two* distinct products namely :

• **a steam–volatile ketone, and**

• **an acetate of a hydroxy ketone (which being non-steam volatile).**

❑ The **'ketone'** has been found to be **isohexylmethyl ketone** $[H_3C.CO.(CH_2)_3.CH (CH_3)_2]$. Therefore, this *ketone* should be the *side-chain* of **cholesterol [A]** with its point of attachement at the C-atom of the respective *'keto'*-moiety.

❑ It is worthwhile to state here that the aforesaid reactions might not be expatiated logically provided the side-chain is duly linked to the '**steroidal nucleus of cholesterol [A];**

❑ Thus, assuming if the *accepted position of the side-chain* is **C-17 of the steroidal nucleus,** the foresaid reactions may be written as stated under :

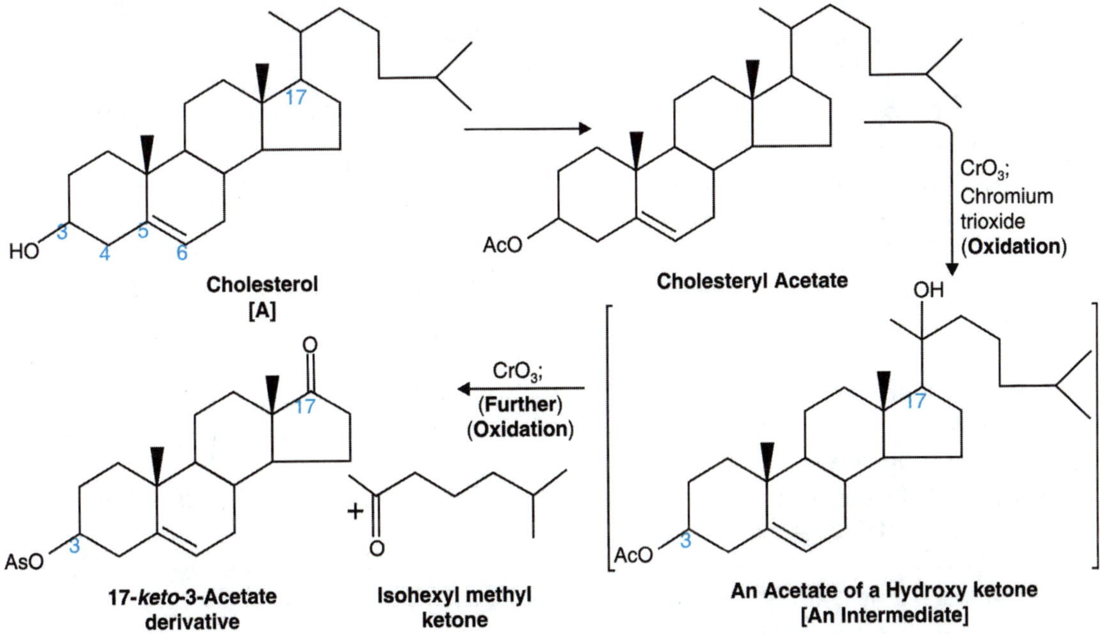

Cholesterol	**Cholesteryl Acetate**
[A]	

17-*keto*-3-Acetate	**Isohexyl methyl**	**An Acetate of a Hydroxy ketone**
derivative	**ketone**	**[An Intermediate]**

An Alternative Method for Establishing the Nature of Side-Chain

The **Barbier-Wieland Degradation*** serves as the *alternative method for establishing the nature of the side–chain in cholesterol.*

It critically explains the stepwise '**carboxylic acid degradation**' of *aliphatic acids* (present specifically in *sterol side-chains*) to the respective **next-lower homolog.** The resulting '**ester**' is duly converted to a **tertiary alcohol** that is dehydrated with a*cetic anhydride*, and the **olefin** oxidized with chromic acid (CrO_3) to a corresponding **lower homologous carboxylic acid.**

Thus, we may have :

$$RCH_2.\overset{\overset{O}{\|}}{C}-OCH_3 \xrightarrow[\text{(li) HX}]{\text{(I)2Ph Mg X}} RCH_2.C.OH.Ph_2 \xrightarrow{Ac_2O} RCH{=}CHPh_2 \xrightarrow{CrO_3} RCOOH + Ph_2CO$$

First and foremost the **cholesterol [A]** gets converted in to **5β-cholestane (or coprostane) (1)** (*i.e,* a stereoisomer of **cholestane (D).** Let us assume that the nucleus of **5β-cholesterol (1)** is duly designed as the *side-chain* as **Cn** and *Barbier-Wieland Degradation* as **BWD.** Thus, we may intelligently formulate the degradation of **5β-cholestane (1)** as stated under :

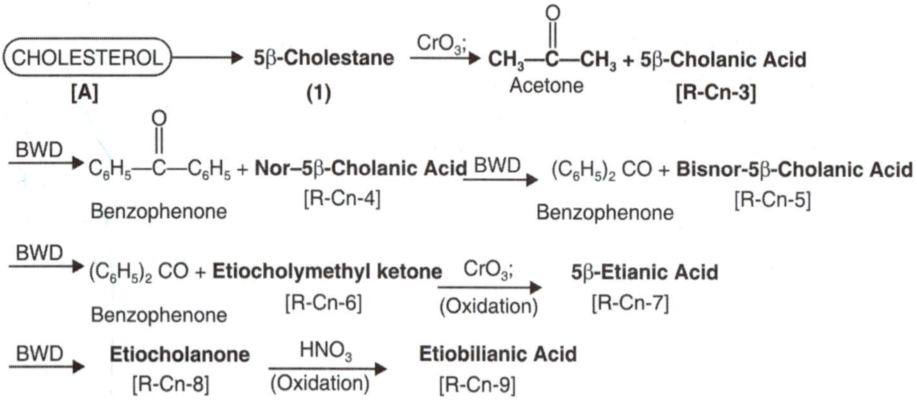

Inferences : Based on the above set of sequential reactions one may critically draw the following **inferences**, namely :

1. Critical formation of **acetone** from **5β-cholestane (1)** suggests that the **side-chain (C_n)** does terminate in an '**isopropyl moiety** ' as shown under :

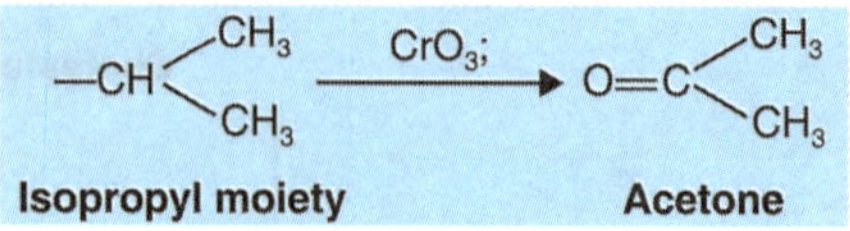

| Isopropyl moiety | Acetone |

2. Specific formation of a *ketone i.e,* **etiocholymethyl ketone** [$R\text{-}C_{n-6}$] right from **bisnor-5β-cholanic acid** [$R\text{-}C_{n-5}$] reveals that an, *alkyl functional moiety* is located strategically at the **α-C-atom'** of the bisnor-**5β-cholanic acid.**

Comment – Had there been *no alkyl moiety at all*, one would have obtained an '**acid**', instead of a '**ketone**', *via* the **BW-degradation** process.

---

* Wieland H :*Ber*, **45**,484,1912; Barbier P *et al.* : *Compt Rend*, **156**:1443.1913.

3. Since, the **etiocholylmethyl ketone** $[R\text{-}C_{n\text{-}6}]$ gets oxidized to **5β-etianic acid** $[R\text{-}C_{n\text{-}7}]$ with the loss of **one C-atom**, thereby ascertaining the fact that the said *ketone* should be a '**methyl ketone**'; and, therefore, the *alkyl moiety* attached to the **a-C-atom** in **bisnor-5β-cholanic acid** $[R\text{-}C_{n\text{-}5}]$ happens to be essentially a **methyl moiety**.

4. Interestingly, in **5β-etianic acid** $[R\text{-}C_{n\text{-}7}]$, the **carboxyl (COOH)** moiety is attached directly to the nucleus, In fact, this observation may be substantiated as given under:

   • **5β-Etianic acid** $[R\text{-}C_{n\text{-}7}]$ on being subjected to **BWD** gives a *ketone i.e*, **etiocholanone** $[R\text{-}C_{n\text{-}8}]$.

   • The resulting compound upon *oxidation* with **HNO₃** gives rise to the formation of a *dicarboxylic acid i.e*, **etiobilianic acid** $[R\text{-}C_{n\text{-}9}]$ without the loss of any **C-atoms**.

Comments – It vehemently suggests that **etiocholanone** $[R\text{-}_{n\text{-}8}]$ should by all means a '**cyclic ketone**' only. Hence, to logically support and account for the **foregoing sequential degradation steps** (*i.e*, **BWDs**), there are *eight* **C-atoms** present critically in the *side-chain.* of **cholesterol [A]** at **C-17**, that should be duly arranged in the following sequence :

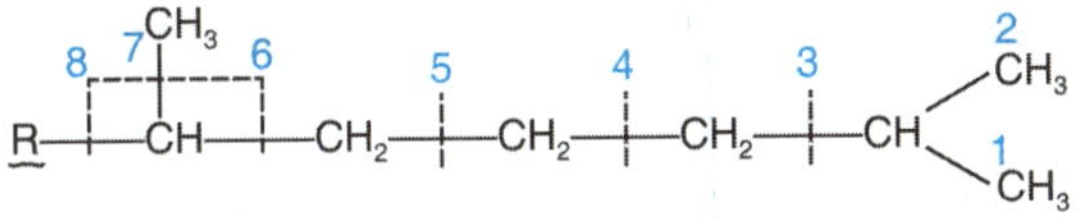

Point of Attachment of the Side-Chain to the Nucleus

The next most crucial problem is to determine the exact position of the *side-chain i.e*, the **point of attachment of the side-chain to the steroidal nucleus (C-17)**. However, the answer to such *intrigue querries* may be duly provided by the following critical findings, namely :

Etiobilianic acid $[R\text{-}C_{n\text{-}9}]$ *i.e*, the dicarboxylic acid, when heated *with acetic anhydride* $[(CH_3CO_2)O]$ it yields the corresponding '**anhydride**'.

Comments: The aforesaid reaction reveals explicitly that the **etiobilianic acid** designates a **1,5-dicarboxylic acid**. Nevertheless, the resulting **1,5-dicarboxylic acid** may be derived from a **5-membered ring** only. Therefore, one may conclude that **etiocholanone** $[R\text{-}C_{n\text{-}8}]$ represents a **5-membered ring ketone**, which ultimately suggests that the *side-chain* must be attached specifically to the **5-membered ring D** (*i.e, cyclopentano ring D in cholesterol*) of the steroidal nucleus.

> **NOTE** However, the above findings still never ascertains the actual '*point of attachment*' of the side-chain to the steroidal nucleus.

❑ Importantly, **cholesterol [A]** upon *dehydrogenation* with **Se** yields the *Diel's Hydrocarbon*. Obviously, this reaction suggests strongly that the *side-chain* (**containing 8 C-atoms**) is located strategically at **C-17** (*or the 5-membered cyclic ring*), because **Se-dehydrogenation** may cause the critical *degradation* of the **side-chain to a methyl (CH₃) moiety**.

❑ **X-ray Diffraction (XRD)** studies pertaining to the *surface-film measurements* do substantiate grossly the **C-17** position of the **8-C side-chain** in **cholesterol [A]**.

❑ The **nor -5β-cholanic acid** $[R\text{-}C_{n\text{-}4}]$ is duly obtained by the *oxidation* of **5β-cholestane (1)** and subsequent **BWD reaction**. Obviously, the aforesaid acid is derived from the oxidation of **deoxychoclic acid** followed immediately by the **Clemmensen reduction*** [*i.e*, reduction of the **carbonyl groups** of *ketones* and *aldehydes* to the corresponding **methylene groups** with Zn-Hg/HCl as stated below :

$$\text{R COR'} \xrightarrow[\text{HCl;}]{\text{Zn–Hg;}} \text{R.CH}_2.\text{R'}]$$

Comment – The above stated methods of preparation for the **nor-5β-cholanic acid** [$R\text{-}C_{n\text{-}4}$] reveal explicitly that the *side-chain* present duly in :

- **cholesterol [A]** , and
- **deoxycholic acid,**

are categorically present in the *same position*.

Further *deoxycholic acid* may also be readily converted into *12 keto -5β-cholanic acid*. However, the latter compound on being heated to 320°C loses a mole each of H_2O and CO_2 to produce *dehydronorcholene*, which subsequently upon *Se-distillation* gives rise to the formation of *20-methyl-cholanthrene*. The chemical structure of the resulting compound is duly ascertained by its oxidation to the *5,6-dimethyl-1,2-benzanthraquinone*, which on further *oxidation* with CrO_3 gives *anthraquinone- 1,2,5,6- tetracarboxylic acid*

NOTE   The chemical structure of *20-methylcholanthrene* may also be confirmed by its *total synthesis.*

In fact, all the above reactions may be duly explained provided the *8-C–side-chain* in **cholesterol [A]** is assumed to be seated at **C-17** position in the **steroidal nucleus** :

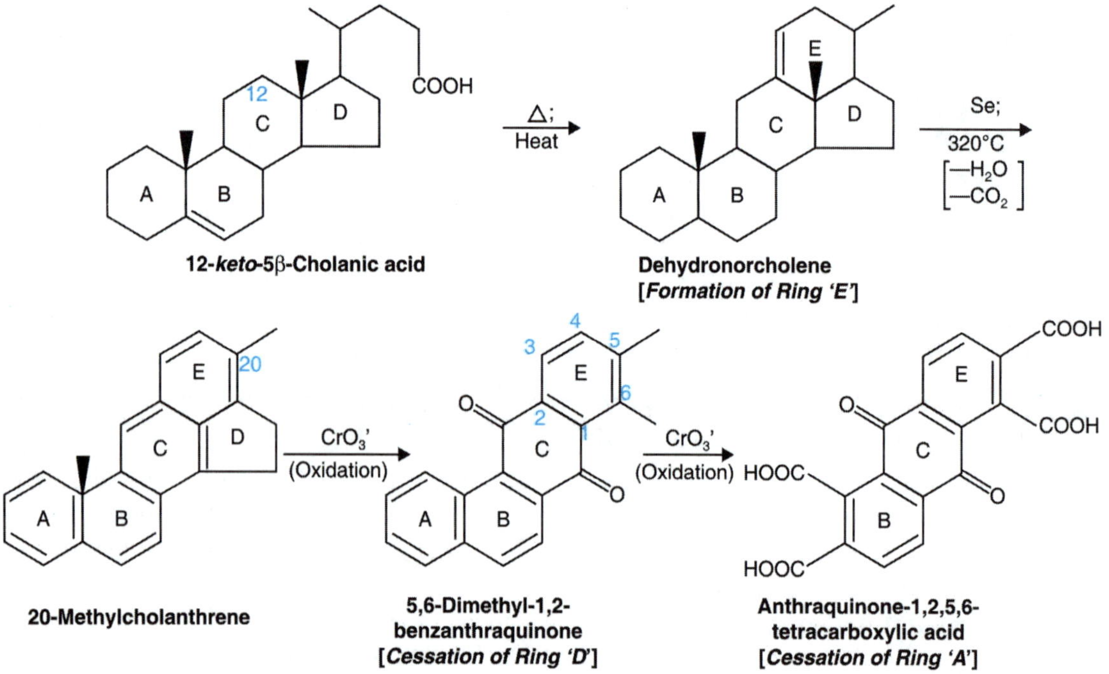

12-**keto**-5β-Cholanic acid

Dehydronorcholene
[*Formation of Ring 'E'*]

20-Methylcholanthrene

5,6-Dimethyl-1,2-benzanthraquinone
[*Cessation of Ring 'D'*]

Anthraquinone-1,2,5,6-tetracarboxylic acid
[*Cessation of Ring 'A'*]

---

*   Clemmensen E : *Ber.* **46**:1837,(1913); **47**: 51,681(1914).

Presence of Cyclopentanophenanthrene Nucleus in Cholesterol [A]

In true sense, the very presence of the so-called **cyclopentanophenanthrene nucleus in cholesterol [A]** may be precisely established based upon the following *two* glaring supportive facts, namely:

(a) **Isolation of 20-Methycholanthrene-** In reality, the critical the Isolation of **20-methycholanthrene** vehemently ascertains that the **cyclopentanophenanthrene nucleus** is duly present in **cholesterol [A].**

(b) **Formation of 1,2-Dimethylphenanthrene-** It has been observed that when **etiobilianic acid** [R-$_{Cn-9}$] being distilled with **Se (at-320°C)** gives rise to the formation of **1,2**-dimethyphenanthrene.

Comments – These essentially include:

1. Formation of **1,2-dimethyphenanthrene provides** an ample evidence for the presence of the so-called **cyclophenanthrene nucleus** in **cholesterol [A].**

2. It also gives *sufficient proof* for the specific location of the **C-13 angular methyl (-CH₃) moiety.**

Thus; we may have the following expression:

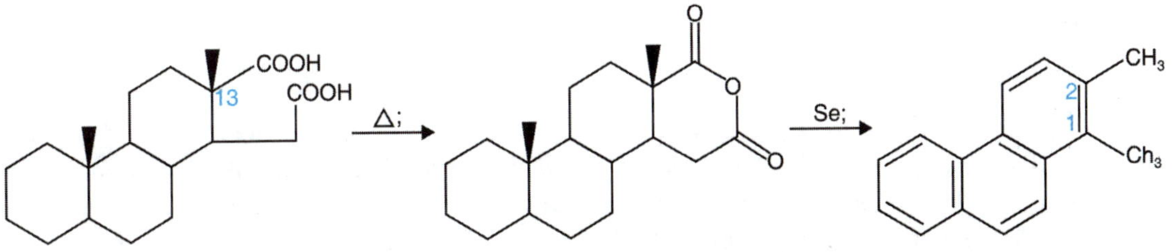

| Etiobilianic Acid [R-C$_{n-9}$] | An Anhydride [Intermediate] | 1,2-Dimethyl-phenanthrene |

[D] Position of Two Angular Methyl Groups

It has been duly observed that the **cyclopentanophananthrene nucleus** logically accounts for **17- C-atoms;** whereas, the side-chain takes care of the 8-C-atoms duly present in the **side-chain** [of **cholesterol (A)**].

However, **cholesterol (A)** precisely contains **27-C-atoms**; and, therefore, [27-(17+18) or **2-C-atoms**] should also be accounted for by all means. Interestingly, these *two* C-atoms have been established to be the *two* angular methyl (CH₃) moieties duly positioned at **C-10 and C-13** respectively.

The aforesaid findings may be further ascertained by the following *three* **explicit explanations.**

(a) *Keto-Acid (J)* (*as obtained in Section 'B'*) When the *keto-acid* **(J)** is being subjected to the **Clemensen reduction** followed immediately by *two* **BW-degradations** successively, one obtains a carboxylic acid that eventually **evolves CO₂** on being heated gently with **H₂SO₄(36M)**; but hardly undergoes *esterification.*

Comments – In fact, such typical characteristic reactions are assumed to be the **special feature of an** 'acid' having a **carboxyl (-COOH) moiety** duly attached to a *tertiary*-C-atom. Therefore, it may be concluded that the **alkyl moiety** should by all means be present at C-10 in the *keto*-acid **(J).** Thus, based upon the aforesaid sequence of *degradation* of **(J)**, one may write the following reactions.

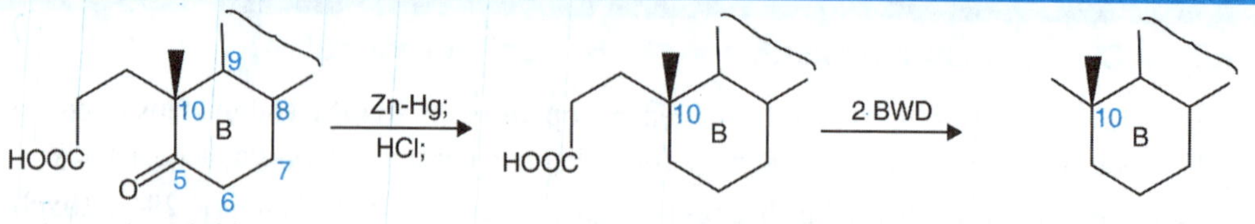

| keto-Acid [J] | A Carboxylic Acid | A Carboxylic Acid |

(b) **Cholesterol[A] yield's Diel's Hydrocarbon and Chrysene :** Cholesterol [A] on being distilled with **Se** gives rise to the formation of **Diel's Hydrocarbon** and **Chrysene** respectively:

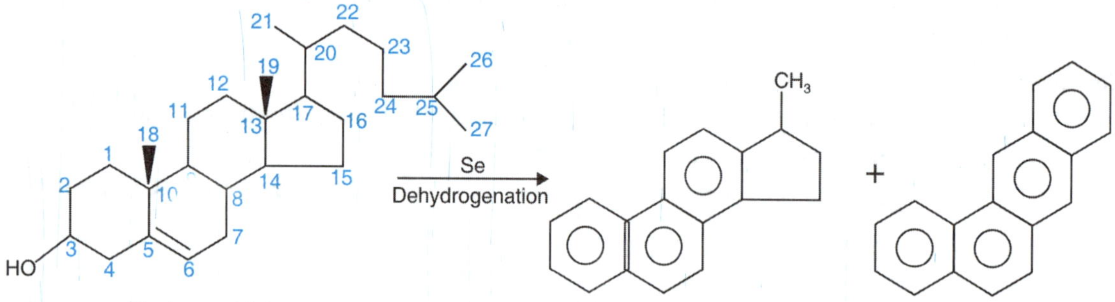

Cholesterol (A)   Diel's Hydrocarbon ($C_{18}H_{16}$)   Chrysene ($C_{18}H_{12}$)

Comments : Thus, one may logically explain the formation of the **Diel's Hydrocarbon** and **Chrysene** by the underlying fact that an '**angular** *methyl* (**CH$_3$**) moiety is located strategically at position **C-13** ; and, therefore, upon *Se-dehydrogenation* the said 'angular methyl group' enters the **5-membered ring D** to form the corresponding **6-membered ring**.

Nevertheless, the aforesaid evidence for the presence of *angular methyl moiety* is certainly not complete in all respects since there would exist *critical* **ring expansion** provided the **methyl moiety is positioned at C-14 instead**.

(E) Formation of 1,2-Dimethylphenanthrene from Etiobilianic Anhydride

The **etiobilianic anhydride** on being subjected to distiallation with Se yields **1,2-dimethylphananthrene**.

Inferences

1. The aforesaid course of reaction suggests strongly that the **methyl (CH$_3$) moiety** is duly present at **C-13 position** only (*i.e*, the **angular methyl group C-19**). Thus, we may have the following expressions :

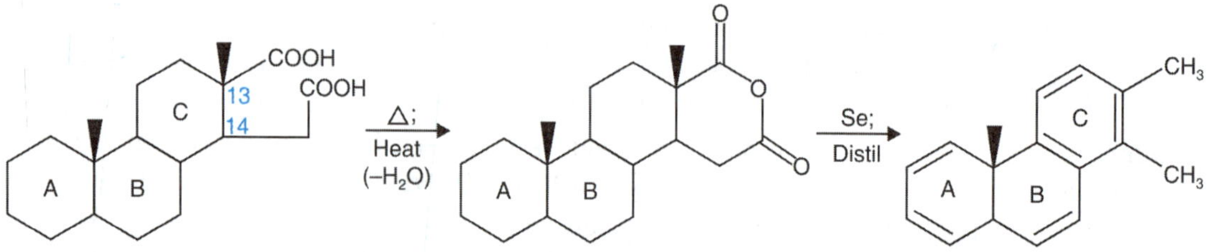

Etiobilianic Acid   Etiobilianic anhydride   1,2-Dimethylphenanthrene

2. Had the said **methyl (CH$_3$) moiety** been located on C-14, one would have obtained exclusively **monomethylphenanthrene,** as given under :

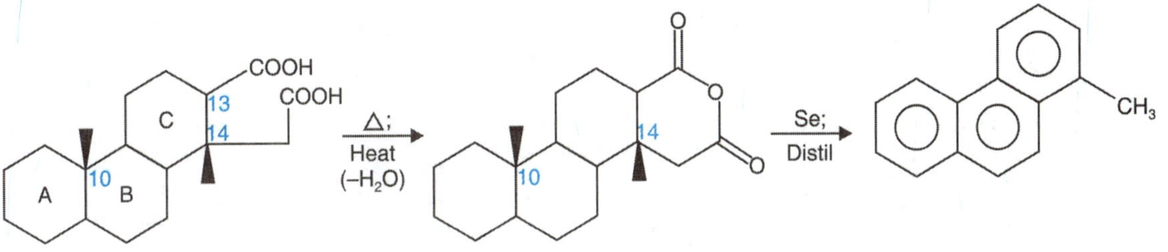

Etiobilianic Acid	Etiobilianic anhydride	Monomethyl-
[Angular methyl group at C-14]	(with angular CH$_3$ group at C-14)	phenanthrene

SYNTHESIS OF CHOLESTEROL [A]

Ultimately, the chemical structure of **cholesterol [A]** has been duly confirmed by its *total synthesis* as put forward by various phytochemists, namely :

- **Woodward (1951),** and
- **Johnson and Robinson (1953).**

Points to Ponder – Following are the *two* major difficulties that are invariably encountered in the *total synthesis* of **cholesterol [A]**, such as :

1. Since, **cholesterol [A]** comprises *eight* **distinct chiral centres** (or **asymmetric C-atoms**), thus it may give rise to *256 optical isomers* (*i.e*, **2n or 28=256**) in reality.

Therefore, each and every step in the **synthesis** shall eventually produce a *new chiral centre* that would actually result in the formation of :

- ➢ a few desired *stereoisomers*, and
- ➢ simultaneous resolution of *'racemic modification'* must also be feasible and practicable.

2. Obviously, another major difficulty usually encountered is to attack at a **specific targeted point in the molecule** thereby *without affecting the other parts of the molecule.* Hence, it may in reality requires the usage of *highly specific reagents only* (*i.e*, **stereoselective reactions**).

There are in all *five* **sequential steps** which are involved in the synthesis of **cholesterol [A]**, namely :

Step -1 : **Preparation of 4-methoxy-2,5-toluquinone (I) from 2-methoxy-*p*-cresol :**

Thus, **2-methoxy –*p*-cresol** undergoes *four* sequential reactions as stated under :

(i) *Methylation* of **2-methoxy-*p*-cresol** with dimethyl sulphate [(CH$_3$)$_2$ SO$_4$] in the presence of KOH to give **3,4-dimethoxy-toluene.**

(ii) **3,4-Dimethoxy toluene** upon nitration with HNO$_3$ yields the respective **3,4-dimethoxy-6-nitro-toluene.**

(iii) Reduction of **3,4-dimethoxy-6-amino toluene** with *Sn/HCl* gives **3,4-dimethoxy-6-amino toluene.**

(iv) Oxidation with FeCl$_3$ of **3,4-dimethoxy-6-amino toluene** gives the desired **compound (I).**

Thus, we may have the following reactions :

| 2-Methoxy-*p*-cresol | Dimethyl Sulphate | 3,4-Dimethoxy-toluene | 3,4-Dimethoxy-6-nitro toluene |

4-Methoxy-2,5-toluquinone [I]   3,4-Dimethoxy-6-amino toluene

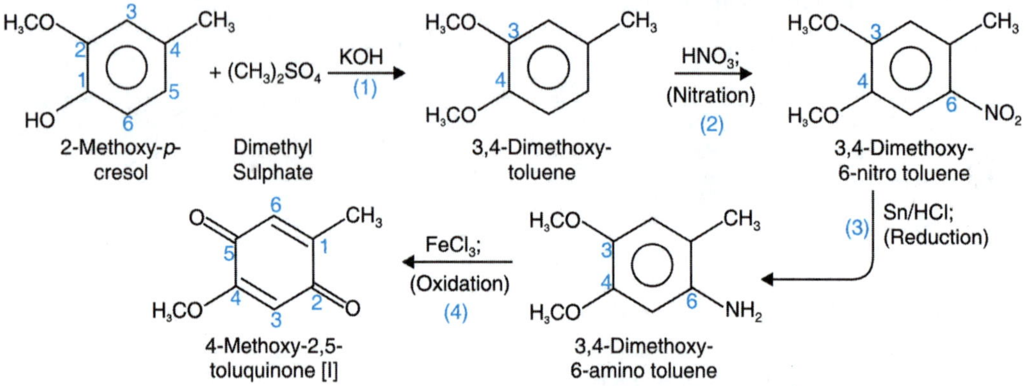

Steps-2 : Formation of 3-*keto*-1,7-diene-4-hydroxy-10-methyl naphthalene (V) from 4-methoxy-2,5-toluquinone (I) [*obtained from Step-1*] :

Following are the various important steps that are involved in the above mentioned reactions :

(i)   Isomerization of **4-methoxy-2,5-toluquinone (I)** to obtain **4-methoxy-3,6-toluquinone,** which upon interaction with *butadiene* yields the *cis*-**3-methoxy-1,4-diketo-2,7-diene-10-methyl naphathalene (II)** (*i.e*, H-atom at C-5 has an (**β-substitution** and viewed as *projecting above the ring plane*).

(ii) Resulting **compound (II)** undergoes geometrical isomerism under the influence of **alkali (OH–)** and **acid (H+)** to produce the corresponding *trans*-**3-methoxy-1,4-diketo-2,7-diene-10-methyl naphthalene (III)** (*i.e*, H-atom at C-5 has an **α-substitution** and viewed as *projecting beneath the ring plane*).

(iii) **Compound (III)** upon treatment with LiAlH$_4$ gives **3-methy 1,4- dihyproxy-2,7-diene-10-methyl naphthalene (IV).**

(iv) Finally, **compound (IV)** on reaction with HCl loses a mole of water to produce **3-*keto*-1,7-diene -4-hydroxy-10-methyl naphthalene (V)** *i.e*, the desired compound.

Thus, we may have the following reactions :

| (I) 4-Methoxy-2,5-toluquinoline | (I) 4-Methoxy-3,6-toluquinoline [**Isomerized product**] | Butadiene | (II) *eis*–3–Methoxy–1,4–diketo–2,7–diene–10–methyl naphthalene |

(V)   (IV)   (III)

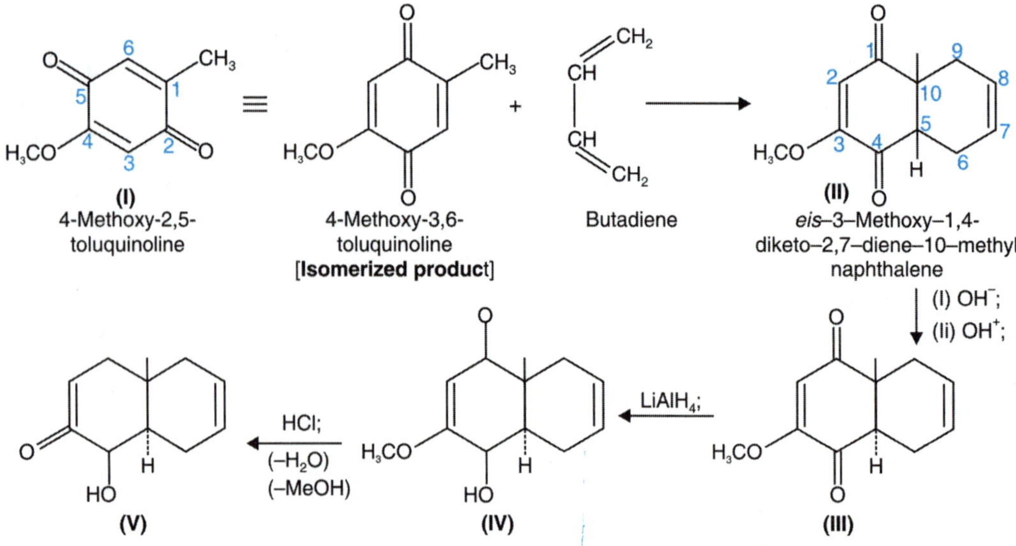

Step -3 : **Preparations of 4-Hydroxy-methylene derivative (XIII) from 3-***keto***-1,7-diene-4-hydroxy-10-methyl naphthalene (V)** :

The following sequential steps are involved in the above mentioned reactions :

(i) Acetylation of **compound (V)** with *acetic anhydride* followed by heating with *Zn in acetic anhydride* gives rise to the formation of **ketol acetate derivative (VI)**.

(ii) The resulting **compound (VI)** on being treated with *ethyl formate (HOCO-C_2H_5)* in the presence of *sodium methoxide* (**CH_3ONa**) undergoes **claisen condensation** to produce the respective **hydroxyl methylene ketone (VII)** derivative.

(iii) **Compound (VIII)** when subjected to treatment with *ethyl vinyl ketone* [**C_2H_5-COCH=CH$_2$**] in the critical presence of **potassium** *tert* -**butoxide** [**(CH$_3$)$_3$COK**] undergoes **Michael condensation** to give **4-aldehyde-4-diethyl ketone derivative (VIII)**.

(iv) **Compound (VIII)** upon treatment with *KOH in dioxan* undergoes **cyclization** almost quantitatively to give the **single product (IX)**.

(v) **Product (IX)** when treated with *osmium tetroxide* [OSO_4], gives *two* **moles of** *cis*-**glycols (X)**. However, one of these *two* **glycols** happens to be *cis*-with respect to the **angular methyl (CH$_3$) moiety;** whereas, the second one has the *trans*-**configuration** (*geometrical isomers*).

(vi) The aforesaid *two* **isomers** (*viz, cis*-and *trans*-) are duly separated; and the *specific isomer* **insoluble in benzene** is treated with *acetone* in the presence of *anhydrous copper sulphate* [$CuSO_4$] to yield the respective **isopropylidene derivative (XI)**.

(vii) The **compound (XI)** on being subjected to reduction with H_2-*Pd* and *strontium carbonate* [$SrCO_3$] gives the corresponding saturated **isopropyl derivative (XII)** by *selective reduction*.

(viii) Finally, **compound (XII)** upon careful condensation with *ethyl formate* [$HCO_2.C_2H_5$] in the presence of *sodium methoxide* [CH_3ONa] gives the desired product **4-hydroxy-methylene derivative (XIII)**.

All the *eight* elaborated sequential reactions may be expatiated as stated under :

Compound (V)   (Acetylation)

Keto acetate derivative (VI)

Hydroxy methylene ketone derivative (VII)

[Michael Condensation]

(I) C_2H_5CO CH$=$CH$_2$
Ethyl vinyl ketone
(Ii) (CH$_3$)$_3$ COK
Pot. *tert*-butoxide

Single product (X)

KOH;
[In Dioxan]
(Cyclization)

4-Aldehyde-4-diethyl ketone derivative (VIII)

(Contd.....)

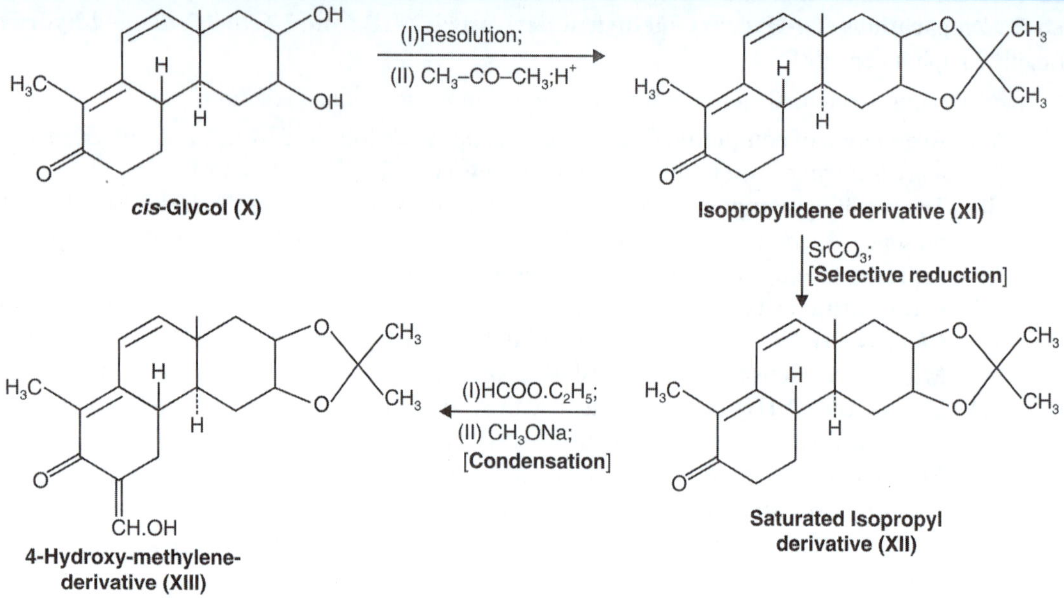

cis-Glycol (X) → (I)Resolution; (II) CH₃–CO–CH₃;H⁺ → **Isopropylidene derivative (XI)**

SrCO₃; [Selective reduction]

4-Hydroxy-methylene-derivative (XIII) ← (I)HCOO.C₂H₅; (II) CH₃ONa; [Condensation] ← **Saturated Isopropyl derivative (XII)**

Step-4 : Preparation of 3-*keto*-4,9,16-triene-17-aldehyde steroid (XX) from 4-hydroxymethylene derivative (XIII) (obtained from Step-3) :

The *six* established sequential steps involved in the synthesis of **compound (XX)** from **compound (XIII)** are duly elaborated as under :

(i) **4-Hydroxy-methylene derivative (XIII)**, as obtained in *Step-3*, on being treated with *methylaniline* [C_6H_5-NH.CH_3] gives **4-methylanilino-methylene isopropyl derivative (XIV)**.

Comment – In fact, the above important step almost proves to be an *absolute necessity* so as to block the so-called **undesired condensation reactions** specifically on this side of the *keto* (>C=O) **moiety** located at *C-3 position.*

(ii) **Compound (XIV)** upon condensation with *vinyl cyanide* (CH_2=CH.CN) and followed immediately by *hydrolysis in an alkaline environment (OH⁻)*, gives an admixture of *two keto*-**acids,** which are separated carefully by *resolution* under *acetic anhydride* [($CH_3CO)_2O$] and *sodium acetate* [CH_3CONa]. Importantly, out the *two keto*-**acids**-the *stereoisomer* (**XV**); wherein, the **methyl (CH_3) moiety** is located *above the plane*, (**solid line**) and the *propionic acid* (CH_3-CH_2-COOH) moiety positioned *beneath the plane of rings* (**dotted line**) gets duly convertd into the respective *enol* **lactone (XVI)**.

(iii) **Compound (XVI)** when reacted with *methyl magnesium bromide* (CH_3-Mg.Br)-a **Grignard Reagent** gives a *keto*-**ethyl methyl ketone isopropylidene derivative (XVII)**.

(iv) **Compound (XVII)** on being treated with an *alkali (NaOH)* undergoes *cylization* to obtain a **cyclic-*keto* isopropylidene derivative (XVII)**.

(v) Subsequently, **compound (XVIII)** on *oxidation* with *periodic acid (HIO₄) in dioxan* gives rise to the formation of a **dialdehyde ketone derivative (XIX)**.

(vi) Finally, on heating **compound (XIX)** in *pure* **benzene solution** in the presence of a small quantum of redistilled (pure) **piperidine** [⟨NH⟩] to produce the desired product *3-keto-***4,9, 16-triene-17-aldehyde steroid (XX)**.

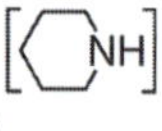

Thus, we may have the following sequential reactions :

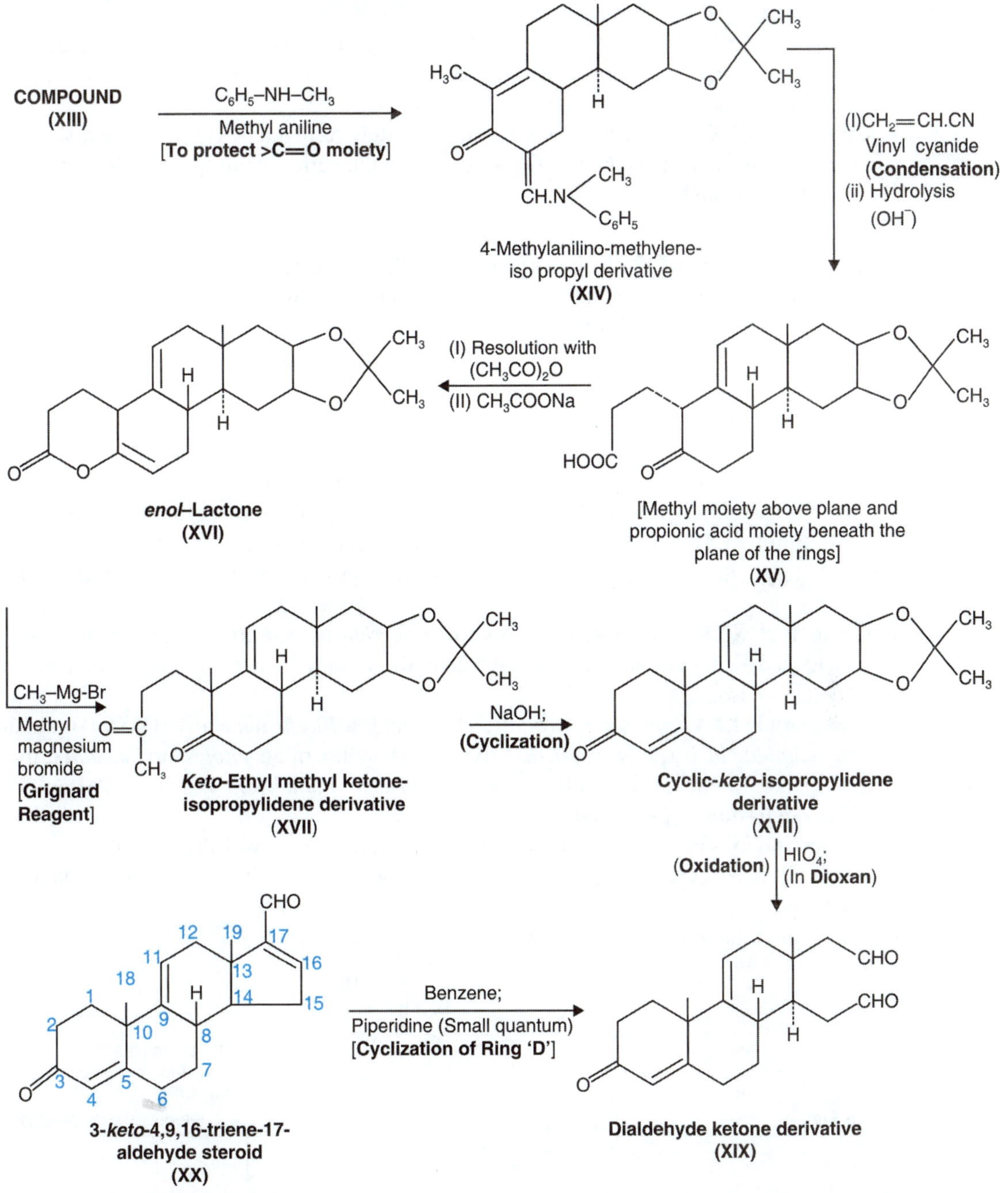

COMPOUND
(XIII)

$\xrightarrow[\substack{\text{Methyl aniline} \\ [\text{To protect} >\text{C}=\text{O moiety}]}]{C_6H_5-NH-CH_3}$

4-Methylanilino-methylene-
iso propyl derivative
(XIV)

(I)CH$_2$=CH.CN
Vinyl cyanide
(**Condensation**)
(ii) Hydrolysis
(OH$^-$)

enol–Lactone
(XVI)

$\xleftarrow[\substack{(II) \text{ CH}_3\text{COONa}}]{\substack{(I) \text{ Resolution with} \\ (\text{CH}_3\text{CO})_2\text{O}}}$

[Methyl moiety above plane and
propionic acid moiety beneath the
plane of the rings]
(XV)

CH$_3$–Mg–Br
Methyl
magnesium
bromide
[**Grignard
Reagent**]

Keto-Ethyl methyl ketone-
isopropylidene derivative
(XVII)

$\xrightarrow[\textbf{(Cyclization)}]{\text{NaOH;}}$

Cyclic-*keto*-isopropylidene
derivative
(XVII)

(**Oxidation**) $\Big| \substack{\text{HIO}_4; \\ \text{(In } \textbf{Dioxan})}$

3-*keto*-4,9,16-triene-17-
aldehyde steroid
(XX)

$\xrightarrow[\substack{\text{Piperidine (Small quantum)} \\ [\textbf{Cyclization of Ring 'D'}]}]{\text{Benzene;}}$

Dialdehyde ketone derivative
(XIX)

**Step-5 : Preparation of Cholestanol [or 5α-Cholestan-3β-ol] (XXVII) from Compound (XX)
(obtained from Step-4) :**

The following *eight* sequential reactions are duly undertaken to obtain the desired **compound (XXVIII),** namely :

(i)   **Compound (XX),** obtained from **Step-4**, on being subjected to *oxidation* by *potassium dichromate* ($K_2Cr_2O_7$) followed by treatment with *diazomethane* (CH_2N_2) yields **3-keto-** 4,9,16-triene-17-methyl carboxylate steroid (**XXI**).

(ii) **Compound (XXI),** being a *racemate mixture*, is duly resolved by careful *reduction so as to convent the* respective **keto (>C=O) moiety** to the corresponding **keto-esters** *viz.,*
   • **(+)-3α-alcohol** and
   • **(+)-3β-alcohol.**

Thus, the resulting **(+)-form of 3β-alcohol** is precipitated almost preferentially by means of **digitonin** (a *known precipitiation substance for steroidal compounds*). The above mentioned form upon *oxidation* gives the desired *stereoisomer* **(+) – (XXI).**

(iii) **Compound (+) – (XXI)** when *reduced catalytically* (**using H₂/Pt**) yields **3-hydroxy-** 17-methyl carboxylate steroid (**XXII.**)

(iv) **Compound (XXII)** upon *Oppenauer Oxidation* in the presence of *aluminium butoxide* [*Al (Bu)₃*] *in acetone* gives a racemate mixture of **3-keto-derivative (XXIII).**

(v)   The racemate, mixture of **compound (XXIII)** is separated reduced by NaBH₄ (sodium borohydride) and hydrolyzed to give **3-hydroxyl 17-carboxylate steroid (XXIV).**

(vi) **Compound (XXIV)** is subsequently *acetylated*, treated with *thionyl chloride (SOCl₂)*, and finally with *dimethyl cadmium [Cd(CH₃)₂]* to produce **acetylated hydroxyketone (XXV).**

(vii) **Compound (XXV)** when treated with *isohexyl magnesium bormide [(CH₃)₂.CH(CH₂)₃.Mg.Br].* a **Grignard Reagent,** yields **3-acetyloxy-20-hydroxy derivative (XXVI),** which being an admixture of **isomers.**

(viii) **Compound (XXVI)** upon *dehydration (-H₂O)* yields **a 20-ene derivative (XXVII),** which upon *catalytic hydrogenation (H₂/Pt)* gives an admixture of **5α-cholestenyl acetates,** that are separated meticulously to obtain the *desired isomer* upon *hydrolysis* **5α-cholestan-3β-ol (or Cholestanol) (XXVIII).**

(ix) **Compound (XXVIII)** is found to be having a *close resemblance* with the naturally occurring '**cholesterol**'. Therefore, it is converted into cholesterol by means of *five* sequential reactions.

Thus, we may have the following series of reactions :

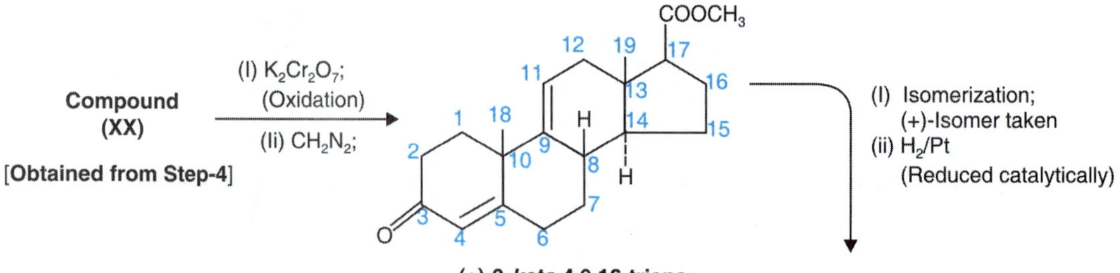

Compound (XX)

[Obtained from Step-4]

(I) K₂Cr₂O₇;
(Oxidation)
(II) CH₂N₂;

(+)-3-*keto*-4,9,16-triene-
17-methyl carboxylate steroid
(XXI)

(I) Isomerization;
(+)-Isomer taken
(II) H₂/Pt
(Reduced catalytically)

(Contd....)

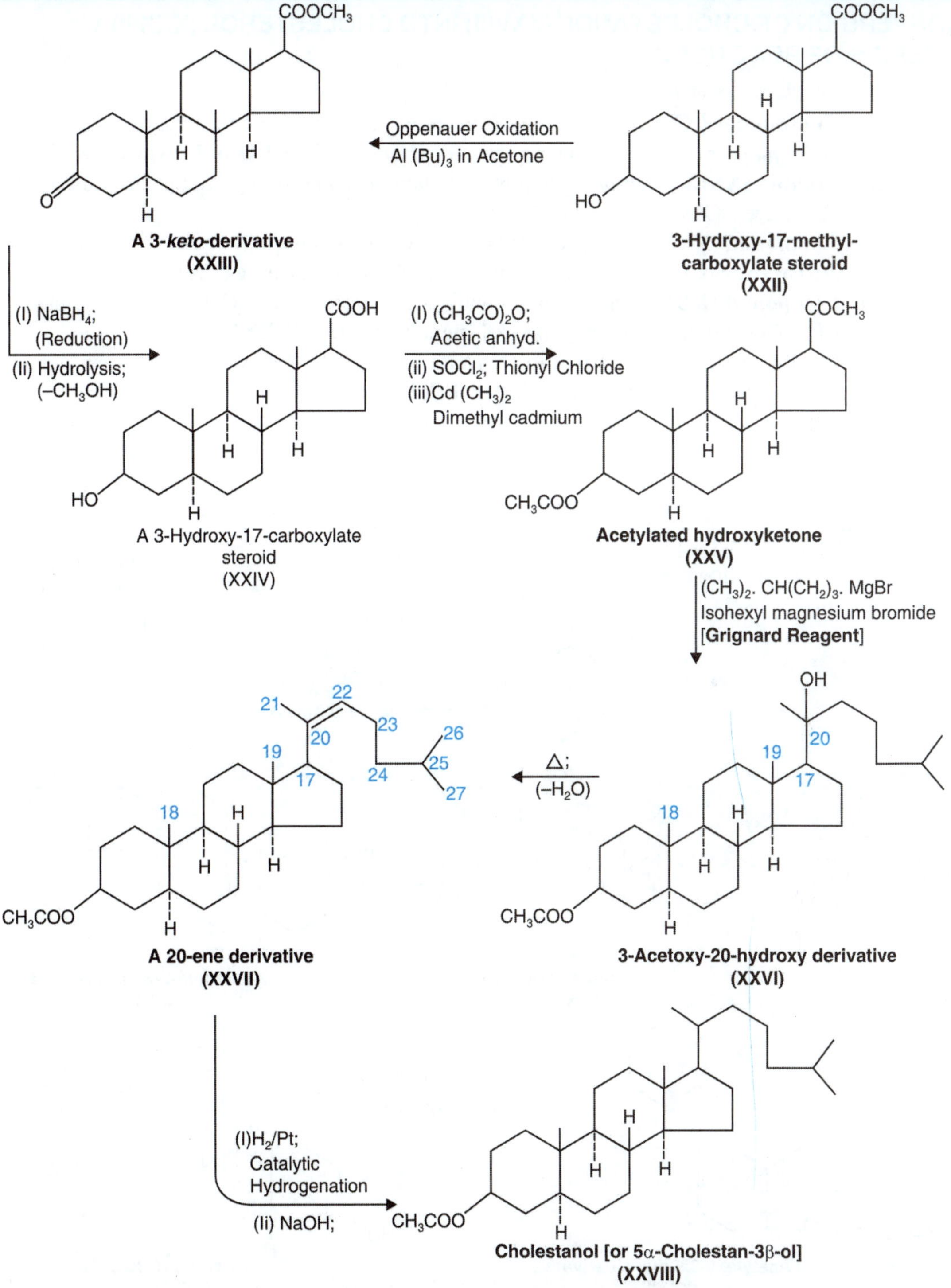

COOCH₃

Oppenauer Oxidation
Al (Bu)₃ in Acetone

A 3-*keto*-derivative
(XXIII)

COOCH₃

HO

3-Hydroxy-17-methyl-
carboxylate steroid
(XXII)

(I) NaBH₄;
(Reduction)
(Ii) Hydrolysis;
(–CH₃OH)

COOH

HO

A 3-Hydroxy-17-carboxylate
steroid
(XXIV)

(I) (CH₃CO)₂O;
Acetic anhyd.
(ii) SOCl₂; Thionyl Chloride
(iii)Cd (CH₃)₂
Dimethyl cadmium

COCH₃

CH₃COO

Acetylated hydroxyketone
(XXV)

(CH₃)₂. CH(CH₂)₃. MgBr
Isohexyl magnesium bromide
[Grignard Reagent]

21   22
20   23   26
19   24   25
17   27
18

CH₃COO

A 20-ene derivative
(XXVII)

Δ;
(–H₂O)

OH

20
19
18   17

CH₃COO

3-Acetoxy-20-hydroxy derivative
(XXVI)

(I)H₂/Pt;
Catalytic
Hydrogenation

(Ii) NaOH;

CH₃COO

Cholestanol [or 5α-Cholestan-3β-ol]
(XXVIII)

CONVERSION OF CHOLESTANOL (XXVIII) INTO CHOLESTEROL (XXXIII) *via* A SERIES OF REACTIONS

It essentially involves *five* sequential reactions, namely

(i) **Cholestanol (XXVIII)**, obtained from **Step-5**, is subjected to *oxidation* with s*odium dichromate (Na$_2$Cr$_2$O$_7$)* in the presence of *H$_2$SO$_4$* to yield **3-*keto* derivative (XXIX)**.

(ii) **Product (XXIX)** on *bromination* in *hydrobromic acid (HBr)* gives the **2α-brome-3-*keto*** derivative (XXX).

(iii) **Compound (XXX)** on reflux with *freshly distilled pyridine* eliminates a mole of HBr (*i.e*, **2α-Br and 5α-H**) thereby producing **3-*keto* 4- ene derivative (XXXI)**.

(iv) **Compound (XXXI)** on acetylation with acetyl chloride (CH$_3$COCl) in acetic anhydride [(CH$_3$CO)$_2$O] yields **3-acetyloxy-3,5-diene derivative (XXXII)**.

(v) Finally, reduction of **compound (XXXII)** with LiAlH$_4$ followed by treatment with *HCl* results into the formation of **cholesterol (XXXIII)**.

All these aforesaid *five* **reactions** may be expatiated as under :

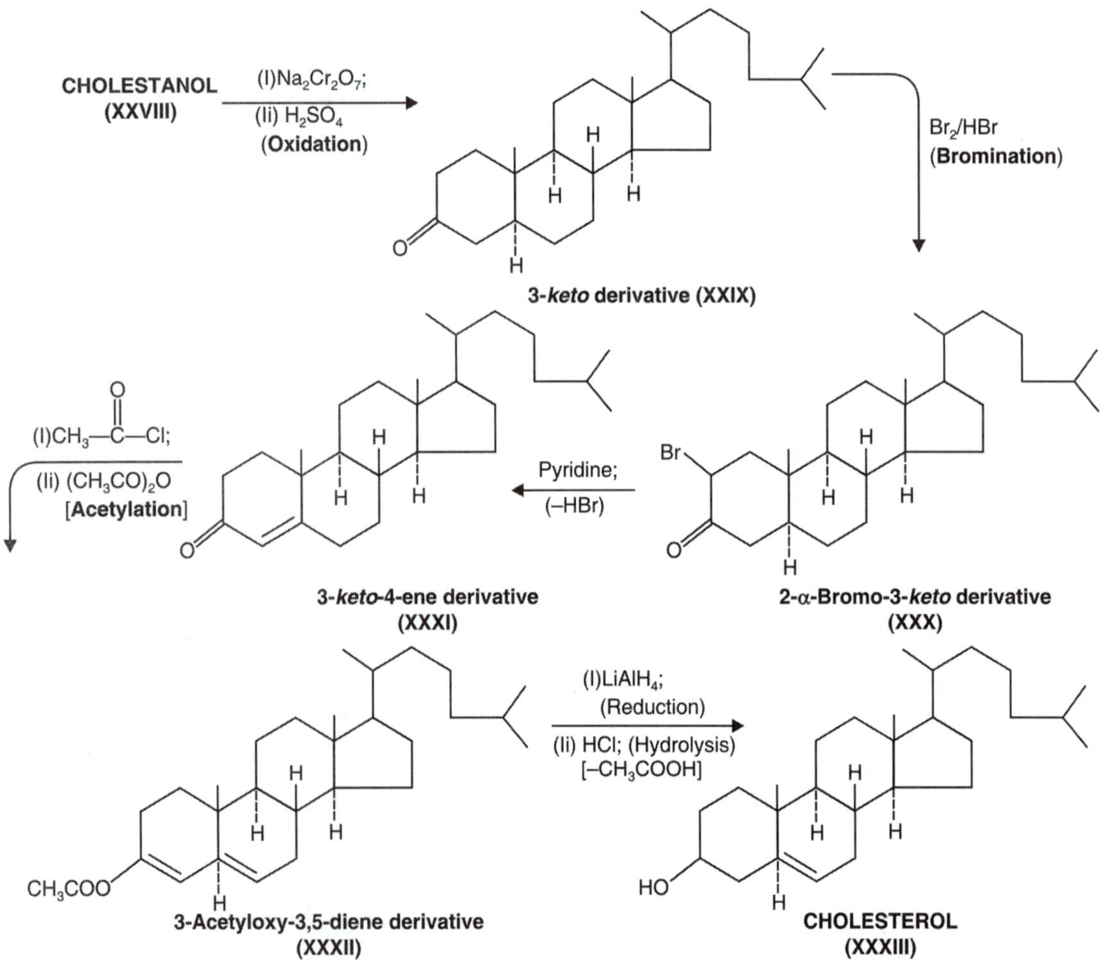

CHOLESTANOL (XXVIII) — (I)Na$_2$Cr$_2$O$_7$; (Ii) H$_2$SO$_4$ (Oxidation) →

3-*keto* derivative (XXIX)

Br$_2$/HBr (Bromination)

(I)CH$_3$—C—Cl; (Ii) (CH$_3$CO)$_2$O [Acetylation]

Pyridine; (–HBr)

3-*keto*-4-ene derivative (XXXI)

2-α-Bromo-3-*keto* derivative (XXX)

(I)LiAlH$_4$; (Reduction) (Ii) HCl; (Hydrolysis) [–CH$_3$COOH] →

CH$_3$COO

3-Acetyloxy-3,5-diene derivative (XXXII)

HO

CHOLESTEROL (XXXIII)

Based on the various chemical reactions involved in **Step-1** through **Step-5** together with the *five* sequential reactions engaged in the *conversion* of **cholestanol (XXVIII)** into **cholesterol (XXXIII)** establishes the *total synthesis* of **cholesterol.**

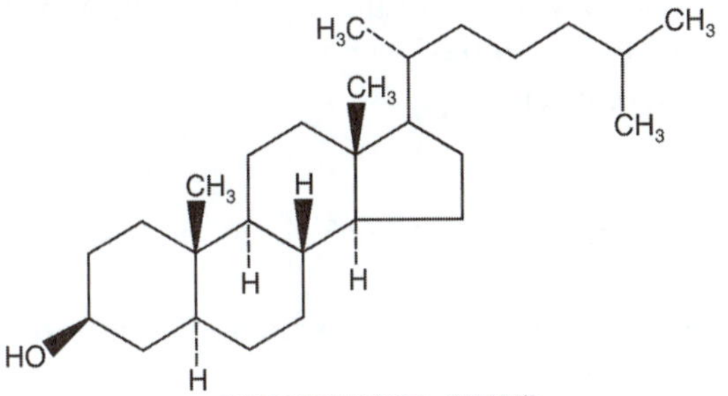

CHOLESTANOL (XXVIII)
[(3β,5α)-Cholestan-3-ol]

- $C_{27}H_{48}O$
- MP 141.5–142°C
- $[\alpha]_D^{22}$ + 24.2° (c=1.3 in CHCl$_3$)
- Precipitated by digitonin
- 1g dissolves in~100mL EtOH
- Freely soluble in hot EtOH, Ether and Chloroform.

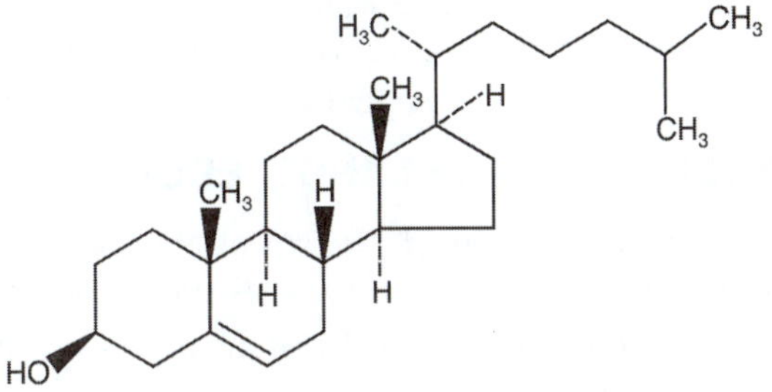

CHOLESTEROL (XXXIII)
[(3β)-Cholest-5-en-3-ol]

- $C_{27}H_{46}O$
- MP 148.5°C (anhydrous product)
- $[\alpha]_D^{20}$ + 31.5° (c=2 in Ether)
- $[\alpha]_D^{20}$ + 39.5° (c=2 in Chloroform)
- 1g dissolves in 2.8mL Ether, 4.5 mL Chloroform, and 1.5mL Pyridine.
- Precipitated by digitonin
- soluble in benzene, Pet. Ether, Oils and Fats

5.2. Lanosterol [$C_{30}H_{50}O$]

Preamble – It has been duly observed that there exists *two* **alcohols** critically present in the so-called *'nonsaponifiable fraction'* of **'wool fat'** together with **cholesterol'** and, therefore, the names do reflect remarkable similarity in the characteristic features to the 'sterols'.

Obviously, **Lanosterol** essentially comprises *two* **distinct double bonds,** namely:

- *one double bond* **(in the side-chain) exhibiting normal reactivity,** and
- *second double bond* **(in the steroidal nucleus) showing absolute invertness to** *hydrogenation reaction.*

Nevertheless, the **second alcohol** known as **'Agnosterol'** designates a *dehydroderivative* of **Lanosterol;** and hence, contains a **typical conjugated** *diene*-system-as given under :

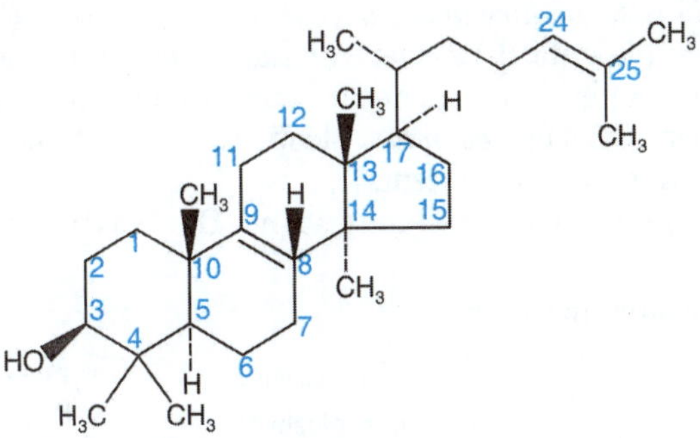

Lanosterol
MP : 139°C
α_D^{20} + 62°

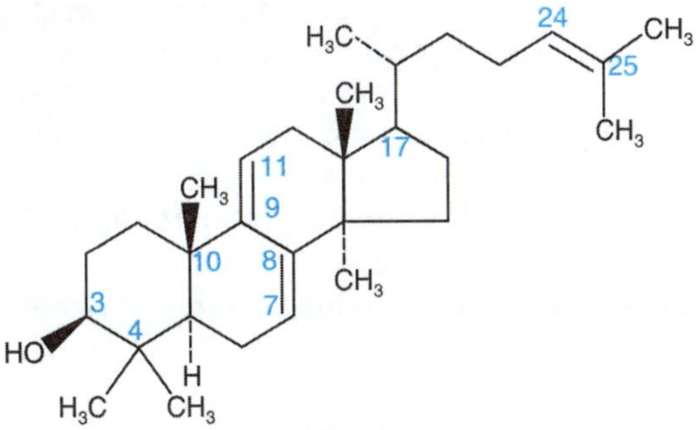

Agnosterol
MP : 165°C
α_D^{20} + 62°; λ_{max} 2400 Å

Interestingly, one important feature of these '**triterpenes**' is that the said *structures* do not necessarily obey the '*isoprene rule*'; and, therefore, these *two* **alcohols** may be rightly regarded as :

4,4,14- trimethyl substituted C27-sterols.

Based on the above findings and observations one may arrive at the logical conclusion that '**Lanosterol**' does have a close and intimate relationship to the '*sterols*' with respect to :

- • **chemical structure, and**
- • **stereochemistry.**

CONSTITUTION OF LANOSTEROL

The **constitution of lanosterol** may be adequately proved and ascertained on the basis of the following *eight* **sequential steps** as stated under :

1. **Molecular Formula** –Based on the *elemental analysis*, the **molecular formula** of '**lanosterol**' is found to be $C_{30}H_{50}O$.

2. **Presence of 2- Double Bonds** – Lanosterol essentially possesses **2-double bonds** of which :
   - • *one* **active** and hence, *reducible by hydrogen (H₂)*, and
   - • *second* **inactive (or inert)**–due to its presence in the **steroidal nucleus** (**precisely at C-8 & C-9**).

3. **Oxidation of Lanosterol produces a mole each of Acetone and Aldehyde-**Thus, the careful *oxidation* of **lanosterol** gives a mole each of **acetone** and **aldehyde**.

Inference – The above reaction indicated that **lanosterol** prominently comprises a **dimethyl** ethylene $[CH=C(CH_3)_2]$ group.

4. **Normal Reactions of Alcoholic (-OH) Moiety Lanosterol** gives the **normal reactions of the alcoholic moiety** present in it.

   Inference- The above observation explicitly shows that *oxygen* is present as an '**alcohol**' in **lanosterol** (*at C-3*).

5. **Formation of Lanostenone (A) from–Lanosterol *via* Reduction followed by Oxidation-Lanosterol** first on *reduction (H₂)* yields **lanostenol** which upon *oxidation (O₂)* gives **lanostenone (A)** as given under:

$$C_{30}H_{50}O \xrightarrow[\text{(Reduction)}]{H_2;} C_{30}H_{52}O \xrightarrow[\text{(Oxidation)}]{O_2;} C_{30}H_{50}O$$
$$\text{Lanosterol} \qquad\qquad \text{Lanostenol} \qquad\qquad \text{Lanostenone}$$

Comments – These essentially include :

   (i) The particular formation of **lanostenone (A)** suggests predominantly that the alcoholic (OH) moiety duly present in **lanosterol** is *secondary in nature*.

   (ii) **IR-Spectral Studies of (A)** – Based on the critical **IR-spectral studies of lanostenone (A)** suggests strongly that this **product (A)** is almost identical to the one obtained from **cyclohexanone**. Thus, it reveals prominently that the *secondary* alcoholic (**-CHOH**) moiety in **lanosterol** is duly present in a **6-membered ring system.**

   (iii) The ketone *i.e*, lanostenone (A) forms a Monobenzylidine Derivative with Benzaldehyde

Thus, we may have the following *three* known sequential reactions :

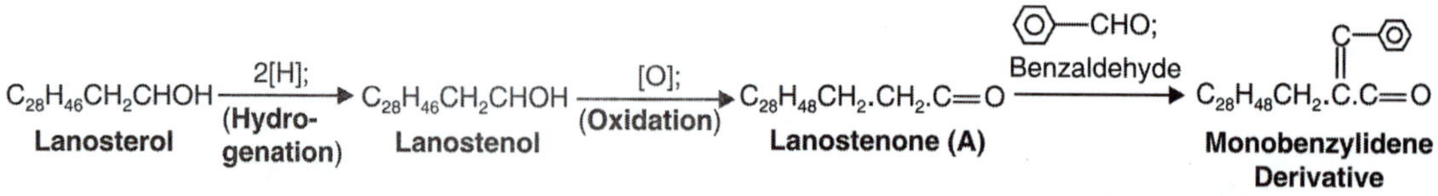

$$C_{28}H_{46}CH_2CHOH \xrightarrow[\text{(Hydro-genation)}]{2[H];} C_{28}H_{46}CH_2CHOH \xrightarrow[\text{(Oxidation)}]{[O];} C_{28}H_{48}CH_2.CH_2.C{=}O \xrightarrow[\text{Benzaldehyde}]{\bigcirc\!-CHO;} C_{28}H_{48}CH_2.C.C{=}O$$
$$\text{Lanosterol} \qquad\qquad \text{Lanostenol} \qquad\qquad \text{Lanostenone (A)} \qquad\qquad \text{Monobenzylidene Derivative}$$

Remarks : The above reactions do reveal that the *ketone* **Lanostenone (A)** contains an **active methylene *keto* moiety (–CH₂CO-)**; and, therefore, **lanosterol** does contain the **methyl hydroxymethyl (-CH₂CHOH-) moiety.**

6. **Conversion of Lanosterol into Isopropylidenecyclopentane (B)** *via* **Nametkin's Rearrangement-** The *cyclohexanol ring* of **lanosterol** on being treated with *phosphorus pentachloride (PCl₅)* gets converted into the corresponding **isopropylidene cyclopentane (B)** *via* the **Nametkin's rearrangement.**

The resulting **compound (B)** when duly treated *with osmium tetroxide [OSO₄]* immediately followed by *oxidation* with **periodic acid [HIO₄]** gives rise to the formation of a mole each of **acetone** and cyclopentanone derivative (C).

Thus, based upon the **Nametkin's rearrangement** one may write the above reactions sequentially as under :

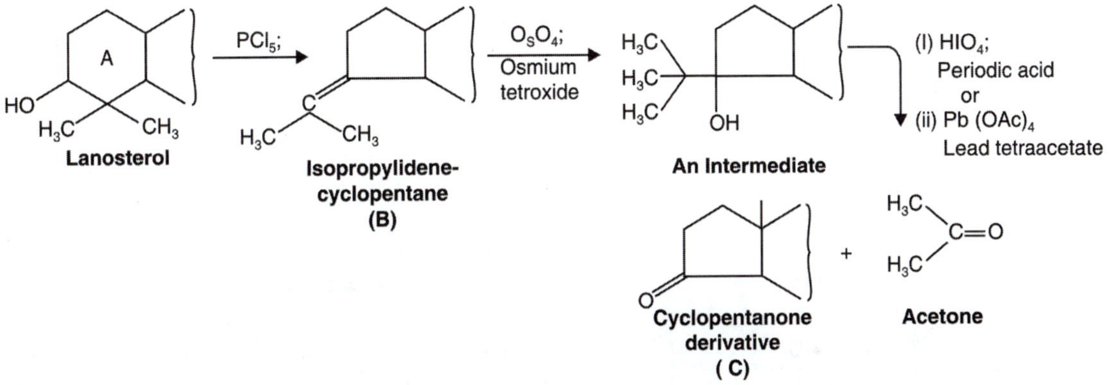

Remarks: These essentially include:

(i) The aforesaid reactions explicitly demonstrate that a *gem*-**dimethyl moiety** is attached strategically on the **C-atom** linked to **secondary alcohol (-CHOH-) group**; and, therefore, the **isopropyl group [–CHOH.C(CH₃)₂]** is duly present in **lanosterol**; and hence, in **lanosterol** it self.

(ii) However, it has already been ascertained in the *earlier segment in Step -6* that the prevailing **alcoholic (OH) group** is duly flanked by **methylene (-CH₂-) moiety.** Hence, it confirms that the **hydroxyl (OH) group** present in **lanosterol** is obviously present as given below :

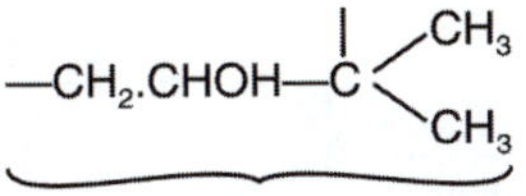

$$-CH_2.CHOH-C\overset{\displaystyle |}{\underset{\displaystyle CH_3}{\overset{\displaystyle CH_3}{<}}}$$

---

* Nametkin SS : *Ann*, 432 : 207 (1923). **Nametkins Rearrangement:** It relates to a special case of **carbonium ion rearrangement** in *camphene hydrochloride derivatives* involving the **migration of a methyl (CH₃) moiety** :

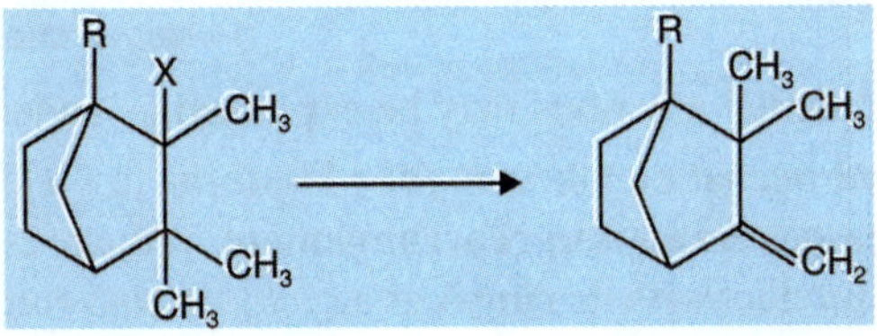

7. Lanosterol and cholesterol Has almost *identical C-skeleton*. In fact, this glaring fact has been duly established on the basis of the following observations namely:

   (a) **Preparation of 14- Methylcholestanol (X) by the conversion of:**
   - **Lanosterol,** or
   - **Cholesterol.**

   (b) **Preparation of Dihydrolanosterol [or Lanostanol] by the conversion of Cholesterol.**

   All the above *three* conversions shall now be treated individually in the sections that follows:

[A-1] Preparation of 14- Methylcholestanol (X) by the Conversion of Lanosterol

Following are the *nine* stepwise reactions starting from **lanosterol** to obtain **14-methylcholestanol (X) :**

Lanosterol   Lanostanol

14-Methyl cholestanol (X)

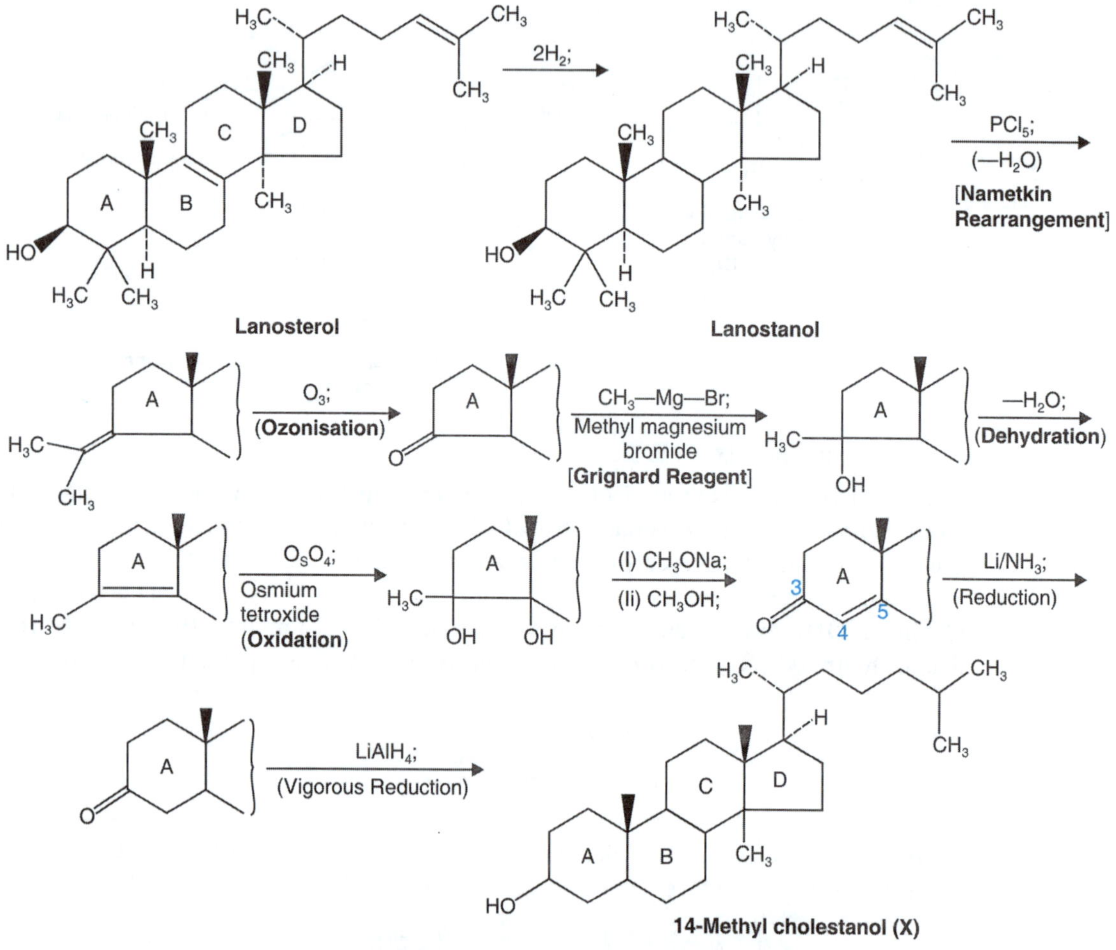

Explanation : All the *nine* sequential reactions may be explained as under :

   (i) **Lanosterol** on hydrogenation takes up *four* **H-atoms** to yield **lanostanol.**

   (ii) **Lanostanol** undergoes **Nametkin rearrangement** in the presence of PCl_5 to lose a mole of water to give a product with its **ring A** as a *cyclopentano ring* and a **dimethyl methylene** group at C-3.

(iii) The resulting product on *ozonization (O₃)* yields a *3-keto derivative*, which upon *Grignardization* with CH₃-Mg.Br gives a **3-methyl-3-hydroxy derivative**.

(iv) The product obtained from (iii) above upon *dehydration (-H₂O)* gives **3-methyl-3-ene derivative**, which on *oxidation* with *osmium tetroxide (OSO₄)* yields a **3-methyl-3,4-dihydroxy derivative**.

(v) The end-product in (iv) when treated with freshly prepared *sodium methoxide (CH₃ONa|)* in *methanol* gives rise to the formation of **3-keto-4-ene-derivative** (and **Ring 'A'** gets converted to a **6-membered ring system.**

(vi) The resulting product undergoes *reduction* with *Li/NH₃* to produce a **3-keto derivative**, which when subjected to vigorous reduction with **lithium-aluminium hydride (LiAlH₄)** yields the targeted product **14-methyl-Cholestanol (X).**

[A-2] Preparation of 14-Methylcholestanol (X) by the Conversation of Cholesterol

Following are the *nine* stepwise reactions that are actually engaged in the conversion of **Cholesterol** into **14-Methyl-cholestanol (X) :**

Explanation : All the *nine* stepwise reactions which are involved in the conversion of **choleseterol** into **14-methyl cholestanol (X)** may be explained as stated under:

(i) **Cholesterol** on being treated with *N-bromosuccinimide* undergoes *bromination* which on reaction with β-*collidine* loses a mole of *HBr* to yield a **5,7-diene derivative.**

(ii) The resulting **5,7-diene** when reacted with *HCl/CHCl₃* helps to migrate one double bond from **C-5/6 to C-14/15 position** by producing a **7,14-diene derivative.**

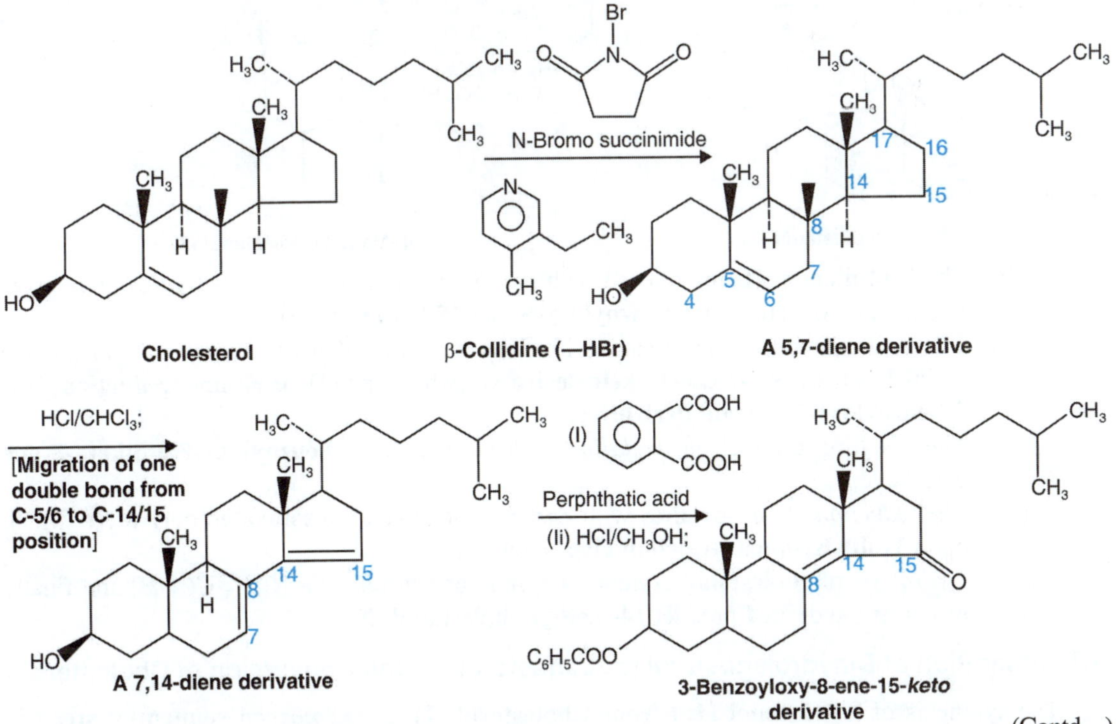

Cholesterol   β-Collidine (—HBr)   A 5,7-diene derivative

A 7,14-diene derivative   3-Benzoyloxy-8-ene-15-*keto* derivative

(Contd....)

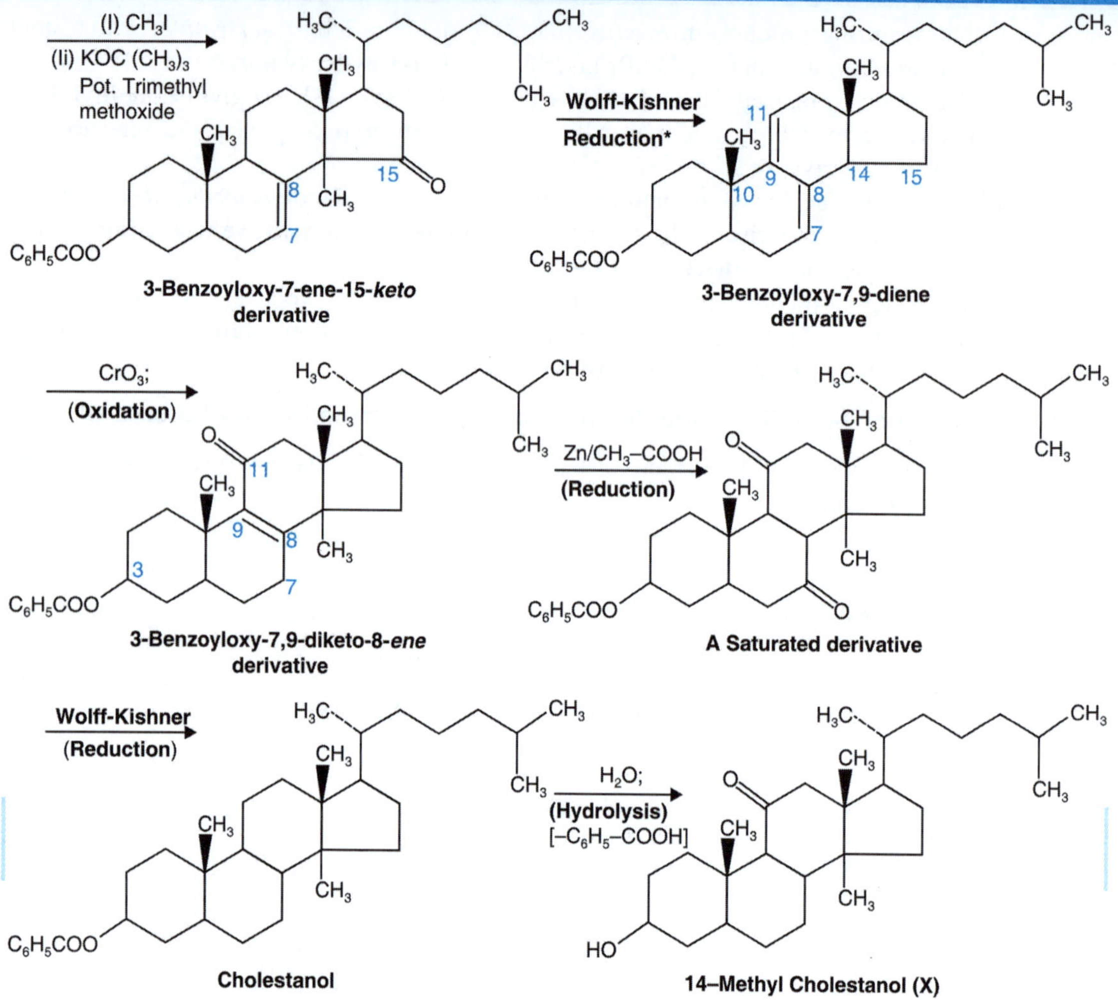

3-Benzoyloxy-7-ene-15-*keto* derivative

3-Benzoyloxy-7,9-diene derivative

3-Benzoyloxy-7,9-diketo-8-*ene* derivative

A Saturated derivative

Cholestanol

14–Methyl Cholestanol (X)

(iii) The **7,14-diene** on being reacted with *perphthalic acid* followed by HCl/CH_3OH gives rise to the production of **3-benzoyloxy-8-ene-15-keto derivative.**

(iv) The resulting product when treated with CH_3I and then with *potassium trimethyl methoxide* yields **3-benzoyloxy-7-ene-15-keto derivative**, which upon **Wolf-Kishner reduction** gives **3-benzoyloxy-7,9-diene derivative.**

(v) The resulting product on *oxidation with CrO_3* gives **3-benzoyloxy-7,9-diketo-8-ene** derivative.

(vi) The *reduction* of **end-product** with Zn/CH_3COOH yields a **saturated derivative**, which upon **Wolff-Kishner reduction** gives **cholestanol.**

(vii) *Hydrolysis* of **cholestanol** eliminates a mole of *benzoic acid* $[C_6H_5\text{-}COOH]$ and finally produces the desired product **14-methyl-cholestanol (X).**

[B] Preparation of Dihydrolanosterol [or Lanosterol] by the Conversion of Cholesterol

The synthesis of **Lanostanol (15)** from Cholesterol **(1)** takes *fourteen* sequential steps as elaborated under :

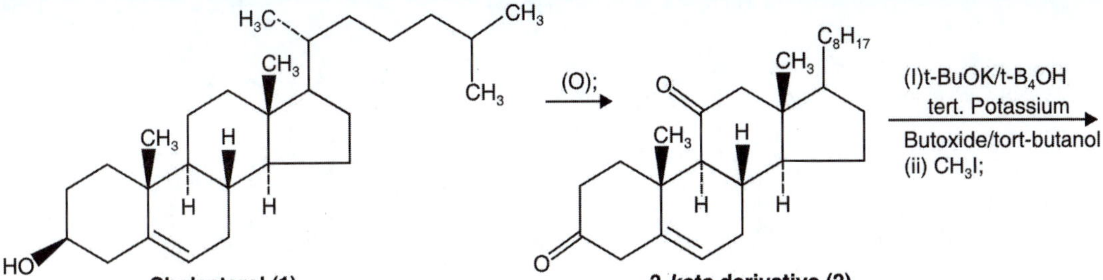

Cholesterol (1)

(O);

3-*keto*-derivative (2)

(I)t-BuOK/t-B₄OH
tert. Potassium
Butoxide/tort-butanol
(ii) CH₃I;

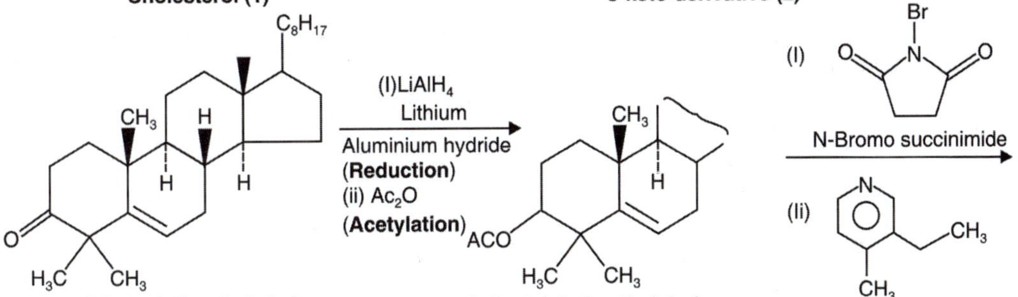

3-keto-4-dimethyl deriv. (3)

(I)LiAlH₄
Lithium
Aluminium hydride
(**Reduction**)
(ii) Ac₂O
(**Acetylation**)

3-Acetyl-4-dimethyl deriv. (4)

(I)

N-Bromo succinimide

(Ii)

(β-Collidine) (–HBr)

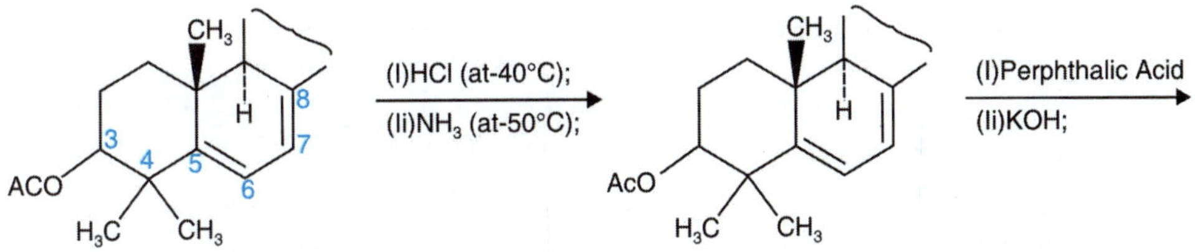

3-Acetyl-4-dimethyl-5,7-diene deriv. (5)

(I)HCl (at-40°C);
(Ii)NH₃ (at-50°C);

7-ene-deriv. (6)

(I)Perphthalic Acid
(Ii)KOH;

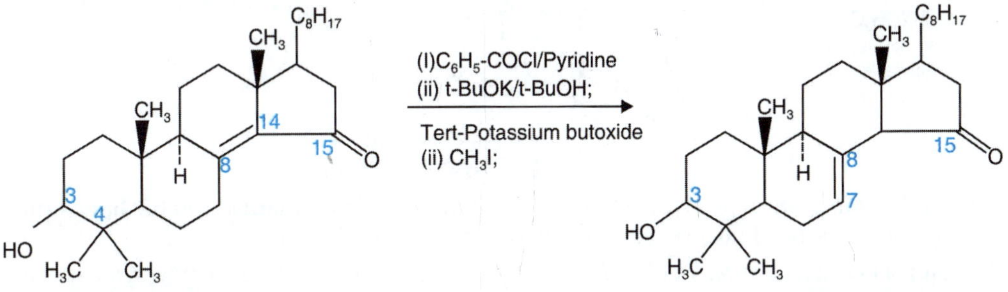

3-Hydroxy-4-dimethyl-8-ene-15-*keto* derivative (7)

(I)C₆H₅-COCl/Pyridine
(ii) t-BuOK/t-BuOH;

Tert-Potassium butoxide
(ii) CH₃I;

3-Hydroxy-4-dimethyl-7-ene-15-*keto* derivative (8)

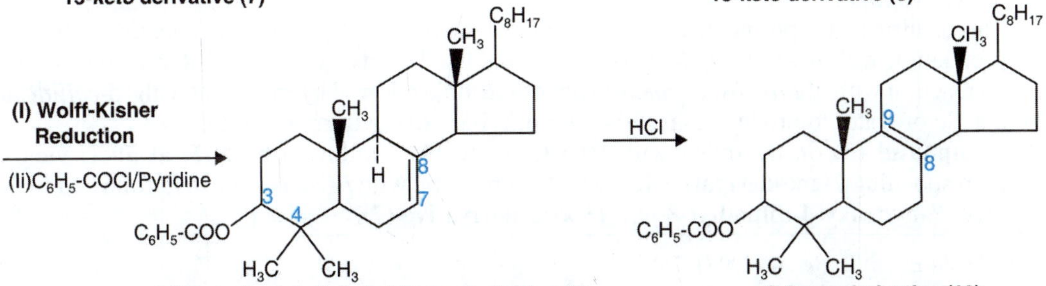

(I) **Wolff-Kisher Reduction**

(Ii)C₆H₅-COCl/Pyridine

3-Benzoyloxy-4-dimethyl-7-ene-derivative (9)

HCl

A 8-ene derivative (10)

(Contd....)

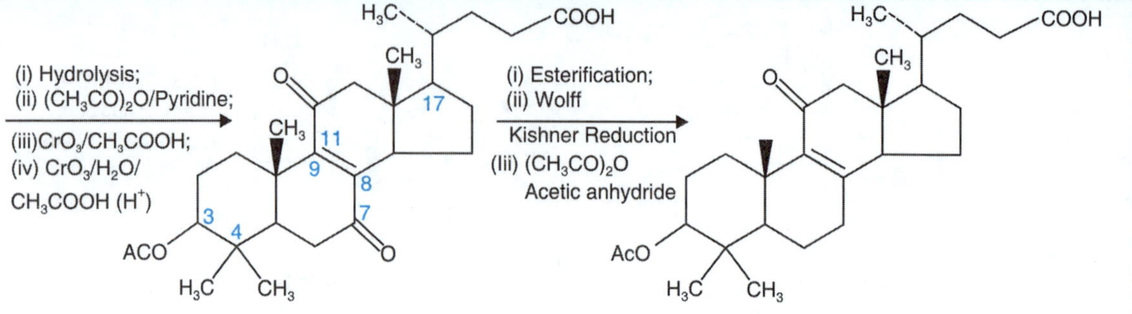

(i) Hydrolysis;
(ii) (CH₃CO)₂O/Pyridine;
(iii)CrO₃/CH₃COOH;
(iv) CrO₃/H₂O/
CH₃COOH (H⁺)

H₃C···COOH
CH₃
17
CH₃
11
9 8
7
3 4
ACO
H₃C CH₃

A 8-ene derivative (10)

(i) Esterification;
(ii) Wolff
Kishner Reduction
(iii) (CH₃CO)₂O
Acetic anhydride

H₃C···COOH
CH₃
AcO
H₃C CH₃

A 11-*keto* derivative (12)

(i) NaOH;
(ii) (COCl)₂
Oxalyl Chloride;
(ii) **Arndt-Eistert
Synthesis***

H₃C···COOH₃
17
9
8

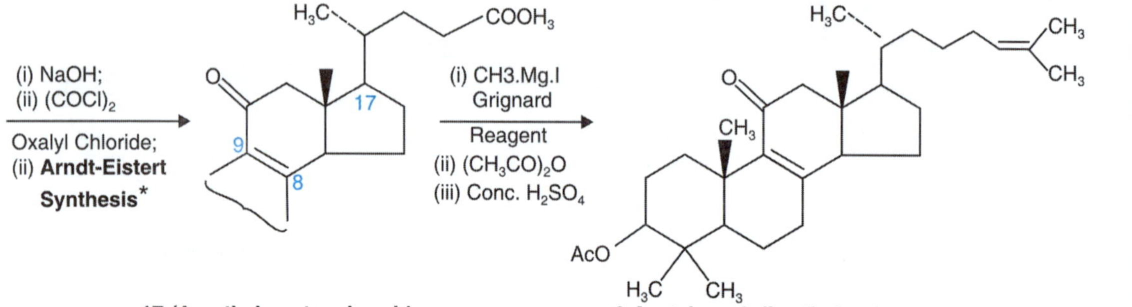

**17-(4-methyl pentanoic acid
methyl ester (13)**

(i) CH3.Mg.I
Grignard
Reagent
(ii) (CH₃CO)₂O
(iii) Conc. H₂SO₄

H₃C···CH₃
CH₃
CH₃
AcO
H₃C CH₃

**3-Acetyloxy-4-dimethyl-
11-*keto* derivative (14)**

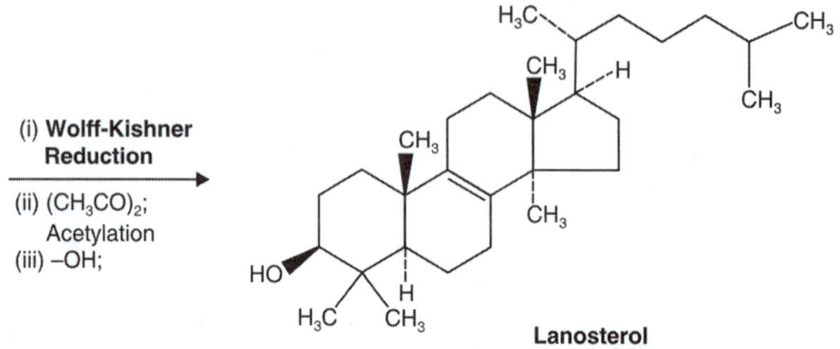

(i) **Wolff-Kishner
Reduction**
(ii) (CH₃CO)₂;
Acetylation
(iii) –OH;

H₃C···CH₃
CH₃ ··H
CH₃
CH₃
CH₃
HO
H
H₃C CH₃

**Lanosterol
[or Dihydrolanosterol] (15)**

Explanation : The *fourteen* steps involved in the preparation of **dihydrolanosterol (or lanostanol)** by the conversion of **cholesterol** (as the starting material) may be explained as under :

(1) **Cholesterol (1)** on *oxidation* yields **3-*keto*-derivative (2)**, which on treatment with *tert*-potassium butanol in *tert*-butanol followed by reaction with **methyl iodide (CH₃I)** gives **3-*keto*-4-dimethyl** derivative **(3)**.

(2) The resulting **compound (3)** on being treated with *LiAlH₄* undergoes *reduction* followed by acetylation with *acetic anhydride* [AC₂O] gives the **3-acetyl-4-dimethyl derivative (4)**, which on reaction with *N-bromosuccinimide* (**bromination**) followed by treatment with (β-*collidie* loses a mole of water to produce **3-acetyl-4-dimethyl-5,7 diene derivative (5)**.

(3) **Compound (5)** on treatment with **HCl (gas) at-40°C** followed by **NH₃ at 50°C** yields the corresponding **7-ene-derivative (6)**, which on reaction with *perphthallilc acid* followed by KOH gives **3-hydroxy-4-dimethyl-8-ene-15-*keto* derivative (7)**.

* Arndt- Eistert B: *Ber*: **68**:200 (1935).

(4) **Compound (7)** when treated with *benzoyl chloride in pyridine*, followed by ***tertiary*-potsassium butoxide** in *tert*-**butanol** and finally with CH_3I gives **3-hydroxy-4-dimethyl-7-ene-15-keto.** **derivative (8),** which upon *Wolff-Kishner reduction* followed by *benzoyl chloride in pyridine* yields **3-benzoyloxy-dimethyl-7-ene derivative (9).**

(5) **Compound (9)** on treatment with HCl yields **8-ene-derivative (10)** which on *hydrolysis, acetylation, oxidation* and *vigorous oxidation in an acidic medium* produces **3-acetyl-4-dimethyl-7,11-diketo 17-(β-methyl butyric acid)** derivative (11).

(6) **Compound (11)** upon *esterification – Wolff-Kishner reduction* and *acetylation* gives a **11-*keto* derivative (12),** which on treatment with an *alkali,* followed by *oxalyl chloride* and finally **Arndt-Eistert synthesis** yields **17-(4-methyl pentanoic acid methyl ester (13) .**

(7) **Compound (13)** when reacted with *methyl magnesium bromimde* (a **Grignard Reagent**), followed by *acetic anhydride in dry pyridine* and finally *conc. H_2SO_4* gives **3-acetyloxy-4-dimethyl-11-keto derivative (14).**

(8) Finally, **Compound (14)** on being subjected to treatment with *Wolff-Kishner reduction, acetylation* and – OH gives rise to the formation of the desired compound **Lanostanol (or** Dihydrolanosterol) (15) .

*Points to Ponde*r : These essentially include :

(i) The critical presence of the *second double bond* has been duly shown to be located between **C-8** and **C-9** in the respective *steroidal nucleus* as could be seen from the oxidation of Lanosterol.

(ii) In addition, the *extra* **methyl (CH₃) moiety** located strategically at **C-14** by the *dehydrogenation* of **lanosterol** to **1,2,8-trimethylphenanthrene, as shown under.**

Therefore, the complete chemical structure of **lanosterol (15)** will be as given below :

Lanosterol
(15)

1,2,8-Trimethyl phenanthrene

SYNTHESIS OF LANOSTEROL

The *total synthesis* of '**Lanosterol**' may be studied elaborately under the following *two* **specific heads,** namely :

❑ **Synthesis of '*Cholesterol*',** and

❑ **Conversion of '*Cholesterol to Lanosterol*'.**

5.3. Stigmasterol [$C_{29}H_{48}O$]

Preamble – Importantly, the '**sterols**' belonging to the *animal origin* (or **Zoosterols**), such as :

- **Cholesterol**   • **β-Cholestanol** and
- **Coprostanol**

do essentially possess almost the '**identical skeletal structure of 27-C atoms**' precisely.

On the contrary, the so-called principal '**plant sterols**' (or '**Phytosterols**'), for instance :

- **Ergosterol** and   • **Stigmasterol**

do critically possess either C_{28}-or C_{29}-**compounds.**

Obviously, the aforesaid naturally occurring substances predominantly inherit the following *typical features*, namely :

- ❑ They have both the **3α-hydroxyl moiety** and the **C-5/C-6 double bond**-characteristic of *cholesterol*;
- ❑ They differ from '**Cholesterol**' in having a *side-chain* with a double bend between **C-22/C-23 (trans-) configuration** (*i.e*, **geometrical isomerism**); and
- ❑ Presence of an '**extra alkyl group : methyl and ethyl moiety**' respectively at **C-24.**

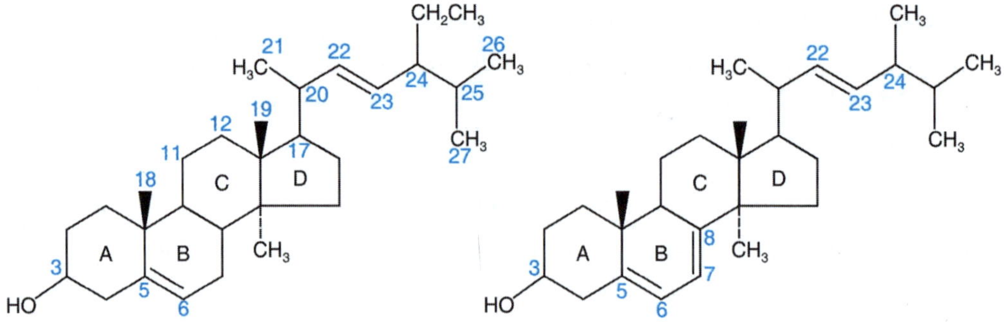

Stigmasterol
[mp : 170°C [α]$_D$–45°]

Ergosterol
[mp : 170° ; [α]$_D$–45°]

Stigmasterol is usually isolated from the **phytosterol mixture*** either from :

- **Calabar Beans**, or   • **Soybeans.**

Besides, **stigmasterol** is widely distributed in plants, which could be present either in the form of :

- **free state**, or   • **glycosides,**

Interestingly, stigmasterol is invariably isolated as the *sparingly soluble* **acetate tetrabromide derivative**. It is mostly available as perfect white crystalline solid, mp 170°C. It is found to be optically active and its specific rotation $[α]_D^{20}$ -55.6° (C=2 in chloroform).

CONSTITUTION OF STIGMASTEROL

The constitution of '**Stigmasterol**' may be deduced in a mehtodical manner and established as given under :

(1) **Molecular Formula** – Based on the analytical data the *molecular formula* of **stigmasterol** has been found to be $C_{29}H_{48}O.$

---

*   Thornton et al.: *J Am Chen Soc*. **62** : 2006, 1940

(2) **Presence of One Hydroxyl (-OH) Moiety** – Since, **stigmasterol** gives rise to the formation of a *'monoacetate'* – it explicitly shows the presence of **one hydroxyl (-OH) Moiety.**

(3) **Nature of the Hydroxyl (-OH) Moiety**– It has been observed that **stingmasterol** upon *oxidation* gives a *'ketone'* – which vividly ascertains that the **hydroxyl (OH) group** is a *secondary* alcoholic in nature (*i.e*, RR'CH.OH).

(4) **Presence of 2 Double Bonds** – *Bromination* of **stigmasterol** yields a **tetrabromide derivative**, thereby it indicates that **stigmasterol** essentially comprises 2- double bonds.

(5) **Presence of the Steroidal Nucleus** – **Stigmasterol** on being distilled with **Se** gives **Diel's Hydrocarbon,** which suggests strongly that a *steroidal nucleus* is duly present in the former. *Two* **Important Observations** – These essentially include :

  ❏ *Hydrogenation of Stigmasterol* – produces **stigmastanol [$C_{29}H_{52}O$]**, which critically confirms the presence of *two* **double bonds** in stigmasterol (*i.e*, **2- double bonds take up 4 H-atoms**).

  ❏ *Molecular Formula of Stigmasterol the parent hydrocarbon* $C_{29}H_{52}$- that corresponds to the *general formula* **CNH_2N-6** *i.e*, for a **'tetracyclic compound'**. Therefore, it ascertains vehemently that **stigmasterol** is a **tetracyclic chemical entity (compound)** *i.e*, it possesses a *four* **ring system** in it.

(6) **Position of the Hydroxyl (OH) Moiety and the Side-Chain**-In fact, the exact **position** and **configuration** (*i.e*, α-or β-**position**) of the *hydroxyl (OH) moiety* was duly confirmed by carrying out the careful *hydrogenation* to the respective **5α-stigmastanol** ; and subsequently, its *acetylation* followed by *oxidation* with Cr_2O_3 *(chromium trioxide)* to **3β-hydroxy-nor-5α-cholanic acid.** Based on this particular reaction one may safely conclude that –'**stigmasterol specifically differs from the so-called 5-cholestan-3β-ol only with respect to the side-chain'.**

Hence, it reveals clearly that **C-3 position** of the *steroidal nucleus* is for the **hydroxyl (-OH) moiety.**

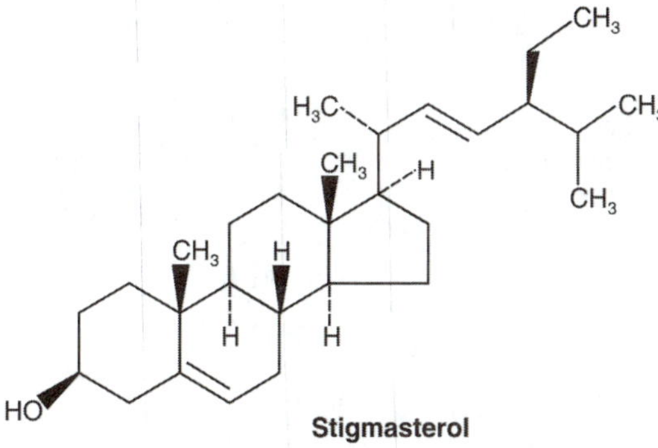

Stigmasterol

Conversion of Stigmastanyl Acetate into Acetate of 3α-Hydroxy-nor-5α-Cholanic acid :

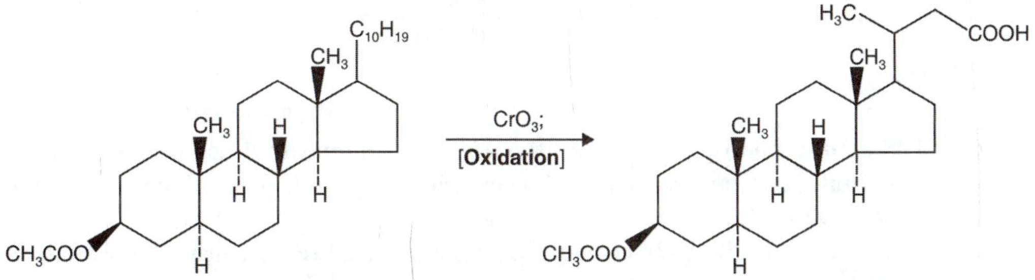

Stigmastanyl Acetate   **Acetate of 3β-Hydroxy-nor-5α-Cholanic acid**

Ozonolysis of Stigmasterol gives Ethyl-isopropyl acetaldehyde (I)

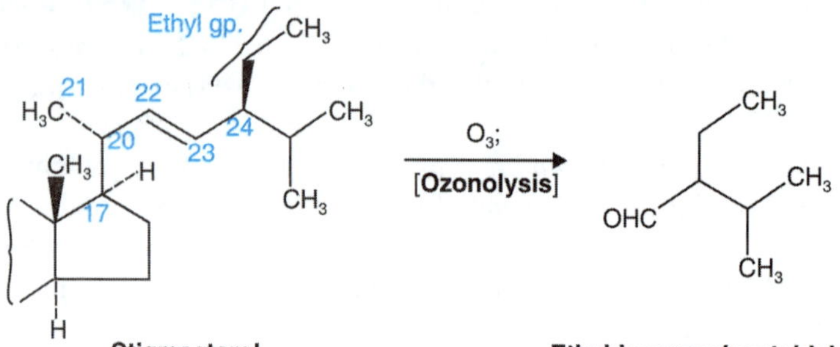

Stigmasterol   Ethyl-isopropylacetaldehyde (I)

Comment – The critical formation of **product (I)** establishes the presence of a **C-24 ethyl (C_2H_5)** **moiety** having a double bond between **C-22/C-23 positions.**

(7) **Exact Position of the Nuclear Double Bond [*i.e*, in the Steroidal Nucleus] :** The so-called 'nuclear double bond' has been proved to be positioned between **C-5/C-6** based on the same procedure as applicable for 'cholesterol' (see *section 5.1*)

Following are the *four* **sequential step**s that are usually followed in ascertaining the exact position of the *nuclear* **double bond** present in 'stigmasterol' as detailed under :

(i) *Hydroxylation* of **stigmasterol** with *hydrogen peroxide (H_2O_2) in acetic acid* gives a **triol**, which further upon *oxidation* with *chromium trioxide (CrO_3)* yields a **hydroxydi ketone (X).** Subsequently, the **compound (X)** on being subjected to *dehydraton* followed immediately by *reduction* with *Zn/CH_3COOH* produces a **dione (Y)**, which eventually *combines* with *hydrazine (N_2H_4)* results into the formation of a respective **pyridazine (Z).**

Thus, we may express all the aforesaid *four* stepwise reactions as given under :

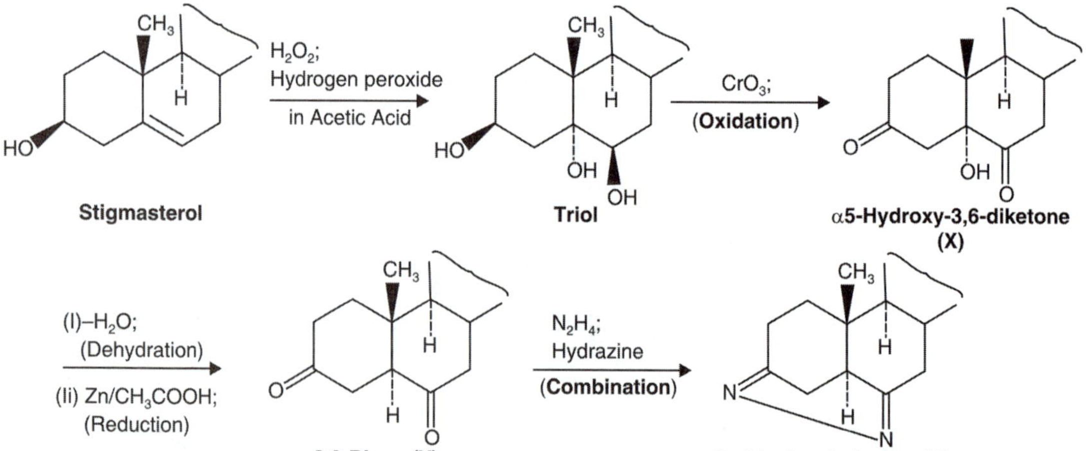

Stigmasterol   Triol   α5-Hydroxy-3,6-diketone
(X)

3,6-Dione (Y)   Pyridazine derivative (Z)

(ii) The **double bond** present in the *side-chain* of **stigmasterol** has been duly assigned a *trans***configuration** (*geometrical isomer*) which is genuinely based upon its **IR-Spectral band at 970m^{-1}.**

(iii) The presence of the **C-24 ethyl (-C_2H_5) moiety** has been adequately proved to be α-*as expatiated below* :

➤ The *ozonized product* gives rise to the formation of an **aldehyde (I)**, which was duly converted by means of *reduction-tosylation-reduction* to the respective *levorotatory* **hydrocarbon (II)**. Amazingly **compound (II)** had been earlier *satisfactorily correlated* with **(+)-isopropylsuccinic acid (III)**. Thus, we may have the following expression :

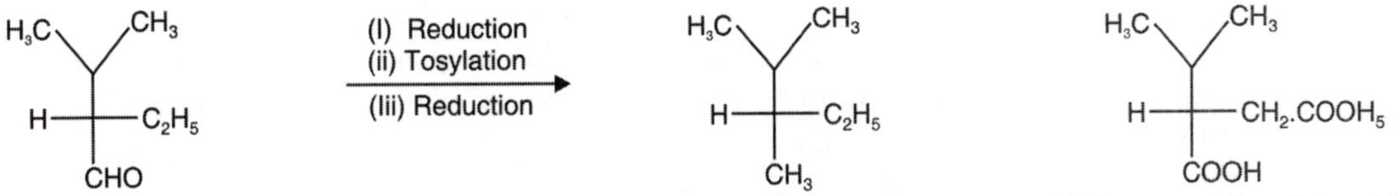

An Aldehyde **(I)**	**A *levo*-rotatory Hydrocarbon** **(II)**	**(+)–Isopropyl-Succinic acid** **(III)**

(I) Reduction
(ii) Tosylation
(Iii) Reduction

Now, based upon the above glaring scientific factual statements 'stigmasterol' has been duly assigned the chemcial structure as given under **item (6)** earlier.

5.4. Ergosterol [$C_{28}H_{44}O$]

Preamble : **Ergosterol** designates the most important member of the **provitamis D**. It is usually obtained from **yeast** which synthesizes it from the *simple sugars e.g,* **glucose**. However, the so-called **damp yeast** yields nearly **2.5%** ergosterol, that eventually depends upon the variety of the yeast which being an extremely important criterion.

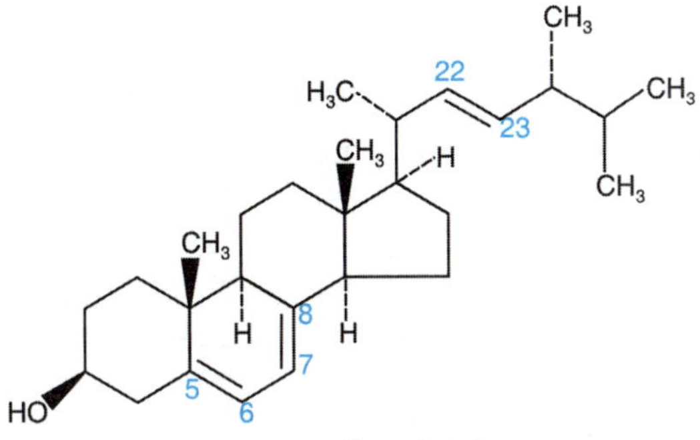

Ergosterol

Important Points – These essentially include :

1. **Ergosterol** possesses a *second* **nuclear double bond** which being critically conjugated with the *first* **nuclear double bond** *i.e,* **C-5/C-6 and C-7/C-8 are conjugated.**

Thus the resulting 'diene system' gives rise to :

- **a characteristic UV-absorption spectrum profile,** and
- **engosteol is duly differentiated from both cholesterol as well as stigmasterol.**

2. **Unsaponifiable Fraction-Ergosterol** is invariably found in the **unsaponifiable fraction from 'yeast'**. Besides, the actual quantum being so *small* and also the substances are generally accompanied by the *related sterols*; and, therefore, to lay hands onto the *'pure material'* is not only cumbersome but also quite expensive.

---

* **Rickets :** A condition due to **vitamin D** deficiency, especially in *infancy, child-hood,* and *disturbance of normal ossification.*

3. **Ergosterol as Provitamin D$_2$** – In 1920$_s$ it was recognized beyond any reasonable doubt that the **rickets*** generally showed a favourable response to the specific treatment either :

- **with direct 'sun-light', or**
- **due supplementation of regular diet with 'Cod Liver Oil,**

However, initially it was probably thought that **cholesterol** could be the '**provitamin**', which was duly converted into the **vitamin D**. Amazinly, it was demonstrated subsequently the '**ergosterol**' happens to be the **provitamin**. Besides, the **provitamin nature of *cholesterol*** was, in fact, caused by virtue of the *contamination* with small quantum of 'ergosterol'. Ultimately, it was overwhelmingly accepted and recognized that '**Ergosterol**' is nothing but a **provitamin D$_2$**.

CONSTITUTION OF ERGOSTEROL

Windaus and co-workers, in Germany, studied **ergosterol** both intensively and extensively in 1927; and named the *pure active product* as **Vitamin D$_2$**.

Irradiation Products from Ergosterol – It has been established that irradiation process leads to a *succession of produces* e.g, **Lumisterol**, **Tachysterol**, and **Vitamin D$_2$** as given in the following expression :

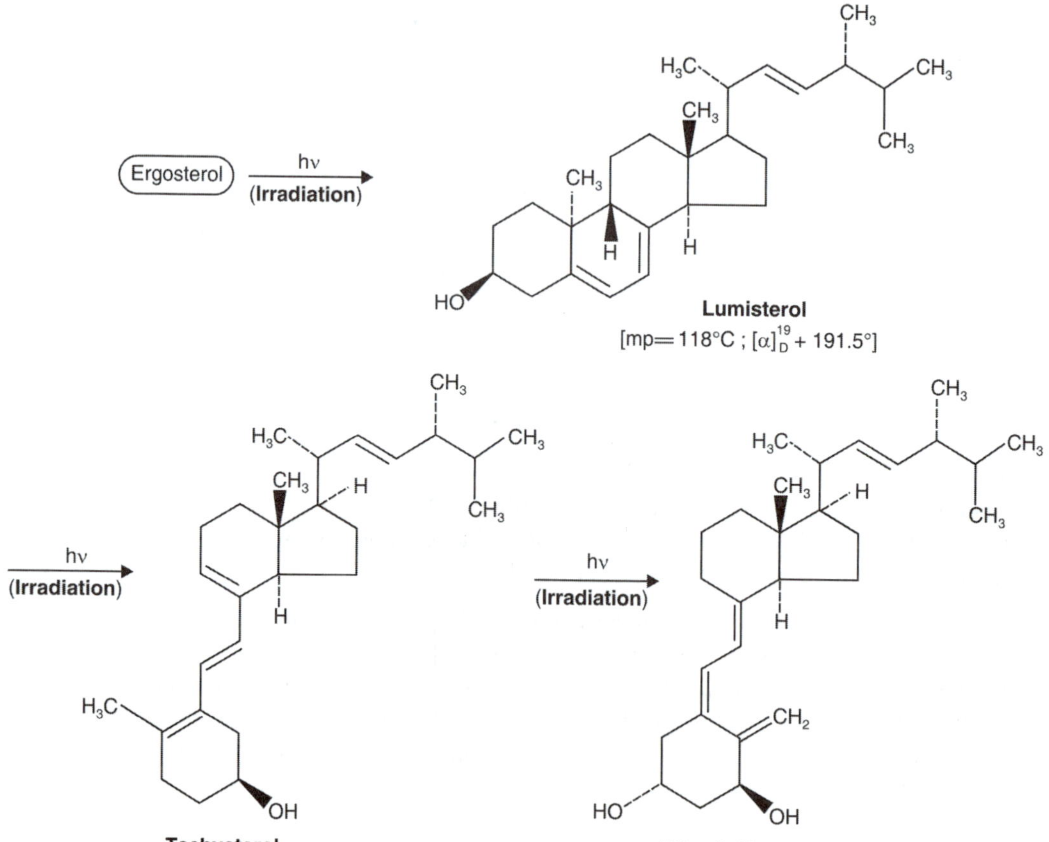

Lumisterol
[mp = 118°C ; $[\alpha]_D^{19}$ + 191.5°]

Tachysterol
$\left[[\alpha]_D^{19} -70° \text{ (24.6mg in 2mL Pet. Ether)} \atop UV_{max} : 280 \text{ nm} \right]$

Vitamin D$_2$
[mp = 115-118°C ; $[\alpha]_D^{25}$+ 82.6°
(c=3 in Acetone]

Importantly, the *constitution* of Ergosterol may be elucidated properly by adopting the following cardinal aspects in a sequential manner :

1. **Molecular Formula** – Based on the analytical data obtained from pure **ergosterol** the *molecular formula* has been found to be $C_{28}H_{44}O$.

2. **Presence of One Hydroxyl (-OH) Moiety** – Ergosterol usually forms the *mono*-esters with *acid derivatives*, such as :

   • **Anhydrides** *viz*, acetic anhydride $[(CH_3CO)_2O]$, and
   • **Acid Chlorides** *viz*, acetyl chloride $[CH_3COCl]$, benzoyl chloride $[C_6H_5COCl]$.

Thus, it strongly reveals the presence of *only* **one hydroxyl moiety** in *ergosterol*.

3. **Presence of 3- Double Bonds in Ergosterol** – **Ergosterol** upon *hydrogenation* with **Pt** gives rise to the formation of 'ergostanol' $[C_{28}H_{50}O]$ *i.e*, it takes up **6 H-atoms (or 3 H_2)**

Ergosterol
$[C_{28}H_{44}O]$

Pt;
[Hydrogenation] →

Ergostanol*

One may note clearly that the **double bond** existing between **C-5/C-6 ; C-7/C-8; and C-22/C-23** in **Ergosterol** have been duly *hydrogenated with Pt*.

4. **Presence of Steroidal Nucleus** – Interestingly, the *dehydrogenation* of **ergosterol** with **Se** gives **Diel's Hydrocarbon** which confirms the presence of a *steroidal nucleus* in it.

5. **Presence of Hydroxyal (-OH) Moiety on the Side-Chain :** In fact, the **acetate of 3β-hydroxynor-5β-cholanic acid (I)** may be obtained from 'ergostanol' (from *previous step*) by *two* **different methods** namely :

   Method -1- Treatment of **ergostanol** with *acetic anhydride* $[(CH_3CO)_2O]$ gives **ergostany acetate**, which upon *oxidation with CrO_3* gives the **Compound (I)** stated above.

   Method -2- Alternatively, **compound (I)** may also be obtained by carrying out the *oxidation* of **5β-cholestanyl-3β-acetate (II)** with *CrO_3*, followed immediately by **Barber-Wieland degradation** of the acetate of **3β-hydroxy-5α-cholanic acid (III)** derived from **(II)**.

Thus , we may express the various sequential reactions starting from **Ergostanol** to obtain **compound (I)** and also the same starting from **compound (II)** *via* compound **(III)**, as stated under :

Comments – These essentially include :

   (i) Based on the avove series of reactions one may conclude that both **ergostanol** and **5α-cholestanyl-3β-acetate (II)** do have essentially. ;
      • **an identical nuclei**.

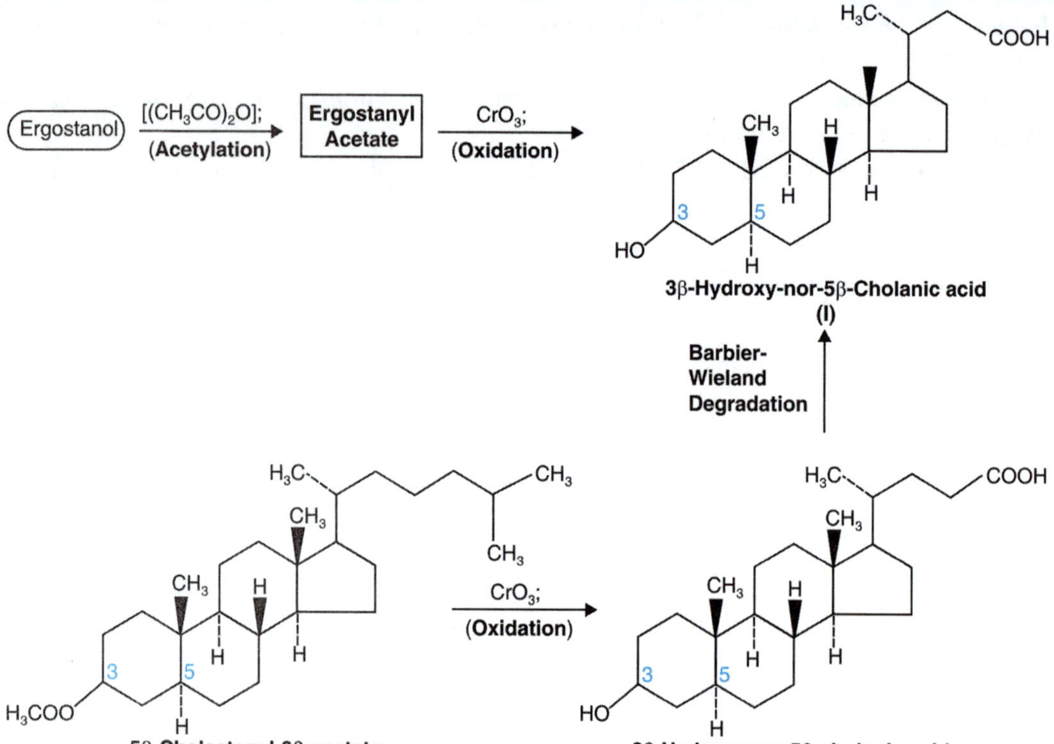

Ergostanol → [(CH$_3$CO)$_2$O]; (Acetylation) → Ergostanyl Acetate → CrO$_3$; (Oxidation) →

3β-Hydroxy-nor-5β-Cholanic acid
(I)

Barbier-Wieland Degradation

5β-Cholestanyl-3β-acetate
(II)

CrO$_3$; (Oxidation) →

3β-Hydroxy-nor-5β-cholanic acid
(III)

- **same position of hydroxyl moiety, and**
- **same position of the side-chain (i.e, at C-17).**

(ii) Hence, it becomes quite evident that **ergosterol (C$_2$H$_{44}$O)** critically possesses *one* **additional C-atom** in its *side-chain* than the **cholesterol (C$_{27}$H$_{46}$O)** (see *section 5.1*) .

(6) Nature of the Side-Chain **Ergosterol** when **ozonyzed (O$_3$)** gives **methyl-isopropyl acetaldehyde (IV)** as one of the products of reaction as given under :

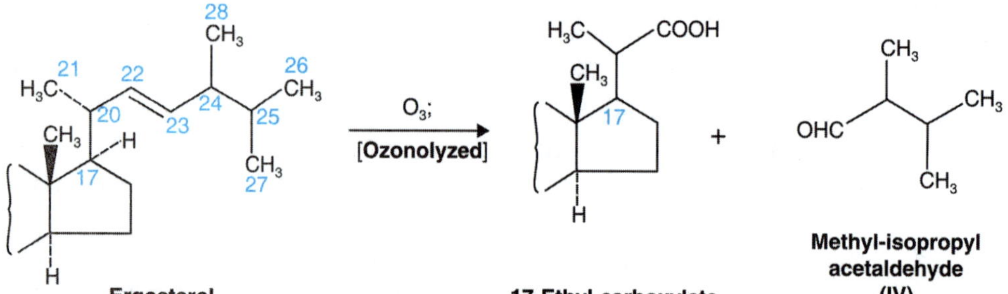

Ergosterol

O$_3$; [Ozonolyzed] →

17-Ethyl-carboxylate

+

Methyl-isopropyl acetaldehyde (IV)

Comments : These essentially comprise :

(i) The formation of **methyl-isopropyl acetaldehyde (IV)** may be explained if the *sidechain* of **ergosterol** is as shown above (*i.e*, **accounting for 28 C-atoms**).

(ii) Formation of **compound (IV)** could be explained clearly only if the *side-chain* of ergosterol is as depicted above.

(iii) Furthermore the ***IR-spectrum*** of **ergosterol** shows explicitly a particular band at ~**970 cm-1**, which categorically reveals that the **C-22/C-23 double band** very much exists in the so called ***trans*-configuration** (*i.e*, **Geometrical isomerism**).

Therefore, based entirely on the above factual observations as well as statements, one may legitimately assign the structure of **ergostanyl acetate (V)** – that eventually explains its *oxidation* with chromium trioxide (CrO₃) to the corresponding *acetate* of **3β-hydroxy-nor-5α-cholanic acid (I).**

Thus, we may have the following expressions :

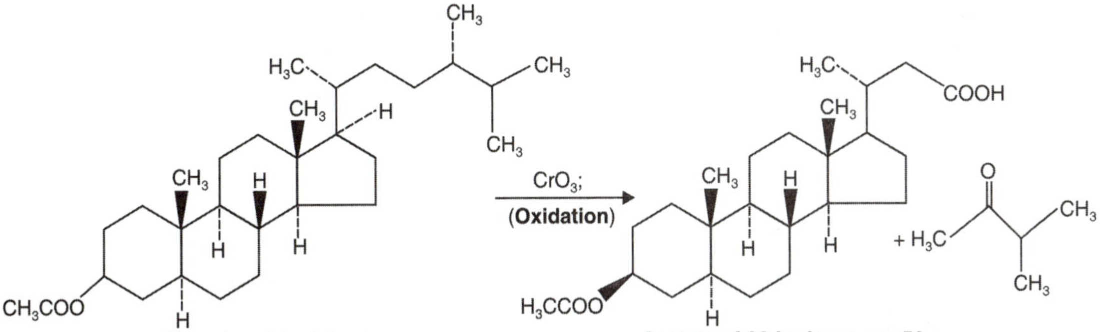

Ergostanyl acetate
(V)

Acetate of 3β-hydroxy-nor-5β-
Cholanic acid (I)

(7) **Positions of the Double-Bonds**-Let us once again consider the *ozonolysis* of **ergosterol** to the corresponding **methyliso-propylacetaldehyde (IV)**-as one of the major products (*see item.6 above*).

Importantly, the aforesaid reaction established the underlying fact that **only one double bond** is duly present in the *side-chain* of **ergosterol**. In case, if it (**ergosterol**) had more than one double-bond located in the side-chain, one would have definitely obtained more than one fragmetend products, such as (IV) being removed upon *ozonolysis*. Besides, the said *double-bond* is present strategically between **C-22/C-23** in **ergosterol** (see *section: 5.4*). However, **ergosterol** essentially comprises *three* **double bonds.**

At this point in time, one is charged with the ***next* problem** of assigning the exact position of the '**remaining *two* double bonds**'. Interestingly, the following *series of reactions* grossly confirms the *exact positions of the double bonds present duly in the* **steroidal nuclear,** such as :

(i) **Presence of Double Bonds in Conjugation Status in Steroidal Nucleus : Ergosterol** on being heated with **maleic anhydride** $\left[\begin{array}{c} H-C-CO \\ \| \\ H-C-CO \end{array} \right> O \right]$ at 135°C, gives an '*adduct*'.

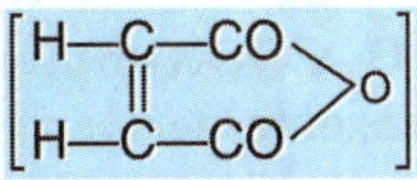

Comment – The above reaction reveals that the **double-bonds** present in the steroidal nucleus of **ergosterol** are in a **conjugation status** to each other.

(ii) **Presence of 2- Double Bonds in the Steroidal Nucleus of Ergosterol in one Ring Only** : It has been duly observed that **ergosterol** shows an **UV-absorption maximum (λ_{max})** at **282nm.** Nevertheless, the so-called **conjugated acyclic** dienes do exhibit an λ_{max} in the **region 220-250nm**; whereas, the **diene system** present duly in a ring system exhibits the λ_{max} in the region varying between 260-290nm.

Comment – The aforesaid observed **UV-absorption maximum values** (λ_{max}) prove beyond any reasonable doubt that the said *two double-bonds,* in the steroidal nucleus of **ergosterol,** are present critically in **one of the rings** (*i.e,* **Ring 'B'** probably).

(iii)   **Conjugated System Present in Ring 'B' of Ergosterol-Ergosterol** on being subjected to **Oppenauer Oxidation** with **aluminium-O-butoxide** [$(CH_3CO)_3$/al CH_3COOH, gives rise to the formation of an (**α,β-unsaturated ketone** (A) with **UV-absorption maximum** (λ_{max} **235nm.**).

Interestingly, one may logically expatiate the formation of the **ketone (A)** by assuming that *one of the double bonds* in **ergosterol** is duly present between **C-4/C-5 position** and eventually moves on to the **C-4/C-5 position** in the process of *Oppenauer oxidation.* Thus, the *second* **double bond** is, therefore, located between **C-7/C-8** so as to be *conjugated* with the one *i.e,* positioned between **C-5/C-6.** Hence, the said **conjugated system** is definitely present in the **Ring 'B'**; and, therefore, the most probably chemical structure of **ergosterol** is as given under-that explains predominanlty the so-called **oppenauer oxidation** as well :

Oppenauer
Oxidation;
[(CH₃)₃CO]₃/
Al-CH₃COOH

Ergosterol   **An α,β-Unsaturated ketone [A]**

Further Evidences of Ergosterol Chemical Structure

Let us consider the following *two* **sequential reactions** starting from **ergosterol** to yield **triol** :

C_6H_5-COOH
Per Benzoic Acid
(Oxidation)

(I)H₂/Pt;
(Reduction)
(Ii) OH⁻;
(Hydrolysis)

Ergosterol   **3β, 5α-dihydroxy-6α-mono**   **Triol**
                                benzoate derivative

The exact position of **double bonds,** as shown in **ergosterol** above, are duly supported by its *oxidation* with *perbenzoic acid* [C_6H_5-**COOH**] to give the corresponding **mono-benzoic acid derivative** (or **3β,5α-dihydroxy-6-monobenzoate derivative**). The resulting prodcuts when *hydrogenated catalytically* with H_2/Pt followed by *hydrolysis* in an alkaline medium yields a **saturated triol.**

The '**saturated triol'** on being subjected to treatment with *lead acetate* [$(CH_3CO)_2Pb$] undergoes '**fission**'

Remarks-(1) In fact, all the foretold reactions could be explained explicitly only if the **2-hydroxyl (-OH) moieties** are duly present specifically in the **vicinal position** only.

(2) Besides, the **saturated triol** (obtained above) may form a **diacetate**; and hence, **one of the hydroxyl (-OH) groups** has got to be *tertiary*-in nature, Thus, it establishes the fact that **one of the double bonds** is certainly present between the **C-5/C-6 position** in **ergosterol**

5.5. Bile Acids

Preamble–Bile acids designate the natural constituent of bile which is essentially an *acidic metabolite of cholesterol.*

It has been observed that the **saponification of bile** usually splits the *glycol-* and *tauro*-cholic **acid components** and yields an *admixture of bile acid*; the most abundant of which from *human* or *ox* **bile** is **cholic acid**, as given under:

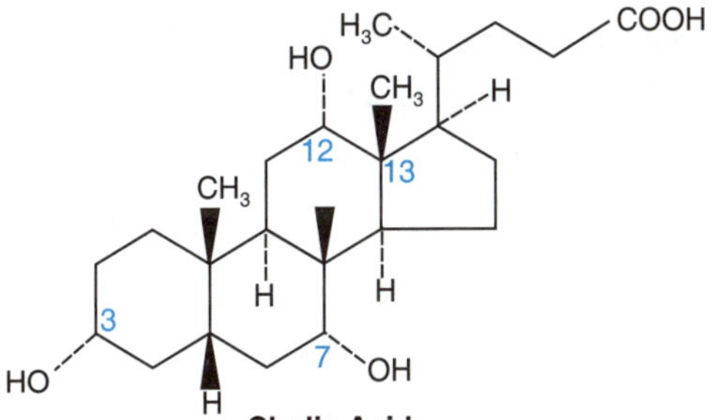

Cholic Acid
[mp=~198°C; $[\alpha]_D^{20}$ + 37° (c=0.6 in alcohol)]

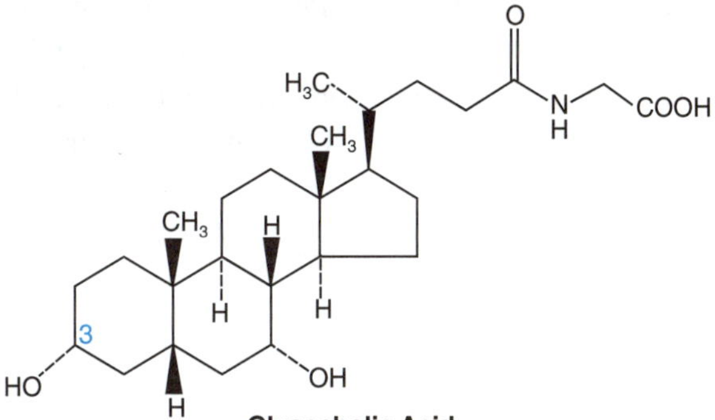

Glycocholic Acid
[mp=~130°C; $[\alpha]_D^{23}$ + 30.8° (c=7.5 in 95% ethanol)]

Points to Ponder- Importantly, the **C-skeleton** has exactly the *same structucre and configuration* as that of 3β-**coprostanol**, except that the prevailing side-chain gets terminated in a **C-24 carboxyl(-COOH) moiety**.

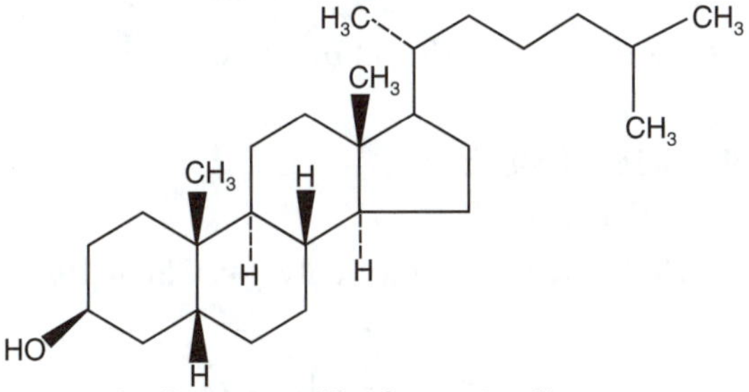

3β–Coprostanol [or Coprosterol]
[mp=~101°C; $[\alpha]_D^{18}$ + 28° (c=1.8 in Chloroform)]

- ❏ All the *three* **hydroxyl (-OH) moieties**, positioned at **C-3, C-7 and C-12** are β-*oriented*;
- ❏ All the **bile acids** (*viz*; **cholic acid**; **glycocholic acid**) do essentially have a **3α- hydroxyl group** (see *section 5.5*); and
- ❏ All the **bile acids are not precipitated by digitonin** (unlike the **steroids**).
- ❏ Importantly, the **3α-hydroxy(-OH) moiety** is *equitorial* in nature whereas, the respective **7α-and 12α-hydroxyls** are located strategically in the *less vulnerable axial orientation*; and, therefore, the **cholic acid** and its respective 'esters' may be **acetylated at C-3 position selectively**.

NOTE   it has been duly established that the so called *3-cathylate* is usually obtainable in high yield specifically.

☐ The observed order of *decreasing susceptibility to oxidation* is found to be $C_7 > C_{12} > C_3$.

☐ Amazingly, the **equatorial β-hydrogen** positioned at **C-12** is observed to be less vulnerable than that at **C-7** which is solely *attributable to hindrance right from the* **β-oriented angular methyl (-CH₃) moiety** located at **C-13** (see *cholic acid structure*).

CHOLANIC ACID

The **cholanic acid** *i.e*; the **parent acid**, is duly obtained by the *pyrolytic dehydration* of **cholic acid** and followed by hydrogenation of the resulting mixture of **cholatrienic acid**. Importantly, it is the product obtained by identical treatment of **deoxycholic acid** (of the isomeric chenodesoxycholic acid), which is nothing but **3α,7α-dihydroxycholic acid**, and also of **lithocholic acid** (generally isolated from the *gall stones*).

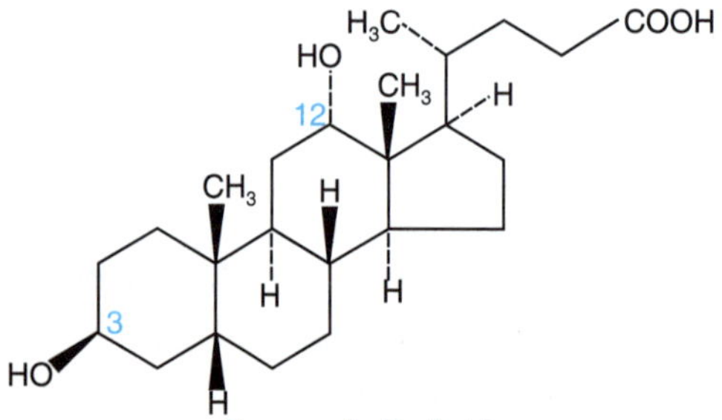

Desoxycholic Acid
[mp=~176-178°C; [α]$_D^{20}$ + 55° (alcohol)]

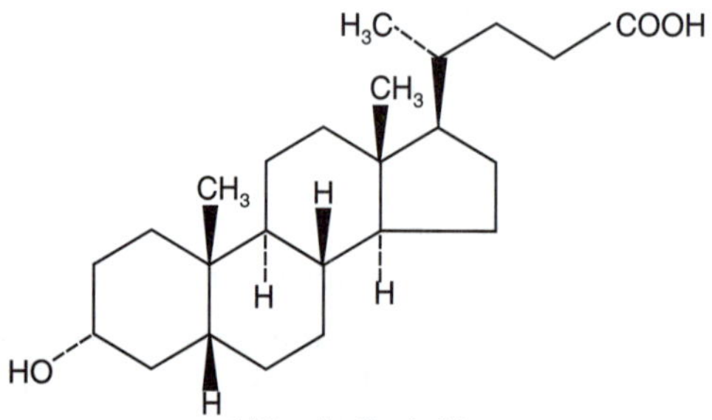

Lithocholic Acid
[mp=~184-186°C; [α]$_D^{20}$ + 33.7° (c=1.5 in absoute ethanol)]

Following are the approximate yield of different **bile acids** as obtainable from '*bile*'

- **Cholic acid**~5-6%;
- **Desoxycholic acid** ~0.6-8%;
- **Lithocholic acid**~0.02%;
- **Chenodesoxycholic acid** ~ in trace only. (or **Chenodiol**)

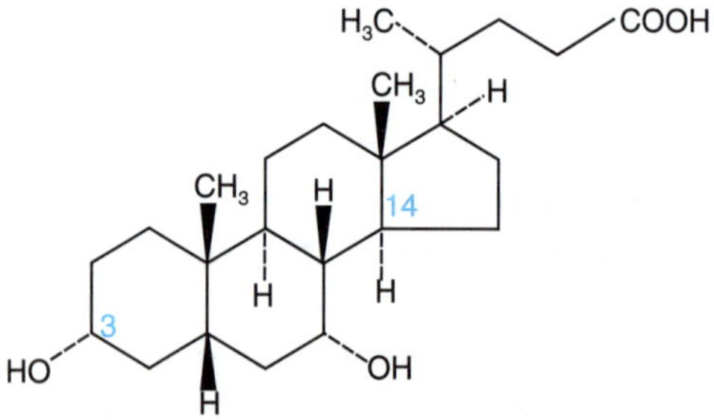

Chenodesoxycholic Acid [or Chenodiol]
[mp=~119°C; [α]$_D^{20}$ + 11.5° (dioxanl)]

It is, however, pertinent to state here that **hyodesoxycholic acid (or hyodeoxycholic acid)** [$C_{24}H_{40}O_4$] is one of the *bile acids* present in **pig bile** (*Greek*; **hys** means swine)- designates the respective **3α,6α- dihydroxy isomer** as shown below :

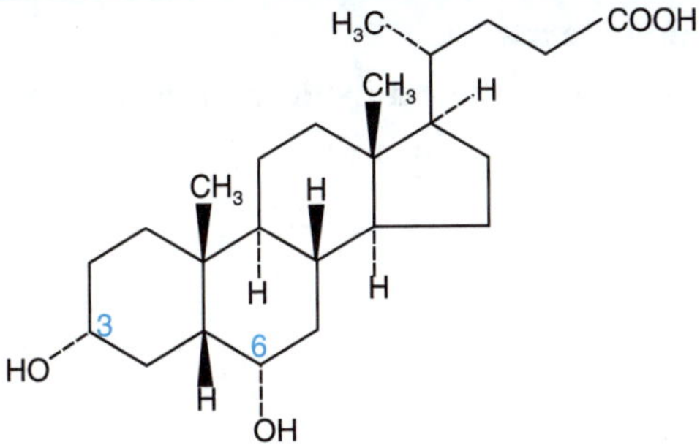

Hyodesoxycholic Acid [or Hyodeoxycholic Acid]
$[mp=196\text{-}197°C; [\alpha]_D^{20} +8° \text{ (alcohol)}]$

Important Features-These essentially comprise:

➤ **Hydroxyl(-OH) moiety at C-6 is oxidized to a carbonyl(>C=0) moiety;**
➤ **Isomerization at C-6 is only possible *via* the formation of '*enol*'; and**
➤ **Ultimate treatment of the *ketone* with an *alkali* (OH) converts tha A/B *cis*-cholanic acid derivative mostly into the more stable *trans*-epimer of the specific *allocholanic acid series*.**

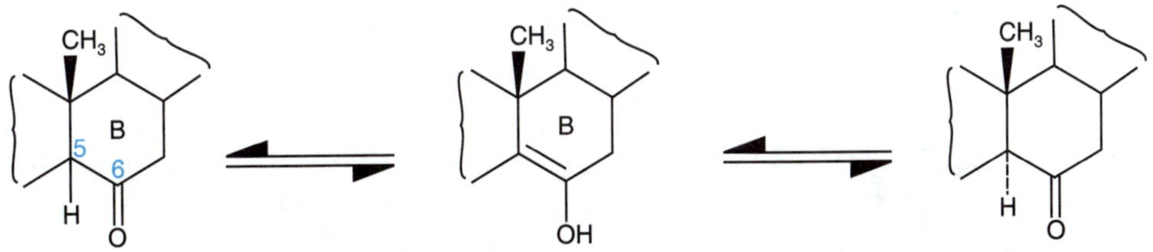

| Cholanic Acid Series | Cholesterol (A) | Allocholanic Acid Series |

Remarks-These essentially include :

(i) **Hyodesoxycholic acid** is fairly convertible by the following *three* processes sequentially:

- **oxidation,**
- **isomerization,** and
- **degradation**

into *allochollanic acid* which critically possesses the so called *trans*-configuration *i.e*; a **characteristic feature of *cholestanol*.**

(ii) Correlation of the '*sterols*' with the respective '**bile acids**' was duly established by **Windaus's discovery (1919)-that** states and establishes the underlying fact- upon *oxidation* with *chromic acid* '**cholestane**' get; duly degraded partially into acetone and **allocholanic acid**; whereas, the corresponding *5-epimer*, **coprostane**, gives **cholanic acid.**

Overview of Bile Acids

Bile acids, as their name indicates, are isolated from the fluids secreted by the concentrated duly in the gallbladder – and eventually powered into the *small intestine via* the so- called **bile ducts**, which aids in *alkalinizing*:

- **the intestinal contents**, and
- **plays a pivotal role in the emulsification, absortion, and digestion of fat.**

However, the major constituents of '**bile**' are, namely:

- **Conjugated Bile Salts**
- **Cholesterol**
- **Phospholipids**
- **Bilirubin** and
- **Electrolytes**

Bile Acids invariably do occur as the respective **sodium salt of the corresponding amides** with either:

- ❏ **Glycine[H₂N.CH₂-COOH], or**
- ❏ **Taurine [H₂N.CH₂ CH₂-SO₃H].**

Examples:

- **Glycocholic Acid**
  [**Glycine+cholic Acid**]

- **Taurocholic Acid**
  [**Taurine+ Cholic Acid**]

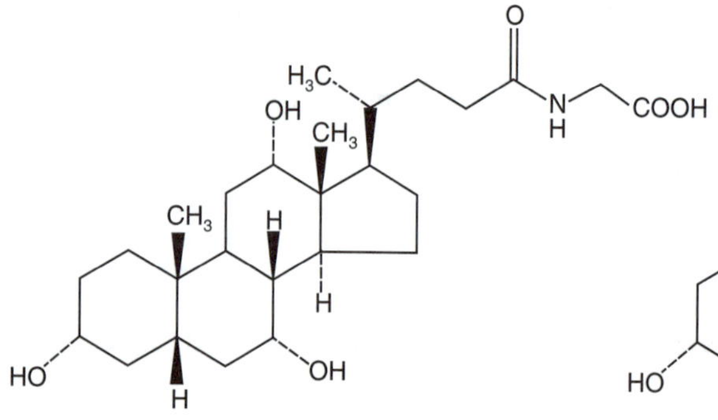

Glycocholic Acid
[mp=~13°C; $[\alpha]_D^{23}$ + 30.8° (c=in 95% EtOH)]

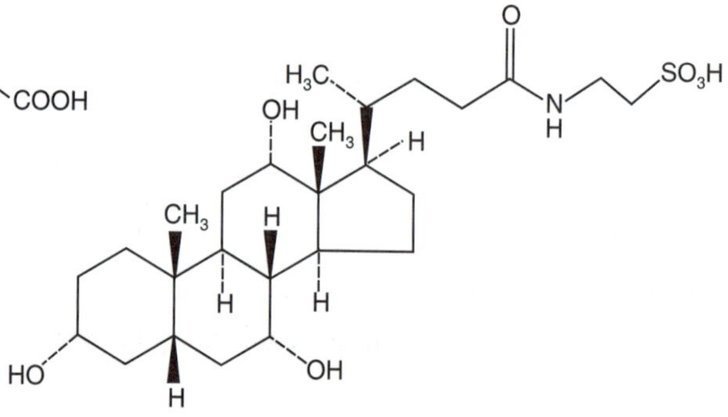

Taurocholic Acid
[Decomp.=125°C; $[\alpha]_D^{18}$ + 38.8° (c=2 in alcohol)]

Function of Bile Acids- The major function of **bile acids** is to augment and facilitate specification the **digestion of fats** since these (*fats*) are almost insoluble in water, but are rendered *soluble by* **bile acids**; and hence, may be absorbed adequately in the **intestinal passage**.

In fact, a large segment of the '**bile acid**' have been found to be the so-called **hydroxyl derivatives** of either:

- **5-β-cholainic acid [or Cholanic Acid], or**
- **5-α-Cholanic acid]or Allocholanic Acid].**

Nevertheless, the **bile acid** on being subjected to *dehydration* by carefully *heating under vacuum*, immediately followed by the **catalytic reduction (H₂/Pt)** may ultimately lead to the production of **5-β-cholanic acid** and **5β-cholanic acid**, as given below:

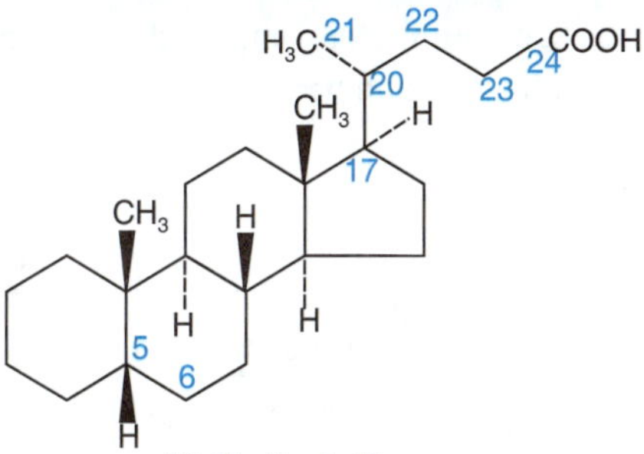

5β-Cholic Acid
[or **Cholanic Acid**]

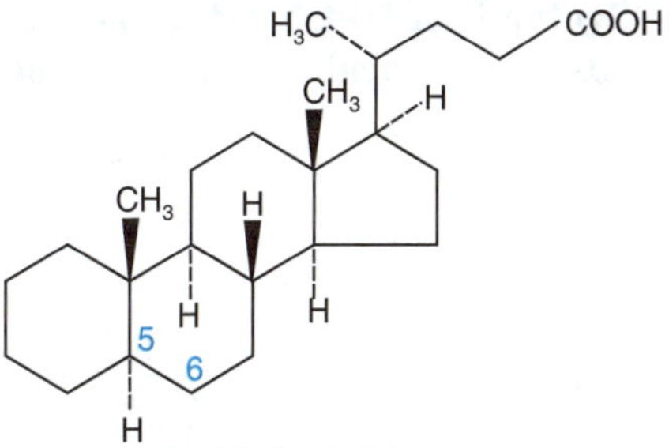

5α-Cholic Acid
[or **Allocholanic Acid**]

Synthesis of 5β-and 5α-Cholanic Acid

The chemical structures of **5β-and 5α-cholanic acid** have been duly established by their *total synthesis* starting from *cholesterol*.

    (a) **Synthesis of 5β-Cholanic Acid**- The various steps involved in the synthesis of **5β-cholanic acid from cholesterol** are as described under:

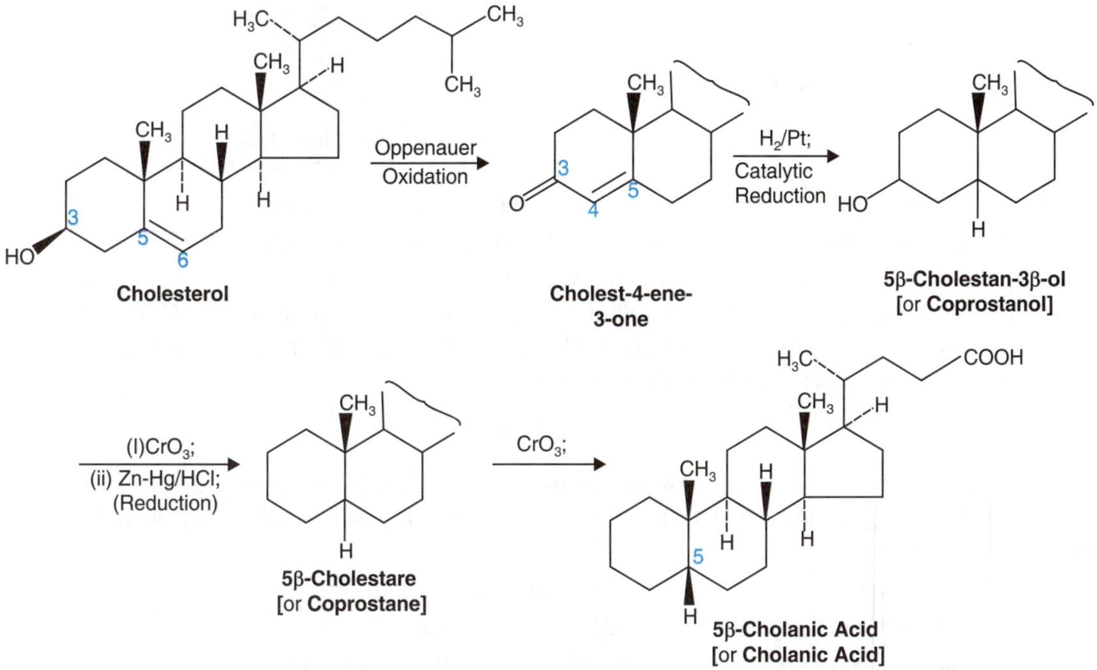

Cholesterol

Cholest-4-ene-3-one

5β-Cholestan-3β-ol
[or **Coprostanol**]

(I)CrO₃;
(ii) Zn-Hg/HCl;
(Reduction)

5β-Cholestare
[or **Coprostane**]

CrO₃;

5β-Cholanic Acid
[or **Cholanic Acid**]

Explanation- The various steps involved above may be explained as under:

   1. **Cholesterol** on *Oppenauer oxidation* yields **cholest-4-one-3-one**, which upon catalytic reduction with *H₂/Pt* gives **5β-choleston-3β-ol** (or **coprostanol**).

   2. **Coprostanol** upon oxidation with *chromium trioxide* followed by *reduction* with *Zn-Hg/HCl* gives **5β-cholestans** (or **coprostane**) which on further oxidation with *CrO₃* yields **5β-cholanic acid (or cholanic acid)**

(b) **Synthesis of 5β-Cholanic Acid-** The various sequential steps engaged in the synthesis of **5α-cholanic acid** from **cholesterol** are as enumerated under:

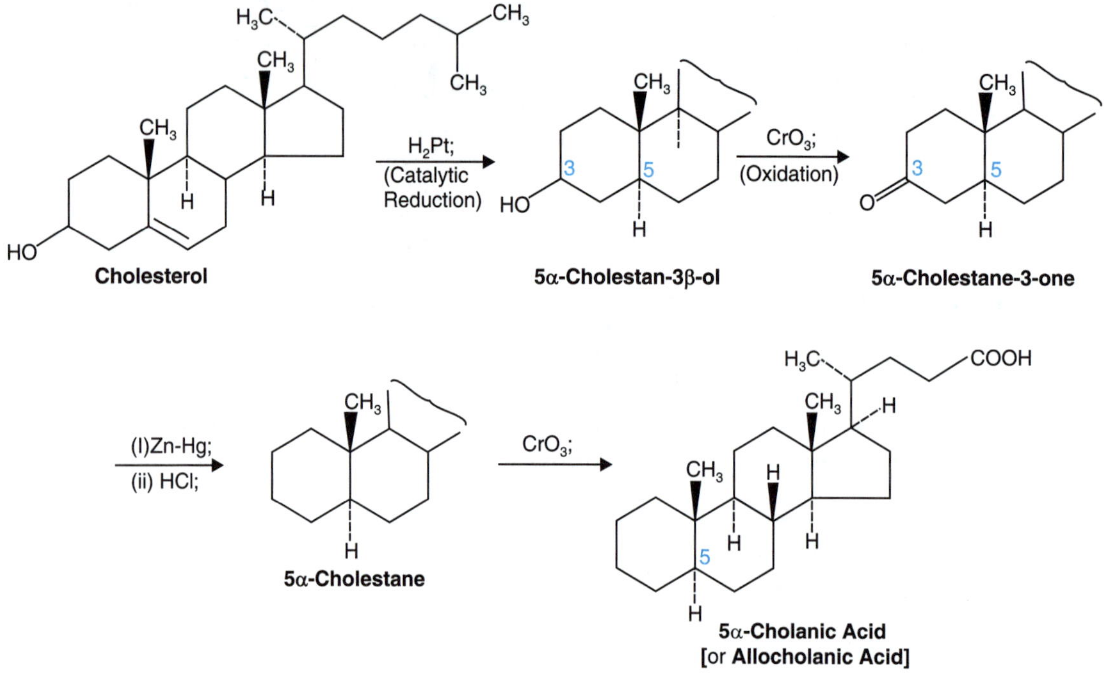

Cholesterol

H_2Pt;
(Catalytic Reduction)

5α-Cholestan-3β-ol

CrO_3;
(Oxidation)

5α-Cholestane-3-one

(I)Zn-Hg;
(ii) HCl;

5α-Cholestane

CrO_3;

5α-Cholanic Acid
[or **Allocholanic Acid**]

Explanation- The different steps engaged above may be explained as under :

1. **Cholesterol** under *catalytic reduction* with **H_2-Pt** gives **5α-cholestan-3β-ol**, which upon *oxidation* with **CrO_3** yields **5α-cholestane-3-one**.

2. The resulting *ketone* on reduction with *Zn-Hg* followed by treatment with *HCl* produces the corresponding **5α-cholestane** which finally on *oxidation* with **CrO_3** gives rise to the formation of **5α-cholanic acid (or allocholanic acid).**

Some Important Naturally Occurring' Bile Acids'

As to date more than ***twenty* natural bile acids** have been duly *isolated purified-characterized* and *identified*. Interestingly, the most vital and abundant **natural bile acids** usually occurring in the ***human bile*** are, namely:

- **Cholic Acid**   : 25-60 % (of *total bile acids*);
- **Deoxycholic Acid**   : 5-25% (of *total bile acids*);
- **Chenodeoxycholic Acids** : 30-35% (of *total bile acids*); and
- **Lithocholic Acid**   : in *small quantum only*.

Point of Difference Amongst Bile Acids – These include:

➢ **Position and number of hydroxyl (-OH) moieties present;**
➢ **Position of the hydroxyle (-OH) group may be positioned at C-3,C-6,C-7,C-12 and C-23 in the bile acids; and**

➤ **Practically all of the '*natural bile acid*'- the hydroxyl and (-OH) groups do essentially have the α–*configurations*.**

Table1.1 records the various important characteristic features of *five* **natural bile acids** including their name, molecular formula, source, exact position of hydroxyl (-OH) group (with actual configuration 'α'-or 'β'), specific optical activity, and melting point

S. NO.	Name of Bile Acid	Molecular Formula	Sources of Bile Acid	Position of Hydroxyl (-OH) Groups	$[\alpha]_D^{20}$	mp(°C)
	Table : 1.1: Characteristic Features of Five Important Natural Bile Acids					
1.	**Cholic Acid** [*TrihydroxycholanicAcid*]	$C_{23}H_{36}(OH)_3 . COOH$	Man; Ox;	3α, 7α, 12α	+37°	195
2.	**Deoxycholic Acid** [*Dihydroxycholanic Acid*]	$C_{23}H_{37}(OH)_2.COOH$	Man; Ox;	3α, 12α	+53°	172
3.	**Lithocholic Acid** [*Monohydroxycholanic Acid*]	$C_{23}H_{38}(OH). COOH$	Man; Ox	3α	+32°	186
4.	**Chenodeoxycholic Acid**	$C_{23}H_{37}(OH)_2 COOH.$	Man; Ox, Hen	3α,7α	+11°	140
5.	**α-Hyodeoxycholic Acid** [*Dihydroxycholanic Acid*]	$C_{23}H_{37}(OH)_2.COOH$	Pig	3α,6α	+8°	197

Isolation of Bile Acids-In general, the '**bile acids**' do have essentially the so-called **peptide bond*** (or **peptide linkage**) [*i.e*, **an amide (-CO-NH-) bond**], which gets subsequently hydrolyzed with alkali. Thus, the liberated '*free bile acids*' from the solution have been carefully isolated either :

- **by crystallization from respective organic solvents, or**
- **by treatment of the *ethereal solution* of the *bile* acids with varying strengths of HCl.**

Examples : Following are a few *typical examples* :

   (i) **Cholic Acid [3α, 7α, 12α-trihydroxy product]** (see Table:1.1) – is extracted duly from the *ethereal solution* by means of HCl (15% v/v).

   (ii) **Dihydroxy Acids** (*viz*, **Deoxycholic Acid** and **Chenodeoxycholic Acid**) are duly extracted from the *ethereal solution* by using HCl (25% v/v)

   (iii) **Monohydroxy Acid** (*viz*, **Lithocholic Acid**) is being extracted with concentrated **HCl (36M).**

CONSTITUTION OF BILE ACIDS

The precise and exact structure of the '**bile acids**' have been duly elucidated based upon the chemical structure of **cholesterol** by virtue of their similarity pertaining to their so-called **structural**

---

 * **Peptide Bond** – It related to the '*chemical linkage*' of amino acis where in the **carboxyl (-COOH) group** of one amino acid forms an **amide (-CO-NH$_2$)** with the **amino (-NH$_2$) moiety** of a *second* amino acid, which in turn may form an **amide bond** with another amino acid, and so on so forth to form a long chain.

relationship. However, it has been duly observed that **all the bile acids** may be duly converted into any one of the *two* **cholanic acids** as stated below :

- **5α-cholanic acid,** or
- **5β-cholanic acid**.

Therefore, one may vividly conclude that the '**bile acids**' do designate the '*hydroxy derivatives*' of the *cholanic acids*'.

1. Conversion of Lithocholic Acid into Cholanic Acid

Thus, **lithocholic acid (A)** (*i.e*, a *monohydroxycholanic acid*) on being subjected to *distillation* under vaccum gives **cholenic acid (B),** which further on *catalytic reduction* (H_2-Pt) yields **cholanic acid (C)**. The resulting **acid (C)** may *alternatively* obtained from **cholesterol (D)**. Compound **(D)** when subjected to *Oppenauer oxidation* followed by *catalytic reduction* (H_2/Pt) yield **coprostanol (E)**. The **compound (E)** on *oxidaotin* with CrO_3 followed by *reduction* with **Zn –Hg/HCl** gives **coprostane (F)** – and this on further *oxidation* with CrO_3 gives rise to the formation of **cholanic acid (C).**

Thus, we may have the following expressions :

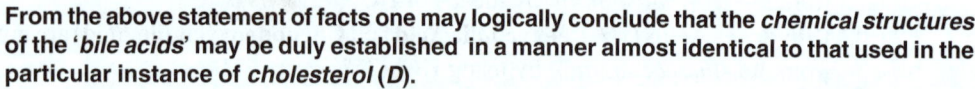

Inferences-Following *two* cardinal **inferences** may be drawn from the above set of reaction, namely :

(i) Amazinlgy, both **lithocholic acid (A)** and **cholesterol (D)** do give the same product **cholanic acid (C).** Therefore, one may vehemently draw the inference that both **(A)** and **(D)** are definitely **related to each other structurally** *i.e*, an *identical nucleus* should be present in **cholesterol (D)** and the **bile acids** [*viz*., **(A)** and **(C)**].

> NOTE   From the above statement of facts one may logically conclude that the *chemical structures* of the '*bile acids*' may be duly established in a manner almost identical to that used in the particular instance of *cholesterol (D)*.

(ii) The actual conversion of **lithocholic acid (A)** into the respective **cholanic acid (C)** explicitly reveals that the **former** (*i.e*, '*A*') is certainly the so-called **hydroxyl derivative** of the **latter** (*i.e*, '*C*').

2. Carbon Skeleton of Bile Acids

Let us take the example of **cholic acid** (a '*bile acid*') which upon *dehydrogenation* with **Se** promptly gives the so-called **1** (see below) as **one of the products of dehydrogenation.**

Likewise, **cholesterol** (a '*zoosterol*') on **Se** *dehydrogenation* yields also **Diel's hydrocarbon** as **one of the products of dehydrogenation.**

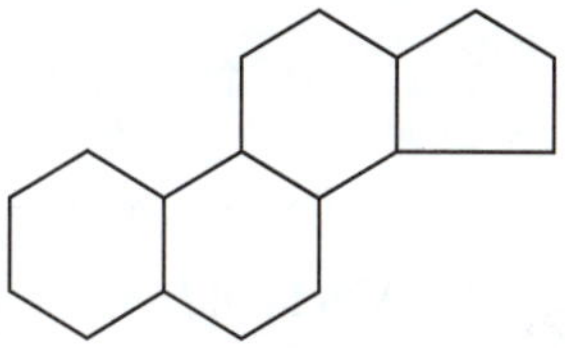

CH₃

17

**Carbon Skeletion of
Bile Acids**

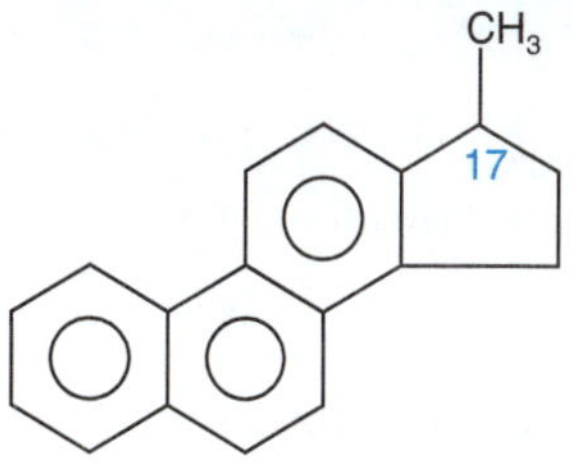

Diel's Hydrocarbon

It is, however, pertinent to observe at this material time that crucial production of **Diel's hydrocarbon** by the *dehydrogenation* of either:

- **bile acids,** or
- **sterols,**

duly establishes the underlying fact that both of them have an absolutely '**identical nucleus**' *i.e*, **bile acids and sterols do comprise a cyclopentanophenanthrene nucleus** (otherwise known as the '*steroidal nucleus*').

Important Features – These essentially include :

❑ **Cholesterol** (*i.e*, **sterol**) glaringly differ from the **bile acids** (*i.e*, **cholanic acids**) in *one prominent aspect* which reveals that he *double-bond in cholesterol nucleus* is **absolutely missing in the bile acids.**

❑ Similarity in the **carbon-skeleton** has been duly confirmed by the formation of :
- **5β-cholanic acid,** and
- **5α-cholanic acid,**

in both *cholesterol* as well as *bile acids*.

Thus, one may represent the **carbon-skeleton of** *bile acids* (as shown above).

3. Side-Chain of Bile Acids

The major points of difference existing premodinantly between **chemical structures** of :
- **Bile Acids,** and
- **Sterols,**

being the ***exact nature*** and *size* of the inherent **side-chain** that may be further *expatiated* and *understood* vividly on the basis of the following *two* glaring facts, namely:

(a) **Cholesterol** upon *oxidation* gives **acetone** exclusively; whereas, the **bile acids** fail to do so. *Comment* – Since, the **8C-side chain in 'cholesterol'** obviously terminates as an '**isopropyl moiety** $\left(-\mathopen{<}^{CH_3}_{CH_3}\right)$', which eventually upon *oxidation* gets duly converted into **acetone**.

On the contrary, the side-chain of bile acids does not terminate in an '**isopropyl moiety**'; and, therefore, the formation of acetone after oxidation is simply not possible at all.

Hence, the **total number of C-atoms** present in the *side-chain* of :
- **bile acids** – *shall be confired to 'five' only,* and
- **cholesterol** – *shall be up to 'eight' only.*

Examples

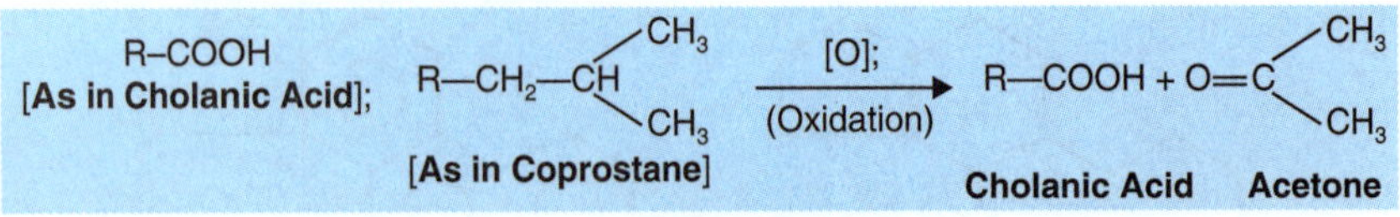

$$R-COOH \quad R-CH_2-CH\begin{smallmatrix}CH_3\\CH_3\end{smallmatrix} \xrightarrow[\text{(Oxidation)}]{\text{[O];}} R-COOH + O=C\begin{smallmatrix}CH_3\\CH_3\end{smallmatrix}$$

[As in Cholanic Acid];   [As in Coprostane]   Cholanic Acid   Acetone

(b) **Nature of the Side-Chain** – Imporantly, the **Barbier-Wieland Degradation** helps to elucidate properly the **nature of the side-chain** in both **bile acids** and **cholesterol**.

Modus Operandi- First of all, the **bile acids** are carefully converted to **cholanic acid**, which on being subjected to **Barbier-Wieland degradation** gives rise to the formation of **etiobilianic acid** – almost in the same fashion as in the particular instance of **cholesterol.**

Remarks – Therefore, the side-chain is attached to the **bile acids** at **C-17 in Ring- 'D'.** Thus,, it is now logically feasible to arrive at the *partial* **chemical structure of bile acid** as given below :

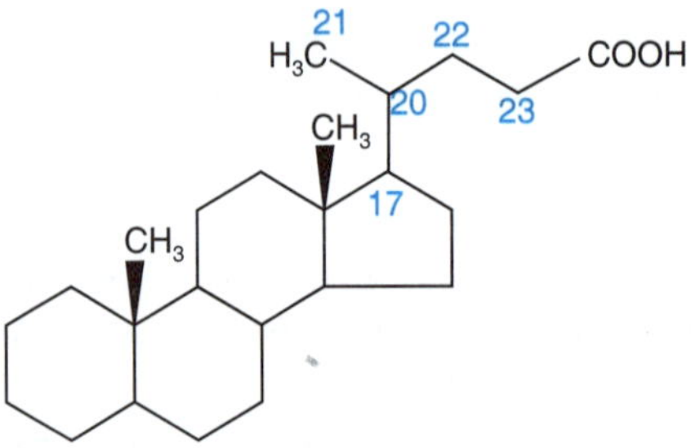

Partical Chemical Structure of a Bile Acid

4. Positions of Angular Methyl (-CH₃) Moieties

Another equally important task is to ascertain precisely the **position of angular methyl (-CH₃) moieties.** Incidentally, these are actually found to have exactly the *same positions* as could be seen in **cholesterol** (*ie.,* **C-10 and C-13 in the steroidal nucleus**). Nevertheless, one may critically ascertain these *two* positions of the angular methyl groups by adopting the same procedure as in **cholesterol.** Thus, it is now quite possible to have the *modified* **partial chemical structure** of 'bile acids' as given under :

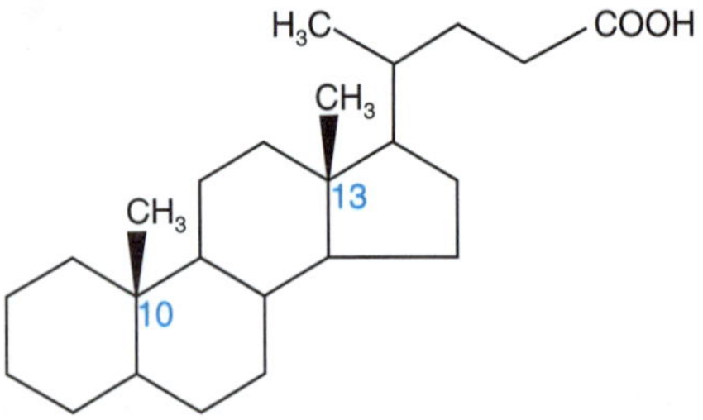

Modified Partial Chemical Structure of a Bile Acid
[Showing 2-angular methyl groups positioned at C-10 and C-13]

5. Position of Hydroxyl (-OH) Group in Bile Acids

In **fact,** the ultimate problem remains to allocate the exact **position of the hydroxyl (-OH) group** in the '**bile acids**'. It has been duly proven and established by carrying out the so-called **oxidative degradation**, such as :

-'the exact position of the hydroxyl (-OH) group present in the *Lithocholic Acid* has been found to be positioned at C-3':

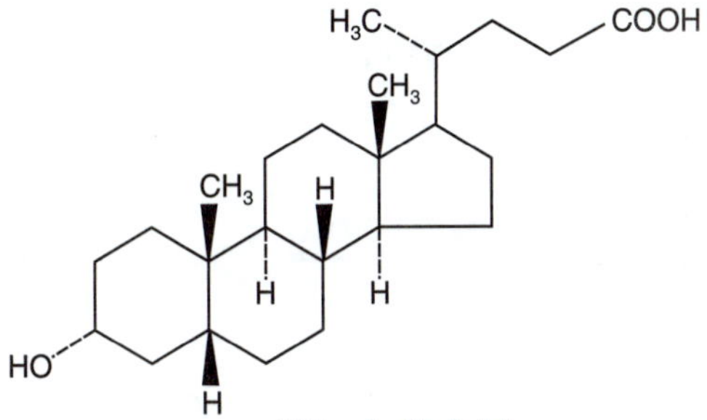

Lithocholic Acid
[A Bile Acid]

The above observations may be further substantiated by means of the following *two* sets of reactions namely:

   (i)   **Conversion of Coprostanol (A) to Lithobilianic acid (C) (a tricarboxylic acid)** *via* a dicarboxylic acid (B); and

  (ii) **Conversion of Lithocholic acid (D) to Lithobilianic acid (C)** *via* dehydrolithocholic acid (E).

Thus, we may have the following reactions :

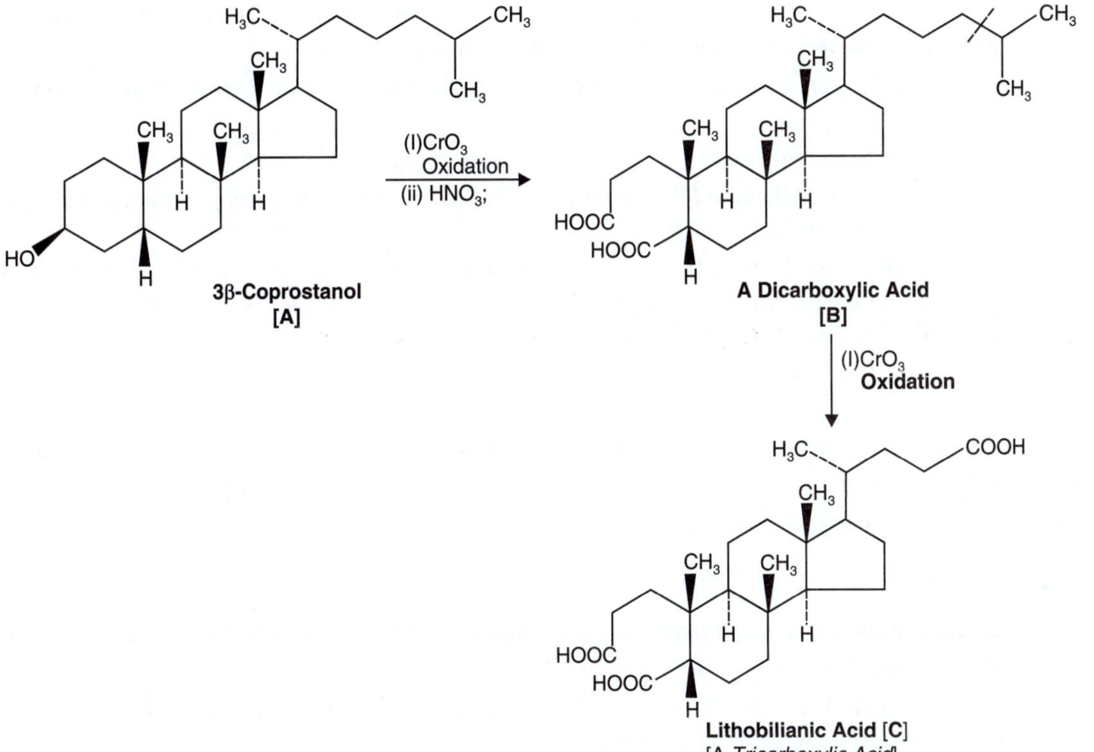

3β-Coprostanol
[A]

(I)CrO₃ Oxidation
(ii) HNO₃;

A Dicarboxylic Acid
[B]

(I)CrO₃ Oxidation

Lithobilianic Acid [C]
[A *Tricarboxylic Acid*]

(Contd........)

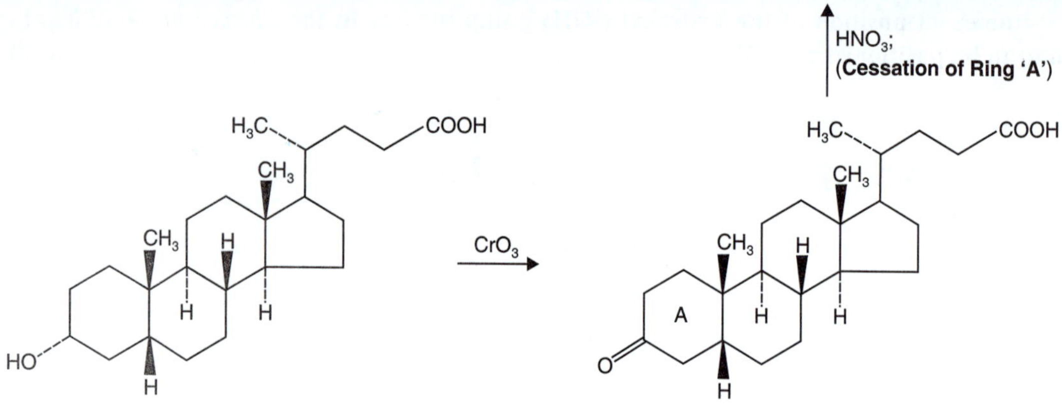

Lithocholic Acid
[D]

Dehydrolithocholic Acid
[E]

Comment-From the above reactions one may conclude that the **hydroxyl (-OH) moiety** in **lithocholic acid (D)** critically occupies the same position as in **5β-cholestan-3β-ol** [or **coprostanol**] *ie.,* **position C-3.**

Further Evidence for Position of Hydroxyl (-OH) Moiety

It is worthwhile to mention at this material time that the foregoing evidences put forward do not really lead us to a *convincing* and *definite* conclusion at all; and this is based on the underlying factual truth that-**'had' the hydroxyl moiety been positioned at C-4 (in** *litocholic acid*)**-even then one, would have still obtained the product** *lithobilianic acid (C);*

In true sence, the careful *oxidation* of **coprostanol (A)** invariably gives *two* **distinct isomeric acids**, namely :

- **Lithobilianic acid (C)** – (*as shown above*), and
- **Dicarboxylic acid (C-1)**-(*which is obtained by cleavage of Ring 'A' between C-2 and C-3 positions*).

Besides, the **dicarboxylic acid (C-1)** on being further oxidized with CrO_3 yields **isolithobilianic acid (D-1)**. Amazingly, the critical oxidation of **lithocholic acid (D)** also gives rise to the formation of an admixture of the *two* acids, namely:

- **Lithobilianic acid (C),** and
- **A Dicarboxylic acid (C-1).**

Inference-Based on the above scientific evidences one may draw a *definite conclusive inference* that the **hydroxyl (-OH) moiety** is present in the said **bile-acid at C-3 position only**.

Let us write the following reaction sequentially:

Configuration of Hydroxyl (-OH) Moiety in Lithocholic Acid [D]

The actual configuration of the hydroxyl (-OH) moiety in **lithocholic acid [D]** is found to be '**α-**' (*i.e., below the ring plane* and *shown by a dotted line*).

The above critical observation has been duly elucidated by performing the *oxidative degradation* of the following *two* chemical entities, namely :

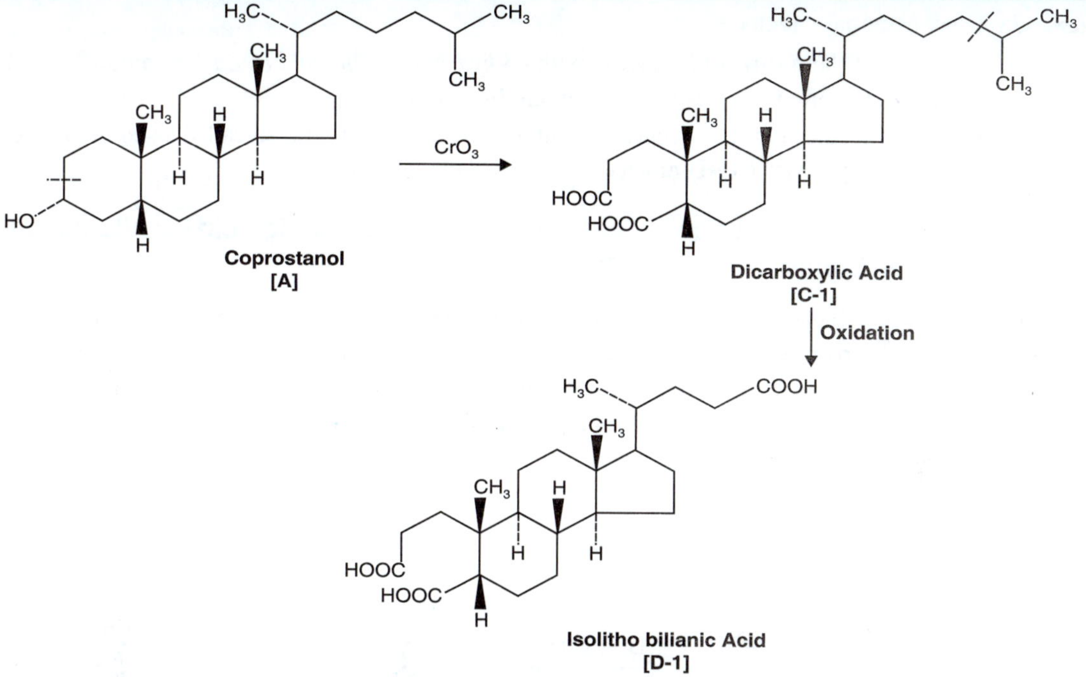

Coprostanol [A] → (CrO₃) → Dicarboxylic Acid [C-1] → (Oxidation) → Isolitho bilianic Acid [D-1]

• **acetates of lithocholic acid (D)**, and • **5β-cholestan-3β-ol(2)** *ie.,* **epicoprostanol**, to the respective **5β-androsterone (1)** *i.e.,* **5-isoandrosterone**). Thus, we may have the following expressions:

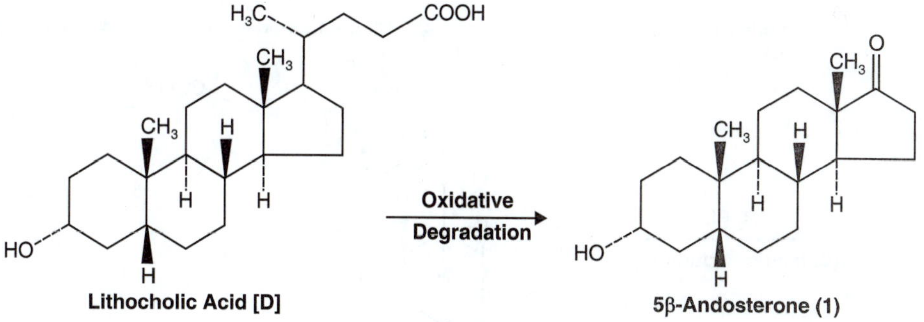

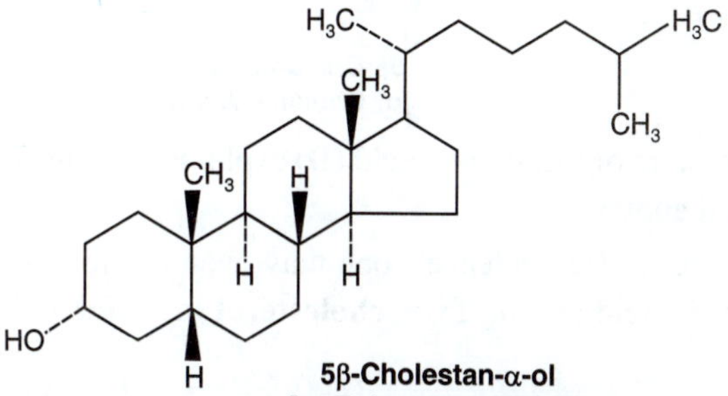

Lithocholic Acid [D] → **Oxidative Degradation** → 5β-Andosterone (1)

5β-Cholestan-α-ol [or Epicoprostanol] [2]

Remarks-These essentially include :

(i) Most of the **natural bile acids** (with (*β-hydroxy-cholic acid as an 'exception'*) may be duly converted into respective **lithocholic acid (D).**

(ii) Therfore, it vehemently suggests that **all bile acids** shall have the so-called '**α-configuration**' for the **hydroxyl (-OH) moiety** at *C-3 position exclusively.*

6. Possibility for the Conversion of Bile Acids into : 5α- Cholanic Acid (or Allocholanic Acid) and 5β- Cholanic Acid (or Cholanic Acid)

Since, practically all the '**bile acids**' may be duly converted into any one of the *two* above mentioned **cholanic acids** (*viz,* **5α-and 5β-cholanic acids**); therefore, the formrr are invariably regarded to be the so-called '*hydroxy-structural analogues*' of the latter. Fieseer and Fieser (1959)* ultimately proved that **cholenic acid** is found to be a mixture of *two* **compounds,** as shown under, wherein **chol-3-enic acid** is identified to be the *major constituent,* which upon *catalytic reduction* with H_2/Pt yields **5β-cholanic acid (or Cholanic Acid)** :

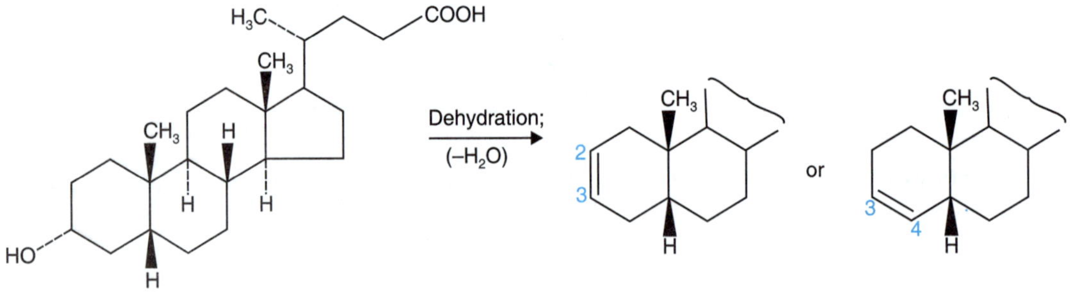

Lithocholic Acid
[D]

2,3-Cholenic Acid

3,4-Cholenic Acid

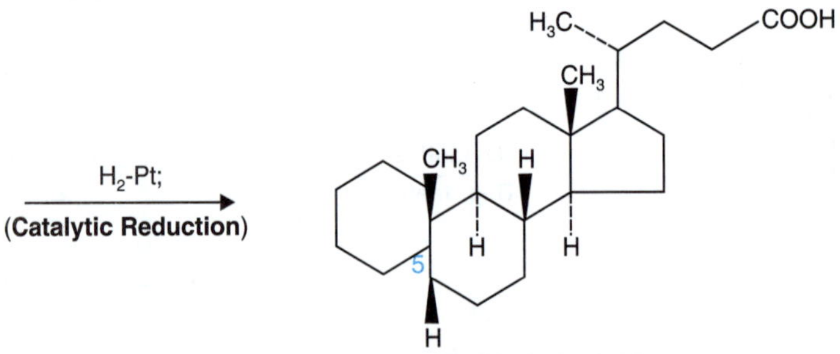

5β-Cholanic Acid
[or Cholanic Acid]

Comment – The chemical structure of **lithocholic acid (D)** explains without any ambiguity the formation of **cholanic acid** as illustrated above.

In addition to the above scientific evidences, one may even put forward another way to produce the **same acid** ie., **5β-cholanic acid** staring from **cholesterol** (a '*zoosterol*') as shown under :

---

* Fieser LF and Fieser M : **Steroids**, Reinhold Publishing Incorp,, New York, 1959.

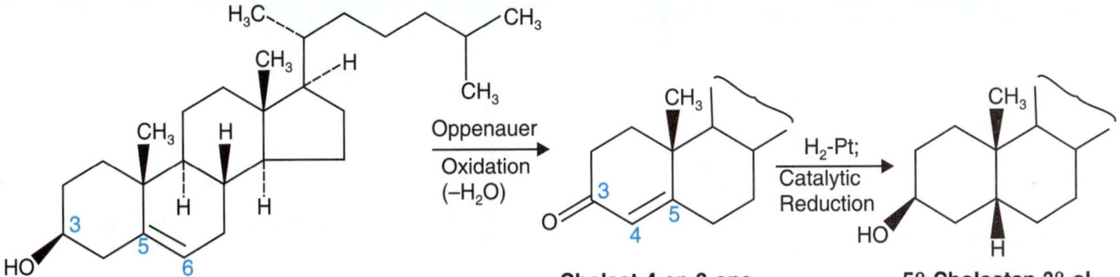

Cholest-4-en-3-one

Cholesterol (1)

5β-Cholestan-3β-ol [Coprostanol] (3)

Explanation – The above reactions may be explained as under :

1. **Cholesterol (1)** on *Oppenauer oxidation* yields **cholest-4-en-3-one (2)** with the loss of a mole of water.

2. The resulting **product (2)** on being subjected to *catalytic reduction* with H_2/Pt gives rise to the formation of **5β-Cholestan-3β-ol (3)**, also called **Coprostanol.**

Conversion of Cholesterol into 5-Cholanic Acid or (or Allocholanic Acid)-Cholesterol (1) under *catalytic reduction* **(H_2/Pt)** gives **5α-cholestan 3β-ol** [4] which on oxidation with **chromium trioxide (CrO_3)** yields **5α-cholestan-3-one (5)**. The resulting **compound (5)** on reduction with *Zn-Hg/HCl* produces **5α-cholestane (6)**, which on further *oxidation with CrO_3* gives the desired product **5α-cholanic acid (7)**.

All these reactions stated above may be expressed as below :

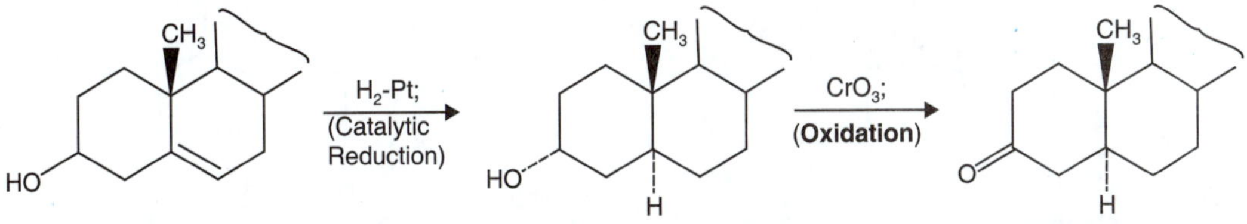

Cholesterol (1)

5α-Cholastan-3β-ol (4)

5α-Cholestan-3-one (5)

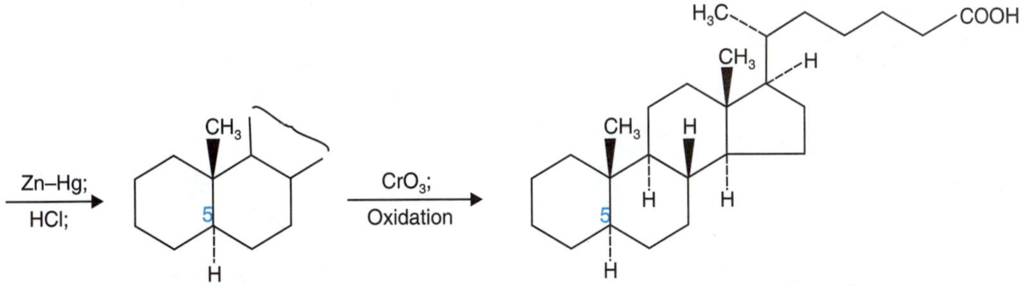

5α-Cholestane (6)

5α-Cholanic Acid* (7)

STRUCTURE OF α-HYODEOXYCHOLIC ACID [or HYODESOXYCHOLIC ACID]

The critical adaptation of the *oxidative degradation* (as described earlier)- one may emphatically ascertain the presence as well as exact position of the *two* **hydroxyl (-OH)** moieties in – hyodeoxycholic **acid (I)** as given under :

---

* That is , **Allocholanic Acid.**

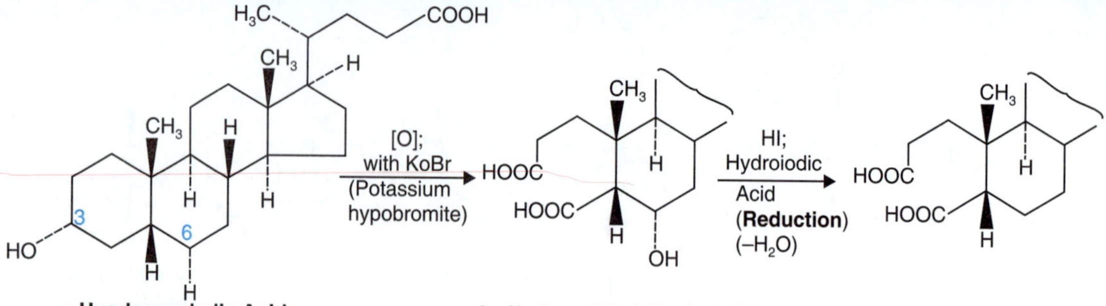

α-Hyodeoxycholic Acid
(I)

6α-Hydroxylithobilianic Acid
(II)

Lithobilianic Acid
(III)

Explanation – The above reactions may be explained as under:

❑ **α-Hyodeoxychoclic Acid (I)** when treated with *potassium hypobromite (KOBr)* causes cleavage of **Ring 'A'** to yield **6α-hydroxylithobilianic acid (II).**

❑ **Product (II)** upon reduction with *hydroiodic acid (HI)* gives rise to the formation of **lithobilianic acid (III)** with the loss of a mole of water (*i.e.*, removal of **6α-hydroxyl group**).

Comments – Based on the above reactions, one point is absolutely clear that the precise position of **one** of the *two* hydroxyl [-OH] group is at C-3 in Ring 'A'; whereas, the exact position of the **other** hydroxyl (-OH) moiety remains to be ascertained by a reported procedure as given under:

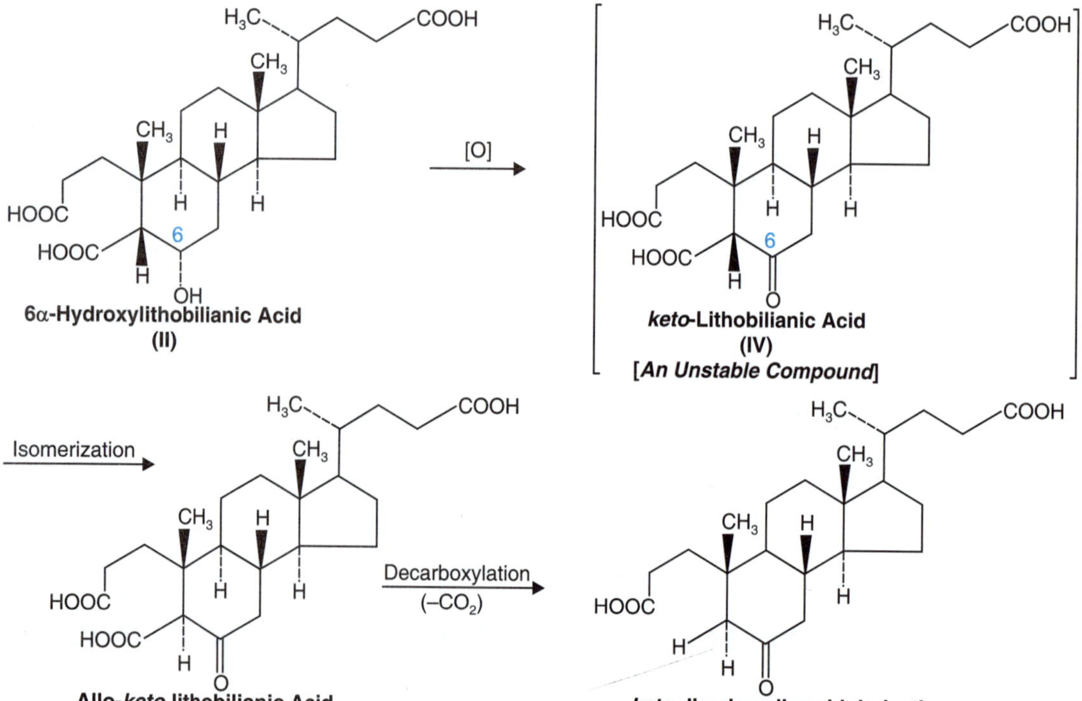

6α-Hydroxylithobilianic Acid
(II)

keto-Lithobilianic Acid
(IV)
[*An Unstable Compound*]

Allo-*keto*-lithobilianic Acid
(V)

keto-dicarboxylic acid derivative
(VI)

Explanation – Various steps involved above are explained as under :

1. **6α-Hydroxylithobilianic acid (II)** on *oxidation* yields an **unstable compound** (*i.e.,* **keto-lithobilianic acid (IV)**, which undergoes *isomerization* to give **allo-*keto*-lithobilianic acid (V)**.
2. The resulting **compound (V)** upon *decarboxylation* gives ***keto*-dicarboxylic acid derivative (VI)**.

STRUCTURE OF CHENODIOL [or CHENODEOXYCHOLIC ACID]

In true sense, the chemical structure of **chenodiol** (or **chenodeoxycholic acid**) is duly elucidated in the same manner more or less as that of ***other bile acids*** (*described earlier by adopting the so-called* **oxidative degradation procedure**).

Important Facts

(1) **Chenodiol** (or **chenodeoxycholic acid**) **(I)** may be duly converted into the *cholanic acid* by *two* sequential steps :
  - *oxidation* to the respective '**diketo acid**', and
  - subsequent treatment by *Clemensen's Reduction*.

(2) Interestingly, the *cholanic acid* may also be obtained directly from **chenodiol** by the following *two* steps, namely:
  - **pyrolysis,** and
  - **hydrogenation.**

Remarks- From the above *two* glaring facts one may arrive at the conclusive observation that **chenodiol (or chenodeoxycholic acid)** happens to be a *structural analogue* (**derivative**) of **cholanic acid.**

Let us take into consideration the **following sequential reaction steps:**

(i) *Oxidation* of **chenodiol (I)** gives **chenodeoxybilianic acid (II)**, which upon *further oxidation* yields **chenocholoidanic acid (III)** *i.e.,* a **pentacarboxylic acid derivative**. Obviously, there are *two* **carboxyl (-COOH) groups** are duly attached to the **same C-atom** (*marked as shown*) in **compound (III)**. Therefore, **chenocholoidianic acid (III)** when subjected to heat promptly **loses one carboxyl (-COOH) moiety** to be eliminated as CO_2–thereby resulting into the production of a **tetracarboxylic acid derivative (IV).**

(ii) However, the **compound (IV)** may also be obtained by the *oxidation* of ***keto*-carboxylic acid (V)** obtained from **hyodeoxycholic acid.**

Thus, we may write all the aforesaid reactions as stated under:

Oxidation →

Chenodiol [Chenodeoxycholic Acid] (I)
[mp=119°C; $[\alpha]_D^{20}$ + 11.5° (dioxane)]

Chenodeoxybilianic Acid (II)

(Contd........)

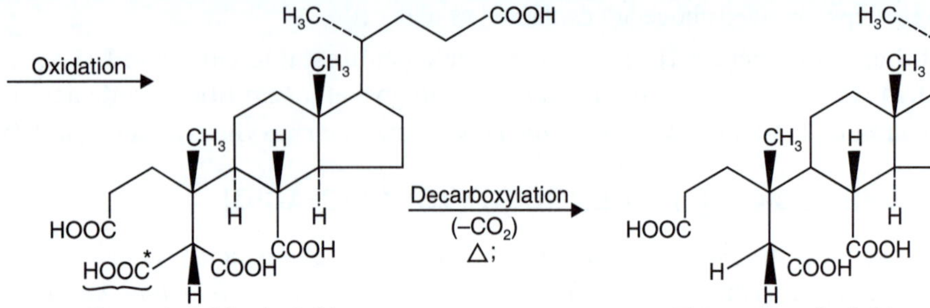

Chenocholoidianic Acid
(III)
(*Decarboxylation)

Tetracarboxylic Acid
Derivative
(IV)

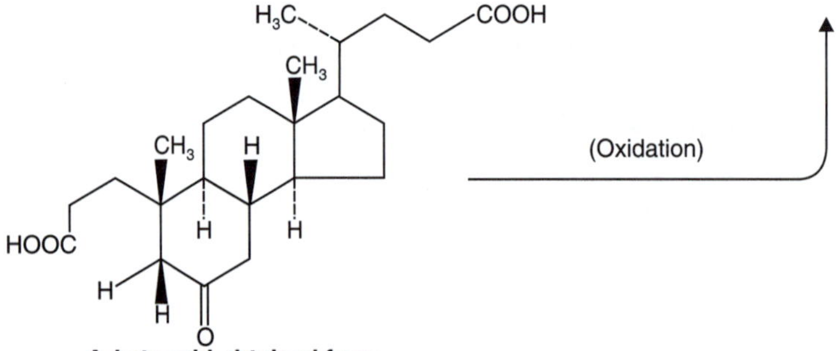

A keto-acid obtained from
Hyodeoxycholic Acid

Comment – The critical presence of *two* **hydroxyl (-OH) moieties** in **chenodiol** (or **chenodeoxylic acid**) (**I**) positioned at **C-3** and **C-7** (as shown above) has been further ascertained by means of the following reaction:

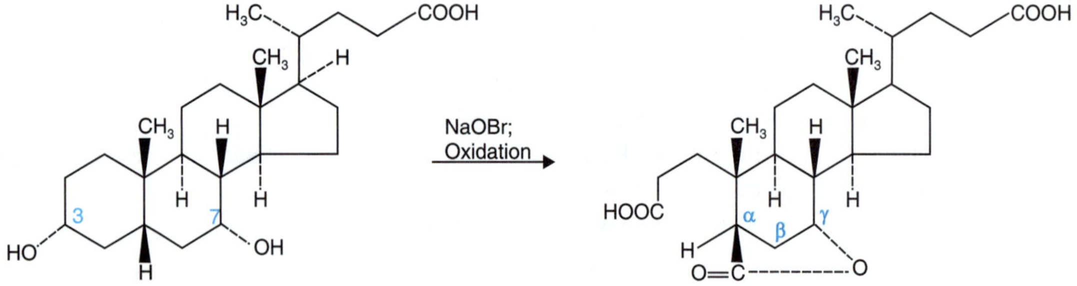

Chenodiol [Chenodeoxycholic Acid]
(I)

γ-Lactone dicarboxylic Acid Derivative
(V)

Explanation – The above reaction may be explained as under:

❑ **Chenodiol (I)** on being *oxidized* with *sodium hypobromite* (**NaOBr**) yields a **γ-lactonedicarboxlic acid derivative (V)**. However, the resulting **lactone analogue (V)** fails to react with either:
  • **hydroxylamine (-NH.OH)**, or
  • **undergo oxidation**.

❑ Therefore, one may infer safely that the *two* **hydroxyl moieties** are located at **C-3** and **C-7** positions in chenodiol (**I**).

Biological Role of the Bile Acids – Following are the six most important **biological role of the Bile Acids**', namely:

1. Interestingly, the so-called '**Bile Salts**' invariably associated themselves with such vital *chemical entities* as :
   - **cholesterol,**
   - **fatty acids**, and
   - **fat-soluble vitamins** (*viz*, **Vitamins A,D,K and E**).

   Incidentally, most of these products are found to be *insoluble in water of digestive juices* (or *secretions*)–and ultimately give rise to the formation of **water–soluble complexes**. Hence, the '**bile acids**' do make it a lot easier for enabling the *critical* absorption of the aforesaid *three* substances (*viz*, **cholesterol, fatty acids, and fat-soluble vitamins**) right into the *circulating blood–stream from the intestinal passage.*

2. **Bile acids** usually facilitate the proper digestion of the *ingested fats* by **emulsifying them first of all** and later on *enhancing the actual surface area* of the material for the ensuing '**pancreatic enzymes**'. Besides, the on-going *emulsifying phenomenon* helps to convert the '**fats**' (that are *water insoluble*) into the respective *water soluble* **chemical entities (compounds)** that may be *duly absorbed in the intestine.*

3. **Bile acids** are also found to **activate** specifically the following *two* enzymes, namely:
   - *cholesterol esterase*, and   • *pancreatic lipase*.

4. **Bile acids** predominantly aid in the overall absorption of **cholesterol** and **fat-soluble vitamins** (*eg.*, **vitamins A, D, and E**) by forming their corresponding **water-soluble complexes**.

5. **Bile acids** do help in a bigway to keep and sustain the **cholesterol in solution**. In a situation, when the prevailing **ratio between bile acids and cholesterol** falls drastically from the *normal level*-then **cholesterol gets duly precipitated** and eventually gives rise to the **formation of gallstones*** in:
   - ❏ **Liver**, and
   - ❏ **Gall Bladder.**

6. It has been duly observed that the '**bile acids**' in the bile entering the *intestinal passage* are absorbed rapidly right into the **circulating blood**-taken **back by the liver**-and **reutilized**.

   Interestingly, the above mentioned phenomenon is usually termed as the '*enterohepatic circulation of bile acids*'. Thus, the so-called **unabsorbed bile acids** are duly attacked by **micro organisms**; and subsequently, get decomposed into several *biochemical degradation products* that are *excreted in faeces* ultimately.

6. SAPONINS AND SAPOGENINS

Saponins- represents a group of '**amorphous colloidal glycosides**' that usually form *soapy-aqueous solutions*.

In fact, the term '**saponin**' (*Latin*: *sapo* means soap) is invariably applied to *two* **groups of plant glycosides** which eventually give rise to the formation of **colloidal, soapy solutions in an aqueous medium.**

---

* **Gallstone :** A calculus duly formed in the **gallbladder** or **bile duct.**

How do the 'Saponins' act in a living system?

1. **Saponins** actually effect **hemolysis of red-blood cells (RBCs)** even at a very high dilution.
Examples :

(1) **Saponins** are quite often used as '**fish poisons**' *i.e.*, fish are dazed by maceration of a *suitable plant* in the water, but amazingly are *not rendered inedible* at all.

(2) **Triterpenoid Glycosides** – It is one of the groups of **saponins** that are found abundantly in nature. In fact, the majority of the **triterpenoid sapogenins** are *pentacyclic in status*, and fall into *three* distinct categories, namely:

- α-amyrin,
- β-amyurin, and
- lupeol.

The **chemical structures** of all the *three* terpenoid sapogenins are entirely based upon the *investigative studies* of the famous *Zurich School* (Ruzicka, Jeger, 1917)*.

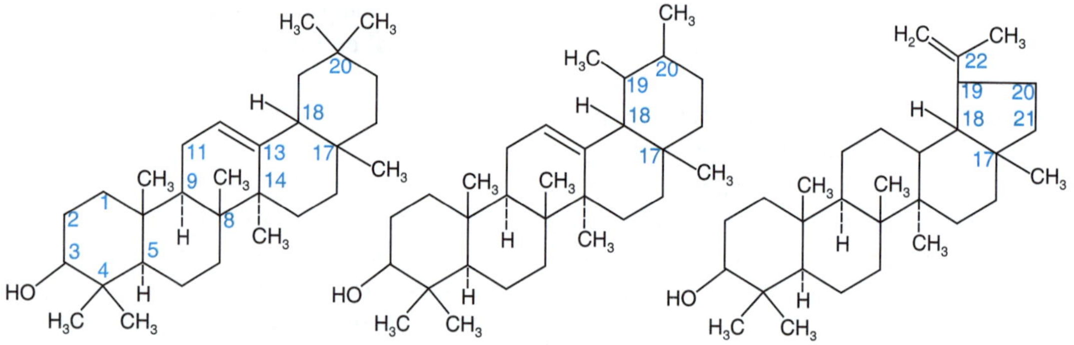

β-Amyrin   α-Amyrin   Lupeol

Reactivity of Double-Bond in 3- Terpenoid Sapogenins – are as given under:

➤ **α-Amyrin** : Markedly inert;

➤ **β-Amyrin** : Somewhat hindered; and

➤ **Lupeol** : Normal reactivity.

Importantly, the *second* **group of saponins** are the so-called '**steroid saponins**'; which being duly exemplified by such **rare glycosides**, such as :

- **Digitonin**
- **F-Gitonin** and
- **Tigonin**

that eventually occur in the *digitalis seeds* usually as an *integral companion* of the **cardiac glycosides**, for instance :

- ❑ **Digitoxin**
- ❑ **Gitoxin** and
- ❑ **Digoxin.**

---

* Okar Jeger, b-1917 Lwow, Poland; D. Sc, ETH Zuruch (Ruzica Prelog); ETH Zurich

Digitonin

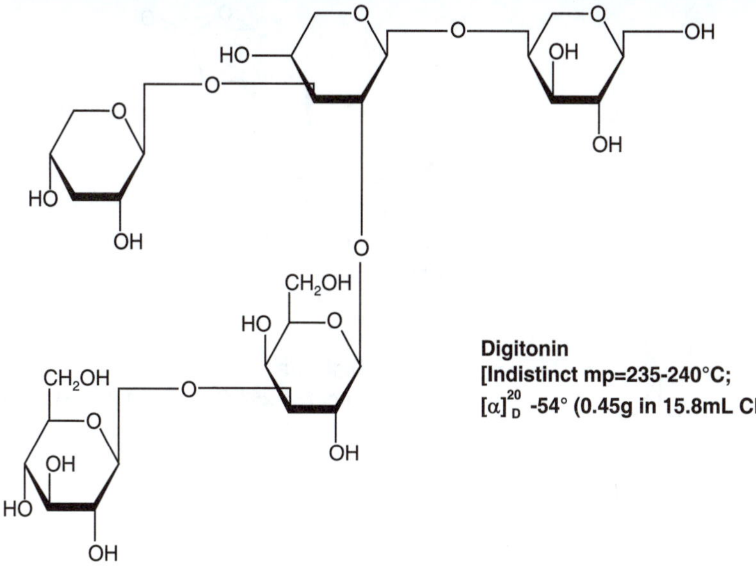

Digitonin
[Indistinct mp=235-240°C;
$[\alpha]_D^{20}$ -54° (0.45g in 15.8mL CH_3OH)]

F-Gitonin

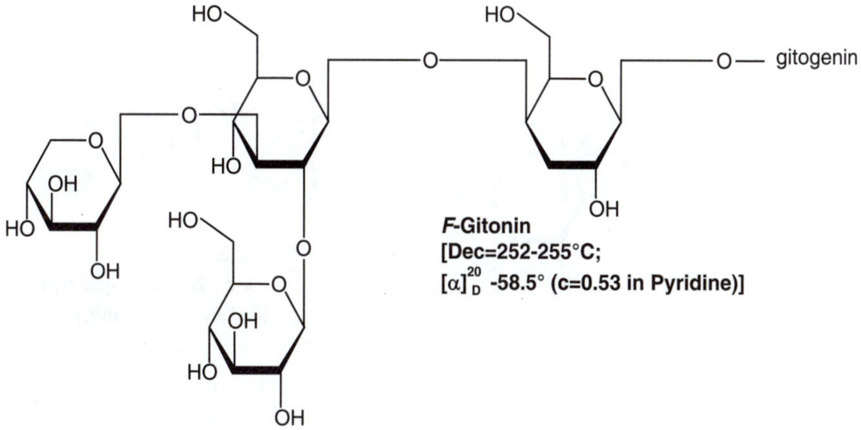

gitogenin

F-Gitonin
[Dec=252-255°C;
$[\alpha]_D^{20}$ -58.5° (c=0.53 in Pyridine)]

Tigonin

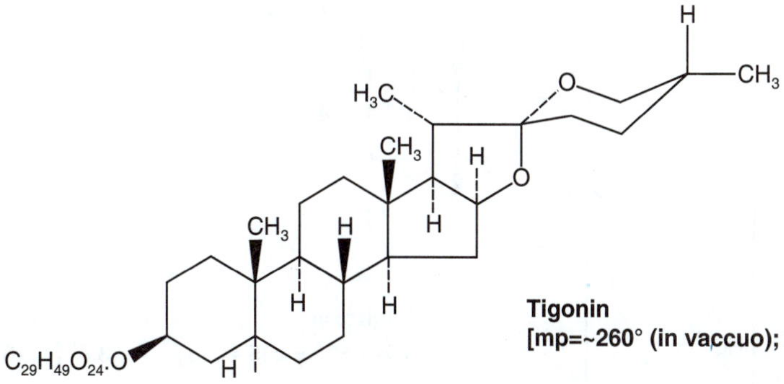

$C_{29}H_{49}O_{24} \cdot O$

Tigonin
[mp=~260° (in vaccuo); Soluble in water]

(Contd....)

Tigonin

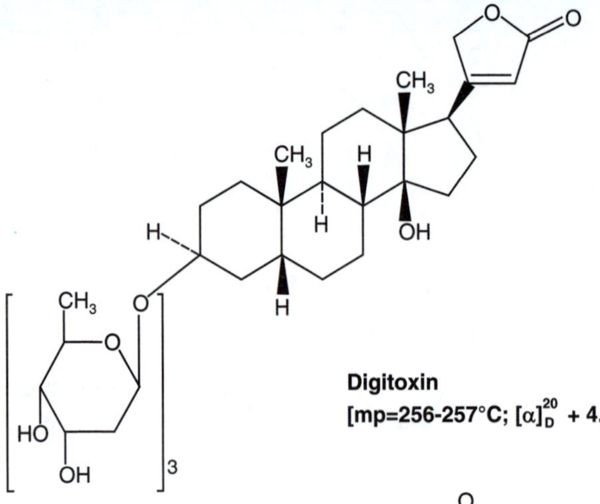

Gitoxin

Digoxin

Digitoxin
[mp=256-257°C; $[\alpha]_D^{20}$ + 4.8° (c=1.2 in dioxana)]

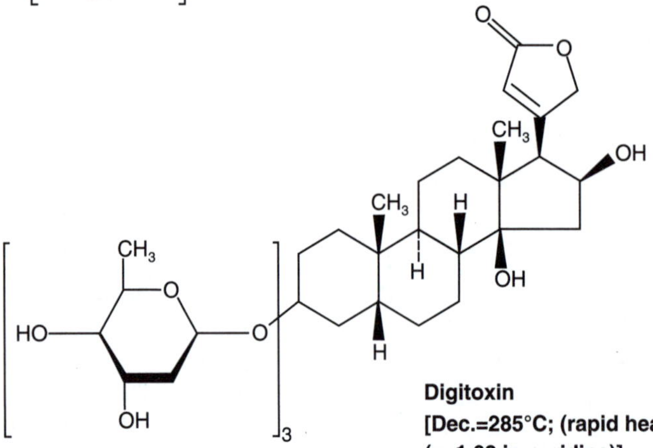

Digitoxin
[Dec.=285°C; (rapid heating); $[\alpha]_{546}^{20}$ + 3.5° (c=1.02 in pyridine)]

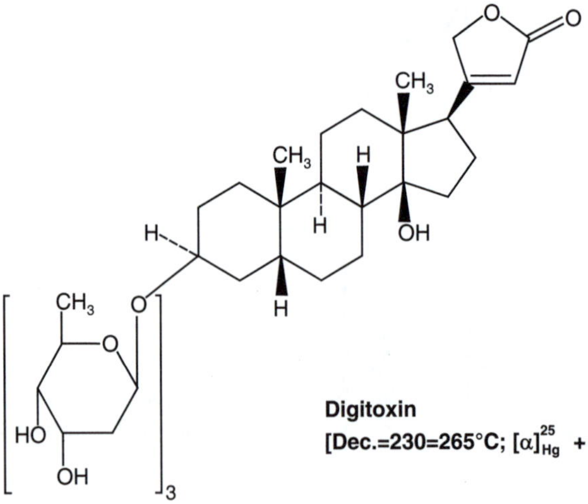

Digitoxin
[Dec.=230=265°C; $[\alpha]_{Hg}^{25}$ + 13.8° (c=10 in pyridine)]

(3) **Properties of Aqueous Solutions of Saponins** – Interestingly, the ingestion of *aqueous solution of saponins* orally prove to be relatively harmless.

Following are *two* **distinct properties of the aqueous solutions of saponins**, namely :

- **cause hemolysis on injection**, and
- notably **possess the unique property of forming insoluble complexes (1:1) critically with the 3β-hydroxyl steroids [***eg., cholesterol + digitonin (ie., the saponin)***].**

(4) **Source of Steroidal Saponins** – The first and foremost known source of the **steroidal saponins** was found to be *Digitalis purpurea*, which gained its immense popularity as a vital source of the 'cardiac glycosides'. However, at a latter stage **Marker (1940-49)** discovered certain *richer sources of steroidal saponins* in a variety of **medicinal plant species**, such as :

- *Liliacease*,
- *Dioscoreaceae*, and
- *Serophulariaceae*.

> **NOTE** The '*saponins*', unlike the '*sapogenins*' are not so-easy to isolate in the purest form; and, therefore, they have been investigated rather scantily. However, as-to-date these studies are being done exhaustively with utmost care and perfection.

(5) **Sugars of Saponins** – In general, the '**saponins**' invariably consist of a

'**linear arrangement of 1 to 6 *hexose* or *pentose* glycoside unites linked to the respective sapogenin aglycon at its C-3 hydroxyl (-OH) moiety**'.

Nevertheless, the most common '**sugars of saponine**' are, namely:

- **D-Glycose** • **D-Xylose** • **L-Arabinose** and • **L-Rhamnose**

(6) **Structure of Saponins** – Importantly, the actual **chemical** structures of a host of naturally occurring '**Saponins**' are duly established by the so-called **conventional degradation procedures**.

Following are the *two* completely established and elucidated structures of '**saponins**', namely:

- **Dioscin**, and • **Sarsaporilloside**.

Dioscin

DIOSGENIN

DIOSGENIN

Dioscin
[Dec.=275=277°C; $[\alpha]_D^{13}$ – 115° (c=0.373 in Ethonal)]

Sarsaporilloside

CH_3

H_3C

CH_3 H

CH_3 H   H

H

H   H

β–D–Glucose

H

H
OH H

α–L–Rhammose–O   H

H   O–β–D–Glucose   **Sarsaporilloside**

Special Remarks – Evident from the survey of literatures it is quite obvious that the present day interest of the 'phytochemists' entirely rests upon the so-called '*Sapogenin Aglycones*' *i.e*, the non-sugar components derived from the *plant glycosides*.

Importantly, a few remarkably important **sapogenin aglycones**, such as :

- **Hecogenin**,   and   • **Diosgenin**,

are indeed found to be useful commercially as the viable *raw materials* in the **exclusive synthesis of the steroids** *eg.*, **Progesterone** (a *female hormone*)- **prepared from 'diosygenin'**.*

SAPOGENINS

Sapogenins may be defined as – 'aglycone of the saponin glycosides and are characterized by the presence of a spiroketal side-chain'.

Based on the nature of the 'aglycone residue' duly present in the *saponin glycosides*, these are broadly classified into the following *two* categories, such as :

- **Tetracyclic triterpenoid saponins (or Steroidal Saponins)**, *eg.*, **Diosgenin, Solasonine, Shatavarin 1** ; and

- **Pentacyclic triterpenoid saponins** *eg.*, **α-Amyrin, β-Amyrin**, and **Lupeol** (*discussed earlier*).

In general, the 'Sapogenins' are recognized to represent a cluster of **27- cholestane derivatives**, which are invariably derived from the respective 'saponins' either:

- **due to** *acid hydrolysis*, or   • **due to** *enzymatic hydrolysis*,

and thus, possess a *unique spiroketal side-chain*.

Marker (1943) discovered **nologenin** and a host of other **sapogenins** through an investigative research of hundreds of plants grown in **Mexico** and **Southern United States** that could be ultimately degraded to **progesterone** (a *female sex-hormone*).

Marker had developed an altogether newer and efficient **3-step procedure** of accomplishing the *critical/specific cleavage* of the **side-chain at C-17 in the steroidal nucleus** as depicted under:

---

* Kar, A: **Medicianl Chemistry,** 6th edn., New Age International, New Delhi, 2015.

□ **Conversion of Sapogenin (A) into 17-methyl ketone derivative (D);** and
□ **Conversion of Diosgenin (E) into Progesterone (F).**
□ **Conversion of Sapogenin (A) into 17- Methyl-ketone derivative (D):**

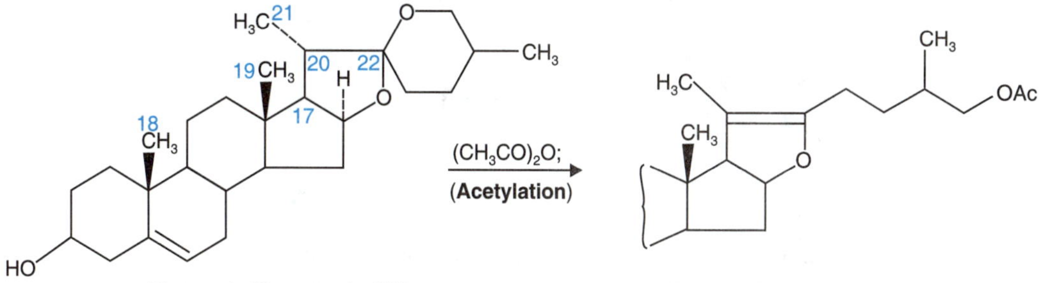

Diosgenin [Sapogenin (A)]
[A Spiroketal Steroid Nucleus]

Pseudosapogenin
[B]

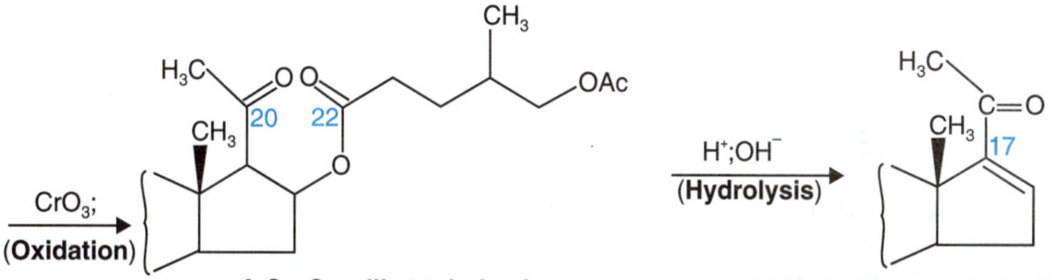

A C_{20}, C_{22}=diketo derivative
[C]

17-Methyl-ketone derivative
[D]

Explanation – The above *three* sequential steps may be explained as under :

1 *Acetylation* of **sapogenin (A)** (*viz*, **diosgenin**) *i.e.*, a *spiroketal steroid nucleus*, yields **pseudosapogenin (B)**, which upon *oxidation* with CrO_3 gives a **diketo (C_{20}, C_{22}) derivative (C)**
2. The resulting **product (C)** when subjected to *hydrolysis* either in an *acidic or alkaline medium* gives rise to the formation of **17-methyl ketone derivative (D).**

Conversion of Diosgenin (E) into Progesterone (G)

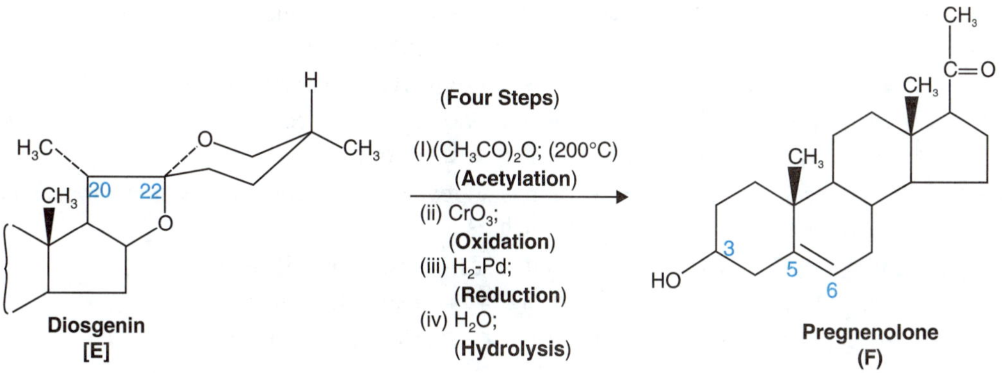

Diosgenin
[E]

(Four Steps)
(I) $(CH_3CO)_2O$; (200°C)
   (Acetylation)
(ii) CrO_3;
   (Oxidation)
(iii) H_2-Pd;
   (Reduction)
(iv) H_2O;
   (Hydrolysis)

Pregnenolone
(F)

(Contd........)

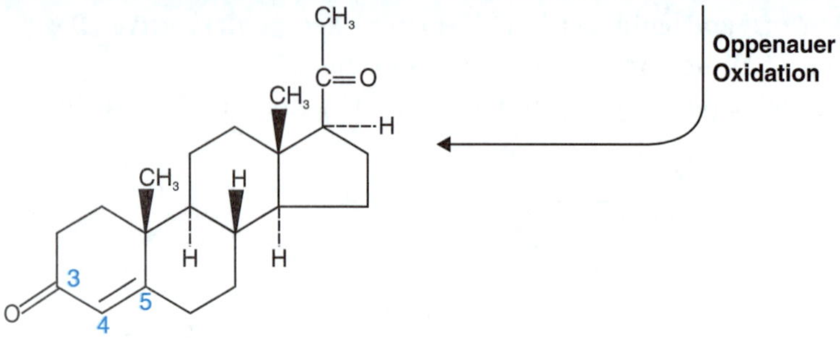

Oppenauer
Oxidation

Progesterone
(G)

Explanation – The above reactions may be further expatiated as under:

(1) **Diosgenin (E)** on being *acetylated* with acetic anhydride (at 200°C) – followed by *oxidation* with CrO_3 – followed by *catalytic reduction* with H_2-Pd-and lastly *hydrolysis* yields **pregnenolone (F).**

(2) The resulting **product (F)** on *Oppenauer oxidation* gives **progesterone (G)** *ie.*, a female–sex hormone.

CONSTITUTION OF SAPOGENINS

The **constitution of sapogenins** may be elucidated in a methodical manner on the basis of the following *nine* **important sequential steps**, namely:

   (i) **C-27 compunds – Authnetication**;
   (ii) **Presence of a Steroidal Nucleus**;
   (iii) **Ring System**
   (iv) **Spiroketal Group**;
   (v) **Iso-Reactions**;
   (vi) **Absolute Configuration at C-25**;
   (vii) **Configuration at C-20 in Naturally Occurring Sapogenins**;
   (viii) **Configuration at C-22**; and
   (ix) **Synthesis of Diosgenin.**

[A] C-27 Compounds : Authertication – It has been duly established that the '**sapogenins**' do represent broadly a group of **C-27 compounds**, which in fact has been reasonabley authenticated and confirmed adequately after the following specified means and ways, namely:

   • **careful purification**, and   • **repetitive spectroscopic analysis*.**

[B] Presence of a Steroidal Nucleus – **Sapogenins** on being subjected to *dehydrogenation* with **Se** yields **Diels Hydrocarbon** plus a '**ketone' having 8 C-atoms.**

Remarks – The aforesaid *two* critical observations do reveal and ascertain the '**sapogenins**' essentially possess the following *two* **definite systems,** such as:

   ❑ **a steroidal Ring System,** and   ❑ **a 8C-atom System.**

[C] Ring System – When either of the following *two* **sapogenins,** namely:

   • **Tigogenin acetate (1A),**   or   • **Sarsasapogenin (1B),**

---

   * That is, **sophisticated characterization methods** *eg.*, FT- IR, **UV**,NMR,MS, HPLC, GLC, HPTLC, RP-chroma-
   tography, CD, ORD, XRD, Electro-chromatography.

wre degraded carefully to the corresponding **etioallobilianic acid (5)** and finally to **eitobilianic acid (7)** respectively.

Remarks – The above scientific evidences explicitly establish the following *three* **genuine and valuable clues and informations,** for instance :

➤ *chemical structure* of the ring system, ➤ *configuration* of the ring system, and
➤ *side-chain* comprises essentially a *tetrahydrofuran ring fused to Ring D.*

Thus, we may have the following set of ractions :

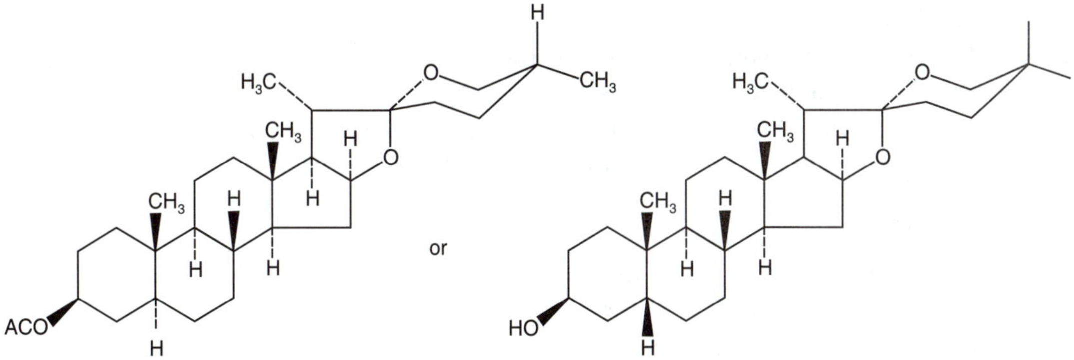

Tigogenin Acetate [1A]
[mp=203°C; $[\alpha]_D^{20}$ −62°]

Sarsasapogenin [1B]
[mp=199-199.5°C; $[\alpha]_D^{25}$ −75° (c=0.5 in CHCl$_3$)]

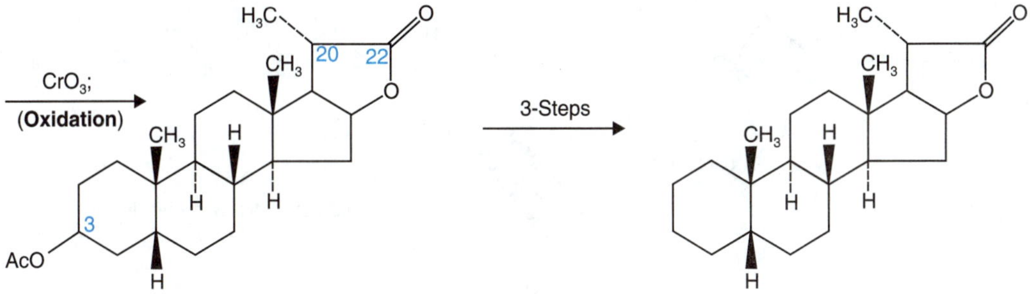

$\xrightarrow[\text{(Oxidation)}]{\text{CrO}_3;}$

3β-Aceto-22-*keto* lactone derivative [2]

$\xrightarrow{\text{3-Steps}}$

22-keto-lactone derivative [3]

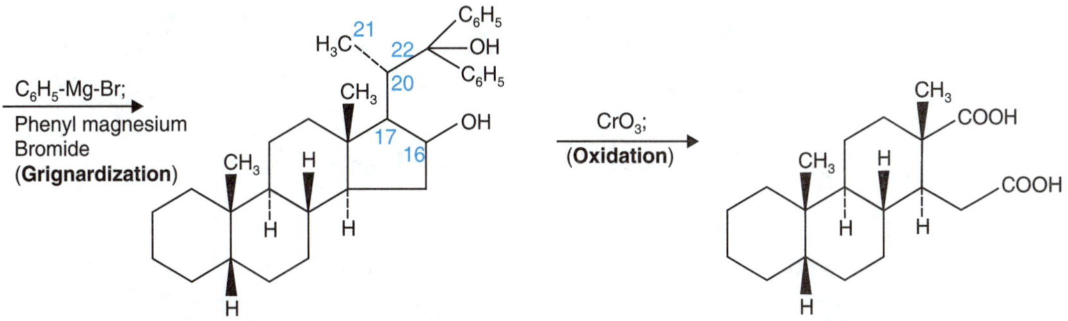

$\xrightarrow[\substack{\text{Phenyl magnesium} \\ \text{Bromide} \\ \textbf{(Grignardization)}}]{\text{C}_6\text{H}_5\text{-Mg-Br;}}$

16,22-Dihydroxy-22-diphenyl derivative
[Cessation of *Lactone Ring*] [4]

$\xrightarrow[\text{(Oxidation)}]{\text{CrO}_3;}$

Etioallobilianic Acid [5]

Explanation – These essentially include :

1. **Tigogenin acetate (1A)** or **Sarsasapogenin (1B)** when *oxidized* with **CrO₃** yields **3β-aceto-12-keto lactone derivative (2)**, which on further treatment gives **22-keto-lactone derivative**. The *three* sequential steps gives the **22-keto-lactone derivative (3)**.

2. The resulting **product (3)** under *Grignardization* **with phenyl magnesium bromide ($C_6H_5.Mg.Br$)** produces the **16, 22- dihydroxy-22-diphenyl derivative (4)** due to the cessation of the 'lactone ring'.

3. Finally, *oxidation* of **product (4)** with *chromium-6-oxide (CrO₃)* gives rise to the formation of **etiallobilianic acid (5)**.

[D] Spiroketal Group–It is, however, pertinent to state here that **Marker (1943)** perceptively observed the following *two* **vital and critical aspects**, for instance:

❑   presence of *two* **O-atoms in the side-chain of the 'sapogenin'**- that are found to be unusually active in an *acidic environment viz.,*

    •   **halogenation,** and
    •   **bromination**; and

❑   **Clemmensen Reduction – that predominantly indicate the specific presence of a carboxyl (>C=O) moiety in the form of a 'spiroketal entity'.**

Let us look into the following sequence of reactions:

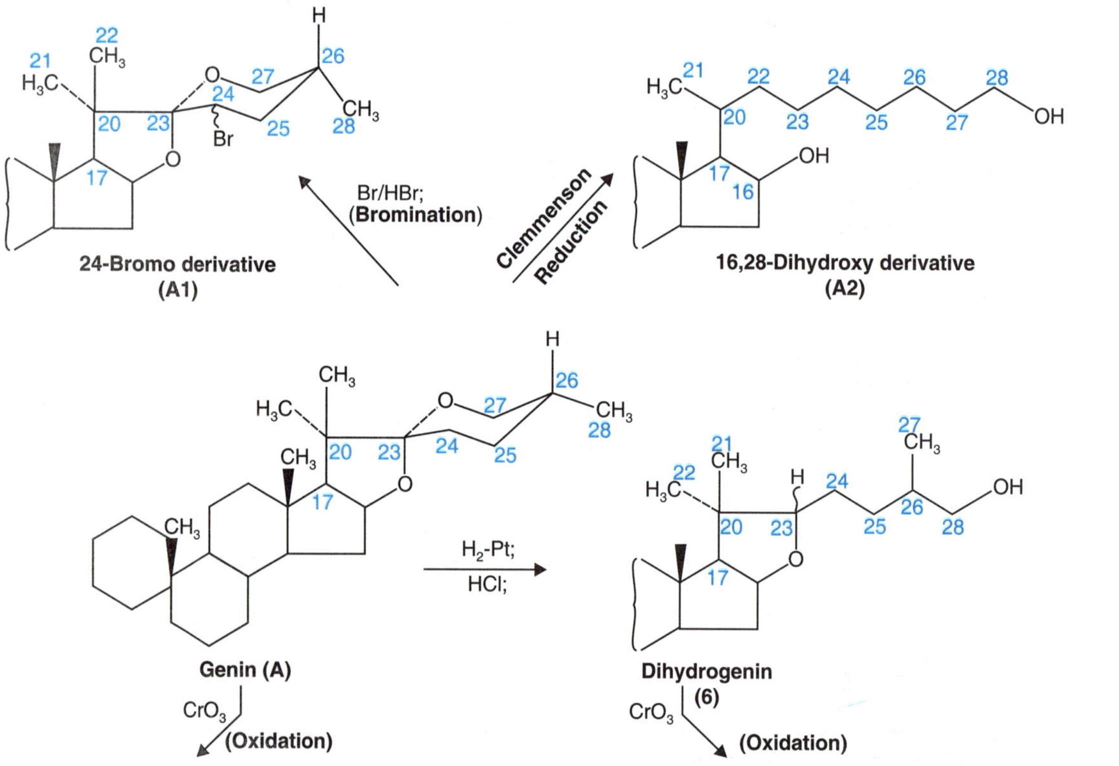

24-Bromo derivative
(A1)

Br/HBr;
(Bromination)

Clemmenson
Reduction

16,28-Dihydroxy derivative
(A2)

Genin (A)

CrO₃
(Oxidation)

H₂-Pt;
HCl;

Dihydrogenin
(6)

CrO₃
(Oxidation)

(Contd........)

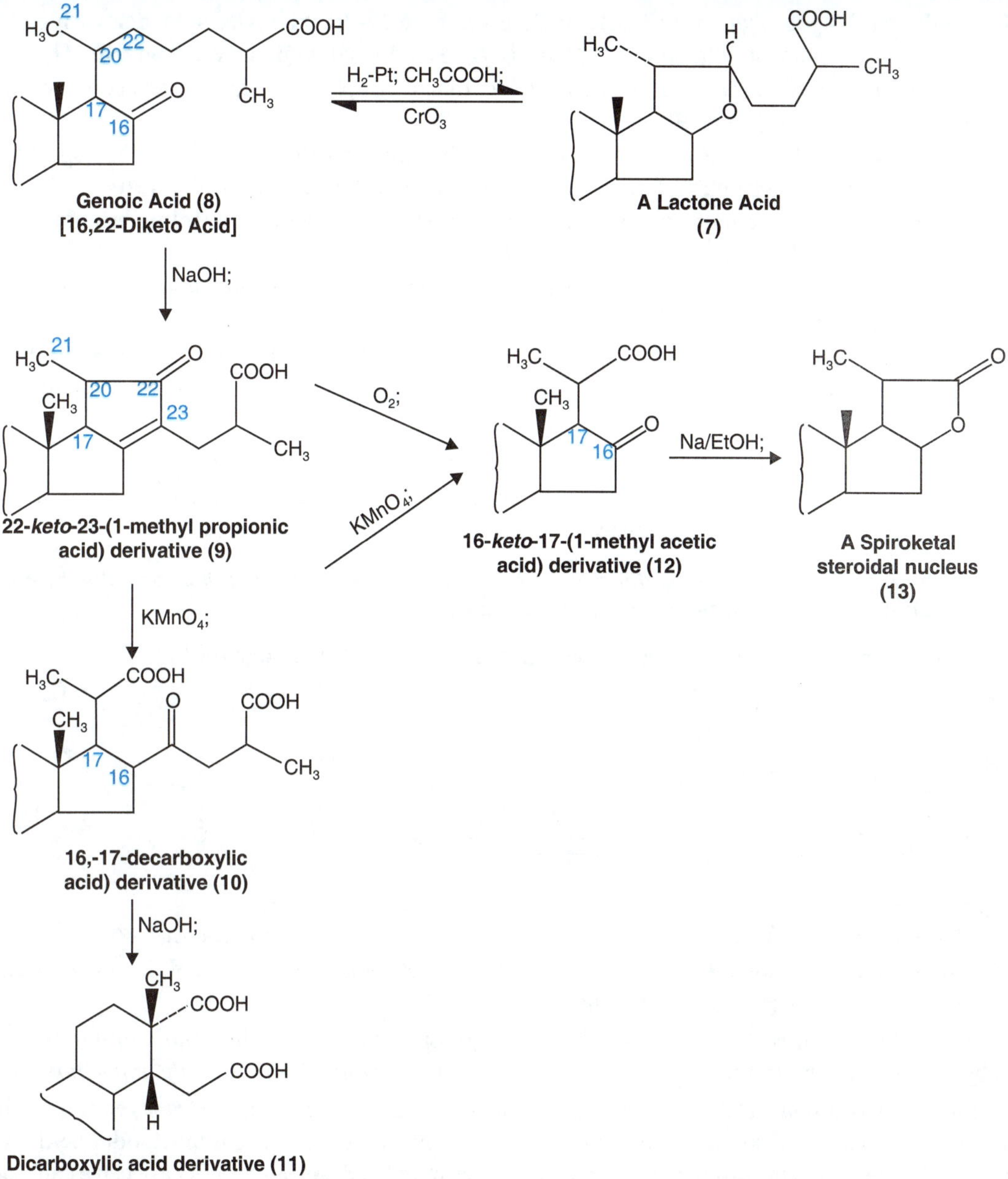

Genoic Acid (8)
[16,22-Diketo Acid]

H₂-Pt; CH₃COOH;
CrO₃

A Lactone Acid
(7)

NaOH;

22-keto-23-(1-methyl propionic acid) derivative (9)

O₂;

KMnO₄;

16-keto-17-(1-methyl acetic acid) derivative (12)

Na/EtOH;

A Spiroketal steroidal nucleus (13)

KMnO₄;

16,-17-decarboxylic acid derivative (10)

NaOH;

Dicarboxylic acid derivative (11)

Explanation – The various reaction steps involved above may be explained as under:

1. **Genin (A)** on *bromination* with Br_2/HBr yields a **24-bromoderivative (A1)**; whereas, the same on **Clemmensen Reduction** gives the **16,28-dihydroxy derivative (A₂)**.

2. **Genin (A)** on catalytic reduction (H_2-Pt) followed by treatment with HCl yields **dihydrogenin (6)**.

3. **Genin (A)** on *oxidation* with chromium 6-oxide (CrO_3) **gives genoic acid (18)** (*i.e.,* a **16,22-diketo acid**), which upon reaction with an *alkali (NaOH)* yields a **22-keto–(1-methyl propionic acid) derivative (9)**.

4. **Product (9)** upon *oxidation* with $KMnO_4$ gives a **16,17-dicarboxylic acid derivative (10)**, which upon treatment with an *alkali (NaOH)* yields a **dicarboxylic acid derivative (11)**.

5. **Product (6)** on *oxidation* with CrO_3 gives a **locatone acid (7)** which shows a reversible reaction between **product (8) and (9)** as shown above.

6. Both **products (9) and (10)**, as obtained above, when treated with O_2 **and** $KMnO_4$ respectively – gives the same **compound (12)** *ie.*, **16-keto-17(1-methyl acetic acid)** derivative.

7. **Product (11)** when being treated with freshly cut pieces of **Na-metal I absolute ethanol** yields the desired sp*iroketal steroidal nucleus (13)*.

[E] ISO- Reactions– Importantly, the so-called naturally occurring '*steroidal sapogenins*' invariably give rise to several '**pairs of stereoisomers**'.

It is, however, pertinent to state here that when the aforesaid **pairs of stereoisomers** are refluxed carefully with an admixture of $HCl/C_2H_5\text{-}OH$, one may distinctly lay hands on *two* **types of sapogenins**, namely :

- ❏ '**Neo-Sapogenins**' – also known as '**normals**' *e.g.*, **Neosapogenin (X)**, and
- ❏ '**Iso-Sapogenins**' – obtained by the conversion of *neo-sapogenin* and are found to be relatively **more stable** *e.g.*, **Iso-sapogenin (Y)**.

NOTE   Importantly, the '*Iso-Reactions*' has been found to be quite *reversible*; and the *overall equilibrium usually slides to one side specifically.*

Following are *chemical structures* of **Neosapogenin (X)** and **Iso-sapogenin (Y)**:

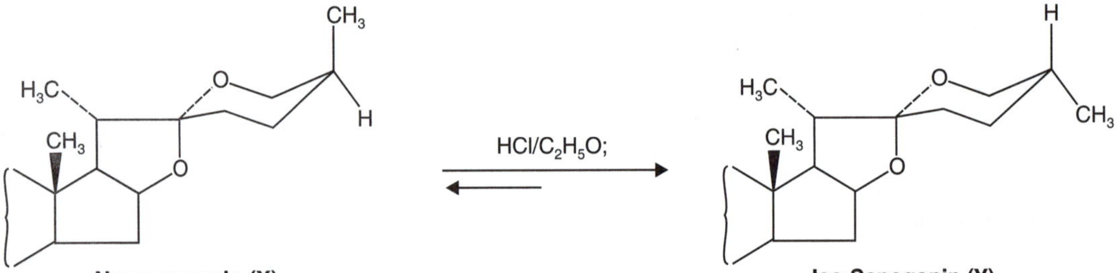

Neosapogenin (X) | Iso-Sapogenin (Y)

(F) Absolute Configuration at C-25– Marker (1943) postulated that both '*neo-*' and '*iso*'- compounds are definitely the *C-22* **isomers** of each other.

Nevertheless, both series *i.e.*, **neo-**and *iso*-compounds do produce the *same* compound (D), thereby ascertaining the fact that they (*ie.*, **X and Y**) *predominantly differ critically at position C-25*.

Besides, the **compound (G)** is obtained duly; and hence, the specific **C-22** (*marked*) configuration should by all means be *identical*-thereby suggesting vehemently that the actual '**antipodal acids (C)**' – are duly obtained from the *two* series. Ultimately, the ensuing **C-25 configuration** has been established justifiably by the degradation of *enone acid (Z)* as given under :

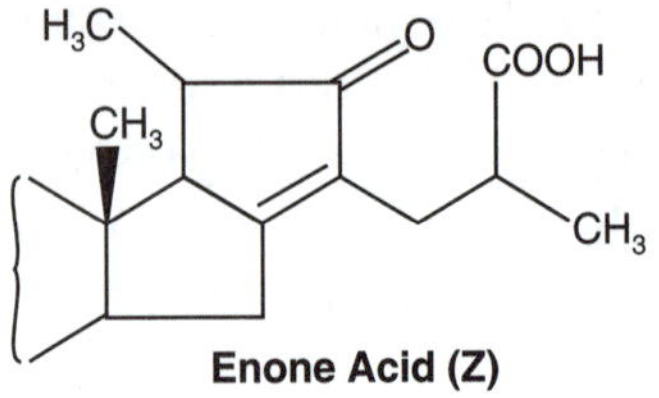

Enone Acid (Z)

In fact, the said **enone acid (Z)** has been derived from the **isosapogenin (A)** [*R-CH₃; R'=H*] to the known **(+)–methylsuccinic acid** as given under :

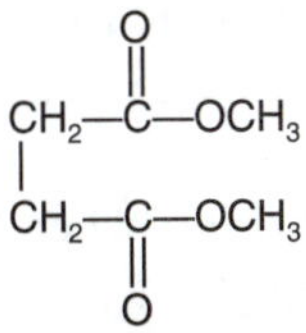

$$
\begin{array}{c}
\text{CH}_2\text{—C—OCH}_3 \\
\quad\quad\; \| \\
\quad\quad\; \text{O} \\
\text{CH}_2\text{—C—OCH}_3 \\
\quad\quad\; \| \\
\quad\quad\; \text{O}
\end{array}
$$

(+)–Methyl Succinate

The various steps involved in the determination for the **absolute configuration at C-25** may be expressed as under :

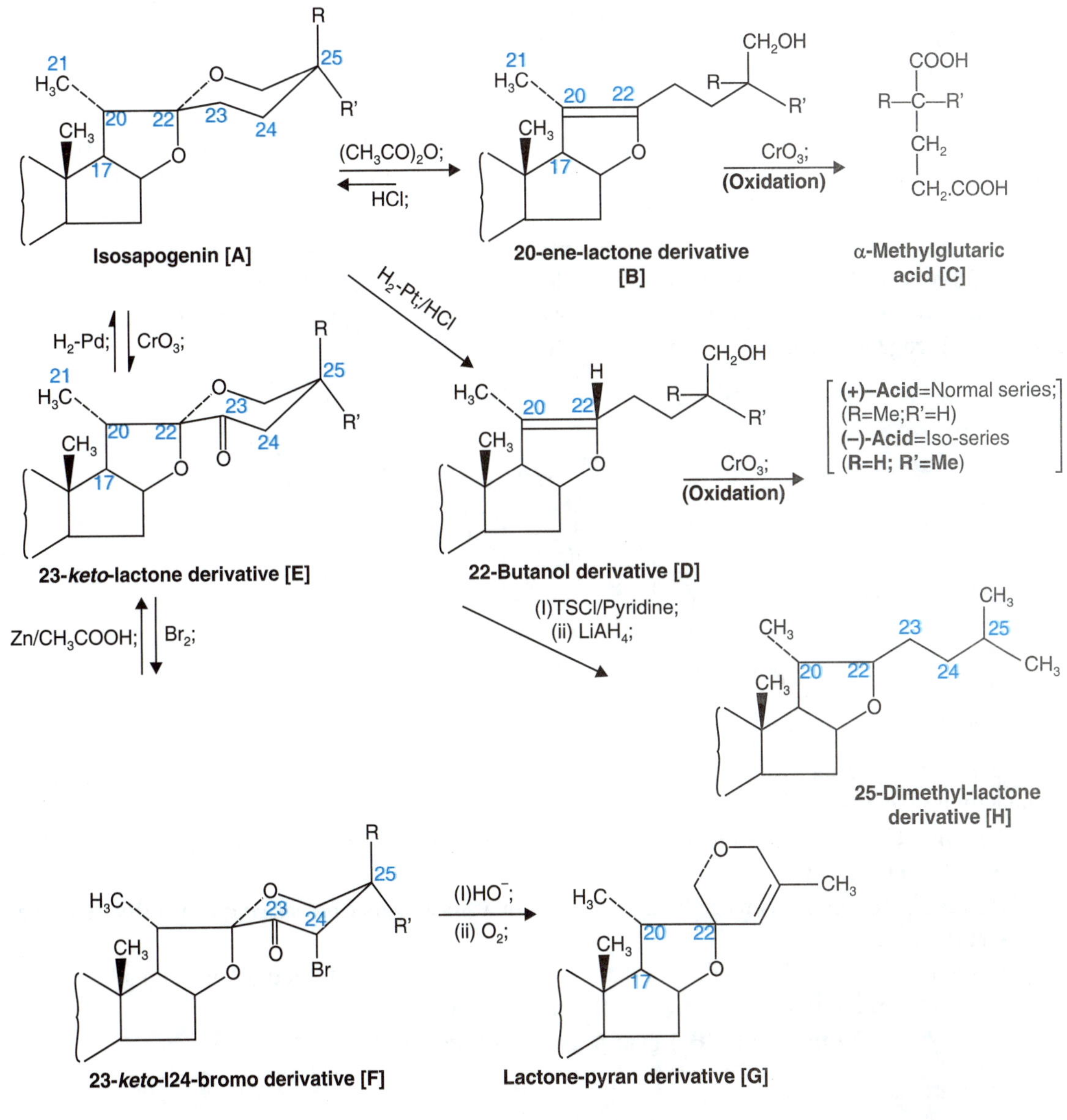

Isosapogenin [A]

20-ene-lactone derivative [B]

α-Methylglutaric acid [C]

23-keto-lactone derivative [E]

22-Butanol derivative [D]

[(+)–**Acid**=Normal series;
(R=Me;R'=H)
(–)-**Acid**=Iso-series
(R=H; R'=Me)]

25-Dimethyl-lactone derivative [H]

23-keto-I24-bromo derivative [F]

Lactone-pyran derivative [G]

Explanation – These essentially include :

1. **Isosapogenin (A)** on *acetylation* yields **20-ene-lactone derivative (B)**, which on treatment with HCl affords a *reversible reaction* to produce **compound (A)**.
2. **Compound (B)** on *oxidation* with CrO_3 gives **α-methylglutaric acid (C)** – that eventually is obtained in *two* **distinct forms**:
   - **Normal series** : (+) – Acid (*ie.*, when R=Me; and R'=H) ; and
   - **Iso series**   : (-) – Acid (*ie.*, when R=H; and R=Me).
3. **Compound (A)** gives on *oxidation* with CrO_3 yields **23-*keto*-lactone derivative (E)**, which affords a *reversible reaction* with **H_2-Pd (catalytic reduction)**.
4. **Product (E)** on *bromination* gives **23-*keto*-24-bromo lactone derivative (F)** and this on treatment with Z_n/CH_3COOH affords a reversible reaction.
5. **Product (F)** on alkaline exposure followed by O_2 gives rise to the formation of a **lactone pyran derivative (G)**.
6. Likewise, **compound (D)** first on treatment with **tosylchloride (TsCl)** and **pyridine**, followed by reduction with **lithium aluminium hydride ($LiAlH_4$)** gives **25-dimethyl-lactone derivative (H)**.

[G] Configuration at C-20 in Naturally Occurring Sapogenis–Based upon the so-called '*biogenetic aspects*', it may be observed critically that-

'the *naturally occurring sapogenins* should by all means possess the same configuration at C-20 as could be seen in most of the other steroids *i.e.*, the C-20 methyl (-CH₃) moiety is found to be quite rare from the *angular methyl (-CH₃) group positioned at C-13*.

Example : Tigogenin Acetate

Tigogenin Acetate [1]

Comments – The glaring evidences mustered in favour of the above stated facts is duly supported by the underlying fact that-

1. **Lactone A (I)** is found to be definitely much more stable in comparison to its corresponding *isomer* **lactone B (J)**.
2. However, **lactone B (J)** when being treated with a 'base' gets *isomerized* quantitatively to the **lactone A (I)**.
3. It is indeed worthwhile to state here that the **inherent vigorous interaction'** prevailing between :
   - 'angular C-13 methyl (-CH₃) moiety, and
   - 'C-20 methyl (-CH₃) group in the lactone ring',

predominantly remains responsible for the apparent instability of **lactone B (J)** with regard to **17-keto-3,16-diacetyl steroid (K)**.

The various steps of reactions involved above may be written as follows;

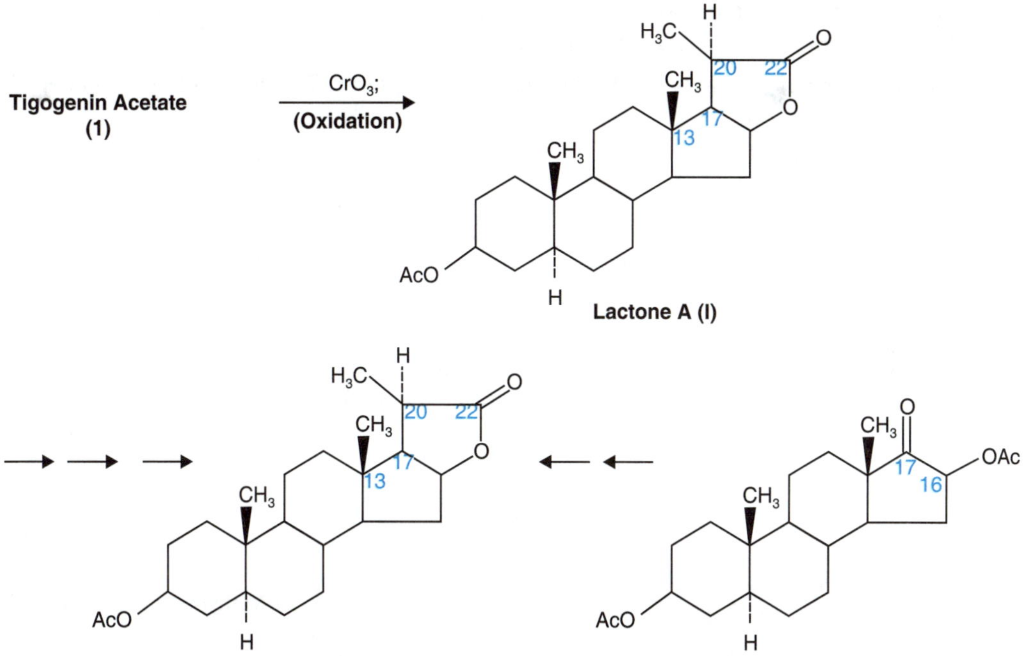

**Tigogenin Acetate
(1)**

CrO₃;
(Oxidation)

Lactone A (I)

Lactone B (J)

**17-keto-3,16-diacetyl steroid
(K)**

[H] Configuration at C-22

Conformation relates to the shape of an organic compound accomplished duly by the critical '*rotation of atoms*' around the single bonds (*i.e.*, covalent bonds). It may be further exemplified by the specific *size and shape of proteins* produced meticulously by:

- **twisting peptide chains,** and /or - **folding peptide chains.**

Therefore, based upon the well known '*conformational analysis*', solely dependent upon the *known greater stability* of the so-called '**iso sapogenins**' ultimately leads to the achieved *configuration* as depicted in :

- **Neosapogenin (X); and** - **Iso-sapogenin (Y).**

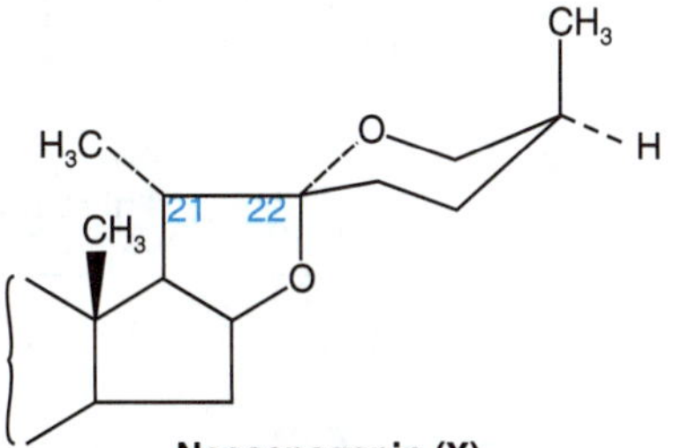

Neosapogenin (X)

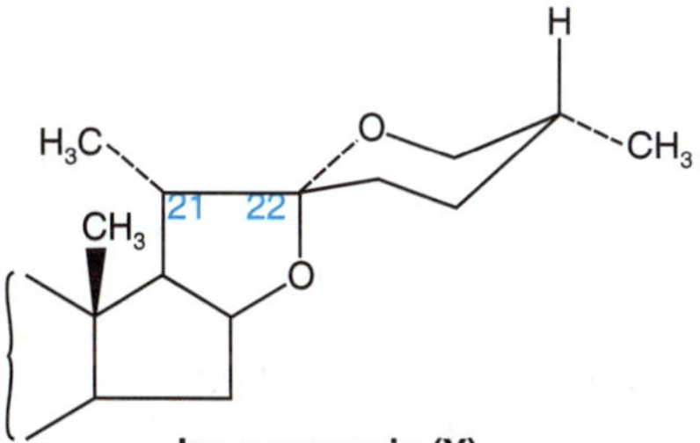

Iso-sapogenin (Y)

[I] SYNTHESIS OF DIOSGENIN

The first and foremost **synthesis of the** *sapogenin i.e.*, **diosgenin***, *via* **non-stereospecific route** was duly put forward by Mazur *et al.* (1960)** and Kessar *et al.* (1966).*** However, the approach adopted by **Kessar** *et al.* proved to be a definitely far superior *stereospecific approach* to the actual synthesis of '**diosgenin**' (a *steroidal sapogenin*).

The various sequential reactions for the synthesis of **diosgenin (T)** may be given as under :

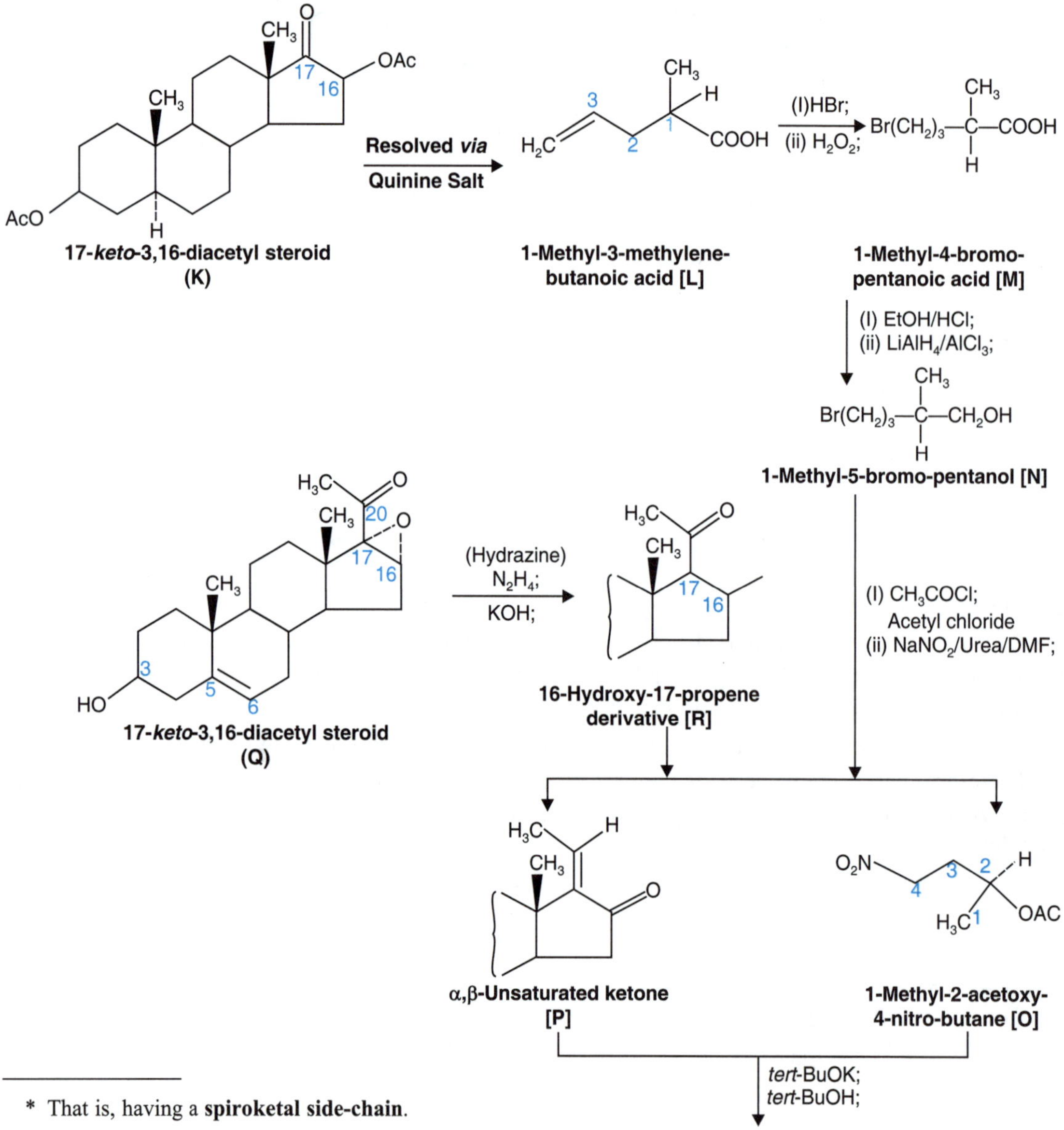

17-keto-3,16-diacetyl steroid
(K)

Resolved *via*
Quinine Salt

1-Methyl-3-methylene-butanoic acid [L]

(I) HBr;
(ii) H_2O_2;

1-Methyl-4-bromo-pentanoic acid [M]

(I) EtOH/HCl;
(ii) $LiAlH_4/AlCl_3$;

$Br(CH_2)_3-\overset{\overset{\displaystyle CH_3}{|}}{\underset{\underset{\displaystyle H}{|}}{C}}-CH_2OH$

1-Methyl-5-bromo-pentanol [N]

17-keto-3,16-diacetyl steroid
(Q)

(Hydrazine)
N_2H_4;
KOH;

16-Hydroxy-17-propene derivative [R]

(I) CH_3COCl;
Acetyl chloride
(ii) $NaNO_2$/Urea/DMF;

α,β-Unsaturated ketone
[P]

1-Methyl-2-acetoxy-4-nitro-butane [O]

tert-BuOK;
tert-BuOH;

---

* That is, having a **spiroketal side-chain**.

** Mazur *et al.* : *J Am Chem Soc*; , **82** : 5889, 1960.

*** Kesar *et al.* : *Tetrahedron Letters*, 4319, 1966.

(Contd........)

Diosgenin [T] ← (I)NaBH$_4$/EtOH; (Δ;3Hr) (ii) H$^+$; ← An Adduct [S]

Explanation -The various steps involved in the synthesis of **diosgenin (T)** may be explained as under :

1. First of all **17-*keto*-3,16-diacetyl steroid (K)** is carefully resolved *via* **quinine salt** to obtain **1-methyl-3-methylene butanoic acid (L),** which upon treatment with *HBr* followed by *mild oxidation* with H$_2$O$_2$ yields **1-methyl-4-bromo-pentanoic acid (M).**

2. **Product (M)** when reacted with *EtOH/HCl* followed by **LiAlH$_4$/AlCl$_3$** gives **1-methyl-5-bromo-pentanol (N),** which on treatment with *acetyl chloride* **(CH$_3$COCl)** followed by a mixture of **NaNO$_2$/urea/DMF** gives *two* products of reactions, namely :
   - **1-methyl-2-acetoxy-4-nitro-butane[O],** and ,
   - **α,β-unsaturated ketone [P].**

3. Besides, reaction of **product [Q]** with *hydrazine (N$_2$H$_4$)* followed by KOH yields **16-hydroxy-17-propene derivative [R],** which upon oxidation with *MnO$_2$* gives **product [P].**

4. The interaction of **product [P]** and **product [Q]** in the presence of **tertiary potassium butoxide** (*tert*-BuOK) and **tertiary butanol** (*tert*-BuOH) gives rise to the formation of an **adduct [S].**

5. Finally, the **adduct [S]** on treatment with *sodium borohydride (NaBH$_4$)* in ethanol followed by exposure in an *acidic medium (H$^+$)* gives the desired product **diosgenin [T].**

7. CARDIAC STEROIDS [or CARDIOTONIC GLYCOSIDES]

Cardiotonic Glycosides or **Cardiac Glycosides** or **Cardenolides** are normally employed for *two* **major functionalities:**

- **slow down the rate of the heart,** and
- **enhance the contractile force of the heart by stimulating the heart muscle.**

Points to Ponder – Intrestingly, *'Digitalis'* (an *age-old pharmaceutical preparation*) duly made from either the **seeds** or **leaves** of *Digitalis purpurea* [*purple fox glove*] is known to possess an inestimable value for the treatment of *heart ailments*. In reality, it is found to be an **admixture of glycoside,** and invariably on hydrolysis a **mixture of aglycones** (*i.e.,* the **non-sugar component**), such as :

- **Digotoxigenin** • **Digoxigenin** and • **Gitoxigenin.**

The above mentioned *three* **chemical entities** (substances) do conform in *skeletal structure* as well as *configuration*. However, there exists a typical exception of **one critical detail,** pertaining to the *'bile acid pattern'*– related to the extent of the **23 C-atoms present in it.**

These noteworthy variants essentially comprise :

- ❑ **having a *cis*-decalin type,**
- ❑ **angular methyl (-CH$_3$) groups located at C-10 and C-13 are β-oriented,**

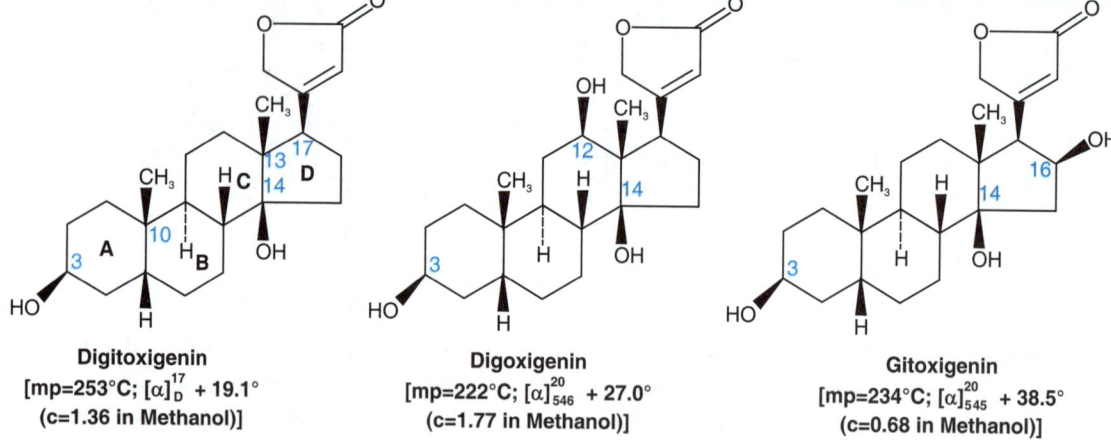

Digitoxigenin
[mp=253°C; $[\alpha]_D^{17}$ + 19.1°
(c=1.36 in Methanol)]

Digoxigenin
[mp=222°C; $[\alpha]_{546}^{20}$ + 27.0°
(c=1.77 in Methanol)]

Gitoxigenin
[mp=234°C; $[\alpha]_{545}^{20}$ + 38.5°
(c=0.68 in Methanol)]

❑   **side-chain positioned at C-17 is also β-oriented,**

❑   **B/C ring juncture is *trans*-in nature,** and

❑   **only apparent difference being that the C/D juncture is *cis*-in nature**

It is, however, pertinent to state here that **C-14** obviously has the so-called **β-configuration,** or the opposite of that in the **sterols** and **bile acids.**

Distinguishing Structural Features – These essentially include :

• **presence of β-oriented hydroxyl (-OH) moiety at C-14,**

• **presence of a 5- membered α,β-unsaturated lactone ring** (*ie.,* a butenolide ring system), and

• **one or more *sugar residues* are attached duly to the hydroxyl (-OH) moiety at C-3 positon; whereas, the respective aglycones (*ie., non-sugar, entity*) do retain/ exhibit *little of the physiological activity.***

An elaborated investigative study with regard to :

➤ **enzymatic hydrolysis,** or

➤ **chemical hydrolysis,**

of the 'cardiotonic glycosides' clearly affords two distinct products :

➤ **genins or aglycones (-the non-sugar moiety),** and

➤ **carbohydrate residues (-one or more).**

The **genins** (or *aglycones*) attribute the effective pharmacological activity; whereas, the **carbohydrate portion(s)** solely imparts :

• **significant favourable solubility,** and

• **distribution profile to the respective glycosides.**

NOTE   Amazingly, the *carbohydrate* duly isolate from the hydrolyzed products of *cardiac glycosides* have not yet been found elsewhere in nature individually.

Cardiac Steroid Variants – Importantly, the **aglycones** present critically in various '*cardiac steroids*' are found to be of *two* distinct types, namely :

❑   **Cardenolides,** and

❑   **Bufadienolides,**

which shall now be discussed briefly in the sections that follows :

(a) **Cardenolides** : The '*cardenolides*' are defined as the :

- '**cardiac glycosides (or cardiac steroids) having an unsaturated butyrolactone ring and form a C-23 steroids'.**

However, the so-called **lactone ring** of the *cardenolides* essentially possesses a *single-double bond* and *is duly attached at the respective C-17 position of the steroidal nucleus.*

Besides, the *cardenolides* do comprise an α,β-**unsaturated γ-lactone ring** thereby exhibiting **UV**$_{max}$ at 220 nm. Following are the *three* commonly available *cardenolides*, such as :

- **Uzarigenin**   • **Strophanthidin**   and   • **Digitoxigenin**

Cardenolides mostly occur in plant species belonging to the order : *Apocynaceae, Liliaceae, Moreaceae, Ranunculaceae*, and *Serophulariaceae*.

NOTE	*Digitalis purpurea* (Fox Gloves) – a particular gener of Digitalis designates the major source of the *active drug substances* bearing the effective therapeutic value .

(b) **Bufadienolides** : The '**bufadienolides**' are defined as the

'**cardiac glycosides having essentially the *six-membered lactone* ring and thereby result into the formation of C-24 steroids; and also the lactone of bufadienolides critically bears *two, double bonds* which are attached at the *C-17 position* of the inherent steroidal nucleus'.**

Uzarigenin	Strophanthidin	Digitoxigenin
[mp=266-270°C; $[\alpha]_D^{20}$ – 27°	[mp=171-175°C; $[\alpha]_D^{25}$ + 43.1°	[mp=253°C; $[\alpha]_D^{17}$ + 19.1°
(c=1.075 in pyridine)]	(c=2.8 in methanol)]	(c=1.36 in methanol)]

In addition, the **bufadienolides** specifically contain a δ-**lactone ring system** that essentially has a *conjugated diene system*. However, a good number of the **bufadienolides** do exhibit λ_{max} at 300 nm. Following are a few typical examples of important **bufadienolides**

- **Bufotalin**
- **Resibufagenin**   and
- **Scillaren A.**

Important Points-These essentially include:

1. **Klyne's Rule-**It is based on the specific *molecular rotation analysis* whereby the **glycosidic linkage** for:

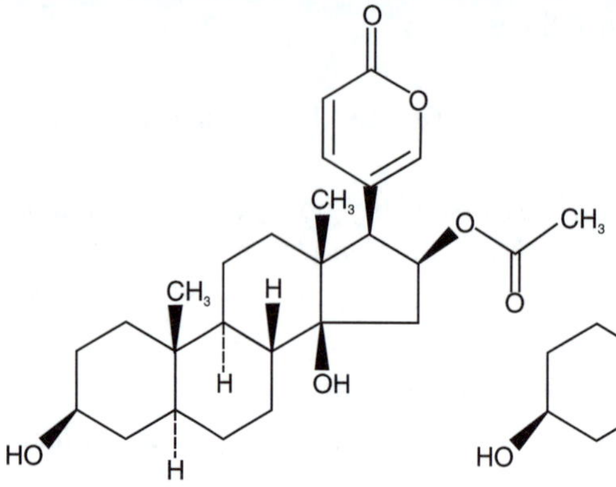

Bufotalin
[Dec.=223°C; λ_{max} 300nm;
$[\alpha]_D^{20}$ + 5.4° (c=0.5 in CHCl$_3$)]

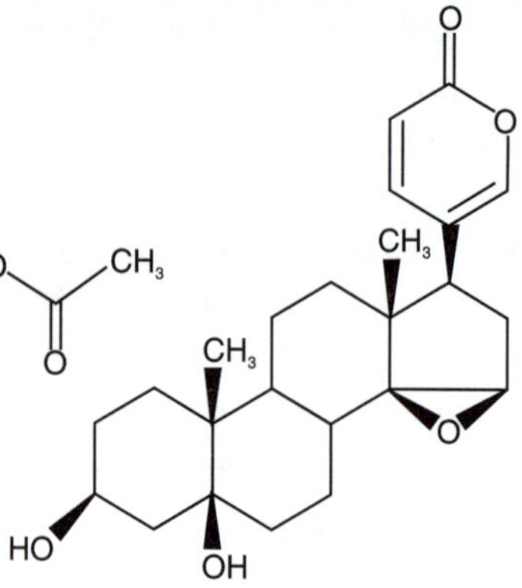

Resibufogenin
[mp=113-140°C;$[\alpha]_D^{23}$ + 7.1°
(c=1.259 in Chloroform)]

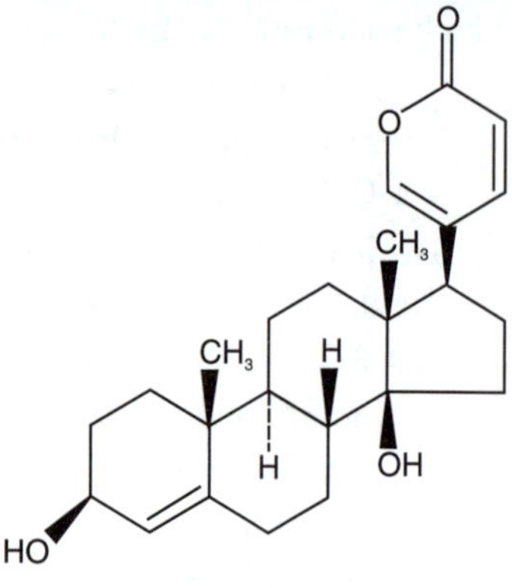

Scillaren A
[mp=184-186°C; $[\alpha]_D^{23}$ −71.9°
(c=1.011 in methanol)]

- **D-sugar** – is found to be a **β-configuration**; and
- **L-sugar** – is found to be an **α-configuration**

2. In general, the **cardiotonic glycoside** comprises prominently of:
   - **several hexose residues, and**
   - **C-3 remains the most preferred linkage between the** *steroidal moiety* **and** *sugar moiety* **(to form the ensuing glycoside).**
3. Importantly, both types of the '**aglycone redisues**'derived either from *cardenolides* or *bufadienolides* essentially have the so-called **normal configuration** prevailing crucially at :

   C-8, C-9, C-10, C-13, and C-17 (*as shown earlier*).

Besides, both **aglycone residues** do have a **C-3 and C-14 hydroxyl (-OH) group with a β-configuration.**
4. Nevertheless, the *two* **aglycone residues** may apparently differ with respect to the following *three* aspects:
   - **configuration (α-or β-) at C-3 and C-5 position,**
   - **presence of double bond (S) in the lactone ring, and**
   - **presence of ketonic function in the lactone ring.**

Brief Account on *k*- Strophanthoside

The **k-strophanthoside** represent a *trioside* (*i.e.*, containing : **cymarose**, and **β-glucose**) and is duly present in the seeds of **Strophanthus kombe** Olive, (*Fam : Apocynaceae*). It is cleared by the *enzyme* known as '**Strophanthobiose**'*. to yield:
   - **cymarin (1), and**
   - **2-moles of glucose.**

Furthermore, the careful *acid hydrolysis* of **cymarin (1)** gives rise to the formation of the following *two* **specific products:**
   ➤ **strophanthidin (2), and**
   ➤ **cymarose (3) (*i.e.*, very rare sugar moiety).**

---

* Duly obtained from the seeds of **Strophanthus courmontii.**

Thus, we may have the following reactions:

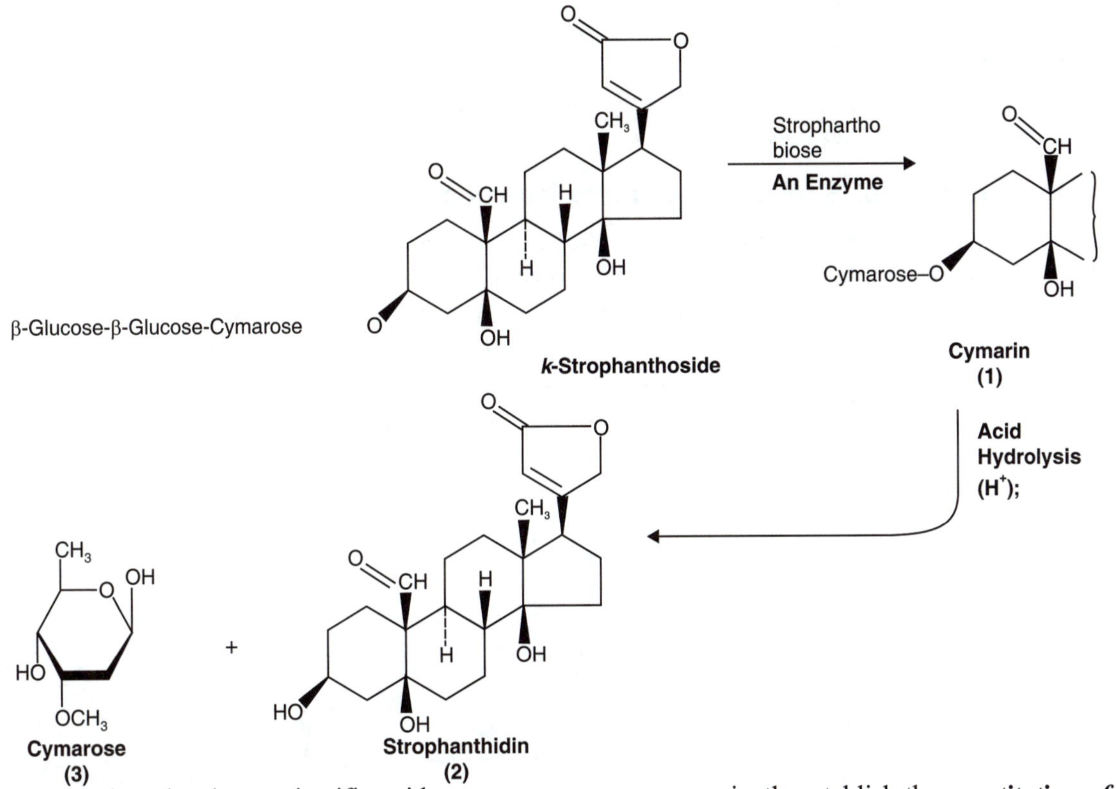

β-Glucose-β-Glucose-Cymarose

k-Strophanthoside

Strophartho biose

An Enzyme

Cymarose–O

Cymarin
(1)

Acid Hydrolysis (H⁺);

Cymarose
(3)

+

Strophanthidin
(2)

Based on the above scientific evidences one may now conveniently establish the *constitution of strophanthidin.*

CONSTITUTION OF STROPHANTHIDIN

Jacobs and Elderfield (1934)* established the **constitution of strophanthidin** on the basis of elaborated scientific and critical evidences as detailed under:

1. Presence of Carbonyl. (>C=O) Functional Moiety

The UV maximum absorption (λ_{max}) recorded for pure **strophanthidin** ascertains the presence of a **carbonyl (>C=O) functional moiety**. Besides, the **carbonyl group** shows the *aldehydic characteristic features* by its direct *oxidation* (**CrO₃**) to the corresponding. **carboxyl (-COOH) moiety.**

2. Formation of Monoacetate or Monobenzoate

It has been duly proved that **strophanthidin (2)** specifically gives rise to the formation of a **monoacetate** or **monobenzoate** only by treatment with *acetic anhydride* or *benzoyl chloride* respectively.

---

* Jacobs and Elderfield, *Science,* **80** :533,1934.

Comment – The above reaction indicates explicitly the presence of a *secondary* hydroxyl (-OH) moiety [-CH(OH)].

3. Positive Legal's Test

Strophanthidin (2) does respond to the **Legal's Test*** thereby ascertaining the presence of the characteristic $\Delta^{\alpha,\beta,\gamma}$-**lactones**.

4. Isomerization of Strophanthidin (2) to Iso-strophanthidine (4)

Isomerization of **strophanthidin (2)** in an *alkaline medium* takes place to produce **iso-strophanthidin (4)** *via* an **intermediate (3)**. Thus, the critical formation of **compound (4)** reveals explicitly that :

- one of the *tert*-hydroxyl (-OH) moiety is duly present at either C-8 or C-14,
- *tert*- hydroxyl group has a *cis*-configuration to the respective *butenolide moiety*, and
- 3,5,14-trihydroxy-10-carboxylate *derivative (5)* on dehydration (-H$_2$O) yields the **monoanhydro strophanthidine (6)**; and this hydroxyl moiety is easiest to eliminate since compound (6) is unable to form the iso-compounds (7).

All these aforesaid reactions may be expressed as under :

5. Strophanthidine has *tert*-Hydroxyl (-OH) Group at C-5 [β] with Respect to *sec*-Hydroxyl (-OH) Group at C-3

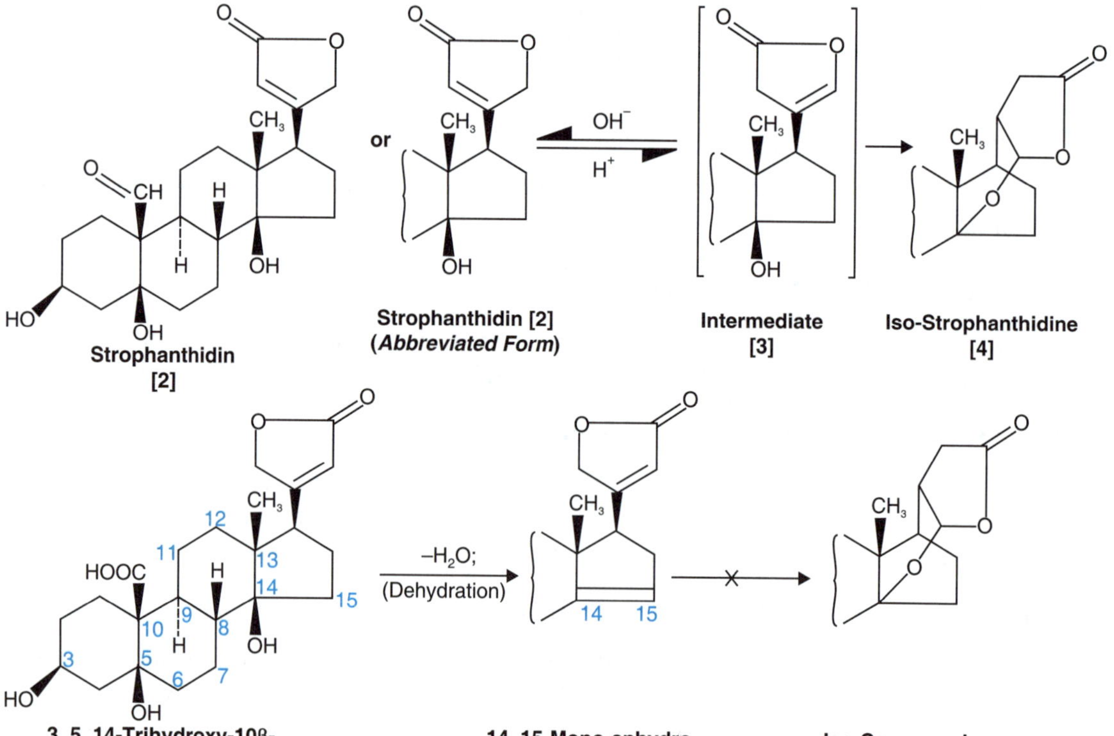

Strophanthidin [2]

Strophanthidin [2] (*Abbreviated Form*)

Intermediate [3]

Iso-Strophanthidine [4]

3, 5, 14-Trihydroxy-10β-carboxylate derivative [5]

−H$_2$O; (Dehydration)

14, 15-Mono-anhydro-strophanthidine [6]

Iso-Compound [7]

---

* **Legal's Test**- The appearance of a deep red colour that becomes violet on dilution by the addition of alkaline sodium nitroprusside solution to **strophanthidin**.

The *oxidation* of the *sec*-hydroxyl (-OH) moiety in **monohydro strophanthidine (6)** helps to activate the remaining *tert*- hydroxyl group at C-5; which may be removed easily by simple heating to give rise to the formation of the respective **enone [8]** *via* the **3-keto-5β-hydroxyl derivative [7]**.

Comment – The above scientific evidences reveal that the *tert*-hydroxyl (-OH) moiety at C-5 has a *β-configuration* only with respect to the **sec-hydroxyl (-OH) moiety** positioned at **C-3**. All these reactions may be written as below :

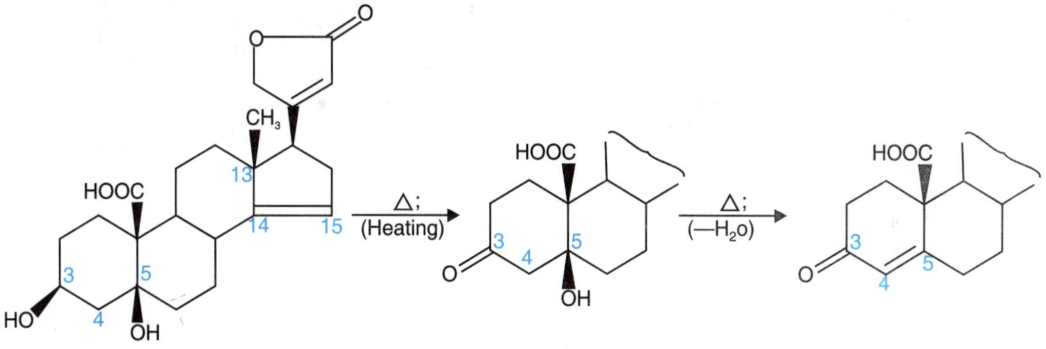

| Monoanhydro (1,4,15)-strophanthidine [6] | 3-keto-5 β-hydroxy derivative [7] | 3-keto-4-ene derivative (An Enone) [8] |

6. Hemiacetal Formation from Strophanthidine [2]

Strophanthidine [2] on being treated with **HCl in cold ethanol (C_2H_5OH)** it gives rise to the formaitno of a *hemiacetal** [9] between the following *two* **functional groups**, namely :

- *aldehyde function* at **C-10**, and
- *secondary* hydroxyl function at **C-3**.

Thus, we may have the following reaction :

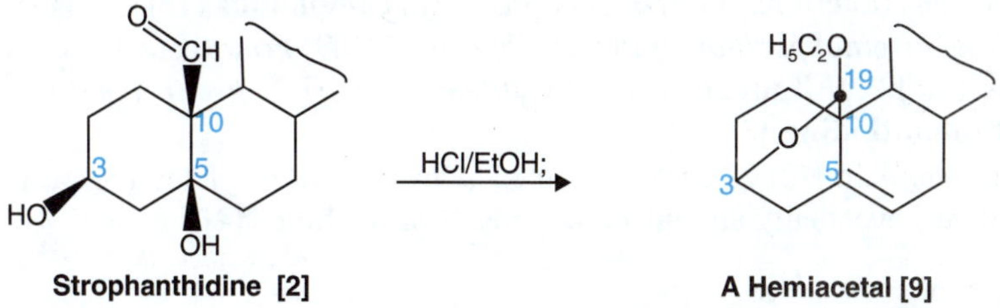

Strophanthidine [2]   A Hemiacetal [9]

Comment – The above reaction justifies the fact that *two* **functional** groups (*viz.*, *sec*-**hydroxy and aldehyde**) are not only close to each other but also have the identical orientation.

7. Specific Structural Information for Ring A in Strophanthidin [2]

The systematic degradation of **strophathidin [2]** is being carried out so as to obtain the exact **structural** information for **Ring A** ; and this could be accomplished by meticulously protecting all the *functional moieties*, but the 'enone moiety' as an *exception*.

---

* **Hemiacetal** : Dissolving either an *aldehyde* or ketone in an alcohol causes the slow establishment of an equilibrium between these *two* **compounds** and an altogether **new compound** called a **hemiacetal**.

Let us look into the following **sequential steps of reactions:**

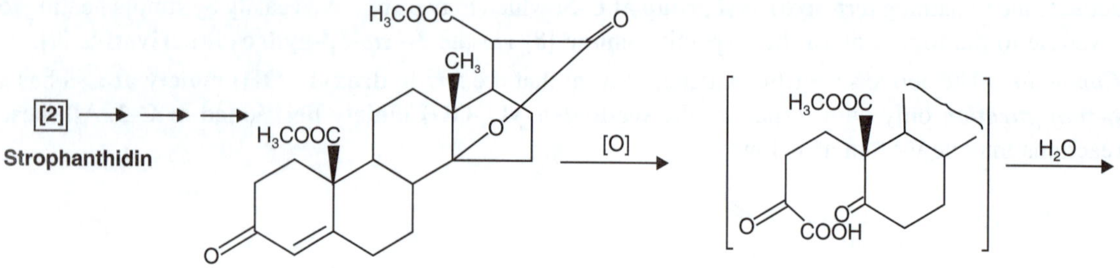

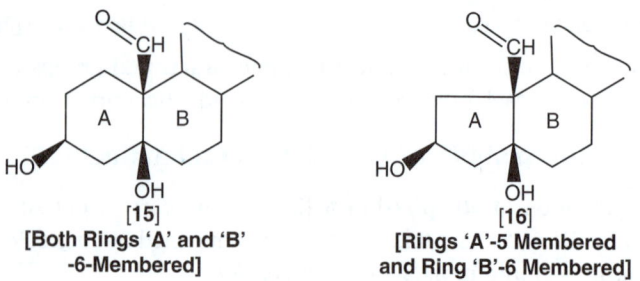

Enone derivatove
[10]

[Intermediate]
[11]

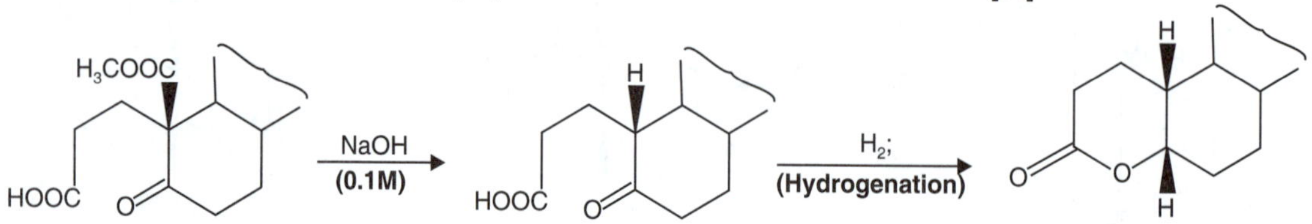

A *keto*-acid ester
[12]

A *keto*-acid
[13]

A *keto*-lactone
[14]

Explanation – The various steps involved above may be explained as under :

1. **Strophanthidine [2]** *via* a series of reaction yields the '**enone derivative [10]**, which upon *mild oxidation* initially gives an **intermediate [11]** (*i.e.*, a di-*keto* acid ester derivative).

2. **Intermediate [11]** on hydration yields the *keto*-acid ester **[12]**, which upon treatment with **0.1M. NaOH** undergoes *decarboxylation* to give a *keto*-acid **[13]**.

3. Finally, **compound [13]** upon *hydrogenation* gives rise to the formation of a *keto*-lactone **[14]**.

Remarks – These essentially comprise:

(i)   The *keto*-**functional moiety** duly present in **compounds [12]** and **[13]** evidently relates to the '*original position of the tert- hydroxyl (OH) group*' ; and, therefore, it essentially has the **β-configuration** to the *aldehyde (-CHO) group* (**seated at C-10**) in the **strophanthidine [2]**.

(ii) The *keto*-**acid [13]** on being subjected *to hydrogenation* yields a respective *hydroxy-acid*, that gets eventually **lactonized** to give a *keto*-lactone **[14]**.

[15]
[Both Rings 'A' and 'B'
-6-Membered]

[16]
[Rings 'A'-5 Membered
and Ring 'B'-6 Membered]

(iii) Hence, based on the logical evidences cited in (**i) and (ii) above** plus the scientific explanations contained in (**5) and (6) above** go a long way in putting a strong defence for either of the following *two* **chemical structures [15]** and **[16]**.

8. Size of Ring 'A' in Strophanthidine [2]

Blanc Reaction-Blanc Rule (1907)* is used to determine precisely the **size of rings** (as in the particular instance of *cholesterol and bile acids*). Importantly, the **Blanc Rule** suggests that the **ring 'A'** of *cardiac glycosides* (or *cardenolides*) happens to be a **6-membered ring** provided the **ring 'A'** occupies the *terminal position*. However, the aforesaid dictum has been duly proved by the help of the *following reactions* :

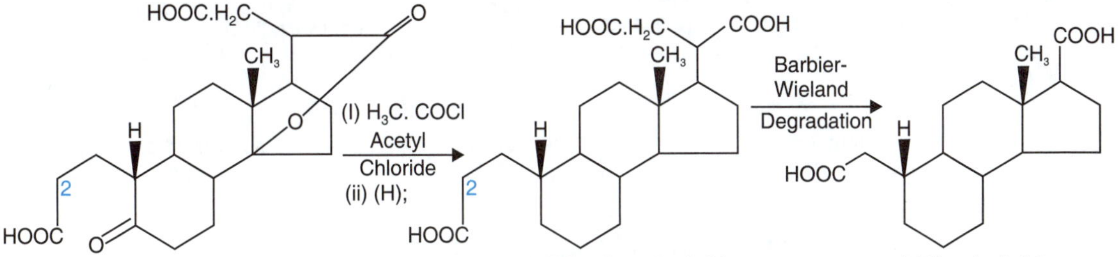

| A *keto*-acid [13] | A Tricarboxylic Acid [17] | A Dibasic Acid [18] |

Explanation – The above reactons are explained as under :

1. The **keto-acid [13]** on acetylation with *acetyl chloride (CH₃COCl)* followed by *hydrogenation* yields a **tricarboxylic acid [17]**.

2. **Compound [17]** on being subjected to **Barbier-Wieland degradation** results into the formation of a **dibasic acid [20]** with **4C-atoms less**.

3. Obviously, out of the **4C-atoms lost**, one may justify the actual manner about their elimination pattern :
   - **3C-atoms** – must have been contributed by the original/inherent '**lactone ring**' of the **keto-acid [13]**; and
   - remaining **1C-aotm** should have been duly provided by the **methylene (-CH₂-) carbon atom** located at *C-2 position in Ring 'A' [as marked in (13)]*.

Remarks – The conclusive remark may be made by the fact that the **Ring 'A'** essentially consists of the moiety :

$$- CH_2\text{-}CH(OH)\text{-}CH_2 -$$

and, therefore, should be present at the *terminal end of Ring 'D'* as shown in [**compound (13)**].

9. Presence of a Terminal 5- Membered Ring

Interestingly, the following set of *degradation reactions* do establish strongly the presence of a so-called terminal 5- membered ring system present in **strophanthidin.**

Thus, **compound [6]** on being oxidized first with $KMnO_4$ and followed by CrO_3 yields a *keto-carboxylate* **[19]**. However, the **ready lactonization of compound [19]** leads to the formation of a **lactonized derivative [20]**.

Comment – The foretold reaction indicates explicitly that one of the **hydroxyl (-OH) group** was definitely located either at **γ-or δ-position** relative to the respective **carboxyl (-COOH) moiety**; and, therefore, the actual '*Ring*' being subjected to cleavage should by all means be a '*5- membered ring*' i.e., **the terminal ring must be a 5- membered ring.**

---

* Blanc HG : *Compt Rend.* 144 :1356,1907

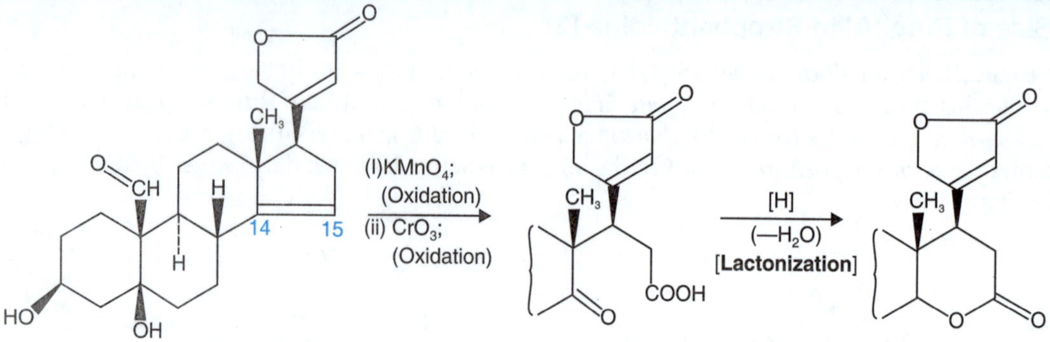

| 14,15-Monoanhydro-strophanthdine [6] | A *keto*-Carboxylate [19] | Lactomized derivative [20] |

10. Presence of Steroidal Nucleus in Strophanthidine [2]

It has been established that **strophanthidine [2]** on being subjected to *dehydrogenation* with **Se** gives a **Diel's hydrocarbon [21].**

Thus, we may have the following reaction :

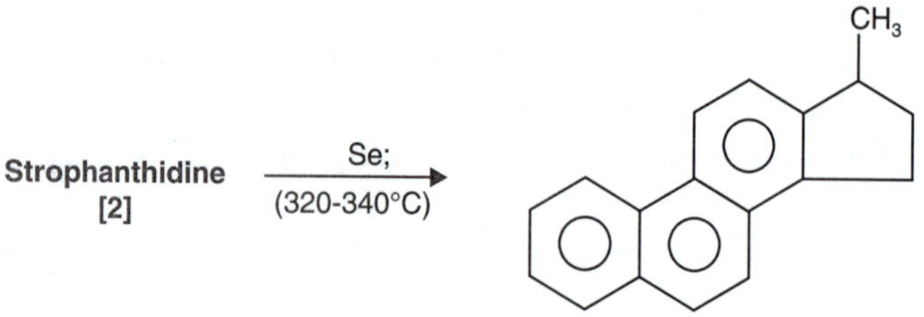

Diel's Hydrocarbon
[21]

Conclusive Remark – Based on the above logically explained facts and statements from item (1) through (10) it is proved beyond any reasonable doubt that the *chemical structure* [2] is the confirmed structure of 'strophanthidine'.

8. TOAD POISONS

Toad poisons (or Toad veroms) relate to the secretion either from the parotid glands or skin of toads designates a rather complex

- **Conjigated Genins** • **Free Genins** • **Bufotanine** • **Epinephrin Sterols** • **Fats** and
- **Tryptamine**.

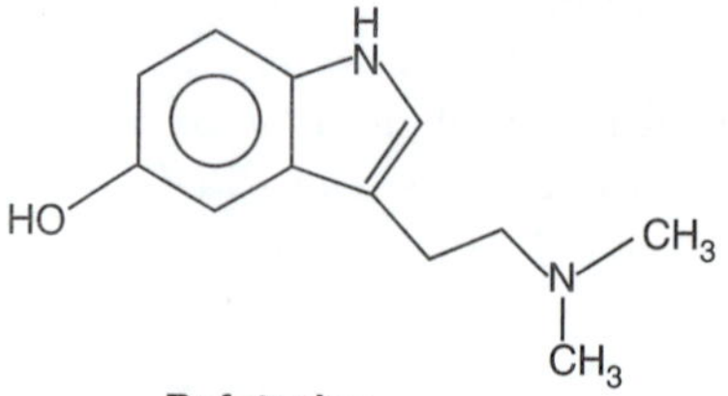

Bufotenine

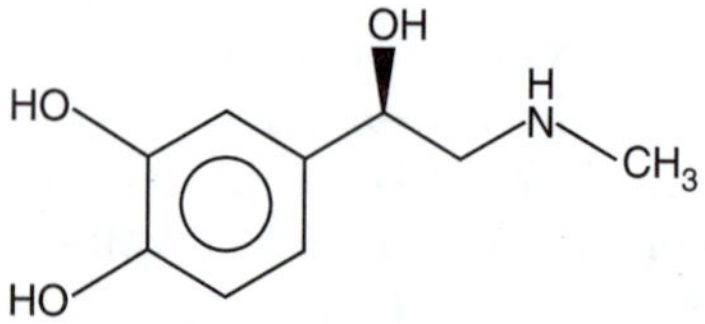

Epinehrine

The **toad poisone** are, in fact, highly toxic to both **frogs** and **mammals**, Importantly, the toad poison do exhibit almoet similar *digitalis-like action* upon the *mammalian frog heart*. The critical physiological action of the **toad poison** is mostly on account of the pressnce of **suberoylarginine**. Obviosly, quit unlike the cardiac poisons (*i.e.*, *digitalis glycosides*) the **toad poisons** happen to be the *esters of suberoylarginine*.

Example: Bufotoxin is the ester of the terminal hydroxyl (OH) moiety of suberoyl-arginine and bufotalin [also known as : Bufotalin -3- suberoyl arginine ester].

Suberoyl–Arginine | **Bufotoxin** | Bufatalin

NOTE The suberoylarginine free moiety is usually termed as Bufagins (*cf.* geninins).

Major Components of Toad Poison – These components having the critical *physiological activity* essentially consist of *two* **major categories**, namely:

❑ **First : (-) Adrenal and Indole – type bases,** such as : **Bufotenine** and **Bufothionine;**

Bufothionine

❑ **Secondly :** a typical group essentially consisting of such compounds as : **Cholesterol, Ergosteroil, 7α- Hydroxycholesterol, β- Sitosterol,** and **Bufodienolides.** *i.e,* both the **Bufogenins** (*aglycon*) and **Bufotoxins** (*suberoylarginine conjugates of Bufogenins*).

NOTE The *Bufotoxins* get rapidly converted *in vivo* to their respective 'aglycons' *i.e,* the non-sugar residues. However, the 'aglycons' and *conjugates* are found to be active physiologically.

Interestingly, almost *twenty* **Bufogenins** have been duly identified from **toads**. The **toad poisons** have long been practiced as the **'Folk Drugs'** in :

• **Chinese System of Medicine (*Chan Su*),** and

• **Japanese System of Medicine (*Senso*)*,**

---

* It is duly obtained from the **'Local Toad'**, *Bufo pargarizans.*

Following are the *three* different variant of '**toad**' along with their respective **genins** :

S.NO.	Name of Toad Variant	Name of Genin
1	*Bufo marinus*	Marino bufotalin
2	*Bufo vulgaris*	Bufotalin
3	*Bufo gargarizans*	Cinobufotalin

Cinobufotalin

Cinobufotalin is duly isolated from the **Chinese** Drug **Ch'an Su** which is prepared meticulously from the **Chinees toads** [*Bufo asiaticus = Bufo gargarizans* Cantor).

9. STEROIDAL ALKALOIDS

Preamble: In true sense, the '**steroidal alkaloids**' are derived solely *via* the **biosynthetic route** using *six* **isoprene units;** and thus, could be classified as **Steroids** or **Triterpenoids**. Nevertheless, they also essentially contain **N-atom** thereby rendering them **basic characteristic features.**

Presence of N- atom in Steroidal Alkaloids – The **N-atom** could be present in the **steroidal alkaloid** either :

- as a part of a ring system *i.e*, duly incorporated at a late stage during biosynthesis phenomenon, or
- as an *N- methyl substituted amino moiety.*

NOTE | The important members of the *steroidal alkaloids* have been duly found in several plant families *viz, Apocynaceae, Buxaceae, Liliaceae,* and *Solanaceae.*

In general, the *N- atom* containing **steroidal alkaloids** are also invariably termed as '**Azasteroids**'. The field of **steroidal alkaloids** have duly acclaimed a wide recognition in the recent past as one of the most prevalent active areas of research in the domain of **steroids.**

Classification of Steroidal Alkaloids – The **steroidal alkaloids** may be classified into *four* **major categories** based solely upon the **parent** *carbon skeleton* **present,** namely:

- **Pregnane Alkaloids,**
- **Solanum Alkaloids,**

- **Veratrum Alkaloids,** and
- **Salmandra Alkaloids,**

which shall now be treated individually in the sections that follows :

9.1. Pregnane Alkaloids

These alkaloids belong to the natural order **Apocynaceae** and **Buxaceae,** that, usually comprise even more than hundred recognized compounds, which are exclusively based upon the *pregnane C- skeleton*. Amazingly, these alkaloids do possess excellent *structural diversity.*

Examples: These include: **Conessine, Cyclobuxine D, Funtumine, and Pachystermine A.**

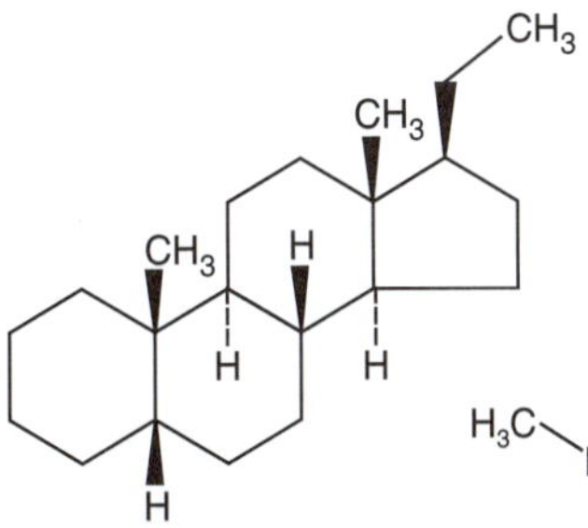

Pregnane
[C-Skeleton]

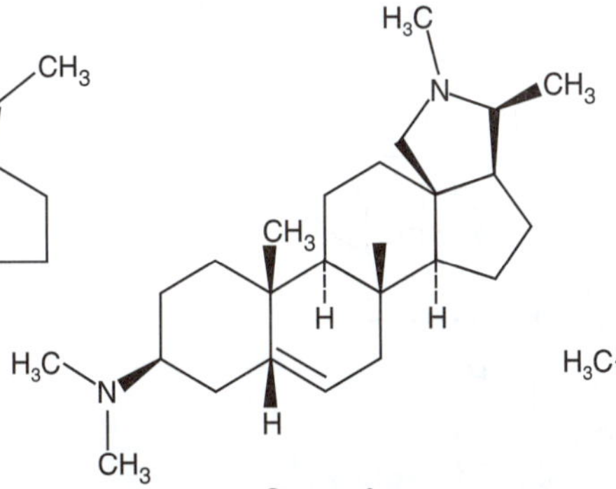

Conessine
Antiemetic, antifungal principle in **kurchi bark** : the bark of *Holarrhena anti-dysentrica* Wall, *Aopcynaceae,* native to India

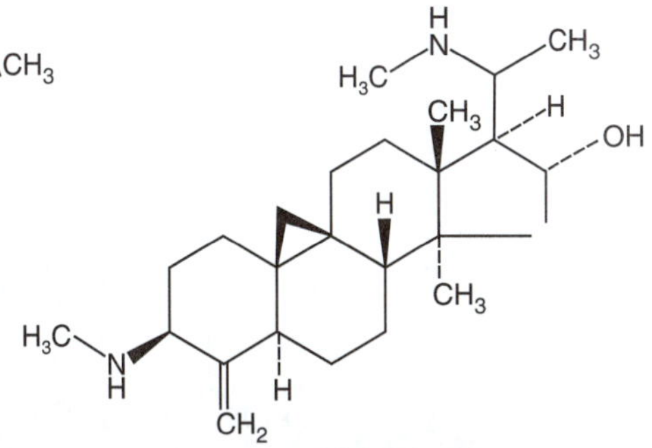

Cyclobuxin D
It is derived from *Buxus Sempervirens* L., *Buxeraceae*

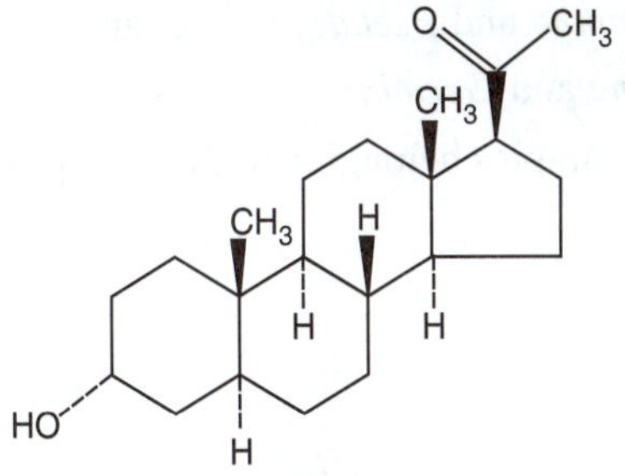

A **steroidal alkaloid** isolated from *Funtumia latifolia* Stapf. *Apocynaceae*

Funtumine

9.2. Solanum Alkaloids

This particular class of **steroidal alkaloids** comprise almost *forty* chemical entities known as *steroidal alkamines* that invariably occur as the **'glycosides'**; and actually isolated form nearly *250 species* of **plant families** of which both *Liliaceae* and *Solanaceae* are distinctly prevalent.

Interestingly, most of these **solanum alkaloids** are the so-called structural analogues of **C-27 cholestane**.

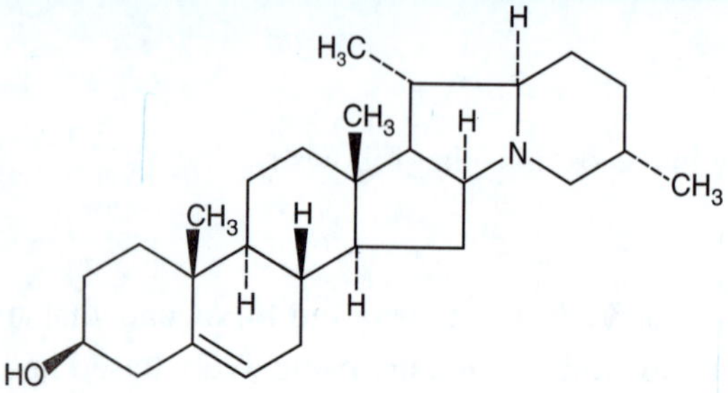

A **steroidal alkaloid** obtained from **Solanum** spp. *Solanaceae.* [Potato sprouts do contain ~**0.008%**].

Solanidine

Examples: **Solanidine** and **Solasondine** are *two* typical examples:

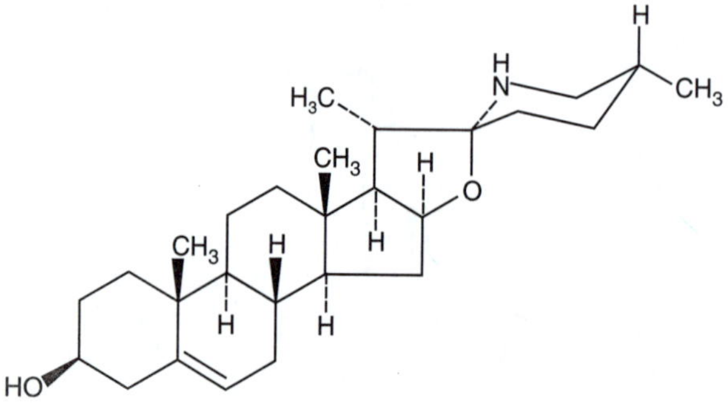

Solasodine

9.3 Veratrum Alkaloids

The **veraturm alkaloids** essentially consist of *two* **distinct C-27 series** duly designated as:

- **Jerveratrum group** as., *Veratramine* and *Pseudoje- vine*, and
- **Ceveratrum group** eg., *Cevadine* and *Germine.*

Jerveratrum Group – The *two* important examples belonging to this group are: **Verteramine and Psuedojervine** :

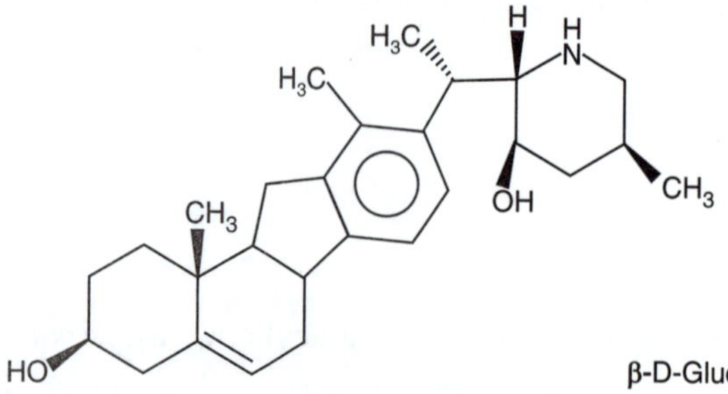

Veratramine

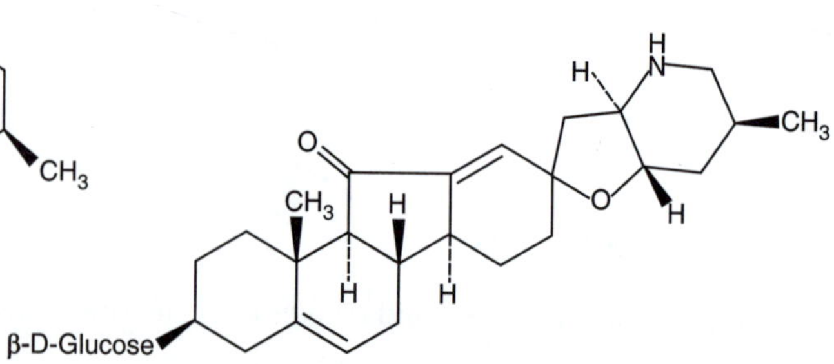

Pseudojervine

Veratramine is the *secondary base* duly obtained from *Veratrum grandiflorum* (Maxim) Loes f. and from *Veratrum viride* Ait, *Liliaceae.*

1

Ceveratrum Group – It mainly comprises esters of the **steroidal bases** (*viz, alkamines*) with *organic acids*, such as:

- **Cevadine and**
- **Germine**

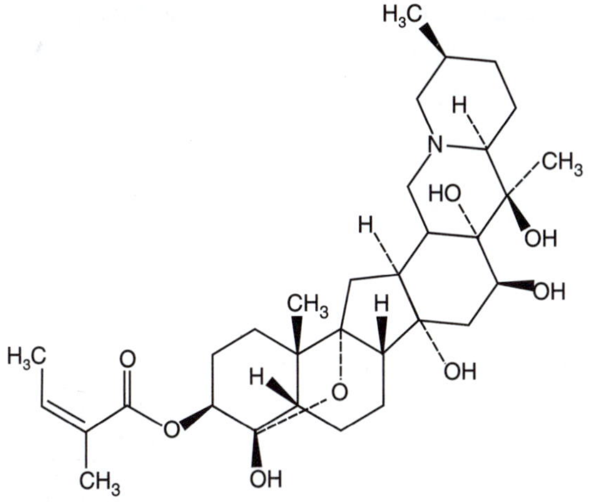

CEVADINE 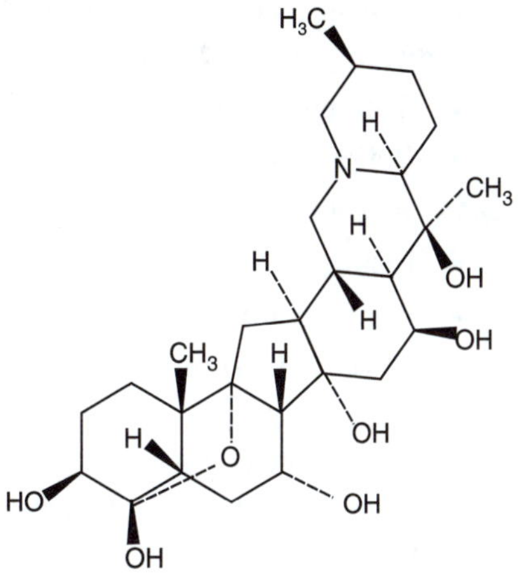   GERMINE

Cevadine : It is obtained from the seeds of *Schoenocaulon officinale* (Schlecht & Cham.) A. Gray (*Sabadilla officina rum* Brandt.), *Liliaceae.*

Germine : The *alkamine* present duly in many **polyester alkaloids** which occur in *Veratrum and Zygadenus* species.

9.4. Salamandra Alkaloids

The **solamandra alkaloids** essentially consist of many extremely toxic chemical components duly isolated from the so-called *salamander species* (an **amphibian** which looks like a lizard) specifically from the **skin glands.**

Examples: **Samandarine** – that critically constitutes together with **somandarone** approximately 75% of the total alkaloid content in the skin glands of the **European Fire** and **Alpine Salamanders.**

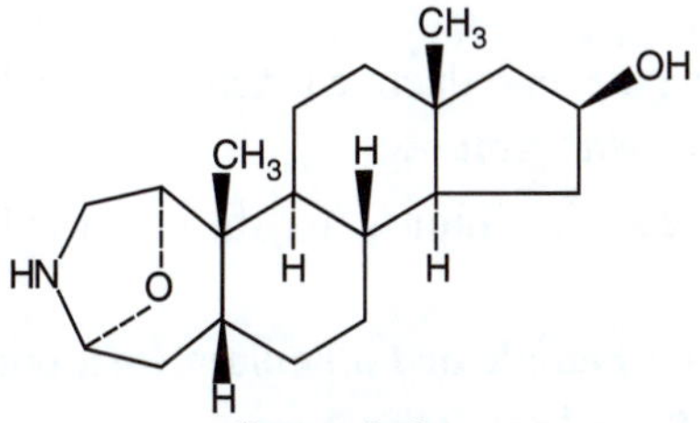

Samandarine

| NOTE | **Samandarine has been duly synthesized from testosterone (male sex-hormone).** |

10. HORMONES

Preamble: **Hormones** refer to the chemical substance being secreted by a *ductless gland* that essentially exhibits physiological effect upon the rest of the parts of the body. In other words, they invariably relate to a product of the prevailing **endocrine system** of the human body which does exhibit *physiological effects* on the entire body. Besides, quite a few of them are even used as **'drugs'** so as the correct and rectify both such anomalies, such as:

- **Abnormal state** and
- **Deficiency Condition**.

Examples: A few typical examples are:

- **Insulin**
- **Sex Hormones** and
- **Adrenal Cortex Hormones**

It may be worthwhile to state here that the exact and precise mechanism(s) by which the **hormones** exert their physiological effects are not yet fully deciphered. Nevertheless, there lies a definite glaring evidence that *certain hormones* show their critical control by acting upon the **DNA of the genes** directly or indirectly.

Points to Ponder – These essentially include :

- **Hormones** are generated into animal body itself, whereas, **vitamins** are duly synthesized in the body.
- The animal body is capable of producing the *hormones independently*; whereas, the **vitamins** are produced in the body of the animal *via* the igestion *plants, foods, the fruits.*

11. SEX HORMONES

Preamble : The **sex hormones** are reported to control and monitor the critical development of *major* and *secondary* **sexual features.** Interestingly, they do regulate the *sex related functions* of a **normal human being.**

Example – A few typical examples are:

- **Production of Sperm and Eggs and**
- **Menstruation,**

The **sex hormones** modulate the reproductive functions; and thus, responsible for the development of *male and female sex characteristic features.*

In a broader perspective, the **sex hormones** may be classified into the following *three* categories, namely :

- ❑ **Oestrogens [*Syn* : Female or Follicular Hormones]**,
- ❑ **Androsterone [*Syn* : Male Hormones]**, and
- ❑ **Gestogens [*Syn* : Corpus Leuteum Hormones]**,

which shall now be discussed with specific examples in the sections that follows:

11.1. Oestrogens [*Syn* : Female or Follicular Hormones]

Preamble : Oestrogens represent the natural or synthetic substances that usually induce the changes in the **Uterus** that precede **Ovolution.** In fact, these are also held responsible for the development of the *secondary* **sex features** in females : that is, the apparent physical changes that invariably take place in a girl at *puberty.*

Example – Enlargement of the breasts, appearance of pubic and axillary hair, and the deposition of fat on the thighs and hips. They are utilized mainly in the management of disturbances of the **menopause,** and also in the treatment of **cancer of the breast.**

Besides, the **oestrogenic hormones** of the ovary are: **Oestradiol** and **Oestrone.** However, that rapid *in vivo* degradation of the so-called *natural oestrogens* limits their use as therapeutic agents. Chemical substitution of the **steroid molecule** as in **ethinyl oestradiol,** or the use of a *non –steroidal synthyetic oestrogen e.g,* **Stilbesterol** – greatly reduces the rate of degradation and increases the *therapeutic action* to a great extent. A further development has been the use of compounds, for instance: those which are not actually oestrogenic themselves, but which are metabolized gradually to the corresponding **oestrogenic substances** *e.g,* **Chlorotrianesene** which are taken up in the *body fat* and then subsequently get released slowely into the circulation.

Ethinyl oestradiol is regarded to be the most potent **oestrogen,** being 20 folds more active than **stilbesterol.** Other usually utilized **oestrogens** are **Dienoesterol and Oestrol**.

> **NOTE** The critical usage of *oestrogens* in the *hormone replacement therapy* (HRT) is dealt with in the entry on the *Menopause.*

Biological Activity Profiles of Oestrogens – They are found to stimulate the female secondary sex hormones *viz, vagina, uterus, mammary glands,* and *fallopian tubes.* They also exhibit biological activities, for instance : effects on the synthesis and release pattern of the **gonadotropic hormones** of the **anterior pituitary** by way of the **hypothalamic area of the brain.**

Oestrogens may be isolated from **urine** (biological fluid), **placenta,** and **reproductive organs of mammals,** such as: ~ **0.07 mg of Oestrone** – is obtained from *1L of urine* collected from a pregnant woman.

Isolation of Oestrogens – The various steps involved in the **isolation of oestrogens** are as stated under:

1. Since the **oestrogens** do occur in combination with **glucuronic acid;** therefore, the *starting material* is first and foremost hydrolyzed with concentrated hydrochloric acid (12M) to release the **hormones** in its *'free state.'*

2. The resulting hydrolyzed product (containing the **hormones**) is extracted subsequently by suitable organic solvents.

3. The combined extracts are then treated with **Grignarad Reagent** [$(CH_3)_3$ N^+CH_2CONH NH_2] Cl⁻ (*i.e,* **Trimethylaminoacetohydrazide chloride**) – that reacts with *carbonyl (-C=O) moieties* of the **oestrogens** to result into the formation of **water soluble semicarbazones** leaving behind the unreacted *non ketonic onces.*

4. The separated **semi–carbozones** are hydrolyzed with HCl so as to regenerate the **Ketonic hormones.**

Important Examples of Oestrogens [Female or Follicular Hormones]

Following are some of the important classical examples of **oestrogens,** namely :

- **Estrone**
- **Estriol**
- **Estrodiol,** and
- **Non-Steroidal oestrogens,**

which shall now be treated individually in a comprehensive manner in the sections that follows :

11.1.1. Estrone [Syn : Oestrone; Hiestrone; Destrone; Kestrone]

Preamble : Estrone is a metabolite of **17β-estrodiol**, possessing considerably less biological activity. It is mostly isolated as the **d-form**. It occurs in the following sources:

- ➢ **Pregnancy urine of women and mares,**
- ➢ **In Follicular liquor of several animals,**
- ➢ **In Human placenta,**
- ➢ **In Urine of Bulls and Stallions,** and
- ➢ **In Palm kernel oil.**

Chemical Structure

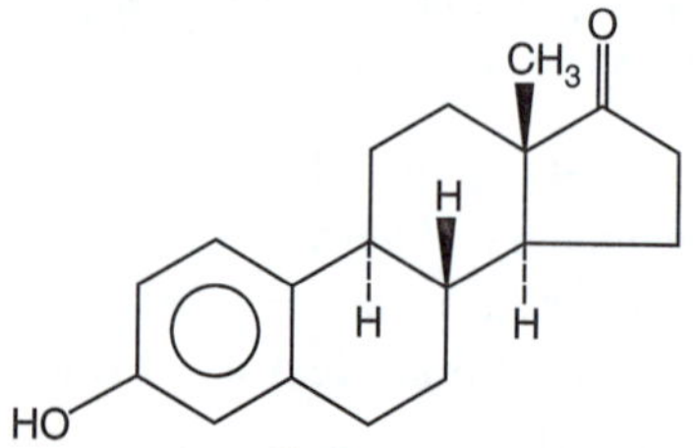

Estrone

3-Hydroxyestra –1,3,5 (10)-trien-17-one;

General Remarks : Estrone is known to be the very first recognized member of the **sex hormones.**

Butenandt (1929)* and Doisy *et al.* (1929)** isolated **estrone** independently from the pregnancy urine of women. Amin *et al.* (1969)*** isolated from the **Moghat roots** and **Date Palm pollen grains.**

Estrone usually occurs as a fairly stable white powdery substance and invariably occurs in *three* **different crystalline** forms, namely;

- **Rhombic metastable form**: mp 254°C;
- **Monoclinic metastable form**: mp 256°C; and
- **Rhombic stable form:** mp 259°C

It has $[\alpha]_D$ + 170° (in dioxane) ad **UV max (in ethanol) 280 nm.**

---

* Butenandt : *Naturwissenchaften,* **17**: 879, 1929.
** Doisy et al: *Am J Physiol,* **90**: 329, 1929.
*** Amin et al :*Phytochemistry,* **8**: 295, 1969.

Constitution of Estrone (Oestrone)

The constitution of **estrone** has been duly elucidated based upon the following glaring **analytical findings** and **synthetic evidences** :

1. **Molecular Formula** – Based on the results of essential analytical date, the **molecular formula of oestrone** has been found to be $C_{18}H_{22}O_2$.

2. **Presence of keto (>CO) functional moiety** – **Estrone** on being treated with **hydroxylamine (HO.NH₂)** yields **Monooxime** $\left[\begin{smallmatrix}R\\R\end{smallmatrix}>R>C=NOH;\right]$ whereas, a similar treatment with **semicarbazide** [H₂NNH CONH₂] it yields a semicarbozane, which indicates the presence of one ketonic functional group in **estrone.**

3. **Presence of a Phenolic (Ar-OH) Group** – **Estrone** is found to be soluble in **alkali** and it forms a **monoacetate** and a **monomethyl ether**. In fact, these chemical reactions do ascertain that **estrone** critically contains one **enolic (or phenolic) hydroxyl group**. Besides, one may further ascertain that the *incumbent hydroxyl group* is **phenolic in nature** due to the fact that **estrone** crucially couples with the respective **diazonium salts** in an *alkaline medium* (which eventually confirms that the ensuing reaction is typical of **phenolic functional moiety**).

4. **Presence of the Steroidal Nucleus** – **Estrone** on being distilled with *Zn dust* yields **chrysene** as given under, which suggests that **estrone** is duly related to the **steroids** :

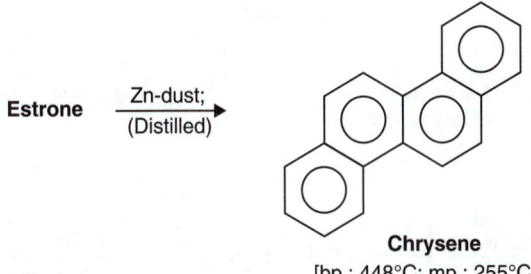

$$\textbf{Estrone} \xrightarrow[\text{(Distilled)}]{\text{Zn-dust;}}$$

Chrysene
[bp : 448°C; mp : 255°C]

The **X-ray diffraction analysis*** ensures the presence of the **steroidal nucleus**; and also ascertains firmly the strategic location of the keto (>C=O) and **phenolic (OH) moieties** at the *two* opposite ends of the said molecule.

5. **Presence of 3-Double Bonds and 1-Benzenoid Ring** – **Estrone** when *hydrogenated catalytically (Pd-Hg)*, it absorbs *four* molecules of hydrogen (H₂) thereby giving rise to the formation of **octahydro estrone** which is a *'diol'* i.e, it contains *two* **hydroxyl (OH) moieties.**

Important Observations – These essentially include :

  ➢ **2 H atoms** are utilized for the critical conversion of the *keto (>C=O)* **group of estrone** into the respective *secondary* **alcoholic moiety [=CH(OH)]** of the *octahydro estrone*, thereby suggesting that the remaining 6H atoms are used to saturate 3-double bonds.

  ➢ If one may assume that the said **3-double bonds** are duly present **in one ring only** i.e, there exists a *benzenoid ring system*, it could be possible to account for the presence of *phenolic (OH) hydroxyl group,* and

---

* Kar A : **Pharmaceutical Drug Analysis** , 3rd edn, New Age International, New Delhi 2013.

➢ In addition, to critical presence of at least **'one benzene ring'** in the **estrone chemical structure,** which has been adequately supported by meticulous measurements of the respective spectral studies, such as:

- **Molecular refraction** (*i.e* **X-ray diffraction**), and
- **UV-absorption spectrum** (λ_{max} **280 nm**).

6. **Exact Position of the Phenolic (-OH) Moiety** – The **monomethyl ether of estrone** on being subjected to the **Wolff- Kishner's Reduction*** it eventually yields a product which on distillation with **dry Se-powder** gives **7-methoxy -1, 2-cyclopentenophenanthrene.**

(A) Indeed the structure of the **resulting product** (A) has been duly confirmed by the synthesis (Cook *et. al.*, 1934)

Synthesis of 7-Methoxy -1,2-Cyclopentenophenanthrene (A)

The various steps involved in the above synthesis are stated below :

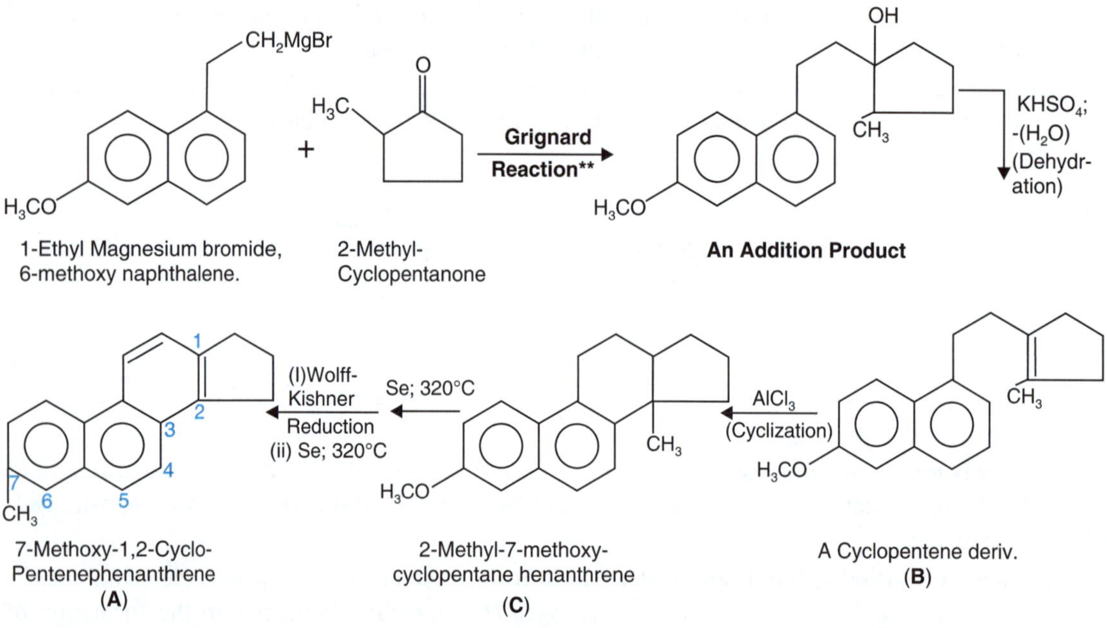

1-Ethyl Magnesium bromide, 6-methoxy naphthalene.

2-Methyl-Cyclopentanone

An Addition Product

7-Methoxy-1,2-Cyclo-Pentenephenanthrene
(A)

2-Methyl-7-methoxy-cyclopentane henanthrene
(C)

A Cyclopentene deriv.
(B)

Explanation – The various steps involved in the above synthesis may be explained as under :

1. **Grignardization** with **1-ethyl magnesium bromide, 6-methoxy naphthalene** and **2-methyl-cyclopentanone** yields an **adduct product,** which upon dehydration with potassium hydrogen sulphate ($KHSO_4$) yields a **cyclopentene derivative (B).**

2. The resulting **product (B)** undergoes **cyclization** with **$AlCl_3$** to give **2-methyl-7-methoxy cyclopentanephenanthrene (C).**

---

* Wolff L : *Ann, 394* : 86, 1912; *Kishner* N : *J Russ Phys Chem Soc.,* **43** : 582 1911.

** **Grignard Reaction**: That is addition of an **alkylmagnesium halide** to a ketone (>C=O).

3. The **product (C)** on being subjected to sequential reactions: first **Wolff-Kishner Reaction**, and secondly treatment with **Se powder at 320°C** yields the desired product **7 methoxy-1,2-cyclopentenephenanthrene (A)** .

Comment: The critical formation of **7-methoxy -1,2-cyclopentene-phenanthrene (A)** indicates explicitly that 'estrone' contains both:

- **steroidal nucleus, and**
- **phenolic (OH) group at C-3,**

and, therefore, the *3-double bonds* has got to be present in **ring 'A'** so as to give a **benezenoid nucleus**. Thus, one may represent the *partial structure of estrone* as given under:

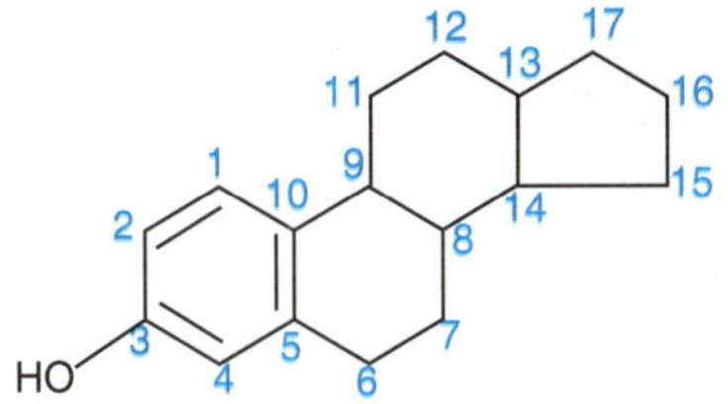

Partial Structure of Estrone

7. **Exact Position of *keto* (>C=O) Group and Angular Methyl Group:** Now, in the above **partial structure of estrone**, there are in all **17C-atoms**; whereas, **estrone** does comprise **18C-atoms**. Thus, we may have to account for the **18th C-atom in estrone** *i.e*, one more C-atom in the said *partial structure*. The exact position of the *keto* (>C=O) group has been shown to be located at **C-17**; whereas, the **extra C-atoms** (*i.e* 18th **C-atom**) should be at **C-13 position**. Obviously, the positions of the two above mentioned moieties have been duly established by the following experiments :

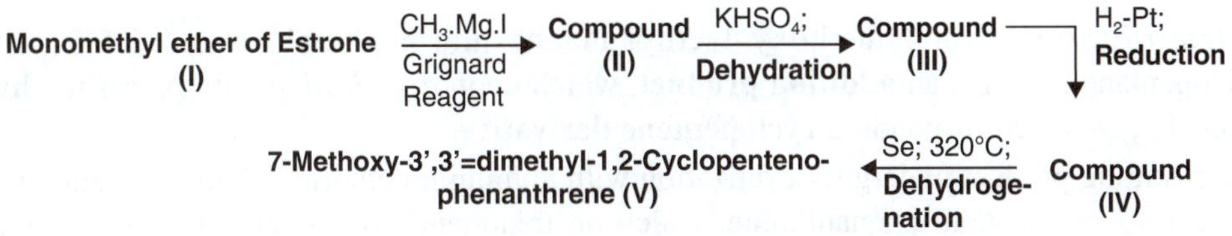

Explanation : The above 4-step sequential reactions may be explained as stated below :

- ❏ The **monomeyl ether of estrone (I)** on being subjected to treatment with **methylmagnesium iodide (CH₃Mg.I)** gives the **compound (II)**, which upon dehydration with KHSO₄ gets duly converted to **compound (III)**.
- ❏ The resulting **compound (III)** on being reduced ecatalytically (**Pd-H₂**) gives rise to the formation **of compound (IV)**, which upon distillation with **Se powder at 320°C** undergoes **dehydrogenation** to yield **compound (IV)**.
- ❏ Finally, **compound (IV)** on distillation with **Se** yields **7-methoxy-3,3-dimethyl -1,2-cyclopentenophenanthrene (V).**

Observation: The critical formation of **compound (V)** may be further expatiated based on the underlying fact that if a **keto (>C=O) moiety** is only present at **C-17 position**; whereas, an **angular methyl (-CH₃) group** is located strategically at **C-3 positions**.

Synthesis of 7- Methoxy-3',3', dimethyl (-1,2-cyclopenteno-phananthrene (Compound V)

Cook *et al.* (1935) proposed the synthesis of **Compound (V)**:

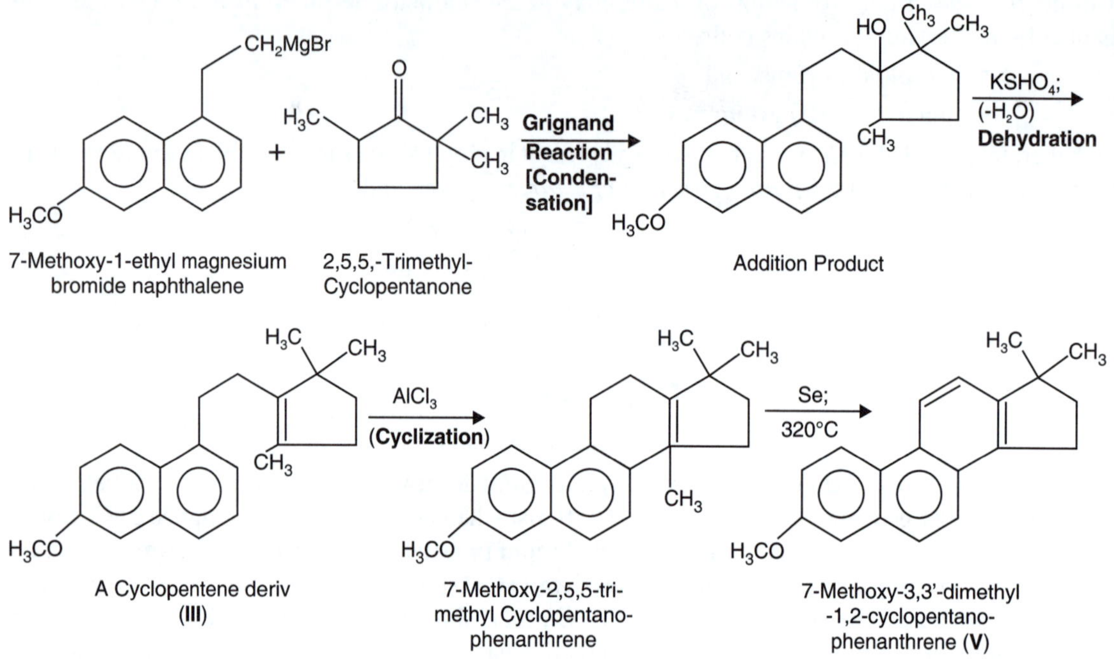

7-Methoxy-1-ethyl magnesium   2,5,5,-Trimethyl-   Addition Product
bromide naphthalene   Cyclopentanone

A Cyclopentene deriv   7-Methoxy-2,5,5-tri-   7-Methoxy-3,3'-dimethyl
(III)   methyl Cyclopentano-   -1,2-cyclopentano-
                                phenanthrene   phenanthrene (V)

Explanation :

1. *Grignardization* with 7-methoxy-1-ethyl magnesium naphthalene and 2,5,5- trimethyl cyclopentanone yields an **addition product**, which upon *dehydration* with potassium hydrogen sulphate gives a corresponding **cyclopentene derivative.**

2. The resulting product undergoes cyclization with aluminium chloride yields 7- methoxy -2,5,5,- trimethyl cyclopentano phenanthrene, which on treatment with Se at 320°C gives rise to the formation of **compound (V).**

Therefore, where the aforesaid **compound (V)** is duly obtained from the respective **monomethyl ether of estrone (I),** the *methyl (CH_3) moiety* gets hooked up at **C-17 position of the steroidal nucleus.**

Observations: These essentially include:

1. The **C- atom** at position **C-17** must possess the **ketonic (>C=O) moiety** which actually is the only critical point of attack for the respective *Grignard reagent* **methyl magnesium iodide ($CH_3Mg.I$).**

2. Importantly, *only one methyl (CH_3) moiety* is usually introduced by $CH_3.Mg.I$; whereas, the **compound (V)** is having *two methyl groups* thereby indicating that the **second methyl group should have already been present as an *angular methyl group* at C-13 position.**

3. However, during the *dehydration* process (with $KHSO_4$) the *methyl groups* gets duly shifted to **C-17 position (Compound III).**

Important Point – Thus, based upon the above logical evidences and observations the following **structure (VI)** may be assigned to **esterone**- that could probably explain most of the above reactions in a plausible manner:

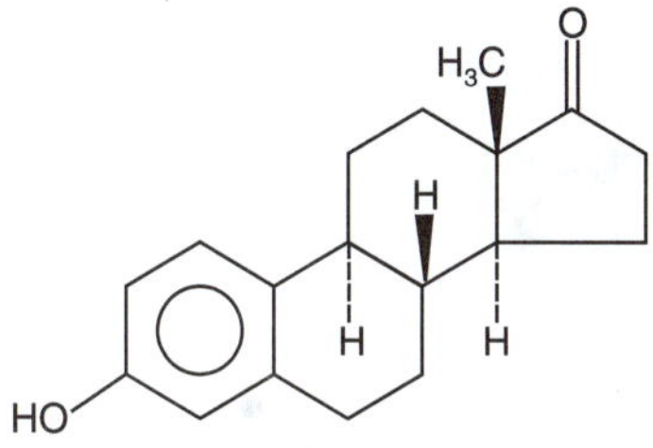
H3C
O
H
H
H
HO

Probable Chemical structure assigned to Esterone.

Structure (VI)

8. Total Synthesis of Esterone (VI)

The **total synthesis** of **esterone (VI)** may be accomplished and confirmed by the ***four* proposed synthetic methods, namely:**

(i) **Anner and Mischer Synthesis (1948)** : Anner and Miescher K: *Helv Chim Acta*, **31**: 2173, 1948;

(ii) **Johnson et al. [1950,1958]** : Johnson WS *et al.* : *J Am Chem Soc* **72** : 1426, 1950; *ibid*,**80** :661, 1958.

(iii) **Hughe Synthesis (1960)** : Hugle *et al.* : *Angew Chem. Int*, **3:** 353,1960.

(iv) **Torgov *et al.* (1960)** : Torgov IV *et al. Dokl Akad Nauk SSSR*, **135**: 73,1960

which shall now be described separately in the sections that follows:

8.1. Anner and Mischer Synthesis (1948)

The various steps involved in the **Anner and Mischer Synthesis** are as follows :

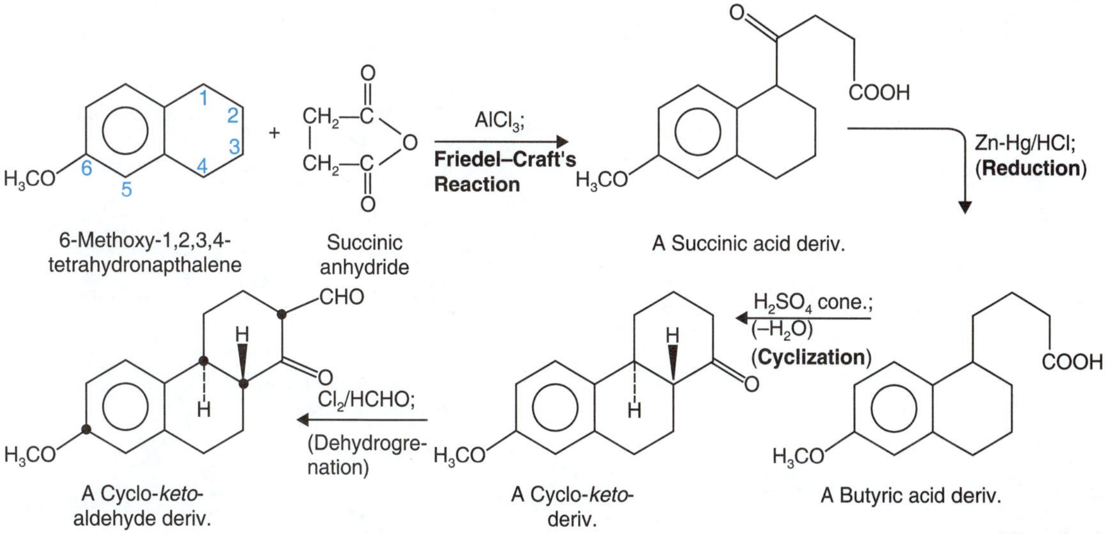
H3CO
1
2
3
4
5
6
6-Methoxy-1,2,3,4-tetrahydronapthalene
+
CH2—C
CH2—C
O
Succinic anhydride
AlCl3;
Friedel–Craft's Reaction
COOH
A Succinic acid deriv.
Zn-Hg/HCl;
(Reduction)
A Butyric acid deriv.
H2SO4 cone.;
(–H2O)
(Cyclization)
A Cyclo-keto-deriv.
Cl2/HCHO;
(Dehydrogre-nation)
CHO
A Cyclo-keto-aldehyde deriv.

(Contd....)

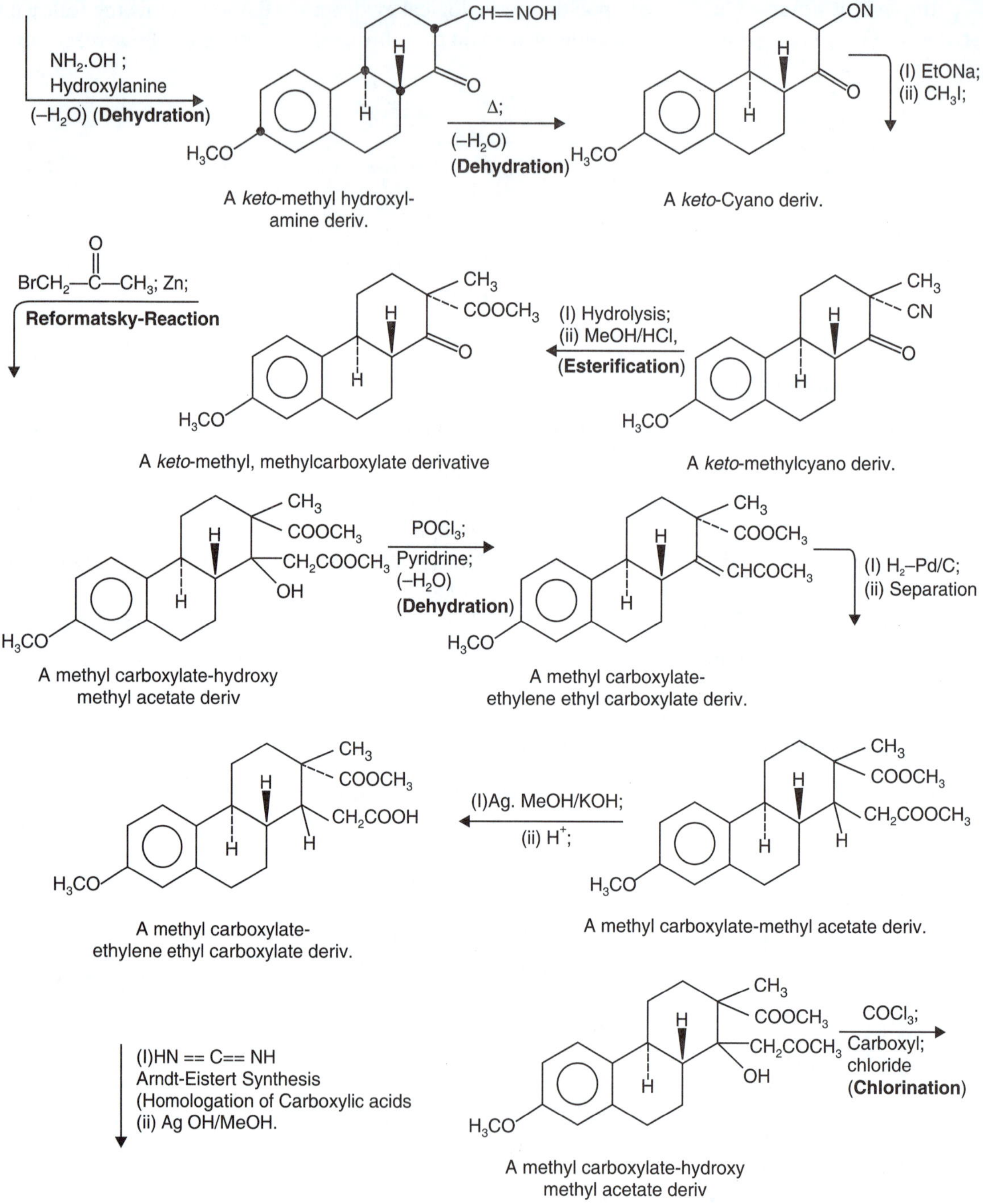

NH$_2$.OH ;
Hydroxylanine
(–H$_2$O) (**Dehydration**)

CH=NOH

A *keto*-methyl hydroxyl-
amine deriv.

Δ;
(–H$_2$O)
(**Dehydration**)

ON

(I) EtONa;
(ii) CH$_3$I;

A *keto*-Cyano deriv.

BrCH$_2$—C—CH$_3$; Zn;
Reformatsky-Reaction

CH$_3$
COOCH$_3$

A *keto*-methyl, methylcarboxylate derivative

(I) Hydrolysis;
(ii) MeOH/HCl,
(**Esterification**)

CH$_3$
CN

A *keto*-methylcyano deriv.

CH$_3$
COOCH$_3$
CH$_2$COOCH$_3$
OH

A methyl carboxylate-hydroxy
methyl acetate deriv

POCl$_3$;
Pyridrine;
(–H$_2$O)
(**Dehydration**)

CH$_3$
COOCH$_3$
CHCOCH$_3$

A methyl carboxylate-
ethylene ethyl carboxylate deriv.

(I) H$_2$–Pd/C;
(ii) Separation

CH$_3$
COOCH$_3$
CH$_2$COOH

A methyl carboxylate-
ethylene ethyl carboxylate deriv.

(I)Ag. MeOH/KOH;
(ii) H$^+$;

CH$_3$
COOCH$_3$
CH$_2$COOCH$_3$

A methyl carboxylate-methyl acetate deriv.

(I)HN == C== NH
Arndt-Eistert Synthesis
(Homologation of Carboxylic acids
(ii) Ag OH/MeOH.

CH$_3$
COOCH$_3$
CH$_2$COCH$_3$
OH

A methyl carboxylate-hydroxy
methyl acetate deriv

COCl$_3$;
Carboxyl;
chloride
(**Chlorination**)

(Contd....)

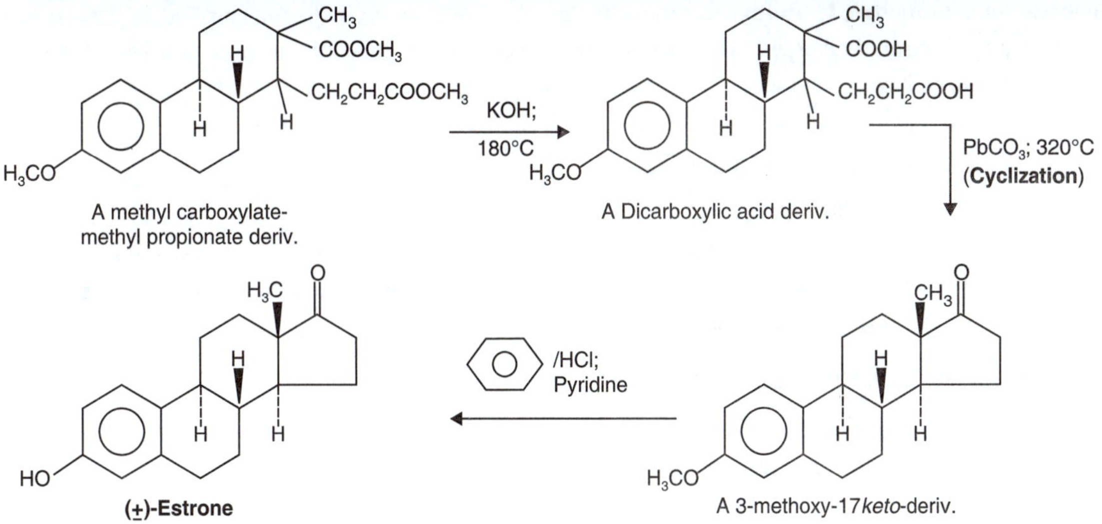

A methyl carboxylate-
methyl propionate deriv.

A Dicarboxylic acid deriv.

A 3-methoxy-17*keto*-deriv.

(±)-Estrone

Explanation: The various steps in the above synthesis may be expatiated as under :

1. **Friedel-Craft's reaction**, between 6-methoxy-1,2,3,4-tetrahydronaphthalene and succinic anhydride yields a **succinc and derivative,** which on reduction with Zn-Hg/HCl gives a **butyric acid derivative**.

2. The resulting product undergoes *cyclization* in the presence of conc. H_2SO_4 to give a **cyclo-*keto*-derivative**,which on *dehydrogenation* with chlorine and formalin yields a *cyclo-keto* aldehyde derivative.

3. The end product obtained above on being subjected to **Reformatsky reaction** with bromomethylacetate and zinc yields a **methyl carboxylate hydroxymethyl acetate derivative**, which upon dehydration with *POCl₃/pyridine* gives a **methyl carboxylate-methylenemethyl carboxylate derivative.**

4. The resulting product with reduction with H_2/Pd/C and subsequent separation gives **a methyl carboxylate methyl acetate derivative,** which on treatment with aqueous MeOH/KOH followed by acidifcation yields a methyl carboxylate–acetate derivative.

5. *Chlorination* of the end product with *carbonyl chloride* ($COCl_2$) yields a methyl carboxylate-acetyl chloride derivative, which on **Arndt-Eistert reaction** with diazomethane affords the so-called *homologation of carboxylic acids* followed by treatment with AgOH/MeOH gives a methyl carboxylate methyl propionate **derivative.**

6. The resulting product on alkalination with KOH at 180°C gives rise to the formation of a **dicarboxylic acid derivative,** which on *cyclization* with $PbCO_3$ at 320°C yields a **3-methoxy-17 keto** derivative.

7. Finally, the reaction of the above end product with **pyridine in HCl** yields the **racemic mixture (dl-) of esterone.**

   Special Remarks- In the aforesaid synthesis of **estrone,** one may not actually encounter with any sort of *complicated stereochemical problems,* since the hormone **esterone** does contain only **4 chiral**

centres (or **asymmetric C-atoms**), as shown by *dark spotted C-atoms* in the final product of the **esterone** synthesis *i.e*, it strongly suggests that *four* **recemates** are vividly possible. Importantly, out of the *four* recemates, *three* have been meticulously isolated and characterized as stated under :

- **one of these has been carefully converted to (±) – esterone (C/D-*trans*);**
- **a stereoisomer (C/D-*cis*); and**
- **(±) – iso-esterone,**

Importantly, all the aforesaid compounds have been duly separated; and the *(+) esterone* has been resolved by the aid of **(-)- methoxyacetic acid**. However, the **(±)–enantiomer** is found to be absolutely identical with the *natural compound*.

8.2. Johnson *et al.* Synthesis [1950,1958]

In reality, Johnson *et al.* have developed intelligently several synthesis of **esterone**; however, we would only describe the **shortest** and the **most stereospecific** synthesis as under:

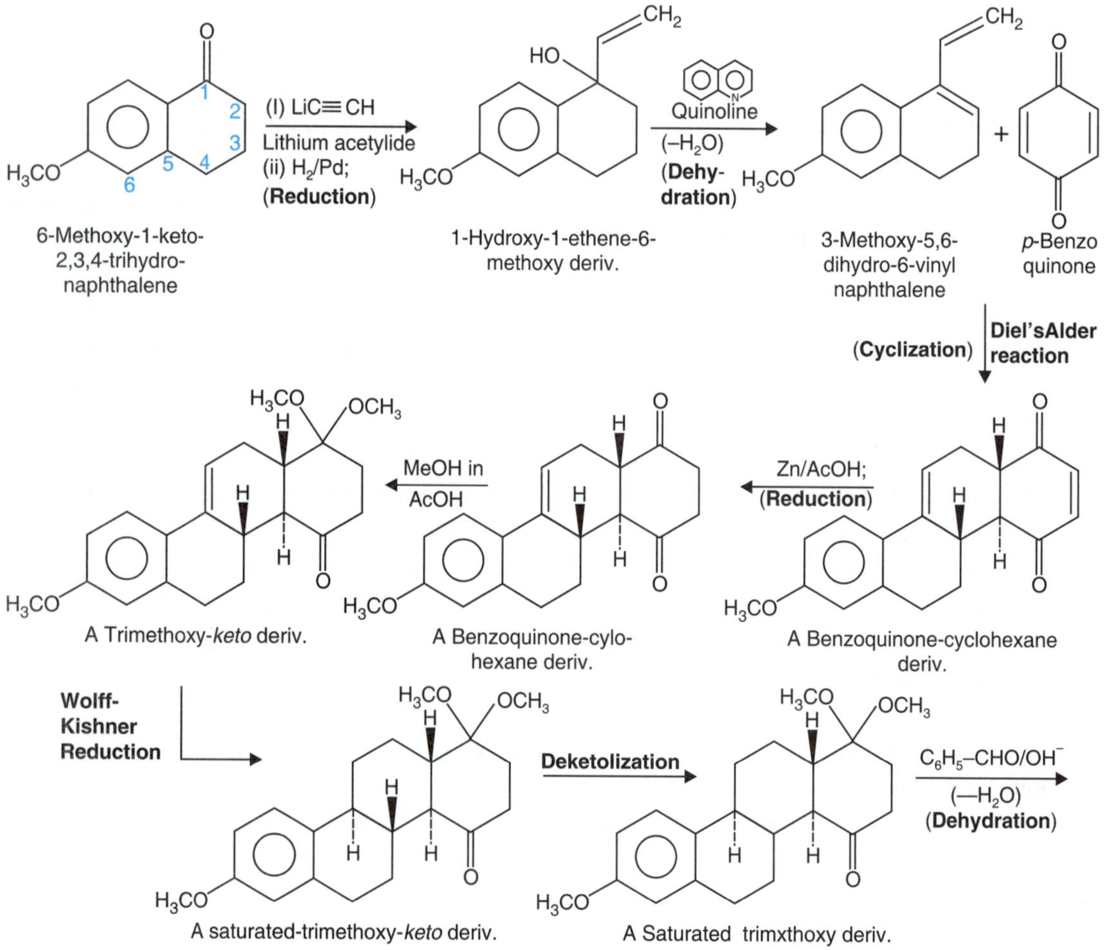

6-Methoxy-1-keto-
2,3,4-trihydro-
naphthalene

(I) LiC≡CH
Lithium acetylide
(ii) H₂/Pd;
(Reduction)

1-Hydroxy-1-ethene-6-
methoxy deriv.

Quinoline
(–H₂O)
**(Dehy-
dration)**

3-Methoxy-5,6-
dihydro-6-vinyl
naphthalene

+

p-Benzo
quinone

(Cyclization) | **Diel'sAlder
reaction**

A Trimethoxy-*keto* deriv.

MeOH in
AcOH

A Benzoquinone-cylo-
hexane deriv.

Zn/AcOH;
(Reduction)

A Benzoquinone-cyclohexane
deriv.

**Wolff-
Kishner
Reduction**

A saturated-trimethoxy-*keto* deriv.

Deketolization

A Saturated trimxthoxy deriv.

C₆H₅–CHO/OH⁻
(—H₂O)
(Dehydration)

(Contd....)

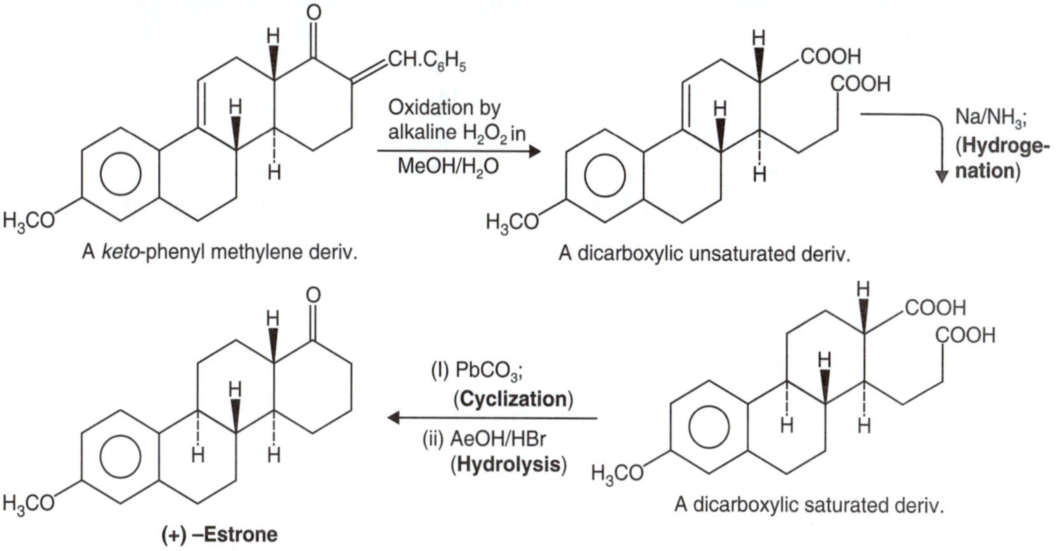

A keto-phenyl methylene deriv.   A dicarboxylic unsaturated deriv.

(+) −Estrone   A dicarboxylic saturated deriv.

Explanation: The various steps involved in the above synthesis may be explained as under:

1. Reaction of **6-mehtoxy-1-keto-2,3,4-trihydronaphthalene** with *lithium acetylide* followed by *catalytic reduction* yields 1-hydroxy-1-ethene-6-methoxy-derivative, which on *dehydration* with quinoline gives the respective **3-methoxy-5,6-dihydro-6-vinyl naphthalene**.

2. The resulting compound on treatment with **p- benzoquinone** undergoes cyclization [**Diel's Alder Reaction**] to give a **benzoquinone cyclohexene derivative,** which on treatment with Zn/ACOH causes reduction to yield a **benzoquinone cyclohexane derivative.**

3. The end product obtained above when treated with *methanol in acitic acid* yields a **trimethoxy-keto-derivative,** which undergoes **Wolf-Kishner reduction** gives a saturated trimethoxy-keto-derivative.

4. The resulting product on *deketolization* gives a **saturated trimethoxy** deriavative, which on *dehydration* with *benzaldehyde in an alkaline medium* yields a *keto-*phenylmethylene derivative.

5. The end product as obtained above on *oxidation* by **alkaline H$_2$O$_2$** in MeOH/H$_2$O yields a **dicarboxylic unsaturated derivative,** which on hydrogenation with Na/NH$_3$ gives rise to the formation of a **dicarboxylic saturated derivative.**

6. Finally, the resulting product *first* on treatment with PbCO$_3$ undergoes *cyclization*; and *secondly* on being subjected to hydrolysis by AcOH/HBr gives the desired compound (+) − esterone.

8.3. Hughe *et al.* Synthesis (1960)

They put forward a rather **simpler** and **efficient total synthesis of esterone,** which essentially involves the use of the following *two* **well known chemical reactions,** namely:

- **Maanich Reaction**, and
- **Michael Condensation.**

The various steps involved in the **Hughe synthesis** are as given under:

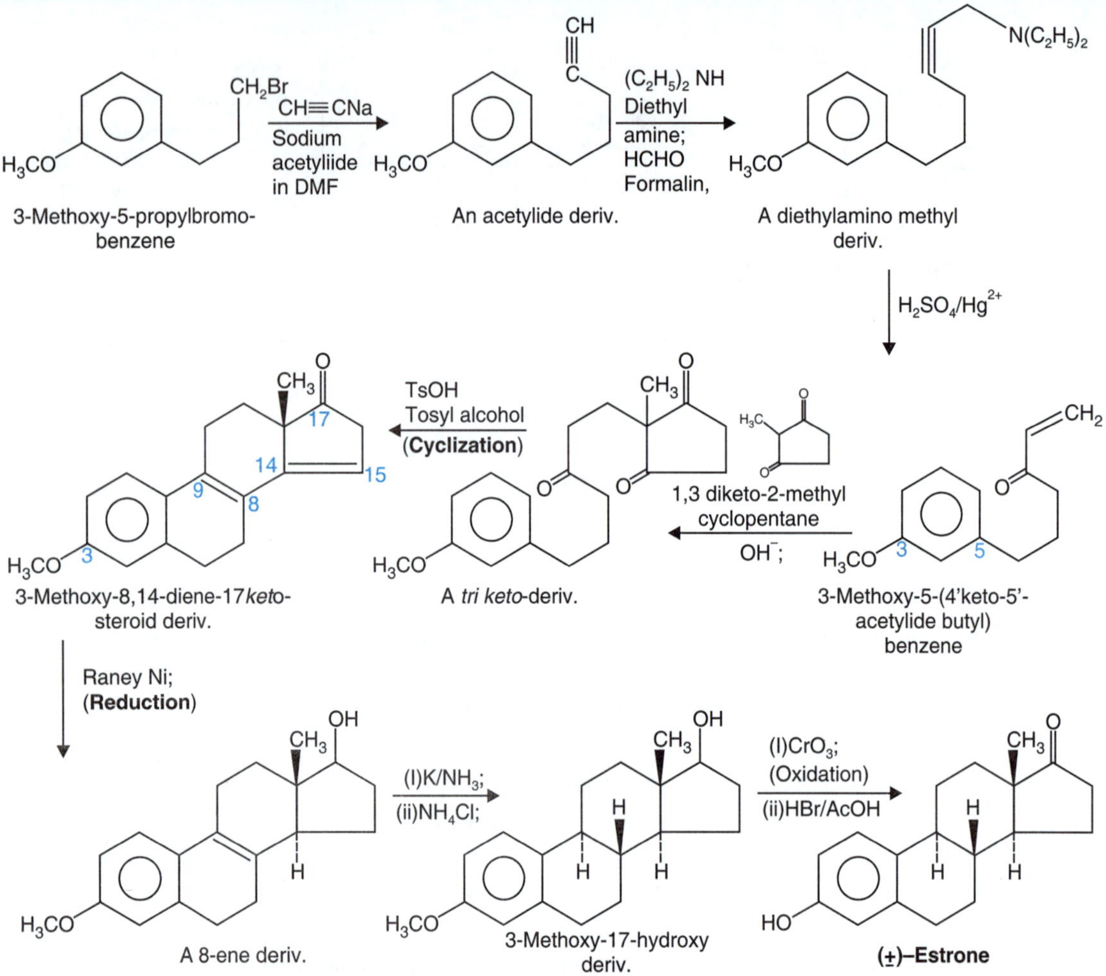

3-Methoxy-5-propylbromo-benzene → An acetylide deriv. → A diethylamino methyl deriv.

3-Methoxy-5-(4'keto-5'-acetylide butyl) benzene

A *tri keto*-deriv.

3-Methoxy-8,14-diene-17*keto*-steroid deriv.

A 8-ene deriv.

3-Methoxy-17-hydroxy deriv.

(±)−Estrone

1,3 diketo-2-methyl cyclopentane

Explanation : The various steps involved in the above synthesis are duly explained as under:

1. Interaction of **3-methoxy-5-propylbromobenzene** with *sodium acetylide* in DMF yields an acetylide derivative which on treatment with *diethyl amine* and then with *formalin* gives a **diethylamino methyl derivative**.

2. The end product obtained above when reacted with **sulphuric acid and Hg^{2+}** gives **3-methoxy-5-(4'-keto-5'-acetylide butyl) benzene,** which on treatment with 1,3-diketo-2-methyl cyclopentane in an alkaline medium yields a **triketo derivative**.

3. The resulting product when reacted with *tosyl alcohol* undergoes *cyclization* to yield **3- methoxy-8,14-diene-17-keto steroid derivative,** which upon reduction with *Raney Ni* gives a **8-ene-derivative**.

4. The end product obtained above when treated with K/NH$_3$ followed by NH$_4$Cl yields **3-methoxy-17-hydroxy derivative,** which upon treatment further with *CrO$_3$ (Chromium-6-oxide)* undergoes *oxidation* and reaction with *HBr/AcOH* gives the desired product **(+)-estrone** *i.e.,* **racemic mixture of estrone.**

8.4. Torgov *et al.* Synthesis (1960)

The proposed **total synthesis of (±)– estrone** by Torgov *et al.* is as stated under:

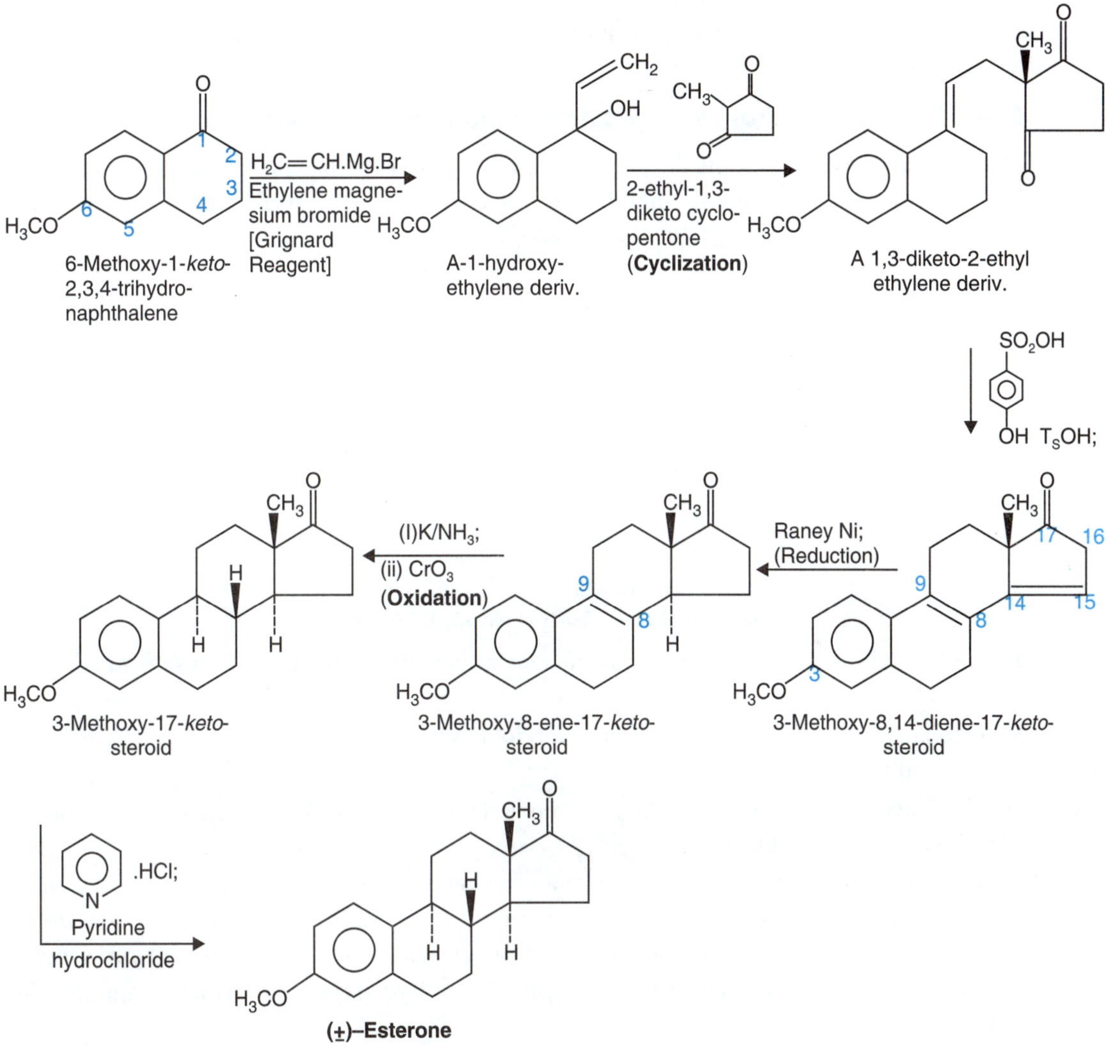

6-Methoxy-1-*keto*-
2,3,4-trihydro-
naphthalene

A-1-hydroxy-
ethylene deriv.

2-ethyl-1,3-
diketo cyclo-
pentone
(Cyclization)

A 1,3-diketo-2-ethyl
ethylene deriv.

3-Methoxy-17-*keto*-
steroid

3-Methoxy-8-ene-17-*keto*-
steroid

3-Methoxy-8,14-diene-17-*keto*-
steroid

(±)–Esterone

Explanation : The various steps involved in the above synthesis may be explained as under :

1. The reaction between **6-methoxy-1-keto-2,3,4- trihydro naphthalene** and **ethylene magnesium bromide** (*Grignard reagent*) yields a **1-hydroxy ethylene derivative,** which undergoes *cyclization* with *2-methyl-1,3-diketo cyclopentone* gives a **1,3-diketo-2-methyl ethylene derivative.**

2. The resulting product on tosylation with TsOH yields **3-methoxy-8,14-diene-17-*keto* steroid,** which on being subjected to reduct with *Raney-Ni* gives **3-methoxy-8-ene-17 keto-steroid.**

3. The end product obtained from above on treatment with K/NH₃ followed by *oxidation* with chromium 6-oxide (CrO₃) gives **3-methoxy-17-keto steroid**, which upon treatment with *pyridine HCl* yields the desired compound **(+) – Esterone.**

Uses

1. The main function of **Estrone** is to induce the state of heat in female partners.
2. It acts in the control of **uterine cycle.**
3. It aids in the development of **female secondary sex characteristic features** (by which they become different from males).

11.1.2. Estriol [Syn : Trihydroyestrin; Follicular Hormone Hydrate;]

Preamble : Estriol is a metabolite of, and considerably less potent than the hormone estradiol. It is invariably the predominant **estrogenic metabolite** found in wine. However, during *pregnancy* the placenta does produce relatively large quantum of **estriol.**

Chemical Structure

Esterone (I)
(16α-17β-)-Estra-1,3,5(10)-triene-3,16,17-triol;

Isolation of Estrial : Marria (1929)* was pioneer in isolating **estriol** from **human pregnancy urine**.

Constitution of Estriol : The **constitution of estriol** has been duly established on the basis of the following analytical results and observations :

1. **Molecular Formula** – The results obtained from various *analytical data*, the **molecular formula of estriol** has been determined to be $C_{18}H_{24}O_3$.
2. **Absence of Ketonic (>C=O) Moiety in Estriol**- The **ketonic (>C=O) moiety** is absent in **estriol.**
3. **Presence of 3- Hydroxyl (-OH) Groups** – The careful *acetylation* of *estriol* (with *acetic anhydride*) gives rise to the formation of a **triacetate**, which indicates that oestriol comprises **3- hydroxyl groups.**
4. **Nature of Hydroxyl (OH) Moieties**- In fact, the *three* hydroxyl **(OH)** moieties are actually shown to be :
   - ❑ **1st hydroxyl moiety** – is critically shown to be **phenolic in nature** since 'estriol' (first similar to *esterogen*) has a tendency to couple with the *diazonium salts* $[C_6H_5-N=N^+Cl^-$, benzene diazonium chloride] in an *alkaline environment (OH⁻)*; and
   - ❑ **2nd & 3rd hydroxy moieties** are of *secondary* alcoholic in nature (>CH-OH), since on being subjected to *oxidation* (Cr_2O_3) yields critically a **diketone analogue.**

     Besides, the **X-ray diffraction analysis** reveals that the *two* alcoholic hydroxyl **(OH) moieties** are located strategically in the *vicinal position i.e*, 1 and 2 respectively.

---

* Marrian : *Biochem* J, **23**: 1090,1929.

5. Exact Position of the 3-Hydroxyl (OH) Moieties – Interestingly, when **estriol** is heated with *potassium hydrogen sulphate (KHSO₄)* it loses a mole of *water* (dehydration) to yield **estrone.** In fact, this reaction indicates vividly that **estriol** does possess exactly the *same carbon skeleton* as that of **estrone**; and therefore, the *two alcoholic groups* are positioned at **C-16 and C-17 of the steroidal nucleus.**

Thus, all the aforesaid facts may be explained adequately and logically provided **structure (I)** (as given in the beginning) is assumed to be the **probable structure of estriol** Of course, it may be further substantiated by the following supportive evidences:

Degradation of Estriol to Respective Phenanthrene Derivative

Assuming **structure (I)** to be the correct structure of **estriol**, its critical degradation to the respective *phenanthrene derivative* may be shown as given under:

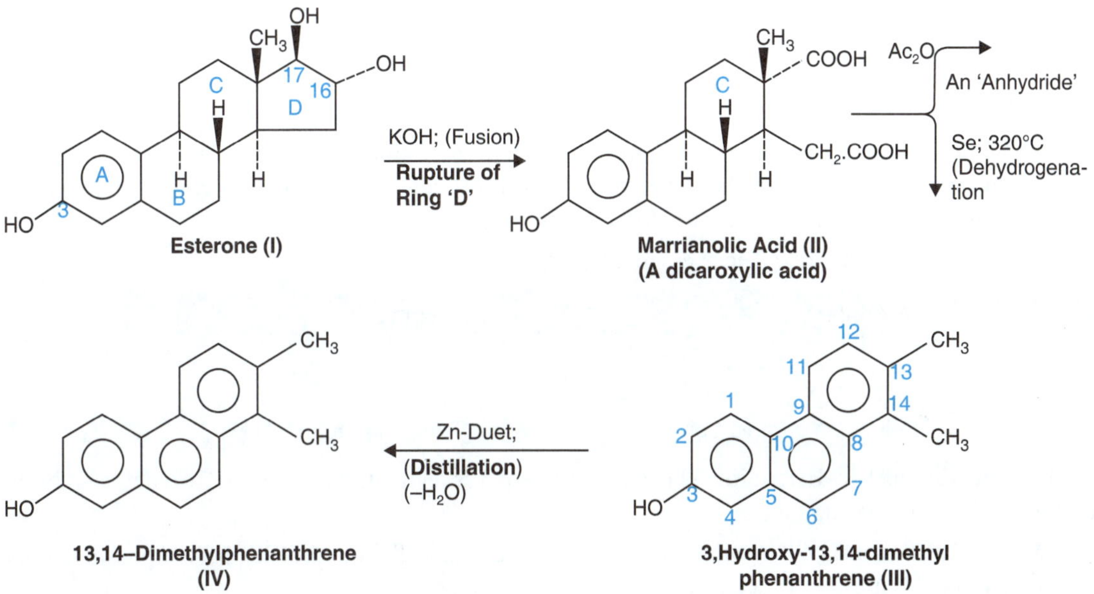

Esterone (I)

Marrianolic Acid (II)
(A dicaroxylic acid)

13,14–Dimethylphenanthrene
(IV)

3,Hydroxy-13,14-dimethyl
phenanthrene (III)

Explanation: The various steps involved in the above degradation from the assumed **structure (I)** of **estriol** to the **phenanthrene derivative** may be explained as under:

1. **Estriol (I)** on being fused with KOH causes rupture of **ring 'D'** to yield **marrianolic acid (II)** *i.e,* a dicarboxylic acid, which on treatment with *acetic anhydride* gives an **anhydride.**

2. **Product (II)** when heated with *Se powder at 320°C* undergoes *dehydrogenation* to give **3- hydroxy-13,14-dimethyl phenanthrene (III).**

3. Finally, the resulting **product (III)** on being subject to distillation with *Zn-dust* loses a mole of water to produce **13,14-dimethylphenanthrene (IV).**

6. **Adjacent Hydroxyl (OH) Moieties in *trans*-Configuration** – Since **estriol** fails to produce an *isopropylidene* with acetone, thereby indicating that the **hydroxyl (OH) moities** must be positioned in the *trans*-**configuration.** In addition, the **total synthesis** of *estriol from estrone*, the configuration of the *hydroxyl moiety at C-17* has been found to be '**β**'

11.1.3. Estradiol [Syn: cis-Estradiol; Dihydrofollicular Hormone; Dihydroxyestrin; Dihydrcthelin;]

Preamble: Estradiol is a *potent mammalian estrogenic hormone* duly produced by the ovary. It critically triggers the production of *gonadotropins* leading to the phenomenon of ***ovulation.*** It is primarily isolated from the follicular fluid of *sow ovaries* (Mac Corquodale *et al.* 1936)

Chemical Structure. It was also isolated from the pregnancy urine of mares (Wintersteiner *et al.,* 1938)

17β-Estradiol (I)

(17β)-Estra-1,3,5 (10)-triene-3,17-diol;

In general, one may come across *two* **stereoisomeric estradiols 'α' and 'β'.** However, of the *two* known **isomers**, the β–isomers is found to be the more active one than **esterone**; whereas, the corresponding α–**isomer** is *much less active physiologically.*

Characteristic Features: These essentially include:

α-Isomer	β-Isomer
mp 178°C	222°C
$[\alpha]_D^{25} + 81°C$	$[\beta]_D^{25} + 54°C$
Soluble in organic Solvents	Soluble in organic solvents

Constitution of 17β-Estradiol: The constitution of **17β-estradiol** has been duly elucidated based upon the following glaring facts:

1. **Molecular Formula** – The *analytical data* obtained for **17β-estradiol** strongly suggests that the molecular formula is $C_{18}H_{24}O_2$.
2. **Presence of 2- Hydroxyl (OH) Groups** – **Estradiol** on *acetylation* (with acetic anhydride) gives rise to the formation of a **diacetyl derivative**, that vehemently suggests the presence of **2-hydroxyl groups**.
3. **Nature of Hydroxyl (OH) Moieties** of the *two* hydroxyl moieties duly present in **estradiol** :
   * **one of the hydroxyl (OH) moieties** is shown to be *phenolic in nature*; and
   * **second hydrodyl (OH) moiety-** happens to be a *secondary* **alcoholic in nature [-CH(OH)]**, since *oxidation* of **estradiol** gives **esterone.**
4. **Presence of a Steroidal Nucleus in Estradiol** – **Estrone** on being subject to *reduction* yields **estradiol**, which strongly suggests that the latter (*estradiol*) essentially possesses exactly the same **C-skeleton (or steroidal nucleus)** as that to estrogen.

Further Evidences for a Phenolic Hydroxyl (OH) Moiety – As it has already been established that **one of the hydroxyl (OH) groups** of estradiol is **henolic;** whereas, the other a *secondary* **alcoholic** is

---

* Mac Corquiodale *et al.* : *J Biol Chem*, **115**:435,1936.

further ascertained by the underlying fact that when **esterone** is subjected to *reduction*, the **keto – moiety located at C-17 gets reduced** *i.e*, the *keto (>C=O)* group at *C-17* yields *sec–alcoholic* **(-CHOH) group**. Therefore, the structure of **estradiol** is **(I)**, which being further *confirmed* by the undermentioned fact:

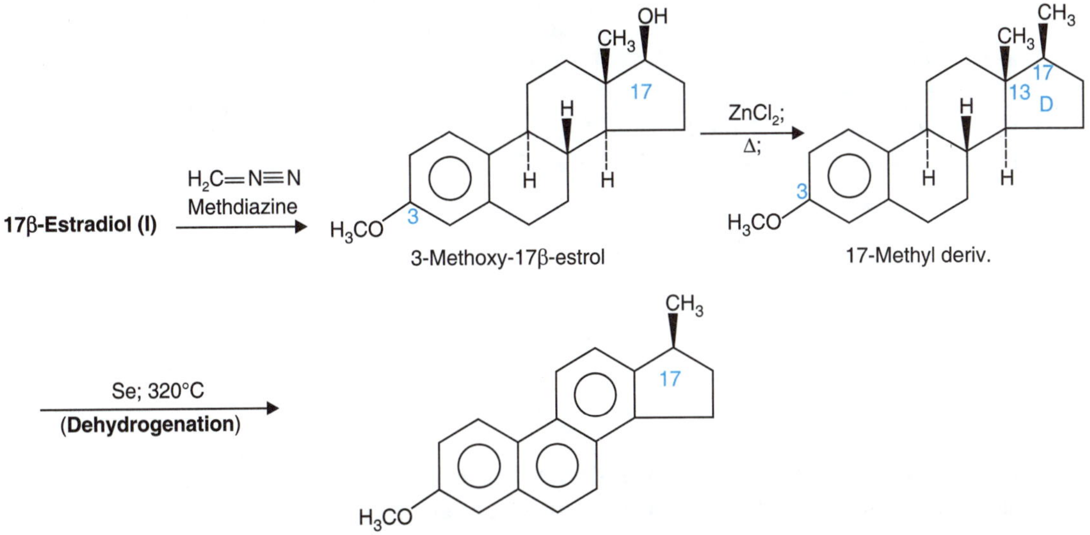

17β-Estradiol (I)

H₂C=N≡N
Methdiazine

H₃CO 3-Methoxy-17β-estrol

ZnCl₂;
Δ;

H₃CO 17-Methyl deriv.

Se; 320°C
(Dehydrogenation)

H₃CO
3-Methoxy-17β-methyl-cyclopentenophnanthrene

Explanation: The above sequential steps in the reaction may be explained as under :

1. **17β Estradiol (I)** when treated with *methadiazine* yields **3- methoxy-17β-esterol**, which on reaction wit $ZnCl_2$ gives the **17-methyl derivative***.
2. The resulting product when heated with *Se powder* at 320°C gives the **3-methoxy-17-methio-cylcopentenophnanthrene***.

 **NOTE :** Interestingly, one comes across *two stereoisomers of estrodiol viz*, α-and β-; of which α– *estrodiol* is more active. Obviously these nomenclatures were based on the *incorrect configuration* at *C-17* , therefore, in order to avoid any kind of confusion, it is always preferred to designate them as the α-and β-*isomers* for instance: 17β- estrodiol and 17α-estrodiol.

The Chemical Relationships: Amongst Estrone, Estriol, and Estradiol

There exists a spectacular **chemical relationship** amongst the estrone, estriol, and estradiol that could be seen explicitly from the following reactions:

❑ **Formation of Estradiol from Estrone** – When **estrone** is subjected to *reduction* either catalytically **(Pd-H₂)** or **aluminium isopropoxide** Al **[OH (CH₃)₂]₃** or **lithium aluminum hydride [LiAlH₄]** it results into the production of **estradiol** as given under:

---

* **Mechanism of Reaction:** A molecular arrangement occurs when the **methylester of estradiol** is heated with **ZnCl₂** whereby the **angular methyl group at C-13** migrates to the *cylopentane ring at position C-17*.

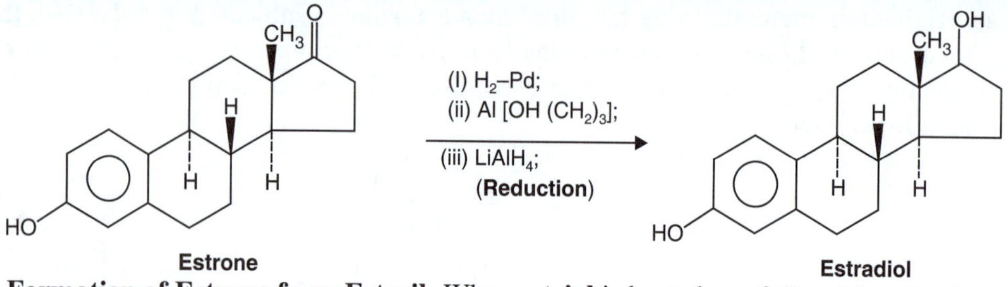

Estrone → **Estradiol**

(I) H₂–Pd;
(ii) Al [OH (CH₂)₃];
(iii) LiAlH₄;
(Reduction)

❑ **Formation of Estrone from Estroil-** When **estriol** is heated carefully with **potassium hydrogen sulphate (KHSO₄)**, it undergoes *dehydration* to give **estrone** as stated below:

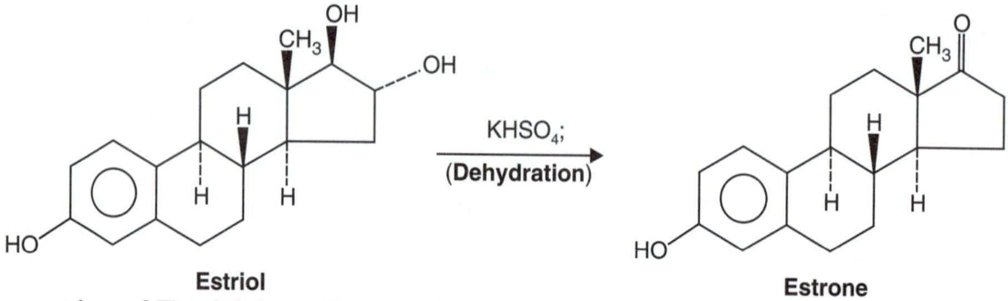

KHSO₄;
(Dehydration)

Estriol → **Estrone**

❑ **Formation of Estriol from Estrone/Methyl Ether of Estrone-** **Estriol** may be synthesized either from **estrone** or **methyl ether of estrogen** as given under:

(a) Estriol from Estrone

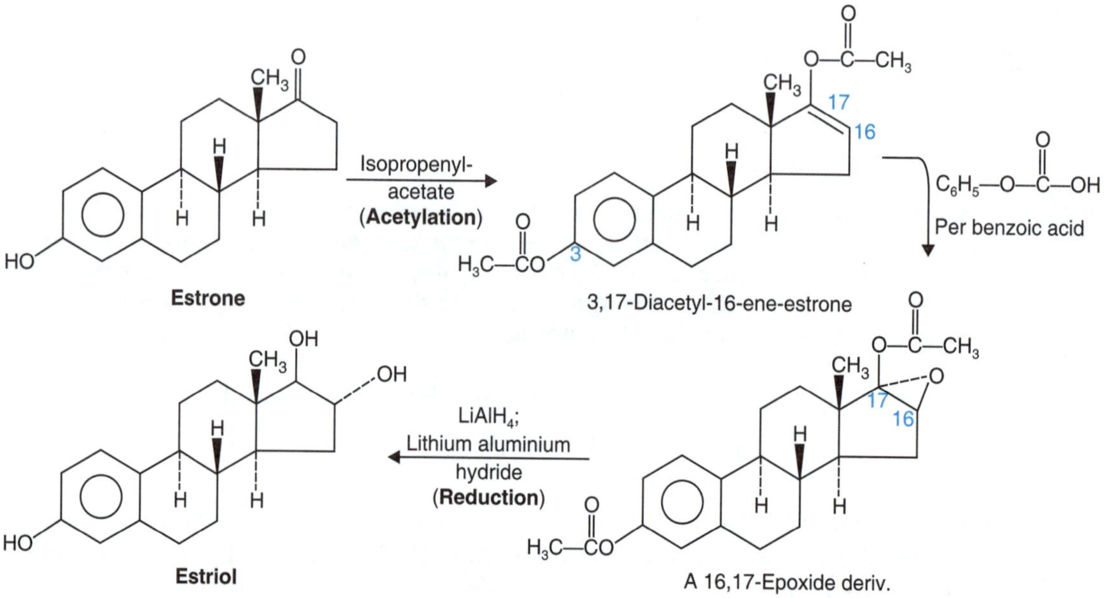

Isopropenyl-acetate
(Acetylation)

Estrone → **3,17-Diacetyl-16-ene-estrone**

C_6H_5—O—C—OH
Per benzoic acid

LiAlH₄;
Lithium aluminium hydride
(Reduction)

Estriol ← **A 16,17-Epoxide deriv.**

Explanation : **Estrone** on *aceylation* with isopropenyl acetate forms 3,17-diacetyl-16-ene-estrone, which on treatment with *perbenzoic acid* undergoes oxidation to produce **a 16,17-epoxide derivative.** This product upon careful reduction with *LiAlH₄* yields **estriol.**

(b) **Estriol from Methyl Ether of Estrone – Estriol** may also be synthesized from **methyl ether of estrone** as given below:

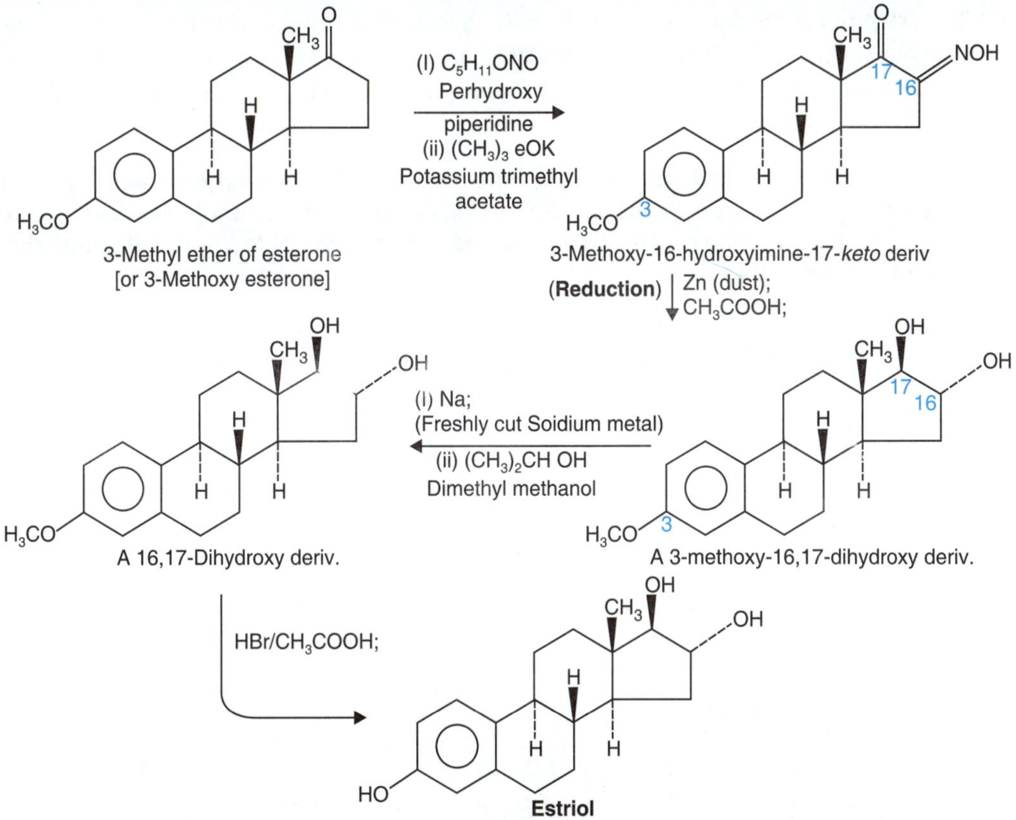

(I) $C_5H_{11}ONO$
Perhydroxy
piperidine
(ii) $(CH_3)_3$ eOK
Potassium trimethyl
acetate

3-Methyl ether of esterone
[or 3-Methoxy esterone]

3-Methoxy-16-hydroxyimine-17-*keto* deriv

(**Reduction**) Zn (dust);
CH_3COOH;

(I) Na;
(Freshly cut Soidium metal)

(ii) $(CH_3)_2CH$ OH
Dimethyl methanol

A 16,17-Dihydroxy deriv.

A 3-methoxy-16,17-dihydroxy deriv.

HBr/CH_3COOH;

Estriol

Explanation : The 3-**methoxy esterone** (*i.e, 3- methyl ether of esterone*) on being *treated with perhydroxy* piperidine followed by *potassium trimethyl acetate* yields **3-methoxy-16-hydroxyimine-17** *keto* **derivative**, which on reduction with **Zn (dust /acetic acid** gives rise to the formation of **3- methoxy 16,17-dihydroxy derivative**. The resulting product when reacted with feshly cut *sodium metal* pieces and *dimethylmethanol* gives a **16,17-dihydroxy derivative**, which finally on treatment with *HBr in* CH_3COOH yields the desired compound **estriol**.

SYNTHETIC ESTROGEN (or Ethinylestrodiol) [*Syn*: 16-Ethinylestradiol; Ethinyl estradiol]

Ethinyl estradiol is a synthetic steroid having an exceptionally high **oral estrogenic potency.**

Chemical Structure

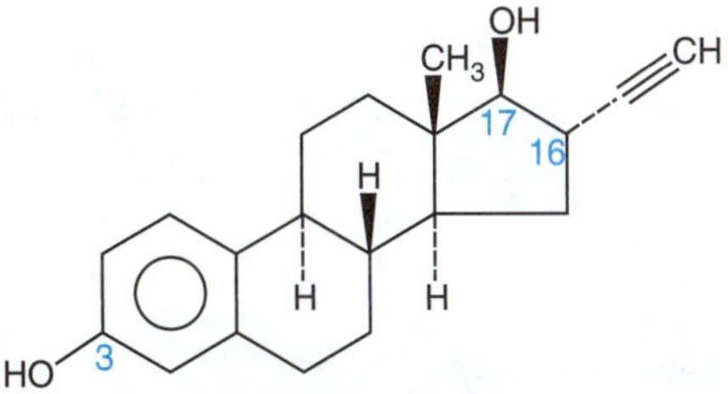

16-Ethinyl estradiol
(17α)-19-Norprogna-1,3,5 (10)-trien-20-yne-3,17-diol;

Ethinyl estradiol represents an extremely *active* **synthetic estrogen.** However, its most vital and important advantage being that it is found to be definitely much more active when administered *orally*. Besides, the **ethinyl (....CH=CH) redical** has shown explicitly to delay the ensuing inactivation of the **estradiol molecule** in the *stomach*, *intestine*, and *liver*.

> **NOTE** It is, however, pertinent to state here that both-*17 α-ethinylestradiol and its structural analogue 3- methyl ether derivative i.e*, **Mestranol, has been used** perceptively as one of the *vial ingridients* in a majority of the presently marketed *oral contraceptives'*.

Mestranol

Preparation from Estrone- Ethinyl estradiol may be prepared by the interaction of *potassium acetylide* (*KC≡CH*) upon **esterone** in a *medium of liquid ammonia* as depicted under:

KC≡CH
Potassium
acetylide
(**Liq. NH₃**)

Esterone   **17α-Ethinylestradiol**

11.1.4. Non-Steroidal Estrogens [Syn; Synthetic Estrogens; Artificial Hormones;]

A good number of **synthetic estrogens** have been prepared that do possess *oestrogenc activity*, but do not have a steroidal nucleus at all. Interestingly, based on the following *two* cardinal features, namely:

- **reasonably cost effective** and
- **noticeable potency when given orally,**

these **non-steroidal estrogens have** been found to be competing much more efficiently to an *overwhelming advantage vis-a-vis* the so called **naturally occurring steroidal estrogen** or their corresponding orally active *structural analogues* :

Following are the *four* important member of the **non steroidal estrogens** such as :

➢ **Stilbesterol,**
➢ **Hexosterol,**
➢ **Dienosterol,** and
➢ **Benzosterol (Chemestrogen),**

which shall now be discussed individually in the sections that follows :

11.1.4.1. *Stilbesterol [Syn: Diethylstilbesterol;]*

Stilbesterol (or **diethylstilbesterol**) may be synthesized by *two* different methods :

- **Dodds and Lawson (1938)*** and
- **Kharasch *et al*. (1943)****.

(a) Synthesis of Stilbesterol from Anisaldehyde (Dodds et al., 1938)

The various steps involved in the synthesis are as follows :

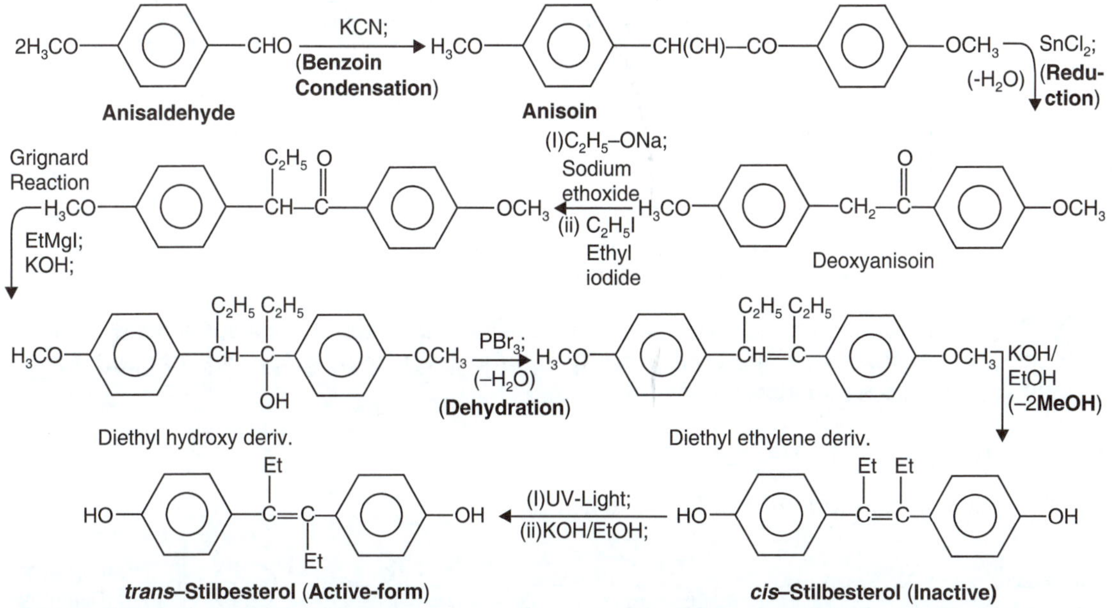

NOTE The actual *cis-* and *trans* configurations have been duly confirmed by *X-ray diffraction analysis* (Crawfoot et al., 1941)

Explanation : The various steps in the above synthesis may be explained as under:

1. Two moles of anisaldehyde undergoes *condensation* with KCN to yield **anisoin**, which upon *reduction* with stannous chloride ($SnCl_2$) gives **deoxyanisoin** with the loss of a mole of water.

2. The resulting product, *first* on treatment with freshly prepared *sodium ethoxide* (C_2H_5-ONa) followed by reaction with *ethyl iodide* (CH_3I) yields **ethyl deoxyanisoin**, which undergoes *Grignard reaction* in the presence of *ethyl magnesium iodide* and *KOH* gives **diethyl hydroxy derivative.**

3. The end product obtained in *step-2* (above) undergoes *dehydration* with **phosphorus tribromide** (**PBr₃**) to produce **diethyl ethlene derivative**, which on treatment with *alkaline ethanol* (EtOH) loss *two* moles of *methanol* to yield the ***cis*-stilbesterol** (*i.e,* **the inactive form**).

4. Finally, the ***cis*-isomer of stilbesterol** on exposure to **UV-light** followed by treatment with *alkaline ethanol* gives the desired product ***trans* - stilbesterol,** the *active form* of the **non-steroidal estrogens**

---

* Dodds *et al.* : *Nature*, **141** : 247, 1938.

* Kherasch *et al.*: *J Am Chem Soe.*, **65** : 1820, 1943

(b) Synthesis of Stilbesterol from Anethole (Kharasch *et al.*, 1943)

The various steps involved in the synthesis are as follows:

Explanation: Anethole on treatment with HBr undergoes **Markownikoff's addition** to yield **anethole hydrobromide,** which on reaction with *sodamide (NaNH₂)* in liquid ammonia gives rise to the formation of an *intermediate*. The resulting *intermediate* on *demethylation* followed by *isomerization* in an *alakline medium* produces the desired product *trans*-**stilbesterol.**

> **NOTE**   *trans*-Stilbesterol shows similar uses and actions to those *estradiol*. It is indicated in the treatment of menopausal symptoms and also in *secondary* amenorhoea due to ovarian insufficiency.

11.1.4.2. *Hexosterol [or Hexesterol]*

The synthesis of *hexosterol* may be expatiated as under:

Explanation: Two moles of *anethole hydrobromide* undergoes **Wurtz-Fitting reaction** in the presence of freshly cut pieces of Na metal to yield a corresponding **1,2-diethylethane derivative,** which upon treatment with *alkaline ethanol* (KOH/EtOH) gives **hexosterol.**

> **NOTE :**   **(1)** Since *hexosterol* contains *two chiral centers* (marked with asteriks), hence it may exist in *diastereomeric forms*. However, the *meso* form (mp:185-188°C) is found to be much more potent than the *lower melting recemic forms* (mp:128-130°C) . It is only the meso form that does have the therapeutic importance.
>
> **(2)** *Hexosterol* may also be prepared from the corresponding *diamino compound i.e,* 4,4-diamino-β-diethyldibenzyl by subjecting to *diazotization* and subsequent treatment with hot water.

11.1.4.3. Dienosterol [or Dienesterol]

It differs from **hexosterol** in possessig *two* **inherent double bonds** in the *side chain position* . Interestingly, it is nearly 1/4th as potent as **stilbesterol**; and also being recongnized as the **most potent amongst the so-called synthetic estrogens.**

Dienosterol may be synthesizd by **different methods.** The synthesis of **Dodds** *et al.* **(1939)*** is described as under:

Chemical Structure

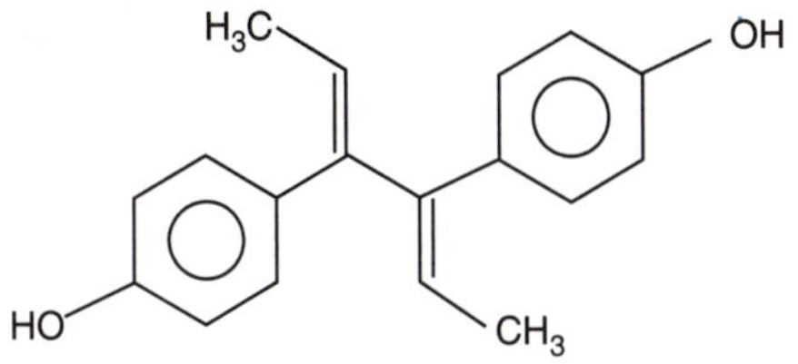

(*trans-trans*)-Dienosterol

Synthesis

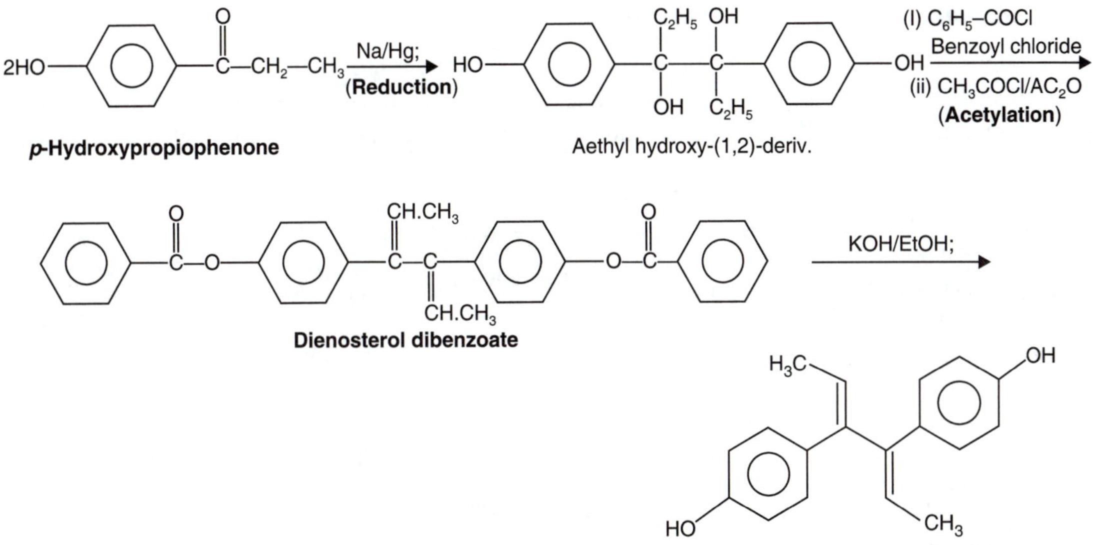

p-Hydroxypropiophenone   Aethyl hydroxy-(1,2)-deriv.

Dienosterol dibenzoate

trans-trans-Dienosterol

Explanation: The *tran-trans*-**Dienosterol** may be prepared by the careful reduction of **p-hydroxy-propiophenone** with *sodium-analgam (Na/Hg)* to give a diethyl (1,2)-derivatie which upon benzoylation followed by acetylation gives the **dienosterol dibenzoate.** The resulting product on being subjected to *KOH in ethanolic medium* produces the desired compound *trans-trans*-**dienosterol.**

Another Method from 2,3-Dibromopropyl-4-methoxy benzene: The alternate synthesis has been duly described as under:

---

* Doods *et al.* : *Proc Roy Soc* ., **127b**, 162,1939.

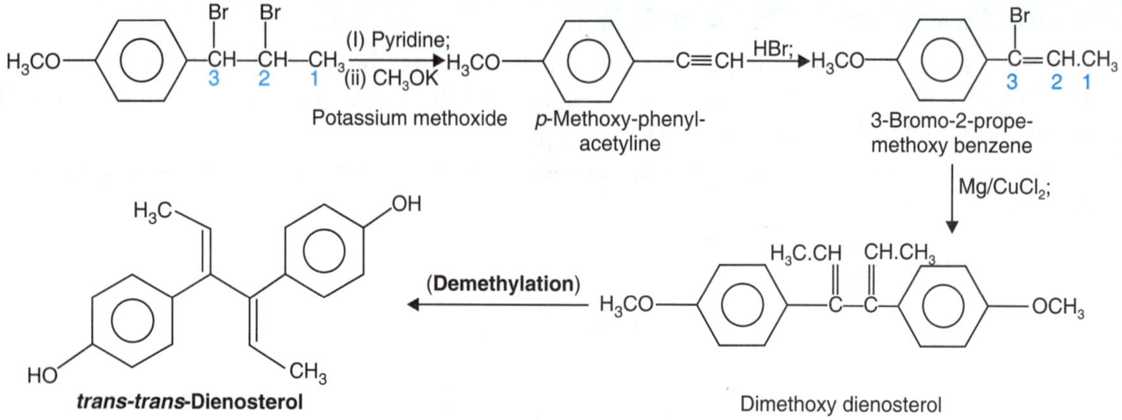

trans-trans-Dienosterol

Dimethoxy dienosterol

Explanation : The treatment of **2,3-dibromopropyl-4-methoxybenze** with *potassium methoxide (CH₃OK) in pyridine* yields **p-methoxy phenyl acetylide** which on reaction with HBr gives **3-bromo-2-propene methoxy benzene.** The resulting compound on treatment with *Mg/CuCl₂* produces the **dimethoxy dienosterol**, which upon *methylation* finally yields the desired compound ***trans-trans-dienosterol.***

11.1.4.4. Benzosterol [or Chemestrogen]

Importantly, it designates the classical example of a purely **synthetic estrogen** which may be derived from **hexosterol** (see *section 11.1.4.2.*) by meticulously lengthening of the **C-Chain** between the *two* **phenyl moieties.** The various steps in the above *synthesis* of **Benzosterol** are as enumerated under:

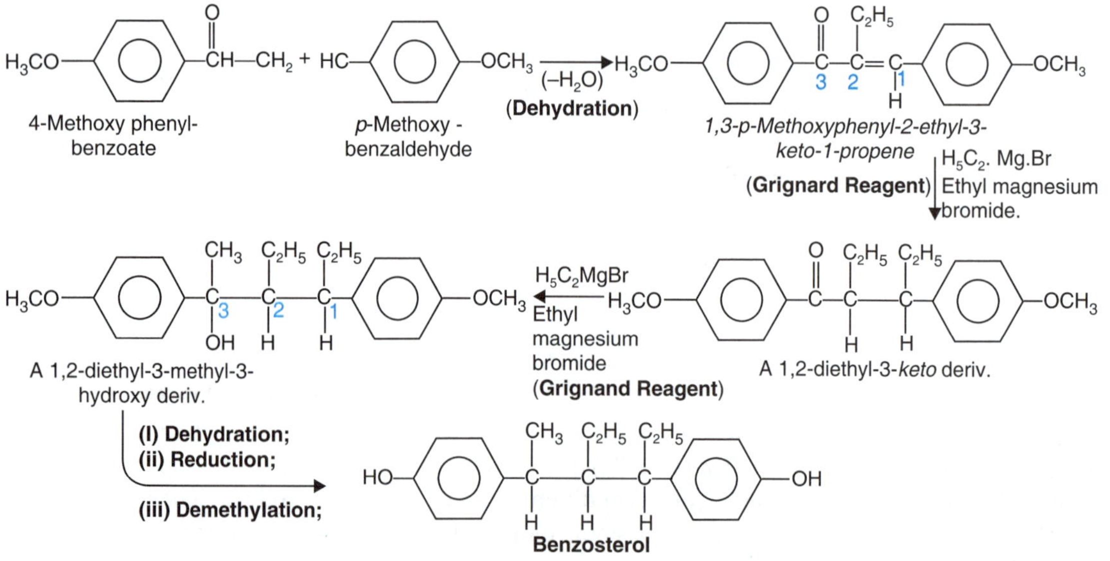

Benzosterol

Explanation: The interaction between **4-methoxy-phenyl benzoate** and *p*-methoxy benzaldehyde results into dehydration to yield **1,3-p-methoxyphenyl-2-ethyl-3-keto-1-propene.** The resulting product makes use of the **Grignard Reagent** (*ethyl magnesium bromide*) in *two* sequential steps to yields a **1,2-diethyl-3-methyl-3-hydroxy derivative.** This product when subjected to the sequential treatment of **dehydration-reduction and demethylation** gives rise to the formation of the desired product, **Benzosterol.**

11.2. Corpus Leuteum Hormones [or Gestogens]

Preamble: Corpus Leuteum refers to the *small yellow endocrine structure* that critically develops within a *ruptured ovarian follicle,* and duly secretes *two* chemical entities namely: **Progesterone** (which shall be discussed in this partucular section), and Estrogen (which has already been described earlier under:

- **section 11.1.3. for Estradiol,** and
- **section 11.1.4.1 for Diethylstilbesterol.**

11.2.1. Progesterone [Syn: Luteohormone;]

Preamble: Progesterone represents an active principle of the ***corpus leuteun, duly*** secreted during the **latter half of the menstrual cycle.** In case, pregnancy continues, the secretion also continues. It ciritcally exerts an **antiovulatory effect** when administered during days 5 to 25 of the *normal menstural cycle.*

Chemical Structure

Progesterone
Pregn-4-ene-3, 20-dione; Δ^4–Pregnene–3,20-dione;

Occurrence: Progesterone was first and foremost obtained from the *corpus leuteum (i.e, gland with internal secretions), adrenal cortex placenta,* and *testis.* However, in the course of *full corpus leuteum activity state*-the *human ovary* does generate nearly **1000mg (or 1g) of progesterone** per day.

Isolation: Various steps involved in the *isolation* of **progesterone** are as given under:

1. The **ovaries** are removed carefully, cleaned, and *minced* in a mincer and then treated with *methanol.*
2. The resulting *methanolic extract* is duly diluted with water and is then *filtered* so as to get rid of the **fatty matter** (*impurities*)
3. The clear filtrate thus obtained is treated with **petroleum ether (40-60°C bp)**-when **progesterone** gets extracted in the solvent *ie.,* **petroleum ether.**
4. The solvent is filtered and concentrated (but *evaporation*) under vacuo; and the residue thus obtained is treated with **ethanol (95%)** followed by water and the **'released progesterone'** is extracted again from this by **petroleum ether (40-60°C bp)**; whereas, *estrone* remains very much in the **alcoholic phase.**

---

* Wintersteiner and Allen: *J Biol Chem,* **107:**321,1934.

5. **Progesterone** is knocked out of the said *alcoholic phase* on being chilled up to **-20°C** in a **deep freezer** as a **crystalline product.**

Characteristic Features: These essentially include:

1. **Progesterone** exists in *two* **crystalline forms** having almost equal physiologic activity profile and which gets interconverted readily.

2. **α-Pregesterone**-is orthorhombic (obtained as prisms from diluted ethanol) having **a: b: c= 0.750:1:0.905)**

3. **β-Progesterone** - is obtained as **orthorhombic (needles)** having **a: b: c=0.563:1:0.275.**

> **NOTE** **The crystals are acicular with *parallel extinction* and *negative elongation*.**

4. **Solubility Profile: Progesterone** is insoluble in water, but soluble in ethanol, acetone, dioxane, conc,H_2SO_4; and sparingly soluble in **Vegetable oils** (*edible grade*).

The various sequential steps given in the above synthesis are indeed absolutely self explanatory.

CONSTITUTION OF PROGESTERONE

The **constitution of progesterone** has been logically established based on various scientific evidences as described under:

1. **Molecular Formula;** The data collected so far from an array *of analytical experiments* do ascertain the **molecular formula of progesterone** as $C_{21}H_{30}O_2$

2. **Presence of 2-*keto* (>C=O) functional moieties:** Since, **progesterone** forms a **dioxime** [>C=NOH)$_2$ with *hydroxylamine,* which clearly shows that it contains *two keto* (C=O) **functional groups**.

3. **Presence of 1-Double Bond:** It has been observed categorically that progesterone on being reduced *catalytically (pd/H$_2$)* – usually takes up **3- moles of hydrogen (H$_2$)** to yield the **dialocohol [$C_{21}H_{36}O_2$].** Obivously, *two* **moles of hydrogen (H$_2$)** are duly utilized for the conversion of *two* keto (C=O) **moieties** directly into the respective *sec*-alcoholic (CHOH) groups; and hence, the **3rd molecule of hydrogen (H$_2$)** is duly added on the **double bond (C=C)** in order to saturate it.

Comment – Therefore, **progesterone** should possess **one double bond.**

4. **Presence of a Steroidal Nucleus** – Since **progesterone** essentially consists of **2-*keto* (>C=O) moieties** plus **1-double bond**; thus, one may conclude that '**parent hydrocarbon of progesterone must be $C_{21}H_{36}$ (ie C_nH_{2N}-6).** Therefore it, shows explicitly that **progesterone** represents- **a tetracylic (D,B,E, of $C_{21}H_{36}O_2$ is 21+1-36/2=4 Rings)** i.e., it must posses a '*steroidal nucleus*'.

> **NOTE** **In addition, the *X-ray diffraction studies* and its subsequent formation from *'Cholesterol'* and *Stigmaterol'* do confirm the fact that *progestrone* critically contains a '*steroidal nucleus*'.**

5. **Presence an α,β-Unsaturated Moiety:** Because, **progesterone** is found to be *quite sensitive to alkalies* (NaOH/KOH), it vividly suggests that it does comprise an ,β-**unsaturated** *keto* (>C=O)

---

* **Vegetable Oils:** These are, in fact, all edible grade double refined vegetable oils such as : **Rice Bran Oil; Olive Oil, Sunflower Oil, Ground nut Oil, Mustard Oil, Soyabean Oil, Seasame Oil, Corn Oil, Coconut Oil, Kurdi Oil, etc.**

group. Besides the UV absorption spectrum (λ_{max} **240nm**) of **progesterone** also confirms fully the presence of the *particular grouping*.

6. **Presence of Acetyl (-CO-CH$_3$) Moiety: Progesterone** when heated with any **halogen** (**Cl$_2$ or Br$_2$ or I$_2$ or F$_2$**) and an **alkali** (**NaOH or KOH**) it gives rise to that formation of (**haloform** *HCX$_3$*) ie, it undergoes the *'haloform reaction.'* Therefore, the *"haloform reaction"** indicates vehemently that **progesterone** essentially consists of an **acetyl(-C-CH$_3$) moiety**.

7. **Exact Position of the Double Bond [>C=]:** It has been stated earlier that **progesterone** may be synthesized either from;

- **Cholesterol** or   - **Stigmasterol**,

which explicitly suggests that the critical position of the *'double bond'* could be either:

- **between C-4 and C-5** , or   - **between C-5 and C-6 ,** (of the *steroid nucleus*).

In addition, the *UV absorption spectrum* of **progesterone** suggests that the exact position of the **double bond** to be between **C-4 and C-5.** Nevertheless, progesterone may be duly synthesized from *'Pregnahediol'*,

That certainly establishes the underlying fact that the **positions of the double bond** in *'progesterone'* is located definitely between **C-4 and C-5** of the *steroid nucleus*.

Probable Structure of Progesterone : Therefore, based on the above discussed facts right from **item(1) through item (7),** one may probably assign the following *structural formula* of **'Progesterone':**

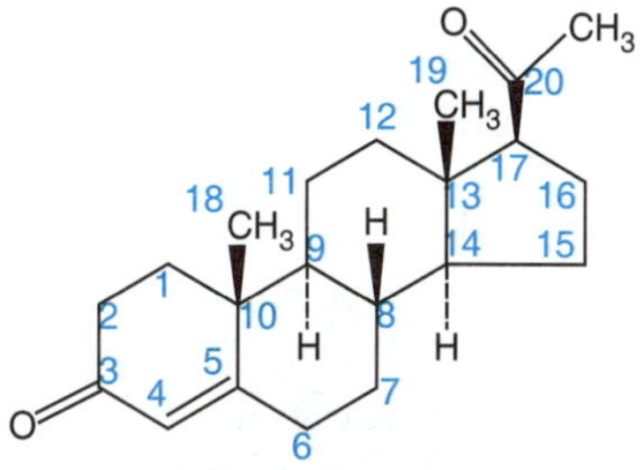

Progesterone

---

* **Haloform Reaction**: In fact, *chloroform, bromoform,* and *iodoform* are duly obtained from a process called the **halo form reaction** (*Chloroform* is so named since the structure is derived from that of **formic acid**, HOOH, by the replacement of all **oxygen substituents** by **chlorine atoms.** The above process consists in the treatments of **acetone** with **bleaching powder** or with a *solution of sodium hypochlorite* ;

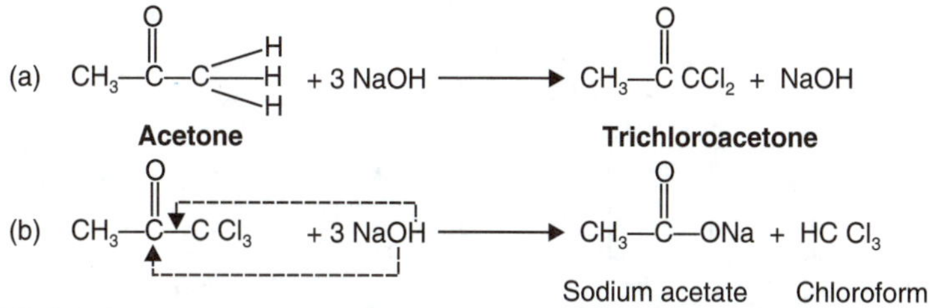

(a)   CH$_3$—C—C—H   + 3 NaOH ⟶ CH$_3$—C CCl$_2$ + NaOH

Acetone   **Trichloroacetone**

(b)   CH$_3$—C—C Cl$_3$ + 3 NaOH ⟶ CH$_3$—C—ONa + HC Cl$_3$

                              Sodium acetate   Chloroform

First phase of reaction – (a) one methyl (CH$_3$) group is fully substituted by chlorine; *Second phase of reaction* (b) alkaline cleavage of **trichloroacetone** takes place with the production *sodium acetate* and *chloroform*.

SYNTHESIS OF PROGESTERONE

The **Progesterone** has been duly synthesized by several different synthetic routes using altogether *different starting materials;* and hence, **five** of them shall be discussed in this particular section since they do have enough commercial importance and utilization. It is, however, pertinent to state here that a majority of the procedures mentioned are more or less those utilized for the preparation of **'Pregnenolone'** since this **particular intermediate** gets oxidized to **Progesterone** rather rapidly and conveniently.

Following are the *five* different **synthetic methods for Progesterone** using various starting materials namely :

   i. **Cholesterol,**
  ii. **Stigmasterol,**
 iii. **Stigmastadienone,**
 iv. **Diogenin,** and
  v. **Ergosterol,**

which shall now treated individually in the sections that follows:

11.2.1.1. *Progesterone from Cholesterol*

Butenandt *et al.* (1934)* were pioneer in putting forward the synthesis of **progesterone** from **cholesterol** as described under :

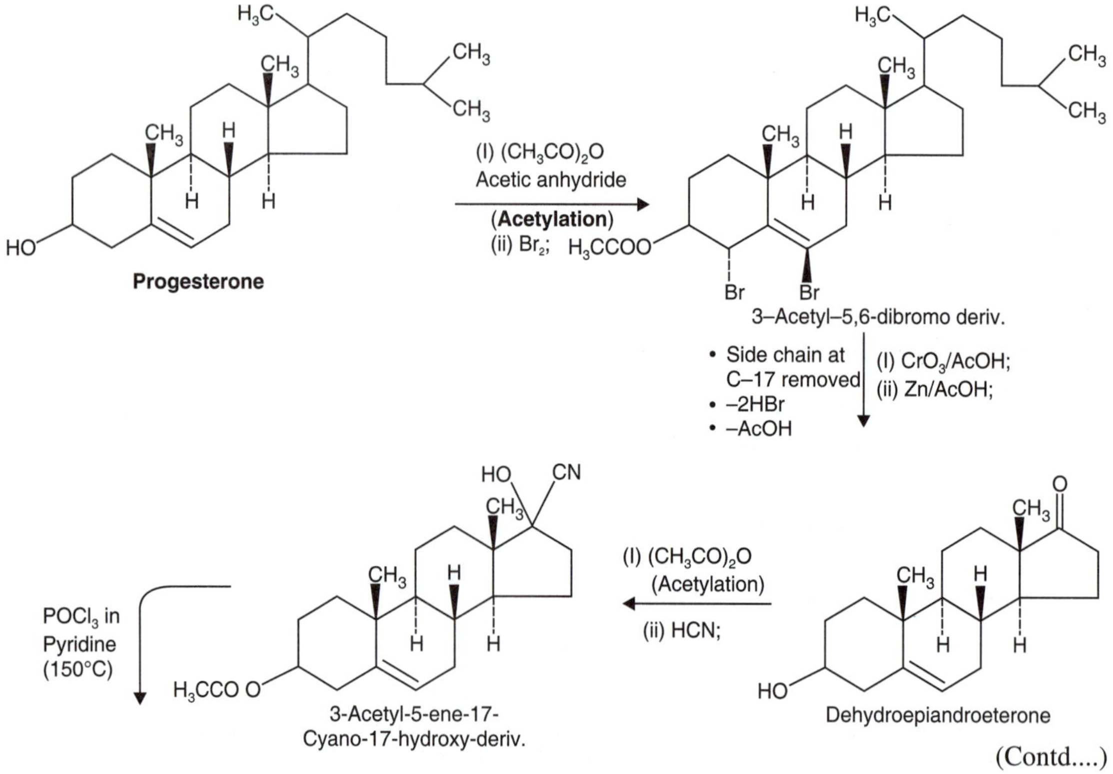

Progesterone

(I) (CH$_3$CO)$_2$O
Acetic anhydride

(Acetylation)
(ii) Br$_2$; H$_3$CCOO

3–Acetyl–5,6-dibromo deriv.

- Side chain at C–17 removed
- –2HBr
- –AcOH

(I) CrO$_3$/AcOH;
(ii) Zn/AcOH;

Dehydroepiandroeterone

(I) (CH$_3$CO)$_2$O
(Acetylation)

(ii) HCN;

3-Acetyl-5-ene-17-Cyano-17-hydroxy-deriv.

POCl$_3$ in
Pyridine
(150°C)

H$_3$CCO O

(Contd....)

---

* Butenandt *et al.* : *Ber,* **67:** 1611,1934.

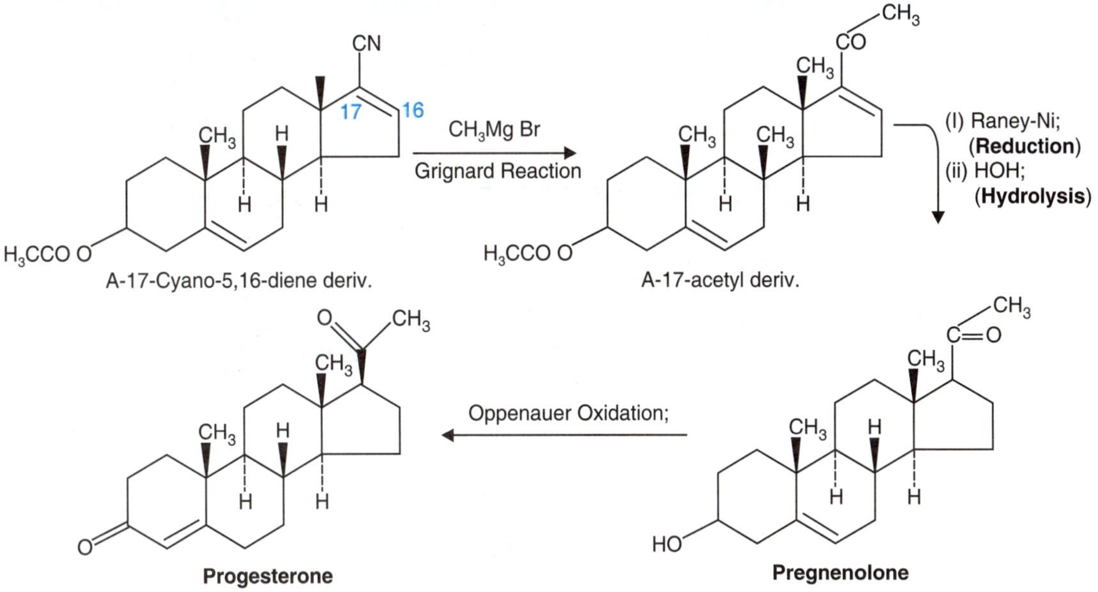

A-17-Cyano-5,16-diene deriv.

CH₃Mg Br
Grignard Reaction

A-17-acetyl deriv.

(I) Raney-Ni;
   (**Reduction**)
(ii) HOH;
   (**Hydrolysis**)

Oppenauer Oxidation;

Progesterone

Pregnenolone

Explanation : The various steps involved in the above synthesis may be explained as under ;

1. **Cholesterol** upon **acetylation** with acetic anhydride followed by **bromination** yields the respective **3- acetyl -5, 6 dibromo derivative**, which on **oxidation** with **chromium 6- oxide in AcOH** and **reduction** with Zn/AcOH gives **dehydroepiandrosterone.**

2. The resulting product on further **acetylation** followed by treatment with **HCN** yields **3- acetyl-5-ene-17-hydroxyderivative,** which on reaction with *Phosphorus Oxytrichloride (POCl₃)* in pyridine at 150°C gives **a 17- cyano-5,16 diene derivative.**

3. The resulting product on **Grignard reaction** with CH₃Mg Br gives a **17-acetyl derivative**, which upon reduction with **Raney Ni** followed by **hydrolysis** yields **pregnenolone.**

4. The end product on being subjected to **Oppenauer oxidation** gives the desired product **progesterone.**

11.2.1.2. *Progesterone from Stigmasterol*

Butenandt and Westphal, (1934)* proposed the synthesis of **progesterone** form **stigmasterol** as detailed below:

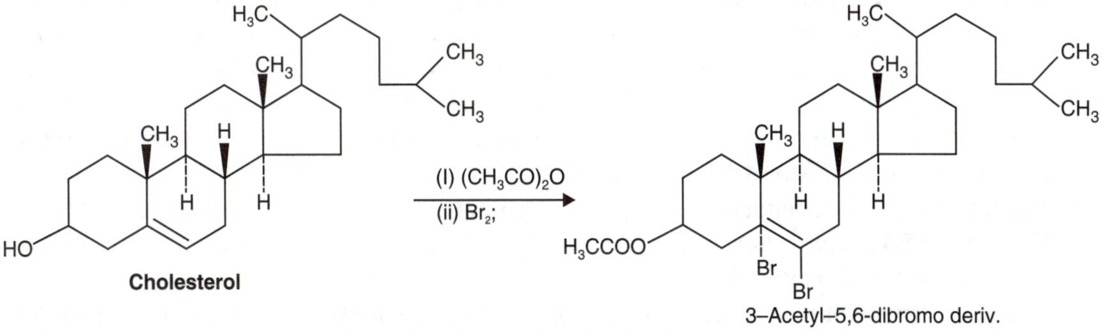

Cholesterol

(I) (CH₃CO)₂O
(ii) Br₂;

3–Acetyl–5,6-dibromo deriv.

---

*   Butenandt and Westphal : *Ber,* **67**: 1440,1934.

(Contd....)

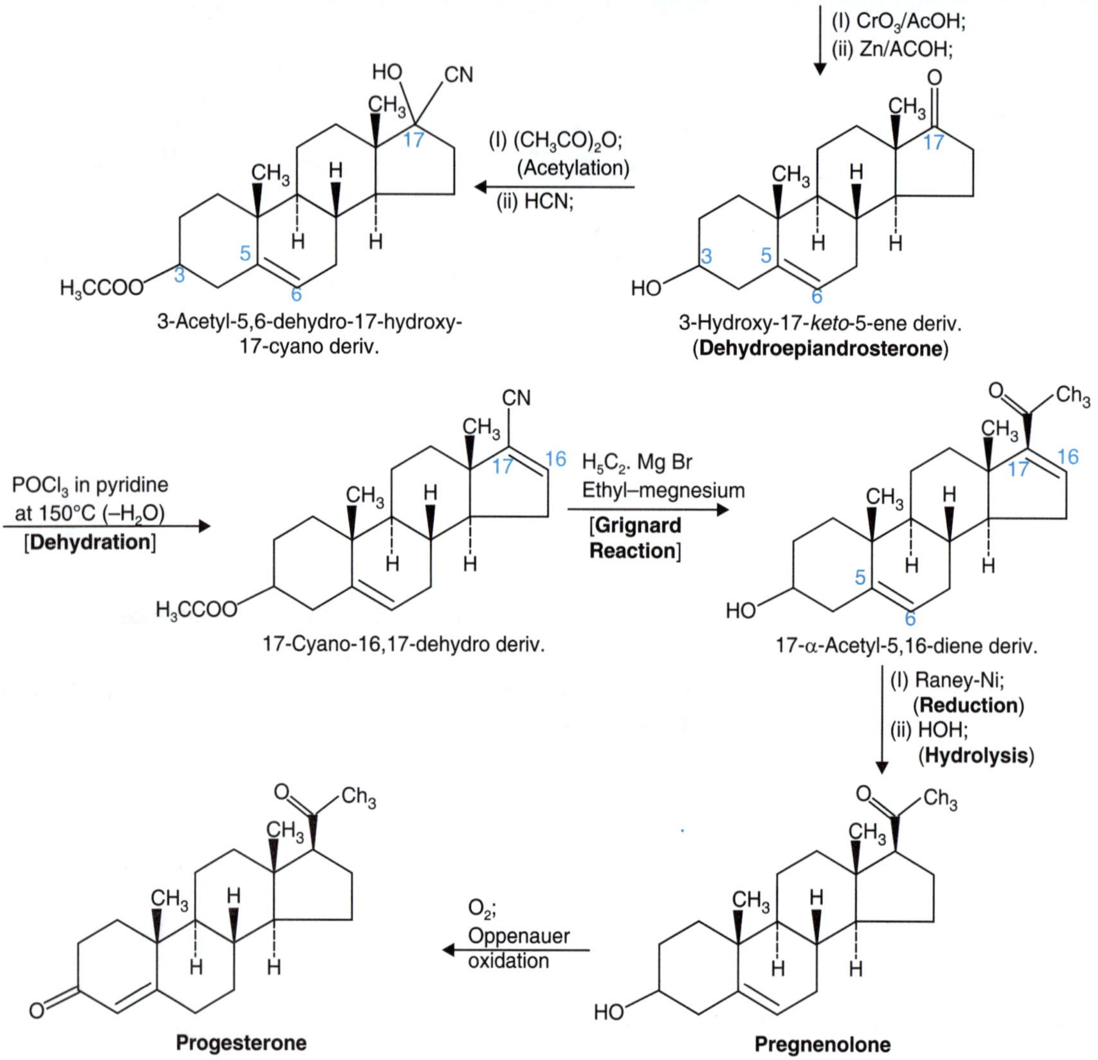

(I) CrO$_3$/AcOH;
(ii) Zn/ACOH;

3-Acetyl-5,6-dehydro-17-hydroxy-
17-cyano deriv.

(I) (CH$_3$CO)$_2$O;
(Acetylation)
(ii) HCN;

3-Hydroxy-17-*keto*-5-ene deriv.
(**Dehydroepiandrosterone**)

POCl$_3$ in pyridine
at 150°C (–H$_2$O)
[**Dehydration**]

17-Cyano-16,17-dehydro deriv.

H$_5$C$_2$. Mg Br
Ethyl–megnesium
[**Grignard
Reaction**]

17-α-Acetyl-5,16-diene deriv.

(I) Raney-Ni;
(**Reduction**)
(ii) HOH;
(**Hydrolysis**)

Progesterone

O$_2$;
Oppenauer
oxidation

Pregnenolone

Explanation: The various steps involved in the above synthesis may be explained as under:

1. **Acetylation** of **cholesterol** followed by *bromination* yields **3- acetyl-5,6-dibromo derivative ,** which on oxidation with *chromium-6-oxide* in acetic acid followed by reduction with *Zn/AcOH* gives the corresponding **3- hydroxyl-17-keto-5-ene derivative**. (*i.e*, **Dehydroepiandrosterone**).

2. The resulting product on *acetylation* and reaction with **HCN** yields **3-acetyl-5,6-dehydro-17-hydroxy-17, cyano derivative,** which undergoes **dehydration** with **POCl$_3$ in pyridine at 150°C** gives **17- cyano-16,17-dehydro derivative**.

3. The resulting product undergoes **Grignard Reaction** with *ethyl magnesium bromide* gives **17-acetyl-15,16-diene derivative,** which upon *reduction* using *Raney-Ni* followed by *hydrolysis* yields **Pregnenolone.**

4. Finally, **Pregnenolone** on being subject to **Oppenauer oxidation** produces the desired product **Progesterone.**

11.2.1.3. *Progesterone from Stigmastadienone*

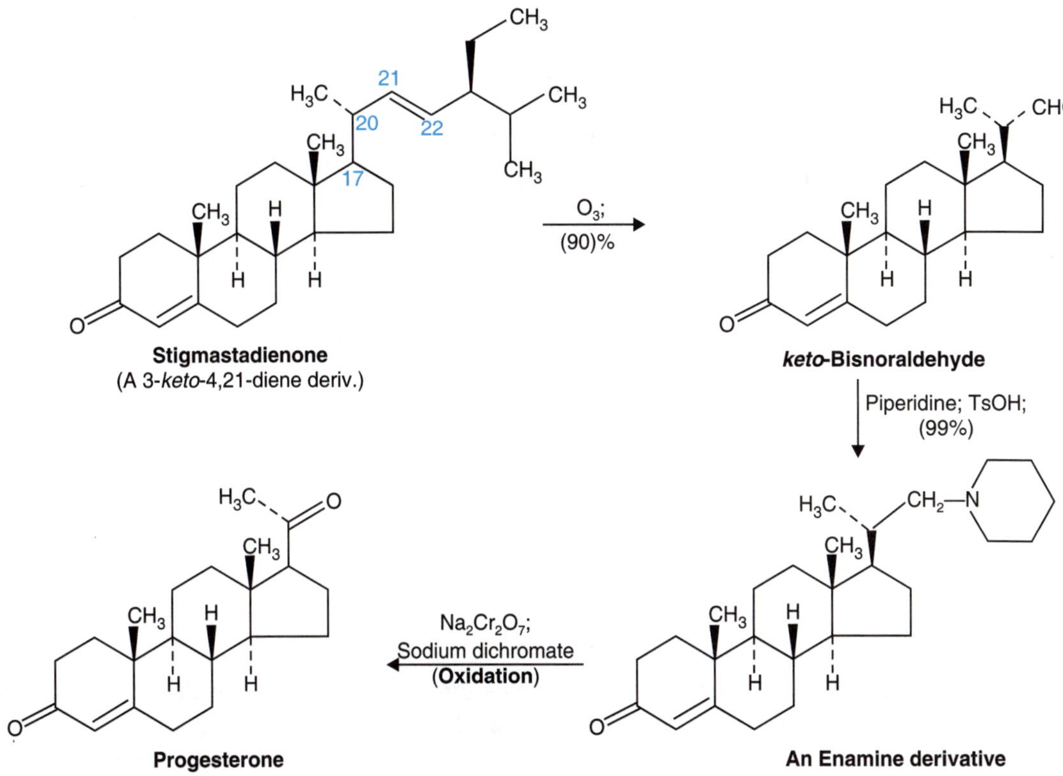

Stigmastadienone
(A 3-*keto*-4,21-diene deriv.)

O₃;
(90)%

keto-**Bisnoraldehyde**

Piperidine; TsOH;
(99%)

Na₂Cr₂O₇;
Sodium dichromate
(**Oxidation**)

Progesterone

An Enamine derivative

Explanation: *Ozomization* of *stigmastadienone* yields the *keto-bisnor aldehyde*, which on treatment with *piperidnie/TsOH (99%)* gives an *enamine derivative*. This product on being subjected to *oxidation* with *sodium dichromate* gives rise to the production of **progesterone**.

11.2.1.4. *Progesterone from Diosgenin*

Progesterone has also been synthesized from **diosgenin** by Marker *et al* (1940-1941) as stated under:

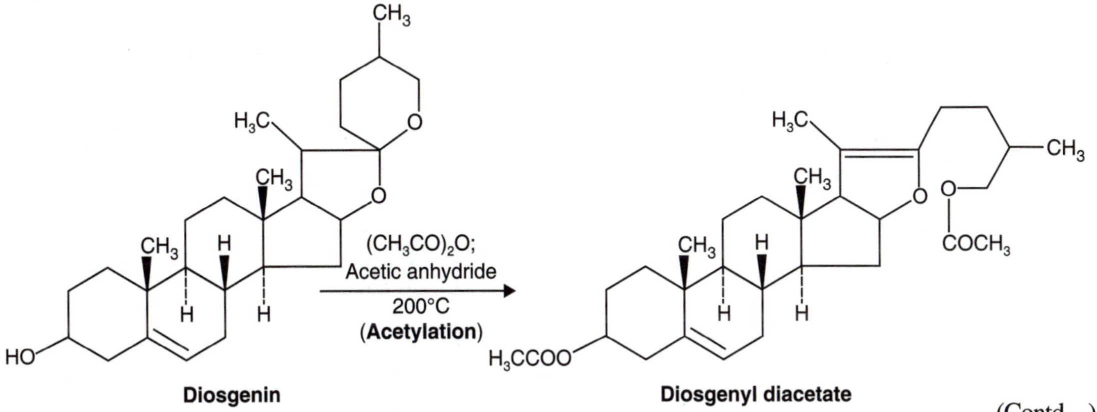

(CH₃CO)₂O;
Acetic anhydride
200°C
(**Acetylation**)

Diosgenin

Diosgenyl diacetate

(Contd....)

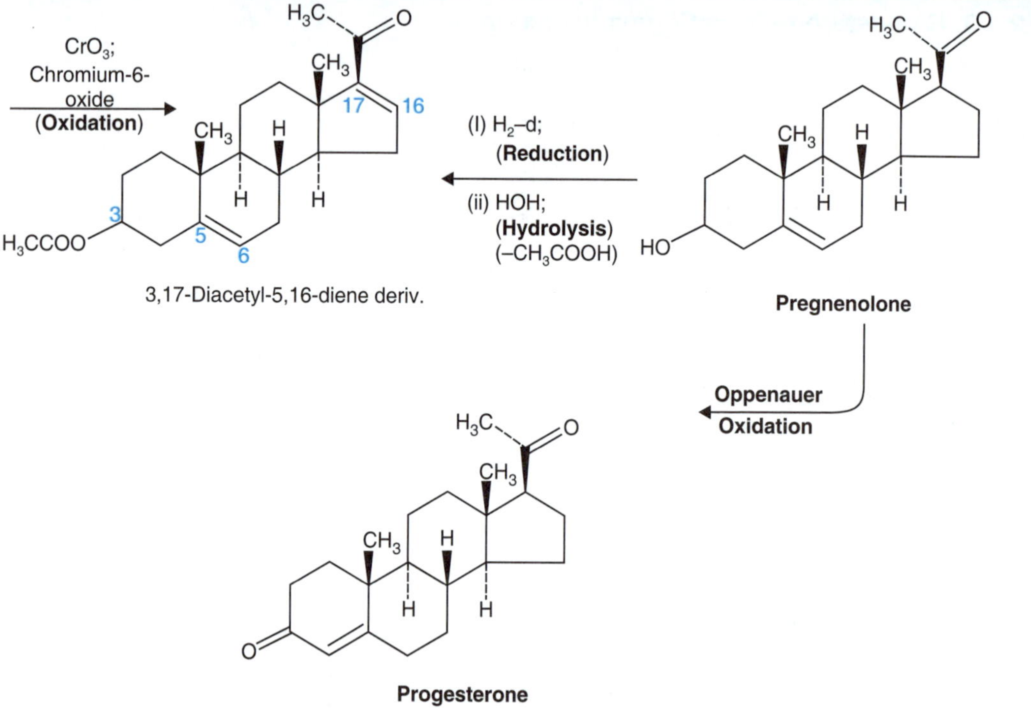

3,17-Diacetyl-5,16-diene deriv.

Pregnenolone

Progesterone

Explanation: *Acetylation* of **diosgenin** at 200°C yields **diosgenyl diacetate** which upon **oxidation** with chromium 6-oxide (CrO$_3$) gives **3,17-diacetyl-5,16-diene derivative**. The resulting product *first* on reduction with *H$_2$/Pd* followed by *hydrolysis* loses a mole of acetic acid to give **pregnenolone,** which on **Openauer oxidation** yields the desired product **progesterone.**

11.2.1.5. *Progesterone from Ergosterol*

Shephard *et al.* (1954) put forward the synthesis of **progesterone** from **ergosterol** (a naturally occurring phytosterol from '**Ergot',** as elaborated below :

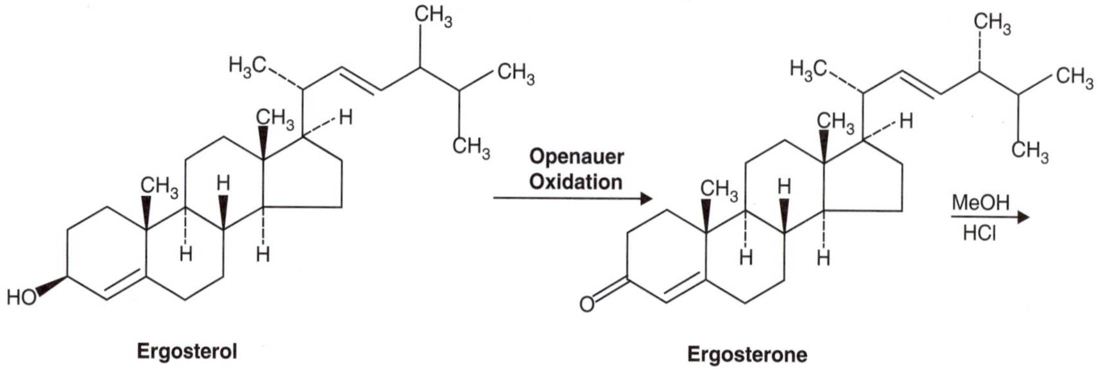

Ergosterol

Ergosterone

(Contd....)

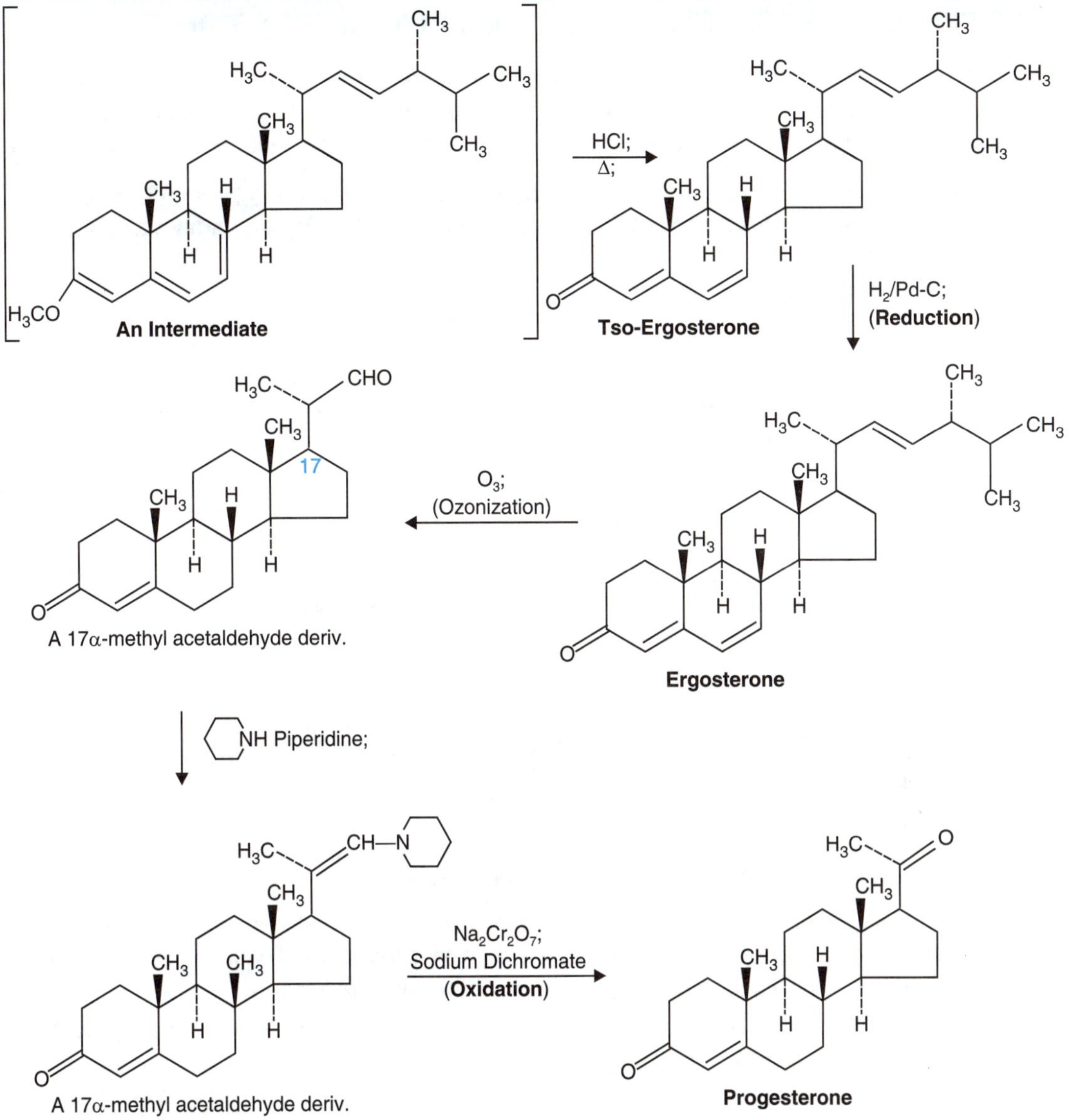

An Intermediate

Tso-Ergosterone

HCl;
Δ;

H₂/Pd-C;
(Reduction)

A 17α-methyl acetaldehyde deriv.

O₃;
(Ozonization)

Ergosterone

NH Piperidine;

A 17α-methyl acetaldehyde deriv.

Na₂Cr₂O₇;
Sodium Dichromate
(Oxidation)

Progesterone

Explanation:

1. **Ergosterol** on *Opppenauer oxidation* yields **ergosterone**, which on treatment with *acidic methanol* gives an **intermediate**.

2. The resulting **intermediate** on heating with HCl gives **iso-ergosterone**, which upon *reduction* with *pd-c* yields **ergosterone.**

3. **Ergosterone** on *ozonization* yields a **17α- methyl acetaldehyde derivative** which being treated with **piperidine** (freshly distilled) gives **α-methyl-ethylene** piperidie derivative.

4. Finally, this product on oxidation with **sodium dichromate** gives the desired product **progesterone**.

OTHER SYNTHETIC PROGESTERONE [17α–ETHINYLTESTOSTERONE]

Evidence from the literature reveals the existence of an important synthetic progestational **chemical entity** (compound), **17α-ethinyltestosterone** (also known as **17α-isopregneninolone; anhydrohydroxyprogesterone; or ethisterone**) which being only related but indirectly to **progesterone.**

> **NOTE**   It is, however, pertinent to state that *17- ethinyl testosterone* has proved to be almost *fifteen folds* more active than progesterone itself by oral administration.

Synthesis of 17α-Ethinyltestosterone

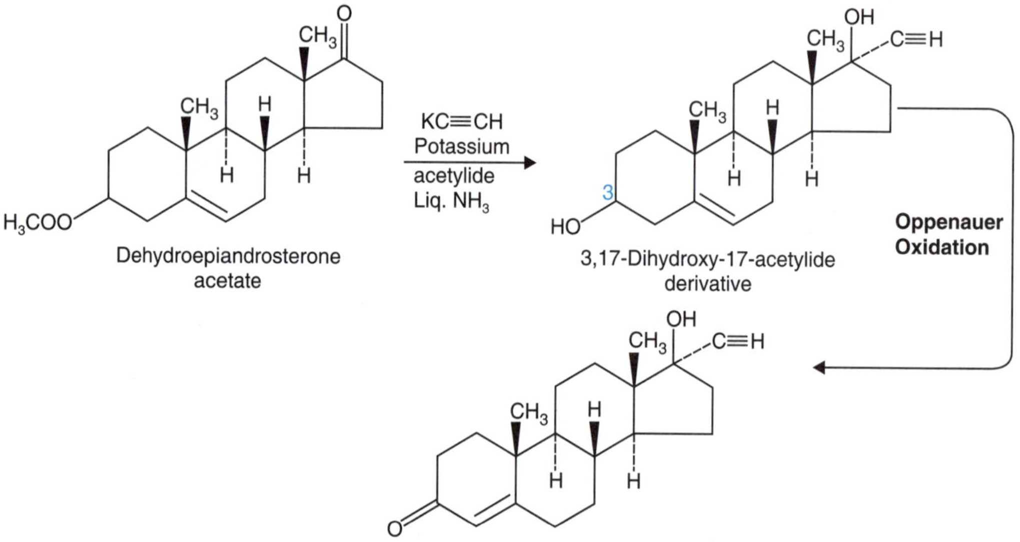

Dehydroepiandrosterone acetate

3,17-Dihydroxy-17-acetylide derivative

17 α-Ethinyltestosterone

The starting material **dehydroepiandrosterone** acetate gets acetylidated with *potassium acetylide in liquid ammonia* to give **3,17-dihydroxy-17-acetylide derivative**. This product on being subject to *Oppenauer oxidation* produces the desired compound **17-ethinyltestosterone.**

Comment: The above *condensation reaction* must be carried out under perfect anhydrous environment in the presence of *liquefied ammonia*;

11.3. Oral Contraceptives

Contraception relates to the means of *avoiding pregnancy* despite the usual **sexual activity.** Nevertheless, there is no *ideal contraceptive*; and hence, the actual choice of technique depends solely upon the so-called balancing considerations **of safety, effectiveness**, and **acceptability**.

Oral Contraceptives- refers to a **'contraceptive'** taken by mouth. It essentially comprises *one or more* **synthetic female hormones,** usually an *estrogen,* that blocks critically the *normal ovulation phenomenon;* and a **progestogen** which influences the *Pituitary Gland*; and, therefore, blocks normal control of the **woman menstrual cycles.**

> **NOTE**   *Progestogens* also make the uterus less congenial for the fertilization of an ovum by the sperm.

Oral Contracetptive Variants : The **oral contraceptives** that are available commercially are of *three* variants, namely:

❑ **Combination of Progestonal Agents and Estrogenic Agents**- In this frequently used combination of 'oral contraceptives, the *estrogenic agents* do help to control the duration of the **menstural cycle and regular menstrual flow.** It has been established beyond any reasonable doubt that these particular type of **contraceptives** invariably exhibit at **least** *three* **points of attack, namely:**

➤ *First* – they do inhibit the phenomenon of *ovulation* by specifically blocking the release of **leutinizing hormone (LH);**

➤ *Secondly*- these render the *cervical n   s hostile* to **sperm penetration process;** and

➤ *Thirdly*- they usually i   uce *endom   al changes* that are regarded to be quite unsuitable for the *implantation phenomenon.*

❑ **Combination of Progesterogen and Sequestering (Chelating) Agents** – In this specific type of combination consisting of **progesterogen** and **sequestering agents** (or *chelating agents*). In fact, these **contraceptives** do have a tendency to inhibit the phenomenon of **ovulation** by suppressing predominantly the release of both **FSH and LH**. In fact, no other *point of interference* with the **menstrual cycle** has been established**.**

❑ **Low- Level Luteal Supplementation**- In this sense, these **contraceptives** do control without causing any sort of **inhibition of ovulation**. Perhaps such an effect is due to the changes induced in the *endometrium* and/or *cervical mucus*, - thereby enabling them **hostile to the sperm penetration**. Besides, it is also possible that such *potential agents* do virtually affect the *mobility of the ovum in the fallopian tube* – in order that the so-called fertilized ovum critically reaches the so-called **uterine-cavity** just prior to the *'time'* wherein the implantation conditions are maximum.

A few important **'Oral Contraceptives'** are given in the following table:

Brand name	Progestrin Content (mg)	Estrogen Content (mg)
Enovid [R]	Norethylnordrel [9.85]	Mestranol [0.15]
Norlestrin[R]	Norethindroneacetate [2.5]	Ethinyl estradiol [0.05]
Orthonovin[R])	Norethindrone [10]	Mestranol [0.00]
Ovulen[R]	Ethynodiol diacetate [1.0]	Mestranol [0.1]
Oracon[R]	Dimethesterone [25]	Ethinyl estradiol [0.1]

Norethynodrel   Mestranol   Norelthindrone

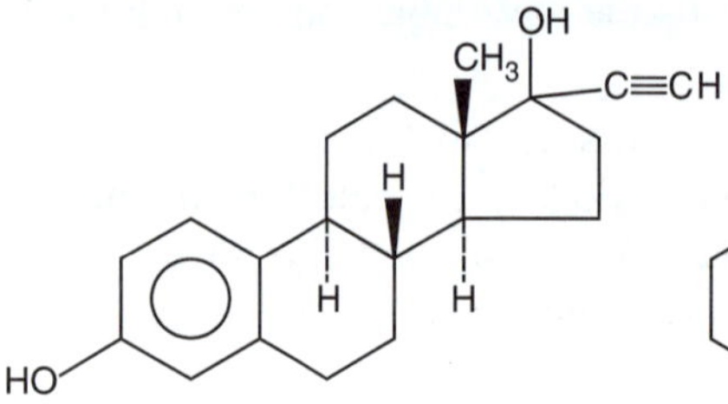

Ethinylestradiol

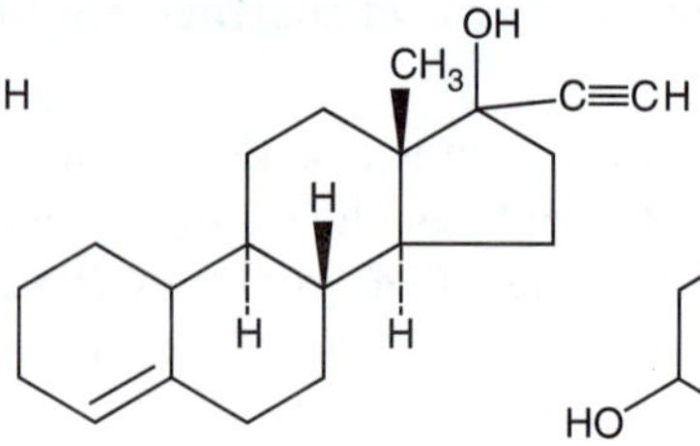

Lynoestrenol

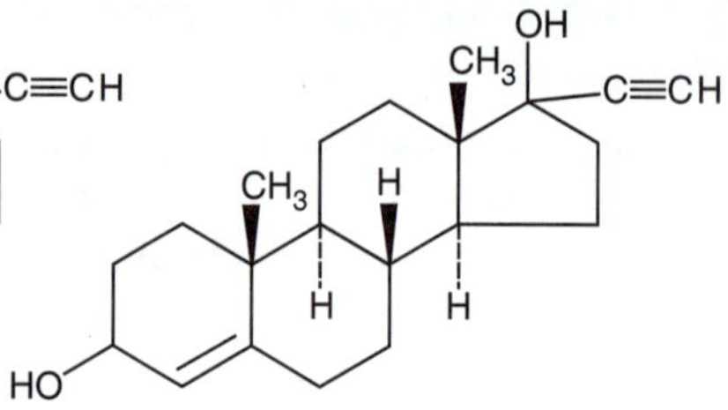

Ethynodiol

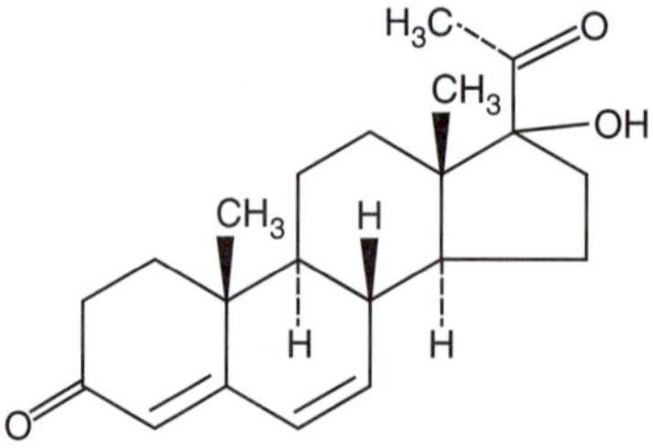

Megestrol

Dosage Regimen:

- Usual recommended dose is *1- Tablet* of the **combination- contraceptive** daily.
- Patient starts taking the **'drug'** on the **5ᵗʰ Day** of her *menstrual cycle* (counting the very first day of the menses as Day l). Thus, she consumes **one Tablet per day** for the next **twenty (20) consecutive days** and stops there.

> **NOTE** The *'medication'* is generally advised to be taken along with the super (*i.e*, evening meal) on a regular basis as stated above.

12. MALE SEX HORMONES

The **male sex hormones** are also known as; **Androgens** or **Male Hormones**. Interestingly, the experiments carried out with the **'testicular extract'**; has more or less enjoyed a **chequered career**. **Vernoff** made a successful attempt in the **transplantation** of *testes* derived from monkeys into *elderly volunteers* and claimed legitimately to **rejuvenate them** to a great extent. Similarly, the specific litgature of *vas deferens*-which is actually known to cause atrophy of **spermatogenic tissue**, and **hypotropy of intestinal tissue** indirectly, which **secrete testosterone**. Besides, another researcher **Steinach** performed almost identical investigative studies by **ligaturing the *vas deferens*** and really obtained similar results.

Following are the various important **'androgens'** or **'male sex hormones'**, namely.

- **Androsterone,**
- **Testosterone,** and
- **Dehydroepiandrosterone,**.

which shall now be discussed individually along with their **'total synthesis'** in the sections that follows:

12.1. Androsterone

Androsterone is a naturally occurring androgen that may be isolated from *male urine* after removal of the respecitve **phenolic estrogen fraction**. It has also been found that **urine** does contain nearly **1 mg of androsterone** per L of wine. It is found as a conjugate with **glucuronic acid** and in **blood** as the **sulphate**.

Chemical Structure

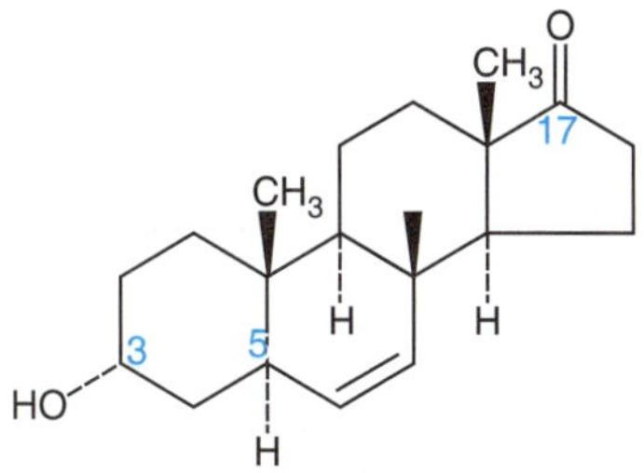

Androsterone
(3α,5α)-3-Hydroxyandrostan-17-one;

Isolation of Androsterone – The various steps involved are as follows:

1. Freshly collected **urine** is duly concentrated under *vacuo*-**hydrolyzed with acid** – and **extracted with chlorofom(CHCl$_3$)**

2. The *CHCl$_3$ layer* comprising **hormone** is treated successively with aqueous KOH solution to get rid of all such impurities as:
   - **acidic impurities**, and
   - **phenolic impurities**.

3. Careful *steam distillation* of the so-called **'neutral fractions'** do help into the removal of additional quantum of the **inactive ingredients**.

4. The **steam distilled fraction** is treated with **Grignard's Reagent,[(CH$_3$)$_3$N$^+$.CH$_2$CONHNH$_2$].Cl$^-$** when the respective hormone **(androsterone)** critically forms a **Grignard derivative.**

5. The resulting product **(Grignard derivative)** is now *extracted hydrolyzed* to give the product **'androsterone'**.

> **NOTE** **Further purification may be carried out by *chromatographic procedures.***

Characteristic Features – These essentially comprise.

1. It is obtained as a *crystalline form* from an admixture of acetone ether.
2. It undergoes **sublimation** in high vacuum.
3. It has $[\alpha]_D^{20} + 94.6°$(C=0.7 in **absolute ethanol**); and $[\alpha]_D^{15} + 87.8$ (C=1.5 in **dioxane**).
4. **Androsterone** in not precipitated by **digitonin.**
5. It is hardly soluble in water, but freely soluble in organic solvents.

---

* Butenandt and Tscherring : *Z Physiol Chem*, **229** : 167, 1934.

** Ruzicka : *Helv Chim Acta*, **17** : 1389, 1934.

*** Marken : *J Am Chem Soc*, **57** : 135, 1935.

CONSTITUTION OF ANDROSTERONE

Butenandt (1934)* first and foremost established the structure of **androsterone** even with relatively small sample of the steroid. Ruzicka (1934)** and Marker (1935)*** gloriously succeeded in synthesizing the steroid, **androsterone,** starting from '**cholesterol**' (a *sterol* derived from **animal sources**) and thereby ascertained the so-called accuracy of the chemical formula proposed by **Butenandt.**

Nevertheless, the **constitution of androsterone** has been duly established on the basis of the following facts :

1. Molecular Formula

Based on the various accrued *analytical data* as well as the **molecular weight determination** the *molecular formula* of **androsterone** has been found to be $C_{19}H_{30}O_2$.

2. Presence of a Hydroxyl (OH) Moiety

Since **androsterone** gives rise to the formation of '**mono-esters**' thereby indicating that *one oxygen atom* is certainly present as a **hydroxyl (-OH) moiety**. In addition, this **hydroxyl (OH) moiety** has been shown to be a *secondary* **alcoholic functional moiety** by virtue of the fact that **aldosterone** on being subject to *oxidation* gives a '**diketone**'.

3. Presence of a *keto* (>C=O) Moiety

In fact, the steroid **androsterone** fails to take up a **mole of bromine (Br_2);** however, it does form:

- **a monoxime**, and
- **a monoacetate**.,

thereby suggesting that it is definitely a **saturated hydroxyl ketone**.

4. Critical Presence of Four Fused Rings

Aldosterone on being subject to *reduction* specifically adds on **one mole of hydrogen (H_2)** thereby prodcing a '**diol**' with a composition of $C_{19}H_{32}O_2$. Hence, the aforesaid reaction reveals explicitly that **androsterone** *does not comprise any double bond at all*. Based on this observation the **parent hydrocarbon of androsterone,** $C_{19}H_{30}O_2$, happens to be $C_{19}H_{32}$. Since this **molecular formula** corresponds to the generalized formula C_nH_{2N-6}, it indicated clearly that the said **androsterone moleculule** is certainly of a *tetracyclic* **in nature**. That is, DBE of $C_{19}H_{30}O_2$=19+1-30/2=5; which implies **one double bond (>C=O)** ; and hence, there are *four* **rings.**

In fact, all these striking evidences do ascertain as well as support the fact that **androsterone** may essentially comprise the *steroid nucleus.*

> **NOTE**   Besides, it represents a *'hydroxyketone'*- which also shows that it may be closely related to oestrone.

5. Chemical Structure of Androsterone

On the basis of the aforesaid *scientific and logical facts,* first of all **Butenandt (1934)** proposed a **structure (I)** for Androsterone; which **Ruzicka (1934)** proved to be *incorrect* and put forward a **structure (II)** that eventually found to be correct as given under :

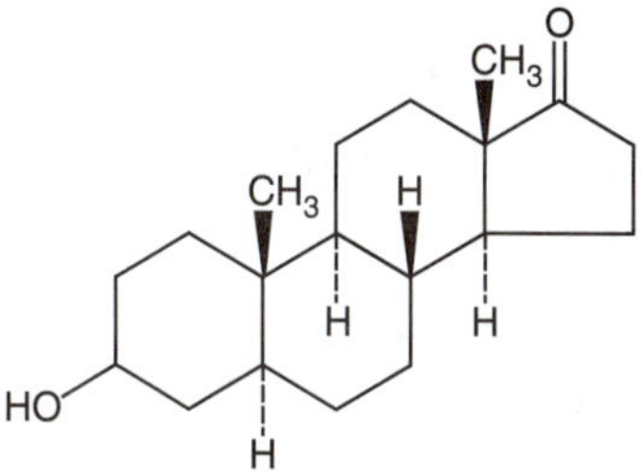

Epiandrosterone
(I)

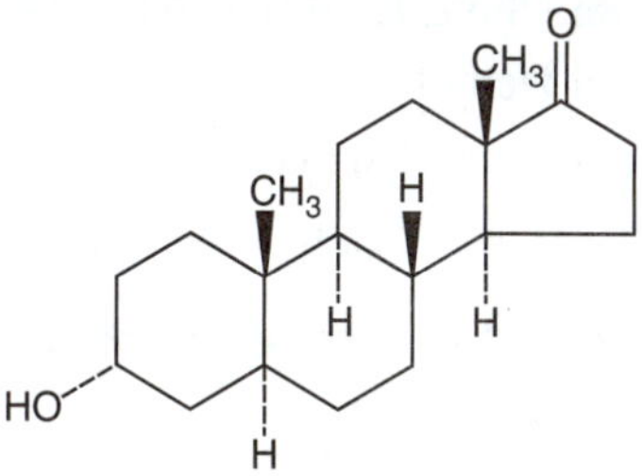

Androsterone
(II)

Further Evidences: **Ruzicka** when subjected **5α-cholestanyl-3β-acetate** with *chromium 6- oxide* in acetic acid it gave **epiandrosterone (I)**, and certainly not **androsterone (II)** [as shown above].

Comment : The above reaction and observation reveals that the **structure (I)** as proposed by **Brutenandt** was *not correct.*

Nevertheless, **5α-cholestanyl-3α-acetate (A)** on being oxidized with *chromium 6- oxide/AcOH* gave **androsterone (II).** Obviously, this specific reaction suggested clearly that the **configuration of the hydroxyl (OH) moiety at C-3 is and α-not β-**, as earlier proposed by **Butenandt.** Furthermore, **Ruzicka** revealed that **epiandrosterone (I)** has exhibited almost **12.5%** of the activity comparable to **androsterone (II).**

The above discourse from **5α-cholestanyl-3β-acetate** to **Androsterone (II)** *via* **epiandrosterone (I)** could be expatiated as under:

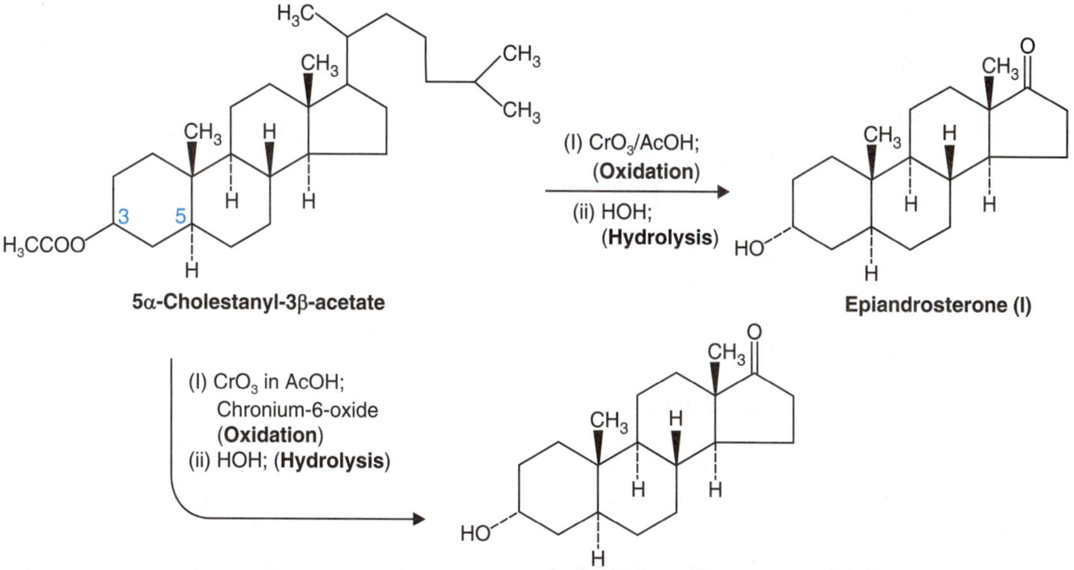

5α-Cholestanyl-3β-acetate

(I) CrO₃/AcOH;
(Oxidation)

(ii) HOH;
(Hydrolysis)

Epiandrosterone (I)

(I) CrO₃ in AcOH;
Chronium-6-oxide
(Oxidation)
(ii) HOH; (Hydrolysis)

Androsterone (I)

6. Synthesis of Androsterone (II)

The **total synthesis** of **androsterone (II),** as concluded in *item (5) above,* may be accomplished by any one of these, *three* methods namely:

- **From Cholesterol (Ruzicka, 1934) ;**
- **From Dehydroepiandrosterone (Caglioti *et al.*, 1964); and**
- **From Epiandrosterone *p*-toluene sulphonate (Sondheiner *et al.* (1954).**

which shall now be treated individually in the sections that follows:

6.1. Synthesis of Androsterone (II) from Cholesterol [Ruzicka, 1934]

The various steps in the above synthesis are as given under in a sequential manner:

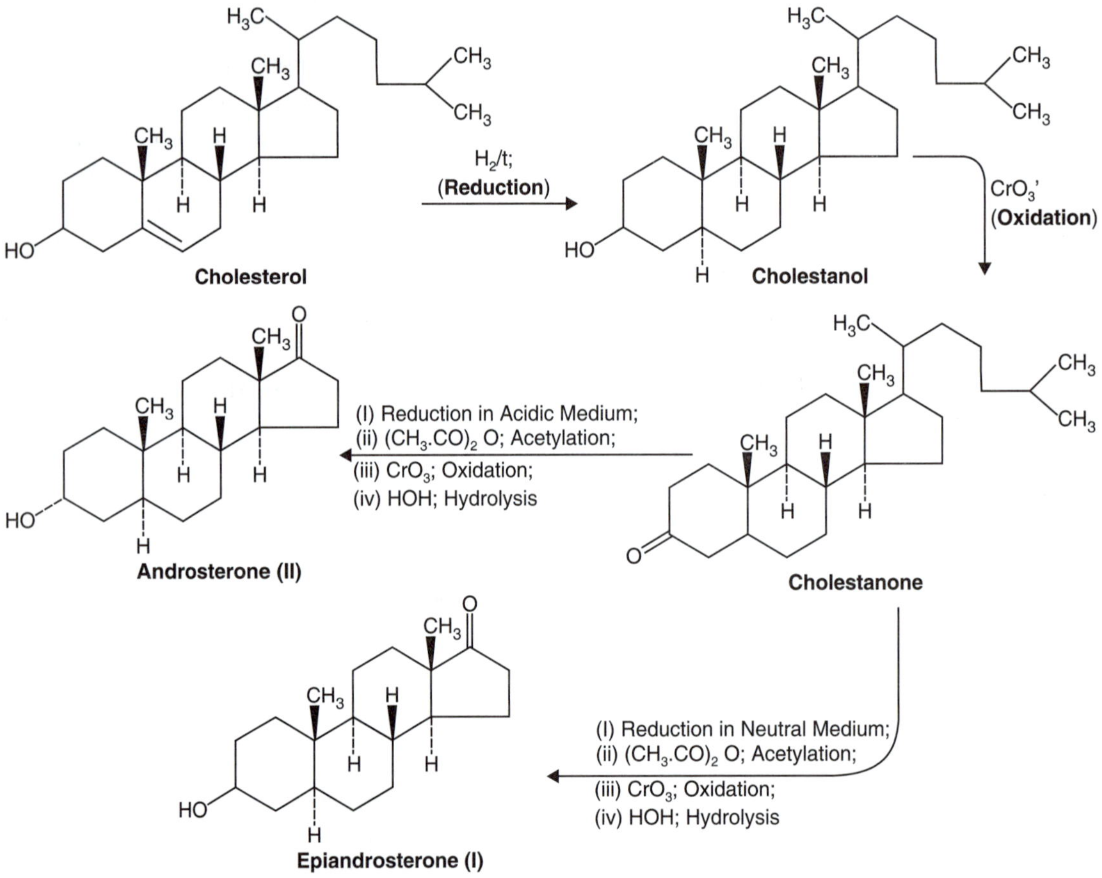

Explanation: *Reduction* of **cholesterol** with H_2-Pt yields **cholestanol,** which on *oxidation* with *chromium 6- oxide* gives **cholestanone**. This product now gives rise to the formation of *two* **different products** as detailed under:

- ❏ **Androsterone (II)** by *reduction* in **ACIDIC** medium *acetylation oxidation* and **hydrolysis;** and
- ❏ **Epiandrosterone (I)** by reduction in **NEUTRAL** medium acetylation oxidation and **hydrolysis.**

6.2. Synthesis of Androsterone (II) from Dehydroepiandrosterone [Calioti *et al.* (1964)]

Calioti *et al.* started the synthesis of **androsterone (II)** from **dehydroepiandrosterone** which essentially consists of almost *ten* **sequential steps** as elaborated under :

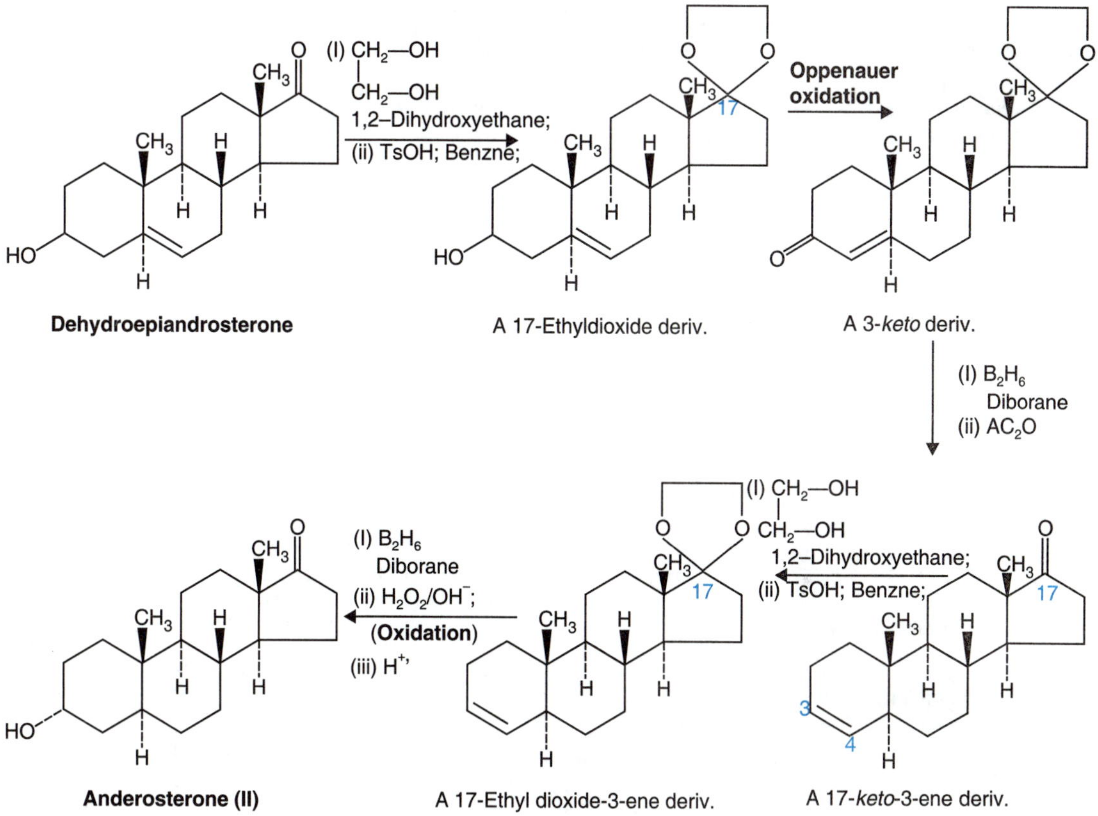

Dehydroepiandrosterone

A 17-Ethyldioxide deriv.

A 3-*keto* deriv.

Anderosterone (II)

A 17-Ethyl dioxide-3-ene deriv.

A 17-*keto*-3-ene deriv.

Explanation : **Dehydroepiandrosterone** on treatment with *1,2-dihydroxyethane* followed by **toluene sulphonic acid (TsOH)** in *benzene* yields a *17-ehthyl dioxide derivative*, which on being subject to **Oppenauer oxidation** gives a **3- keto derivative**. The resulting product on reaction with **diborane (B_2H_6)** followed by *acetylation* with acetic anhydride gives a **17-keto -3-ene derivative**, which on treatment *with 1,2-dihydroxyethane* followed by *TsOH/Benzene* yields a **17-ethyldioxide-3-ene derivative.** Finally, the resulting product on reaction with **diborane-oxidation** with H_2O_2/OH – and *acidification* results into the production of the desired product **Androsterone.**

6.3. Synthesis of Androsterone (II) from *para*-Toluene Sulphonate [Sondheimer *et al.* (1954)]

The synthesis commences by using **epiandrosterone** *para* **toluene sulphate** as the starting material. The various steps involved in the *synthesis* are as described below:

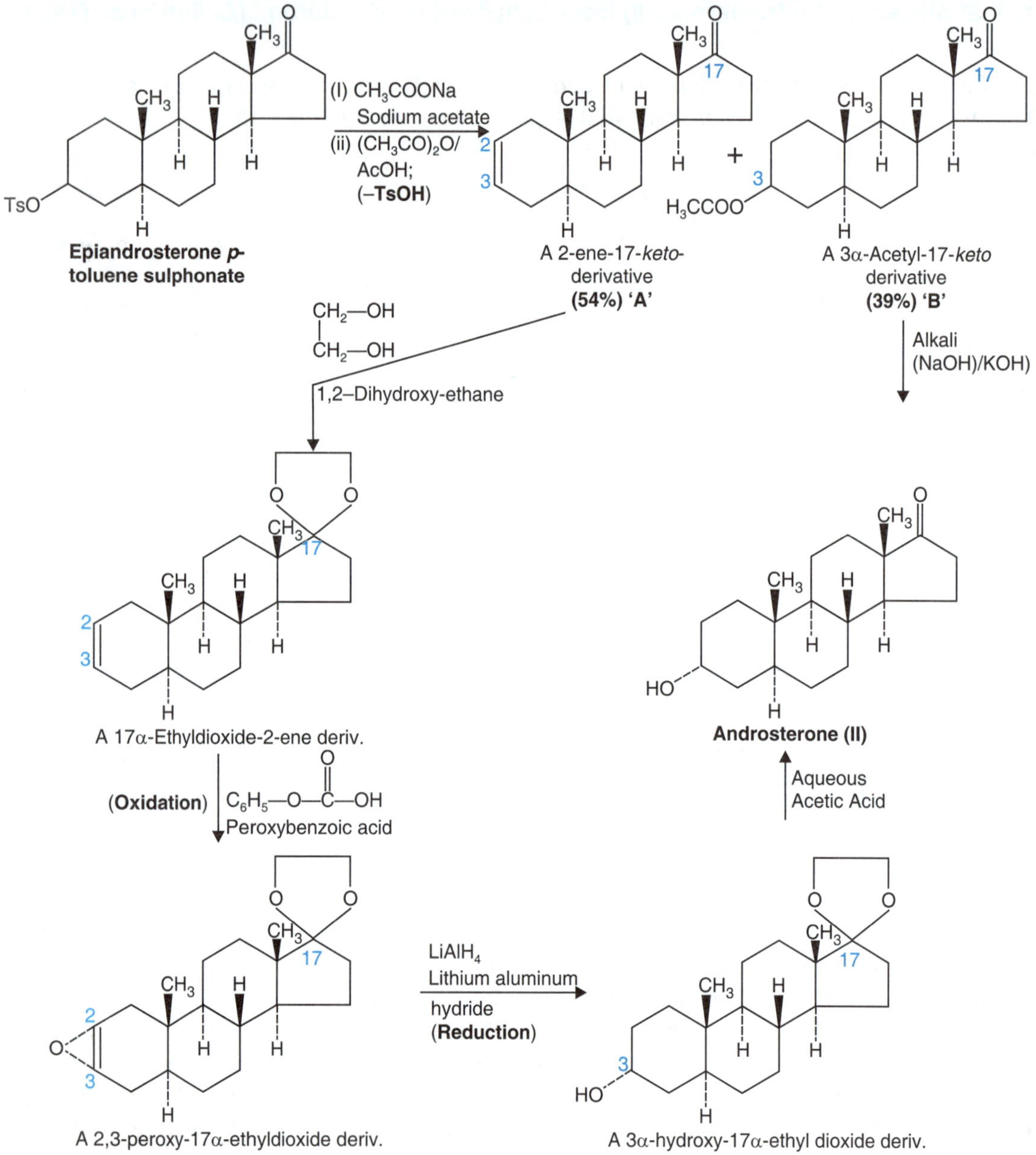

Epiandrosterone p-toluene sulphonate

(I) CH₃COONa Sodium acetate
(ii) (CH₃CO)₂O/ AcOH; (–TsOH)

A 2-ene-17-*keto*-derivative **(54%) 'A'**

+

A 3α-Acetyl-17-*keto* derivative **(39%) 'B'**

CH₂—OH
CH₂—OH
1,2–Dihydroxy-ethane

A 17α-Ethyldioxide-2-ene deriv.

(Oxidation) C₆H₅—O—C—OH Peroxybenzoic acid

A 2,3-peroxy-17α-ethyldioxide deriv.

LiAlH₄ Lithium aluminum hydride **(Reduction)**

A 3α-hydroxy-17α-ethyl dioxide deriv.

Alkali (NaOH)/KOH

Androsterone (II)

Aqueous Acetic Acid

Explanation : The various steps involved in the above synthesis are explained as under:

1. **Epandrosterone p-toluene sulphonate** when treated with *sodium acetate* and *Ac₂O/AcOH* loses a mole of **TsOH** and yields a **2-ene-17-*keto* derivative 'A'** (54%) plus a **3α-acetyl -17 *keto* derivative 'B'** (39%)

2. **Product 'A'** when reacted with *1,2-dihydroxyethane* gives a **17α-ethyloxide-2-ene derivative,** which on treatment with **perbenzoic acid** undergoes *oxidation* to yield **2,3-peroxy-17α-ethyldioxide derivative**.

3. The resulting product on *reduction* with LiAlH$_4$ gives a **3α-hydroxy17α-ethyldioxide derivative,** which on treatment with aqueous acetic acid yields **Androsterone (II)** .

4. Furthermore, **product 'B'**, obtained earlier, when reacted with an **alkali** (NaOH/KOH) also gives **Androsterone (II).**

12.2. Testosterone

Preamble: Testosterone is the **principal hormone** of the *testes*, and produced by the **interstitial cells**. However, it is a major *circulating androgen*, duly converted by the enzyme **5α-reductase** specifically in **androgen dependent target tissues** to the respective **5α-dihydrotestosterone**- that is required predominantly for *'normal male sexual differentiation'*.

> **NOTE** *Testosterone* gets usually converted by aromatization to *estradiol*.

Chemical Structure

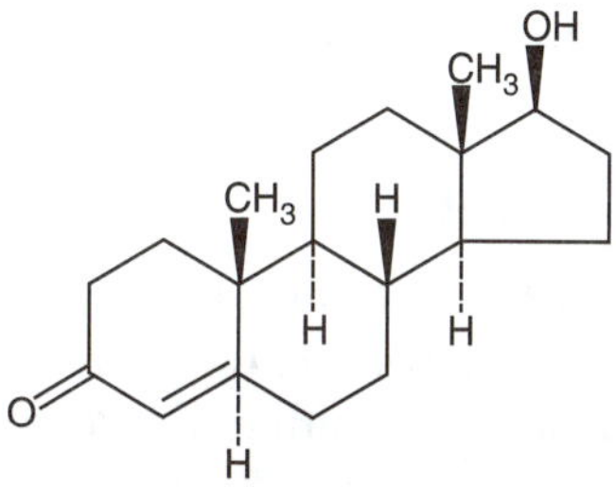

Testosterone

(17β)-17-Hydroxyandros-4-en-3-one;

Isolation: David *et al.* (1975) isolated **testosterone** for the first time from the testes of oxes. They isolated nearly **10 mg of testosterone** from *100 kg fresh testes*.

Characteristic Features. : These essentially include:

1. **Testosterone** is obtained as needles from dilute acetone.
2. It shows **mp 155°C**
3. It has $[\alpha]_D^{24}$ + 109 (C=4) in ethanol).
4. It exhibits UV absorption (UV$_{max}$) 238nm .
5. *Solubility Profile* –It is insoluble in water, but soluble in *alcohol, ether*, and *other organic solvents*.

CONSTITUTION OF TESTOSTERONE

The **constitution of testosterone** has been duly established on the basis of the following scientific and logical explanations:

1. Molecular Formula

Based upon various analytical data, it was duly inferred that the **molecular formula** of *testosterone* is **C$_{19}$H$_{28}$O$_2$**.

---

* David *et al* : *Z Physiol Chem,*. **233**:281, 1935.

2. Presence of a Steroidal Nucleus

The usual *standard tests* do reveal and ascertain that **testosterone** essentially comprises :

- **1 double bond (C=C)**
- **1 *keto* (>C=O) moiety,** and
- **1 *sec*– alcoholoic (>CHOH) group,**

In addition, the **molecular formula** of the *parent hydrocarbon* corresponds to the *'general formula'* having a **tetracyclic system.**

3. Presence of α,β-Unsaturaed *keto* (>C=O) Moiety

Since **tstosterone** is found to be exremely sensitive tto *alkali* (**NaOH/KOH**), it reveals that **testosterone** vehementy contains an α,β- **unsauraed keto (>C=O) moiety**]. However, the presence of this particular group could be further substantiated by its **UV- spectrum (λ_{max}=240 nm)**

Comment -The striking evidences that *UV spectrum* as well as its *marked sensitivity to alkali* predominantly supports the fact that **testosterone** is intimately and structurally related to **progesterone** *(see section 11.2.1.1)*

4. Oxidation of Testosterone to a Diketone

David (1935)* carried out the *oxidation* of **testosterone** to form a *diketone* **'androst-4-ene-3,17-dione'(A)** *i.e*; a chemical entity with a known **chemical structure**. However, the *diketone* thus obtained **'A'** may also be obtained by the **Oppenauer oxidation** of *dehydroepiandrosterone.*

Importantly, the formation of the **diketone compound (A)** could only be explained logically if **(I)** is assumed to be the probable structure of **testosterone.**

Thus, we may have the following set of reactions:

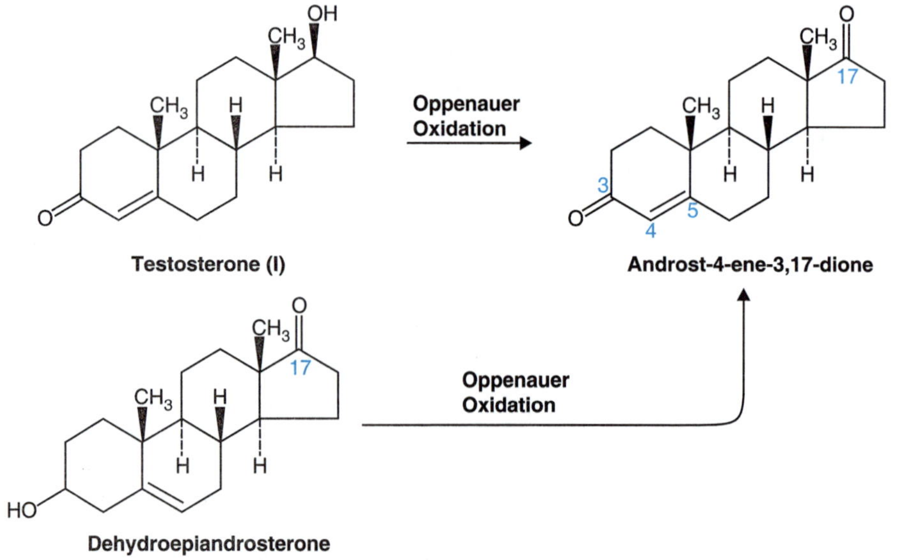

Testosterone (I)   Androst-4-ene-3,17-dione

Dehydroepiandrosterone

---

* David *et al* : *Z Physiol Chen,.* **233**:281, 1935.

5. Synthesis of Testosterone (I)

The *total synthesis* of **Testosterone** has been put forward by :

- **Butenandt A and Hamisch G (1935):** *Ber,* **68:** 1859; and
- **Mamoli (1938)**

5.1. Testosterone From Cholesterol [Butenandt and Harisch (1935)]

The various steps involved in the synthesis are as stated under :

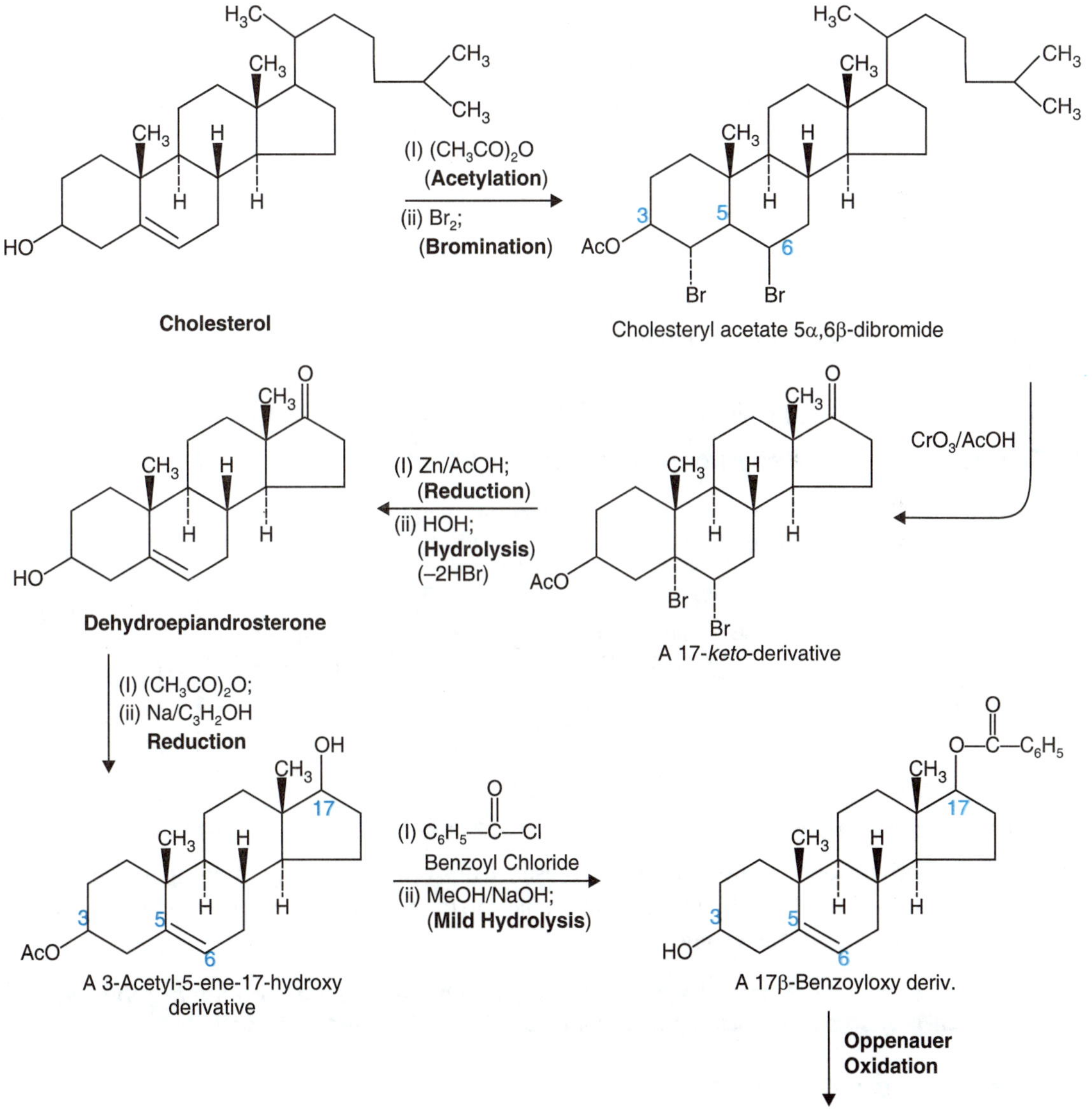

Cholesterol

(I) $(CH_3CO)_2O$
(**Acetylation**)

(ii) Br_2;
(**Bromination**)

Cholesteryl acetate 5α,6β-dibromide

$CrO_3/AcOH$

A 17-*keto*-derivative

(I) Zn/AcOH;
(**Reduction**)
(ii) HOH;
(**Hydrolysis**)
(−2HBr)

Dehydroepiandrosterone

(I) $(CH_3CO)_2O$;
(ii) Na/C_2H_2OH
Reduction

A 3-Acetyl-5-ene-17-hydroxy derivative

(I) $C_6H_5\text{—}\overset{O}{\overset{\|}{C}}\text{—Cl}$
Benzoyl Chloride
(ii) MeOH/NaOH;
(**Mild Hydrolysis**)

A 17β-Benzoyloxy deriv.

Oppenauer Oxidation

(Contd.....)

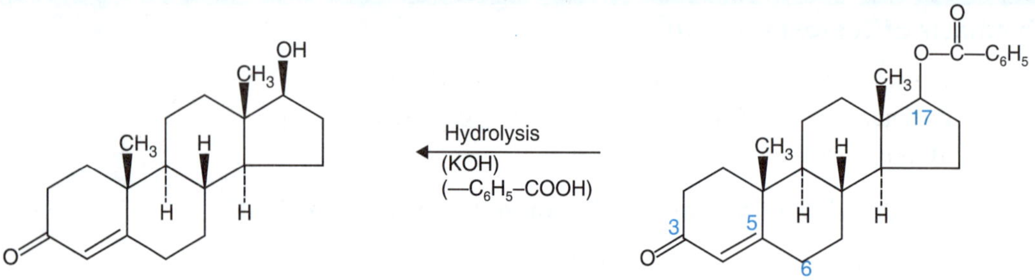

Testosterone (I) ← Hydrolysis (KOH) (—C₆H₅—COOH) — A 3-*keto*-4-ene deriv.

Explanation

1. **Cholesterol** on **acetylation** with acetic anhydride followed by **bromination** yields **cholesterol acetate 5α,5β-dibromide,** which upon *oxidation* with chromium-6-oxide in the presence of AcOH gives a **17-keto derivative**.

2. The end product from above on *reduction* with Zn/AcOH followed by *hydrolysis* loses **2 moles of the HBr** to give **dehydroepiandrosterone,** which on *acetylation* with acetic anhydride followed by reduction with *Na/propanol* yields a **3-acetyl-5-ene-17-hydroxy derivative.**

3. The resulting product on being subject to *benzoylation* followed by *mild hydrolysis* with **MeOH/KOH** yields a **17β-benzoyloxy derivative,** which on *Oppenauer oxidation* gives a **3-keto -4-ene derivative**

4. This product on *hydrolysis* loses a mole of *benzoic acid* to produce **testosterone (I).**

5.2. Testosterone from Dehydroepiandrosterone [Mamoïi (1938)]

Mamoli (1938) put forward the following sequential reactions for the synthesis of **testosterone** from **dehydroepiandrosterone:**

Dehydroepiandrosterone → Oxidising Yeast (Oxygen) → Androst-4-ene-3,17-dione → Fermenting Yeast (Hydrogen) → Testosterone

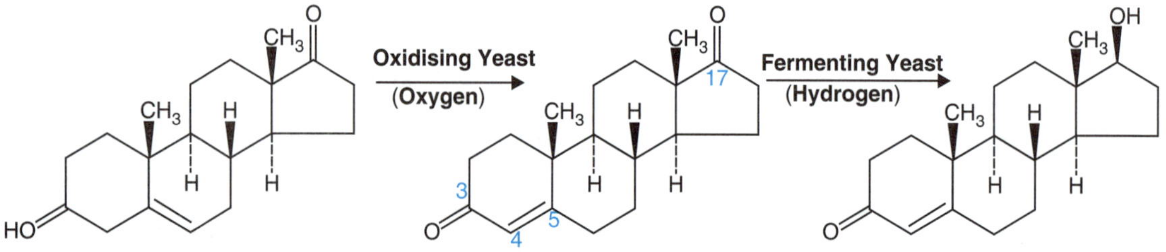

Explanation : Dehydroepiandrosterone in the presence of *oxidizing yeast* (**providing oxygen**) yields **androst -4-ene-3,17-dione.** This product on being subject to the *fermenting yeast* (**providing hydrogen**) gives the desired product **testosterone.**

> **NOTE** Instead of the *fermenting yeast* one may also make use of *sodium borohydride* (NaBH₄) so as to reduce *androst-4-ene-3,17-dione* to testosterone.

6- Stereochemistry of Testosterone

Interestingly, except the *configuration* of the **hydroxyl moiety** at **C-17**, the **Stereochemisty of testosterone**- has been adequately established and ascertained by its **methodical preparation from**

cholesterol. Nevertheless, the ensuing particular configuration of the **hydroxyl (OH) moiety** located strategically at **C-17** has been proved to be of *(β-type* by means of the following *two* highly specialized determinations, namely:

- **Molecular Rotation Measurements;** and
- **Estimation of Rates of Hydrolysis pertaining to various testosterone esters.**

SYNTHETIC ANDROGENS

Importantly, the **synthetic compound,** *17α-methyl testosterone*, designates a controlled *'anabolic steroid'* and falls very much within the therapeutic category-**'androgen'**. In fact, it is slightly less active in comparison to **'testosterone'** when administered parenterally (*i.e, via* **injection),** but it is indeed found to be more potent than the **testosterone** itself when taken orally.

Synthesis of 17α-Methyltestosterone from Androstenolone

Androstenolone   **17α-Methyl-17β-hydroxy deriv.**   **17α-Methyltestosterone**

The *Grignardization* of **androstenolone** with *methyl magnesium bromide* yields **17α-methyl-17β-hydroxy derivative** which upon *Oppenauer oxidation* gives the desired product **17α-methyltestoterone**

13. ADRENOCORTICAL HORMONES

Preamble: The **adrenocortical hormones,** also known as **adrenocortical steroids** or **corticosteroids,** and possessing essentially **21C- atoms.**

In general, the **adrenal glands** are located strategically close to the *kidneys* and comprise *two* **distinct segments,** namely:

- **Medula** and   • **Cortex.**

The **medulla** gives rise to the formation of **catecholamines** *i.e,* **(-)-adrenaline;** whereas, the **cortex** synthesised citically a number of *steroidal hormones*. Interestingly, the production of these *adrenocortical hormones* (or **corticoids**) has been duly controlled and modulated by the *hormone* termed as **'adrenocorticotrophic hormone (ACH)',** which is produced meticulously inside the **anterior lobe of the pituitary gland.**

Physiological Activity Profile The major **physiological funcitionaries** of the **corticoids** are as stated under :

- ❑ **Critical control of Crabohydrate and Protein Metabolism**; and
- ❑ **Excellent control of balance of Electrolytes and Water in humans**.

Following are a few important **'Adrenocortical Hormones'**, namely:

- **Cortisone**   • **Hydrocortisone**   • **Aldosterone**   • **Cortexolone**
- **Corticosterone** •   **11-Dehydrocorticosterone**-whose *chemical structures* are given below :

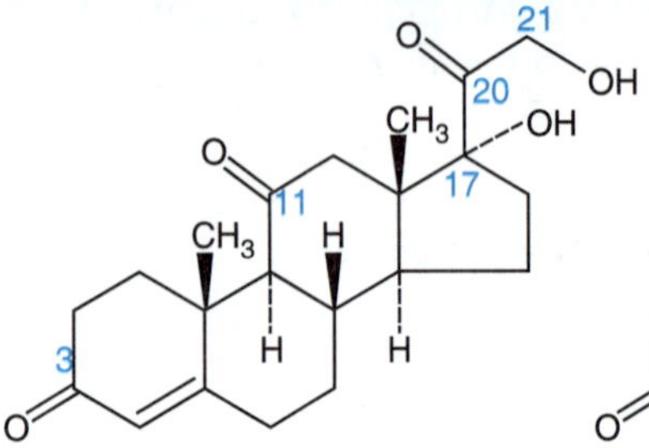

CORTISONE
[(17,21)-Dihydroxypregn-4-ene-
3,11, 20-trione;]

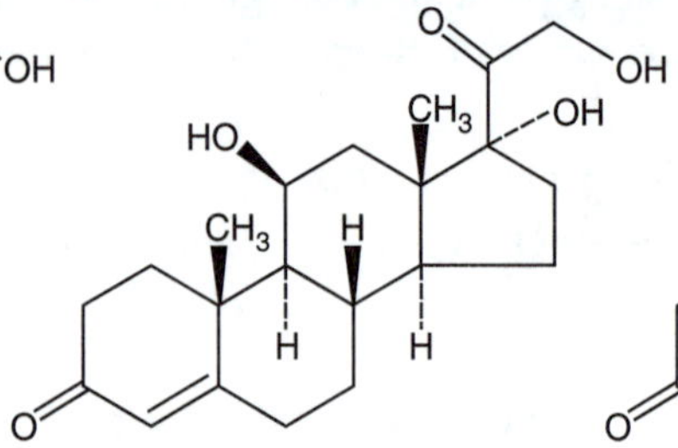

HYDROCORTISONE [CORTESOL]
[(11β)-11,17,21-Trihydroxypregn-
4-ene-3,20-dione]

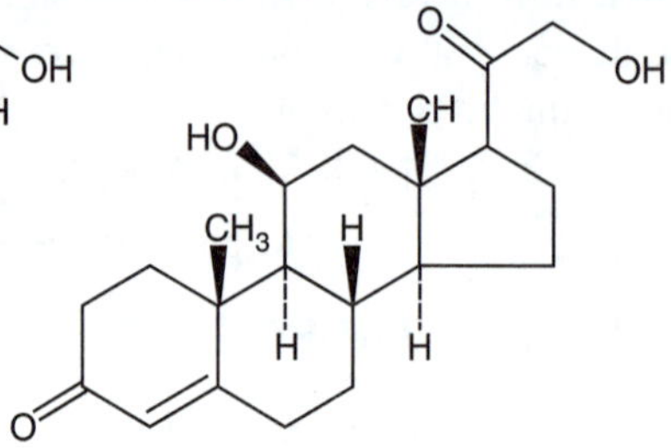

ALDOSTERONE
[(11β)-11,21-Dihydroxy
dioxopregn-4en-18-al]

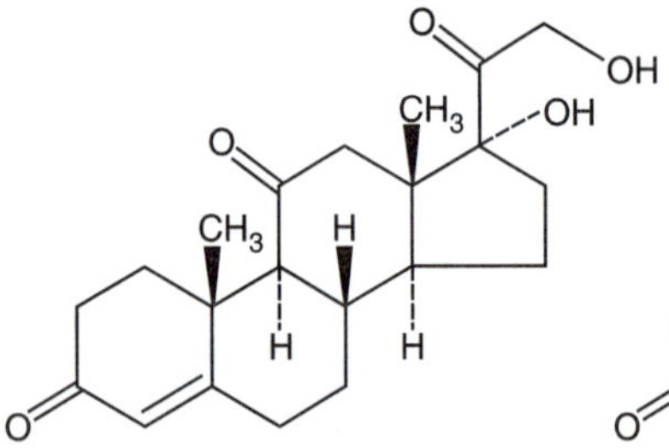

CORTEXOLONE
[11-Desoxy-17-hydroxy-
corticosterone]

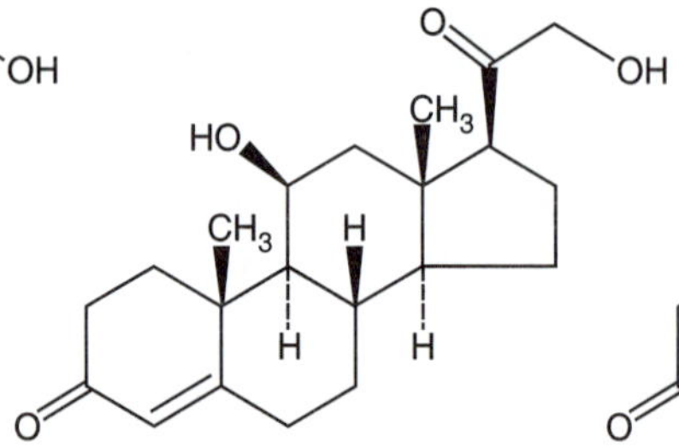

CORTICOSTERONE
[(11β)-11,21-Dihydroxy-
pregn-4-ene-3,20-dione]

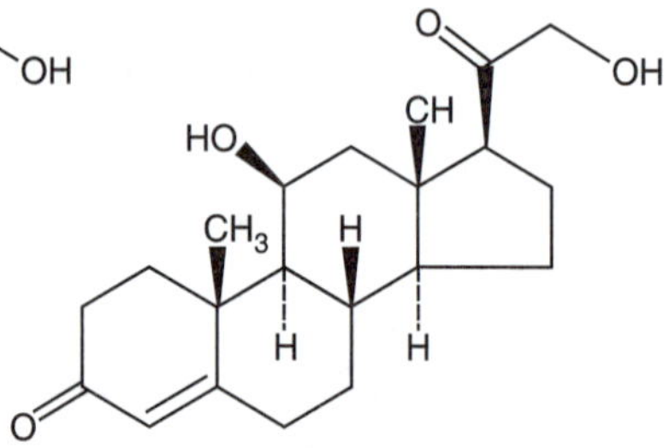

11-DEHYDROXY CORTICOSTERONE
[21-Hydroxy-pregn-4en-
3,11,20-trione]

Evidence from the literature reveals that various researcher, namely: Pfiffner (1935), Kendall (1946), and Reichstein (1943) isolated from the **adrenal glands** nearly *forty* steroids; however, only *six* of them are proved to be reasonably active in exerting specific physiological reactions in humans. It is, however, pertinent to state here that :

- **all of these *adrenocortical hormones* do have the same *steroidal nucleus* (*i.e.,* cyclopentanophenanthrene structure);**
- **show their difference with respect to the *oxidation level variance* at C-3, C-11 C-17 ,C-18 ,C-20, and C-21 positions; and**
- **above all the so-called 'physiologically active' corticoids are all having the – Δ⁴enone structures.**

The following *three* vital and important members of the **'adrenocortical hormones;,** namely:

- **Cortisone,**
- **Hydrocortisone (Cortisol),** and
- **Aldosterone**

will be disscused at **length** along with their respective **'total synthesis'** in the sections that follows:

---

*   P fiffner *et al.* : *Biol Chem.,* **111***:* 585,1935.

**   Kendall *et al. J Bil Ceh.,* **166**:.345,1946.

*** Reichstein *et al.* : *Helv Chim Acta,* **26:**562,1943.

13.1. Cortisone [*Syn* : Kendall's Compound E]

Preamble : Cortisone was first derived form the *suprarenal glands*. **Cortisone** exhibits prominently a **good systemic anti-inflammatory activity** and a **low to moderate salt retention activity** soon after its appropriate *in vivo* conversion to the respective hydrocortisone acetate. Importantly, this conversion is duly mediated by **11β-hydroxysteriod dehydrogenase** *(an enzyme).* It really exerts a host of *therapeutic advantages,* such as :

- **Addison's diseases,**
- **Collagen disease,**
- **Allergic reactions,**
- **Acute shock,** and
- **Several other indications.**

> NOTE **Cortisone has a plasma half life ranging between 1.5-3.0 hours.**

Characteristic Features: These essentially include :

1. It occurs as **rhombohedral platelets** from ethanol (95%).
2. It has mp 220-224°C
3. It shows $[\alpha]_D^{25}$+209 (C=1.2 in 95% ethanol); $[\alpha]_{546}^{25}$ + 269°(C=O.125 in benzene); $[\alpha]_{546}^{25}$ + 248°(C=0.1 to 0.2 in alcohol)
4. **Cortisone** exhibits UV$_{max}$: 237nm (ε1.4x10)
5. **Solubility Profile**:
    - Fairly soluble in : Cold methanol ethanol, and acetone
    - Much less soluble in : Ether, Benzene, Chloroform.
    - Slightly soluble in water (28mg/100mL at 25°C)
6. It reduces **Benedict's solution** on heating.
7. It gives orange red solution with *intense green fluorescence* with conc. H_2SO_4.
8. The aqueous solution is neutral to litmus.

CONSTITUTION OF CORTISONE

The **constitution of cortisone** has been duly elucidated based upon the host of scientific and logical evidences as detailed under:

1. Molecular Formula

The various analytical accrued data, the **molecular formula of cortisone** has been found to be $C_{21}H_{20}O_5$.

2. Presence of an α,β-Unsaturated *keto* (>C=O) Moiety

Since cortisone is shown to be extremely sensitive to *alkali* (KOH/NaOH), it explicitly indicates the critical **presence of α,β unsaturated keto moiety**. In fact, its presence has also been further ascertained by *the UV- absorption of spectral studies.*

3. Important

It is based on the *standard well defined tests* the **cortisone** has been found to contain:

- **1- Double Bond**
- **2-*keto* Moieties**
- **1-*pri*-Alcoholoic Moiety**
- **1-*tert*-Alcoholic Moiety.**

4. Presence of an α keto –Moiety

Since **cortisone** causes the *reduction* of both:

- **alkaline solution of silver nitrate ($AgNO_3$), and**
- **Fehlings Solutions,**

it indicates strongly that **cortisone** does contain one **α-*keto*-(-CO.CH$_2$.OH) moiety**.

5. Oxidation of Crotisone

Cortisone on being *oxidized* with *chromic acid (Cr_2O_3)* yields a compound that was identified to be **adrenosterone**. Interestingly, when the **UV absorption spectral studies** of **Cortisone** and **Adrenosterone** were carried out the exact location of a ***keto* (>C=O) functional moiety at C-3** was revealed and confirmed.

Remarks: Since **Cortisone** and **Adrenosterone** also happen to be, **α,β-unsaturated ketones** it offers a vivid indication of the underlying fact that the **double bond** is located strategically between **C-4 and C-5**

NOTE	The above factual observation has also been duly substantiated by doing a comparative study with *Testosterone* and *Progesterone*.

6. Structural Relationship of Cortisone to Androsterone

The *hydrogenation* of *adrenosterone* yields a ***triketone*** structural analogue having a predetermined structure with a **mp 178°C.** It has, however, been observed that the resulting '*triketone* analouge' happens to be **strongly androgenic in nature**.

Thus, one may genuinely conclude that it is certainly a derivative of '**androstane-3,17-dione**'. Nevertheless, the above act may be further ascertained based on the observation that '**androstane**' is duly obtained by carrying out the *reduction* of the said '*diketone*' by:

- **Clemmensen reduction,** and
- **Reduction by hydrogen (H$_2$).**

Inference: Therefore, one may logically infer that '**cortisone**' is structurally related to '**androstane**' positively.

7. Exact Position of the Second Hydroxyl (OH) Moiety

Cortisone on being *oxidized with periodic acid (HIO_4)*, gives rise to the formation of a **17-hydroxy acid,** which vehemently indicates the **exact position of the second hydroxyl (OH) moiety.**

Therefore, based exclusively upon the *scientific deliberations* and *logical explanations*, one may propose and assign the following structure of '**cortisone (I)**' – which might justifiably explain all the *foregoing reactions* :

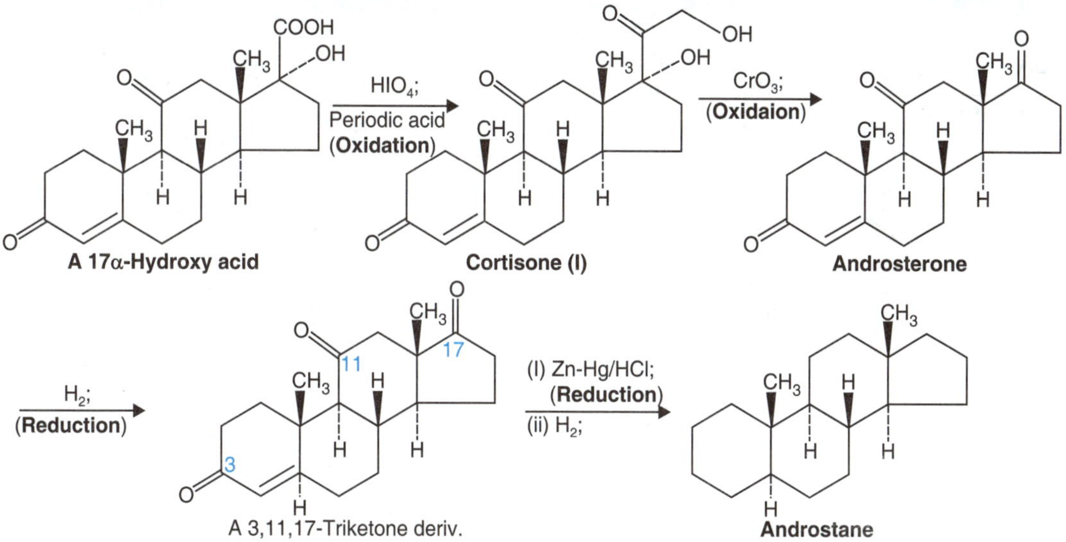

A 17α-Hydroxy acid → HIO₄; Periodic acid (**Oxidation**) → **Cortisone (I)** → CrO₃; (**Oxidaion**) → **Androsterone**

H₂; (**Reduction**) → **A 3,11,17-Triketone deriv.** → (I) Zn-Hg/HCl; (**Reduction**) (ii) H₂; → **Androstane**

The **17α-hydroxyl acid** on being oxidized with *periodic acid (HIO₄)* yields **cortisone (I)** which on further oxidation with *chromium -6-oxide (CrO₃)* gives **androsterone.** The resulting product on *reduction* gives a **3,11,17, triketone derivative,** which upon reduction at *two* **stages:**

- *first* **with Zn-Hg/HCl,** and   • *secondly,* **with** *hydrogenation,*

yields **androstane**.

8. Synthesis of Crotisone

The *total synthesis* of **cortisone** may be accomplished from different starting materials as stated under:

   i   **Cortisone from 3α, 21-diacetoxypregnane-11,20-dione: Sarett (1946)*;**
   ii **Total Synthesis of Cortisone: Sarett et al. (1952-1953)**; and**
  iii **Synthesis of Crotisone from 6- Methoxytetralone: Leon Velluz, (1960),**

which shall now be discussed individually in the sections that follows :

8.1. Cortisone, (I) From 3α, 21-Diacetoxypregnane-11,20-dione (Sarett,1948)

The synthesis commences from **3α,21-diacetoxy- pregnane-11, 20-dione** as described undersequentially:

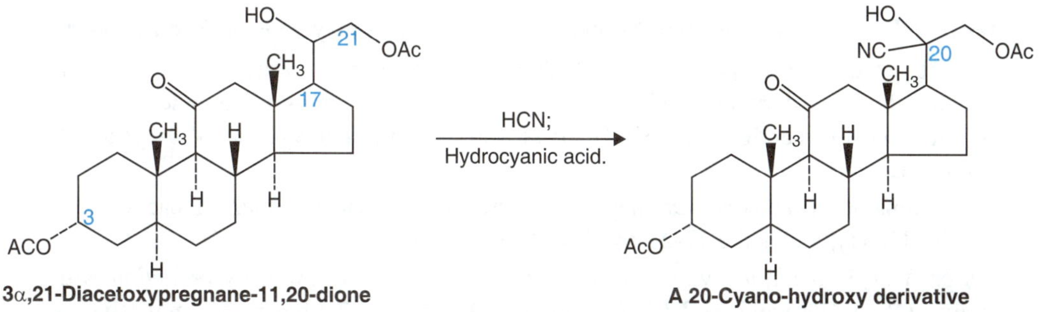

3α,21-Diacetoxypregnane-11,20-dione → HCN; Hydrocyanic acid. → **A 20-Cyano-hydroxy derivative**

---

  *   Sarett: *J Biol Chem.,* **162**:601,1946.
 ** Sarett:*J Am Chem Soc.* **74**:1871,1952.

(Contd.....)

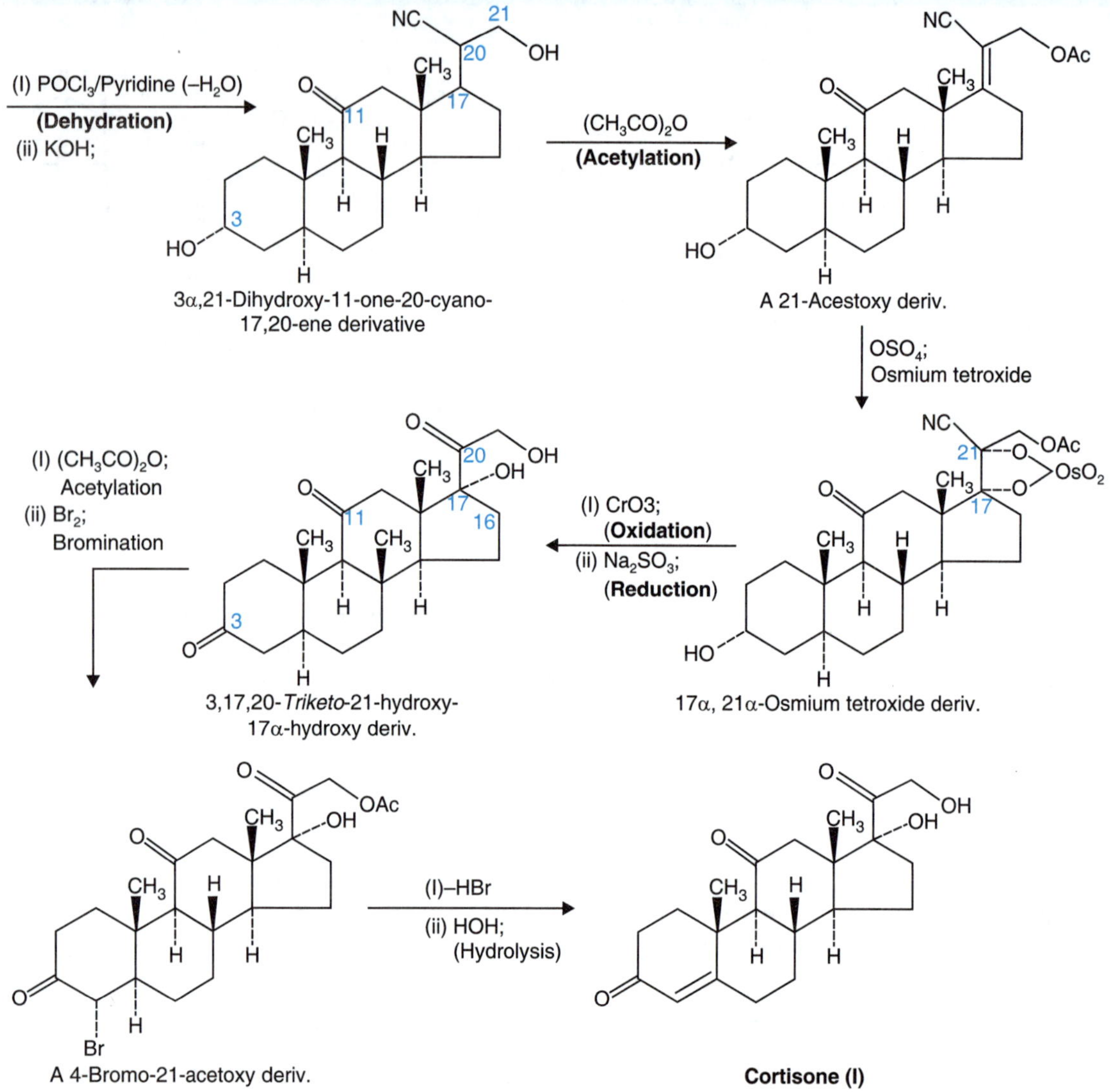

(I) POCl₃/Pyridine (–H₂O)
(Dehydration)
(ii) KOH;

3α,21-Dihydroxy-11-one-20-cyano-
17,20-ene derivative

(CH₃CO)₂O
(Acetylation)

A 21-Acestoxy deriv.

OSO₄;
Osmium tetroxide

17α, 21α-Osmium tetroxide deriv.

(I) CrO3;
(Oxidation)
(ii) Na₂SO₃;
(Reduction)

3,17,20-*Triketo*-21-hydroxy-
17α-hydroxy deriv.

(I) (CH₃CO)₂O;
Acetylation
(ii) Br₂;
Bromination

A 4-Bromo-21-acetoxy deriv.

(I)–HBr
(ii) HOH;
(Hydrolysis)

Cortisone (I)

Explanation : The various steps involved in the above synthesis may be explained as under:

1. When **3α, 21-diacetoxypregnane-11,20-dione** is treated with *hydrocyanic acid (HCN)* it yields a 20- cyano-hydroxy derivative, which on reaction with *POCl₃ in pyridine* undergoes dehydration and followed by reaction with KOH gives a **3α,21-dihydroxy-11-one-20-cyano-17-20-ene derivative.**

2. The resulting product on *acetylation* gives a **21- acetoxy derivative** which on reaction with *osmium tetroxide* **[OsO₄]** yields the **17-21-osmium tetroxide derivative**.

3. This product on *oxidation* with chromium 6-oxide (CrO₃) followed by *reduction* with *sodium sulphite (Na₂S₂O₃)* yields a **3,17,20-triketo-21-hydroxyl-17α-hydroxy derivative**, which on *acetylation* followed by *bromination* gives a **4- bromo-21-acetoxy derivative**.

4. Finally the end product obtained above loses a mole of *hydrobromic acid (HBr)* and followed by *hydrolysis* gives the desired product **cortisone (I).**

8.2. Cortisone From 3- Ethoxypenta-1,3-diene (1) and *para*-Benzoquinone (2) [Sarett *et al.* (1952)]

The synthesis involves the following sequential steps of reactions:

Step-1: 6-Ethoxy-5-ene-7-methyl-3,8-dihydroxy naphthalene(5) from 3- Ethoxypenta-1,3-diene (1) and p-Benzoquinone (2) The various reactions that are involved in the synthesis proposed by *Sarett et al.* (1952) are as detailed below:

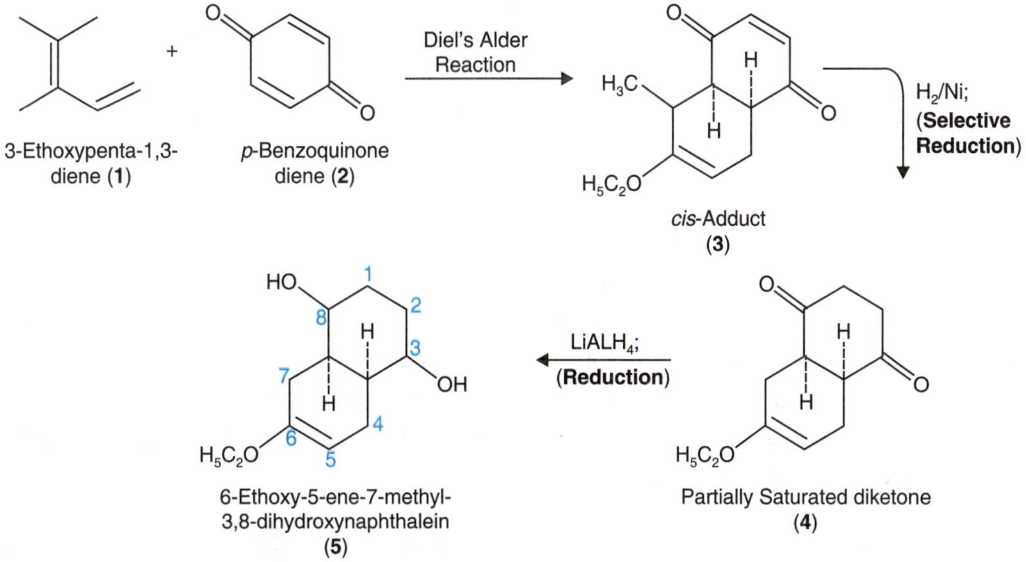

3-Ethoxypenta-1,3-diene (1)   *p*-Benzoquinone diene (2)   Diel's Alder Reaction   *cis*-Adduct (3)   H₂/Ni; (Selective Reduction)

6-Ethoxy-5-ene-7-methyl-3,8-dihydroxynaphthalein (5)   LiALH₄; (Reduction)   Partially Saturated diketone (4)

Explanation: The **Diel's Alder reaction** between **3- ethoxypeta-1,3-diene**

(1) and **p-benzoquinone (2)** yields a *cis*-adduct **(3),** which on *selective* reduction with raney Ni gives partiahy saturated diketone (4).

(2) The product (4) on further reduction with **lithium aluminium hydride (LiARH₄)** gives the product **6-ethoxy-5-ene-7-methyl-3,8-dihydroxy naphthalein(5).**

Step-2 : Conversion of Product (5) to a Tricyclic ketone (9)

The various sequential steps involved in **step-2** are enumerated as under:

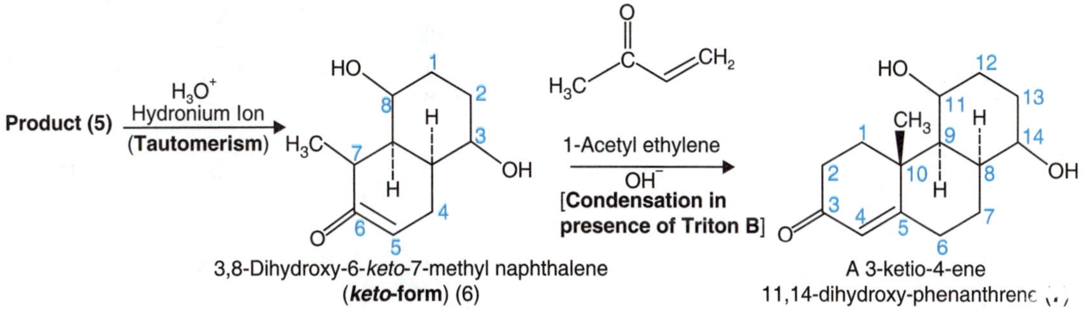

Product (5) H_3O^+ Hydronium Ion (Tautomerism)   3,8-Dihydroxy-6-*keto*-7-methyl naphthalene (**keto**-form) (6)   1-Acetyl ethylene   OH⁻ [Condensation in presence of Triton B]   A 3-ketio-4-ene 11,14-dihydroxy-phenanthrene (7)

(Contd.....)

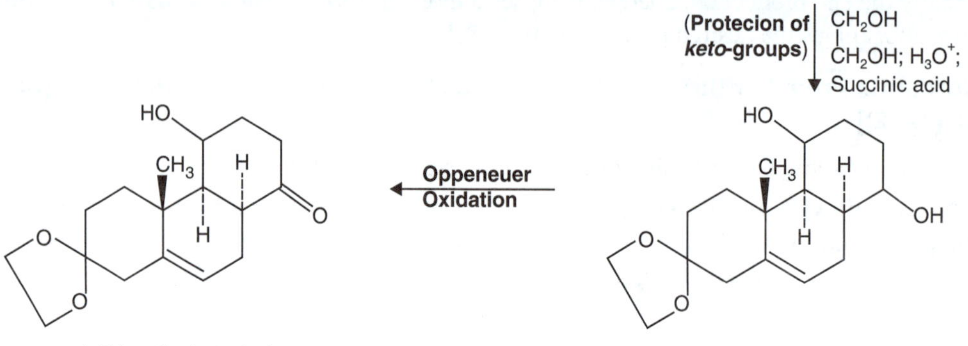

(Protecion of keto-groups) | CH_2OH
CH_2OH; H_3O^+;
Succinic acid

Oppeneuer Oxidation

A Tricyclic *keto* deriv
(9)

A Tricyclic dihydroxy deriv
(8)

The **product (5)**, obtained from **Step-1** undergoes *keto-enol* **tautomerism** in the presence of *hydronium ion* (H_3O^+) to give **3,8-dihydroxy-6-*keto*-7-methyl naphthalene** (*keto*-form **(6)**, which on treatment with *1- acetyl ethylene* and alkali gets condensed in the presence of **Triton B** to yield **3- *keto* -4-ene-11,14-dihydroxy phenanthrene (7)** . The resulting **product (7)** on tratment with *succinic acid* plus *hydronium ion* gives a **tricylic dihydroxy derivative (8)**, which finally on being subject to **oppenauer oxidation** yields a **tricyclic *keto* derivative(9)**.

Steps-3 Conversion of compound (9) into Cortisone : The sequential various steps involved in the conversion of **compound (9) to cortisone** are as given under:

The various steps involved in the above synthesis are self-explanatory.

The synthesis of cortisone (I) from 6-methoxytetralone (1) was duly put forward by Leon Velluz (1960) that esentially includes the following sequel of reactions:

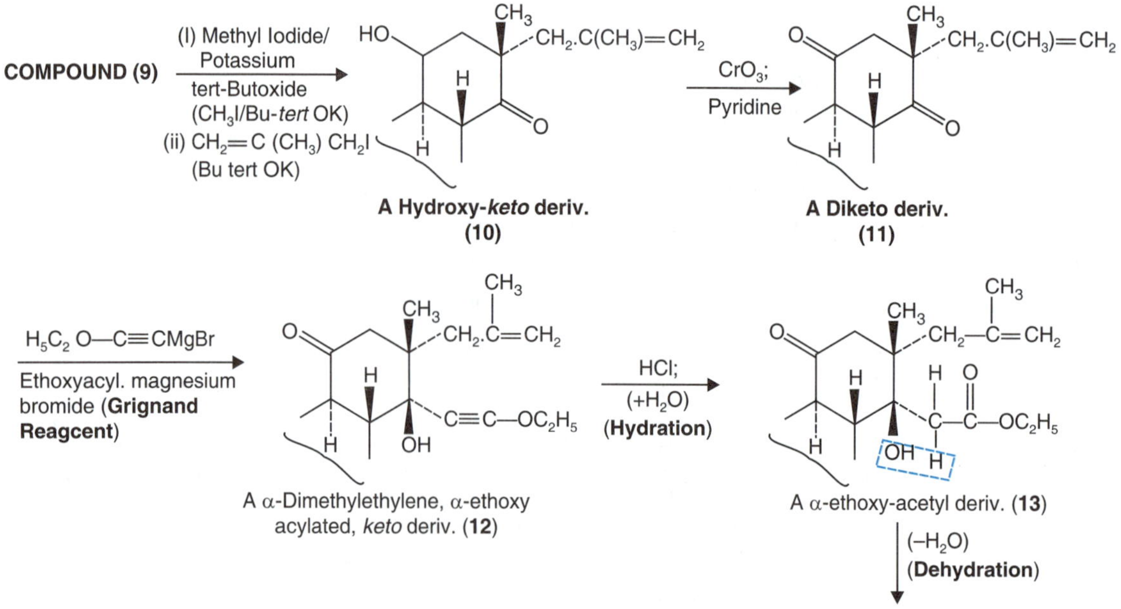

COMPOUND (9)

(I) Methyl Iodide/ Potassium tert-Butoxide (CH_3I/Bu-*tert* OK)

(ii) CH_2=$C (CH_3) CH_2I$ (Bu tert OK)

CrO_3; Pyridine

A Hydroxy-*keto* deriv.
(10)

A Diketo deriv.
(11)

$H_5C_2 O$—C≡CMgBr

Ethoxyacyl. magnesium bromide (**Grignand Reagcent**)

HCl; (+H_2O) (**Hydration**)

A α-Dimethylethylene, α-ethoxy acylated, *keto* deriv. (12)

A α-ethoxy-acetyl deriv. (13)

(−H_2O) (**Dehydration**)

(Contd.....)

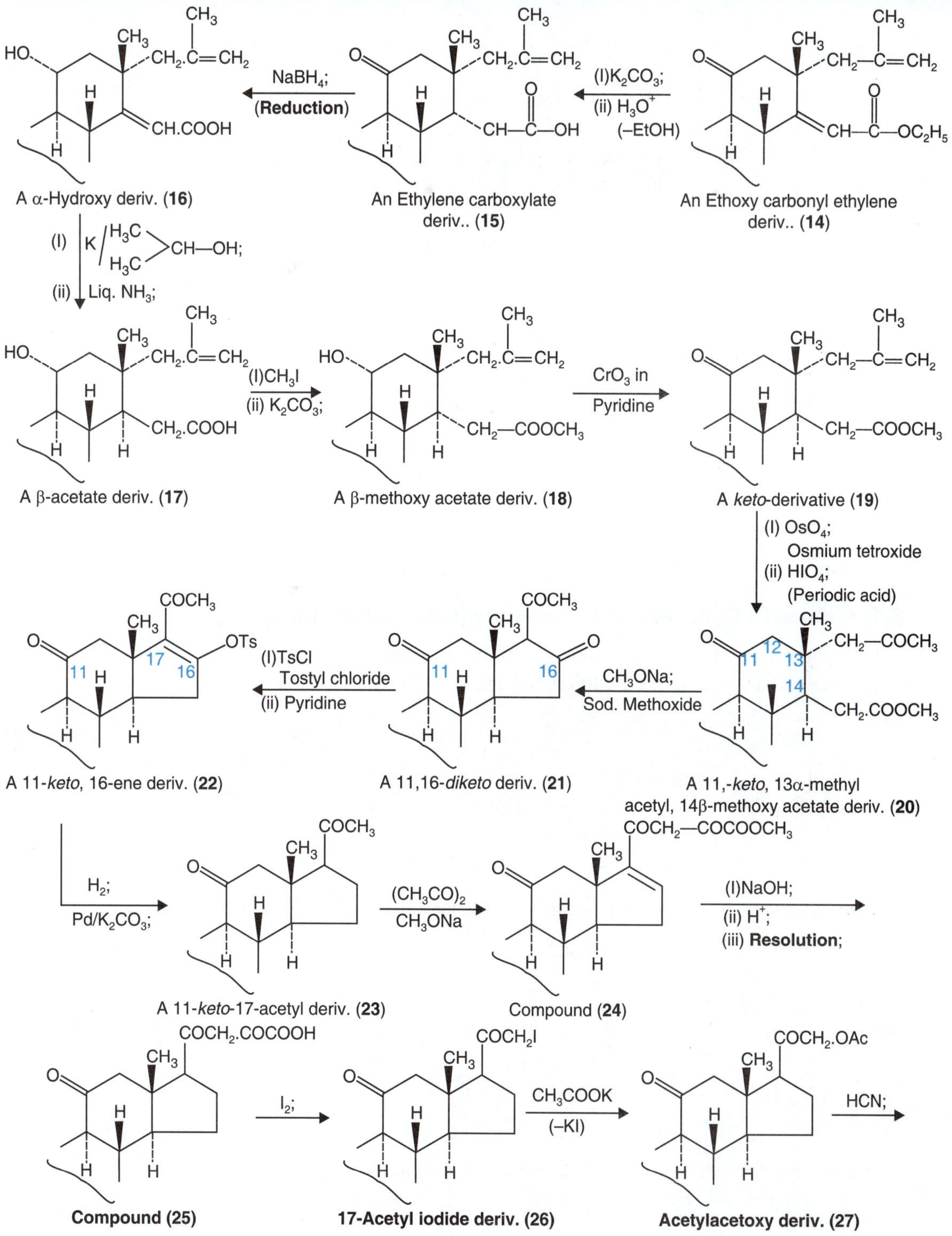
NaBH$_4$; (Reduction)
(I)K$_2$CO$_3$; (ii) H$_3$O$^+$ (–EtOH)
A α-Hydroxy deriv. (16)
An Ethylene carboxylate deriv.. (15)
An Ethoxy carbonyl ethylene deriv.. (14)
(I) K / H_3C>CH—OH; H_3C
(ii) Liq. NH$_3$;
(I)CH$_3$I (ii) K$_2$CO$_3$;
CrO$_3$ in Pyridine
A β-acetate deriv. (17)
A β-methoxy acetate deriv. (18)
A keto-derivative (19)
(I) OsO$_4$; Osmium tetroxide
(ii) HIO$_4$; (Periodic acid)
(I)TsCl Tostyl chloride (ii) Pyridine
CH$_3$ONa; Sod. Methoxide
A 11-keto, 16-ene deriv. (22)
A 11,16-diketo deriv. (21)
A 11,-keto, 13α-methyl acetyl, 14β-methoxy acetate deriv. (20)
H$_2$; Pd/K$_2$CO$_3$;
(CH$_3$CO)$_2$ CH$_3$ONa
(I)NaOH; (ii) H$^+$; (iii) Resolution;
A 11-keto-17-acetyl deriv. (23)
Compound (24)
I$_2$;
CH$_3$COOK (–KI)
HCN;
Compound (25)
17-Acetyl iodide deriv. (26)
Acetylacetoxy deriv. (27)

(Contd.....)

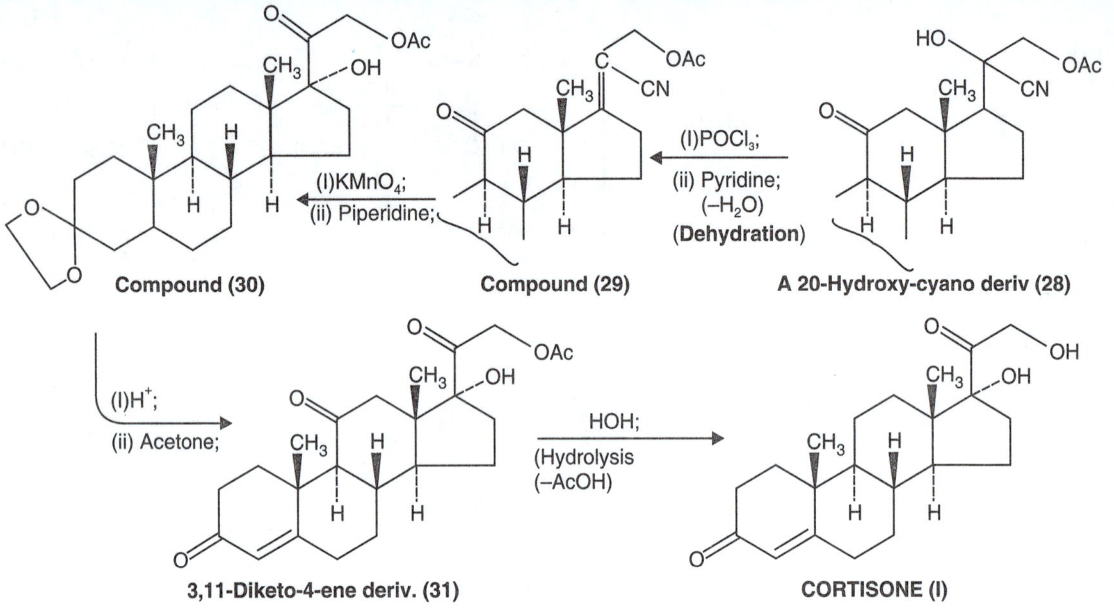

Compound (30)

(I)KMnO$_4$;
(ii) Piperidine;

Compound (29)

(I)POCl$_3$;
(ii) Pyridine;
(−H$_2$O)
(Dehydration)

A 20-Hydroxy-cyano deriv (28)

(I)H$^+$;
(ii) Acetone;

3,11-Diketo-4-ene deriv. (31)

HOH;
(Hydrolysis
(−AcOH)

CORTISONE (I)

The various steps involued in the above synthesis are self-explanatory.

8.3. Cortisone (I) From 6-Methoxytetralone (Leon Velluz, 1960)

The synthesis of cortisone (I) from 6-methoxytetralone (1) was duly put forward by Leon Velluz (1960) that essantially includes the following sequel of reactions :

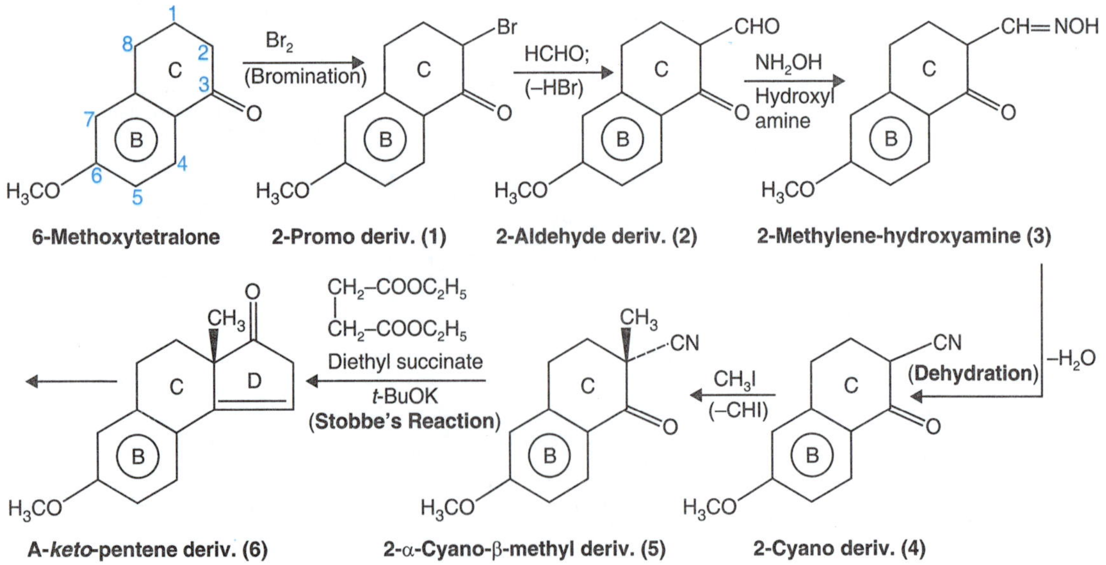

6-Methoxytetralone

Br$_2$
(Bromination)

2-Promo deriv. (1)

HCHO;
(−HBr)

2-Aldehyde deriv. (2)

NH$_2$OH
Hydroxyl
amine

2-Methylene-hydroxyamine (3)

A-*keto*-pentene deriv. (6)

CH$_2$–COOC$_2$H$_5$
CH$_2$–COOC$_2$H$_5$
Diethyl succinate
t-BuOK
(Stobbe's Reaction)

2-α-Cyano-β-methyl deriv. (5)

CH$_3$I
(−CHI)

2-Cyano deriv. (4)

(Dehydration) −H$_2$O

(Contd.....)

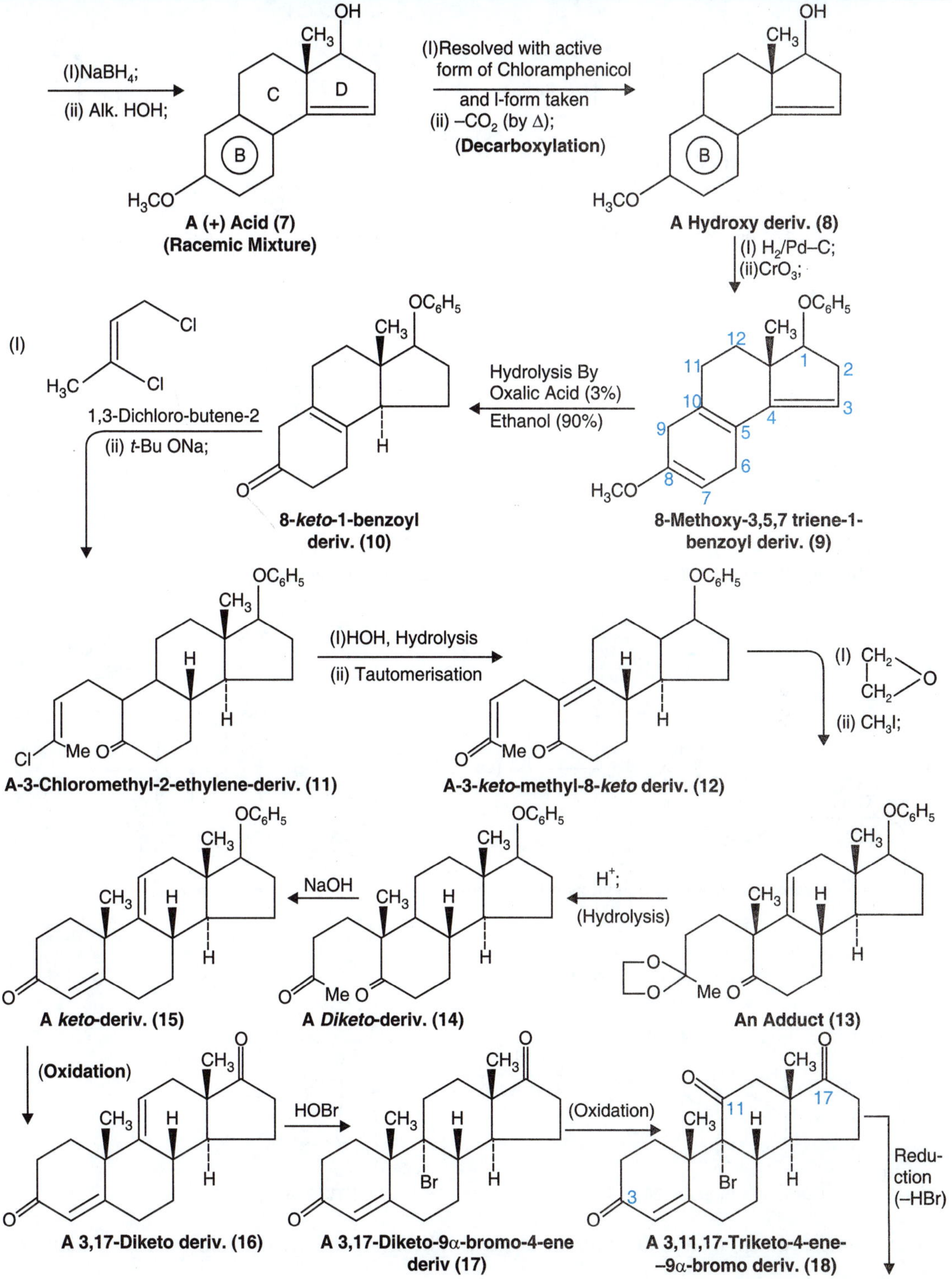

(I)NaBH₄;
(ii) Alk. HOH;

A (+) Acid (7)
(Racemic Mixture)

(I)Resolved with active
form of Chloramphenicol
and I-form taken
(ii) –CO₂ (by Δ);
(**Decarboxylation**)

A Hydroxy deriv. (8)
(I) H₂/Pd–C;
(ii)CrO₃;

1,3-Dichloro-butene-2
(ii) t-Bu ONa;

Hydrolysis By
Oxalic Acid (3%)
Ethanol (90%)

8-keto-1-benzoyl
deriv. (10)

8-Methoxy-3,5,7 triene-1-
benzoyl deriv. (9)

A-3-Chloromethyl-2-ethylene-deriv. (11)

(I)HOH, Hydrolysis
(ii) Tautomerisation

A-3-keto-methyl-8-keto deriv. (12)

(I) CH₂—CH₂—O
(ii) CH₃I;

A keto-deriv. (15)

NaOH

A Diketo-deriv. (14)

H⁺;
(Hydrolysis)

An Adduct (13)

(Oxidation)

A 3,17-Diketo deriv. (16)

HOBr

A 3,17-Diketo-9α-bromo-4-ene
deriv. (17)

(Oxidation)

A 3,11,17-Triketo-4-ene-
–9α-bromo deriv. (18)

Redu-
ction
(–HBr)

(Contd.....)

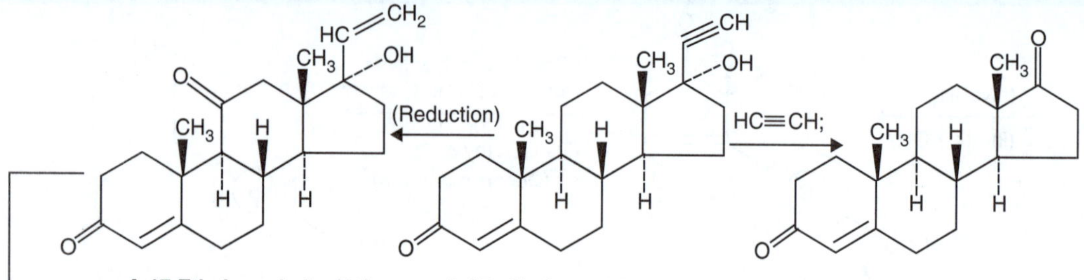

A 17-Ethylene deriv. (21)

A 17α-Hydroxy-17-acetylene deriv (20)

A 3,17-Diketoo-4-ene-deriv. (18)

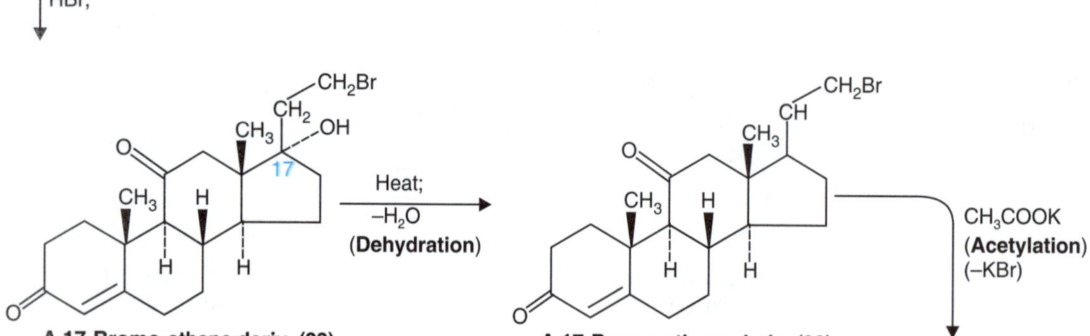

A 17-Bromo-ethane deriv. (22)

A 17-Bromo-ethane deriv. (23)

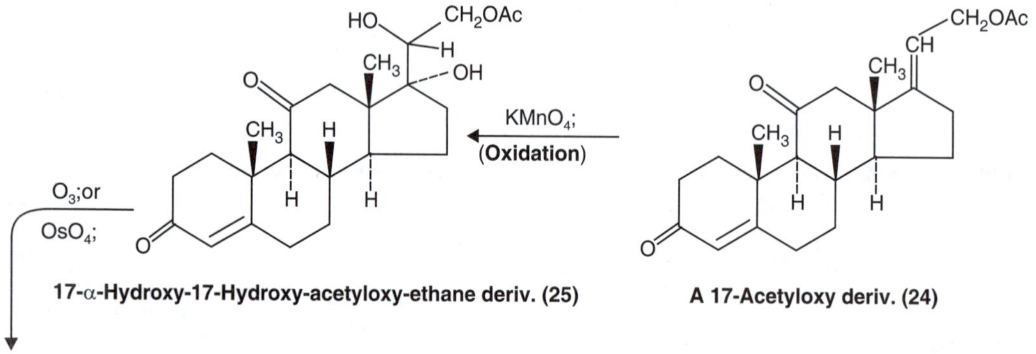

17-α-Hydroxy-17-Hydroxy-acetyloxy-ethane deriv. (25)

A 17-Acetyloxy deriv. (24)

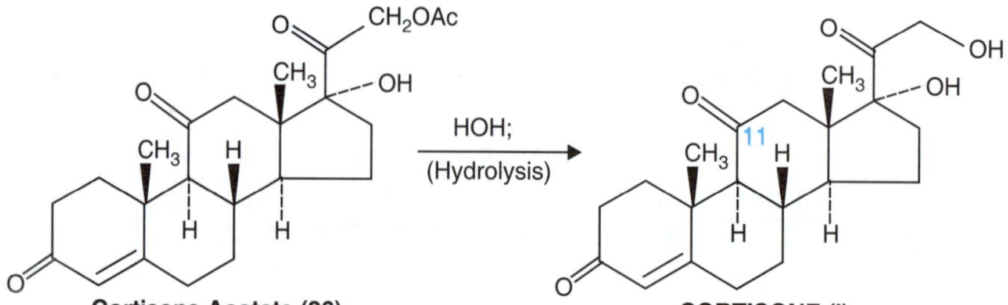

Cortisone Acetate (26)

CORTISONE (I)

The various sequential steps given in the above synthesis are indeed absolutely self-explanatory.

13.2. Hydrocortisone [Cortisol; Anti-inflammatory Hormone; Kendall's compound F; Reichestein's Substance M]

Preamble: Hydrocortisone (Cortisol) designates the principal *glucocorticoid hormone* produced by the **adrenal cortex**. The *biosynthesis* of **hydrocortisone** is duly stimulated by **ACTH** *(i.e. , adrenocorticotropic hormone)*. It usually circulates in *plasma* and primarily bound to the so-called **corticosteroid binding globulin (or transcortin)**, and also to **albumin.** It is isolated from the **adrenal glands**.

Besides, **cortisol** is found to be mose active physiologically than **cortisone** (see *section 13.1*). It has been established that both **Cortisol** and **Corticosterone** are secrated critically by the *zona fasciculata* of the 'adrenal cortex' actually in response to **ACTH – stimulation** from the respective **anterior pituitary gland.** Importantly, the **ACTH secretious** are meticulously controlled and regulated by the respective **hypothalamic secretion** of a *'peptide'* solely in response to *varying levels* of these *two* steroids in the blood.

Chemical Structure

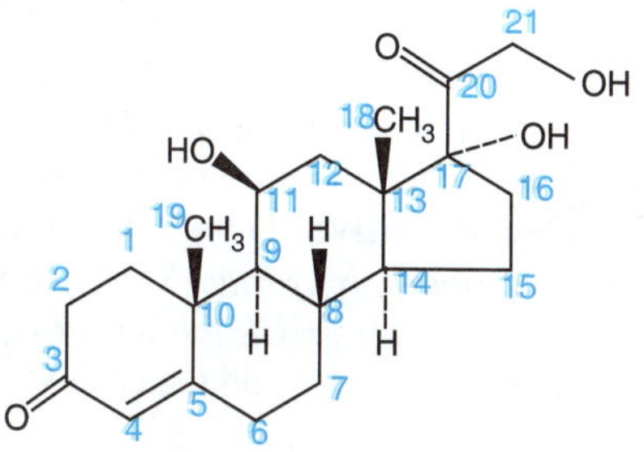

Hydrocortisone [Cortisol]

(11β)-11,17,21-Trihydroxypregn-4-ene-3, 20-dione;

CONSTITUTION OF HYDROCORTISONE

The only *glaring difference* between the structures of **Cortisol** and that of **Cortisone** being the presence of a **β-hydroxyl (OH) moiety** at C-11 in the former instead of a **carbonyl (>C=O) group** in the latter. Therefore, based upon such prominent similarities the **constitution of hydrocortisone (cortisol)** may be elucidated by adopting more or less the some sequential course of experimentation profile as already described for 'cortisone'.

Synthesis of Hydrocortisone from Cortisone Acetate [Wendler *et al* (1950)]

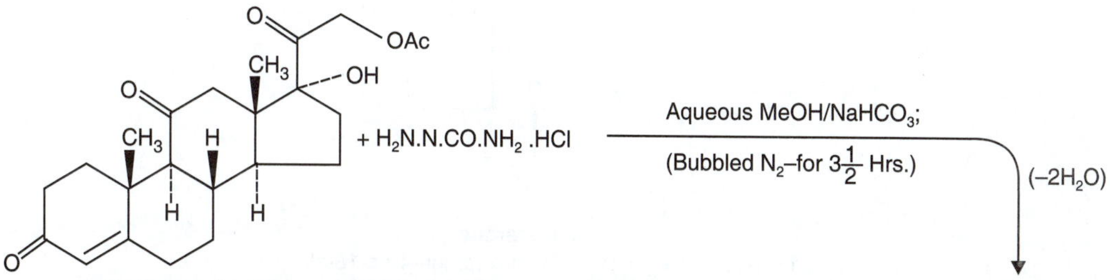

Cortisone Acetate   **Semi-Carbazide hydrochloride**

$+ H_2N.N.CO.NH_2 .HCl$

Aqueous MeOH/NaHCO$_3$;

(Bubbled N$_2$–for $3\frac{1}{2}$ Hrs.)

$(-2H_2O)$

(Contd.....)

HCl.H$_2$N.CO.NH.N

20 21 OH

CH$_3$ OH

HO

11 17

H

H H

KBH$_4$
Pot. borohydride

in Tetrahydro-
furan (THF)

HCl.H$_2$N.CO.HN.N

A 11β, 17α, 21-Trihydroxy deriv. (2)

HCl.H$_2$N.CO.NH.N

20 21 Oac

CH$_3$ OH

O

17

CH$_3$ H

3

H H

HCl.H$_2$N.CO.HN.N

A 3,20-semi-carbazide deriv. (1)

(I)HCl;

(ii) Excess of NaNO$_2$;

O

OH

CH$_3$ OH

HO

11

CH$_3$ H

3

H H

O

Hydrocortisone [Cortisol]

Explanation: The various steps involved may be explained as under:

1. Interaction of *cortisone acetate* with *semi-carbazide* HCl in aqueous methanol and sodium bicarbonate and passing through the reaction medium a constant pakage of **N$_2$-gas** (dry) for $3^{1/2}$ hours yields **3,20-semi-carbazidededrivative (1)** with the loss of *two* moles of water.

2. **Product (1)** on being treated with *potassium borohydride* (KBH$_4$) in **THF** gives a **11β 17α, 2l trihydroxy derivative (2),** which on further reaction with *HCL followed by excess of sodium of sodium nitrite* (*NaNO$_2$*) **(producing nitrons acid at 0-4 ºC)** to yield the desired product **hydrocortisone (cortisol).**

13.3. Aldosterone

Preamble: Aldosterone represents an *adrenocortical steroid* that exerts a *regulatory influence* upon the metabolism of **electrolytes and water**. In fact, **aldosterone** is the *mineralocorticoid* that was first and foremost isolated in a *crystalline form* by Mattox *et.al.* (1953)*, - who obtained only *45mg* from the **adrenal Cortex gland.**

NOTE   Soon after its isolation from the *adrenal cortex gland*- the compound was named as *'Electrocortion'*; however, its name was later changed to *'aldosterone'*, since its structure comprised of an *aldhyde and ketonic moieties.*

Chemical Structure

O

O

CH

20 OH

18

HO

11 17

CH$_3$ H

3

H H

O

4

Aldosterone
(11β)-11,21-Dihydroxy-3,20-dioxopregn-4-en-18-al;

* Mattox et al : *J Am Chem Soc.*, **75** : 4869, 1953.

Chemical Characteristics: These essentially include:

1. **Aldosterone** is obtained as hydrated crystals from dilute acetone.
2. Its mp ranges between 108-112 °C; when anhydrous mp : 164 °C.
3. It has $[\alpha]_D^{23}$ anhydr: CO_2 in acetone); and $[\alpha]_D^{23}+161°$ (C=0.1 in chloroform).
4. It has UV_{max} : 240nm (log ϵ 4.20 for the monohydrate, ϵmol 15,000 for the anhydrous).
5. **Aldosterone** predominantly exists in solution as an *equilibrium mixture* comprising the **18-aldehyde** and **11, 18-hemiacetal forms.**
6. Interestingly, the synthetic *racemic mixture (i.e.,* (+)-**aldosterone**) essentially consists of *two* **products,** namely:
   - **aldosterone;** and   •   **17-isoaldosterone**

   However, this is found to be absolutely devoid of any **physiological activity**.
7. The mp of **aldosterone -21- acetate** actually melts at **198°C;** whereas $[\alpha]_D^{23}$ +123 °C in ethanol.

SYNTHESIS OF ALDOSTERONE [JOHNSON *et.al.* (1958, 1963)

The survey of literature would reveal that a good number of *partial* and *total* **synthesis of aldesterone** has been reported; however, the synthesis of **Johnson** *et al.* **(1958, 1963)** is being discussed in detail as under perhaps due to the following *two* **cardinal reasons,** namely:

   - **the stereospecific nature of the 'total synthesis'; and**
   - **the starting material being a** *tetracyclic ketone(I)*

Thus, the **synthesis of aldosterone** as put forward by Johnson *et al.* (1958, 1963) critically comprise *five* **sequential steps** as enumerated under:

Step-1: Conversion of Tetracyclic Ketone (I) into 3α-ol (IV)

The various steps engaged in the conversion of the **tetracyclic ketone (I)** into **3α-ol (IV)** are as given under:

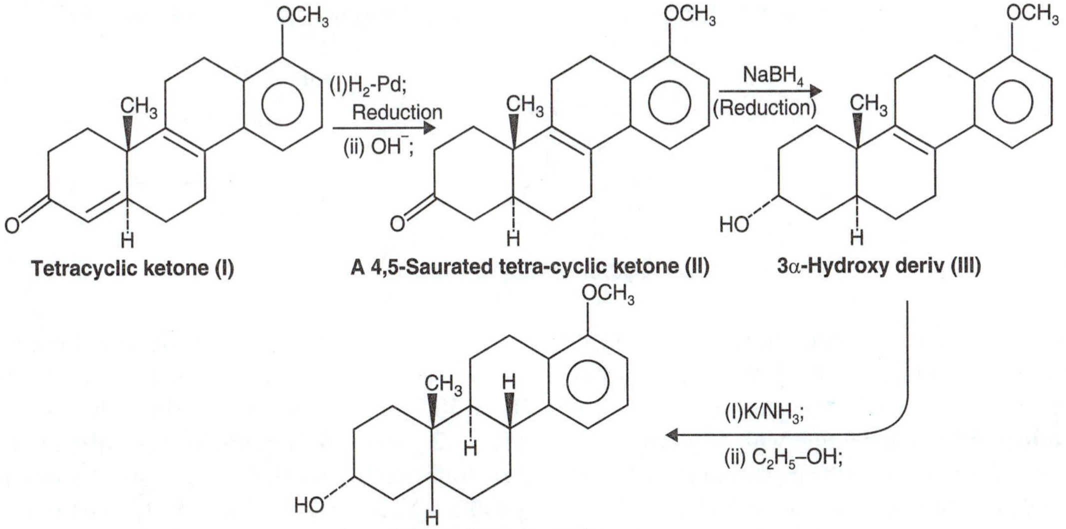

Tetracyclic ketone (I)   A 4,5-Saurated tetra-cyclic ketone (II)   3α-Hydroxy deriv (III)

A 3α-ol derivaive (IV)

* Johnson *et al.* : *J Am Chem Soc.*, **80** : 2585, 1958; *ibid*, **85** : 1409, 1963.

The **tetracyclic ketone (I)** on *reduction,* with H_2/Pd in an alkaline medium yields a **4,5-saturated tetracyclic ketone (II),** which on further reduction with *sodium borohydride (NaBH$_4$-)* gives 3α- **hydroxyl derivative (III).** The **end product (III)** on treatment with *K/NH$_3$* followed by ethanol gives the product **3α-ol derivative (IV).**

Step-2: Conversion of 3α-ol(IV) to a *keto* hydroxy phenyl methylene derivative (X)

The various sequential steps engaged in the aforesaid conversion are as detailed under:

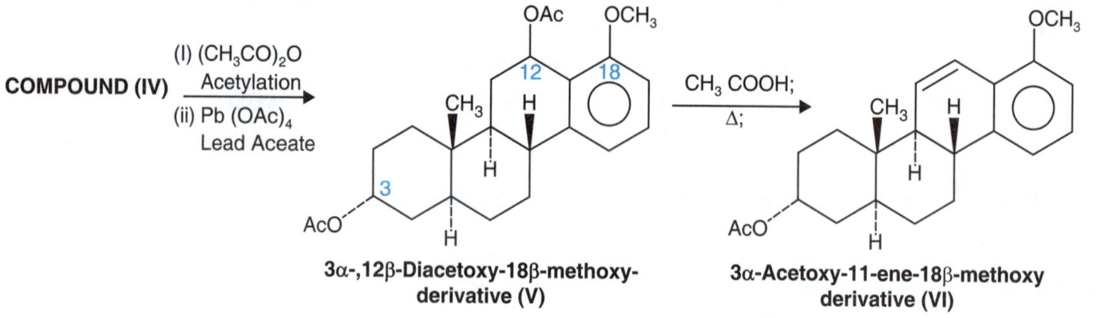

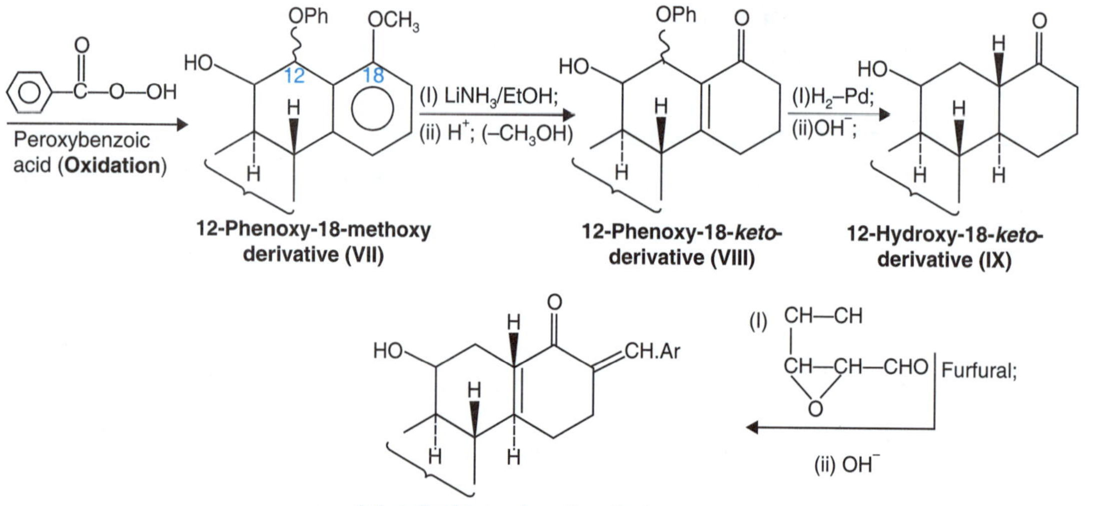

**3α-,12β-Diacetoxy-18β-methoxy-
derivative (V)**

**3α-Acetoxy-11-ene-18β-methoxy
derivative (VI)**

**12-Phenoxy-18-methoxy
derivative (VII)**

**12-Phenoxy-18-*keto*-
derivative (VIII)**

**12-Hydroxy-18-*keto*-
derivative (IX)**

**A *keto*-hydroxy-phenyl methylene
derivative (X)**

Explanation: The **compound (IV)** on being subject to *acetylation* with acetic anhydride followed by treatment with *lead acetate* yields **7α, 12β-diacetoxy-18α-methoxy derivative (V),** which on further treatment with acetic acid heating gives **3α acetoxy-11-*ene*18β-methoxy-derivative (IV).** The resulting **product (VI)** on *oxidation* with *peroxybeuric acid* gives **12-phenoxy-18-methoxy derivative (VII),** which on treatment with an admixture of *Li/NH$_3$/ EtOH* followed by an acid loses a mole of *methanol* to *yield* **12-phenoxy-18-*keto* derivative (VIII).** This **product (VIII)** on reduction with H_2 – Pd followed by *alkaline* treatment yields **12-hydroxy-18-*keto* derivative (IX),** which on reaction with *furfural* and by *alkali* gives a *keto*-**hydroxyphenyl methylene derivative (X).**

Step-3: Conversion of Compound (X) [*i.e.*, keto-hydroxy-phenyl methylene derivative] to a *bis*-methy ethyl ketone derivative (XV)

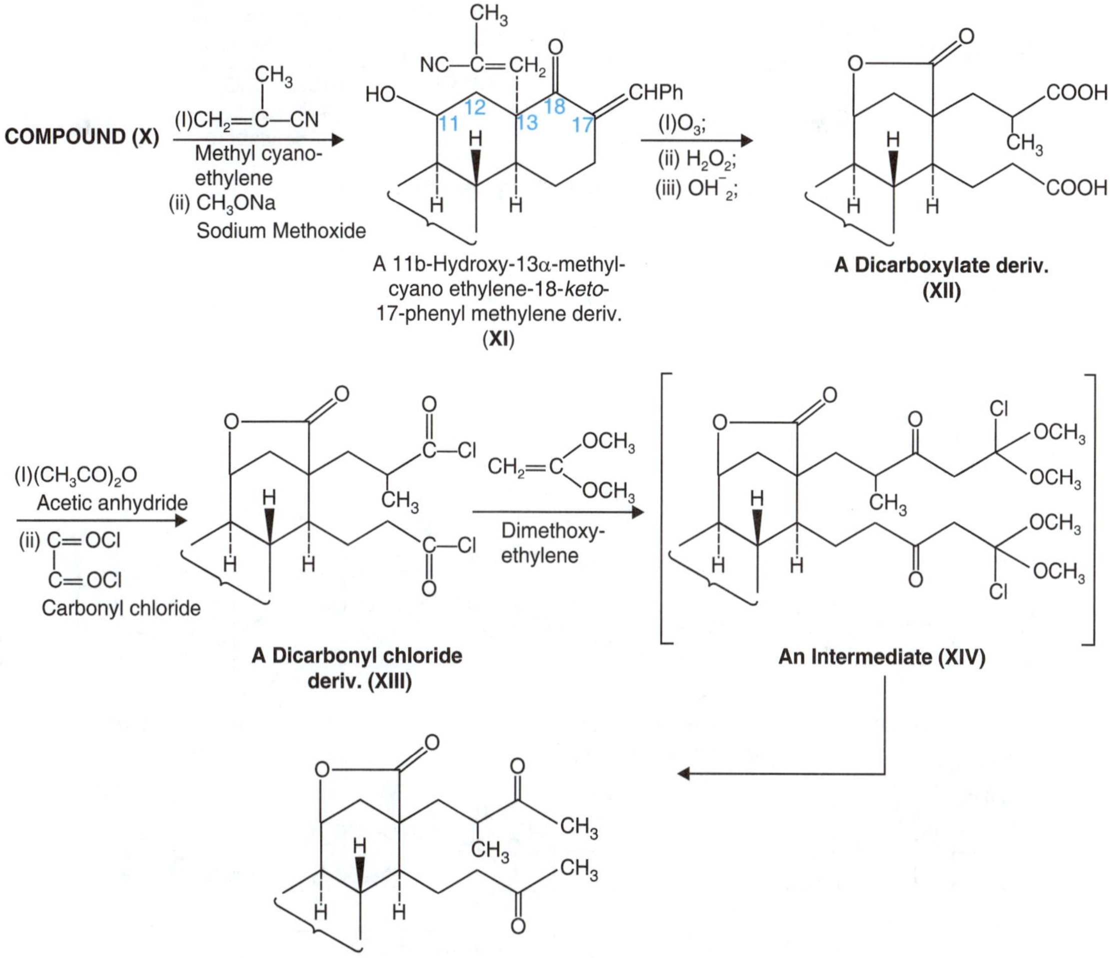

A 11b-Hydroxy-13α-methyl-
cyano ethylene-18-*keto*-
17-phenyl methylene deriv.
(XI)

A Dicarboxylate deriv.
(XII)

A Dicarbonyl chloride
deriv. (XIII)

An Intermediate (XIV)

A Diacetyl deriv. (XV)

Explanation: The various steps involved are explained as under:

1. **Compound (X)** when reacted with *methyl cyano ethylene* followed by *sodium methoxide* (**freshly prepared**) gives a **11-hydroxy-13α-methyl-cyano ethylene 18-*keto*-17-phenylmethylene derivative (XI),** which on *ozonization-oxidation-*and *alkalizaion* yields a **dicarboxylate derivative (XII).**

2. The resulting **compound (XII)** on *acetylaiton* with acetic anhydride followed by treatment with *carbonyl chloride* [**COOCl)$_2$**] yields a **dicarbonyl chloride derivative (XIII),** which on treatment with *dimethoxy ethylene* (CH$_2$=C(OCH$_2$)$_2$) gives an **intermediate (XIV).**

3. Finally, the **intermediate** on *hydrolysis* yields a **diacetyl derivative (XV).**

Step-4 : Conversion of Compound (XV) to a 17β-Acetyl Acetoxy Derivative (XXV)

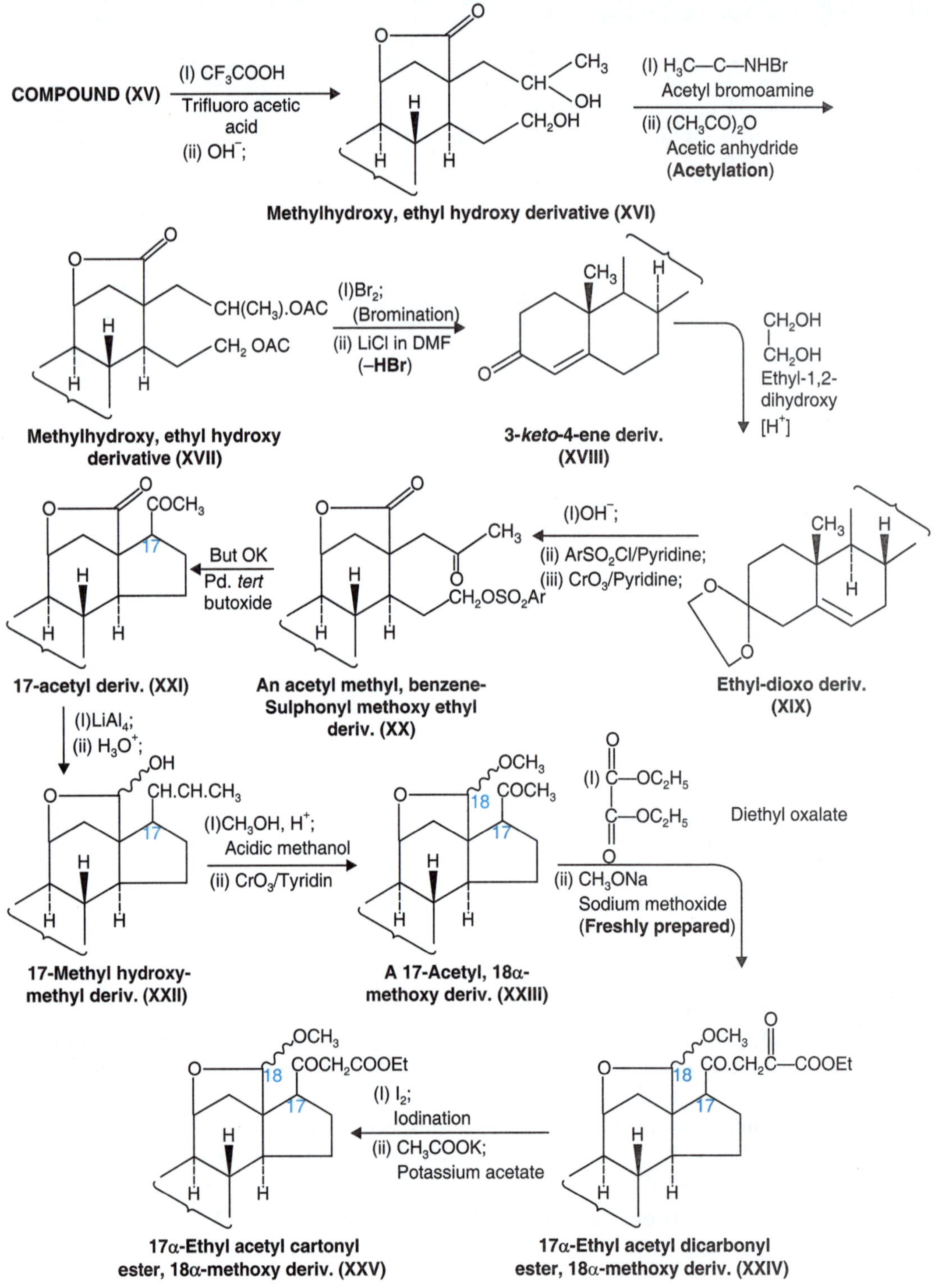
COMPOUND (XV)
(I) CF3COOH
Trifluoro acetic acid
(ii) OH⁻;
CH3
CH
OH
CH2OH
(I) H3C—C—NHBr
Acetyl bromoamine
(ii) (CH3CO)2O
Acetic anhydride
(Acetylation)
Methylhydroxy, ethyl hydroxy derivative (XVI)
CH(CH3).OAC
CH2 OAC
(I)Br2;
(Bromination)
(ii) LiCl in DMF
(–HBr)
Methylhydroxy, ethyl hydroxy derivative (XVII)
3-keto-4-ene deriv. (XVIII)
CH2OH
CH2OH
Ethyl-1,2-dihydroxy
[H⁺]
COCH3
But OK
Pd. tert butoxide
CH3
CH2OSO2Ar
(I)OH⁻;
(ii) ArSO2Cl/Pyridine;
(iii) CrO3/Pyridine;
17-acetyl deriv. (XXI)
An acetyl methyl, benzene-Sulphonyl methoxy ethyl deriv. (XX)
Ethyl-dioxo deriv. (XIX)
(I)LiAl4;
(ii) H3O⁺;
OH
CH.CH.CH3
(I)CH3OH, H⁺;
Acidic methanol
(ii) CrO3/Tyridin
OCH3
COCH3
(I) C—OC2H5
C—OC2H5
Diethyl oxalate
(ii) CH3ONa
Sodium methoxide
(Freshly prepared)
17-Methyl hydroxy-methyl deriv. (XXII)
A 17-Acetyl, 18α-methoxy deriv. (XXIII)
OCH3
COCH2COOEt
(I) I2;
Iodination
(ii) CH3COOK;
Potassium acetate
OCH3
CO.CH2C—COOEt
17α-Ethyl acetyl cartonyl ester, 18α-methoxy deriv. (XXV)
17α-Ethyl acetyl dicarbonyl ester, 18α-methoxy deriv. (XXIV)

1

NOTE The various sequential steps that are involved in the critical conversion of *compound (XV)* through *compound (XXV)* are quite self-explanatory in nature.

Step-5: Conversion of Compound (XXV) to Aldosterone (XXVI)

The following reaction comes into play:

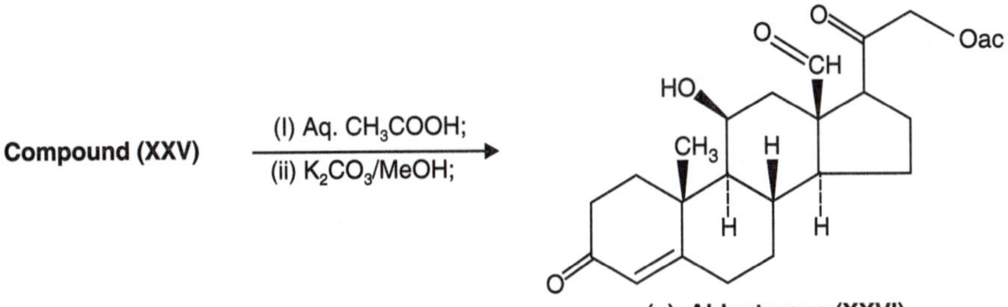

Compound (XXV) $\xrightarrow{\text{(I) Aq. CH}_3\text{COOH;}}_{\text{(ii) K}_2\text{CO}_3/\text{MeOH;}}$

(±)–Aldosterone (XXVI)

Explanation: When the **compound (XXV)** is carefully heated with *aqueous acetic acid* in order to carry out the *hydrolysis* of the **ether moiety,** followed by treatment with **methanolic carbonate** so as to cause the *hydrolysis* of the **acetate (-COCH$_3$) moiety.** Besides, it specifically helps in the *isomerisation* of the existing **17α-side chain** in **compounds (XXIV) and (XXV)** to the respective-**configuration i.e., 17-configuration,** thereby giving rise to the production of (±)-**Aldosterone (XXVI).**

NOTE : The aforesaid structure of 'aldosterone' (XXVI) has been duly ascertained by a host of researchers based on scientific evidences as stated under:

- Structure : Tait *et al.* : *Experientia,* **10,**132, 1954;Tait et al. : *Helv Chim Aeta,* **37:** 1200, 1954;
- Crystal Structure and Molecular Conformation: Duax and Hauptmann, *J Am Chem Soc.,* **94:**5467,1972.
- ^{13}C-NMR Spectrum : Gerand P : *Org Magn Reson,* **11:**478,1978

14. PLANT GROWTH HORMONES

The **plant growth hormones** are solely responsible for the growth of plants. In general, one may come across with *two* types of **plant hormones,** namely:

(a) **Plant growth hormone-** such as:
- **Auxin** *viz,* Indolylaectic Acid,
- **Cytokinins** *viz,* Zeatin,
- **Ethylene** *viz,* Abscisic Acid, and
- **Gibberellins** *viz,* Gibberillic Acid.

Following are the *chemical structures* of these *four* compounds:

(b) **Wound Hormone-** Responsible for the healing of *damaged tissue,* for instance:

Traumatic Acid

$$\text{HOOC.CH=CH (CH}_2)_8.\text{COOH}$$

Traumatic Acid

In a broader perspective, the **genetic makeup** and **growth regulators** do invariably constitute vital and important *internal factors* which predominantly regulate both:

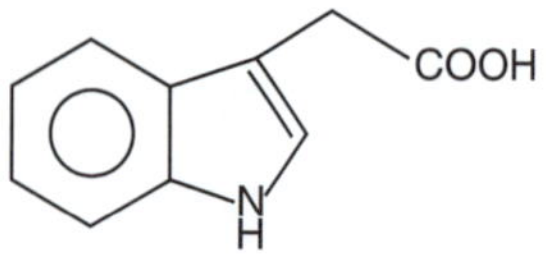

Auxin

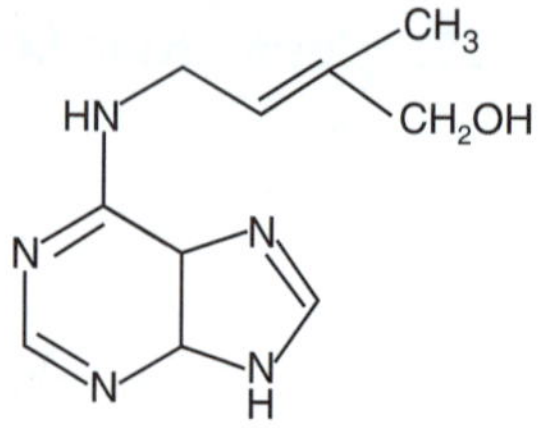

Zeatin

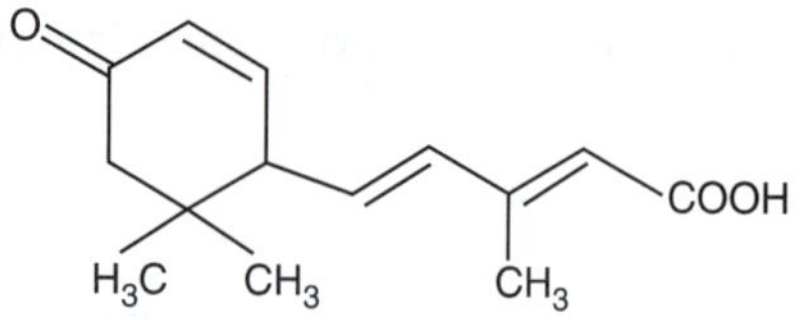

Abscisic Acid

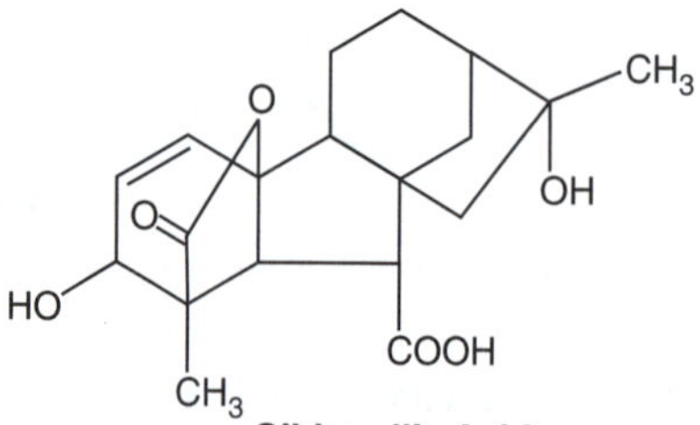

Gibberellic Acid

- **plant growth, and**
- **development**

NOTE : In plants, these chemical entities invariably occur in small amounts and are able to control efficiently several physiological processes involved in plant development.
The following *three* commonly used 'plant growth hormones', namely:
- **Auxins,**
- **Gibberllic Acid [Gibberellins], and**
- **Cytopinins**

shall now be discussed separately in the sections that follows:

14.1. Auxins

Auxins represent the so called **natural plant growth hormones**. Interestingly, the subsequent release of **tissue cells** and **tissue fragments** from not so easily crumbling **callus masses,** one may essentially afford the *critical maintenance* of a significant extent of *cell separation.* Thus, it may sustain a relatively high level of **auxin** concentration in the *prevailing liquid medium.*

It is, however, pertinent to state here that amongst the **natural plant growth hormones**, the 'auxins' were recognized to be the very *first*- studied extensively. Went (1928) duly proved that its ultimate effect upon the plant growth by carring out certain experiments using the *cat coleoptile.*

However, Kogl *et al.* (1933) reported the spectacular *isolation* of *two* most vital and important chemical entities (compounds):

- **auxin-α :** isolated from '**human urine**'; and
- **auxin -β :** Isolated from '**maize-green oil**',

which were considered to be *'active'* in the perfect **regulatory mechanism** pertaining to the **plant growth.** The structure of **auxin- α** and **auxin-β :**

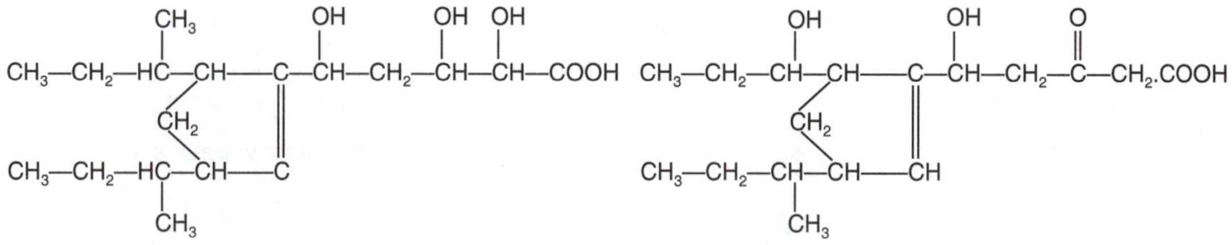

Auxin-α (I)   Auxin-β (II)

NOTE | **Since, both Auxin-α and Auxin-β could not be isolated from the growing plant tissue; and hence, these two compounds are merely of historical importance.**

HETEROAUXIN [or INDOLEACETIC ACID]

Heteroauxin (or Indoleacetic Acid) is a *plant hormone*, being recognized as the **'principal auxin'** of higher plants. It may be prepared by the reaction of *indole* with *potassium glycolate* at 250°C.

The *growth promoting substances* were first investigated in a systematic manner in 1931 by the **Dutch scientists**, who succeeded in the isolation of *two* **growth-regulating acids (auxin-α)** and **(auxin-β),** obtained respectively from:

- **Human urine** and
- **Cereal Products**.

The **phytochemicals** observed subsequently that both of these *derived auxins* almost exhibited **identical characteristic features** to **indole-3-acetic acid (IAA)**. Amazingly, **IAA** is now regarded to be the chief and prevalent **auxin of plants;** and found extensively and specifically in the so called **'actively growing tissues'**.

A survey by literature would reveal that quite a few *similar acids-* that proved to be *potential precursors of IAA,* have also been documented as **'Natural Plant Products'**, such as:

- **Indoleacetaldehyde**
- **Indoleacetonutrile** and
- **Indolepyruvic acid**

Nevertheless, the amino acid **'Tryptophan'** being the original source of **these aforesaid compounds** and **IAA**

The typical effects of **auxins** are as stated under:

- ➤ **cell elongation thereby are as stated under:**
- ➤ **inhibition of root growth,**
- ➤ **adventitious root generation,** and
- ➤ **fruit-setting in the absence of pollution.**

Examples: Following are a few typical examples of *synthetic auxins*:

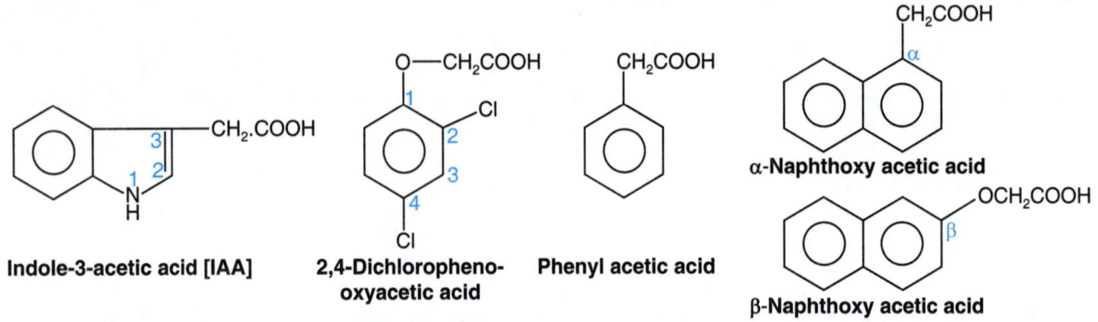

Indole-3-acetic acid [IAA]　　2,4-Dichloropheno-　　Phenyl acetic acid
　　　　　　　　　　　　　　　oxyacetic acid

α-Naphthoxy acetic acid

β-Naphthoxy acetic acid

It has been reported that the **heteroauxins** are duly synthesized in various plant segments, such as:

- **Shoot Tips**
- **Developing Seeds**
- **Young Leaf Primorida**

–and right from there they usually migrate to the *region of elongation* gradually.

Functionalities of Heteroauxins: They essentially include:

1. They help particularly in the *promotion of cell elongation.*
2. They also play a major role in such activities as:
   - **initiation of roots formation,**
   - **critical developments of fleshy (pulpy) fruits,**
   - **delaying the phenomenon of 'leaf drop' and**
   - **enhancing the process of 'fruit drop'**
3. The **'heteroauxins'** in conjunction with **'cytokinins'** do play a predominant role in the phenomenon of *cell division.*

Synthesis of 3-Indole-Acedic Acid (IAA)-Interestingly, (IAA)-is produced in the natural process in the course of *decomposition of proteins e.g.,* **putrefaction of meat protein in dead animals;** and *decomposition of fish-proteins.*

The *chemical synthesis* of **IAA** starts from **indolyl magnesium iodide (Grignard Reagent)** and **methylchlorocyanate** as detailed under:

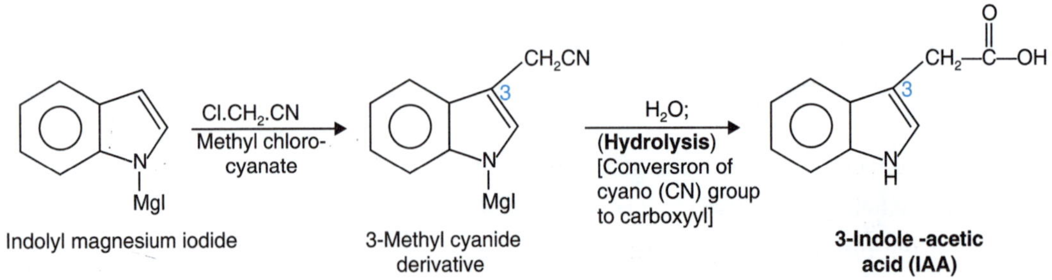

Indolyl magnesium iodide　　　3-Methyl cyanide　　　　　　　　　　3-Indole -acetic
　　　　　　　　　　　　　　　derivative　　　　　　　　　　　　　　acid (IAA)

The interaction of **indolyl magnesium iodide** and **methyl chlorocyanate** yields a **3-methylcyanide derivative,** which upon *hydrolysis* helps in the conversion of the *cyano groups* into a *carboxyl (COOH) group* to gives **3- indole acetic acid (IAA).**

Synthesis of IAA from Ethyl iodide acetate [King *et al.* (1934)] It is an alternate method for the synthesis of **IAA** from *ethyl iodide acetate* and the various steps involved are as enumerated under:

3-Indole acetic
acid (IAA)

Indolyl-2-carboxylete-
3-acetate (3)

Indolyl-2-ethyl carboxylate
-3-ethylacetate (2)

Explanation: A **double ester (1)** is formed by the interaction *of ethyl iodide acetate* and *monosodium ethyl acetoacetate* by the climination of a mole of sodium iodide (NaI). The resulting **product (1)** on treatment with *phenyl diazonium chloride in aqueous ethanol* followed by heating with HCL/EtOH at 100°C yields **indolyl-2-ethyl carboxylate-3-ethyl acetate(2),** which on *hydrolysis* with alkaline water loses *two* moles of ethanol to yield **indolyl-2 carboxylate-3-acetate(3).** Finally, the **product (3)** undergoes *decarboxylation* in the presence of freshly distilled anhydrous *quinoline* and *Cu powder* at 195°C to produce **IAA.**

14.2. Gibberellic Acid [or Gibberellins]

Gibberellic acid [or Gibberellin] designates a **plant hormone**, which is regarded to be the most outstanding of the **plant-growth promotion metabolites** of *Gibberella fujikuroi [Fusarium moniliforme]*

Historical Evidences: Infact, it was reported as early as **1898,** there prevailed an *uncommon* and *anomalous growth* of the so-called *rice seedlings* invariably termed as- **'baka-nae'** (*i.e., stupidly overgrown seedling*) disease of plants. Almost a gap of *three* decades (**1928**) the said *'plant disease'* was correctly *attributed* and *identified* due to a **'toxin'** generously produced by a *fungus,* that was subsequently recognized as a *Gibberella fujukuroi*.

Yabuta and Hayashi (1936) succeeded in isolating a crystalliney, sample of the active material which they called '**Gibbrellin**' Amazigly even after a decade (*i.e.,* upto 1946) one could not appreciate the significance of these excellent research observations and findings intimately associated with '**gibbbrellin**', perhaps due to the following glaring facts:

- Preoccupation with the development of *'auxins'* by the *Western Plant Physiologists*;
- Obvious existence of *'language barriers'* between the Weet and the East (Japan); and
- The crucial advent of World War II

Concerted Research Efforts: These efforts were made in a concerted manner by **Japan, United States,** and **Great Britain** to substantiate further investigation of these *wonderful compounds*, which were shown to have **amazing overall effects when applied to plants.** Obviously, it paved the way to establish and ascertain that *a complete range of gibberellins was actually involved,* that are as-to-date being distinguished as: **GA$_1$, GA$_2$,** and **GA$_3$**. Importantly, '**GA$_3$**' is presently referred to as '**Gibberellic Acid'**; and produced commercially by the unique process of *'fungal' cultivation'*.

Occurrence of Gibberellins: The '**gibbrellins'** (or **gibbrellic acids**) are duly synthesized in the leaves; and get subsequently accumulated in comparatively huge quantum within the so-called *'immature seeds and fruits of certain plant sources'*.

Extraordinary Apparent of Gibbrellins: In reality, such effects may be observed explicitly by their application to the **short -mode plant species:**

Examples- These essentially include:

❑ **Digitalis; Hyoscyamus**	: *i.e.,* such plants producing *rosettes of leaves,* when bolting and howering induced; and
❑ *Dwarf Varieties of Several Plants*	: *i.e.,* when such plants are treated with **gibberellins,** they eventually grow to almost the *same height* as the **taller varieties.**
❑ *Initiation of Synthesis of Various* Enzymes	: *i.e.,* the synthesis of various *proteolytic* and *hydrolytic* enzymes upon which the *seed germination* and *seedling establishment* depend exelusively.

Important Functions of Gibberellic Acids : The different vital and important functions of **Gibberellic Acids** (**Gibberelins**) are started under:

1. **Stem Elongation**- *Gibberellins* do cause stem elongation predominantly.
2. **Genetically Dwarf Plants**-The *genetically dwarf plants* on being treated with **gibberellic acid** tend to become tall and big.
3. **Negate Buds/Tubers Dormanoy-Gibbrellic acid** critically negates the dormancy of *buds* and *tubers* prominently.
4. **Promote seed Germination-Gibbrellin** (*Monsanto Brand*)[R] specifically promotes the *seed germination phenomenon.*

Example: Malting Process of Barley. *i.e.,* the germination of barley seed is enhanced progressively by spraying an aqueous dilute solution of *'gibberellin'* on the sprouted grains at intervals periodically at a maintained **moisture level, temperature,** and **O$_2$-level** in aerated germination beds (*with perforated bottom plates*).

5. **Promotion of Long-Day Flowering Plants**- **Gibberellins** under the *noniductive environmental parameters* do promote the *long day flowering plants.*
6. **Critical Induction of Partheno carpy-Gibberellic acids** also helps in the critical induction **seedless fruit production**, such as:
   • **seedless tables grapes,**
   • **seedless strawberrys** etc.
7. **Gibberllins,** in general, enhance the **leaf and flower size** of several plant species.

CONSTITUTION OF GIBBERELLIC ACID (GA₃)

The **constitution of gibberellic acid (GA₃)** may be elucidated based on the following scientific evidences and logical explanations:

1. **Molecular Formula**: The data obtained from *chemical analysis* lead one to the *molecular formula* of *gibberellic acid* to be $C_{19}H_{22}O_6$.

2. **Presence of 2-Ethylenic Double Bonds:** The gibberellic acid on being subject to **microhydrogenation** reveals the presence of *two* **ethylenic double bonds**.

3. **Presence of 2-Hydroxyl (OH) Moieties:** *Acetylation* of **gibberellic acid** yields as *diacetate* which vividly indicates the presence of *two* **hydroxyl (OH) moieties**.

    Additional Evidences – Gibberellic acid duly forms a *monoaectate promptly*, but not a *diacetate;* thereby suggesting that one of the **hydroxyl groups** happens to be a *tertiary* **alcoholic group [(CH₃)COH]**. Besides, it has been observed that the **reductions products** duly obtained from **gibberellic acid** may be *oxidized easily to a ketone*, thereby ascertaining that the *second* **alcoholic (OH) moiety** is a *secondary* alcoholic group [CH₃ CH₂ CH-CH₃] (see-butanol)

4. **Presence of a Carboxyl (COOH) Moiety:** Since, **gibberellic acid** on **esterification** (with methanol) yields a **monomethyl ester**; it indicates the presence of a **carboxyl (CODH) moiety**.

5. **Presence of a Lactone Ring:** Since, **gibberellic acid** takes up *two* **equivalents of alkali**, it suggests vehemently that it essentially contains a '**lactone ring**'. However, it has been duly demonstrated; and hence, confirmed to be a **ψ-lactone** due to the presence of specific bond **~1780 cm⁻¹** in the **IR-spectrum**.

6. **DBE* of Gibberellic Acid stands at 9 [*i.e.*, 19+1-22/2] :** Intereslthgly **gibberellic acid** essentially contains:

    *   **2-Ethylenic >Double Bonds,** and
    *   **2-Carbonyl> Double Bonds.**

    Therefore, the total number of 'rings' in **gibberellic acid** should be **(9-4=) 5 rings.** Of these **5 rings,** one is the so called '**lactone rings**'. Based on the above evidences one may safely conclude that **gibberellic acid** comprises: "**Four carbocyclic rings**" *,i.e.*, it is indeed a *'tetracyclic chemical entity'*.

7. **Probable and Proposed Structure of Gibberellic Acid(I):** The careful acid hydrolysis of **gibberellic acid(I)** gives *two* **hydrolyzed products**: **Gibberic acid(II)** and **Allogibberic Acid(III),**

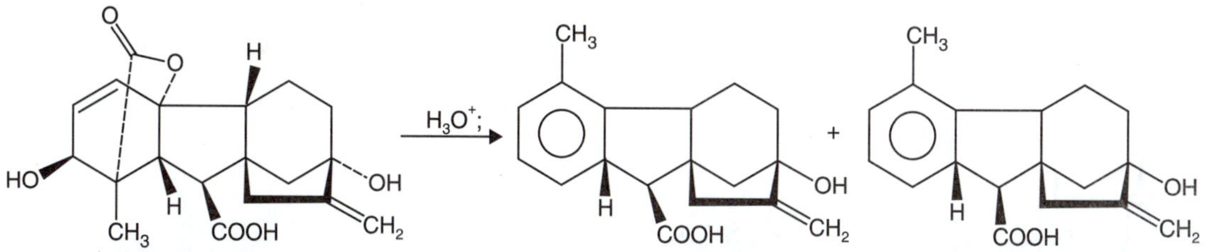

| Gibberellic Acid [I] | Gibberic Acid [II] | Allogibberic Acid [III] |

Besides, both **gibberic acid (II)** and **allogibberic acid (III)** on being subject to *hydrogenation* (with **Se powder at 315⁰C**) yields **gibberine, C₁₅H₁₄(IV)**, as given under:

---

* **DBE** : Double Bond Equivalent.

Remarks: It has been found that **gibberine (IV)** on being exposed to **UV-spectrum** proved to be a *'fluorene derivative'*. Furthermore, its *respective degradation* and *subsequent synthesis* showed it to be **1,7- dimethyl fluorine** i.e.,

Inference: The above scientific evidences and logical explanation do suggest strongly that all the *three* proposed compounds **(I),(II), and (III)** predominantly possess **the same C-skeleton** as that of **(IV)** *i.e.,* **Gibberene.**

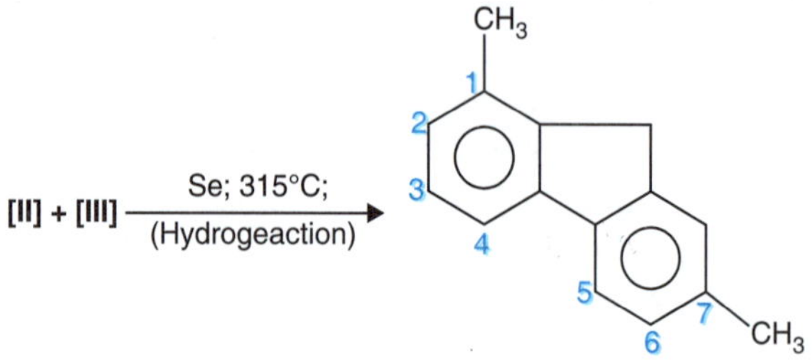

Gibberene (IV) [or 1,7-Dimethyl fluorene]

8. **Gibberic Acid (II) Contains a Carbonyl (>C=O) Group:** Based on the prescribed standard chemical test methods, the **gibberic acid (II)** has been shown to comprise:

- **1 Carbonyl (>C=O) Moiety** – Since it forms an *oxime(=NOH)*,
- **1 Carboxyl (COOH) Moiety**- Since it forms an *ester(-C-OR)*.

Besides, the **IR- Spectrum** of **compound (II)** exhibits a critical bond at **1744 cm^{-1}**, -which designates strongly a *typical characteristic, bond* pertaining to a *keto (>C=O) moiety* in **cyclopentenone.** Moreover, the **UV spectrum of compound (II)** explicitly exhibits that it contains a **phenyl ring λ_{max}: 265, 274 nm].**

9. **Presence of a *Hexahydrofluorene nucleus* in Gibberric Acid (II):** The following sequential *degradations* from **compound (V) through compound (VIII)** are indeed self- explanatory:

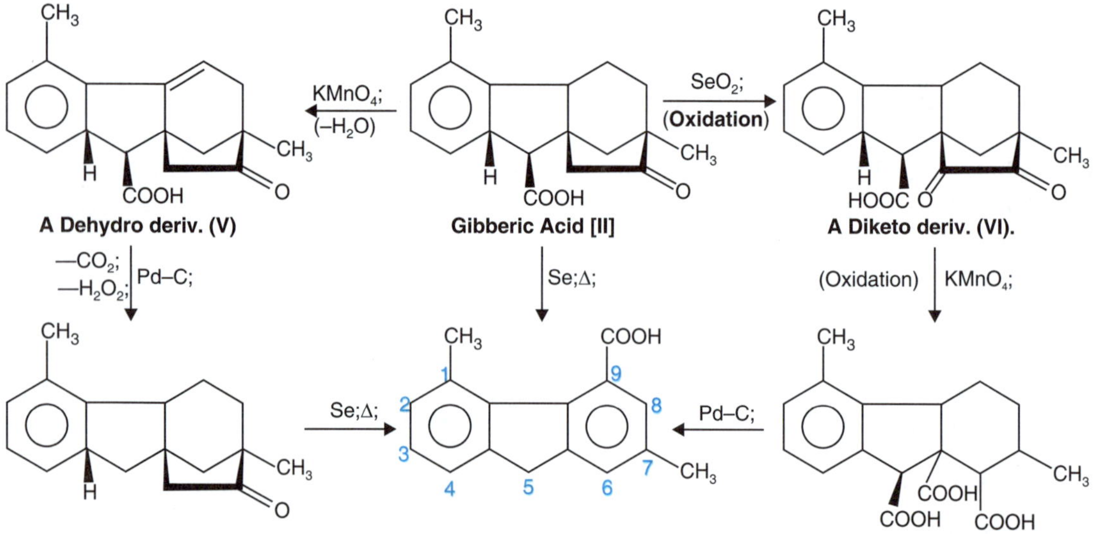

A Dehydro deriv. (V)　　　Gibberic Acid [II]　　　A Diketo deriv. (VI).

A monoketo deriv. (VI)　　　A Dimethyl deriv. (VII)　　　A Tri-Hydroxy deriv. (VIII)

Explanation:

1. The chemical structures of **Gibberone (VI)** and **Gibberic Acid (VIII)** have been duly confirmed and ascertained by their respective *'total synthesis'* (Loewenthal *et al.*, 1960, 1962)

2. The critical **UV-spectrum** of the *diketo* derivative **(VI)** shows the absence of the *ionizable hydrogen,* which explicitly reveals the substitution pattern of the repective **'cyclopentanone ring fragment'**.

3. Besides, the careful *dehydrogenation* of **compound (VI)** to the corresponding **1,7-dimethyl fluorine-9-carboxylate (VII),** establishes beyond any reasonable doubt the exact positions of the **carboxyl (COOH) group**.

10. Conversion of 1,7-Dimethyl fluorene-9- carboxylate (VII) to Diastereoisomer (XII)

Let us look into the various sequences of reactions from **compound (VII)** to the respective **diastereoisomer (XII):**

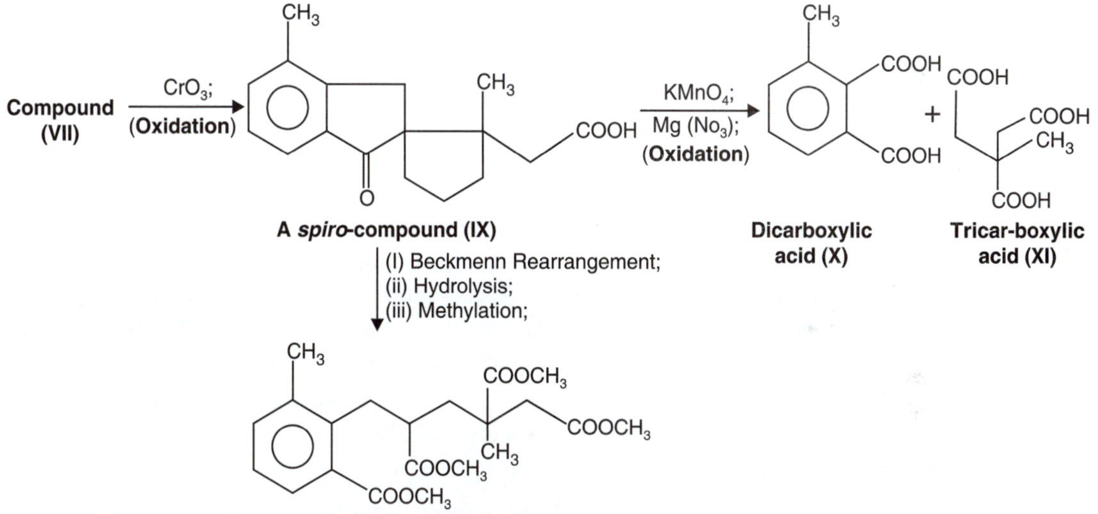

A *spiro*-compound (IX)

(I) Beckmenn Rearrangement;
(ii) Hydrolysis;
(iii) Methylation;

Dicarboxylic acid (X)   **Tricar-boxylic acid (XI)**

A *tetra*-Methylcarboxylate (Deiastereoisomer) (XII)

Explanation : Compound **(VII)** on *oxidation* with *chromium 6- oxide (CrO₃)* forms a *spiro* compound **(IX),** which on further *oxidation* with KMnO₄ /Mg(NO₃)₂ yields *two* **products : Dicarboxylic Acid (X)** and **Tricarboxylic Acid (XI)**. The *spiro* –compound **(IX)** on being subject to : **Beckmann Rearrangement-Hydrolysis**, and-**Methylation** gives rise to the formation of a **tetra-methylcarboxylate (XII)** due to the rupture of *two* **cyclopentane ring system in (IX)**.

11. Confirmaiton of Structure of Allogibberic Acid (III) by its Degradation to 1-Methyl-7-hydroxy fluoren (XV) *via* (XIII) and (XIV):

The various steps involved in the above reactions are as given under :

The **UV spectrum** analysis provide the following critical **max values:**

- **For allogibberic acid (III): 266 and 274nm-** indicating that the structure contains a **'benzene ring'** system.

Further the **ozonolysis** of **compound (III)** gives a mole of **formalin** and also a respective *keto*-**derivative (XIII),** which vividly shows that **compound (III)** essentially contains an **oxocylic methylene moiety.**

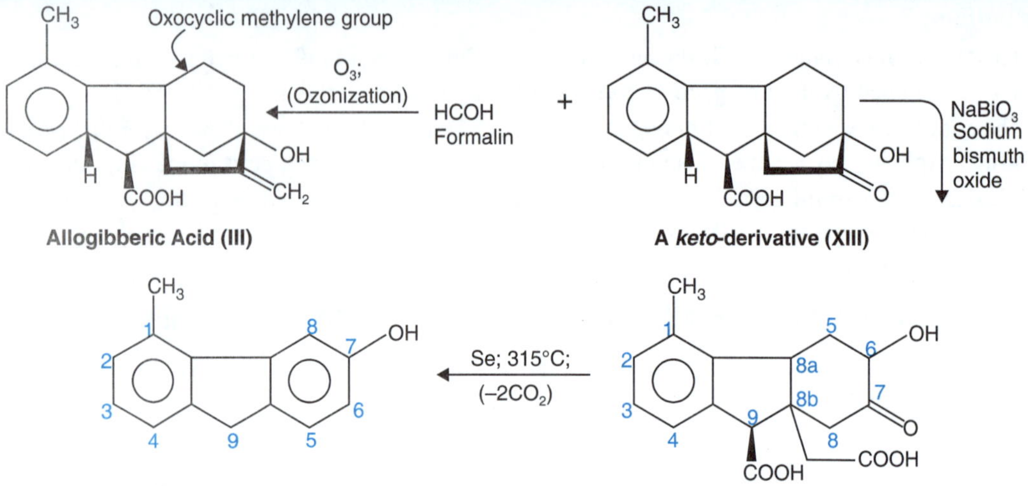

Allogibberic Acid (III)

A *keto*-derivative (XIII)

1-Methyl-7-hydroxy-fluoren (XV)

1-Methyl-7-*keto*-9-carboxy-8a-methyl carboxylate derivative (XIV)

Observations

- It is, however, pertinent to state at this point in time that the respective '*acid*' (**XIV**) eventually *isomerizes* the corresponding **methylester of allogibberic acid to methyl gibberate.** Therefore, it clearly ascertains the exact positions of the respective **carboxyl (COOH) moiety** is the same in **both acids.** However, the *third* O-atom in **allogibberic acid** (**III**) has been emphatically thrown to be *hydroxylic in nature* (λ_{max} **3460cm^{-1}**)

- Besides, this particular moiety may also be assumed to be *tertiary* **in character** – perhaps by virtue of the difficulty encountered in these aspects, namely:

    - **acetylation process,** and
    - **failure to cause oxidation of dihydroallogibberic acid to the corresponding** '*ketone*'.

- **Compound (XIV)** has been duly proven to be a *keto*-**carboxylic acid,** wherein the **carbonyl (C=O) moiety** is critically located in a **cyclohexane ring system.**

- **Compound (XIV)** on being subject to *dehydrogenation* on **Se power** (at 315°C) gives **compound (XV)** ie; **L-methyl-7-hydroxy flouoren.** Thus, the above reaction establishes firmly the **exact position** of the **carbonyl (>C=O) moiety.**

> **NOTE** Based on the above scientific evidences and logical explanations it may be concluded that *structure (III)* is the proposed structure of allogibberic acid.

12. Conversion of Methyl Ester of Gibberellic Acid [I(a)] to Dimethyl Ester of Compound (IV) [XIV (a)]

The following sequence of reactions come into play:

Methylester of gibberellic acid [1(a)] on being subjected to *ozonolysis (O_3)* followd by *oxidation* yields a **methyl gibberellate [I(b)],** which upon *methylation* gives a **dimethyl ester [XIV(a)].**

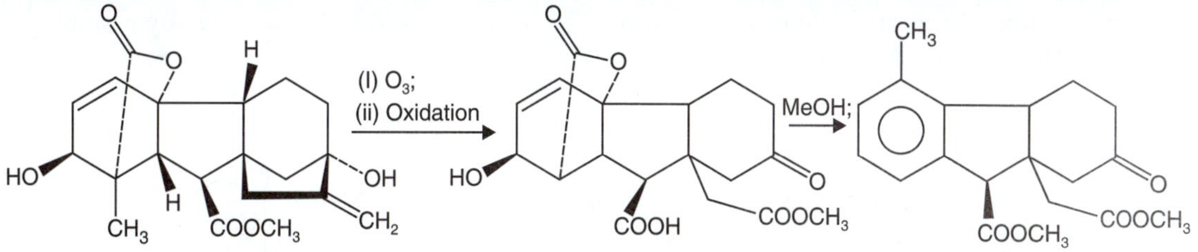

| Methyl ester of Gibberellic Acid [I(a)] | Methyl Gibberellate [I(b)] | A Dimethyl Ester [XIV (a)] |

Remarks : In fact, these reactions do reveal that **gibberellic acid (II)** and **allogibberellic acid (III)** actually differ distinctly with respect to **ring 'A'**, - which should undergo *'aromatization'* in the *former* to produce the *latter*. Therefore, one may infer that the respective **ring 'A'**, in (II) does possess essentially *one ehtylenic double bond* one , *sec*-hydroxyl moiety, and a **lectone ring.**

13. Position of Hydroxyl (OH) Moiety

The exact position of the **hydroxyl (OH) moiety** has been duly established by the underlying fact that *two* **ketones** obtained from **Gibberellin A$_1$** on being subject to **Se-dehydrogenation**, gives *two* **products**, namely:

- **1-methyl fluoren-2-ol (X),** and
- **1,7-dimethyl fluoren-2-ol (Y),**

which evidently ascertains that the **hydroxyl (OH) moiety** is located strategically quite close to the **methyl (CH$_3$) group** in ring 'A'.

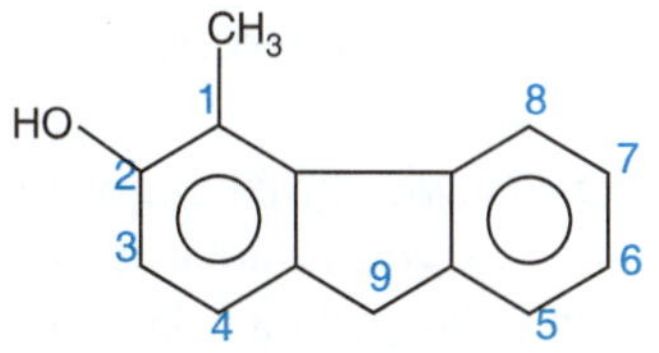

Compound (X) 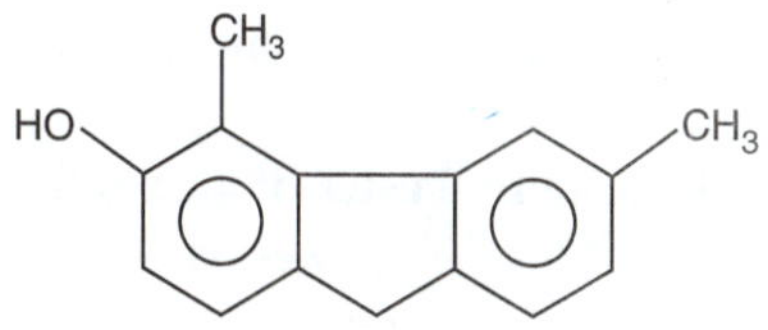   Compound (Y)

14. Exact Position of Double Bond

Since the *oxidation* of **methyl ester of gibberellic acid (or methyl gibberellate)** with MnO$_2$ gives an **α,β-unsaturated ketone (XVI),** having *UV-absorptions* (λ_{max} 228 nm) , it establishes the **position of the double bond.**

15. Allylic Position of 'Lactone Ring'

The precise and exact position of the **'lactone ring'** has been established based upon the following scientific facts:

❑ *First,* the **methylester of gibberellic acid (or methyl gibberellate)** on being subjected to *catalytic reduction* with **H$_2$/pt** gets converted into the corresponding *carboxylic acid (XVII)* in a substantial quantity; and

❑ *Secondly,* a **heteroannular diene** *i.e,* **gibberellinic acid (XVIII)** (having λ_{max}: 253nm) being formed from **gibberellic acid (I)** in an **aqueous medium.**

Importantly , the above statement of facts may be further proved and expatiated by assuming the **structure [I(a)] for methyl gibberellate** as given under:

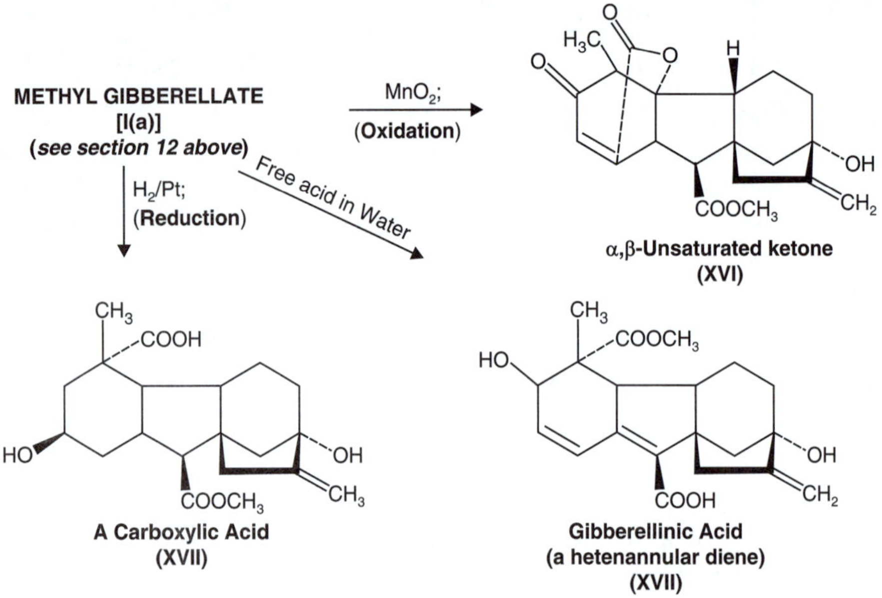

METHYL GIBBERELLATE
[I(a)]
(*see section 12 above*)

MnO₂;
(Oxidation)

H₂/Pt;
(Reduction)

Free acid in Water

α,β-Unsaturated ketone
(XVI)

A Carboxylic Acid
(XVII)

Gibberellinic Acid
(a hetenannular diene)
(XVII)

16. Stereochemistry of Gibberellic Acid (XXII)

Let us look into the following important aspects with respect to the **stereochemistry of gibberellic acid (XXII):**

❑ When the **dimethyl ester of gibberellic acid [XIV (a)]** is hydrolyzed, it gives **gibberellic acid (I)** plus its **C-9** *epimer*. However, when the *latter compounds* are heated with **acetic anhydride**, they are duly converted into the *same anhydride* that on subsequent *hydrolysis* gives exclusively the **dimethyl ester of gibberellic acid [XIV(a)]** *i.e;* there is no formation of the *C-9 epimer.*

   Inference: Therefore, one may conclude logically that the **substituents** at **C-9 and 5α– apositions** should have essentially the *cis*-**configuration**. Besides, the *carboxyl (COOH) moiety* positioned at **C-10** and the respective **C-8/C-9 bridge** should be also *cis*-**configuration** in **allogibberic acid (III).**

❑ Importantly, because the **catalytic hydrogenation (Pt/H2)** of the respective **dehydroallaogibberiic acid (XX)** is likely to occur possibly at the *less-hindered segment of the molecule,* it indicates explicity that the **H-atom located strategically at C-4b** shall be on the exact opposite side to the corresponding **C-10 carboxyl (COOH) moiety** and also the respective **C-8/C-9 bridge.**

   Nevertheless, this partucualr reduction indeed brought forth the *original stereochemistry at C-4b.*

   Inference : Thus, one may safely draw a conclusion that the **rings B/C** should by all means have *trans* **fused in the allogibberic acid (III).** However, it has already been proved that gibberellic acid **[XIV(a)]** actually differs from the respective **allogibberic acid (III)** only with respect to **ring 'A'**. Hence, it is very much expected that **ring fusion B/C in the gibberellic acid [XIV(a)]** would be *trans* configuration only.

❑ A close examination of the *CD- curve* of the *ketone* analogue **(XXI),** that is eventually derived from **compound (XIX),** strongly reveals and ascertains that the *configuration* of the **C-4b hydrogens** to be *ie,* the **rings B/C** are *cis-* **fused.** Hence it would be the **correct configuration** in **Gibberellic acid [XIV (a)].** From the above deliberations and evidences one may safely infer that in the *conversion* of **Gibberellic Acid [XIV (a)]** into the respective **allogiberellic acid (III)** to **C-4b** gets **epimerized** ultimately.

Compound	Dehydroallogibberic Acid	A *keto* analogue
(XIX)	**(XX)**	**(XXI)**

Important Points to Ponder: These essentially comprise:

1. Since, one may observe a *prompt isomerization* specifically at the *C-2 hydroxyl (OH) moiety* of **compound (XIX),** which indicates clearly that the **C-2 hydroxy group** is *axial* in nature **(XIX);** and, therefore, it designates a *quasi axial* in **gibberelllic acid (i.e; unstable axial stable equitorial).**

2. Based on the inherent chemical behavior of **compound (XVII),** one may draw a conclusion that the *'lactone ring'* in the **gibberellic acid [XIV (a)]** should have the **β-*configuration*.** Nevertheless, the **X-ray diffraction analysis** reveals explicitly that the *'lactone ring'* does possess an **α-configuration.**

3. Therefore, bearing in mind all the *scienntific evidences* and *logical explanations* pertaining to the **stereochemistry of gibberellic acid** is as depicted in **structure (I).**

15. CYTOKININS [KININ]

The **'Cytokinin'** (or **Kinin**) relates to the growth substances which primairily stimulate the *cell division*; and the overall effect takes place only in association with the **'auxin'.** Different proportions of **'auxin'** and **'cytokinin'** may predominantly induce altogether different types of **meristematic activity.**

The **cytokinins** (or *phytokinins*) are the so called **'purine derivatives'**; and hence, regarded as the degradative products of **adenine** *(i.e;* a component of **nucleic acids** *viz.,* **DNA and RNA**). Stock et al. carried out an intensive and extensive research on the **cytokinin.** Evidence from the literature reveals that the **'endosperms'** of the *seeds* and *roots* are the most preferred primary sites of the **synthesis of cytokinins.** However, the ultimate effect of **cytokinins** upon the cell division invariably comes into being especially in conjunction with the **'auxin'.**

❑ **Kinetin** - is the degradation product of **DNA** and fails to occur in nature as such.

❑ **Zeatin**-designates the most extensively studied and explored naturally occurring **cytokinin**, that is being isolated from the *unripe maize grains* (**sweet corn kernels**) *Zea mays* L., *Graminiae*. It is, in fact, a **plant- growth hormone**.

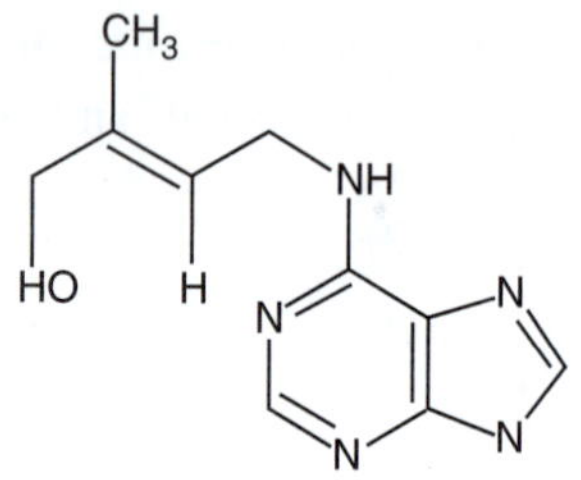

Zeatin

The major utility and function of '**cytokinins**' is the cell division. Importantly, the '**cytokinins**' are also intimately involved in the **cell expansion** and **retardation of aging of leaves.**

FURTHER READING REFERENCES

Andrews T	:	**A Bibloigraphy on Herbs, Herbal Medicine, 'Natural' Foods and Unconventional Medical Treatment,** Libraries Unlimited, Colorado, Lillteton (USA), 1992.
Duke JA	:	**Handbook of Phytochemical Constituents of GRAS, Herbs and Other Economic Plants,** CRC-Press, Boca Raton, Florida (USA), 1992.
Principe PP	:	**The Economic Significance of Plants and Their Constituents as Drugs:** *Economics and Medicinal Plants Research, Vol. 3.,* Wagner H et al. [Eds], Academic Press, London (UK), 1989.
Tyler VE	:	**Herbs of Choice; The Therapeutic Use of Phytomedicinals,** Pharmaceutical Products Press, New York, 1994.

REVIEW QUESTIONS

1. Enumerate briefly the **Stereochemistry of Steroids.** Give suitable examples in support of your answer.
2. Describes the **Constitution of Cholesterol** and explain briefly the various sequences of reactions incurred.
3. Give a comprehensive account on **Lanosterol** and discuss its **synthesis.**
4. How would you elaborate on the **Constitutions of Stigmasterol?** Explain with proper examples.
5. Discuss '**Ergosterol**' with specific references to the '**Constitution of Ergosterol**'.
6. What are **Bill Acids?** Elaborate on the **Constitutions of Bile Acids** and support your answer with appropriate brief explanations whenever necessary.

7. Write a brief account any **one** of the following:
   - (a) **Saponins**
   - (b) **Sapogenins**
   - (c) **Synthesis of Diosgenin**
8. How would you elaborate comprehensively the **'Cardiac Steroids'** [or **Cardiotonic Glycosides**]?
9. Explain any **ONE** of the following:
   - (a) **Constitution of Strophanthidin**
   - (b) **Classification of Steroidal Alkaloids**
   - (c) **Solanum Alkaloids**
   - (d) **Salamandra Alkaloids**
10. Give a detailed account of **Sex Hormones.**
11. Explain the Synthesis of **(+)-Estrone** by **Johson's Synthesis?**
    Elaborate various steps briefly.
12. Write explanatory notes on any **one** of the following:
    - (a) **Estriol**
    - (b) **Chemical Relationships amongst Estrone, Estriol, and Esteradiol.**
    - (c) **Synthesis Estrogen (or Elthinylestradiol)**
13. Give a brief account on any **one** of the following:
    - (i) **Benzosterol**
    - (ii) **Progesterone from Diosgenin**
14. Discuss in details the **'Oral Contraceptives'**
15. Explain the following:

    Male Sex Hormones   OR

    Adrenocortical Hormones

16. Write descriptive account on any **ONE** of the following:
    - (a) **Constitution of Hydrocootisoue**
    - (b) **Aldosterone**
    - (c) **Plant Growth Hormones**
    - (d) **Constitutions of Gibberellie Acid**

Contents at a Glance

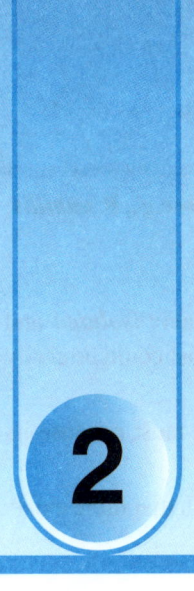

2

Antibiotics

1. INTRODUCTION

Vuillemin (1889) was pioneer in introducing the term **'antibiotic'** and defined it as —'the *active component* duly involved in the process of 'antibiotics' or to the opposition of one living microorganism to another.'

However, another school of thought defines **'antibiotics'** as– 'substances that are nothing but the *'microbial metabolites'* that could in relatively high dilution may inhibit the growth of microorganisms.

Waksman (1942–44)[*] proposed the widely cited **definition** that – 'an antibiotic or an antibiotic substance is a substance produced by the microorganisms, which has the capacity of inhibiting the growth and even of destroying other microorganisms'.

> **NOTE** **Importantly, *selective toxicity* happens to be the 'key concept' amongst the antibiotics.**

Obviously, at a relatively very low concentrations the antibiotics do perceptively bring about their respective antibiotic activity. They are also removed to as the potent *'chemotherapeutic agents'*.

Some Ardent Satisfying Parameters – Following are some of the **Ardent Satisfying Parameters** of the **'antibiotics'** that are duly summarised as under:

1. The original concept of an **'antibiotic'** relates to a *product of metabolism* even though it might have been duly *synthesized*.

2. In a situation, when the **'antibiotic'** is a purely *synthetic product*, it must be a **structural analogue** of the actual **naturally occurring antibiotic**.

3. An **'antibiotic'** should be effective at a relatively **low concentrations only**.

4. The **'antibiotic'** should essentially possess the ability to antagonize either:
   - **the growth**, or   • **the survival**,

   of one or more than one **species** *of the microorganisms*.

Antibiotic as a Therapeutic Agent – Importantly, a substance to be known as – **'antibiotic effective as a therapeutic agent'**, may have to cater for the following conditionalities, namely;

---

[*]   Waksman SA: 1888 Priluka, Russia, Ph.D., Calif: Rutgers Irst Microbiol, Nobel Prize, 1952.

- It should be effective against a pathogenic (*disease – producing*) *microorganism* **e.g.,** **B. subtilis**, **S. aureus**, **E. coli** and the like,

- It must not cause an appreciable *toxic side–effects*,

- Its *stability profile* should be significantly high in order that it may be suitably *isolated* and *transformed* into the appropriate forms of **drug products** (or **dosage forms**) that could be easily **administered and absorbed** *in vivo*,

- It should withstand reasonably long and good storage profile in order that it must not lose its activity significantly,

- The rate of *detoxification* and *elimination* from the human body should be such that there exists enough time interval prevailing between:

  ➤ *two* **successive dosage forms**, and

  ➤ **during the period a proper concentration level has been duly accomplished**, and

- The **'antibiotic'** must be eliminated as completely as possible from the *human-body system* soonafter its administration has been terminated completely.

1.1. History of Discovery

It has often been observed that in **mixed cultures of microorganisms** one specific type may be suppressed by another. Thus, one such example was duly reported in 1929, when it was found that a certain strain of the mold **Penicillium notatum** Westling does critically produce the so-called *bacteriostatic agent –* **'Penicillin'** (Fleming). However, the observation **passed almost unnoticed**. Nearly, a decade later: **Florey and Chain**, at the **Oxford University** started a vigorous investigative research of the aforesaid **'antibiotic agent'**. They eventually succeeded in obtaining the *crude preparations* that had remarkably high activity against several **Gram-positive organisms.**

Importantly, the very first **'clinical results (1941)'** were indeed found to be **highly successful**. Besides, one advantage of **penicillin** over the *previous chemotherapeutic agents* is its **remarkably low toxicity**; however, the disadvantages are that it loses much of its activity when administered orally and also that it gets **excreted rapidly**. In addition, it is found to be active only against **Gram-positive organisms** (*viz.*, *Staphylococci*, *Streptococci*, *Pneumococci*, *Meningococci*, and *Gonococci*) and also against certain **spirochetes**; certain strains are *less sensitive* in comparison to other strains.

Interestingly, there are **organisms** that occasionally develop *fastners to penicillin*. It has been observed that although the penicillin has a **broad-chemotherapeutic spectrum**, the **penicillin** is relatively ineffective against **Gram-negative bacteria**. The so-called commercial preparations do comprise only or mainly **Penicillin G**; and *P chrysogenum* is the most **preferred organism**. However, the intelligent addition of **phenlacetic acid** to the *fermentation tank* certainly enhances the actual yield of **Penicillin G** significantly. The most *useful derivatives of penicillin* are the respective **'amine salts'**, specifically the one formed with *procaine* (*i.e.*, **Procaine Penicillin**).

The **Anglo-American** collaborated research mustered altogether remakable chemotherapeutic characteristic features critically during the **World War II** – which eventually resulted into the pure crystalline product, the chemical structure was duly established, and the underlying problems of manufacture of the highly-labile antibiotic substance on a commercial scale was solved.

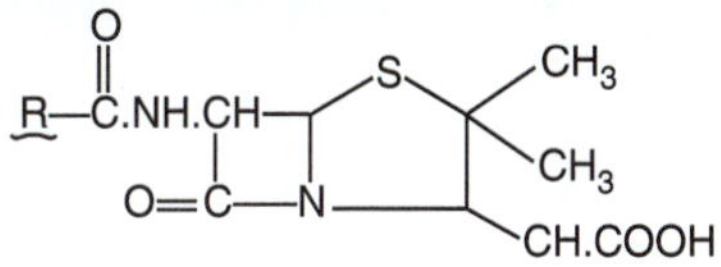

Penicillin

when,

R = – CH$_2$ – CH = CH CH$_2$CH$_3$: Pent–2–enylpenicillin

R = – CH$_2$ – C$_6$H$_5$: Bencylpenicillin

R = – (CH$_2$)$_6$. CH$_3$: n-Heptylpenicillin

R = – CH$_2$ – C$_6$H$_4$ – OH (p) : p-Hydroxybenzylpenicillin

1.2. Importance of Antibiotics

In the recent past, a good number of **'antibiotics'** have gotten the due well-deserved recognition as vital and important **therapeutic agents**. In a broader perspective many **antibiotics** have been demonstrated to be effective clinically in these *two* **kinds of infections**:

- **Protozoan infections,** and
- **Fungal infections**.

Besides, quite a few **'antibiotics'** are proven to be **'carcinostatic'** *i.e.,* will inhibit or arrest the development or continued growth of *cancer, carcinomas,* or *malignant tumors,*

- ❑ **Antitumor Antibiotics –** A sincere, intelligent, and methodical search for the **antitumor antibiotics** has been initiated with wonderful results – that eventually established their potential usage in the control and management of the so-called **malignant diseases**. Following are a few well-established **antitumor antibiotics** which have been recognized to be active against the **experimentally induced tumors** in *laboratory animals,* such as:

  - **Azaserine** • **Actinomycin-D** • **Mitomycin-C** • **Actinobolin** • **Streptovitacin**-A • **DON [6-diazo-5-oxo-norleucine]** • **Psicofuranine** • **Streptonigrin and** • **Refuin (Anthramycin)**

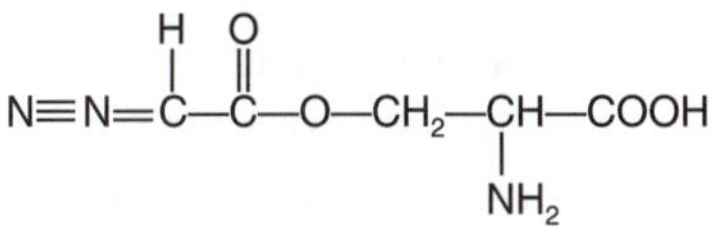

Azaserine

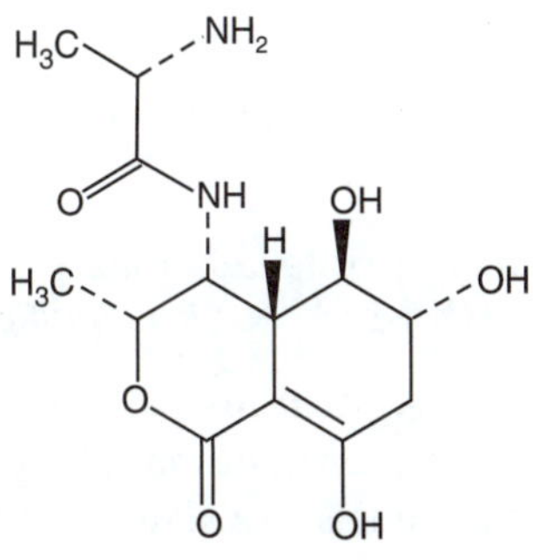

Actirobolin

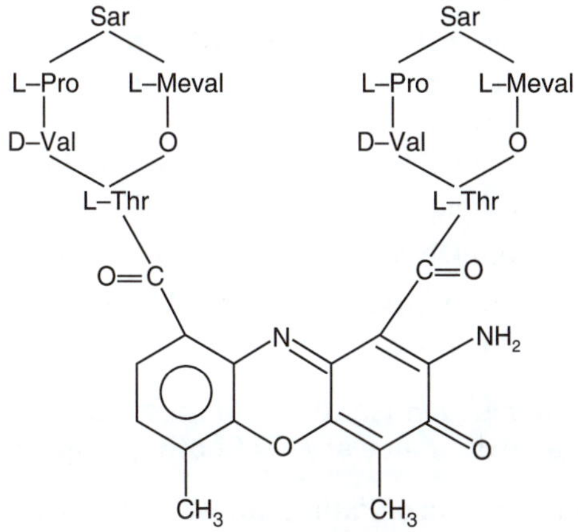

Actinomycin D [Dactinomycine]

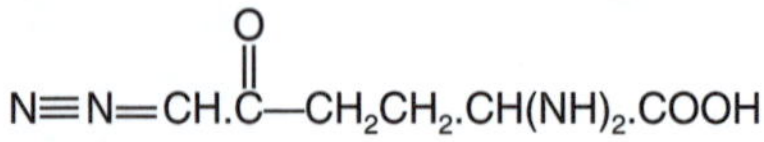

$$N\equiv N=CH.\overset{\displaystyle O}{\overset{\|}{C}}-CH_2CH_2.CH(NH)_2.COOH$$

**6-Diazo-3-oxo-L-norleucine
(DON)**

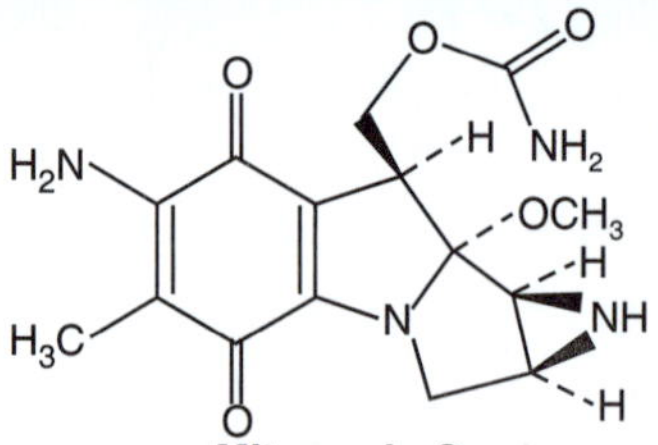

Mitomycin C

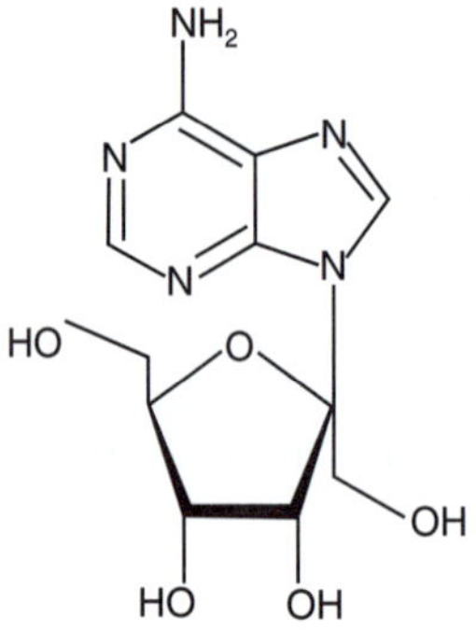

Psico furanine

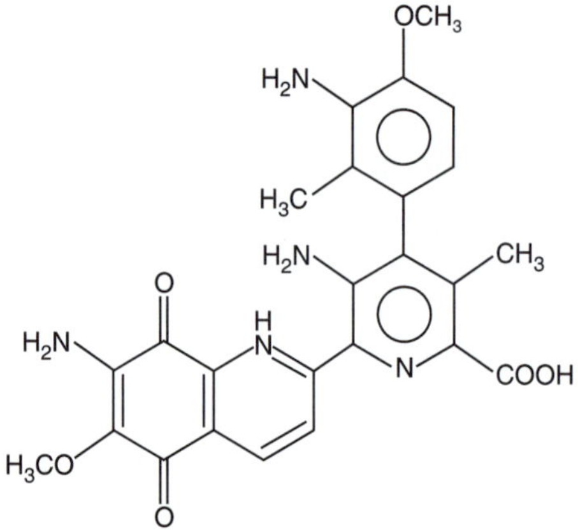

Streptonigrin

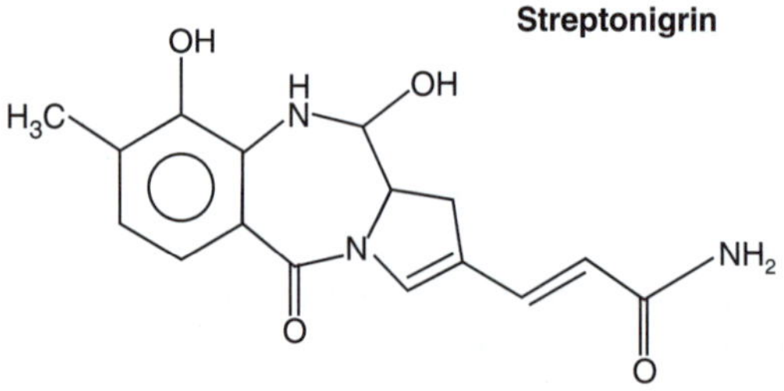

Anthramycin [Refuin (oldname)]

❑ **Biochemical Cellular Mechanisms** – It has been well-established that the **antibiotics** have become more or less an indispensible tool for performing the study of **biochemical cellular mechanisms**. In 1960, the spectacular discovery with respect to the underlying inherent ability of the '**actinomycin D (or dactinomycin)**' to *complex with DNA* opened an entirely new technique of investigating **DNA dependent synthesis** in such entitles as:

• **from DNA synthesis, or**   • **from protein synthesis,**

by the respective **antibiotic** unless and until the corresponding synthesis in the host cells is fully suppressed.

NOTE	**In this manner, the '*antibiotic*' has been employed quite judiciously to carry out an exhaustive study of both: • *messenger RNA* and • *virus multiplication.***

❑ **Suppression of Pathogenic Microbes** – In the field of '**Agriculture Sciences**' the **antibiotics** are used abundantly so as to suppress appreciably the growth of the pathogenic microbes, which ultimately lowers the normal growth-rate of animals to a significant level. Besides, 'antibiotics' are also used by the 'genetist' to cause **mutations in molds**. These are also employed effectively to supplement animal feed and preservation of seafood and poultry products.

2. CLASSIFICATION OF ANTIBIOTICS

The survey of literature reveals that the **'antibiotics'** may be classified in the following *three* ways, namely:

- ❏ **Broad based classification;**
- ❏ **Classification based on the type of bacteria attacked,** and
- ❏ **Classification based on their chemical structures.**

2.1. Broadbased Classification

It categorizes the **antibiotics** into the following *two* variants, namely:

(a) **Broad-Spectrum Antibiotics –** These refer to such **'antibiotics'** that are being employed extensively as *therapeutic agents* against many dreadful human diseases.
   Examples: **Chloramphenicol, Penicillins**, and **Tetracyclines**.

(b) **Narrow-Spectrum Antibiotics –** They are invariably very specific in their so-called *curative activities.*
   Examples: **Bacitracin, Nystatin** etc.

2.2. Classification Based on the Type of Bacteria Attacked

Importantly, this specific and highly acclaimed classification was introduced for the first time by *Christian Gram*; and hence, is broadly known as the ***Gram-Staining Technique***.

Specific Details – In this particular method, the carefully prepared **'fixed bacterial smear'** is being treated:

- **first** – with a **Crystal Violet** solution (a **dye solution**), and
- **secondly** – with an **Iodine solution**.

Subsequently, the fixed and **dye-treated bacterial smear** is washed with ethanol slowly and repeatedly; and thus, the colour duly attained by bacteria divides them into *two* **different kinds:**

- ❏ **Gram-Positive Bacteria –** those which **retain the colour of the 'crystal violet'** and usually appear as **'deep violet'**; and
- ❏ **Gram-Negative Bacteria –** those which **lost the violet colouration** and get counter-strained by *safranin* and usually appear **red in colour** are invariably termed as *Gram*-negative bacteria. Following are a few **'typical examples'** of both *Gram*-positive and *Gram*-negtative bacteria.

Gram-Positive Bacteria	*Gram*-Negative Bacteria
Diphtheria bacillus	*Colt bacillus*
Leprocy bacillus	*Gonococcus*
Pneumococcus	*Meningococcus*
Staphylococcus	*Plague bacillus*
Streptococcus	*Typhoid bacillus*
Tubercle bacillus	*Vibrios cholerea*

2.3. Classification Based on Their Chemical Structure

Following are a few typical examples of **'antibiotics'** whose classification is entirely based upon their inherent chemical structures, such as:

1. **Aminoglycosides:** They do comprise a **sugar residue** (*e.g.*, *D-Ribose* or *D-glucosamine*)– that are linked glycosidically to an amino compound.
   Examples: Streptomycin; Neomycin C;

2. **Antitubercular Antibiotics:** These prominently possess *antitubercular activity profile*.
   Examples: Cycloserine; Viomycin Sulphate.

3. **Antineoplastic Antibiotics:** These antibiotics are employed particularly for the *control and management* of *cancerous growth of tumors*.
   Examples: Dactinomycin; Mitomycin.

4. **Chloramphenicol/Synthetic Structural Analogues:** Such as:
   Examples: Chloramphenicol Succinate; Chloramphenicol Palmitate;

5. **Lincomycins:** They are **S-containing antibiotics**, wherein the **S-atom** is *not present in the ring-system*
   Examples: Lincomycin; Clindamycin.

6. **Macrolides:** They specifically contain a *large lactone-ring system.*
   Examples: *Erythromycin*; *Oleandomycin*.

7. **Penicillins:** These are solely derived from the Amino Acids.
   Examples: Phenoxymethylpenicillin; Penicillin-G; Cephalosporins;

8. **Polypeptides:** They critically contain a polypeptide chain.
   Examples: Bacitracin; Tyrothyricin.

9. **Polyenes:** These have a *conjugated polyene system.*
   Examples: Amphoterian; Nystatin.

10. **Tetracyclines:** They do comprise of *four six-membered fused ring systems.*
    Examples: Oxytetracycline; Chlorotetracycline; Aureomycin; Terramycin.

11. **Unclassified Antibiotics:** These include any **'antibiotic'** that are not covered under the above categories from (1) through (10):

Examples: Griscofulvin; Fulvicin; Vancomycin (or **Vancomycin Hydrochloride**).

It may be worthwhile to state here that the following *five* **categories of antibiotics**, namely:

- **Penicillins,**
- **Cephalosporius,**
- **Streptomycin,**
- **Chloramphenicol, and**
- **Tetracyclines,**

Shall now be discussed at length in the sections that follows; along with their brief introduction, constitution, total synthesis, and therapeutic applications wherever applicable.

3. ANTIBIOTICS

3.1. The Penicillins

The **β-lactam antibiotics** do comprise a *beta-lactamase* i.e., one of a group of enzymes (duly produced by various *Gram*-negative bacteria and *Gram*-positive bacteria) that catalyzes the hydrolysis of the **β-lactam ring**. However, the **β-lactam antibiotics** may be further sub-divided into *two* **major groups,** namely:

- **Penicillins** and
- **Cephalosporins.**

Penicillin is the name assigned to the mixture of certain naturally occurring compounds having the *molecular formula* $C_9H_{11}O_4N_2$ **SR** and differing only in the **nature of 'R':**

R—C.NH.CH
O=C—N
S
CH₃
CH₃
COOH

A = Thiazolidine Ring;
B = β-Lactam Ring;

Penicillin

Naturally Occurring Penicillins: Following are the *six* **naturally occurring penicillins**, whose *chemical name*, *other names* and *attachment equivalent to* **'R'** are as given under:

S.No.	Chemical Name	Other Names	R
1.	**Benzylpenicillin**	Penicillin–G or H	—CH₂—⬡
2.	**Pent-2 enylpenicillin**	Penicillin I or F	– CH₂ CH == CH CH₂ – CH₃
3.	**4-Hydroxybenzylpenicillin**	Penicillin III or X	—CH₂—⬡—OH
4.	**n-Heptylpenicillin**	Penicillin IV or K	– (CH₂)₆ – CH₃
5.	**n-Amylpenicillin**	Dihydro-F-Penicillin	– (CH₂)₄ – CH₃
6.	**Phenoxymethylpenicillin**	Penicillin V	—CH₂—O—⬡

 NOTE : **(1) The critical modification of the '*prosthetic moiety* R*',-duly attached to the *6-aminopenicillinic acid segment* of the respective penicillin molecule, determines the actual relative stability profile of the form of *penicillin to penicillinase.***
(2) It has been observed that the *more complex nature* of the *prosthetic moieties* would import a greater extent of *stability to hydrolysis by penicillinase.*

Commercial Production of Penicillin – The **'Penicillin'** is mostly produced on a *commercial scale* by one of the following *three* known methods:

(a) **Bran Method** – In fact, the use of **moist bran** (*i.e. husks of cereal grain*) is regarded to be a good substrate for the growth of mold; and hence, the resultant penicillin may be duly extracted either:
- **in a liquid medium,** or
- **the penicillin-containing bran** (used directly)

---

* **Prosthetic Moiety:** It relates to the *nonprotein portion of a conjugated protein* which is essential to the *physiological activity* of the protein (often a **complex organic molecule**).

Demerits – The *two* major (drawbacks) of the **Bran Method** are as follows:

- **Bran** being a *'bad conductor of heat'* poses a great difficulty to sterilize it due to the *long duration involved* for the internal medium to attain the desired stipulated temperature; and

- Heat is generated by the mold undergoing development, which is ultimately difficult to dissipate; and hence, partially deteriorates the goodness of penicillin.

(b) **Liquid Surface-Culture Method** – This method makes use of an *aqueous solution of molasses** as a medium for the growth of the **microorganisms**, since it duly comprises:

- **Sucrose (8 to 10%)**   •   **Mineral salts**   •   **Nitrogenous materials**;

    and, therefore, it tends to be an ideal medium for the growth of mold. The pH of molasses is duly adjusted between 7 to 8. Thus, in a few days duration, the growth of the *microbes* commences briskly and then after a period of 6-7 days, the prevailing concentration in the *fermentation both* of **penicillin** ranges between **0.3–0.4 mg.mL^{-1}**

Comment – The **surface culture method** is indeed a simple but a definitely *laborious method* that essentially entails:

- **High cost** and   •   **Lengthy process**

    and hence, is not usually adopted as a method for the **commercial production of penicillin**.

(c) **Submerged Culture Method –** It is, in fact, regarded to be the most preferred method that is being employed for the general **commercial production of penicillin**. This method makes use of the strain *Penicillium chrysogenum* (as the starting **culture**) importantly, the culture medium actually being employed for this **100 parts** predominantly contains the following **vital components**, namely:

(i)	**Lactose:** 20 parts;	(ii)	**Corn Steep Solids:** 20 parts;
(iii)	**Sodium Nitrate:** 3 parts;	(iv)	**Dipotassium Phosphate:** 0.05 parts;
(v)	**Magnesium Phosphate:** 0.125 parts;	(vi)	**Calcium Carbonate:** 1.8 parts;
(vii)	**Phenyl Acetie Acid (Precursor):** 0.5 parts.		

*Modus Operandi***:** The various steps involved are as follows:

1. Both the **'culture'** and **'culture medium'** are taken in a thoroughly cleaned-sterilized 55 large tanks.

2. The solution is agitated by a stream of **'sterile air'** at the temperature of the *culture broth* is specifically maintained at about **24°C for 48–72 hours**.

3. After this, the **'Fermentation Vessel'** is duly sealed and a small positive pressure is maintained precisely inside the **'fermentation vessel'** in order that the so-called contagious *harmful material* from the atmospheric exposure may not gain an entry into the vessel.

4. Under such **stringent parameters** of *agitation – aeration* and incubation allows the proper mold growth throughout the **main bulk of the liquid** – as globular pellicles comprising mainly of mycelium.

Isolation of Penicillin – The **isolation of penicillin** from the fermentation broth is duly accomplished through extraction procedures. It is, however, pertinent to state here that such extraction procedures not only pose certain *intricate problems* but also do require close control of *several parameters*, such as:

** **Molasses:** A byproduct from a sugar industry containing almost 8–10% of residual non-crystallizable cane sugar (i.e., disaccharide)

1. **Penicillin** gets decomposed rapidly by stronger acidic/alkaline solutions.

2. It also gets readily decomposed by **heavy metal ions** e.g., Pb^{2+}, Hg^{2+}.

3. An *elevated temperature* is quite detrimental to the **stability of penicillin**; and hence, one must totally avoid an **elevated temperature** so as to obtain the **final solid product** from the *concentrated liquid extract*. Obviously, such a severe **constrain** would ultimately evolve the *most sophisticated* and *scientifically based* **'Freeze-Drying Methods'**.

4. **Yield of Penicillin from the Concentrated Culture Filtrate** – It quite often gets affected by the so-called **microbial contamination,** which eventually comes into being due to the fact that **certain bacteria** (i.e., streams of *Escherichia coli*) critically secretes an **'enzyme penicillinase'** which predominantly attacks the *basic* **penicillin structure**. Therefore, such organisms (**E. coli**) must be completely eradicated from the *fermentation* and *working area*, by the *stringent maintenance of perfect asceptic environment*.

Isolation:

❑ After fermentation the entire contents of the vessel need to be **chilled rapidly** by passing liquid ammonia through a cooling coil for several hours.

❑ **Mycelium** is filtered subsequently on a '*sterile rotary filter*'.

❑ pH of the filtrate is adjusted suitably. The precise magnitude of pH solely depends upon the nature of solvent extraction.

  Example: In case, **Amyl Alcohol** is employed as an extracting solvent, the pH is adjusted to 2 and 3; and if instead **Butanol** is used the pH is adjusted at **6.4** by adding $(NH_4)_2SO_4$.

Solvent Extraction: Having adjusted the **pH of the filtrate**, the **solvent extraction** is carried out by using either:

  • **Amyl Alcohol** or   • **n-Butyl Alcohol**.

➢ **Petroleum Ether (bp: 40–60°C)** is now added to the **pre-cooled** solvent extract, which is subsequently shaken with *diluted NaHCO$_3$ aqueous solution* (to obtain the respective sodium salt of penicillin). The pH of the resultant mixture is carefully adjusted between **6 to 7**; and then subjected to rapid **freeze-evaporation** so as to obtain the respective **sodium salt**, as a final product in a perfect **pure and dry state**. Thus, the *sodium salt of penicillin* is subject to drying in a **high vacuum**.

NOTE   Importantly, throughout the aforesaid sequenced operations the *solutions of penicillin* should be maintained at a temperature very close to the '*freezing-point*' so as to check and prevent the '*inactivation phenomenon*' as far as possible.

Separation of Resultant Admixture of Penicillins: The **penicillin** obtained above is an *admixture of penicillins*, which need to be separated individually by the careful addition of an *appropriate chemical compound* right into the **culture medium**. Thus, it really becomes a lot easier to enhance both:

• **yield of the the desired penicillin**, and

• obtain *newer types of penicillins* (or penicillin variants).

Examples: Following are a few typical glaring examples:

1. **Benzyl Penicillin** – When **2-phenylacetamide** $[C_6H_5–CH_2–CO–NH_2]$ is added to the *fermentation vessel* as a precursor – **benzyl penicillin** is produced instead of **pent-2-enylpenicillin**.

2. **Increased yield of Benzyl Penicillin** – When **2-phenylethylamine [c₆H₅–CH₂–CH₂–NH₂]** is being incorporated as a precursor, the final yield of **benzyl penicillin** is increased significantly.

3. *Three*-**Fold increase in yield of penicillin** – When one adds either **0.4% phenyl acetic acid [C₆H₅ – CH₂ – COOH]** or 0.1% **cysteine** (or **cystine**) as *precursors* the yield of **penicillin** get increased upto **3-folds**.

4. **Fluoropenicillin** – When **2-*para*-fluoro phenylethyl arrive [F–C₆H₄–CH₂–CH₂–NH₂]** is duly added as a precursor, it results into the formation of **fluoropenicillin** (*i.e.*, a *new type of penicillin*).

   Characteristic Features of Penicillins: These essentially include:

1. **Pure penicillin** is usually obtained as white or slightly creamy white crystalline powder –having an **unpleasant taste**.

2. Most natural **penicillins** do occur as *dextro*-**rotatory**.

3. **The Penicillins** are sparingly soluble in water. However, their corresponding **sodium and potassium salts** are fairly soluble in water.

 NOTE   Some of these penicillin salts are found to be '*hygroscopic*' in nature; and hence, may have to be stored in *perfect sealed containers*.

4. **The Penicillins** are mostly soluble in *organic solvents*.

5. In general, the penicillins are reasonably *strong carboxylic acids* having a pKa value of 2.8.

6. However, **the penicillins** available in the form of **'Free Acids'** have not yet been accomplished as a *crystalline form*; and hence, they usually undergo *rapid decomposition* in the **presence of moisture**.

7. **Penicillins** have a tendency to undergo **hydrolysis rather swiftly** and the nature of the resulting product so formed solely depends upon the *nature of the hydrolyzing agent*.

Examples: Following are a few typical examples:

❑ **Alkaline hydrolysis of penicillin** – yields **penicilloic acid** which eliminates a mole of CO_2 to give **penilloic acid**;

❑ **Acidic hydrolysis of penicillin** – affects the amide-side chain thereby causing cessation of the **β-lactam ring** structure – that causes *loss of activity*. Perhaps it could be the possible viable reason – **'why the natural penicillins'** are *not quite effective* when administered by mouth.

   Important Points – However, the aforesaid *serious* **'drawback'** has been overcome partially by the critical introduction of an **'*electron attracting moiety*'** – specifically in the **α-position** in the **amide (-CO–NH) side chain** as could be seen in **Penicillin V**, which being *more acid-resisting in nature*:

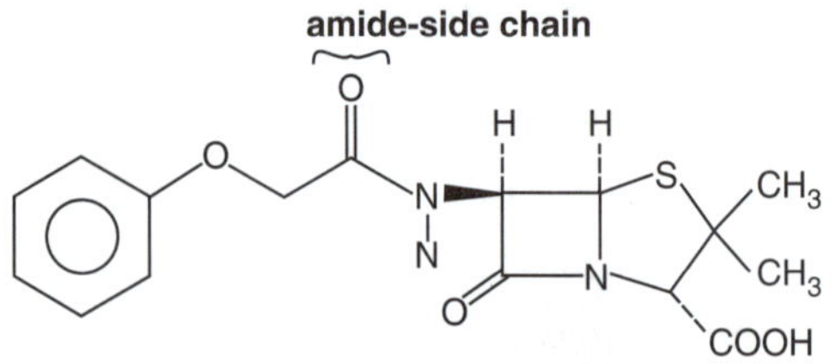

amide-side chain

Penicillin V

❑ Besides, the **gastric juices** (*stomach acids*) do contain the **enzymes penicillinases**, which subsequently cause *hydrolysis* of **the penicillins** thereby giving rise to the cessetion of β-**lactam ring system** followed by the *loss of activity*.

Activity Profile of the Penicillins: These essentially comprise:

❑ **Active against *Gram*-positive strains;**

❑ **Inactive against *Gram*-negative strains;** and

❑ **Quite often *certain organisms* do develop *critical resistance* to the Penicillins**.

Side-Reactions of the Penicillins: These side-reactions of the *penicillins* may sometimes be referred to as the '**toxicity**' of the penicillins. It may however, be stated that **the penicillins** do exhibit relatively '*low toxicity*' profile *vis-a-vis* the '**sulpha-drugs**'. Importantly, **the penicillins** invariably causes:

- **Allergic reactions** and • **Diarrhoea.**

NOTE   **Hence, Penicillin, is usually injected ater a 'Test Prick' on a patient.**

How do the Penicillins exert their Physiological Activity? These essentially include:

- By causing inhibition of the so-called *metabolic functions* vital to the bacterial *cell-wall*;
- **Penicillins** could act as *bacteriostatic* or *bactericidal* which solely depends upon the actual '*age*' of the microorganism as well as the nature of the environment;
- The '**young fast-growing microbial-cells**' are indeed found to be definitely *more susceptible*; whereas, the respective **mature cells are more resistant to the penicillins;** and
- At a specific point when the *rate of multiplication of the microorganisms* reaches the maximum level, the *bacteriolytic activity* attains the **peak level**.

Constitution of Penicillins

The **constitution of penicillins** has been duly elucidated based on the basis of various scientific and logical explanations as stated below in a sequential manner:

1. **Molecular Formula:** On the basis of various analytical data obtained one may assign the general **molecular formula** of **penicillins** as $C_9H_{11}N_2O_4$ SR.

2. **Presence of 1 carboxyl (–COOH) moiety:** Since the **penicillins** yield *monosalts*, which ascertains that these essentially comprise **a carboxyl (–COOH) moiety.**

3. **Absence of either Free Amino (–NH$_2$) or Thiol (–SH) Moiety:** Various confirmation tests have shown that the **penicillins** do not contain either free **amino group** or **thiol moiety.**

4. **Hydrolysis of the Penicillins:** The **penicillins** on being subject to *hydrolysis* in the presence of *hot dilute mineral acids* (*viz.*, HCl, H_2SO_4) actually affords a loss of one C-atom in the form of carbon dioxide (CO_2); besides, they are duly *degraded* to the **equimolar proportion** of:

- **an amine:** *Penicillamine*; and
- **an aldelyde:** Penilloaldelyde.

Importantly, almost all the **penicillins** get adequately degraded ultimately to the • **same amine;** but • **different adelydes,** since the *retrieved fragment* resides in the **aldelyde portion.** Thus, we may have:

$$C_9H_{11}N_2O_4SR + 2H_2O \longrightarrow CO_2\uparrow + C_5H_{11}NO_2S + C_3H_4NO_2\widetilde{R}$$

| | Carbon dioxide | Pericillamine | Penillo-aldehydes |

Fragment '*R*' – Since, the **fragment '*R*'** very much resides in the *aldelyde* portion of most **penicillins** – it yields the **same amine** logically, but certainly *different* 'aldehydes'.

5. **Probable Structure of Penicillamine [$C_5H_{11}NO_2S$]:** The probable structure of **penicillamine** may be elucidated on the basis of the various sequential underlying facts:

   (i) **Molecular Formula** – Based on various analytical data, the molecular formula of penicillamine was found to be $C_5H_{11}NO_2S$.

   (ii) **Presence of Thiol (–SH) Moiety** – Since **penicillamine** responds to the specific **indigo colour reaction** with *ferric chloride [FeCl$_3$]* solution and also gives **colour reaction** with **sodium nitroprusside**, $Na_2[Fe(CN)_5 NO]$, which indicates that **D-penicillamine** essentially comprises a **thiol (–SH) moiety**.

 NOTE In fact, these colour reactions do reveal that D-penicillamine is presumably a substituted cysteine (an amino acid).

   Besides, the *oxidation* product of **penicillamine** also shows the presence of a **thiol (–SH) group**.

   (iii) **Three pKa values Given by D-Penicillamine:** Based on the **'Electrometric Titration'** it has been ascertained that **D-penicillamine** essentially shows *three* **distinct pKa values: 1.8, 7.9 and 1.05** that critically corresponds to:

   • **Carboxylic** • **α-Amino Acid and** • **Thiol Moiety** – respectively.

   Hence, the presence of these *three* **functional groups** do reveals that **D-penicillamine** is most probably a **substituted cystein.**

   (iv) **D-Penicillamine Gives Similar Chemical Reactions to Cysteine:** It has been duly observed that **D-penicillamine** gives almost similar chemical reactions to **cysteine**.

Example: These essentially include:

   1. Both **D-penicillamine and cysteine** react with *acetone* to give an **isopropylidene derivative**. However, the latter does not contain either a **free amino (–NH$_2$) group** or a **thiol (–SH) group**; and hence, is reconverted to **D-penicillamine** on *hydrolysis*. Therefore, these aforesaid reactions do reveal that these *two moieties* (*viz.*, **–NH$_2$** and **–SH**) are intimately attached to the so-called **adjacent C-atoms**.

   2. Furthermore, oxidation of **D-penicillamine** with *bromine* (Br_2) *water* gives rise to the formation of **sulphonic acid** (*e.g.*, **benzene sulphonic acid: C_6H_5–SO_3H**), which reaction is characteristic of a **'thiol'**.

   (v) **Kuhn-Roth Method for the Determination of Methyl Side-Chains to D-Penicillamine:** The *methyl side-chain* in **D-penicillamine** when determined by **Kuhn-Roth method** – it yields an extremely *low value* (~ **0.2 moles**), which eventually reveals that the former essentially consists of an **'ispropyl end-group'** and *not a methyl end-group*.

   Comment: Based on the foregoing evidences – it follows that the **D-penicillamine** has a *β, β-dimethylcysteine structure*, that has been duly ascertained by its **synthesis** starting from **DL-valine** and **chloroacetylchloride** as given under:

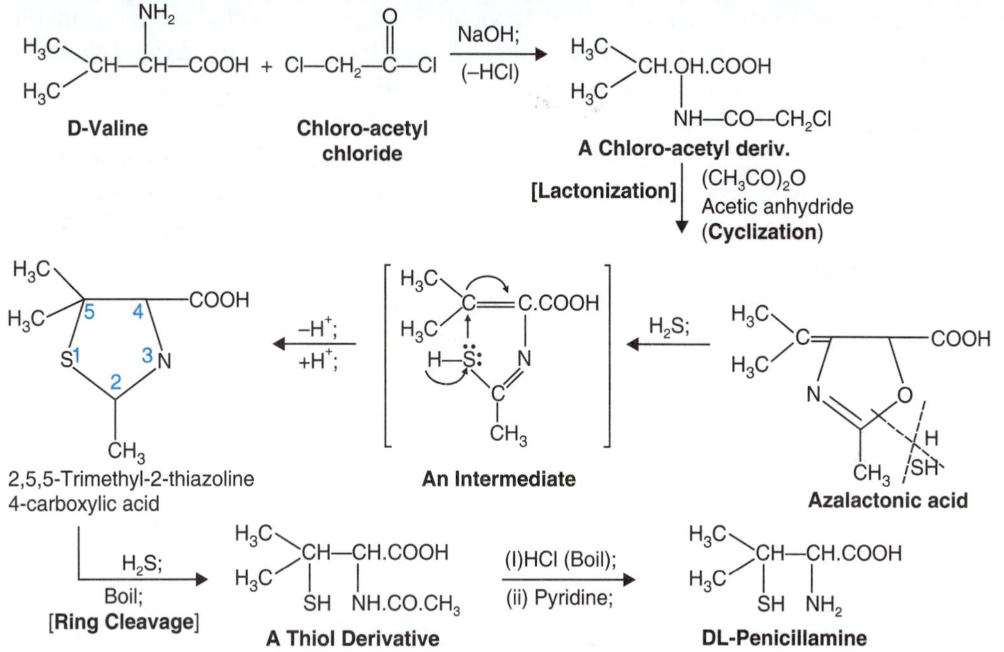

H₃C–CH–CH–COOH + Cl–CH₂–C–Cl $\xrightarrow[(-HCl)]{NaOH;}$ H₃C–CH.OH.COOH

D-Valine   **Chloro-acetyl chloride**   **A Chloro-acetyl deriv.**

NH₂ (on D-Valine); O (on chloroacetyl chloride); NH–CO–CH₂Cl (on product)

[Lactonization] $\downarrow$ $(CH_3CO)_2O$ Acetic anhydride **(Cyclization)**

2,5,5-Trimethyl-2-thiazoline 4-carboxylic acid $\xleftarrow[+H^+;]{-H^+;}$ **An Intermediate** $\xleftarrow{H_2S;}$ **Azalactonic acid**

| | $\downarrow$ H₂S; Boil; [Ring Cleavage] |

H₃C–CH–CH.COOH (SH NH.CO.CH₃) **A Thiol Derivative** $\xrightarrow[(ii) Pyridine;]{(I)HCl (Boil);}$ H₃C–CH–CH.COOH (SH NH₂) **DL-Penicillamine**

Explanation: The various steps involved above may be explained as under:

1. The interaction between **D-valine** and **chloroacetylchloride** in the presence of NaOH, loses a mole of HCl and yields a chloroacetyl derivative, which on treatment with acetic anhydride undergoes *lactonization* (or cyclization) to give **azalactone.**

2. The **azalactone** on reaction with H_2S–*gas* gives an **intermediate**, which on treatment with an acid yields **2,5,5-trimethyl-2-thiazoline 4-carboxylic acid**.

3. The resulting product on further treatment with H_2S followed by boiling causes the *ring cleavage* to give a ***thiol*-derivative**, which on boiling with HCl followed by *pyridine* yields the desired **DL-penicillamine** in a *dl*-mixture.

Resolution of the Racemic (±) Mixture of DL-Penicillamine – It is accomplished by treatment with *formic acid* (HCOOH) to obtain the respective **formyl derivative**, which is subsequently resolved by the aid of brucine (an **alkaloid**); and ultimately the *formyl moiety* is removed by **hydrolysis** – as given below:

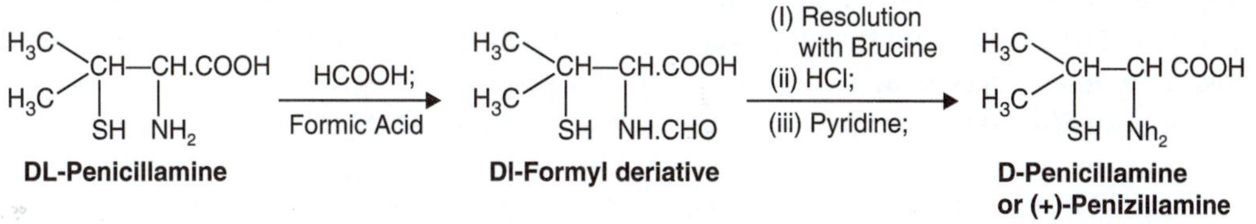

H₃C–CH–CH.COOH (SH NH₂) **DL-Penicillamine** $\xrightarrow{HCOOH; Formic Acid}$ H₃C–CH–CH.COOH (SH NH.CHO) **Dl-Formyl deriative** $\xrightarrow[(iii) Pyridine;]{(I) Resolution with Brucine \\ (ii) HCl;}$ H₃C–CH–CH COOH (SH NH₂) **D-Penicillamine or (+)-Penizillamine**

Comments: The **D-Penicillamine** as obtained from the aforesaid means of resolution of the *racemic mixture* has been found to be quite identical with the **natural penicillamine.**

Diazomethane Treatment of Penicillin: It has been duly established that when **penicillin** is treated with **diazomethane** [$CH_2=N^+=N^-$] it gets converted into the **methylester**, which on being treated with an aqueous solution of **mercuric chloride ($HgCl_2$)** yields the respective **methyl ester of penicillamine**. Therefore, these above cited reactions do indicate vehemently that the **carboxyl (–COOH) moiety** present in penicillamine happens to be the **carboxyl (–COOH) moiety** present duly in the penicillin molecule itself.

6. **Chemical Structure of Penilloaldehyde:** The **chemical structure of penilloaldehyde** has been duly elucidated based upon the following *four* cardinal scientific evidences, namely:

   (i) **Molecular Formula** – Based on the various *analytical data* the **molecular formula** of **penilloaldehyde** has been found to be $C_3H_4NO_2R$.

   (ii) **Penilloaldehyde Hydrolysis** – When **penilloaldehyde** is subject to *vigorous hydrolysis* – it yields a *substituted acetic acid* and an *aminoacetaldehyde*, as given under:

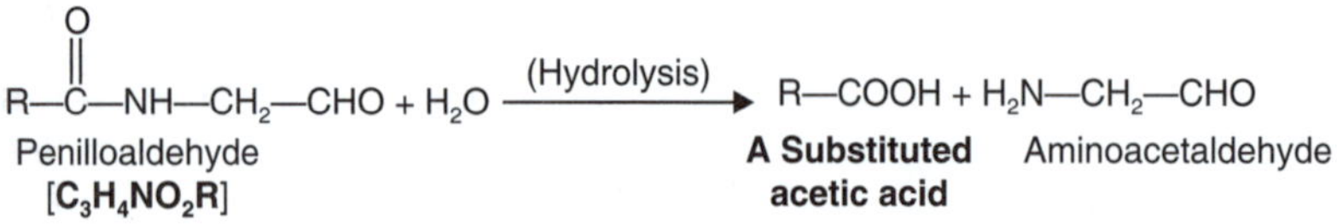

Penilloaldehyde $[C_3H_4NO_2R]$ + H_2O → (Hydrolysis) → R—COOH **A Substituted acetic acid** + H_2N—CH_2—CHO Aminoacetaldehyde

   (iii) **Synthesis of Penilloaldehyde** – It may be duly confirmed by its synthesis as shown below:

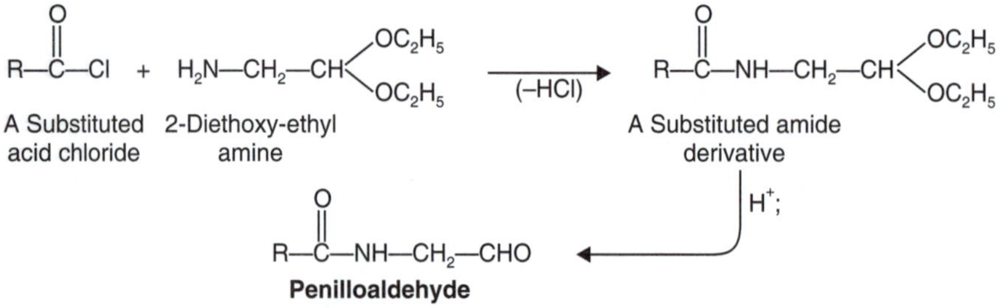

A Substituted acid chloride + 2-Diethoxy-ethyl amine → (–HCl) → A Substituted amide derivative → H^+; → **Penilloaldehyde**

That is, the reaction between a *substituted acid chloride* and *2-diethoxyethyl amine* loses a mole of HCl to yield a **substituted amide derivative**, which on exposure to an acidic environment gives the desired product **penilloaldehyde**.

   (iv) **Formation of a CO_2-Molecule:** In item (4) under **constitution of the penicillin**, it has already been stated that the respective *acid hydrolysis of penicillin* yields a mole each of:
   - **Penicillamine**
   - **Penilloaldehyde** and
   - **Carbon dioxide**

Importantly, the critical formation of a mole of **carbon dioxide (CO_2)** vehemently suggest that certain *highly unstable acid* gets duly generated as an 'intermediate', – which subsequently undergoes **rapid decarboxylation** (*i.e., loss of CO_2*), to produce a mole of **carbon dioxide**. However, such an 'acid' could be a β-keto acid; and, therefore, the formation of the so-called penaldic acid *i.e.*, the **penilloaldehyde-carboxylic acid** must have been formed as an 'intermediate' in the course of **hydrolysis of penicillin**. Thus, we may have the following expression:

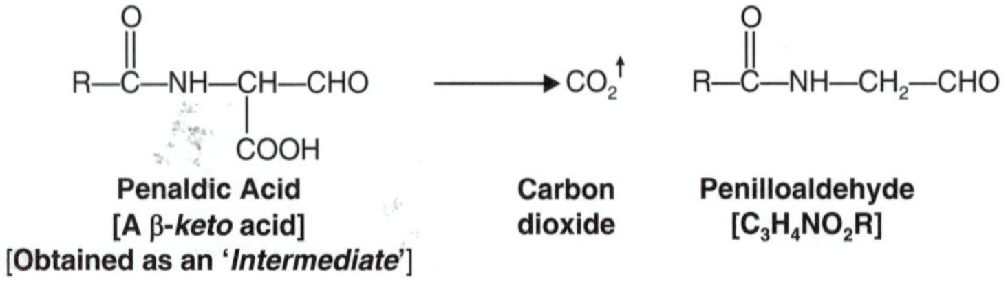

Penaldic Acid [A β-*keto* acid] [Obtained as an '*Intermediate*'] → $CO_2\uparrow$ **Carbon dioxide** **Penilloaldehyde** $[C_3H_4NO_2R]$

7. **Mode of Linkage Between Penicillamine and Penilloaldehyde:** Importantly, at this point in time the most *critical* and *vital* issue comes into being with respect to the exact **mode of linkage between penicillamine and penilloaldehyde** in the molecule of '*penicillin*' in order that it may be explained explicitly the formation of **Penaldic Acid** (*i.e.*, a *keto-acid*), as given in (6) above. This eventually expatiates clearly the formation of CO_2. Thus, one may scientifically explain the linkage of these *fragments and* establish the aforesaid dictum.

(a) **Hydrolysis of Penicillin** – The penicillin on being hydrolyzed carefully either:
   - **using the enzyme penicillinase;** or   •   **with diluted alkali,**

   it yields **penicilloic acid** (a *dicarboxylic acid*), and hence, eliminates rapidly a mole of **carbon dioxide (CO_2)** to produce **penilloic** acid (a *monocarboxylic acid*).

   Comments: It indicates strongly that in **penicilloic acid** *one of the* **carboxyl (–COOH) groups** is located strategically at the β-**position** with respect to an **electron-attracting moiety**.

(b) **Probable Structure of Penilloic Acid** – The chemical structure of **penilloic acid** has been duly proved and established on the basis of the fact that when it is subject to hydrolysis with an aqueous solution of $HgCl_2$, it yields *two* products, namely:
   - **Penicillamine** and   •   **Penilloaldehyde.**

Inference: In fact, this kind of **hydrolysis** is characteristic of, such compounds essentially possessing a '**thiazolidine ring system**'. However, the presence of the so called **thiazolidine nucleus in penicillin** has been duly ascertained by the fact that it possesses: neither:
   - **a free amino (–NH$_2$) moiety;** nor
   - **a free thiol (–SH) moiety.**

Therefore, one may probably assign the following structure for **penilloic acid (I)**, since it would ultimately lead to the production of the designed products, as given under:

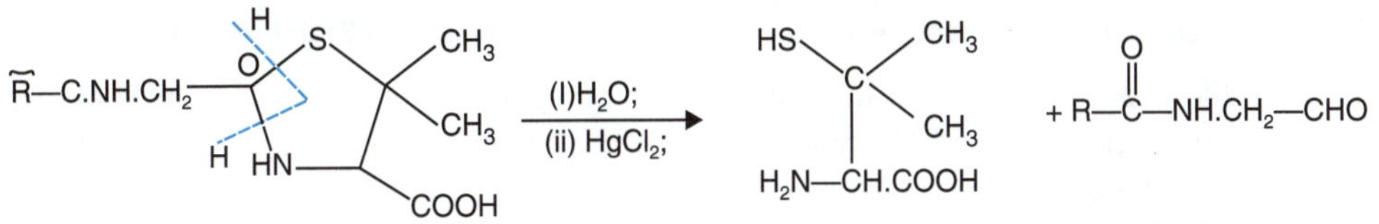

Penilloic Acid (I)   Penicillamine   Penilloaldehyde

Penilloic acid (I) on *hydrolysis* in the presence of *mercuric chloride* yields a mole each of **penicillamine** and **penilloaldehyde**.

Now, if **(I)** is the structure of **penilloic acid**, one may assign **(II)** as the probable structure of **penicilloic acid**.

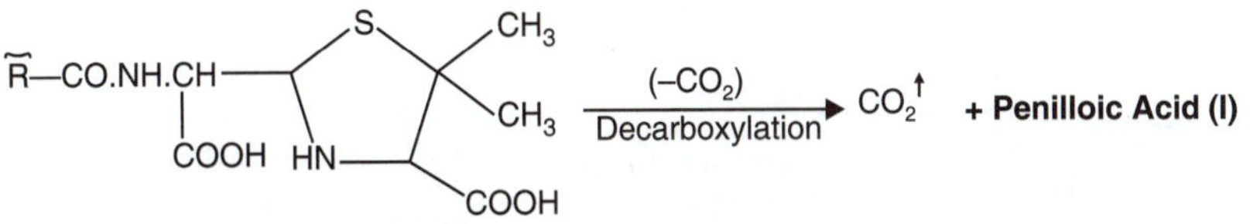

Penicilloic Acid (II)

Confirmation of Penicilloic Acid (II) Structure: Importantly, **the penicilloic acid (II)** structure may be duly confirmed by the underlying fact that **penicillin** on being treated with methanol gives rise to the formation of **methyl penicilloate (B)**, which on *hydrolysis* with **aqueous mercuric chloride (Hg Cl₂)** yields *two* different products, namely:

- **Methyl penaldate (C)** and • **Penicillamine (D)**

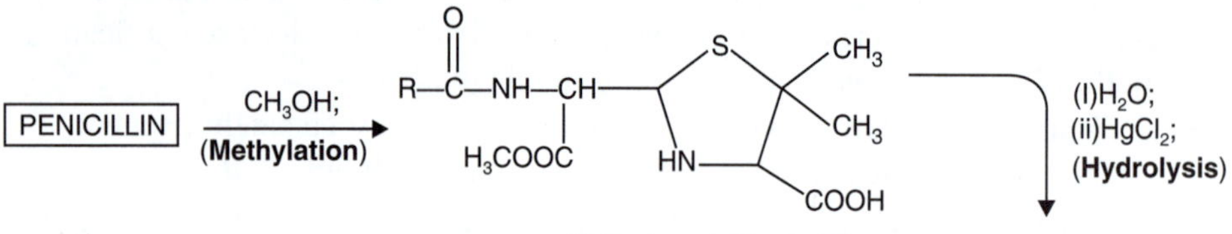

Methyl penicilloate (B)

Methyl Penaldate (C)   **Penieillamine (D)**

8. **Structure of Penicillin:** Based on the above mentioned scientific evidences and explanations *two* probable structures:

- **one based on the Oxazolone Structure (III); and**
- **second based on the β-Lactam Structure (IV),**

were proposed, as given below:

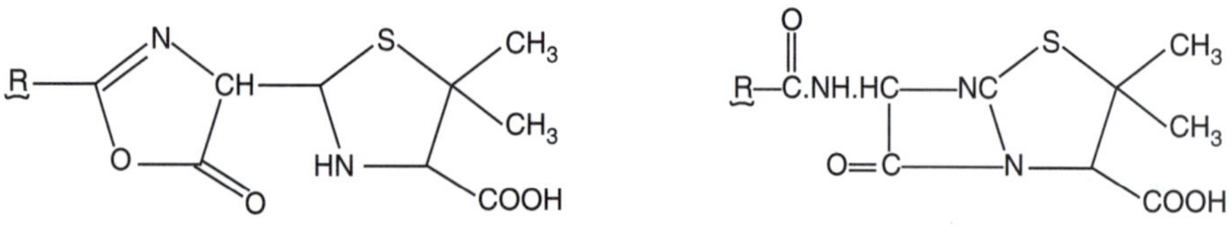

Oxazolone Structure (III)   **β-Lactum Structure (IV)**

However, the treatment of penicillin with dilute mineral acid affords a **molecular rearrangement** promptly to give **penillic acid (A)**; and, therefore, the ensuing *chemical evidence* could not decide with regard to the correct structure either as **(III)** or **(IV)** of **penicillin**.

9. **Definite Evidence of β-Lactam Ring in Penicillin Structure (IV):** This could be accomplished by *desulphurization* of the **benzylpenicillin** with **Raney–Ni** to produce **desthiobenzylpenicillin (V),** which upon *hydrolysis* in an **acidic medium (H⁺)** yields **desthiobenzylpenicilloic acid (VI)**. Subsequently, the **product (V)** thus obtained on being boiled with *benzylamine in dioxane* solution gives rise to the production of the **dibenzylamide of desthiobenzylpenicilloic acid (VII)**. Thus, we may express the following sequence of reactions:

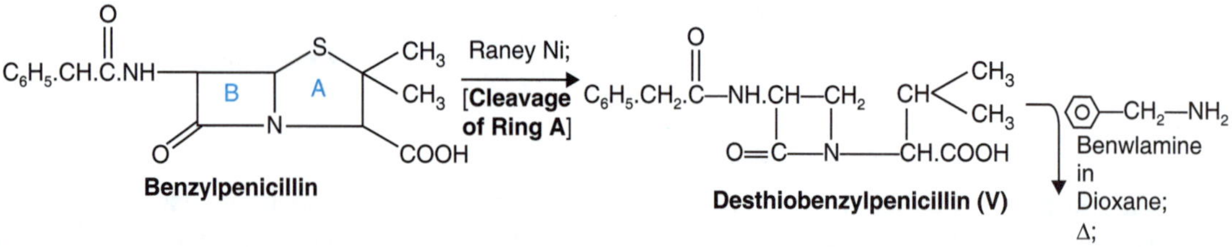

Benzylpenicillin   **Desthiobenzylpenicillin (V)**

Contd.....

H$^+$;

(Hydrolysis)

C$_6$H$_5$—CH$_2$CO.NH.CH—CH$_2$ CH$<$CH$_3$/CH$_3$

HOOC NH——CH.COOH

Desthibenzylpinicilloic Acid
(VI)
[Due to cleavage of Ring B]

C$_6$H$_5$—CH$_2$.CO.NH.CH——CH$_2$$<CH_3$/CH$_3$

C$_6$H$_5$.CH$_2$.NH.C=O NH.CH.COOH

Dibenzylamide of desthiobenzyl-
penicilloic acid (VII)

NOTE Importantly, one may also obtain the compounds (VI) and (VII) by performing carefully the *desulphurization* of *benzylpenicilloic acid* with Raney-Ni and subsequently to its dibenzylamide derivative respectively.

10. **More Concrete Evidences Based on Physico-Chemical Methods:** At this point in time, it could not be possible to decide for certainty which of the *two* **possible structures of penicillins** (*i.e.*, **III or IV**) to be the *correct* – based on the **chemical evidences** alone, which might be due to the **extreme sensitive inherent features of penicillin**, since it undergoes **molecular rearrangement** on treatment with *dilute mineral acid* to give **penillic acid (A)** (*see section-8 also*):

Thus, we may have the following expression:

PENICILLIN $\xrightarrow{\text{HCl;}}$

HOOC—

S—CH$_3$/CH$_3$

N N

COOH

R

Penillic Acid (A)

The *two* most suitable *physico-chemical methods of analysis,* namely:

*   **IR-Spectral Studies** and
*   **X-ray Diffraction Studies,**

    have been employed to ascertain precisely whether the proposed **β-Lactam Structure (IV)** or the **Oxazolone Structure (III)** is the correct structure for the **penicillin** molecule.

10.1. IR–Spectral Studies

The **IR–spectral studies** of the **methyl ester and sodium salt of Benzylpenicillin**, that essentially comprised various important segments of **structures (III) and (IV)** have been recorded duly as given below; and from this one could easily establish an acceptable and logical correlation between:

*   **different bands** and
*   **functional moieties,**

that are present in the penicillin molecule.

Nevertheless, the various observations and findings could be understood by taking into consideration the **methyl ester and sodium salt of benzylpenicillin** that might exhibit explicitly the *different observed maxima* (*i.e., the characteristic features of practically all the penicillins in these regions–as depicted in the above **tabular presentation***).

Infra-Red Maxima of Methyl Ester and Sodium Salt of Benzylpenicillin

IR-Maxima of Methyl Ester (cm^{-1}) of Benzylpenicillin	Assignment Given	IR-Maxima of Sodium Salt (cm^{-1}) of Benzylpenicillin	Assignment Given
3333	NH Group (str.)	3333	NH Group (str.)
1770	β-Lactam Ring	1770	β-Lactam Ring
1748	Carbonyl (> C== 0) (str.)	1613	Carbonyl (>e=0 (str.)
1684 ⎱	The Secondary	1681 ⎱	The Secondary-
1506 ⎰	–amide structure	1515 ⎰	–amide structure

Explanation of the Recorded IR-Spectra: Based on the aforesaid **IR-spectral data** pertaining to:

* **methylester of Benzylpenicillin**, and
* **sodium salt of Benzylpenicillin**,

the following vital and important *conclusions* may be drawn:

(i)   IR-spectral band at **3333 cm^{-1}** in the *two compounds* stated above may be duly assigned to the **imino (–NH–) moiety** (str.)

(ii) The *two observed* bands at:
* **1748 cm^{-1}** – of the *ester*, and   • **1623 cm^{-1}** – of the *salt*,

have been assigned to the respective **carbonyl (>C==0) moiety** (str.) duly present in the **carboxyl (–COOH) moiety** (as *ester* or *salt*).

(iii) Interestingly, the **IR-spectra** of the '*model oxazolones*' are duly recorded which categorically show *two* **typical characteristic bands, namely:**
* **1825 cm^{-1}** – *due to the carbonyl (>C==0) group*, and
* **1675 cm^{-1}** – *due to cyano (or nitrile) (–CN) group.*

However, the **former band** (*i.e.,* at 1825 cm^{-1}) is totally absent, but the latter band (*i.e.,* at 1675 cm^{-1}) is distinctly present in the **benzylpenicillin derivatives**.

Obviously, this fails to provide any definite clue in the **correct assignment** either to the proposed **structure (III)** or **structure (IV)** for '**penicillin**'.

Critical IR-Spectral Investigative Studies – Therefore, when the **IR-spectral determinations** of a good number of '**thiazolidines**' are being *thoroughly investigated* particularly in the region of **double band** down to 1470 cm^{-1}, one could only find the bands at ~ **1748 cm^{-1} and 1613 cm^{-1}** due to the presence of **carbonyl (>C==0) bond**. Besides, the IR-spectra of an array of '**amides**' are also recorded. Hence, one may clearly observe a band at ~ **1670 cm^{-1}**, which may be duly attributed to the respective **carbonyl (>C==0) moiety**.

Furthermore, the presence of the *primary*-**amides** may also exhibit a critical band at ~ **1613 cm^{-1}**; and the so-called *secondary*-**amides** do also show a distinct band at ~ 1515 cm^{-1}.

Comments: The observed **IR-spectral data** predominantly *indicate* the presence of a *secondary*-**amide (–NH–CO–)** strucutre for the 'penicillin' *i.e.,* the assigned *structure* (IV) (**or the β-lactam structure for penicillin**), which fact could be further substantiated due to:

* **Secondary amide band** at 1670 cm^{-1} (**1684**), and **1681 cm^{-1};**
* *band* at **1515 cm^{-1}** (**==1506**), and **1515 cm^{-1}.**

In this manner, out of the *five* **IR-bands** cited above, at least *four* **bands** have been expatiated logically. Thus, we may have:

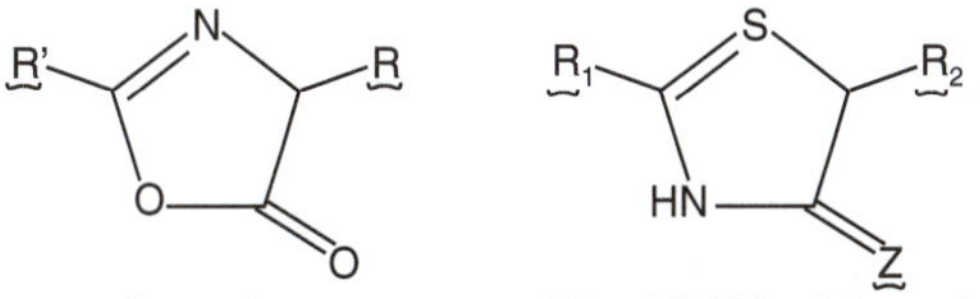

Oxazolone   [Z=COOMe, CO₂ etc] Thiazolidines

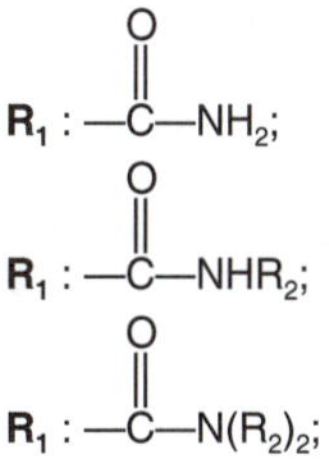

$R_1 : -\overset{\overset{\displaystyle O}{\|}}{C}-NH_2;$

$R_1 : -\overset{\overset{\displaystyle O}{\|}}{C}-NHR_2;$

$R_1 : -\overset{\overset{\displaystyle O}{\|}}{C}-N(R_2)_2;$

(iv) Ultimately, an attempt was duly made to determine and record the **IR-spectra** of several compounds.

- **β-Lactams and** • **Fused Thiazolidine-β-lactams.**

Observation: The **β-lactams** fail to show a band at ~1770 cm⁻¹; whereas, the **fused thiazolidine - β-lactams** exhibit a distinct band at ~**1770 cm⁻¹**; and this vehemently explains the *fifth band* thereby ascertaining the fact **structure (IV)** *i.e.*, the **β-lactam structure** is the **most probable** and **correct** structure for **penicillins**.

10.2. X-Ray Diffraction Studies [or XRD-Studies]

X-ray chromatographic radiation emitted from an atom when an orbital electron changes to a lower energy level. Hence, the resulting *wavelengths* are **longer** in comparison to *Gamma*-rays and *shorter* than **ultraviolet** light rays and falls very much within the range of **10⁻⁸ meter**.

The elucidation of the structure of an array of **naturally occurring products** *e.g.*, **benzylpenicillin** (*antibiotic*), **semi-synthetic drug substances** are exclusively determined by **XRD-studies** in their respective powder form. Based on the underlying fact that the observed **XRD-pattern** seems to be absolutely unique for each and every crystal sample; and, therefore, provides a typical *'fingerprint identification'* of each pure sample could be possible.

Importantly, when the **XRD-analysis** of the *sodium, potassium,* and *rubidium salts of benzylpeni-cillin* is performed, it reveals the presence of a *β-lactam ring system* thereby supporting the structure (IV) of penicillin, as proposed earlier.

Various chemical reactions may now be formulated squarely as shown under:

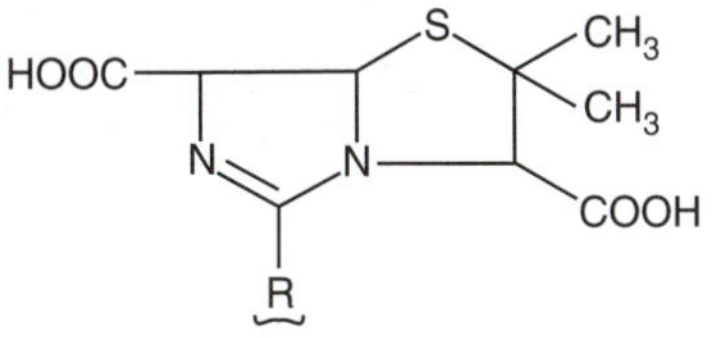

Penillic Acid (A)

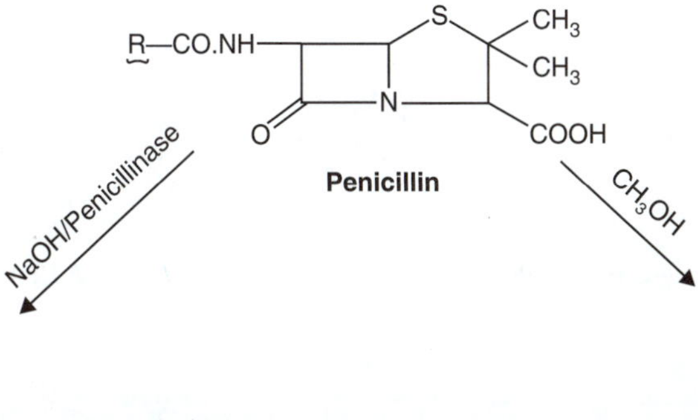

Penicillin

(Contd.....)

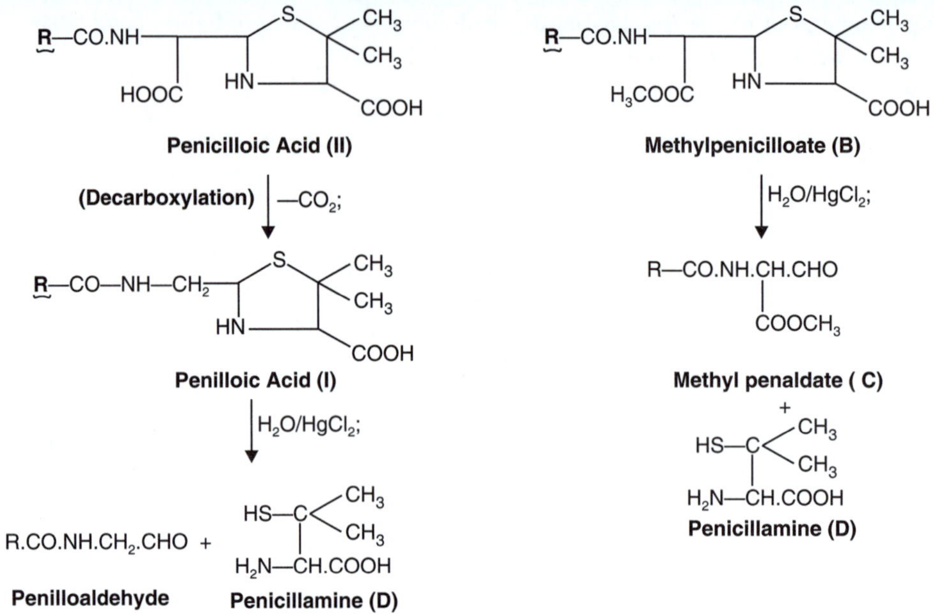

Penicilloic Acid (II)

(Decarboxylation) | $-CO_2$;

Penilloic Acid (I)

| $H_2O/HgCl_2$;

$R.CO.NH.CH_2.CHO$ +

Penilloaldehyde

Penicillamine (D)

Methylpenicilloate (B)

| $H_2O/HgCl_2$;

$R-CO.NH.CH.CHO$
$|$
$COOCH_3$

Methyl penaldate (C)

Penicillamine (D)

11. Synthesis of Penicillins

The *total synthesis* of **Penicillin** may be accomplished by any *one* of the *three* reported procedures, namely:

  (a) **From Methylphenaceturate:** Vigneaud *et al.* (1946);
  (b) **From DL-Valine and Chloroacetyl Chloride:** Sheehan *et al.* (1957); and
  (c) **From Compound (X)** [*obtained in (b)*]: Sheehan *et al.* (1962),

which shall now be treated individually in the sections that follows:

11.1 Synthesis of Penicillin from Methyl Phenaceturate [Vigneaud *et al.* (1946)]

The various sequential steps involved in the synthesis are as enumerated under:

Part-a

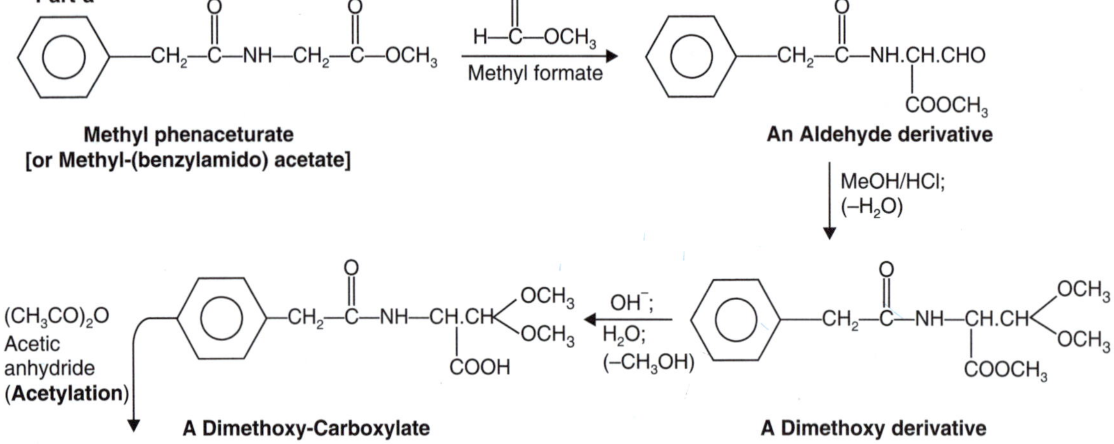

Methyl phenaceturate
[or Methyl-(benzylamido) acetate]

An Aldehyde derivative

| MeOH/HCl;
  $(-H_2O)$

A Dimethoxy-Carboxylate

A Dimethoxy derivative

$(CH_3CO)_2O$
Acetic
anhydride
(Acetylation)

Contd.....

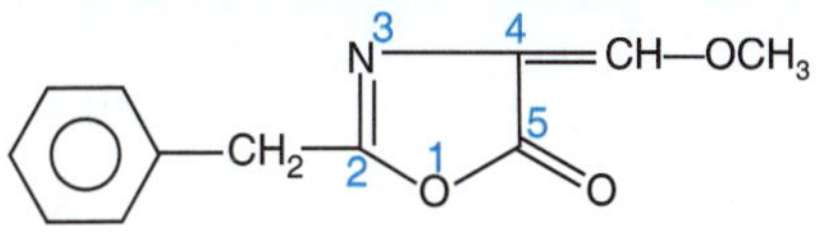

**2–Benzyl-4-methoxy-
methylene-5-oxazolone
(VIII)**

Part–b :

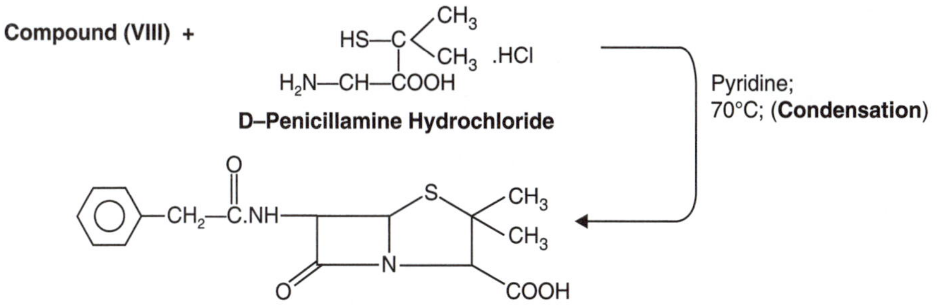

Compound (VIII) +

D–Penicillamine Hydrochloride

Pyridine;
70°C; **(Condensation)**

**Benzyl penicillin
(Yield <0.1%)**

Explanation: The various steps involved above may be explained as under:

1. In **Part-a**, Methyl phenaceturate (*i.e.*, methyl-benzylamido acetate) is made to react with *methyl formate* to yield an *aldehyde derivative*, which upon dehydration on treatment with *acidified methanol* gives a **dimethoxy derivative**.

2. The resulting product on *hydrolysis in a mild alkaline medium* gives a **dimethoxy carboxylate** with the loss of a mole of *methonal* which upon acetylation undergoes cyclization to yield **2-benzyl 4-methoxy-methylene-5-oxazolone (VIII)**.

3. In **Part-b**, the resulting **product (VIII)** is made to react with **D-penicillamine hydrochloride** in **pyridine** at an elevated temperature of *70°C* to afford a *condensation reaction* to yield **benzylpenicillin** (having a yield *less than 0.1%*).

NOTE	Importantly, the '*penicillin*' synthesized by the method suggested by Vineaud *et al.* (1964) possesses a physiological activity almost 50% to that of the *natural product*.

11.2. Synthesis of Phenoxymethyl penicillin [or Penicillin V] from DV-Valine and Chloroacetylchloride [Sheehan *et al.* (1957)]

The above synthesis is accomplished in *three* **different sequential steps**, as given under:

(a) DL-Penicillamine from DL-valine and Chloroacetyl chloride:

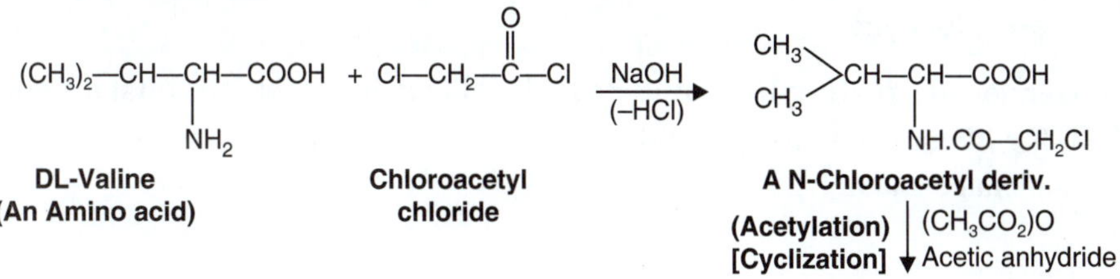

**DL-Valine
(An Amino acid)**

**Chloroacetyl
chloride**

A N-Chloroacetyl deriv.

(Acetylation) $\Big|$ $(CH_3CO_2)O$
[Cyclization] $\downarrow$ Acetic anhydride

Contd.....

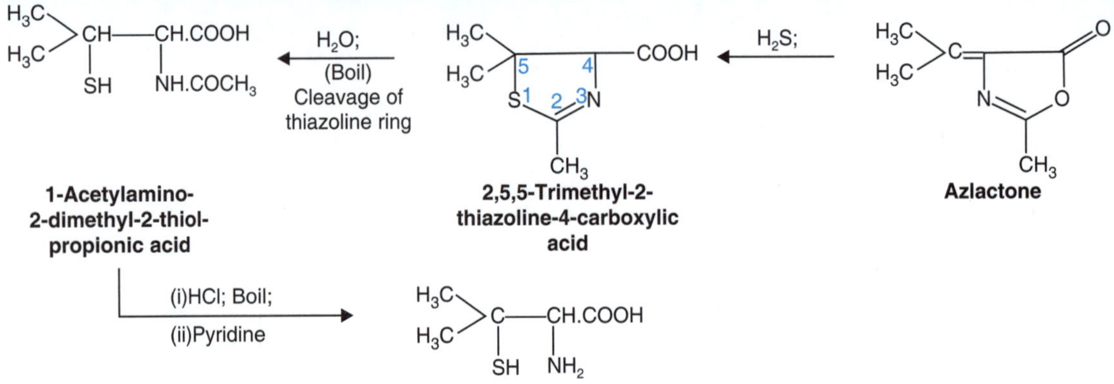

The reaction scheme (top of page, left to right):

1-Acetylamino-2-dimethyl-2-thiol-propionic acid

$$\underset{\substack{H_3C\\H_3C}}{>}CH\text{---}CH.COOH$$ with SH and NH.COCH$_3$ groups

$\xleftarrow{\begin{array}{c}H_2O;\\(Boil)\\ \text{Cleavage of}\\ \text{thiazoline ring}\end{array}}$

2,5,5-Trimethyl-2-thiazoline-4-carboxylic acid

$\xleftarrow{H_2S;}$

Azlactone

Below:

$\xrightarrow{\begin{array}{c}(i)HCl;\ Boil;\\(ii)Pyridine\end{array}}$

$$\underset{\substack{H_3C\\H_3C}}{>}C\text{---}CH.COOH$$ with SH and NH$_2$ groups

DL-Penicillamine

Explanation:

1. Interaction between **DL-valine** and **chloroacetyl chloride** in an *alkaline medium* yields a **N-chloroacetyl derivative** (with the loss of a mole of HCl), which on acetylation with *acetic anhydride* undergoes cyclization to give **azlactone**.

2. The resulting product when treated with H$_2$S forms **2, 5, 5-tri-methyl-2-thiazoline-4–carboxylic acid**, which on boiling with water causes a cleavage of **thiazoline ring** to produce **1-acetyl-amino-2-dimethyl-2-thiol propionic acid**.

3. The end-product when treated with *HCl/Boil/*followed by *pyridine* dissolution yields the desired racemic **DL-penicillamine**.

   (b) *tert-*Butylphthalimidomalonaldehyde **(X)** *via Gabriel Synthesis:* This synthesis commences from the starting material phthalimide as stated under:

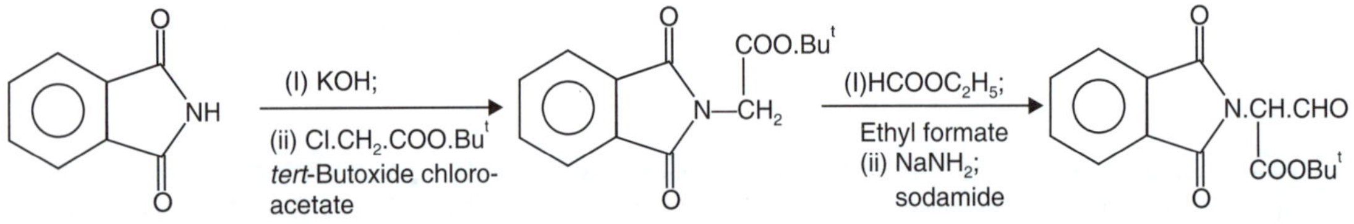

Phthalimide

$\xrightarrow{\begin{array}{c}(I)\ KOH;\\(ii)\ Cl.CH_2.COO.Bu^t\\ \textit{tert-}Butoxide\ chloro-\\acetate\end{array}}$

tert-Butoxide chloroacetate

$\xrightarrow{\begin{array}{c}(I)HCOOC_2H_5;\\ Ethyl\ formate\\(ii)\ NaNH_2;\\ sodamide\end{array}}$

tert-Butyl-phthalmido-malonaldehyde (X)

NOTE : All the aforesaid reactions take place at room temperature only when phthalimide is treated with an *alkali* and made to react with *tert-*butoxide chloroacetate it yields *tert-*butylphthalimido malonal, which on further treatment with *ethyl formate* followed by *sodamide* gives the desired product: tert-butylphthalimido-malonaldehyde.

   (c) **Conversion of** *tert–* **Butylphthalimido malonaldehyde (X) into the final product Penicillin (VII) [or Phenoxy-methyl penicillin]:** The various steps involved in the above synthesis are as enumerated under:

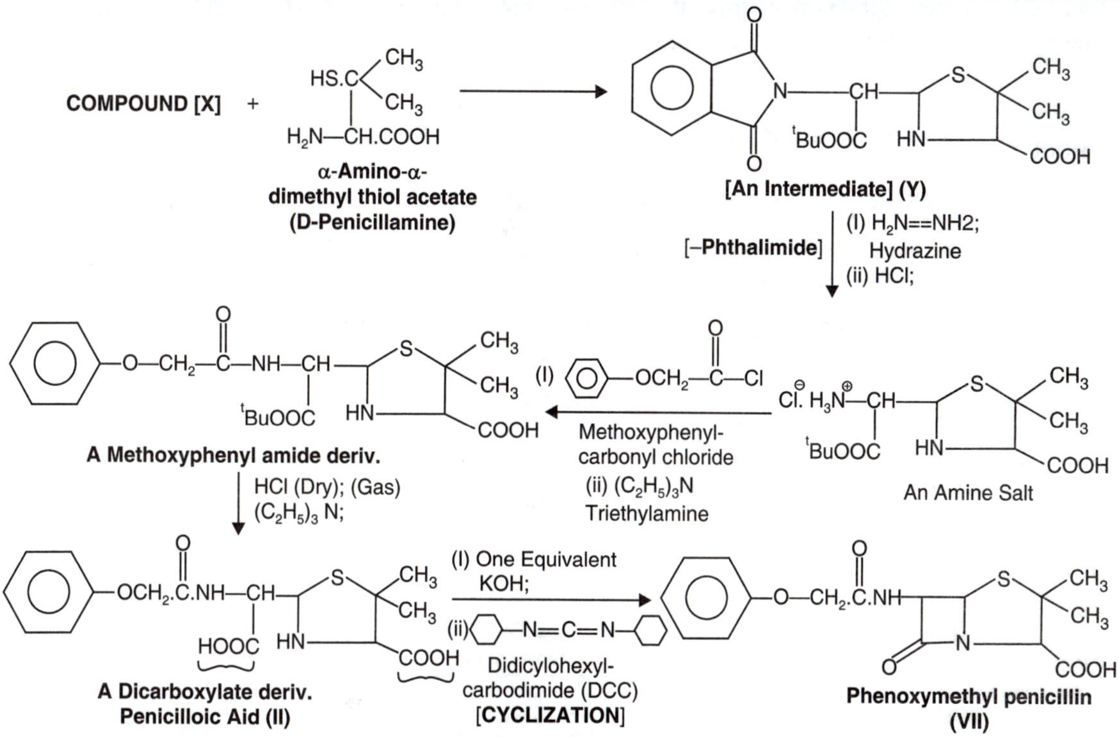

Explanation:

1. The **compound [X]** [*from step. (b)*] on being treated with *α-amino α-dimethyl thiol acetate* yields an intermediate **(Y)**, which on treatment with *hydrazine* followed by HCl gives an **amine salt**.

2. The resulting product on reaction with *methoxyphenyl-carbonyl chloride* followed by *triethylamine* gives a *methoxy-phenylamide derivative*, which on being treated with **dry HCl gas followed by triethylamine** yields a **dicarboxylate derivative (II)** or **penicilloic acid**.

3. Finally, the *resulting product* on being reacted with an equivalent of KOH, followed by treatment with **dicyclohexyl carbodimide** gives **phenoxymethyl penicillin (VII)**. due to the *cyclization* of the **β-lactam ring system**.

Further Deliberation on item (3) above

In the ultimate critical conversion of the respective **dicarboxylate derivative** to the **phenoxymethyl penicillin (VII)** – one may observe the *ring closure* due to the presence of **dicyclohexyl carbodimide**, which represents a *mild reagent*, that eventually forms the **amide bond**. In fact, the so-called **phthalimido moiety** has been duly utilized to protect the amino (–NH_2) group exclusively. Besides, one of the **carboxyl (–COOH) groups [Marked – A]** *i.e.*, the one involved in the *ring closure* – has been solely protected by the *tert*-butyl ester formation. Hydrazine ($H_2N = NH_2$) helps particularly to affect *two* changes:

- to leave the '*ester moiety*' **intact**, and
- to eliminate the rather bulky '*phthalimido group*'.

Cyclization *i.e.*, *the* final step has been performed on the corresponding '**K-salt**' of the **penicilloic**

acid (**II**). However, the **potassium salt of phenoxymethyl penicillin** (**VII**) could be duly purified by means of the *counter-current distribution* using

- **an isobutyl ketone**; and   • *two* successive **phosphate buffers**.

> **NOTE**   The final yield of *Phenoxymethyl (penicillin (VII)* is found to be *5.4%*.

11.3. Synthesis Proposed by Sheehan *et.al.* (1962)

The above synthesis of penicillins were intelligently improved upon by preparing 6-Amino penicillanic Acid from **Compound (Y)** – as obtained in (**C**) above.

The various steps involved in the above *synthesis* are as described under:

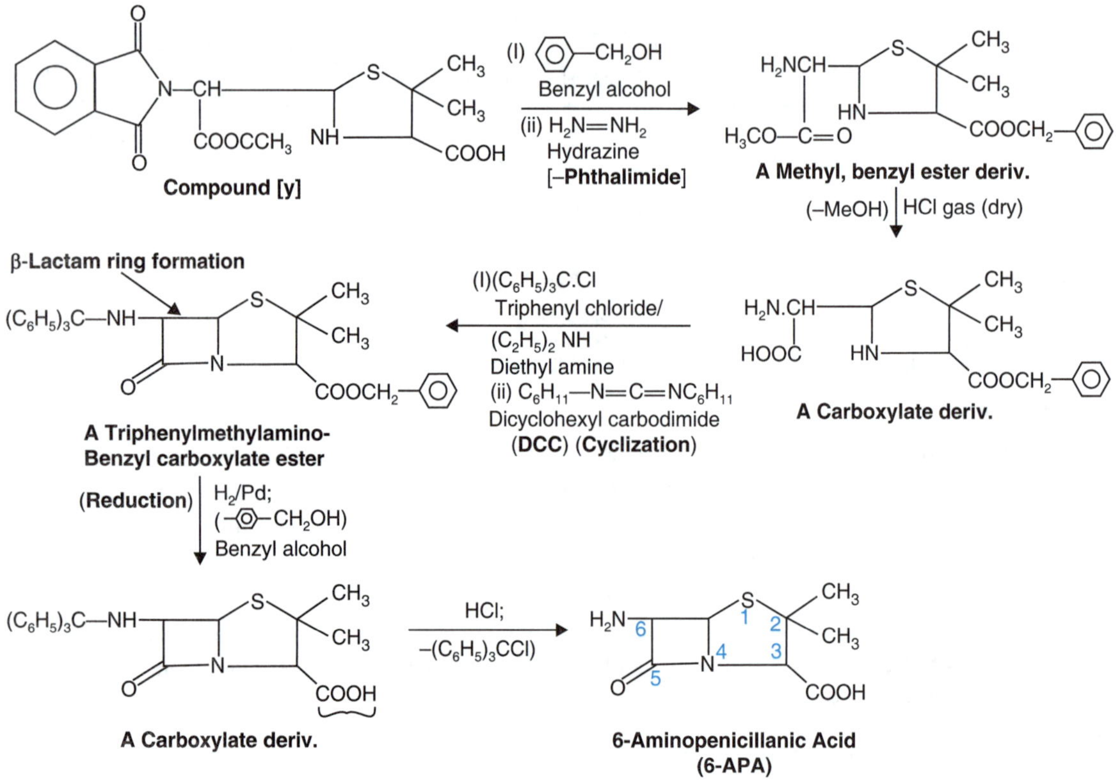

Compound [y]

A Methyl, benzyl ester deriv.

A Triphenylmethylamino-Benzyl carboxylate ester

β-Lactam ring formation

A Carboxylate deriv.

A Carboxylate deriv.

6-Aminopenicillanic Acid (6-APA)

Explanation:

1. The **compound (Y),** obtained as an '**intermediate**' in (e) above, when treated with *benzyl alcohol* followed by *hydrazine* gives a **methyl, benzyl ester derivative** with the loss of a mole of **phthalmide.**

2. The resulting product on treatment with *dry HCl gas* yields a **carboxylate derivative** and eliminating a mole of *methanol,* which on reaction with *triphenyl chloride/diethyl amine* followed by **dicyclohexylcarbodimide (DCC)** undergoes '*cyclization*' to produce a triphenylmethylamino – benzylcarboxylate ester (*i.e., a β-lactam ring is formed duly*).

3. Reduction of the end-product with H_2/Pd loses a mole of *benzyl alcohol* to yield a **carboxylate derivative,** which on treatment with HCl loses a mole of *triphenylmethyl chloride* to give **6-APA**.

THE PENICILLIN VARIANTS

Following are the *five* important **penicillin variants**, namely:

- **Penicillin G**;
- **Procaine Penicillin**;
- **Benzathine Penicillin G**;
- **Penicillin V** ; and
- **Staphicillin [or Methicillin Sodium]**,

which shall now be described in the sections that follows:

[A] Penicillin G [*Syn*: Benzylpenicillin; Benzylpenicillinic Acid; Penicillin II;]

Preamble: Penicillin G designates a '*natural penicillin*' and is regarded to be one of the most versatile and effective *antibiotic* for the critical therapeutic treatment of a plethora of *bacterial infections* in comparison to any *other antibiotic* (*except solely for such patents who are allergic to penicillin reaction*).

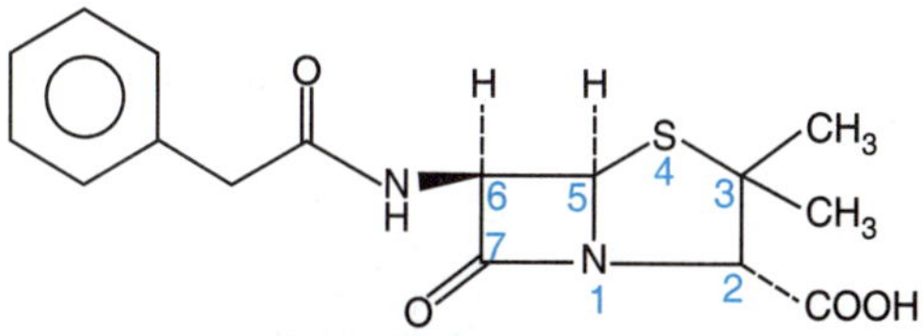

Penicillin G

(2S, 5R, 6R) -3,3-Dimethyl-7-oxo-6 [(phenylacetyl) amino]-4-thia-1 azobicyclo [3,2,O] heptane - 2-carboxylic acid;

Salient features:

1. It is reasonably a cost effective **antibiotic**.
2. Penicillin G is a *bacteriocidal drug* which is *active* against such **vital infections** caused due to:
   - **Pneumococcus**
   - **Streptococcus**
   - **Gonococcus**
   - **Treponema**
   - **Anthrax**
   - **Clostridia**
   - **Penicillinase Negative Staphylococcus and**
   - **Actinomycosis (fungus).**
3. In a broader sense, the action of **Penicillin G** is found against a large segment of *Gram*-**positive microorganisms** and relatively fewer *Gram*-**negative bacteria (cocci).**

Characteristic Features: These essentially include:

1. It is mostly obtained as the water-soluble **Na⁺ or K⁺-salts**
2. The aqueous solution (*water-for-injection*) is invariably used for both **IV/IM - injections**.
3. Interestingly, soonafter an IM-injection, the absorption is rather *rapid* and eventually ~ *60%* gets excreted in wine almost as such (*i.e.*, unchanged).
4. **Penicillin G**, when administered orally, gets destroyed partially by the *gastric juice* (HCl) present in the stomach.

NOTE Therefore, an excess dose ranging between *4–5 times the normal dose* is usually recommended when *Penicillin G* is to be given to a patient *via* oral route.

5. The **aqueous Penicillin G salt solutions** are fairly stable at a temperature of +10°C for a duration of *3-weeks without significant loss of activity*.

Caution: The clinical usage of *Penicillin G* has been largely confined due to the following two cardinal reasons:

- **allergic reactions (skin-rashes)**; and
- **analphylactic shock**.

[B] Procaine Penicillin G [*Syn* Penicillin G Procaine]

Penicillin G procaine (or **Procaine Penicillin G**) represents the *procaine salt of penicillin G*. It may be synthesized by the interaction of equimolar proportion of *sodium salt of benzyl penicillin G* and *procaine hydrochloride*, as given under:

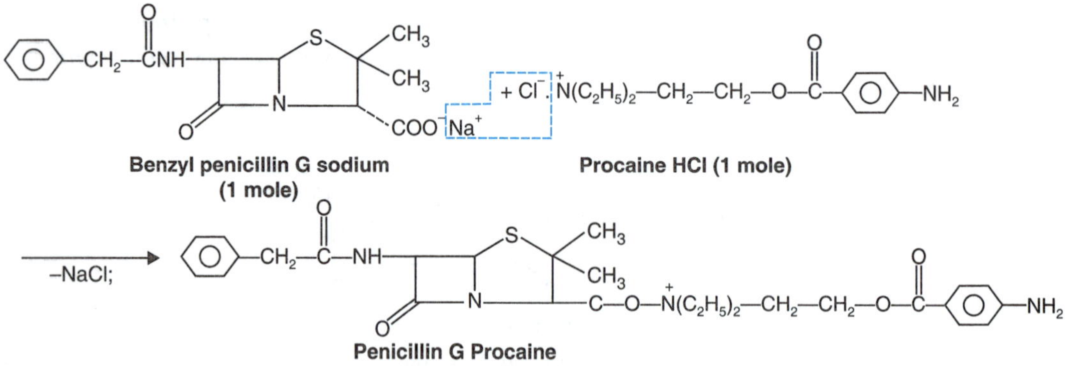

Benzyl penicillin G sodium
(1 mole)

Procaine HCl (1 mole)

Penicillin G Procaine

Highlights

- ❑ **Penicillin G procaine** being the *insoluble salt* is found to be quite useful in the treatment of **respiratory disorders**.
- ❑ A unique *combination* of **Penicillin G procaine** and **Aluminium stearate** has proved to be an **excellent long-acting penicillin** specifically useful in the *prolonged treatment of syphillis* (a **dreadful vinereal disease**).
- ❑ It is not significantly affected either by *light* or *air*.
- ❑ Its aqueous solutions are **dextrorotatory** in nature.
- ❑ **Solubility Profile:** In water 6.8 mg.mL^{-1} at ~28°C; MeOH >20; Isopropanol 6.5; Benzene 0.075; Toluene 1.05; Ethyl acetate 3.35;

[C] Benzathine Penicillin G [Penicillin G Benzathine]

Benzathine penicillin G (or Penicillin G Benzathine) designates a **dibenzyl penicillin** that finds its abundant usage in the treatment, control, and management of *chemoprophylaxis* against the so-called **streptococcal infections** in such patients suffering from the **rheumatic fever**. The usual dosage varies between **1 to 2 million IU** *via* IM every 2 to 4 weeks gap.

Penicillin G Benzathine is normally synthesized by the interaction of *two moles of benzyl penicil-lin* and *one mole of N', N-dibenzyl ethylene diamine*, as given under:

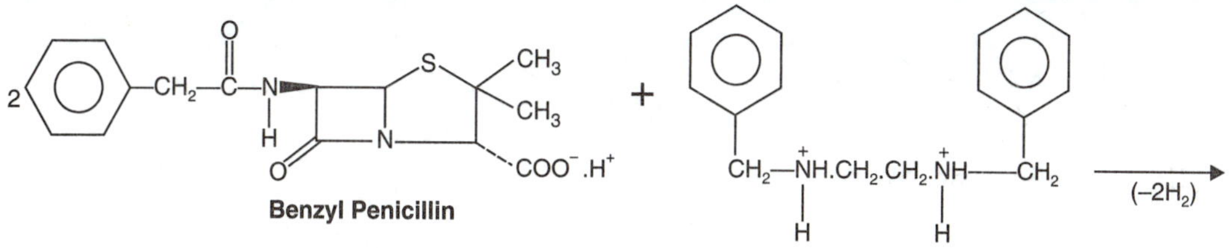

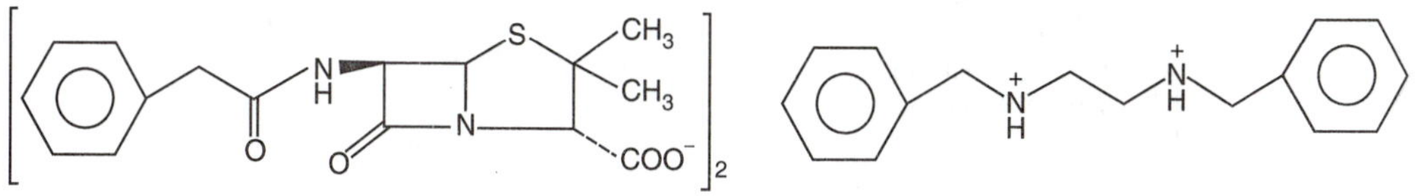

Benzyl Penicillin

+

(−2H₂)

Penicillin G Benzathine

Characteristic Features

1. **Penicillin G benzathine** is not only highly insoluble in an aqueous medium, but also is more stable together with a prolonged duration of action.
2. It is fairly stable at **gastric pH** (*acid pH*).
3. The pH of a saturated aqueous solution is ~ 6.
4. Solubility Profile: In water 0.315; methanol 16.9; and ethanol 15.4.

[D] Penicillin V [Phenoxymethylpenicillin; Phenoxymethylpenicilline Acid]:

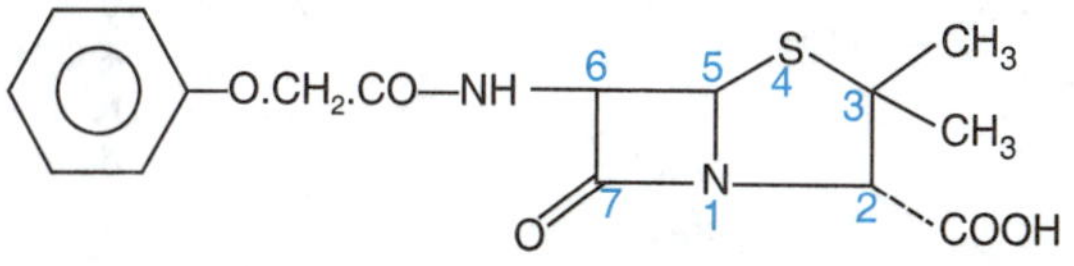

Penicillin V

(2S,5R,6R)-3,3-Dimethyl-7-OXO-6-[(phenoxyacetyl) amino]–4-thia]-azabieyelo [3,2,0] heptane – 2-carboxylic acid;

Synthesis: In the **section 11.1(c)**, the synthesis of **Penicillin V** has been duly accomplished starting from *tert*-**butylphthalimidomalonat dehyde (X)** and α-amino-α-dimethyl thiol acetate.

Alternative Synthesis of Penicillin V: This synthesis commences from the *condensation* of:

- *tert*-**Butyl-a-phthalimidomalonaldelyde**; and   • **D-Penicillamine,**

and the various steps involved in the above synthesis are as given under:

Synthesis

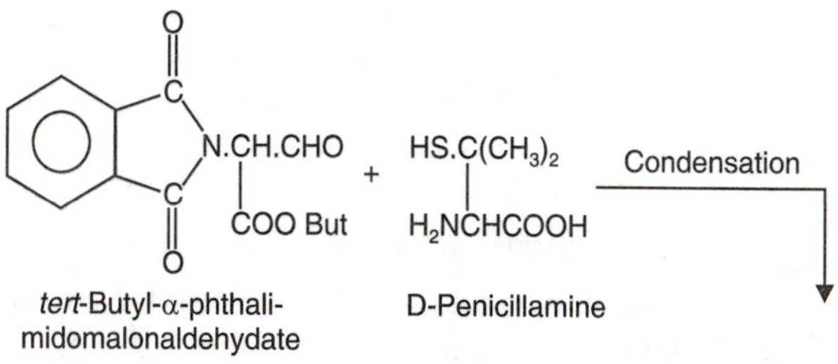

tert-Butyl-α-phthali-
midomalonaldehydate

D-Penicillamine

Contd.....

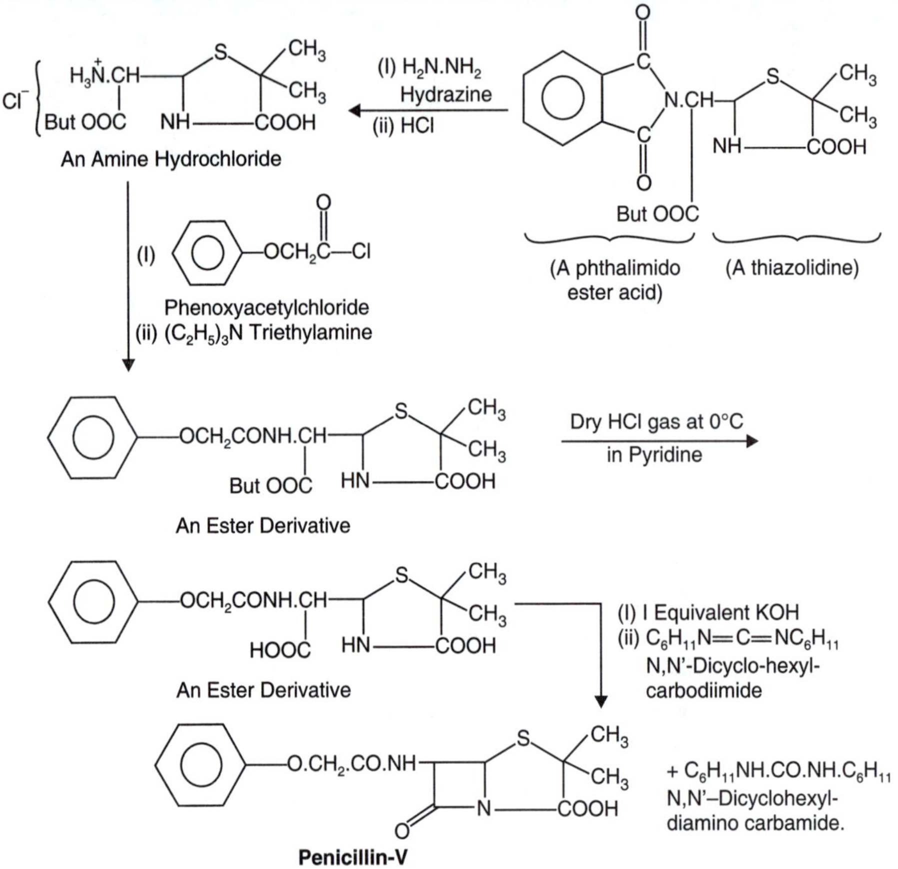

Penicillin-V

Penicillin V is found to be very effective in the control and management of infections caused due to the *Gram*-positive organisms: • Streptococcal • Staphylococcal • Pneumococcal and • Clostridium Infections. It is also the drug of choice in the treatment of a good number of *Gram*-positive bacteria e.g., Gonococcal and *Meningococcal infections*. In addition, it is also indicated solely in the critical treatment of **Pneumonia** and **Respiratory tract infections** caused profusely by such organisms as:

• **S. aureus**   • **B. anthracis** and   • **S. pyrogenes.**

NOTE   **Penicillin V is now used exclusively in the treatment of both Syphillis and Gonorrhea.**

[E] Staphicillin [Methicillin Sodium; Dimethoxyphenecillin Sodium]:

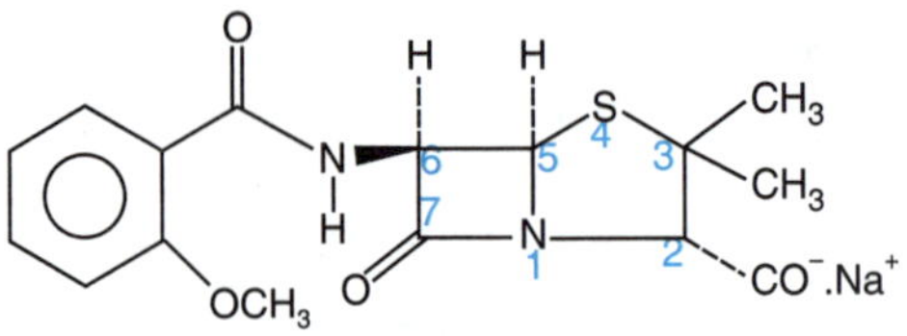

Methicillin Sodium

(2S,5R,6R)-6-[(2, 6–Dimethoxybenzoyl) amino] -3,3-dimethyl-7-oxo-4 thia-1-azabicyclo [3, 2,0] heptane -2-carboxylic acid monosodium salt;

Synthesis

   Methicillin sodium is prepared by the interaction of **2,6-dimethoxybenzoyl chloride (I)** with **6-aminopenicillanic acid (6APA) (II)** in the presence of *HCl* followed by *NaOH* (in **stoichiometric proportion**), as given under:

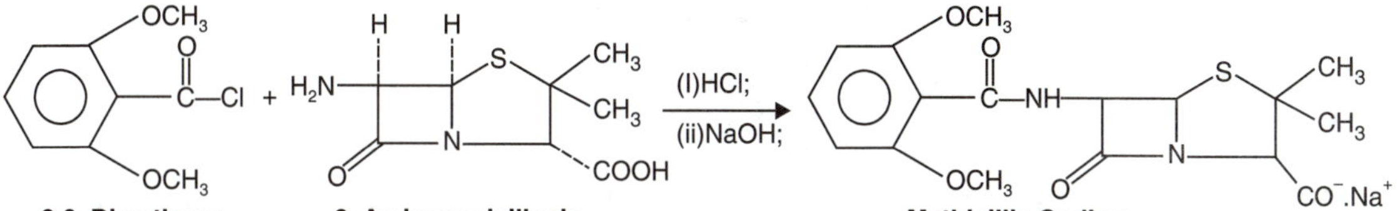

| 2,6–Dimethoxy-benzoyl chloride (I) | 6–Aminopenicillanic Acid (II) | | Methicillin Sodium |

Uses

1. **Methicillin sodium** is the drug of choice for treating infections of a serious nature invariably caused by *staphylococci*, which are usually resistant to **Benzylpenicillin [see (A) above]**.
2. However, the drug is required to be used under the close supervision of a **physician only** since it may cause **severe allergic reactions**.

3.2. The Cephalosporins

   The **cephalosporins** represent a group of *bactericidal antibiotics* that critically block the final cell of cell wall *biosynthesis in microorganisms* by inhibiting *transpeptidase* (**an enzyme**). It designates a type of compound wherein a **β-lactam ring** is meticulously fused to :

"dihydro – 1, 3 – thiazine ring system"

   In reality, the **'cephalosporins'** are *antibiotics* duly obtained from a *species of fungus* known as **cephalosperium**; and also by the **semi-synthetic procedures**. There are *three* different compounds which have been duly isolated from **cephalosporin**, namely:

- **Cephalosporin P₁** • **Cephalosporin N** and • **Cephalosporin C**
- ❏ **Cephalosporin P₁** – The antibiotic substance produced by a *cephalosporium* sp. cultivated from the sea near **Sardinia**. Interestingly, the *erude cephalosporin P* contains at least **5 known components** *viz.*, **P₁, P₂, P₃, P₄, and P₅**; the major *active* substance being **cephalosporin P₁**.

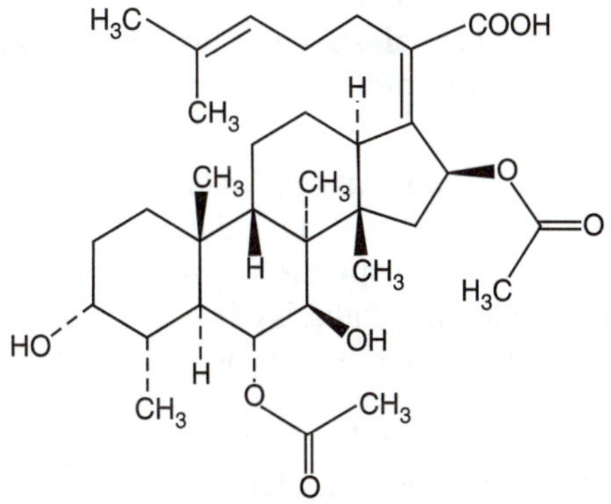

Cephalosporin P₁

It possesses relatively **low antibacterial activity profile**.

❑ **Cephalosporin N** – The antibiotic substance produced by *cephalosporun spp.* found usually in the *sewage outpours*.

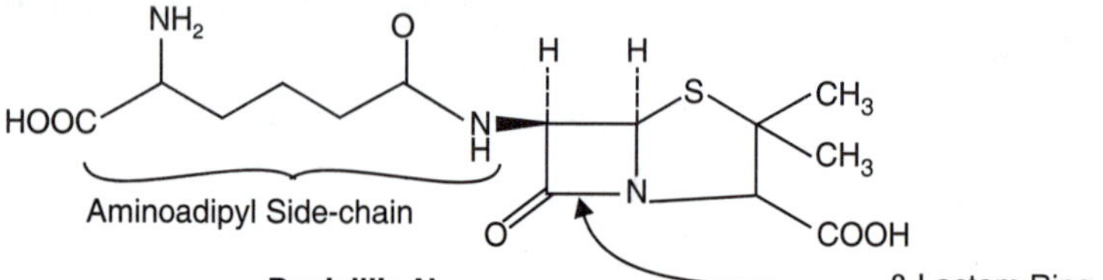

Penicillin N

From the above structure one may observe explicitly that the **β-lactam ring** is duly fused with the **dehydrothiazin ring 'A'**. However, it is found to be less effective against the so-called *Gram*-**positive microbes** in comparison to *other penicillins*, but definitely more active than **penicillin G**.

❑ **Cephalosporin C** – It designates a natural antibiotic duly produced by the fungus *Cephalosporium acremonium*.

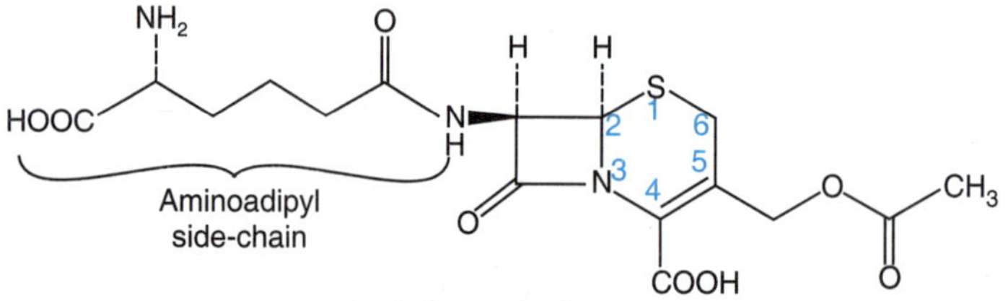

Cephalosporin C

The antibacterial spectrum of activity of **cephalosporin C** is very much identical to that of penicillin N, but it is comparatively **less active**. However, it is found to be quite useful as a *starting material* for the preparation of a series of the so-called *'penicillinase-resistant antibiotics'*, thereby showing an appreciable activity against *Gram*-**positive bacteria** and also *Gram*-**negative bacteria** almost to the same degree.

Historical Evidences

Giuseppe Brotzu's epoch making discovery, in 1945, in the species **cephalosporium fungi** obtained from *C. acremonium* showed a remarkable inhibition in the growth of a rather wide spectrum of both *Gram*-positive and *Gram*-negative organisms. Abraham and Newton (1961) at Oxford for the first time not only isolated successfully but also characterized **Cephalosporin C.*** However, the confirmation of its structure was ascertained by X-ray crystallography.**

Inspite of the glaring evidence that **cephalosporin C** was resistant to *S. aureus* β-lactamase, besides its prevailing antibacterial activity was inferior in comparison to **penicillin N** and other penicillin structural analogues.

It has been observed critically that the **natural products** usually exhibit a relatively lower level of **antibacterial activity**. Therefore, the articulate and judicial 'cleavage' of the amide bond of the **aminoadipyl side-chain** present in **cephalosporin C** provides **7-amino-cephalosporanic acid (7-ACA)**, which is most ideally suitable for the synthesis of a wide range of semisynthetic cephalosporins *via* acylation of the C(7)-amino functional moiety*** as depicted under:

---

    *   EP Abraham and GGF Newton, *Biochem J.*, **79**, 377, (1961)

  **   DC Hodgkin and EN Masien, *Biochem. J.*, **79**, 393, (1961).

***   Fechtig B *et al., Helv. Chim Acta.* **51**, 1108, 1968.

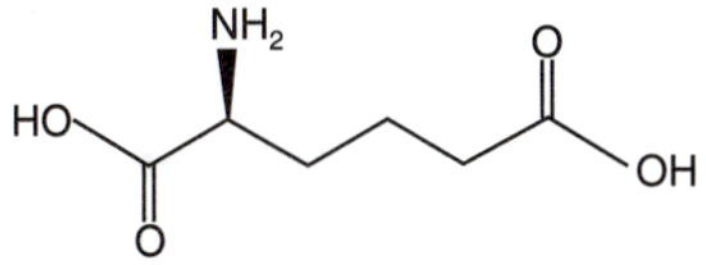

Aminoadipyl Side-chain
(Present in Cephalosporin C)

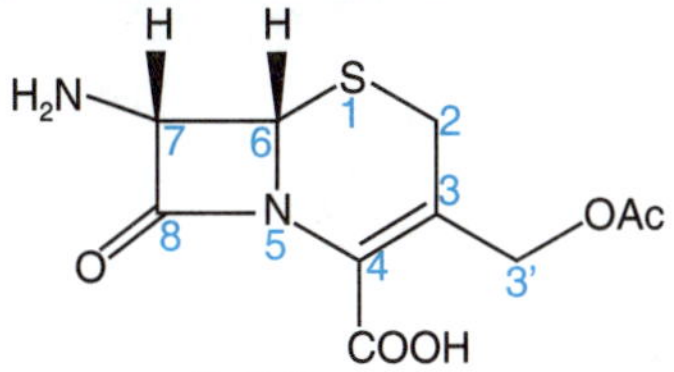

7-ACA

3.2.1. *Classification of Cephalosporins*

The **'cephalosporins'** may be classified under the following *four* categories:

(a)   First generation (Staph, some enteric *Gram*-negative, Bacilli)

(b)   Second ageneration (more active *Vs Gram*-negative, some active *Vs H. Influenzae* and anaerobes)

(c)   Third generation (best *Gram*-negative spectrum, β-lactamase resistant, poor *Vs.* Staph.)

(d)   Fourth generation.

A few typical examples of **'cephalosporins'** belonging to each of the above four generation shall now be discussed more explicitly in the sections that follows:

3.2.1.1. *First Generation Cephalosporins*

The following *three* drugs belonging to this class of compounds shall be treated in an elaborated manner, namely: **Cefazolin Sodium** and **Ceptradine.**

[A] Cefazolin Sodium

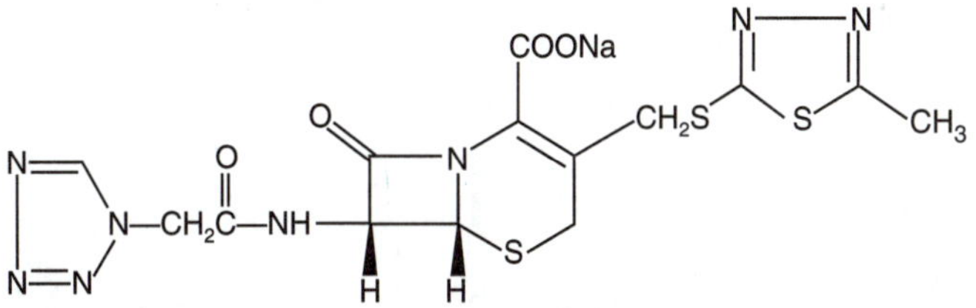

5-Thia-1-azabicyclo[4,2,0]oct-2-ene-2-carboxylic acid, 3-[(5-methyl-(6R-trans)-1,3,4-thiadiazol-2-yl) thio] methyl]-8-oxo-7-[[(1H-tetrazol-1 yl]-amino]-monosodium salt;

Synthesis

The **acylation** of the sodium salt of 7-**aminocephalosporanic acid** (*i.e.*, **7-ACA**) with 1H-tetrazole-1-acetyl chloride gives rise to the formation of an intermediate with the elimination of a mole of HCl. The resulting product on being treated with 5-methyl-1, 3, 4-thiadiazole-2thiol affords the displacement of the acetoxy moiety which upon treatment with an equimolar concentration of NaOH yields the desired compound.

It is a **first-generation cephalosporins** given **IM** or **IV**. The **'drug'** may be employed to treat infections of the skin, bone, soft tissues, respiratory tract, urinary tract, and endocarditis and septicemia caused by susceptible organisms. It has been observed that amongst UTIs, cystitis responds much predominantly and better in comparison to *pyelonephritis*.* It is regarded to be the preferred **cephalosporin** for most surgical prophylaxis due to its inherent long half-life

---

*   Inflammation of *kidney* and *renal pelvis*

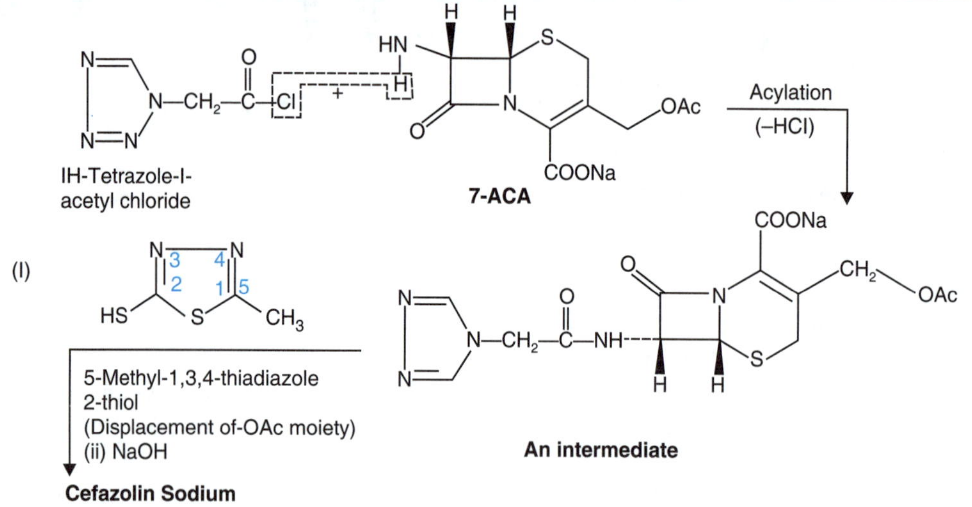

IH-Tetrazole-l-
acetyl chloride

7-ACA

Acylation
(–HCl)

(I)

5-Methyl-1,3,4-thiadiazole
2-thiol
(Displacement of-OAc moiety)
(ii) NaOH

An intermediate

Cefazolin Sodium

Explanation: The acylation of 1H-tetrazole-1-acetylchloride and 7-ACA loses a mole of HCl to yield an **intermediate**, which affords the *displacement of acetyloxy (–OAc) moiety* in the presence of **5-methyl –1, 3, 4-thiadiazole-2-thiol** followed by treatment with NaOH to produce the desired product, **cefazolin sodium**.

[B] Ceptradine

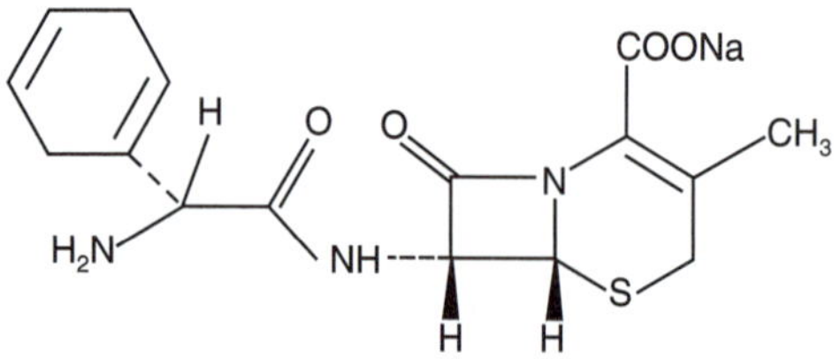

5-Thia-1-azabicyclo [4,2,0] oct-2-ene-2-carboxylic acid [6R-[6α, 7b-(R*)]]-[(amino-1,4-cyclohexadien-1-ylacetyl] amino]-3-methyl-8-oxo-;

It is recognized as a short-acting **first-generation cephalosporin** administered IM or IV. Recommended usually for the treatment of UTIs and respiratory tract infections.

3.2.1.2. *Second Generation Cephalosporins*

In this particular class of compounds the following typical examples shall be treated individually in details *e.g.* **Cefamandole, Cefoxitin,**

[A] Cefamendole Nafate

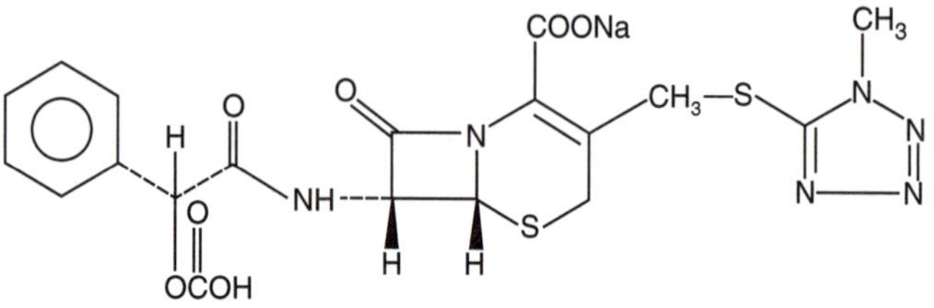

5-Thia-1-azabicyclo [4,2,0] oct-2-ene-2-carboxylic acid, [6R-[6a,7b (R*)]]-7-(formyloxy) phenylacetyl] amino]-3-[[(1-methyl 1H-tetrazol[-8-oxo-, monosodium salt;

[B] Cefoxitin Sodium

It is invariably used as an **'alternative drug'** for the treatment of intra-abdominal infections, colorectal surgery or appendectomy and ruptured viscus because it is active against most enteric anaerobes including the organism *Bacteroides fragilis*. It is also indicated in the management and treatment of bone and joint infections produced by *S. aureus*, gynecological and intra-abdominal infections caused by *Bacteroides* species together with other common enteric anaerobes and *Gram*-negative bacilli; lower respiratory tract infections produced by *Bacteroides* species, *E.coli, H. influenzae, Klebsiella* spp. *S. aureus* or *Streptococcus* spp. (except *enterococci*); septicemia caused by *Bacteroides* spp. *E. coli, Klebsiella* spp., *S. aureus* or *Strep pneumoniae*; skin infections produced by *Bacteroides* spp., *E. coli*, Klebsiella spp., *S. aureus* or *epidermidis* or *Streptococcus* spp. (except *enterococci*) or UTIs by *E. coli, Klebsiella* spp. or indole positive Proteus, and for preoperative prophylaxis.

3.2.1.3. *Third Generation Cephalosporins*

Though there are several drugs that are approved and marketed belonging to the **'third generation cephalosporins'**, but only *two* such compounds shall be discussed in this particular section, namely: **Cefixime, Ceftazidime Sodium**.

[A] Cefixime

It is a well-known orally active **third generation cephalosporin** having superb and excellent therapeutic profile against a plethora of *E. coli, Klebsiella, H. influenzae, Branhametla catarrhalis, N. gonorrhoeae* and *meningitidis, besides* including β-*lactamase producing strains*. It is found to be active against certain common **streptococci** spp. whereas **staphylococci** are genuinely resistant. It is recommended for the respiratory infections*, Otitis media and uncomplicated UTIs; however, its actual therapeutic role is yet to be understood exhaustively.

---

* Infections due to acute bronchitis, *pharyngitis* and *tonsitlitis*.

[B] Ceftazidime Sodium

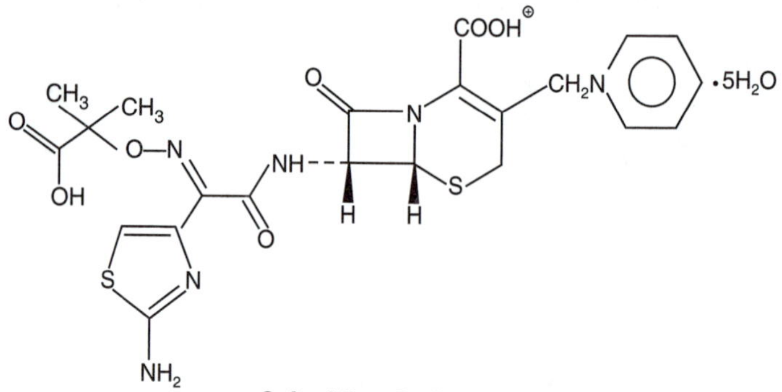

Ceftazidime Sodium

The 'drug' displays its special interest due to its inherent high activity against the Pseudomonas and Enterobacteriaceae but bails to do so for enterococci. It is well recognized widely as an 'alternative drug' specifically for the management and treatment of hospital-acquired Gram-negative infections. However, a combination with **amikacin** in the treatment of infections in immunocompromised patients when *Ps aeruginosa* happens to be a causative organism.

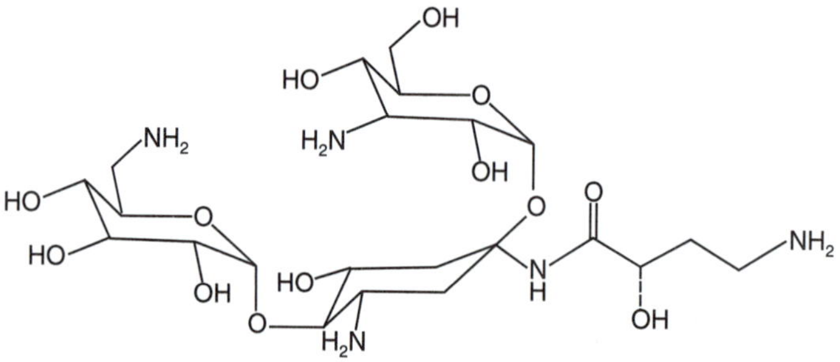

Amikacin

Ceftazidime is profusely recommended for use in the treatment of bone and joint infections, CNS-infections, gynecological infections, lower respiratory tract infections, spticemia, skin and UTIs.

3.2.1.4. *Fourth Generation Cephalosporins*

The only approved drug substance that belongs to the **fourth generation cephalosporins** is **cefepime** which will be discussed as under:

[A] Cefepime Hydrochloride

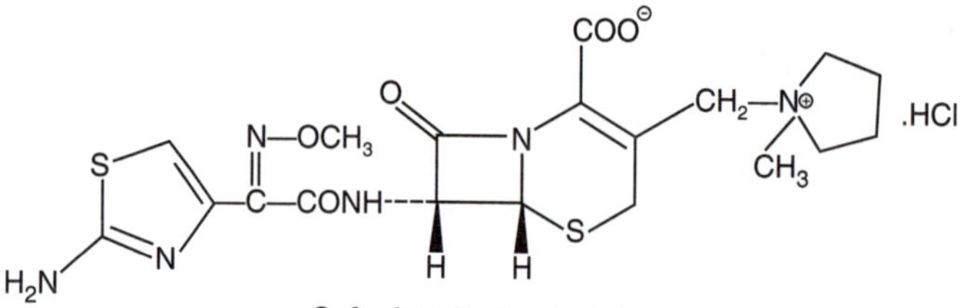

Cefepime Hydrochloride

It is profoundly recognized as an altogether new approved **fourth-generation cephalosporin** which essentially possesses an extended *Gram*-negative spectrum against *Gram*-negative aerobic bacilli usually covered by **cefotaxime** and **ceftazidime** including certain strains that are found to be resistant to these **third-generation cephalosporins**.

It is, however, pertinent to state here that cefepime definitely exhibits an improved antibacterial profile against *Streptococcus pneumoniae* and *Staphylococcus aureus* in comparison to the **third-generation cephalosporins**. Interestingly, its specific activity against *P. aeruginosa* is found to be variable just like other antibiotics; and the profile of activity resides between that of **ceftazidime** and **ceftoxime**.

The **'drug'** gets excreted mostly in the urine having a half-life of 2.1 hours. It is found to be bound almost minimally to the plasma proteins.

NOTE **Cefepime HCl may be administered IV or IM for the treatment of UTIs, pneumonias and skin infections.**

THE CEPHALOSPORIN C

Preamble: First and foremost **cephalosporin N** was isolated which was produced by a species of *Cephalosporium*. Abraham (1955)* duly reported that this antibiotic was indeed nothing but **penicillin** wherein the value of **'R'** is found to be:

$$–H_2C – CH_2 – CH_2 – CH (NH)_2 – COOH$$

Later on, Abraham (1956)** meticulously isolated *another* **antibiotic** from the *crude* **cephalosporin N** and subsequently, named it as **cephalosporin C.** Abraham was able to ascertain as well as prove that **cephalosporin C** was having the following *three* **cardinal characteristic features,** namely:

- **shows distinct antibacterial activity,**
- **possesses more stability to acid *vis-a-vis* cephalosporin N**, and
- **exhibits critical resistance to the action of 'enzyme' from penicillinase (*i.e.*, difference from the 'penicillins').**

Constitution of Cephalosporin C

Abraham virtually established the exact constitution of **cephalosporin C** based on the following scientific and logical evidences:

1. **Molecular formula** – Its molecular formula, based on various analytical data was found to be $C_{16}H_{21}N_3O_8S$.

2. **Presence of an α-Amino Acid** – It gives a distinct positive test with **Ninhydrin Reagent** thereby confirming the presence of an α-amino acid in **cephalosporin C.**

3. **Presence of an Aminodicarboxylic Acid** – Based on the **electrometric titration** it was inferred that **cephalosporin C** possesses an **aminodicarboxylic acid**. In addition, it also inherits *three* ionizable functional moieties that critically attribute *three* **distinct pKa values** at: **2.6, 3.1**, and **9.8** respectively.

4. **IR-spectral Analysis** – It shows a definite *IR-band* at **1783 cm^{-1}**, which is duly ascribed to the presence of a carboxyl (>C=0) moiety located strategically in the β-**lactam ring system** (*we must recall that in penicillin a similar IR-band at 1770 cm^{-1}*).

---

* Abraham, *Nature*, **175**: 548, 1955.
** Abraham, *Biochem J.*, **62**: 651, 1956.

Remarks: Cephalosporin C on being subjected to hydrolysis does not yield *penicillamine* (which is unlike the 'penicillin'). Thus, it ascertains emphatically that the so-called – 'structures duly attached to the β-lactam ring' in *cephalosporin C* and *penicillin* are found to be altogether different.

5. **Acid Hydrolysis of cephalosporin C** – It yields one mole each of these *three* products:
    *   **1 mole of D-a-aminoadipic acid (I);**
    *   **1 mole of carbon dioxide (CO_2); and**
    *   **2 moles of ammonia (NH_3)**

6. **Reduction followed by Hydrogenolysis of cephalosporin C** – The reduction of cephalosporin C with Raney – Ni followed by hydrolysis (hydrogenolysis) gives rise to the formation of:
    *   **D-a-aminoadimpic acid (I),**
    *   **L-alanine (II), and**
    *   **some quantum of DL-valine.**

7. **Controlled Hydrolysis of Cephalosporin C-**It yields a **dipeptide (III)** together with **DL-aminoadipic acid (I)** plus α, β-**diamino-propionic acid**.

$$HOOC.CH(NH_2).(CH_2)_3.COOH$$
D-α-Adipic Acid (I)

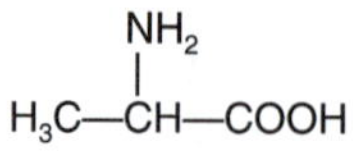

$$\underset{\overset{|}{NH_2}}{H_3C-CH-COOH}$$
L-Alanine (II)

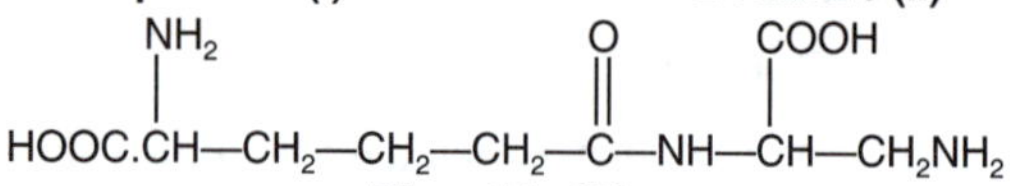

$$\underset{\overset{|}{NH_2}}{HOOC.CH-CH_2-CH_2-CH_2-}\overset{\overset{O}{\|}}{C}-NH-\underset{\overset{|}{COOH}}{CH}-CH_2NH_2$$
Dipeptide (III)

8. **Hydrolysis of Cephalosporin C in Neutral Aqueous Medium** – It usually yields at 37° C D-2-(4-amino-4-carboxybutyl) thicazole 4-carboxylic acid (IV):

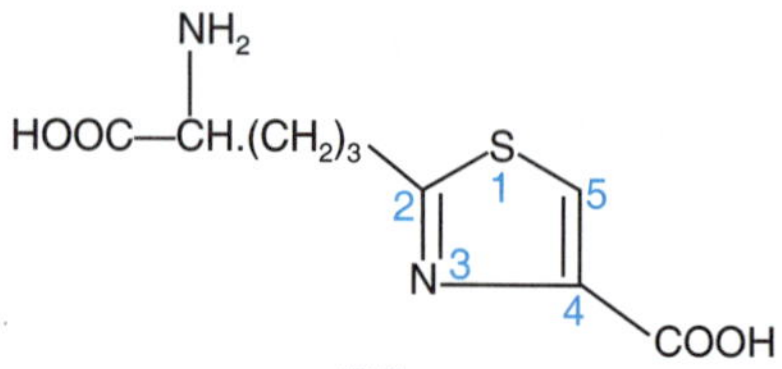

(IV)

❑   **Electrometric Titration** – Thus, when **compound (IV)** is subjected to **electrometric titration**, it showed the critical presence of a basic moiety (**pKa = 9.9**); and also *two* **acidic moieties** having **pKa values** 2.6 and 4.0 respectively.

❑   **UV-Spectral Studies** – This reveals the presence of *two* different peaks (**compound IV**):
    *   at λ_{max} **237 nm : for H_2O; and**
    *   at λ_{max} **233 nm : for NH_4Cl,**

    which is found to be exactly the same as that of **2-(1-amino - 2 methylpropyl) thiazole-4-carboxylic acid**

❑   The precise and exact nature of the *side-chain* is duly ascertained by the careful isolation of **D-α-aminoadipic acid (I).**

9. **Probable Structure (V) Assigned to Cephalosporin C** – Based on the above *scientific data and spectral analysis data* one may assign the following **probable structure (V)** for **Cephalosporin C:**

$$\begin{array}{c} \text{NH}_2 \\ | \\ \text{HOOC} - \text{CH.(CH}_2)_3.\text{CO.NH} \end{array}$$

(β-lactam ring system with S, N and $C_7H_8O_4$)

(V)

10. **Hydrolysis of Cephalosporin C with H_2SO_4** – It results into the production of **one mole of acetic acid (CH_3COOH)**, that shows clearly the presence of an **acetoxy (–O – COCH$_3$) moiety** in the cephalosporin C (also explained due to the revelation of the IR-band at 1738 cm^{-1}). However, the respective IR-band at 1031 cm^{-1} may be attributed due to the **O–C (str.)** in the ensuing group $CH_3CO-O-C$.

   Remark: Importantly, the presnce of the **acetoxy [–O–COCH$_3$] moiety** in the **cephalosporin C molecule** helps to reduce the nature of the **remaining fragment.** in **compound (V)** to **C–5** ultimately (**see numbering in cephalosporin C earlier**).

11. **Hydrogenolysis of cephalosporin C with Raney–Ni**–It critically yields these *two* chemical entities, namely:

   - **DL-valine** and   • α–oxo–isovaleric acid.

   It is worthwhile to state at this point in time that when **'penicillin N'** is subjected to **hydrogenolysis** it resulted D-valine (right from the penicillamine fragment).

   Obviously, neither **cephalosporin C** nor **the penicillins** fails to yield *penicillamine* upon hydrogenolysis, suggesting thereby that

   "structures of the corresponding fragment attached to the β-lactam ring system in the *two* aforesaid antibiotics are absolutely different from each other by all means".

> **NOTE** The above important findings and observations have been further confirmed by carrying out the NMR-spectral studies of *cephalosporin C* thereby showing the absence a typical signal at τ 7.9 (*i.e.*, a specific signal caused due to the presence of a gem*-dimethyl moiety)

12. **Formation of Lactone Cephalosporin C** – Dissolution of cephalosporin C carefully in **0.1M HCl at room temperature (25°C)** gives rise to the formation of a **lactone cephalosporin C$_c$**, with the complete elimination of the attached **acetoxy (–O–COCH$_3$) moiety**.

   Subsequent hydrogenolysis of the resulting cephalosporin C$_c$, in the critical presence of Raney – N$_i$ yields α–amino–β-methyl butenolide (VI), – which eventually upon *hydrogenation* [Pt – Pt O$_2$] gives the respective γ-hydroxyvaline lactone (VII). The structures of above cited products: (VI), (VII) are given below:

[VI]   [VII]   [VIII]

---

* gem (means *geminal*) – *i.e.*, dimethyl moieties on the basis of the basis of the attachment of **two methyl groups** to *two* adjacent C-atoms and one C-atom respectively. [e.g., CH_3–CH (CH_3)$_2$ Dimethyl ethylidene (gem.)]

Remarks – The specific formation of **compound (VI)** from the **cephalosporin C** may be explained explicitly based on the underlying fact that the latter must be having the *inherent grouping* as depicted in **structure (VIII)** above.

13. **Exact Position of the Double Bond (C=C) in Structure (VIII) –** In reality, the exact position of the **double bond** (or *olefinic bondage*) in **structure (VIII)** appears to be very much consistent with respect to the **isolation** of: '2, 4 – dinitrophenyl hydrazone of hydroxyacetone as obtained due to the '*ozonolysis*' of *cephalosporin C* followed by reduction using Raney – Ni'

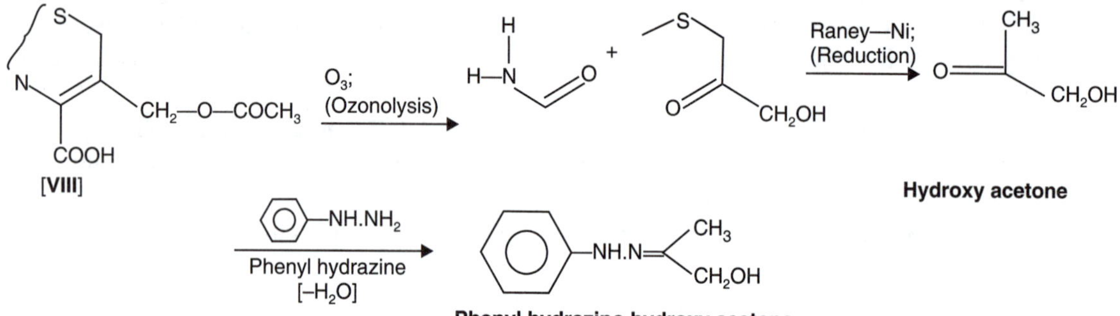

Hydroxy acetone

Phenyl hydrazino hydroxy acetone

14. **Proposed Structures of Cephalosporin C (IX) and Cephalosporin C$_c$ (X) by Abraham** *et al.* **(1955, 1956)**

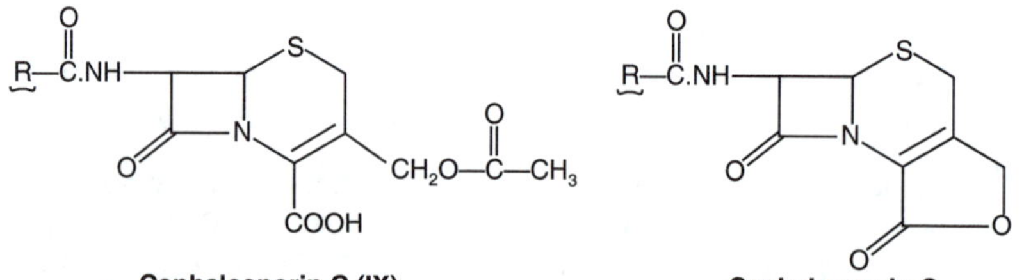

Cephalosporin C (IX)

Cephalosporin C$_c$
[Deacetyl cephalosporin C Lactone] (X)

In fact, the above *two* structures **(IX)** and **(X)** are duly put forward on the basis of the aforesaid evidences and results as well.

15. **Cephalosporin C (IX) Proposed Structure Expatiates the Various Degradative Reactions Satisfactorily-** Thus, the underlying **mechanism** involved particularly for the actual conversion of **cephalosporin C** to D-2-(4-amino-4-carboxybutyl) thiazole-4-carboxylic acid (IV) critically takes the following sequential steps:

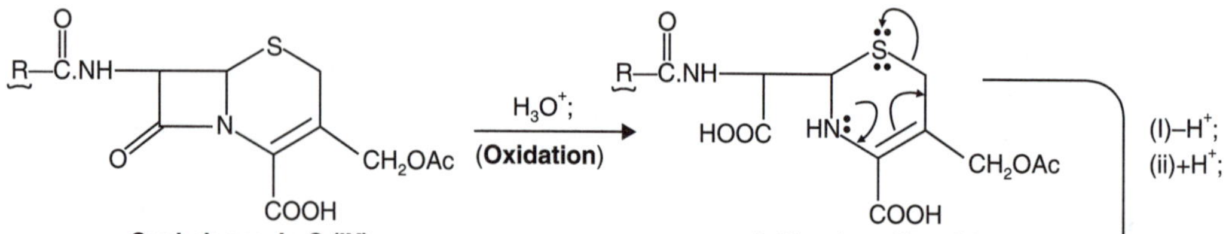

Cephalosporin C (IX)

A Dicarboxylic acid derivative

Contd.....

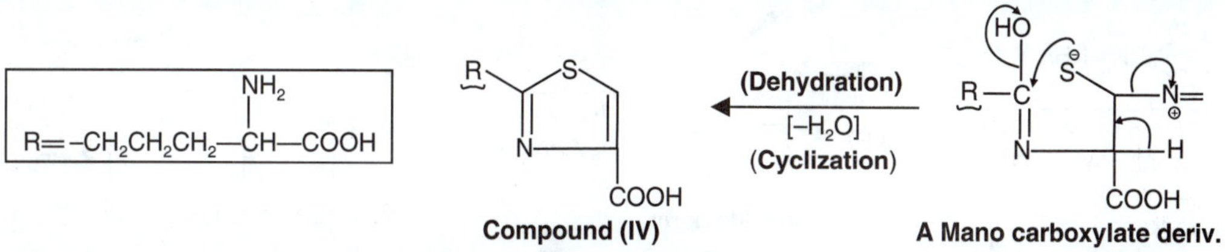

Compound (IV)   **A Mano carboxylate deriv.**

(Dehydration)
[–H₂O]
(Cyclization)

R = –CH₂CH₂CH₂–CH–COOH with NH₂

NOTE **It must always be understood clearly that:**

• **Penicillins – designate β-lactam thiazolidines; and**
• **Cephalosporins – represent β-lactamdihydrothiazines**

---

16. **Confirmation of the Proposed Structure of Cephalosporin C by means of Synthesis–** Woodward *et al.* (1966)* was pioneer in putting forward the *total synthesis* of cephalosporin C starting from L(+)–cysteine by making use of certain highly stereospecific reactions**, as stated under:

Synthesis

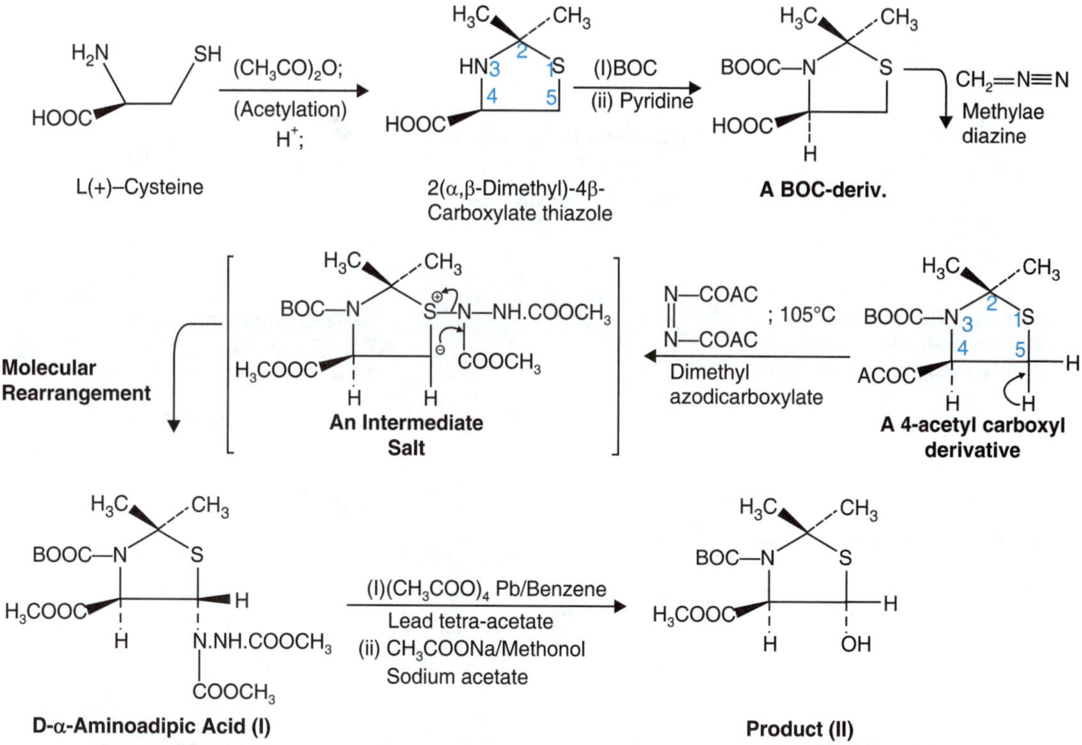

L(+)–Cysteine

(CH₃CO)₂O;
(Acetylation)
H⁺;

2(α,β-Dimethyl)-4β-
Carboxylate thiazole

(I)BOC
(ii) Pyridine

A BOC-deriv.

CH₂=N=N
Methylae
diazine

**Molecular
Rearrangement**

**An Intermediate
Salt**

Dimethyl
azodicarboxylate

; 105°C

**A 4-acetyl carboxyl
derivative**

D-α-Aminoadipic Acid (I)

(I)(CH₃COO)₄ Pb/Benzene
Lead tetra-acetate
(ii) CH₃COONa/Methonol
Sodium acetate

Product (II)

---

*   Woodward *et al. Am. Chem Soc.*, **88**: 852, 1966.

**   **Stereospecific reactions:** Refer to such reaction which is said to be stereospecific, if the starting materials are do differ from its product **only in the configuration**.

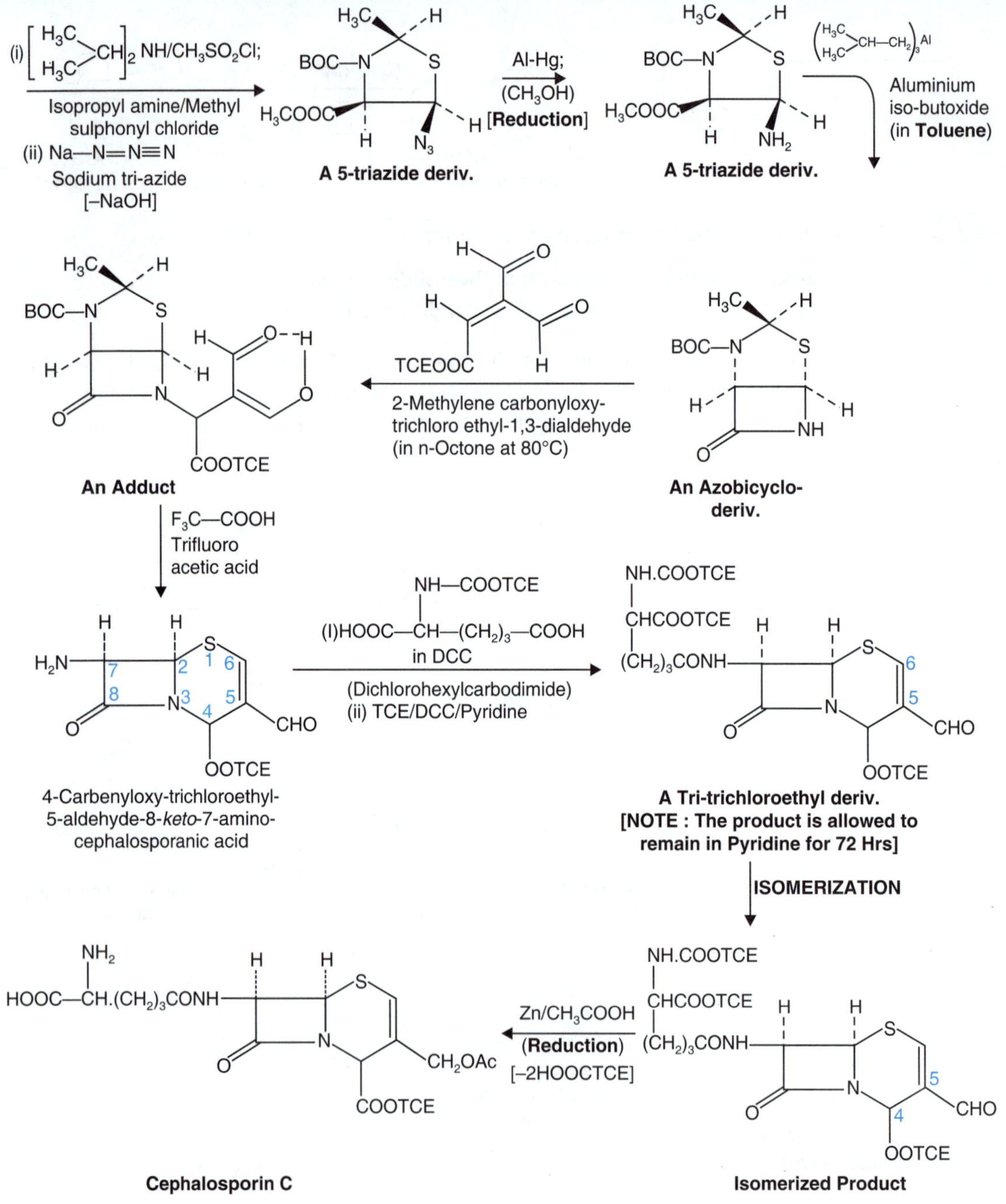

A 5-triazide deriv.

A 5-triazide deriv.

Isopropyl amine/Methyl sulphonyl chloride

(ii) Na—N≡N≡N
Sodium tri-azide
[–NaOH]

Al-Hg;
(CH₃OH)
[Reduction]

Aluminium iso-butoxide (in Toluene)

An Adduct

2-Methylene carbonyloxy-trichloro ethyl-1,3-dialdehyde (in n-Octone at 80°C)

An Azobicyclo-deriv.

F₃C—COOH
Trifluoro acetic acid

4-Carbenyloxy-trichloroethyl-5-aldehyde-8-*keto*-7-amino-cephalosporanic acid

(I)HOOC—CH—(CH₂)₃—COOH
in DCC
(Dichlorohexylcarbodimide)
(ii) TCE/DCC/Pyridine

A Tri-trichloroethyl deriv.
[NOTE : The product is allowed to remain in Pyridine for 72 Hrs]

ISOMERIZATION

Zn/CH₃COOH
(Reduction)
[–2HOOCTCE]

Cephalosporin C

Isomerized Product

Decoding of Abbrevations :
BOC = tert-Butyl oxycarbonyl;
TCE = b,b,b-Trichloromethyl ester;
DCC == Dichloro hexylcarbodimide;

Explanation: The various steps involved above may be explained as under:

1. *Acetylation of* **L(+)-cysteine** with acetic anhydride yields **2 (α,β–dimethyl)–4β-carboxylate thiazole**, which on treatment with ***tert*-butyloxy carbonyl (BOC)** followed by *pyridine* gives the respective **BOC-derivative**.

2. The resulting product on interaction with methylene diazine gives a **4-acetyl carbonyl derivative**, which on being subject to treatment with *dimethyl azo-carboxylate* (at 105°C) yields an intermediate salt.

3. This product on *molecular rearrangement* forms **D–amino–adipic acid (I),** which on subsequent treatment with *lead tetraacetate* in benzene followed by *sodium acetate* (*in methanol*) yields **Product (II)**.

4. **Product (II)** on being treated with *isopropyl amine* (in *methyl sulphonyl chloride*) followed by *sodium thiazide*, loses a mole of NaOH to give a **5-triazide derivative**, which on reaction with *Al-Hg* (in MeOH) undergoes reduction to yield a **5-amino derivative**.

5. The end product obtained above when reacted with *aluminium iso-butoxide* (in toluene) gives an **aza-bicyclo derivative**, which on treatment with *2-methylene carbonyloxy-trichloroethyl–1, 3-dialdehyde* (in n-octane at 80°C) yields an **'adduct'**.

6. The resulting **'adduct'** on reaction with *trifluoroacetic acid* yields **4-carbonyloxy trichloroethyl-5-aldehyde-8-keto-7-amino-cephalos poranic acid**, which on treatment with a *TCE derivative* (in *DCC*) followed by *TCE/DCC/pyridine* gives a **tri-trichloroethyl derivative**.

7. The end-product undergoes isomerization at C–4 and C–5, which on *reduction* with *Zu/AcOH* loses *two* **moles trichloromethyl carboxylate** to produce the desired cephalosporin C.

Important Salient Features in the Synthesis of Cephalosporin C (Woodward et al., 1966)

Following are the *four* **important salient features** in the above described synthesis of cephalosporin C, namely:

(a) **Use of *tert*-Butyloxycarbonyl (BOC) Moiety – as an Amino-(–NH₂) Protecting Group**

In actual practice, the aforesaid '**BOC moiety**' [*i.e.*, (CH₃)₃ – C.OCO–] in introduced meticulously with ***tert*-butyloxy-carbonyl azide** and promptly removed (eliminated) by subsequent treatment with cold **tri-fluoroacetic azid** [F₃C–COOH]–as given below.

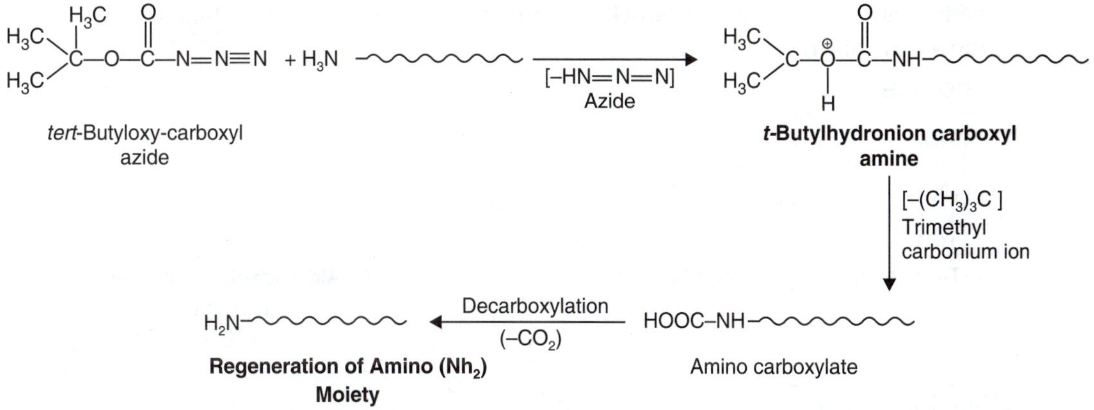

All the above reactions are self-explanatory.

(b) Conversion of D-α-Aminoadipic Acid (I) and Product (II)

Following are the various sequential steps that are involved in the conversion of (I) and (II):

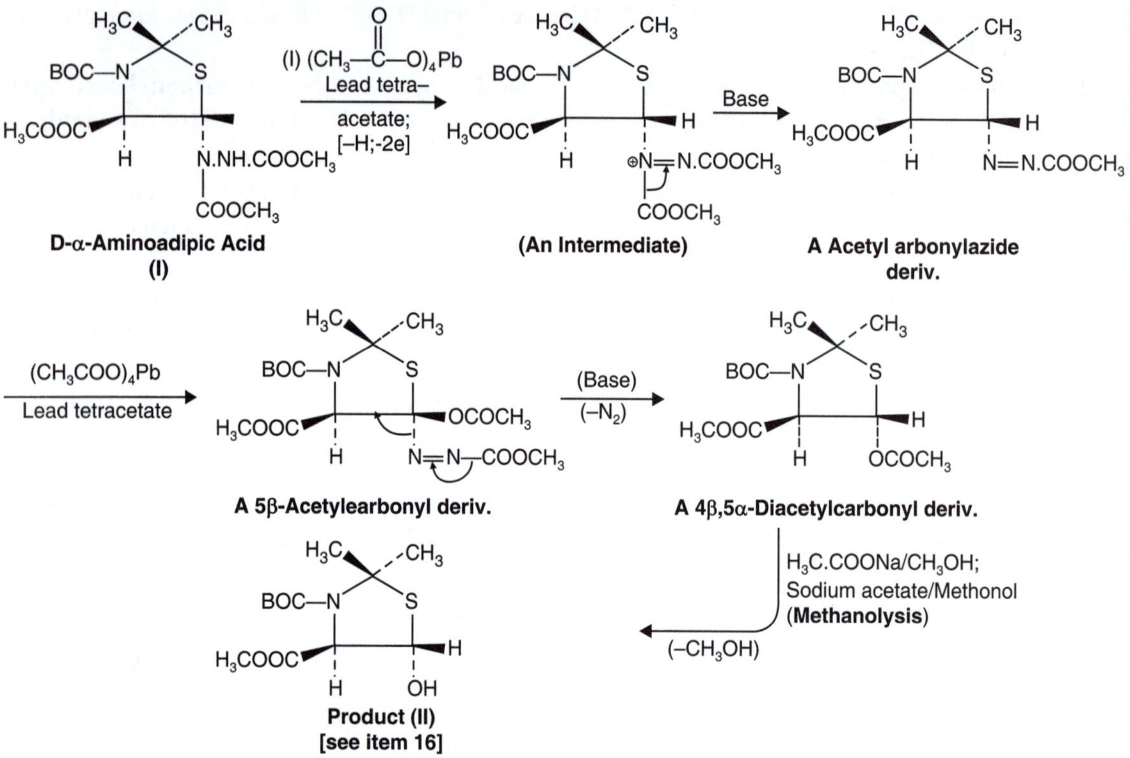

D-α-Aminoadipic Acid
(I)

(An Intermediate)

A Acetyl arbonylazide
deriv.

A 5β-Acetylearbonyl deriv.

A 4β,5α-Diacetylcarbonyl deriv.

Product (II)
[see item 16]

NOTE   The '*methanolysis*' takes place only at *C-5* of the *thiazole ring* due to its critical
electronic environment (and not at C–4).

(c) Formation of β-Lactam Ring System [$\boxed{N}$] in *Cephalosporin C* Accomplished due to a Novel Dialdehyde (X)

The formation of the **Novel Dialdehyde (X)** may be carried out using *D-tartrate ester* in the following manner:

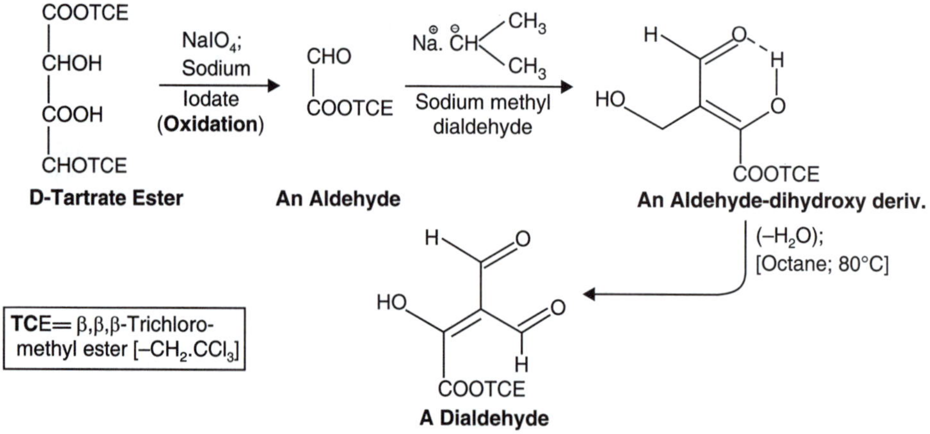

D-Tartrate Ester

An Aldehyde

An Aldehyde-dihydroxy deriv.

A Dialdehyde

TCE= β,β,β-Trichloro-
methyl ester [–CH$_2$.CCl$_3$]

(d) Removal of the Protecting Carboxylic (–COOH) Moieties of Tartaric Acid

It is, however, pertinent to state at this point in time that the respective inherent carboxyl (–COOH) moieties present in **tartaric acid** were carefully protected as the **β, β, β-trichloroethyl esters (TCE)** by means of:

'**reaction of Tartaric Acid with Trichloroethanol in the presence of Dichlorohexylcarbodimide (DCC)**'.

Thus, the *protecting moiety* may be removed by causing reduction with Z_n-dust in acetic acid (90%) at 0°C – as given below:

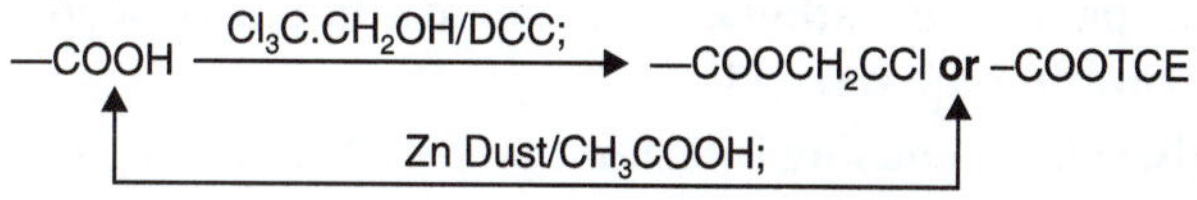

$$—COOH \xrightarrow[\text{Zn Dust/CH}_3\text{COOH;}]{\text{Cl}_3\text{C.CH}_2\text{OH/DCC;}} —COOCH_2CCl \text{ or } —COOTCE$$

Remark: The compound **D–2–(4–amino–4–carboxybutyl)–thiazole–4–carboxylic acid** [*i.e.,* **compound (IV)**] is duly acylated with *N–β, β, β–trichloroethyloxyacarbonyl–D–α–amino–adipic acid*.

STRUCTURAL ANALOGUES OF CEPHALOSPORIN C AND THEIR ACTIVITY PROFILE

Since it has been established duly that '**cephalosporin C**' does exhibit *sufficiently weaker antibacterial activity profile*; and hence, it was thought worthwile to attempt the synthesis for comparatively stronger structural analogues of *cephalosporins*. Nevertheless, the ensuring antimicrobial activity of the **cephalosporin C** could be enhanced largely by carrying out the following *two sequential reactions*, namely:

- **mild acidic hydrolysis to 7-aminocephalosporanic acid (7-ACA)**; and
- **Conversion to N-acetyl derivatives.**

The **7-ACA** may be synthesized in *two* different ways as given under:

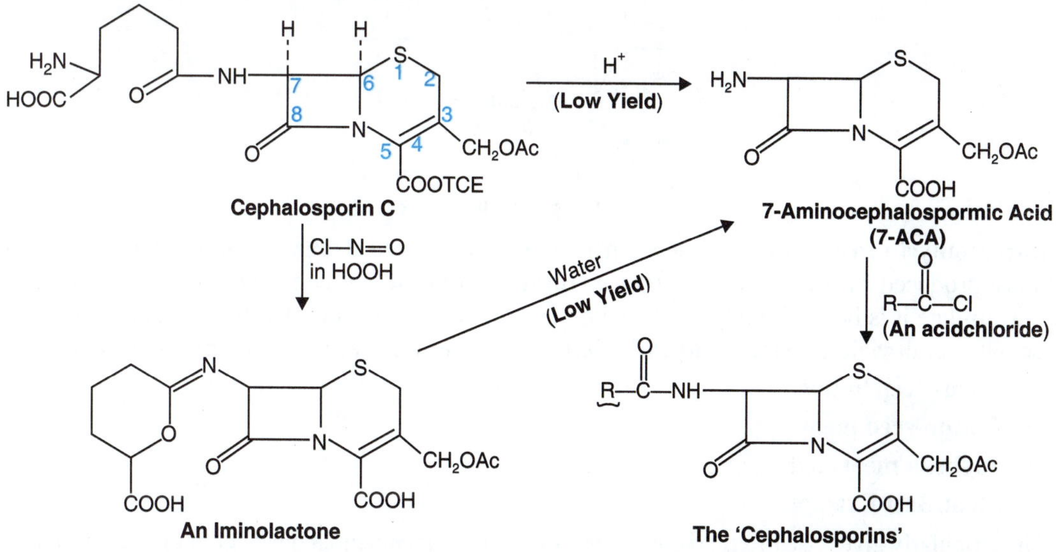

Cephalosporin C

7-Aminocephalospormic Acid (7-ACA)

An Iminolactone

The 'Cephalosporins'

3.3. Streptomycin

Streptomycin designates one of the **aminoglycoside antibiotics**. The respective sulphate and hydrochloride salts are crystalline in nature. It is, in general, is found to be effective in inhibiting the growth of both:

- *Gram*-positive microorganisms; and   • *Gram*-negative microorganisms.

Amazingly, **streptomycin** is also found to be effective to a great extent against the **mycobacteria**.

Points to Ponder: There are *two* extremely viable features:

(i) **It may be prepared biosynthetically from an admixture of carbohydrate components of the fermentation media; and**

(ii) **It invariably acts by causing interference with the *reading* of the genetic code.**

Streptomycin was the first even *antibiotic* having enormous **practical importance** and **significance,** and was isolated primarily from the class of organisms usually termed as ***Actinomycetaceae***. The usefulness of the *antibiotic* centres around the specific treatment of:

- **Meningitis** • **Pneumonia** and • **Tuberculosis**

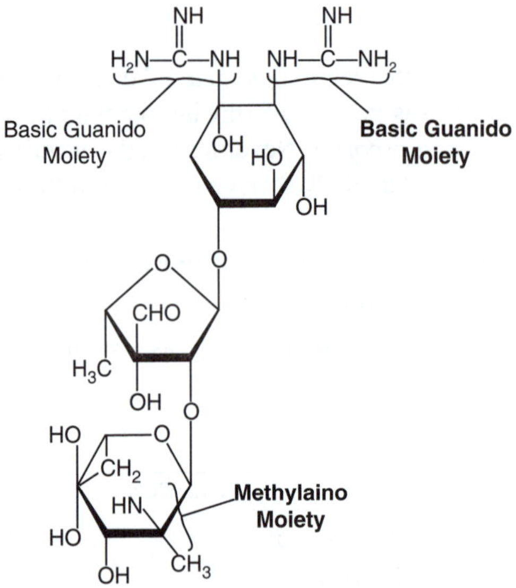

Streptomycin

Streptomycin Production: Evidence from the literature reveals that **streptomycin** was first and foremost produced by the so-called **'surface-culture technique'** (a relatively *older procedure*); however, as-to-date it is being, largely produced by making use of the **'submerged cultures'**. Obviously, the actual overall yield of **'streptomycin' solely depends upon the** *medium being employed*, namely:

- **Corn-steep liquor,**
- **Cotton-seed meal,**
- **Soybean meal**, and
- **Meat/Beef Extract,**

which particularly gives rise to a marked and pronounced enhancement in the level of ultimate production of the *antibiotic*.

Addition of Glucose [1% (w/v)] – The addition of **glucose** also largely influences and favours the **streptomycin production** to an appreciable degree.

Composition of the Medium Used:

- **Glucose (1%)**   • **Peptone (1%)**   • **Meat Extract (0.3%)/or Corn Steep Liquor**
  (1–2%) and   • **Aqueous**   • **NaCl (0.5%).**

Modus Operandi

1. The *culture medium* is transferred to a *pre-cleaned* and live-steam sterilized fermentation vials (*capacity 2.5–5 kL*) where the growth of the microorganism duly commences at **24–28°C**.
2. Sterile compressed air is made to pass through the culture medium periodically (*to agitate the contents uniformly*) and the maximum yield is accomplished after a duration of **3 to 5 days**.
3. Interestingly, the careful irradiation of the resulting '**culture medium**' either:
   - **By UV-light** or
   - **By X-ray radiation** definitely enhanced the final yield of the antibiotic.

Isolation of Streptomycin – The isolation of **streptomycin** from the **fermented broth** may be carried out as follows:

1. The fermented broth is treated suitably to separate both waste and mycelium and filtered to collect the filtrate.
2. The '**streptomycin**' is duly extracted from the filtrate either:
   - **upon base-exchange resins**, or
   - **by adsorption upon activated charcoal**.
3. The extracted streptomycin is now eluted either from the column by the help of either:
   - **diluted aqueous mineral acids**, or
   - **alcoholic mineral acids**.
4. The '*acidic eluate*' collected in the manner from the column could be purified adequately by passing it *via* an **ion-exchange resin**.
5. The so-called **pure form of streptomycin** may be isolated either as:
   - **sulphate salt**; or
   - **crystalline double-salt of calcium chloride ($CaCl_2$)**.

 NOTE   **It is worthwhile to state at this point in time that absolute '*aseptic parameters*' should always be maintained both in the course of *production* and *isolation* of streptomycin.**

Further Purification of Streptomycin – In order to obtain the perfectly sterile streptomycin, the crystalline product obtained above is carefully subject to dissolution to yield a 25% (w/v) solution – that is eventually forced from *undesirable* impurities viz., **colouring matter, heavy metals (Pb, Hg)** by making it to pass through:

   - *first, via* a **seitz filter**; and
   - *secondly*, subjecting it to *Freeze Drying*.

Finally, the resulting *Freeze Dried* product (powder is transferred) aseptically into small cleaned/ sterilized glass vials.

Characteristic Features: These essentially include:

1. **Streptomycin** is usually available as the *trihydrochloride, trihydrochloride–calcium chloride double salt*, phosphate, or *sesquisulphate*, that occur as **granules** or **powder**.
2. It is odourless, or nearly so, with a slightly bitter taste.
3. Most salts are hygroscopic and deliquesce upon exposure to air, but are not affected at all either by **air** or **light (UV)**.
4. **Solubility Profile** – The **salts of streptomycin** are quite soluble in **water**; but almost insoluble in **solvent-ether** and **chloroform**.
5. The **solutions of streptomycin** are laevorotatory.
6. The **streptomycin molecule** essentially comprises *two* **strongly basic guanido moieties** (*see structure given earlier*) plus **one weakly basic methylamino moiety**.

Constitution of Streptomycin

Based on the various analytical data, scientific and logical explanations evidences the **constitution of streptomycin** may be eleborated as under:

1. **Molecular Formula** – From various critical analytical data, the molecular formula of streptomycin has been found to be $C_{21}H_{39}N_7O_{12}$.
2. **Nature of Seven N-Atoms** – Since the **streptomycin** specifically forms a **trihydrochloride salt** respectively, it glaringly indicates that the *three* **N-atoms** present in the molecule must be *strongly basic in nature*.

    Thus, we may have:

$$C_{21}H_{39}N_7O_{12} + 3HCl \longrightarrow C_{21}H_{39}N_7O_{12} . 3HCl$$

Streptomycin	**Streptomycin trihydro-chloride (Crystalline Product)**

3. **Hydrolysis of Streptomycin** – The **hydrolysis of streptomycin** in a *strongly acidic solution* clearly yields *one mole each of:*
    - **streptidine** $(C_8H_{18}N_6O_4)$; and
    - **streptobiosamine** $(C_{13}H_{23}NO_9)$*.

    Thus, we may have:

$$C_{21}H_{39}N_7O_{12} + H_2O \xrightarrow[\text{(Hydrolysis)}]{\textbf{Mild Acid}} C_8H_{18}N_6O_4 + C_{13}H_{23}NO_9$$

Streptomycin	**Streptidine**	**Streptobiosamine**

- **Alkaline Hydrolysis** – of *streptomycin* gives rise to the formation of a γ-pyrone structural analogue, **Maltol**, which is regarded to be an outcome from *one of the fragments (i.e., streptose)* of **streptobiosamine** (as obtained above).

Maltol

---

* Folkers KA: Decature III; Ph.D., Wisconsin (Adkins)

Elucidation of the Structure of '*Streptidine*' and '*Streptobiosamine*':

At this *critical juncture* an attempt shall be made to elucidate the exact constitution of '**streptidine**' and '**streptobiosamine**' individually as under:

- ❏ **Structure of 'Streptidine'** – The following steps are involved in establishing precisely the structure of streptidine:

  (i) Molecular Formula – Based on the different *chemical* data the **molecular formula** of '**streptidine**' has been found to be $C_8H_{18}N_6O_4$.

  (ii) Presence of Two Guanido $\underset{(-NH.\overset{\displaystyle \|}{C}-NH-)}{\overset{NH}{}}$ **Moieties** – The **streptidine** on being subject to *oxidation* with **KMnO$_4$** – gives *two* moles of guanidine [$H_2N-CO-NH_2$], thereby showing the presence of *two* '**guanido moieties**' in streptidine.

  (iii) Alkaline Hydrolysis of Streptidine – When **streptidine** is subjected to careful *alkaline hydrolysis* – it gives *two* products *viz.*, **streptamine** and **ammonia (NH$_3$)** (Brink and Folker, 1947)*.

  Thus, we may have:

$$C_8 H_{18} N_6 O_4 + 4H_2O \longrightarrow C_6 H_{14} N_2 O_4 + 4NH_3\uparrow + 2CO_2\uparrow$$

  Streptidine   **Streptamine**

Furthermore, it has been established that with the interaction of *benzoyl chloride* and *pyridine* the streptamine yield a **hexabenzyl derivative**. The *periodate oxidation* of **streptamine** results into the formation of a **diaminotetrahydroxy cyclohexane**. Besides, when **N, N–dibenzoyl streptamine** is oxidized with *periodic acid (HIO$_4$)*, the former takes up *two* molecules of **HIO$_4$** and gives a **dialdehyde**, which on further *oxidation* with bromine (Br$_2$) water yields **2, 4–dibenzamido 3-hydroxy glutaric acid**.

Comments: These essentially include:

1. The above sequence of reactions do reveal explicitly that the **2-amino (NH$_2$) moities** duly present in **streptamine** are not present on the *two* **adjacent C-atoms** but are located strategically at the **alternate C-atoms**.

2. Therefore, one may legitimately conclude that streptamine should be all means be:

   "1, 3 – diamino–2, 4, 5, 6–tetrahydroxy cyclohexane"

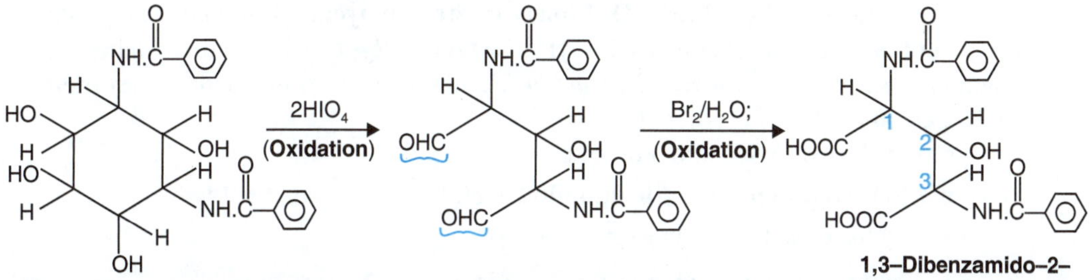

1,3(N,N–Dibenzoyl streptamine)   A Dialdehyde   1,3–Dibenzamido–2–hydroxy glutaric acid.

  (iv) Probable Chemical Structure Assigned to Streptidine – Since one may critically obtain **streptamine** by the **alkaline hydrolysis** of streptidine, the following *probable chemical structure* may be duly assigned to '**streptidine**'.

---

* Brink and Folkers: *J AM Chem Soc.*, **69**: 1234, 1947.

Thus, we may have the following expression:

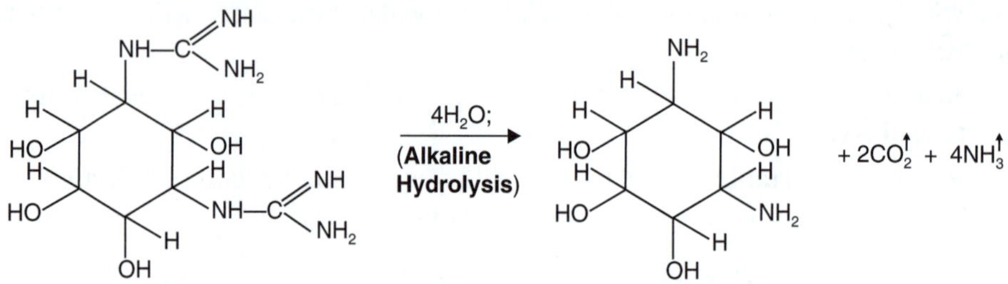

Streptidine → **Streptamine** $+ 2CO_2\uparrow + 4NH_3\uparrow$

$4H_2O$;
(Alkaline Hydrolysis)

Both CO_2 and NH_3 escape from the reaction mixture.

(v) **Confirmation of Streptamine Structure of synthesis proposed by Wolform** *et al.* **(1948).**

(vi) *Meso*-**configuration of streptidine with two guanido moieties located strategically at the *cis*-position.**

The above *meso*-**configuration** may be shown as under:

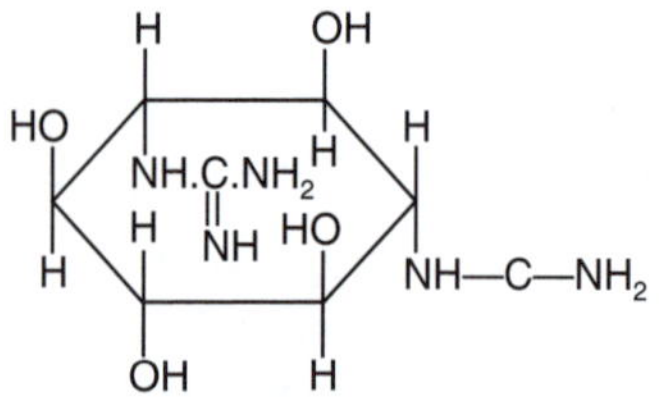

Streptidine

❑ **Structure of 'Streptobiosamine'**

(a) **Molecular Formula** – Based on the *analytical data, the* **molecular formula** of *streptobiosamine* has been found to be **$C_{13}H_{23}NO_9$**.

(b) **Formation of N-Methyl – L-Glucosamine (X) from Streptomycin** – when **streptomycin is made to react with** *methanolic hydrogen chloride* (*i.e.*, methanolysis), followed by hydrolysis, and ultimately subject to *acetylation* (with **acetic ahydride**) – it results into the formation **of N-methyl – L – glucosamine pentaacetate**. The resulting product upon careful hydrolysis yields **N-methyl–L-glucosamine (X)**.

(c) **Presence of One Carbonyl (>C=O) Moiety in Streptomycin** – It has been duly established that **streptomycin** contains **one carbonyl (>C=O) moiety**, which eventually takes part in the reaction when the *streptomycin hydrochloride* is subjected to interaction with **methanolic hydrogen chloride** (CH_3OH/HCl) to give:

 • **streptidine dihydrochloride**, and

 • **methyl streptobiosaminide dimethyl acetal**.

Thus, we may have the following expression:

$$C_{21}H_{39}N_7O_{12} + 3CH_3OH \longrightarrow C_8H_{18}N_6O_4 + 2HCl + H_2O$$

Streptomycin hydrochloride

Streptidine dihydrochloride
+ $C_{13}H_{22}NO_7(OCH_3)_3$. HCl
Methyl streptobiosaminide dimethyl acetal

Two Most Critical Observation: These essentially comprise as stated under:

❑ **Drastic Alkaline Hydrolysis** – When **methyl streptobiosaminide dimethyl acetal** is treated under *drastic alkaline hydrolysis*, it gives '**Methyl acetal**' which ascertains the presence of a **methylamino (CH_3–NH_2) moiety; and

❑ **Moderate Drastic Degradation** – Likewise, when the *moderate drastic degradation* of methyl **streptobiosaminide dimethyl acetal** is carried out with **acetic anhydride** and HCl, it yields *two* products of reaction:

• **N-Methyl glucosamine** and • **Fragment of a Hexose Sugar*.**

Inference: These observations strongly suggested that the respective '**Streptobiosamine**' must be:

• **a Glycoside of N-methylglucosamine**; and
• **the L-Streptose (a sugar which is the part of a** *streptomycin molecule*).

❑ **STRUCTURE OF N-METHYL GLUCOSAMINE**

The structure of **N-methylglucosamine** may be duly ascertained by virtue of the *four* **important sequential steps**, namely:

Step-1: The interaction of **N-methylglucosamine** with **phenyl hydrazine** [–Figure–] gives rise to the formation of **phenyl hydrazone** [Ph – CH = N.NH_2]. The resulting product may be easily converted into a **phenylosotriazole** that eventually shows the same **mp** plus an *equal but oppositive specific rotation* as that of **D-glucose phenylosotriazole** as shown below:

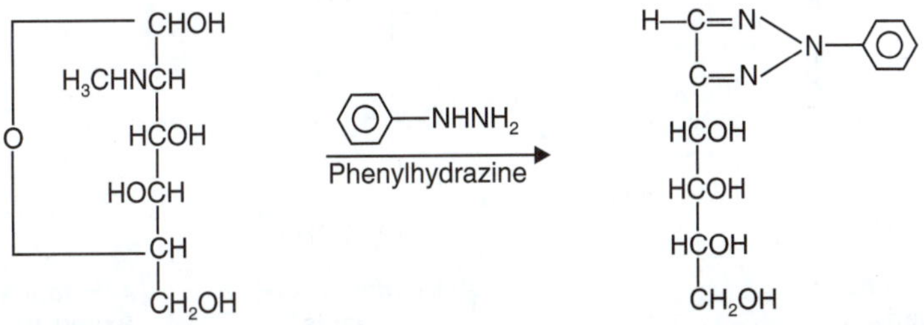

N–Methyl–L–glucosmine   L–Glucose phenyl ostriazole

Step-2: Interestingly, when **N-methylglucosamine** is carefully subjected to *oxidation* by **mercuric oxide (HgO)**, it produces an '*acid*' – whose **mp** very much corresponds to that of **N-methyl-D-glucosamic** acid; however, whose **specific optical rotation** is found to be just opposite to that of **N-methyl-D-glucosomic acid**.

Inference–Based on these critical observations it is quite evident that the **N–methyl glucosamine** duly derived from the **streptomycin** is certainly, *not* **N-methyl-D-glucosamine** but '**N-methyl-L-glucosamine**'– as shown under:

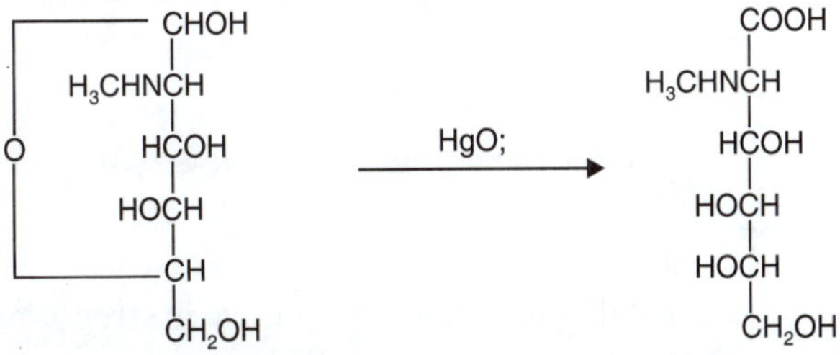

N–Methyl–L–glucosamine   N–Methyl–L–glucosamine acid

* It was subsequently termed as – '**Streptose**'.

Step-3: **N-Methyl-L-glucosamine** does possess a '*pyranose ring*' (just like '*glucose*') that could be reaffirmed by the following set of *chemical reactions:*

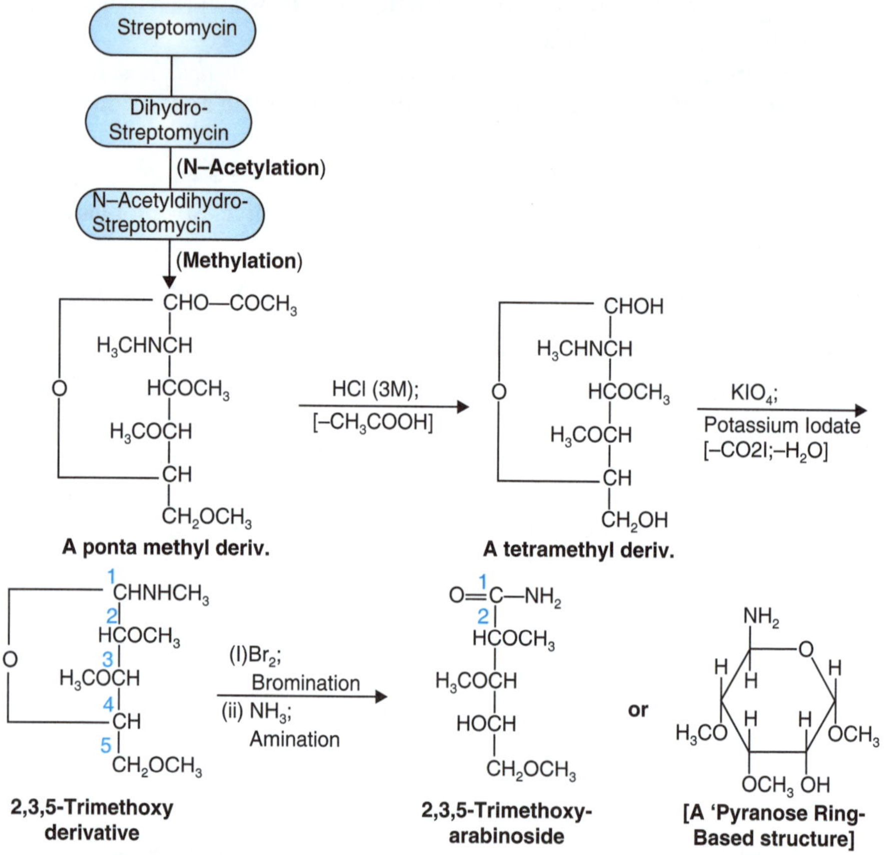

Step-4: Very much akin to '**glucose**' – the N-methyl-L-glucosamine has been proven duly to have a '*Pyranoside Ring*'. Thus, one may ultimately ascertain and confirm the structure of **N-methyl-L-glucosamine** based on its actual synthesis commencing from **L-arabinose** as given below:

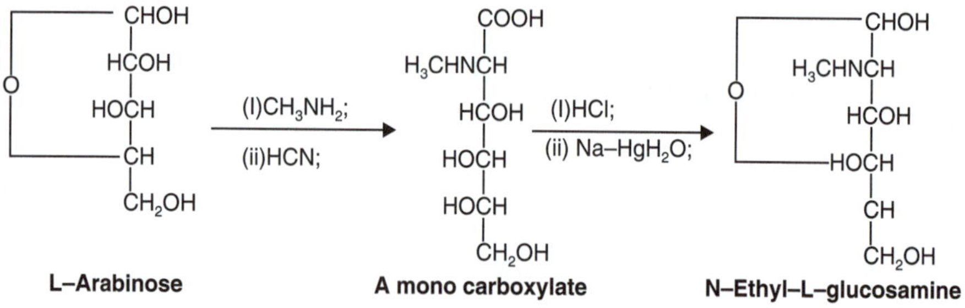

□ STRUCTURE OF STREPTOSE

It is, however, pertinent to state here that the '*streptose unit*' present in **streptomycin**, could not be isolated from it by *degradation procedure*, since it is highly unstable in nature.

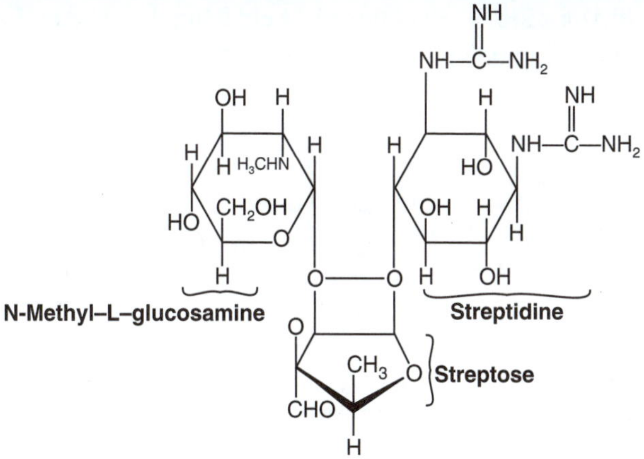

Streptomycin

Nevertheless, the precise *chemical structure* of '**streptose unit**' may be elucidated justifiably by means of an array of **degradative known procedures**.

Examples: These essentially include:

❑ **Hydrolysis of Streptomycin** – The **alkaline hydrolysis** of *streptomycin* gives **maltol**, which critically involves the expansion of a '*furanose ring system*' (*i.e.*, the '**streptose unit**' – *see structure above*) present in **streptomycin** into the corresponding **γ-pyrone**.

❑ **Formation of Maltol** – It could only be possible an *aldeyde (–CHO) moiety* is positioned strategically at C–1 in the **streptose** moiety gets solely invoved in the **glycosidic linkage**; whereas, the respective **carbonyl (>C=O) moiety** present at **C–3** remains free potentially.

❑ **Dihydro Streptomycin Fails to Form 'Maltol'** – In fact, the **dihydro streptomycin** having a primary (–CH$_2$OH) alcoholic functional moiety, instead of an aldehyde (–CHO) moiety, fails to form '**maltol**' under similar experimental parameters.

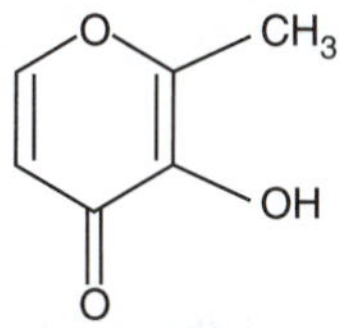

Maltol

Supportive Chemical Evidences for Streptose Moiety – The various supportive chemical evidence for establising the structure of streptose are as enumerated under:

Streptobiosamine when subjected to oxidation with bromine (Br$_2$), it critically gives a product that upon **hydrolysis**, yields **two** products respectively:

• **Streptosonic Acid**
• **Monolactone (C$_6$H$_8$O$_6$)**.

In fact, the **monolactone** gets duly converted into an **amide (–CONH$_2$)** that specifically takes up *two* **moles of periodic acid (HIO$_4$)**.

Thereby suggesting vehemently the presence of *three* **hydroxyl (–OH) moieties**. Thus, based upon these reactions, the proposed structures of:

- **Streptosonic Acid Diamide** • **Streptosonic Acid Monolactone and** • **Streptose,**

may be duly written as given under:

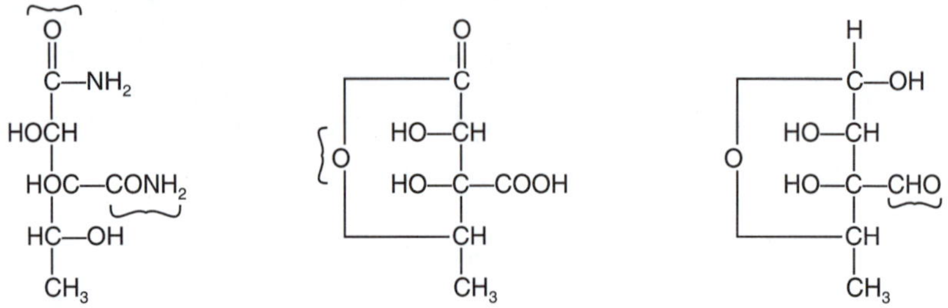

| Streptosonic Acid Diamide | Streptosonic Acid Monolactone | Streptose |

Comment: Importantly, all the above *three* chemical entities have been duly confirmed by **FTIR–studies**.

Based upon the above elaborated and confirmed chemical structures, namely:

- **N-methyl-L-glucosamine,**
- **streptose,** and
- **streptobiosamine (a *disaccharide*),**

One may assign the following *two* structure for '**streptobiosamine**':

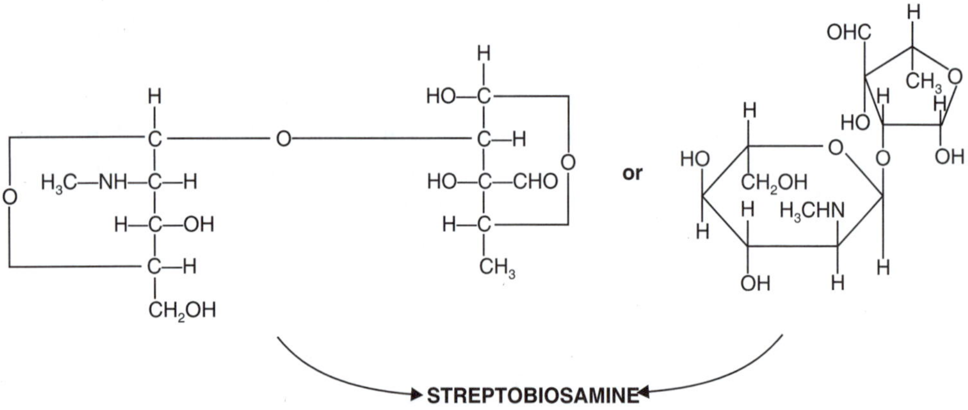

→ **STREPTOBIOSAMINE** ←

(d)   Point of Linkage Between 'Streptidine' and 'Streptobiosamine' –

At this point in time, a sincere intelligent attempt was made to establish the precise and exact point of linkage between the *two* **chemical entities**:

  • **Streptidine**   and   • **Streptobiosamine**.

In fact, the concerted efforts of the group of chemists (at the famous *Merck Laboratories, USA*) carried out the hydrolysis of the so-called completely '*benzoylated streptomycin*' with utmost care without causing any possible disturbance to the ensuring '**ester linkage**' to produce an *optically active* **heptabenzoyl streptidine**. Furthermore, the resulting product has a *hydroxyl (–OH) moiety* that is being duly converted to:

- **Methyl derivative** • **Iodo compound** • **Dihydro derivative and**
- **Ultimately to Deoxy Derivative.**

The end-product (*i.e.*, the **deoxy derivative**) on being subject to hydrolysis so as to get rid of the *ortho*-**benzoyl moieites** gives **N, N′–dibenzoyl deoxy streptamine**. However, the latter product when subjected to the respective *Periodate (HIO₄) oxidation yields α, γ-dibenzamido-β-hydroxy-adipaldehyde*. Thus, we may have the following expression:

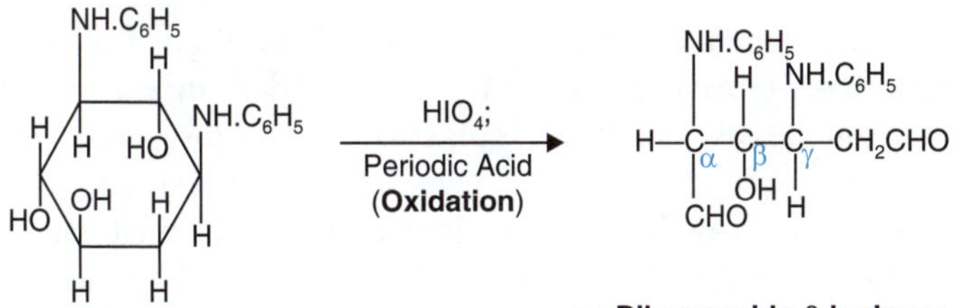

N,N′—Dibenzoyl deoxy-
streptadine

α,γ-Dibenzamido-β-hydroxy
adipaldehyde

Comments: These essentially include:

1. Importantly, the *product of oxidation* establishes explicitly the chemical structure of the **dibenzoyl derivative** that in turn reveals that the *hydroxyl (–H) moiety*, which is located strategically adjacent **to one of the C-atoms** attached to the respective **guanido functional groups of streptidine** remains involved in the ensuring linkage between:

    • **Streptidine** and **Streptobiosamine**.

2. Furthermore, the so-called **C-1 atom of the streptose unit** is intimately involved in the structure of **streptobiosamine**, which is duly ascertained due to the underlying fact that:

    ❏ *Maltol* – **is critically produced by the molecular rearrangement upon the** *alkaline degradation* **of** *streptomycin*; **and**

    ❏ *Maltol* – **is NOT generated from both** *tetra*-**acetyl and N-acetyl** *streptobiosamine* **upon alkaline degradation.**

Comments:

   • The latter *chemical entities* do not have *any substituents* attached to **C-1 atom**.

   • Hence, it clearly indicates that the *substituents should by all means by present for the maltol production.*

   • Therefore, the *C-1, atom of streptose* is attached with a **'glycosidic linkage'** to the C-4 of the *streptidine fragment* in order to give the following chemical structure to '**streptomycin**' (see structure of 'streptomycin' under structure of streptose earlier)

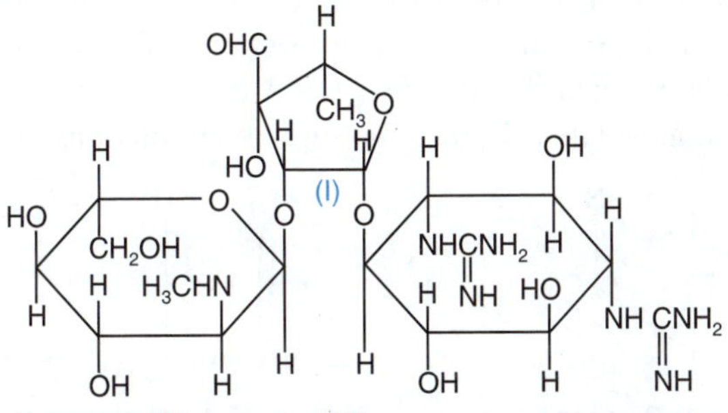

---

* Lemieu X, Wolform : *The Chemistry of Streptomycin*, In: Pigman WW and Wolform ML, *Adv. Carbohydr. Chem.*, **3** : 337–384, 1948.

HUDSON'S ISOROTATION RULES

Hudson's Rules – Comparison of isomers that differ in configuration at each of two contiguous asymmetric centers is vitiated by an as yet undeterminable vicinal effect of each center on the rotatory contribution of the other. This *vicinal action* is illustrated by the comparison given in Table 2.1 of the molecular rotations of *pairs of pyranoses* that differ only in the **configuration at C_1**. The **isomer of the α-configuration** is invariably **more dextrorotatory** than the **Iβ-epimer**, but the difference amounts to an average of + I77 units in the case of **glucose, galactose,** and **lactose,** and to an average of only + 86 units in the case of **mannose, talose,** and 4-glucomannose. The difference in the averages (+ I77 and + 86) affords an indication of the magnitude of the vicinal effect concerned. **Hudson*(1909)** expressed the relationship illustrated in Table 2.1 by a **first rule** stating that the rotatory contribution of C_1 is affected in only a minor degree by changes in the structure of the remainder of the molecule provided the changes are not on *contiguous atoms*.

Table 2.1 Comparison of C_α- and C_β-Sugars		
Sugar	M_D	$M^\alpha{}_D - M^\beta{}_D$
	α β	
Glucopyranose Type		
Glucose	+ 202.1 + 33.7	+ 168.4
Galactose	+ 271.5 – 95.1	+ 176.4
Lactose (disaccharide)	+ 306.3 + 119.5	+ 186.8
Glucopyranose Type		
Mannose	+ 52.8 + 30.6	+ 83.4
Talose	+ 122.5 – 23.8	+ 98.7
4-Glucomannose	+ 53.0 – 22.0	+ 75.0

Hudons's second rule states that changes at the glucosidic carbon atom (C_1), for example the change from C_1-OH to C_1-OR, affect in only a minor degree rotation of the remainder of the molecule. That methyl α-D-glucoside is distinctly *more dextrotatory* than α-**D-glucose** can be interpreted as meaning that the bulkier OCH_3 increases dissymmetry of C_1; and hence, accentuates the destorotatory contribution of C_1. **Methyl β-D-glucoside** is correspondingly more levorotatory than the *parent β-D-glucose* (Table 2.2). Increase in size of the alkyl group from methyl to benzyl results in progressive, smaller changes in the same direction. If the rotatory contribution of C_1 is +A in the α-form and –A in the β-form and the contribution of the rest of the molecule is B, then the sum of the rotations of any pair

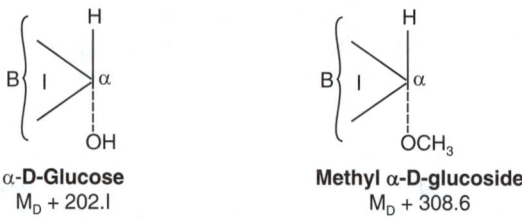

α-D-Glucose
M_D + 202.I

Methyl α-D-glucoside
M_D + 308.6

*   Claude S. Hudson, 1881–1952; b. Atlanta; Ph.D. Priceton; U.S. Public Health Service; J. Chem. Soc., 4042 (1954)

Table 2.2. D-Glucopyranosides

C_1-Substituent	Iα	Iβ	Sum
OH (D-Glucose)	+ 202.1	+ 33.7	+ 235.8
OCH_3	+ 308.6	− 66.4	+ 242.2
OC_2H_5	+ 313.6	− 69.5	+ 244.1
OC_3H_{7-n}	+ 312.9	− 77.6	+ 235.3
$OCH_2C_6H_5$	+ 354.1	− 150.2	+ 203.9
OC_6H_5	+ 463.3	− 181.9	+ 281.4

of I-epimers should be a constant quantity, 2B. The data of Table 2.2 show that $(M^\alpha_D + M^\beta_D)$ is indeed substantially a constant; whereas M^α_D varies from + 202 to + 463, the sum of the molecular rotations deviates from the average of + 240 to the extent of only ±15 (average).

The data of Tables : 2.1 and 2.2 include several illustrations of the general relationship that in a pair of glycosidic anomers, the isomer with the glycosidic hydroxyl or alkoxyl group in the α-orientation (down, to the rear) is invariably more dexrorotatory than the α-epimer. The rule holds for the pentoses, D-lyxose, D-xylose, and L-arabinose, for the molecular rotation differences $(\alpha − \beta)$ are all positive.

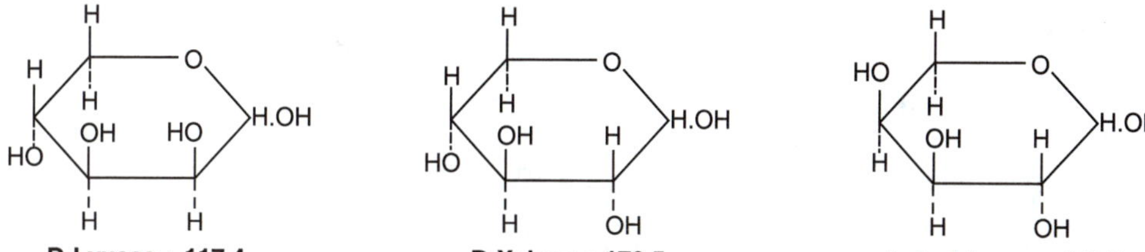

D-Lyxose + 117.4   D-Xylose + 170.5   L-Arabinose + 170.5

$$M_D^{\alpha} - M_D^{\beta}$$

The example of **L-arabinose** and the further example of **L-rhamnose**, a natural methylpentose, show that in the **L-series** as in the D-series, the **glycosidic carbon makes a greater dextorotatory**

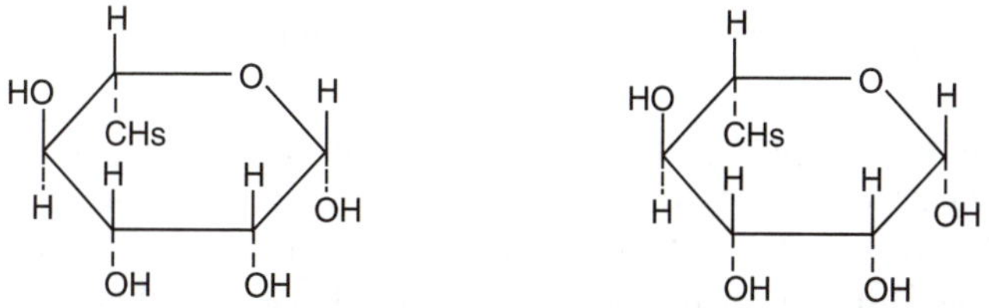

α-L-Rhamnose MD + 63 β-L-Rhamnose MD + 15.7

contribution when the *hydroxyl* is oriented **below the plane of the ring than when above the plane**. D-Fructose forms a pair of **methylpyranosides epimeric at C_2**; the one with the **glycosidic methoxyl oriented below the plane of the ring** is defined as the α-**isomer** and this isomer is the **more dextorotatory member of the pair**.

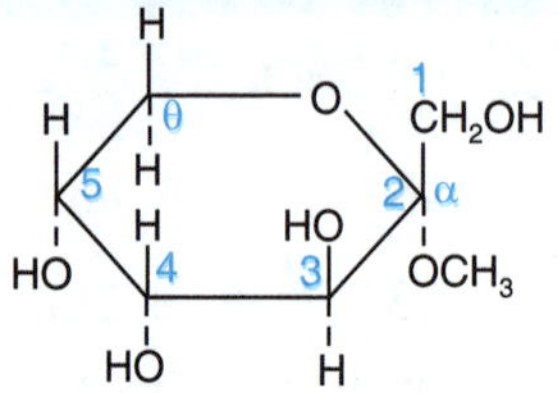

Methyl α-D-fructopyranoside
$M_D + 85$

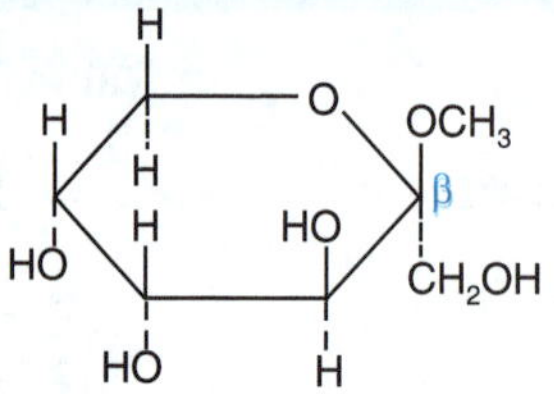

Methyl α-D-fructopyranoside
$M_D + 334$

Wolform *et al.* (1950)* duly proposed the following *specific linkages* in *streptomycin*:

(a) **β-L-Glycosidic Linkage** – is found to be present between the *streptose* (which designates the **L-isomer**) and *streptidine* (III); and

(b) **β-L-glycosidic Linkage** – is known to be present between the **N-methyl-L-glucosamine (II)** and **streptose**.

> **NOTE** The aforesaid linkages have been duly accepted for quite sometime and these are explicitly depicted in the above structure of *streptomycin* (under structure of 'streptose')

Reinehart *et al.* (1965) made his spectacular observations based solely on the **NMR-spectral studies of streptomycin** and ultimately came to the conclusion that the *precise linkage* between (I) and (III) (under **section-'d'**) is found to be α-L; and hence, they vehemently assigned the following structure for '**streptomycin**':

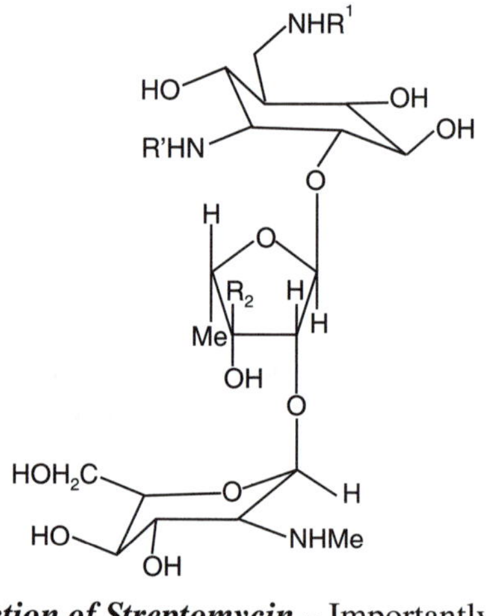

[III] $R_1 = -\overset{\overset{\displaystyle NH}{\|}}{C}-NH_2$;

[I] $R_2 = CHO$; α-Linkage

[II] α-Linkage

Mechanism of Action of Streptomycin – Importantly, the exact underlying *mechanism of action of* **streptomycin** against various microbes still remain a mystery – a good number of meaningful investigative studies have duly expatiated that **streptomycin** predominantly interferes with the *normal cellular division*. In fact, **streptomycin** vividly intervenes in the **RNA-metabolism** by causing a blockade in the chemical reaction just preceding the formation of **mononucleotides**. Furthermore, Gale (1963) has demonstrated explicitly the inhibition pertaining to the typical *oxidation of RNA* with certain microorganisms exclusively.

---

* Wolform *et al.*, *J Am Chem Soc.*, **76**: 3675, 1950.

Umbreit *et al.* showed that **streptomycin** largely interfers with the ongoing interaction between pyruvic acid and oxaloacetic acid particularly in the **cell-metabolism**. However, the aforesaid analogy is found to be equally important in:

- **Bacterial Cells** and **Animal Tissues**.

Besides, the above '*concept*' has been duly substantiated by the *vital observations* that:

❏ **Streptomycin** – *does prevent the critical generation of an altogether newer metabolic intermediate*; *and*

❏ *Resulting products formed normally in microorganisms only when: Pyruvic Acid and* **4 C-dicarboxylate** *are duly present*.

It is, however, pertinent to state here that **streptomycin**, exerts its ensuring activities upon the so-called '*tubercle bacilli*' when they are in a '**growing state**' rather than in a '**resting phase**'. Therefore, one may safely conclude that the undergoing *phenomenon of division* taking place due to its *rapid synthesis of bacterial cell constituents* **– remains virtually blocked by the drug substance particularly.**

NOTE The foretold crucial observation is very much in accordance with the recognised *insufficient response of streptomycin* to *clear old dormant TB lesions –* whereupon the bacilli are usually present in a '*resting state*'.

Uses

1. Streptomycin serves as a '**bactericidal drug**'.
2. It is extensively employed in the treatment of an array of human ailments, such as:
   - **Pulmonary Tuberculosis** • **Plague** and • **Tularaemia.**
3. **Streptomycin** is also found to be quite useful in the **infections** produced by a vari of **pathogenic microbes**, such as:
   - *E. Coli* • *P. pyocyaneus* • *H. influenzae* • *Proteus vulgaris* and • *Brucella abortus*.
4. Besides, **streptomycin** in conjunction with **penicillin** in the specific treatment of:
   '**Subacute bacterial endocarditis duly produced by** *Streptococcus faccalis***, resistant staphylococci, and postoperative infections**'.
5. **Streptomycin** is also a drug of choice for the treatment of:
   - **Pneumonia** and
   - **Urinary Tract Infections (UTIs).**

NOTE 1. Oral administration of *streptomycin* is not so-effective in the specific treatment of '*systemic infections*'
2. Unlike *penicillin* – streptomycin is not inactivated by the so-called *microbial enzymes.*

3.4. Chloramphenicol

Chloramphenicol (chloromycetin) is a **lavorotatory broad-spectrum antibiotic** orginally produced from a host of **Streptomyces** spp. namely:

S. venezualae; and *S. phacochromogenes var. chloromycetieus*.

Interestingly, it has been reported to be the drug of choice for the specific treatment of *Typhus* and Typhoid **fever**.

Paramount Importance of Chloramphenicol – The following three reasons are of **paramount importance of chloramphenicol**, namely:

❑ It is a naturally occurring aromatic nitro compound of which there are only one previously recorded example of *hiptagin* duly obtained from the root bark of *Hiptage madablota* Gartn.

❑ It is quite capable of exerting its effect against the *viral diseases* and also those caused due to the **microbial invasion**; and hence, opens up the entire field of chemotherapy of *virus and ricketsial* infections in humans including: **Typhus** and **Undulant Fever**. Besides, it is also indicated in an array of other ailments:

- *Salmonella septicaemia*   • **Whooping cough**   • **Typhoid**
- **Paratyphoid** and   • *Lymphogranuloma inguinale*.

NOTE	**As-to-date, the chloramphenicol-fast strains have not yet been isolated**. and

❑ It is amenable to synthesis on a commercial scale.

Large-Scale Production of Chloramphenicol: The large-scale (or commercial) production of chloramphenicol may be accomplished by adopting the following *three* sequential steps:

Step-1: The first and foremost task is to undertake the preparation of culture medium by mixing together:

- **Maltose** (or *glycerol*);   • **Wheat gluten** (or *meat products*);
- **Sodium carbonate (Na_2CO_3); and**   • **Sodium chloride (NaCl),**

in order to *adjust the pH* with the addition of **anti-forming agents.**

Important Observation – It has been duly ascertained that the *'fermentation residue'* obtained from molasses perceptively helps in the stimulation of the **production of chloramphenicol**.

Step-2: Importantly, *Streptomyces venezuelae i.e.,* the organism responsible for producing **chloramphenicol**, is duly obtained from the *soil of Venezula (a country)*. The master cultures (or the *mother culture*) of *Streptomyces venezuelae* are preserved meticulously upon the 'soil' and the 'slope cultures' are subsequently prepared from time-to-time as per need. Now, by making use of these (*slope cultures*) – the requisite suspensions are carefully prepared in a so-called '*soapy solution*'. The resulting prepared solutions are preserved under refrigerated conditions (in **sealed units**) and put into actual usage as per requirement.

Step-3: The actual progressive growth of *Streptomyces venezuelae* is most diligently performed in a **strict step wise manner** as elaborated under:

❑ **First** – the sub-cultures of *S. venezuelae* [prepared as in (step-2) above] are carefully transferred to the respective culture medium aseptically [prepared duly as in a (**Step-1**) above] in a **200 L tank**, in which the normal growth is allowed to take place for **24 hours**.

❑ **Secondly** – the resulting treated culture medium, the total contents of the 200L tank (**SS**) are transferred to another pre-sterilized SS–2000L tank–wherein a further growth of culture is allowed for a duration of 24 hours.

❑ **Thirdly** – the entire contents of 2000 L tank capacity are duly transferred to the **20,000 L** pre-sterilized SS-tank (*i.e.,* **fermentor**) – wherein, the ultimate growth of culture is allowed for a span of **72 hours**.

Isolation of Chloramphenicol: In fact, the isolation of chloramphenicol is much simpler in comparison to that of penicillin perhaps due to its relatively higher stability profile and also its least dependency on **pH adjustment**.

- The 20,000 L of the medium (obtained from 3rd stage above) in the fermentor is duly concentrated. Subsequently, the concentrated culture is extracted with either:
- **Ethyl Acetate** or **Amyl Acetate** by means of the principle of counter-current.
- The *ethylacetate* or *Amylacetate* (along with the '*culture*') is duly subjected to separation by distillation process especially under Vacuo so as to preserve the santity and authenticity of the chemical entity. Later on, the *distillate* is carefully concentrated under reduced pressure and washed several times with H_2SO_4, **NaOH solutions, and DW finally**. The resulting product is evaporated slowly when one may lay hands on to the crystals of chloramphenicol.
- The *crude product* is further recrystallized either :
- **Ethylene Dichloride [Cl–CH$_2$–CH$_2$–Cl]**, or
- **DM-water containing activated charcoal**.

Characteristic Features

1. **Chloramphenicol** occurs as needless or elongated plates from *water* or *ethylene dichloride* (**mp:** 150.5–151.5°C)
2. It sublimes in high vacuum.
3. **Solubility Profile** – It is soluble in **water** at 25°C; (2.5 mg.mL^{-1}) **propylene glycol** (150.8mg.mL^{-1}); very soluble in methanol, ethanol, butanol acetone, and **ethyl acetate;** and its solubility in 50% (w/v) **acetamide ~ 5%**.
4. Its aqueous solutions are found to be '**neutral**'; besides, the *neutral* and *acidic solutions* are fairly stable on heating.

Chloramphenicol

2,2-Dichloro-N[(1R,2R)-2-hydroxy-1-(hydroxymethyl)-2-(4-nitrophenyl)-2-amino-1,3-propanediol;]

Constitution of Chloramphenicol

Based on various scientific logical evidences and analytical data the **constitution of chloramphenicol** has been duly elucidated by:

- **Contronlis *et al.* (1949):** *J AM Chem Soc.,* 2463, 1949; and
- **Long and Troutman (1949):** *J Am Chem Soc.,* 2469, 949.

1. **Molecular Formula** – The essential analytical data suggests and ascertains the *molecular formula* *of* chloramphenicol to be $C_{11}H_{12}N_2O_5Cl_2$.
2. **Presence of One Nitro (–NO$_2$) Moiety** – Based on the critical UV-absorption spectrum of *chloramphenicol* resembling to that of **nitrobenzene** suggests strongly that the former does contain

a **nitro (NO$_2$) functional moiety**. Furthermore, the presence of the **nitro moiety** was confirmed by the specific reduction of **chloramphenicol** with Sn/HCl, followed immediately by *diazodization* and finally coupling with β-napthtol to obtain an **orange-coloured precipitate**.

Further Confirmation of Nitro (NO$_2$) Group in Chloramphenicol – may be accomplished by the fact that the drug on being reduced catalytically with *Pd* gives a product whose **UV-absorption spectrum** matches similar to that of *para*-**toluidine [– Figure –]** and whose solution does contain *ionic chlorine*.

Remarks – The aforesaid observations do reveal and confirm that **chloramphenicol** designates a *para*-**nitrobenzene substituted chemical entity** and also its '**chlorine atom**' is located only in the **side-chain**.

3. **Absence of Free-Amino (–NH$_2$) and Carbonyl (>C=O) Moieties** – Based on the *normal tests* available for *amino*-and *carbonyl* moieties, it has been shown that **chloramphenicol** contains **neither free amino group not carboxyl group**.

4. **Presence of 2-Hydroxyl (OH) Groups** – **Chloramphenicol** on being subject to *acetylation* with **acetic anhydride [(CH$_3$CO)$_2$O]** in pyridine gives a diacetyl derivative, which clearly shows that it containing *two* **hydroxyl groups**.

5. **Hydrolysis of Chloramphenicol** – **Chloramphenicol** when hydrolysed with either acids or alkalis, it gives rise to the formation of **dichloroacetic acid** and an '*optically active*' basic entity, as given under:

$$C_{11}H_{12}Cl_2N_2O_5 + H_2O \longrightarrow C_9H_{12}N_2O_4 + Cl_2.CH.COOH$$

Chloramphenicol	**Base**	**Dichloroacetic acid**
	(Optically Active)	

Remarks: At this point in time, it would be necessary to establish the '*constitution of chloramphenicol*' – by elucidating **first** and foremost the **structure of the base**.

STRUCTURE OF 'BASE': The following *five* **sequential steps (a)** *through (e)* may be adopted so as to ascertain the structure of the '*base*' (**optically active in nature**):

(a) It is found to be an **optically active base having a molecular formula: C$_9$H$_{12}$N$_2$O$_4$**.

(b) The usual *known tests* – it has been proven duly to have only one *primary*–**amino (–NH$_2$) moiety**.

(c) The '*base*' on being **acetylated** with *acetic anhydride inpyridine* gives a **triacetyl derivative**– which vividly confirms the presence of:

   • **2-hydroxy (–OH) groups in the parent compound (*i.e.*, chloramphenicol); and**

   • **one pri-amino (–NH$_2$) moiety originally derived by hydrolysis.**

NOTE Besides, the '*base*' fails to produce a distinct colouration with *FeCl$_3$* **solution,** – which indicates that the *hydroxyl moieties* are '*alcoholic*' *in nature* **but not** '*phenolic*'.

(d) When it is subject to treatment with **methyldichloroacetate** it forms distinctly a **dichloroacetamide**, that is found to be absolutely identical with **chloramphenicol**.

(e) **Presence of Propyl (–CH$_2$–CH$_2$–CH$_3$) Moiety** – The '*base*' when duly oxidized with **periodic acid (HIO$_4$)**, it usually takes up two **equivalents of HIO$_4$** to give rise to the formation of:

   • **p-Nitrobenzaldehyde**

   • **Formaldehyde and**

   • **Ammonia.**

Comment: The above reaction reveals explicity the critical presence of a 'propyl moiety' that is located strategically para-to the incumbent **nitro (–NO₂) group** with an **amino (–NH₂) group** duly present on to the **second C-atom**.

Thus, we may have the following reactions:

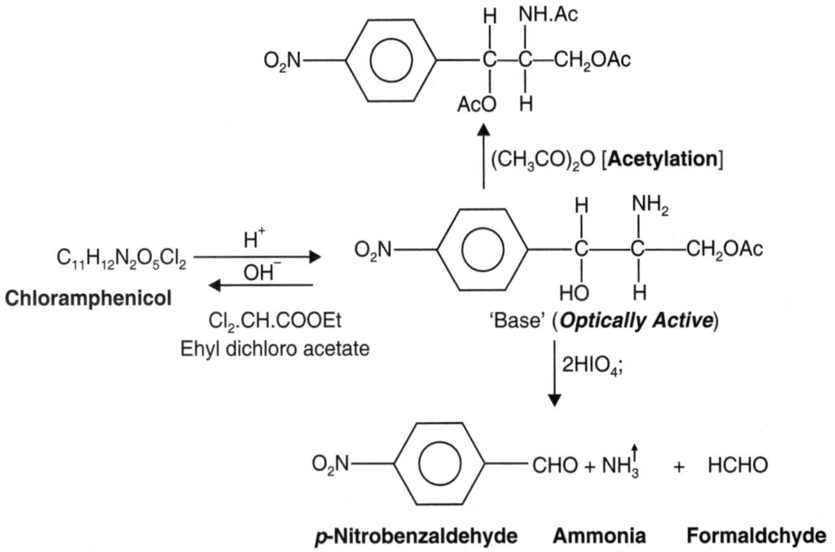

$C_{11}H_{12}N_2O_5Cl_2$
Chloramphenicol

$Cl_2.CH.COOEt$
Ehyl dichloro acetate

$(CH_3CO)_2O$ [**Acetylation**]

'Base' (*Optically Active*)

$2HIO_4$;

p-Nitrobenzaldehyde   Ammonia   Formaldchyde

Conclusion–The actual formation of all the *products of reactions* as depicted above may be expatiated only when the **formula of the '*base*'** is assumed to be 2-amino-1-p-nitrophenyl-propane-1,3-diol.

Thus, we may have:

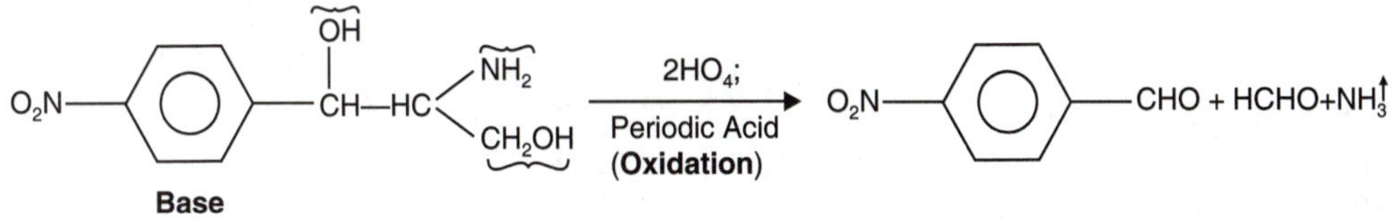

$2HO_4$;
Periodic Acid
(**Oxidation**)

Base

6. **Blockade of Free Amino (–NH₂) Group** – Since **chloramphenicol** fails to react with *periodic acid (HIO_4)*, an oxidizing agent, it indicates clearly that the **free amino (–NH₂) group in the former** gets blocked totally; and therefore, **chloramphenicol** should by all means be:

'**D-(–)-threo-2-dichloroacetamido-para-nitrophenyl propane-1, 3-diol**' and thus is found to be very much consistent with the *factual statements* raised in [5(d)] above.

Thus, we may have the following expression:

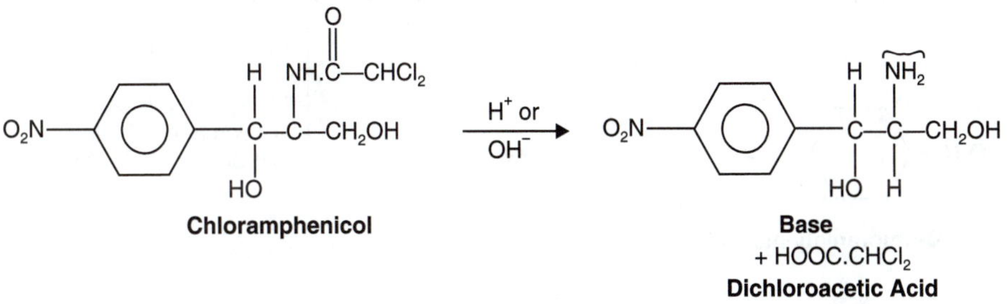

Chloramphenicol

H^+ or
OH^-

Base
+ $HOOC.CHCl_2$
Dichloroacetic Acid

7. **Probable Structure of Chloramphenicol** – Since **chloramphenicol** does not interact with *periodic acid (HIO₄)* i.e., undergoes the phenomenon of *oxidation*, it strongly suggests that the former is virtually devoid of vicinal hydroxyl (–OH) moieties. Thus, based on this observation, the **chloramphenicol** must be **D-(–)-Threo-2-dichloroacetamido-*para*-nitrophenylpropane – 1, 3-diol,** which may be represented as under:

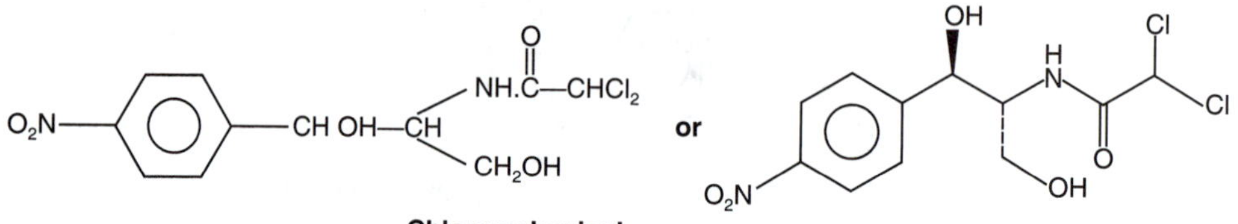

Chloramphenicol

8. **Synthesis of Chloramphenicol** – The **total synthesis** of '*chloramphenicol*' may be accomplished by either of the *two* methods:

- **Long *et al.* (1949):** From p-Nitroacetophenone; and *J Am Chem Soc.*, **2469, 2473**
- **Controulis *et al.* (1949):** From **Benzaldehyde** and *p*-Nitroethano *J Am Chem Soc.*, **2463.** l.

(a) **Synthesis of Chloramphenicol by Long *et al.* (1949):** The synthesis of **chloramphenicol** as put forward by Long *et al.* is as described below in a sequential manner:

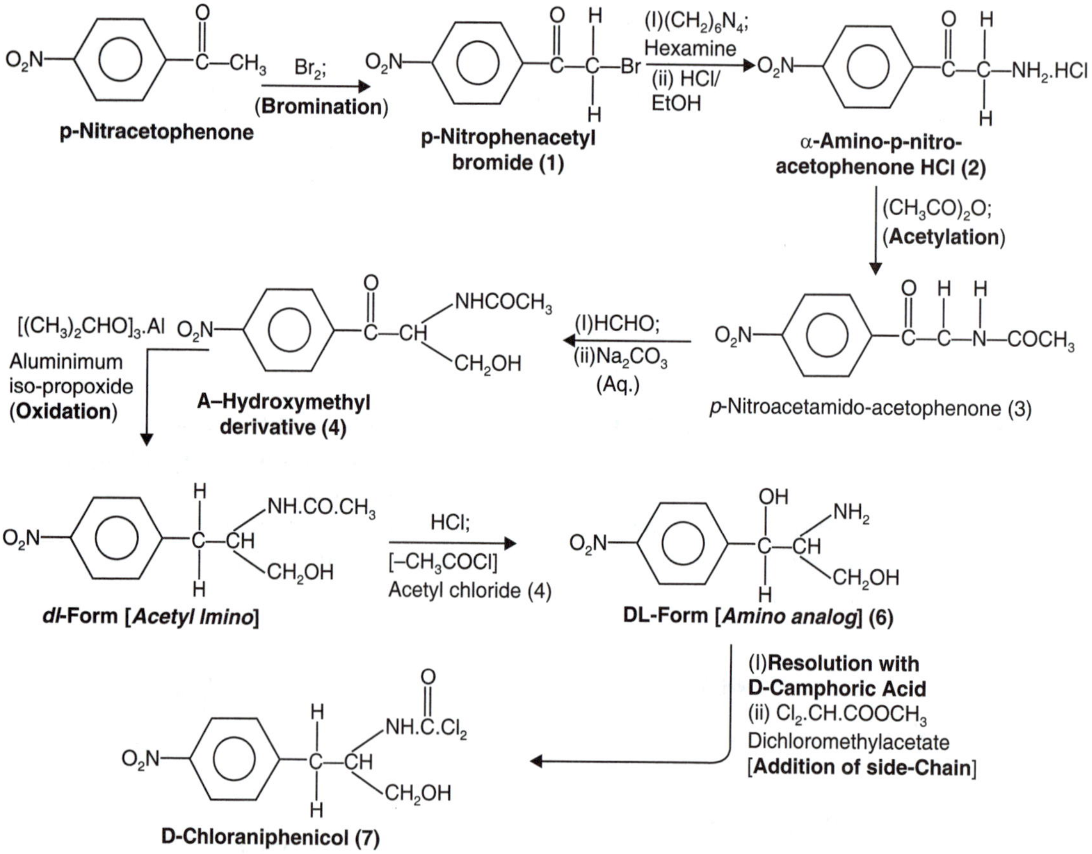

Explanation: The various steps involved in the synthesis may be explained as under:

1. Bromination of *p*-nitroacetophenone yields *p*-nitrophenacetyl bromide (1), which on treatment with *hexamine* followed by *acidified ethanol* gives **α-amino-*p*-nitroacetophenone hydrochloride (2)**.

2. The resulting **compound (2)** on acetylation gives ***p*-nitro-acetamido acetophenone (3)**, which on further treatment with *formalin (HCHO)* followed by *aqueous sodium carbonate yields a* **hydroxymethyl derivative (4)**.

3. The **end product (4)** on being subject to *oxidation* with *aluminium iso-propoxide* yields **dl-form (5)**, which on reaction with *HCl* gives the respective *amino analog* in **DL-form (6)** (with the loss of a mole of **acetyl chloride**).

4. The resulting **product (6)** is resolved with **D-camphoric acid** followed by interaction with *dichloromethyl acetate* yields **D-chloramphenicol (7)** (due to the addition of side-chain.)

   (b) **Synthesis of Chloramphenicol by Controulis *et al.* (1949):** The various steps involved in the synthesis of **Chloramphenicol**, as proposed by **controulis *et al.***, starting from **benzaldehyde** and **β-nitroethanol** are as stated below:

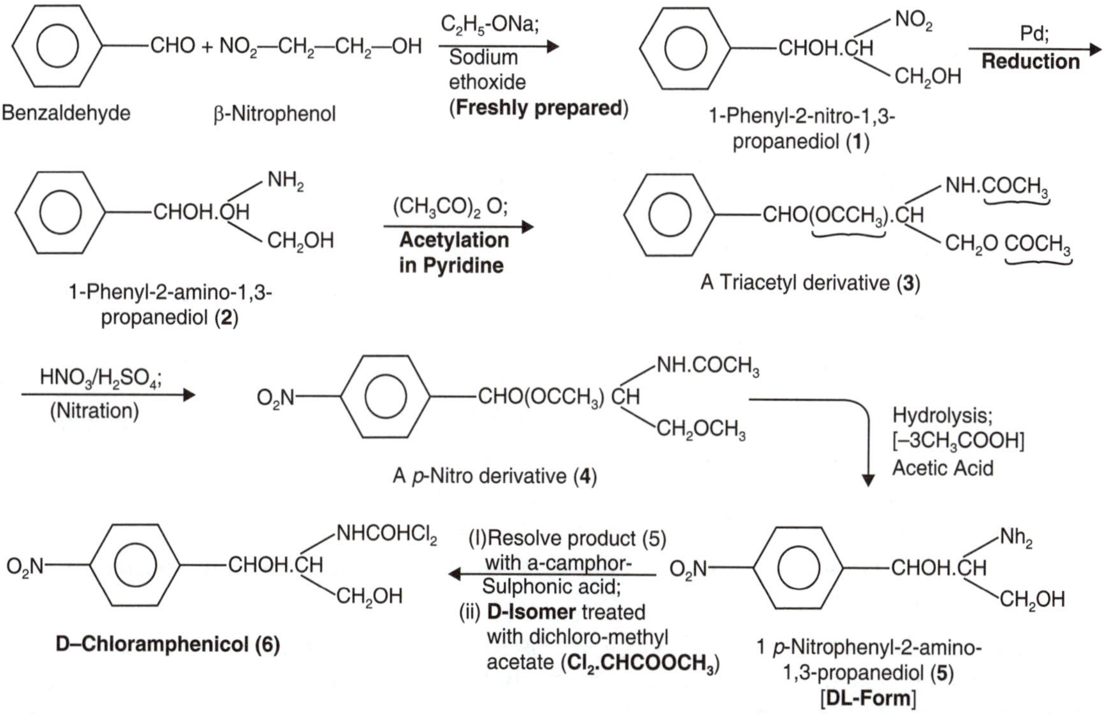

Explanation: The various steps involved in the synthesis stated above may be further explained as under:

1. Interaction of freshly distilled *benzaldehyde* and *β-nitrophenol* in the presence of *sodium ethoxide* yields **1-phenyl-2-nitro-1, 3-propanediol (1)**, which upon *reduction* with **Pd** gives **1-phenyl-2 amino-1, 3-propanediol (2)**.

2. The resulting **product (2)** when *acetylated* in pyridine gives a **triacetyl derivative (3)**, which upon *nitration* (with HNO$_3$/H$_2$SO$_4$) gives a **p-nitro derivative (4)**, which on being subject to *hydrolysis* loses *3 moles of acetic acid* to produce **1-p-nitrophenyl-2-amino-1,3-propanediol (5)** as the **DL-form**.

3. The **end-product (5)** when resolved with **α-camphor**, followed by treatment of the **D-isomer** with *dichloro methyl acetate* gives the desired product **D-chloramphenicol (6)**.

9. **X-Ray Diffraction Studies of Chloramphenicol** – The structure of **chloramphenicol** has been duly confirmed by **X-ray diffraction (crystallographic) studies** by Dunitz *et al.* (1952).

10. **Configuration of Chloramphenicol** – Since it has been proven beyond any reasonable doubt that both:

- **Chloramphenicol** and - **Base**,

essentially comprises *two* **chiral centres** (or *asymmetric C-atoms*), it distinctly indicates that there exists *two* **probable pairs** of '*enantiomers*'. Therefore, when the characteristic features of:

- **Norephenephrine** - **Nor-ψ-Ephedrine and** - **Base**,

are carefully compared, it predominantly reveals that the configuration of the '**base**' is similar to that of 'nor-ψ-ephedrine'

Based on the above scientific revelations it is interesting to observe that '**Chloramphenicol**' is: **D-(–)-threo-2-Dichloroacetamido-1-p-nitrophenylpropane-1,3-diol**.

Points to Ponder: Following are the *two* most critical and prevalent unique features that are duly contained in the structure of '**chloramphenicol**', such as:

- a nitro (–NO$_2$) functional group (first amongst the naturally occurring duly (substance); and
- presence of a dichloromethane (–CHCl$_2$) moiety.

Mechanism of Action – Chloramphenicol possesses the distinct advantage for inheriting the ability to penetrate right into the **CNS (Central Nervous System)**; and, therefore, it is still regarded to be an important **alternative therapy** for the treatment of **meningities**. However, the major course for the metabolism of **chloramphenicol** essentially involves the critical formation of the 3–O–glucuronide. Obviously, the *minor* reactions necessarily comprise the following four vital aspects:

❑ reduction of inherent *p*-nitro group to the respective '**amine (–NH$_2$) function**';

❑ **hydrolysis of the amide (–CONH$_2$) group**;

❑ **hydrolysis of α-chloroacetamido group**; and

❑ **reduction to give α-hydroxyacetyl analogue***.

Modus Operandi of Chloramphenicol – The various modalities by which **chloramphenicol** exerts its *pharmacological actions* are as stated under:

1. It gets absorbed very rapidly from the **GI-tract** with a *bioavailability* of ~ 20%. Besides, almost 60% of the *drug* in blood remains intimately bound to **serum albumin**; and gets ultimately biotransformed in the '*liver*' within a range of 85–95% . However, the volume of distribution v_d^{55} stands at 0.7 mL.g^{-1}.

2. Its plasma **half-life ($t_{1/2}$)** ranges between 1.5 to 5 hours, except over 24 hours in **neonates** (1–2 days old), and *10 hours* in in infants (10–16 days old).

3. The '*drug*' may cross the **placental barrier** and subsequently the **foetus**; and hence, it must not be recommended to a pregnant woman.

 NOTE | The '*prodrug*' of chloramphenicol *e.g.,* chloramphenicol palmitate **(USP)**, which is a tasteless product, is solely intendended for *pediatric usage* commonly, since the parent drug has a distinct *bitter taste*.

Structure-Activity Relationship (SAR) – The **chloramphenicol** critically inherits *two* **chiral (asymmetric) C-atoms** duly present in the '*acylaminopropanediol chain*' as illustrated under:

NO$_2$ NO$_2$

HO—C—H   H—C—OH

H—C—NH.CO.CHCl$_2$ H—C—NH.CO.CHCl$_2$

CH$_2$OH   CH$_2$OH

(*Threo*-Form)   **(*Erythro*-Form)**

Therefore, there are *two* possible pairs of 'enantiomorphs'. Based on the *biological activity profile* solely attributed to the respective:

- **D-*Threo*–Isomer**; (active') and
- **L-*Threo*, D-and L-*Erythro*-isomers remain '*inactive*' virtually.**

Since '**chloramphenicol**' as its **O,N-diacetate derivative (X)** critically undergoes the '*acyl migration*'; and hence, it was duly assigned the *threo*–configuration as given below:

CH$_3$

O$_2$N— —CH—CH—CH$_2$OAc

OH NH

COCH$_3$

C$_2$H$_5$

Intermediate in N→O migration   **O,N-Diacetate Derivative (X)**

---

* Glazko A: *Antimicrob Agents chemother*, 655, 1966.

Structural-Analogues of Chloramphenicol – An attempt has been made to prepare an array of structural analogues of chloramphenicol that are duly based on the following '*themes*', *namely:*

- **removal of the chlorine atom,**
- **transference of chlorine atom to the '*aromatic nucleus*',**
- **transference of nitro (–NO₂) group to the ortho-or meta-position,**
- **esterification of the hydroxyl (–OH) group (S),**
- **replacement of *phenyl ring* with *furyl, naphthyl,* and xenyl rings respectively,**
- **addition of alkyl or *alkoxy* substituents to *arylring*,** and
- **replacement of inherent *nitro* (NO₂) *group* by a *halogen atom*.**

Remarks: None of the structurally modified analogues exhibited an overall activity approaching to that of '**chloramphinicol**' towards *Shigella paradysenteriae*.

3.5. The Tetracyclines

Preamble: The epoch-making discovery of **chlortetracycline (aureomycin)** in 1947 by **Duggar** virtually paved the way for a number of structural analogues used as **broad-spectrum antibiotics** which belong to the **tetracycline family**. Later on, in 1950, **Findlay** and his co-workers could grow successfully in *culture broth* a *new actinomycete*, known as *Streptomyces rimosus*, right from a soil sample that could yield an antibiotic terramycin. Subsequently, in 1952, the most exceptional chemical structure of the *two* **foretold antibiotics** were duly established and thus, a 3rd compound was prepared that eventually possessed **antibacterial characteristic features.** In a generalized way, these '*antibiotics*' were collectively termed as the '**Tetracyclines**'. With the passage of time a host of **structural analogues of the tetracyclines** have been prepared *synthetically* to the advantage of **newer therapeutic drugs.**

Tetracycline [A Generalized Structure]

The '**Tetracyclines**' that are found to be effective therapeutically have been enlisted in the table given below:

Name of Compound	Official Status	Brand Name(s)	R_1	R_2	R_3	R_4	R_5
Tetracycline	BPC (1973); USP;	Tetracyn(R) (Pfizer); SK-Tetracycline(R) (SK & F)	H	OH	CH_3	H	H
Oxytetracycline	USP;	Terramycin(R) (Pfizer)	OH	OH	CH_3	H	H
Chlortetracycline HCl	BP; USP; Eur. P.; Int. P.; Ind. P.;	Aureomycin(R) (Lederle)	H	OH	CH_3	Cl	H
Demeclocycline HCl	BP, USP; Eur. P.;	Ledermycin(R) (Lederle, UK)	H	OH	H	Cl	H
Methacycline HCl	BP (1973); USP;	Rondomycin(R) (Wallace)	OH	CH_2	CH_2	H	H
Doxycycline	USP;	Vibramycin(R) (Pfizer)	OH	H	CH_3	H	H
Rolitetracycline	USP:	Syntetrin(R) (Bristol)	H	OH	CH_3	H–CH_2–N⃞	

Nomenclature

Based on the above 'conventional numbering' of various C-atoms and subsequent labelling of *four* aromatic rings (*viz.*, A, B, C, and D) present in the **tetracycline nucleus** is duly designated on the *chemical basis* as–

"4-Dimethylamino – 1,4, 4a,5,6α,8,11,12α-octahydro–3,6,10,12,12a-penta-hydroxy–6–methyl–1,11-dioxo-2-naphthaceue-carboxamide".

Importantly, some other members of the 'tetracycline family' may be named conveniently as stated under:

- **Methacycline** : 6-Methylene-5-oxytetracycline;
- **Doxycycline** : α-6-Deoxy-5-Oxytetracycline;
- **Rolitetracycline** : N-(Pyrrolidinomethyl)-tetracycline.

General Characteristics of Tetracyclines

Following are the **general characteristics features** of all the members of the **tetracycline** family:

(a) The **tetracyclines** are obtained by fermentation procedures from **Streptomyces species** or by the chemical transformations of the **natural products**.
(b) The important members of this family are essentially derivatives of an **octahydronaphthacene**, *i.e.*, a hydrocarbon made up of a system of four-fused rings.
(c) The **antibiotic spectra** and the chemical properties of these compounds are quite similar but not identical.
(d) The **tetracyclines** are amphoteric compounds, *i.e.*, forming salts with either acids or bases. In neutural solutions these substances exist mainly as **Zwitter ions**.

(e) The acid salts of the tetracyclines that are formed through protonation of the dimethylamino group of C-4, usually exist as crystalline compounds which are found to be very much soluble in water. However, these **amphoteric antibiotics** will crystallize out of aqueous solutions of their salts unless they are duly stabilized by an excess of acid.

(f) The corresponding hydrochloride salts are used most commonly for oral administration and are usually encapsulated owing to their bitter taste.

(g) The water soluble salts are obtained either from bases such as sodium/potassium hydroxides or formed with divalent/polyvalent metals, *e.g.*, Ca^{++}. The former ones are not stable in aqueous solutions, while the latter ones, *e.g.*, calcium salt give tasteless products that may be employed to prepare suspensions for liquid oral dosage forms.

(h) The unusual structural features present in the **tetracyclines** afford *three* **acidity constants (pKa values)** in aqueous solutions of the acid salts. The **thermodynamic pKa values** has been extensively studied by Lesson *et al.* and discussed in the chapter on **'Physical-chemical factors and Biological Activities'**.

(i) An interesting property of the **tetracyclines** is their ability to undergo epimerizaton at C-4 in solutions having intermediate pH range. These isomers are called **epitetracyclines**.

The **four** *epi*-**tetracyclines** have been isolated and characterized. They exhibit much less, activity than the corresponding **'natural' isomers**; thus accounting for an apparent decease in the therapeutic value of aged solution.

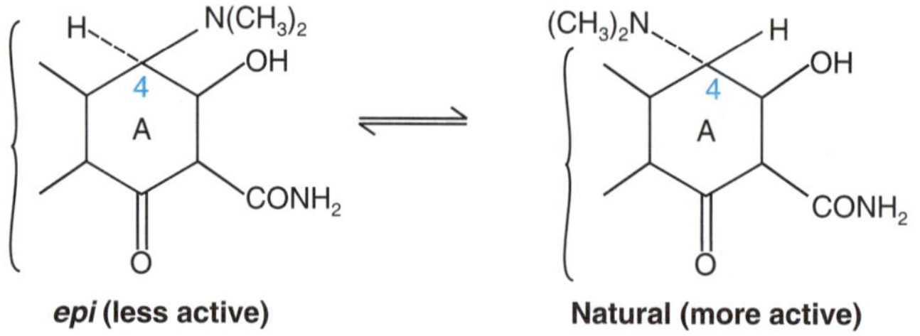

epi (less active)   Natural (more active)

(j) It has been observed that the strong acids and bases attack the tetracyclines having a hydroxy moiety at C-6, thereby causing a considerable loss in activity through modification of the C-ring as shown below:

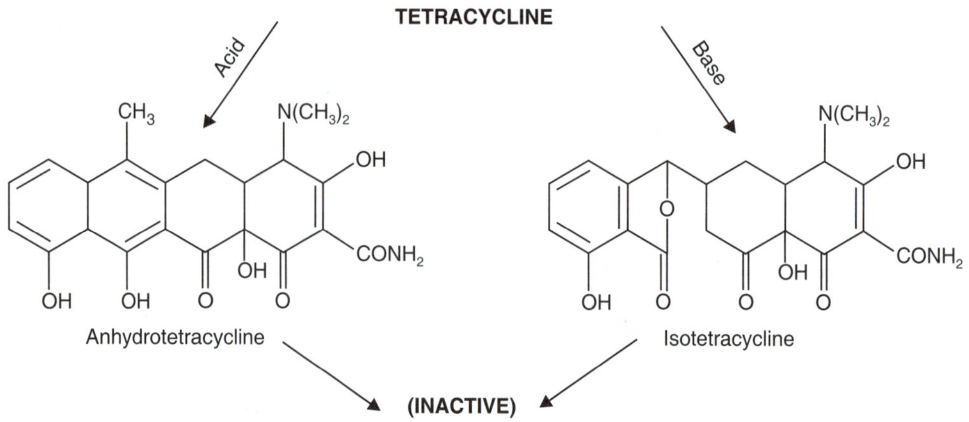

TETRACYCLINE

Anhydrotetracycline   Isotetracycline

(INACTIVE)

Strong acids produce a dehydration through a recuction involving the OH group at C-6 and the H atom at C-5a. The double bond thus generated between positions C-5a and C-6 induces a shift in the position of the double bond between the carbon atoms C-11 and C-11a thereby forming the relatively more energetically favoured resonant system of the naphthalene group found in the **inactive anhydrotetracyclines**.

The strong bases on the other hand promote a reaction between the hydroxyl group at C-6 and the carbonyl moiety at C-11, thereby causing the bond between C-11 and C-11*a* atoms to cleave and eventually form the lactone ring found in the **inactive isotetracyclines**.

(k)   The **tetracyclines** form stable chelate complexes with many metals, e.g., Ca++, Mg++, Fe++, etc.

A few typical examples of the tetracyclines shall be dealt with in the sections that follows:

3.5.1. *Tetracycline*

2-Naphthacenecarboxamide [4S-(4α, 4aα, 5aα, 6β, 12aα)]-4-(dimethylamino)-1, 4, 4a, 5, 5a, 6, 11, 12a-octahydro-3, 6, 10, 12, 12a-pentahydroxy-6-methyl-1,11-dioxo;

Achromycin(R); Cyclopar(R); Panmycin(R); Tetracyn(R);

It is the durg of choice in the treatment of chloera, relapsing fever, granuloma inguinale and infections produced by rickettsia, *Borrelia, Mycobacterium fortuitum and marinum,* and *Chlamydia psittaci* and *trachomatis* (except pneumonia and inclusion conjunctivitis).

It may be employed as an '**alternative drug**' in the following *two* situations, namely:

(a)   For silver nitrate in the prevention of neonatal ocular prophylaxis of chlamydial and gonococcal cojunctivitis, and

(b)   For treatment of actinomycosis, anthrax, chancroid, mellioidosis, plague, rat-bite fevers, syphilis and yaws.

It has also been reported to be beneficial in the treatment of *toxoplasmosis*.

Mechanism of Action. The mechanisms of action of its combination with other agents have been established adequately, such as:

Tetracycline + $MgCl_2.6H_2O$–Panmycin (R)–Enhances the rate and peak of plasma concentration.

Tetracyclines + Aluminium/Calcium gluconates–Observed enhanced plasma levels in experimental animals

From the above two cited examples one may evidently conclude that the **tetracyclines** may form table complexes with bivalent metal ions (e.g., Mg^{2+}, Ca^{2+};) that would appreciably minimize the absorption from the GI-tract. In reality, these '**adjuvants**' seen to compete with the tetracyclines for substances present in the GI-tract which might otherwise be free to complex with these antibiotics, and ultimately retard their absorption significantly. Of course, there is no concrete evidence which may suggest that the metal ions (Mg^{2+}, Ca^{2+}) *per se* serve as '**buffers**', a theoretical explanation quite often forward in the literature.

3.5.2. *Minocycline Hydrochloride*

2-Napththacenecarboxamide, [4S-(4a,4aa, 5aa, 12aa)]-4,7-*bis* (dimethylamino)-1,4,4α,5α,6,11,12α-octahydro-3,10,12,12α-tetrahydroxy-1,11-dioxo-, monohydrochloride ; USP : Minocin[(R)]; Vectrin[(R)];

The 'drug' is found to be 2-4 folds as potent as **tetracycline**; however, it essentially shares an equally low potency against Enterococcus fecalis. Besides, it is observed to be 8 times more potent gainst *Streptococcus viridans*, and 2-4 times against *Gram*-positive organisms in comparison to tetracyclines. It is the drug of choice for the treatment of infections caused by Mycobacterium marinum remarkably differs from the other structural analogues of **tetracyclines** wherein the obsereved bacterial resistance to the drug stands at a *low ebb and incidence*; it is particularly true for *Staphylococci*, in that prevailing cross-resistance is only upto 4%.

Minocycline has been indicated for the management and treatment of chronic bronchitis and the **upper respiratory tract infections (URTIs)**. Though it essentially possesses comparatively low renal clearance, which is partially compensated for by means of its high serum and tissue levels, it has been duly recommended for the treatment of **urinary tract infections (UTIs). The 'drug'** has been equally useful in the virtual erradication of N. meningitidis in specific asymptomatic carriers.

Mechanism of Action. The 'drug' is usually absorbed by the oral route upto 90–100%. However, its absorption is predominantly diminished to a small extent by milk and food intake; and appreciably by the presence of 'iron preparations' and 'nonsystemic antacids'. It is protein-bound in plasma between a range of 70-75%. The volume of distribution v_d^{ss} stands at 0.14–0.7 mL. g^{-1}. The plasma half-life ranges between 11–17 hours. It gets excreted unchanged in urine upto 10%; however, its biological half-life is usually prolonged chiefly in the incidence of renal failure.

Structure Activity Relationship (SAR)

The **structure activity relationsip** amongst the various members of the **tetracycline** family has been studied extensively.

The high level of antimicrobial activity of tetracycline established earlier reveal that the substitutions on the C-5 and C-7 were not an essential requirement.

The activity of 6-dimethyltetracycline (**demecycline**) and **demeclocycline** has established that the methyl function at C-6 may be replaced by hydrogen.

The activity of **deoxycycline** and **6-deoxy-6-demethyltetracycline (minocycline)** shows that the presence of hydroxy moiety at C-6 is not essential either.

The **6-deoxy-6-methylenetetracyclines** and their corresponding **mercaptan adducts** possess typical characteristics tetracycline activity and illustrate further the level of modification feasible at C-6 with the possible retention of biologic activity.

It is, however, interesting to observe that the subsequent removal of the 4-dimethylamino function affords a loss of about 75% of the antibiotic effect of the **parent tetracyclines**.

The **X-Ray Diffraction (XRD)** Studies reveal that the following stereochemical formula repre- sents the orientations, as observed in the **natural tetracyclines**:

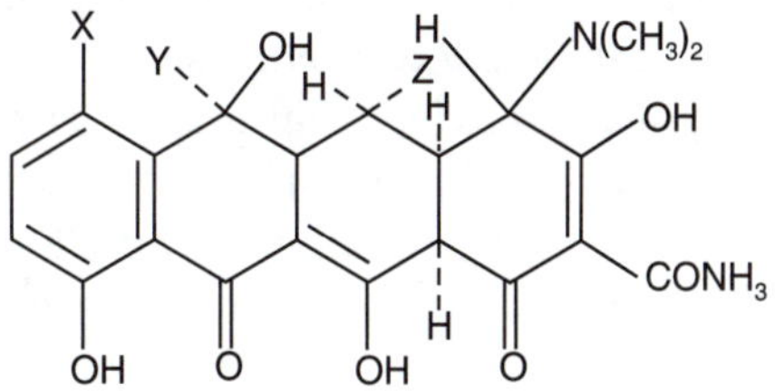

Tetracycline : Z = H ; Y = CH3;

Chlortetracycline : X = CI ; Y = CH3; Z = H ;

Oxytetracycline : X = H ; Y = CH3; Z = OH ;

Demeclocycline : X = CI ; Y = Z = H :

X-Ray Diffraction (XRD) studies further reveal that the 4-dimethylamino function is placed in a transorientation rather than the cis-form as infered earlier by chemical investigations. It further establishes the presence of a conjugated system existing in the structures of **tetracycline** from C-10 through C-12.

NEWER TETRACYCLINES

Since 1992, several newer breeds of '**tetracyclines**' have emerged that were exclusively based on the recent researches focussed on the following aspect, namely:

(a) superb broad spectrum antimicrobial profile of the '**tetracyclines**', and

(b) recent astronomically broad emergence of bacterial **genes** and **plasmids** encoding tetracycline resistance.

Therefore, keeping in view of the stringent limitations imposed on the '**tetracyclines**' as a class has caused the researchers at the Lederle Laboratories to augment extensive and intensive studies to rediscover SARs of tetracyclines with strategical substitutions in the **aromatic ring 'D'** in a meaning- ful and sincere effort to lay hand on to certain newer breeds of tetracyclines that might give rise to such drug substances which are specifically effective against the resistant strains.*

The concerted efforts ultimately gave birth to a **few newer** tetracyclines as illustrated below:*

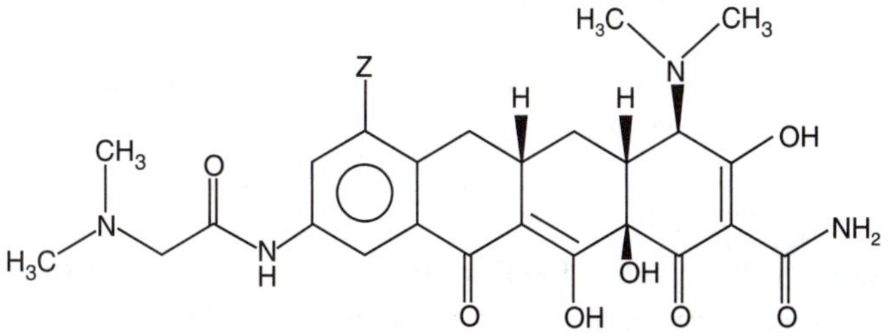

---

*   Tally FT *et al. J Antimicrob. Chemother*, **35**, 449, 1995.

Examples

    (a) 9-(Dimethylglycylamino) minocycline: [DMG-MINO]; $Z=N(CH_3)_2$;

    (b) 9-(Dimethylglycylamini)-6-demethyl-6-deoxytetracycline [DMG-DMDOT]; Z = H;

    Salient Features. The salient features of the '**glycylcyclines**' are as stated under:

    (i) retain essentially both potency and broad spectrum profile as displayed by the '**parent tetracyclines**' against specifically the **tetracycline-sensitive microbial strains**, and

    (ii) exhibit predominantly maximum activity against bacterial strains which show tetracycline resistance either through the ribosomal protecting determinants or afford mediation by efflux.

    The future prospects of a possible '**second generation tetracyclines**' are almost written on the wall provided the meaningful and fruitful clinical trials of the ongoing **glycylcyclines** do emerge both favourable *pharmacokinetic* and *toxicological* profiles for such '**medicinal compounds**' in the near future.

Oxytetracycline (or Terramycin)

    Oxytetracycline was primarily isolated from the so-called *elaboration products* of the **actinomycete**, *Streptomyces rimosus* grown duly on a suitable medium. *Hochstein et al.* (1952)* first and foremost established the structure of **oxytetracycline** as given under;

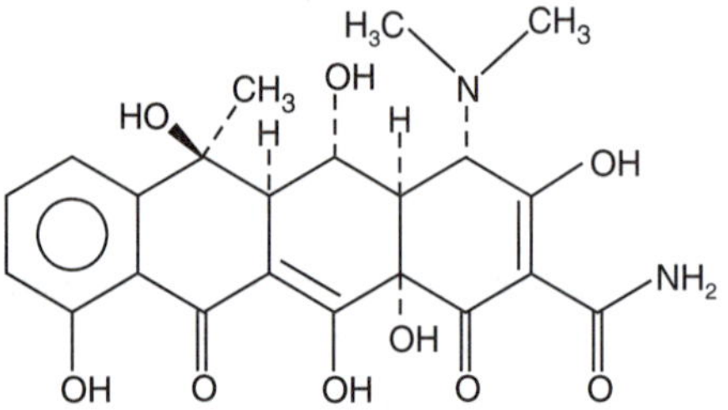

Oxytetracycline

Characteristic Features – These essentially include:

1. **Oxytetracycline** is obtained as a yellow, odourless, crystalline amphoteric substance.
2. **Solubility Profile** –
    • **Oxytetracycline** – 1g in 2L of water.   • **Oxytetracycline HCl** – 2g in 2mL of water.
3. The solutions of the '*base*' and '*HCl-salt*' are not found to be stable at **pH < 2**; and are destroyed rapidly by **alkaline (NaOH/KOH) solutions**.

CONSTITUTION OF OXYTETRACYCLINE

    The *constitution of oxytetracycline* has been duly established based on the *various analytical* and *scientific evidences* as detailed under:

1. **Molecular Formula** – From the essential analytical data, the *molecular weight* of *oxytetracycline* has ben found to be $C_{22}H_{24}N_2O_9$.
2. **Presence of 2-Enolic (–OH) Acidic Moieties** – It has been observed that **oxytetracycline** possesses *three* ionizable functional moieties having *three* distinct pKa values:
    • **3.5** • **7.6 and 9.2.**

---

* Hochstein *et al.* : *J Am Chem Soc.*, **74**: 3708, 1952.

However, the latter *two* values (i.e., **7.6** and **9.2**) explicitly show the presence of '**acidic moities**' in **oxytetracycline,** which are further confirmed by the *critical formation of **dimethyl oxytetracycline*** by treatment with ***diazomethane*** [CH$_2$=N$^\circ$N]. In addition, there stands a clear-cut indication that the so-called '**acidic moieties present**' are certainly of '***enolic* in nature**', since oxytetracycline distinctly gives a *positive test* with **ferric chloride [FeCl$_3$] solution**.

> **NOTE** Oxytetracycline has been found to be completely devoid of any *carboxy* (–COOH) moiety.

3. **Absence of unconjugated aldehyde (–CHO), keto (>CO), carboxyl ester (–COOR), lactone moieties by Spectrophotometric Screening** – The UV-spectrum and IR-spectrum of oxytetracycline are obtained they fail to sow any *critical* and *specific* **bands or peaks** – that could justifiably ascertain the presence of such characteristic moieties as:
   * **Unconjugated aldehyde (CHO)** • *Keto* **(>C=O)** • **Carbonyl Ester (–COOR)** and • **Lactones.**

4. **Presence of 2-Alcoholic Hydroxyl (–OH) Moieties** – *Acetylation* of **oxytetracycline** with **acetic anhydride** yields the respective **diacetyl derivative**, which still possesses *two* **acidic groups**. Thus, it vehemently suggests the presence of *two* **alcoholic groups in oxytetracycline**.

5. **Presence of *One* Amide and *One* Dimethyl Amino Groups** – The *oxytetracycline* on being subjected to *hydrolysis* in the presence of *an alkaline medium* – it gives rise to the formation of *one mole each of:*
   * **Ammonia (–NH$_3$)**;   and   • **Dimethylamine [–NH(CH$_3$)$_2$],**

   thereby indicating strongly that oxytetracycline contains **one amide (–CONH$_2$) moiety** and **one dimethylamino [–NH(CH$_3$)$_2$] group**.

6. **Presence of One Methyl (–CH$_3$) Group** – Based on the usual standard analytical tests the presence of one **methyl (–CH$_3$) group** in **oxytetracycline** has been duly ascertained.

   Therefore, all the above chemical evidences cited above from (1) through (6) – the structure of **terramycin/oxytetracycline** may be confidently expressed as under:

$$C_{18}H_9O_4 \left\{ \begin{array}{l} \text{—C(CH}_3) \\ \text{—20H (enolic acidic)} \\ \text{—2OH (alcoholic)} \\ \text{—CONH}_2 \\ \text{—N(CH}_3)_2 \end{array} \right.$$

7. **Degradation Studies** – The complete chemical structure of the **oxytetracycline** has been established by performing the thorough *degradation studies* under the following three different parameters.
   * *alkaline* degradation, • *acidic* degradation, and • *reductive* degradation.

❑ *Alkaline Degradation*

(a) When fused with *alkali* (**NaOH/KOH**) oxytetracycline gives the *four* carboxylic acids *viz.,* **acetic acid (1), succinic acid (2), salicylic acid (3),** and **m-Hydroxybenzoic acid (4).**

| C$_{12}$H$_{24}$N$_2$O$_3$ | $\xrightarrow[\text{Fusion}]{\text{Alkali}}$ | CH$_3$COOH + | CH$_2$–COOH
$\mid$
CH$_2$–COOH | + | (3) | + | (4) |

Oxytetracycline   (1)   (2)   (3)   (4)

Now, based on the specific formation of aforesaid *four degradation products* [(1) through (4)] – it clearly indicates that the **oxytetracycline** probably has the *following type of structure (I)*:

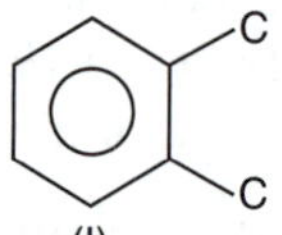

(I)

(b)   On being treated with *aqueous alkali (NaOH/KOH)*, **oxytetracycline** gives the following *two* **carboxylic acids** – as the major products:

*   **Terracinoic Acid (II)**; and

*   *Iso–decarboxyterracinoic acid* **(IIa)**.

In fact, these *two* products of **alkaline degradation** are vehemently assumed to be formed by the so-called *fundamental managements* in the **oxytetracycline molecule**, following the *strategic cessation* of the critical *C–C bonds* located in the **rings B and C** of the **oxytetracycline**;

Thus, we may have the following expression:

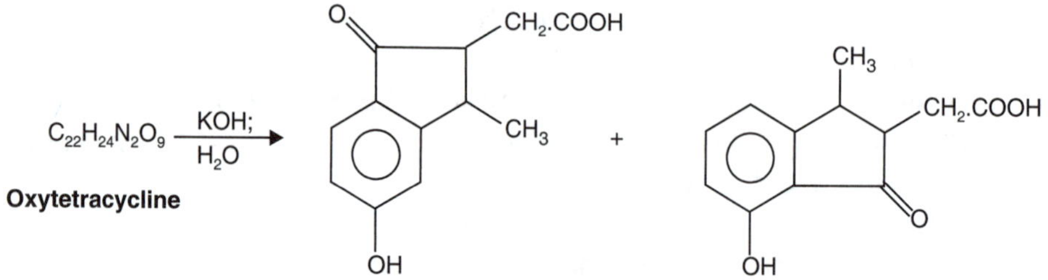

Terracinoic Acid
(II)

Iso-decarboxy-
terracinoic acid
(IIa)

(c)   When subjected to treatment with *caustic alkali (NaOH/KOH)* in the presence of Zn, **oxytetracycline** yields exclusively the following *two* products of alkaline degradation, namely:

*   **Terranaphthol (III)**; and

*   **Methylphthalid (IIIa)**

Thus, we may have the following expression:

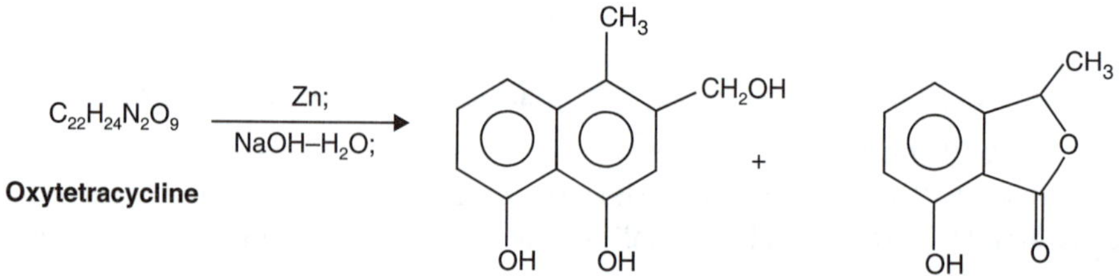

Terranaphthol
[III]

Methyl phthalide
[IIIa]

Remarks – Thus, based upon the *two* **aforesaid alkaline** degradation products [II] and [III], as obtained in (b) and (c), the respective **compound [IV]** was meticulously synthesized as a '*Model*' solely designed for the **UV-spectroscopic examination**.

Compound (IV)

Compound (V)
[As the '*partial structure*' of
Oxytetracycline]

Inference – The resulting observed **UV-spectrum** of **compound (IV)** (*i.e.*, the **prepared '*Model*'**), has been found to have a close resemblance to that of '**oxytetracycline**' itself; and, therefore, it has predominantly led us to the *belief* and *most affirmative suggestion* that compound (V) may justifiably represent as the '**partial structure of oxytetracycline**'.

❑ *Acidic Degradation:* It is, however, pertinent to state at the very outset that '*acid*' is quite unable to afford any root of **plausible degradation** to the **oxytetracycline molecular** thereby leading to the desired **extensive degradation** – but is definitely able to **slight perceptive changes in it**. Nevertheless, a *series of four distinct products* have been isolated meticulously by rendering the ensuring **parameters of the hydrolysis** rather *more vigorous*.

NOTE These resulting products derived from vigorous *acidic degradations* are more or less related simply to their corresponding *precursor(s)*.

(a) **Treatment of Oxytetracycline with Diluted Acids** – In this instance, when **oxytetracycline** is made to react with diluted acids, it gets duly *dehydrated* to give an *anhydro*-**oxytetracycline (VII)**. The resulting **compound (VI)** when further subjected to treatment with either an '*acid*' or a '*base*', it gets duly converted into an admixture of *two* **interchangeable α-and β- apo– oxytetracyclines (VII)**. Obviously, the **α- and β-apo-oxytetracyclines** do designate as the *stereoisomers*,-that actually differentiate with respect to the *orientation* **of the following** *two aspects:*

• **enolised β-carbonyl system**; and
• **attached** *amide (–CONH₂) moiety.*

In addition, one may also obtain the said *two* **isomeric α- and β-apo-oxytetracyclines (VII)** more or less directly, when the *parent* **oxytetracycline** is carefully treated with *diluted HCl* in the presence of **nitrogen (N₂)** – an **inert gas**.

NOTE Importantly, the apo-oxytetracycline when fused with alkali (NaOH/KOH) –

It gives rise to the formation of 2, 5-dihydroxybenzoquinone

(b) When either the *parent* **oxytetracycline** or the **aforesaid compounds (VI) and (VII)** are heated very carefully upto **60°C** along with **0.5M HCl** in a *stream of air* (aeration) at a stretch for a duration of *nine* **days**, then one would clearly observe the elimination of '**dimethylamino**

moiety' duly present in **compound (VII)** thereby producing **terrinolide (VIII)**. The resulting **compound (VIII)** on being refluxed with pure concentrated HCl (12 M) in an environment of N_2 – **gas** it gets duly converted into the respective *decarboxamido terrinolide* (**IX).**

Thus, we may express the various *acid-degradative products* viz., **(VI), (VII), (VIII),** and **(IX)** as given under:

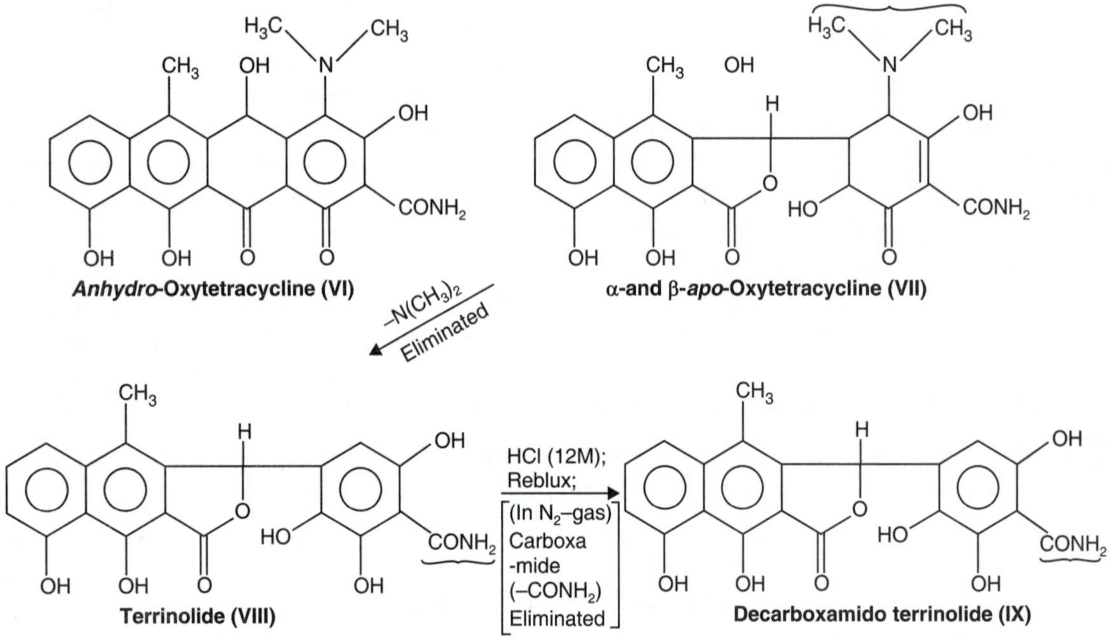

Anhydro-Oxytetracycline (VI)

$-N(CH_3)_2$
Eliminated

α-and β-*apo*-Oxytetracycline (VII)

Terrinolide (VIII)

HCl (12M);
Reblux;
(In N_2–gas)
Carboxa
-mide
(–$CONH_2$)
Eliminated

Decarboxamido terrinolide (IX)

Remarks: Based upon the above products obtained by the so-called *alkaline degradation* – one could possibly and logically undertake the further *partial structure of oxytetracycline* as given under:

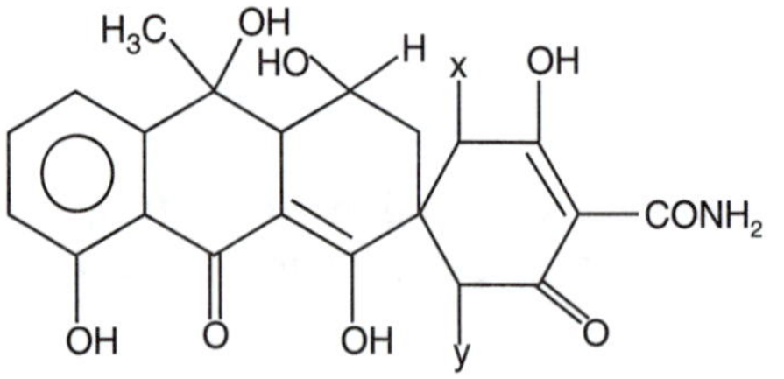

where : x or y $=$.OH or–N(CH$_3$)$_2$

❑ **Reductive Degradation** – It is, however, pertinent to state at this point in time that further information with respect to the complete structure of **Oxytetracycline (or Terramycin)** may be duly accomplished by the help of '**reductive degradation**'.

Thus, when **oxytetracycline** is subjected to *reduction* with *glacial acetic acid* plus Z_n-*dust* in a relatively milder experimental parameters, it critically loses the **dimethylamino** moiety [$-N(CH_3)_2$]; and thereby yields the **desdimethylamino-oxytetracycline (X)**. The resulting compound (X) when duly reduced under *more severe conditions* eventually loses one more '**oxygen atom**' to give rise to the formation of *deoxy-desmethylamino-oxytetracycline (XI)*. [$C_{20}H_{19}O_8N$]. This **product (XI)** on Z_n-*dust distillation* gives **naphthalene**, as given under:

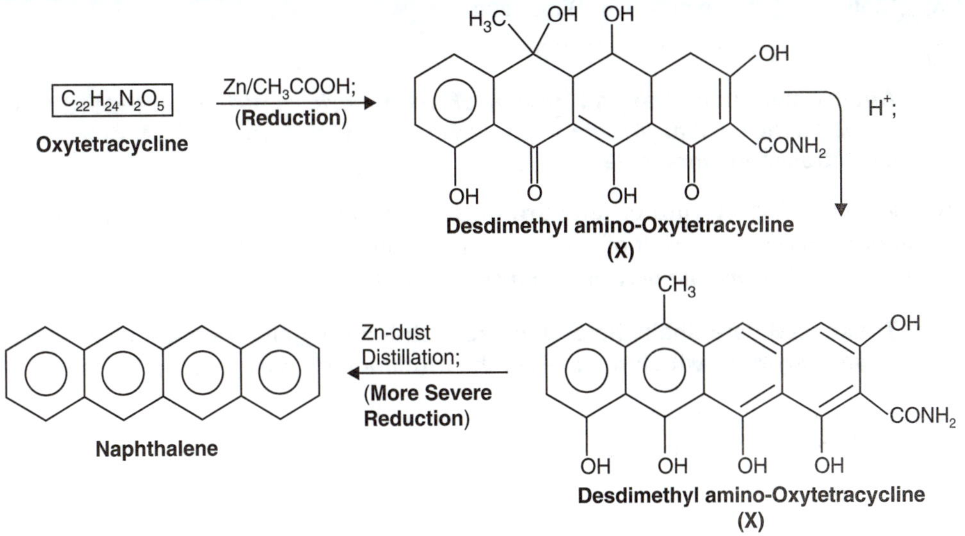

Desdimethyl amino-Oxytetracycline
(X)

Naphthalene

Desdimethyl amino-Oxytetracycline
(X)

8. **UV-Data in Support of the Structural Problem of Oxytetracycline** – The actual strength of the *inherent* **UV-data** (*i.e.,* the *spectroscopic technique*) gallantly and efficiently solve the intricate issues related to **oxytetracycline** structural problems.

Thus, we may have the following *four* **supportive structures A, B, C, and D:**

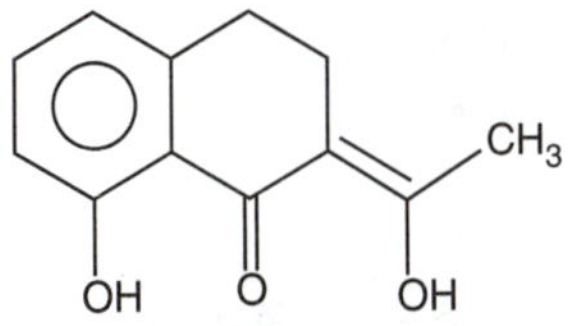

[A]

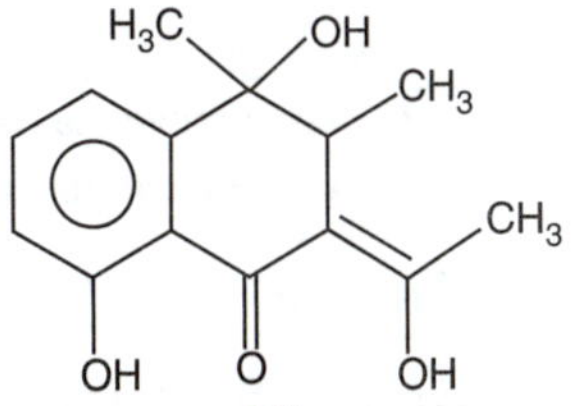

[B]

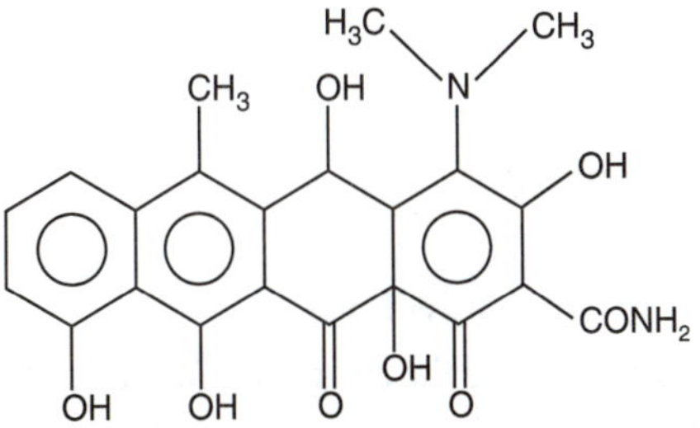

Anhydro-Oxytetracycline [C]

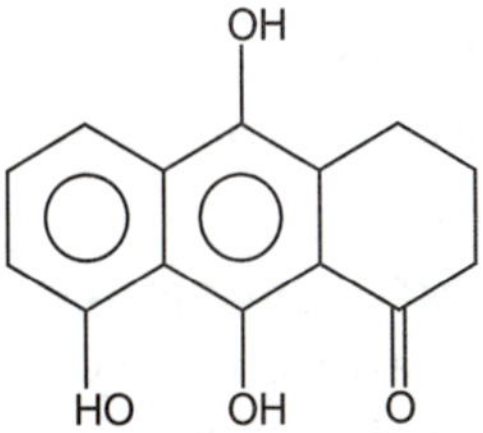

***tetra*-Hydroanthracene [D]**

Important Points – The following *two* **important points** need to be considered with utmost integrity:

☐ The **UV-spectrum of '*Oxytetracycline*'** explicitly exhibits *two* **distinct bands** at:

• λ_{max} **267**   :   (ε max 2,100) nm; and
• λ_{max} **357**   :   (ε max 12,500) nm.

In fact, these bands also appear in the so-called '**synthetic model compound (A)**' having the following *two* UV-spectrum bands as stated below:

- λ_{max} 260 : (ε max 5,700) nm; and
- λ_{max} 345 : (ε max 12,500) nm.

> **NOTE** **Amazingly, such vital and crucial *chromophoric contribution* indeed mostly revealed the structural moiety (B)– which ultimately forms an integral part of the oxytetracycline molecule.**

☐ Besides, the **UV-spectrum** of **anhydro-oxytetracycline (C)**, that predominantly designates the *major product* as a result of acidic treatment – that compared extremely close particularly with the *model compound **tetra*-hydroanthracene (D)**.

> **NOTE** **These vital data finally helped in ascertaining the dehydration phenomenon of *oxytetracycline*, which resulted in the formation of 'naphthalenoid systems'.**

Therefore, based on the above scientific facts and logical explanations, it could be possible to assign the **complete chemical structure** of Oxytetracycline (or **Terramycin**) as given below:

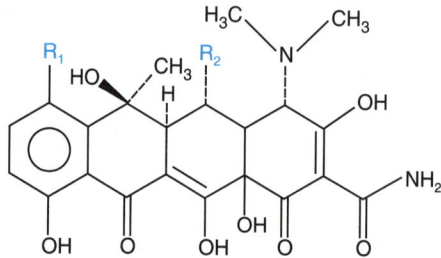

Oxytetracycline [terramycin] :R_1=H; R_2=OH;
Aureomycin [terramycin] :R_1=Cl; R_2=H;

THERAPEUTIC [CLINICAL] ACTIVITY PROFILES

In a broader perspective, the 'Tetracyclines' are found to be active against a good number of:

- *Gram*-positive organisms;
- *Gram*-negative organisms;
- **Spirochaetes**;
- **Rickettsial infections** (viz., Q-Fever, Typhus);
- **Mycoplasma**; and
- **Lymphogranulomaprittacosis group of virus infections.**

Drugs of Choice – The **tetracyclines** are regarded to be the *drugs of choice* for the critical as well as overall treatment of a variety of *dreadful human ailments*, such as:

- **Brucellosis**
- **Lymphogranuloma venereum**
- **Psittacosis**
- **Q-Fever**
- **Typhus** and
- **Whooping cough.**

In addition, the **tetracyclines** are also found to be of *immense utility* in the treatment of:

- **Respiratory-tract microbial infections** *e.g.*, **the specific exacebations of chronic bronchitis**; and
- **Urinary-tract Infections (UTIs).**

Points to Ponder – These essentially include:

1. **Tetracyclines** being *amphoteric compounds* (*i.e.*, forming salts with acids or bases in *neutral solutions; and hence, they exist mostly as the Zwitter ions. These amphoteric antibiotics to have a* tendency to *crystallize out of the aqueous solutions of their respective salts unless they are duly stabilized by an excess of acid.*

2. **Tetracyclines** do have a tendency to **chelate** with divalent ions *e.g.*, Ca^{2+}, Zn^{2+} etc.; and, therefore, its use should be avoided in *pregnant mothers* (who are put on extra Ca-dosage for the healthy growth of the baby in mother's womb).

3. **Epimerization of Tetracyclines** – The **tetracyclines** have an inherent ability to undergo *epimerization* at **C-4 in solutions** (*with an intermediate pH range*). Hence, these specific isomers are known as the '**epitetracyclines**'.

   The *four* **epi-tetracyclines** have been duly isolated and characterized; and interestingly, are found to exhibit much less activity in comparison to the corresponding '**natural isomers**'. In this manner, the '**aged solution**' of tetracyclines have an apparent minimized therapeutic activity profile.

Thus, we may have:

epi **(less active)**   **Natural (more active)**

FURTHER READING REFERENCES

Abraham DJ (Ed)   : **Burger's Medicinal Chemistry and Drug Discovery**, 6th edn., John Wiley & Sons, Inc., New York, 2007.

Berdy J *et al*. (Eds.)   : **Handbook of Antibiotic Compounds, Vols. 1–8**, CRC-Press, Boca Taton, EL(USA), 1982.

Cloute JE   : **Manual of Antibiotics and Infectious Diseases,** 8th edn., Williams and Wilkins, Baltimore (USA), 1995.

Davies JE   : **Antibiotics in Laboratory Medicine**, 3rd. edn., Williams & Wilkins, Baltimore (USA), 1991.

Lancini *et al*.   : **Antibiotics: An Interdisciplinary Approach** 3rd edn., Plenum Press, New York, 1995.

Mandell GL *et al*. (Eds)   : **Principles and Practice of Infectious Diseases**, **Vol. 1,** 4th edn., Churchill–Livingtone, New York, 1995.

Wise EM: : **Antibiotics (Peptides) Kirk–Othmer Encyclopedia of Chemical Technology**, **Vol. 3** 4th edn., Wiley, New York, 1992.

REVIEW QUESTIONS

1. What do you mean by '**Antibiotics**'? Elaborate on the following aspects briefly:
   (i) **Antibiotic as a Therapeutic Agent**
   (ii) **History of Discovery of Antibiotics**
   (iii) **Importance of Antibiotics.**
2. What are **Antitumor Antibiotics**? Give the chemical structure of:
   (a) **Psicofuranine**
   (b) **Anthramycin**
   (c) **Azaserine**
   (d) **Mitomycin C**
3. **Antibiotics** may be *classified* in *three* different ways:
   (i) **Broadbased classification**
   (ii) **Based on the type of Bacteria Attacked**
   (iii) **Based on their Chemical Structures**
   Give suitable examples to support your answer.
4. Write a comprehensive account on the '**Penicillins**' with particular reference to:
   (a) **Naturally Occurring Penicillins**
   (b) **Commercial Production of Penicillin**
   (c) **Isolation of Penicillin**

5. Discuss the 'Constitution of Penicillins'. Give appropriate examples to expatiate each and every step being discussed.

6. Write notes on the following:
   (i) **X-Ray Diffraction Studies of Penicillin**
   (ii) **Synthesis of Penicillin V**

7. Describe any **Two** most common **Penicillin variants:**
   (a) **Procaine Penicillin**
   (b) **Staphicillin**
   (c) **Penicillin G**

8. What are Cephalosporins? Discuss one important example from each of the following groups:
   (i) **First Generation Cephalosporins**
   (ii) **Second Generation Cephalosporins**
   (iii) **Third Generation Cephalosporins**
   (iv) **Fourth Generation Cephalosporins**

9. Give a detailed account on **Streptomycin** including the **constitution of Streptomycin**.

10. Describe the **chloramphenicol** with reference to its:
   (a) **Isolation**
   (b) **Synthesis of Chloramphenicol by Controulis** *et al.* **(1949)**
   (c) **SAR of chloramphenicol**

11. Discuss the constitution of Oxytetracycline and provide all necessary and relevant details wherever required.

Contents at a Glance

3

Lipids

1. INTRODUCTION

Lipids (*fixed oils*, *fats*, *and waxes*) do designate the esters of such typical *chemical entities* as:

- *esters of* **long-chain fatty acids;**
- **alcohols**; and
- **closely related structural analogues**.

The major differences that invariably occur between these substances are mainly due to the following cardinal factors, namely:

> ➤ **type of alcohol present;**
> ➤ **nature of fixed oils and fats;**
> ➤ **glycerol combines with respective fatty acids; and**
> ➤ **'alcohol' in waxes–has a definite higher molecular weight** *viz.*, **cetyl alcohol.**

In a broader perspective, the **'Lipids'** comprise a distinct *heterogeneous group* of compounds that are critically found to be possessing these **elegant characteristic features:**

- ❑ **sparingly water soluble;**
- ❑ **exceedingly soluble in organic solvents** (*viz.*, **ether, benzene, chloroform, hot ethanol, and petroleum ether**); and
- ❑ *hydrolysis yields* **'fattly acids' that are being utilized by the living organisms** *in vivo*.

 NOTE An extensively used 'universal' lipid solvent is an admixture of *chloroform + methanol* [2:1] (v/v)

Presence of Hydrophobic Nature of Hydrocarbon Structure in Lipids

It has been observed that most of the **'lipids'** do have a tendency to undergo dissolution in a variety of **organic solvents** (as stated earlier), which vehemently shows that the *'hydrophobic nature of the hydrocarbon structure'*. A few **'lipids'** may also possess certain inherent *'hydrophilic moieties'*.

Besides, the so-called *hydrophobicity of lipids* (*i.e.*, water-insolubility of lipids) is perhaps solely based on the fact that the **'polar moieties'** they usually possess are found to be *much smaller* in comparison to their *alkane-like* (*nonpolar*) *segments*. Thus, these **nonpolar segments** actually provide the **water-repellant** (or *hydrophobic*) characteristic properties.

1.1. Occurrence of Lipids

The 'lipids' are broadly distributed in plants and animal products normally typified by esters of higher fatty acids but embracing certain other:

- **oil-soluble substances** and   •   **water-soluble substances**.

*Isolation:*The isolation of pure, and individual fatty acids really poses an exceedingly difficult task. Nevertheless, *a trditional method* indulging the separation of acids of appreciably different **molecular weights** is performed by *fractional vacuum distillation* of the corresponding '*methyl esters*'*

Thus, it is possible to segregate the *saturated* and *unsaturated* acids could be achieved by careful *crystallization* of the '**lead**' or '**lithium**' salts from *acetone* or *ethanol*. In this way, the **saturated salts are less soluble and crystallize out first**. The '**polyunsaturated fatty acids** are usually isolated as the **polybromides**', even though one may visualize and record changes taking place predominantly in:

- **Structure or Configuration**

with regard to their respective **bromination** or **debromination**.

Low-temperature crystallization (at < –60°C) –It could be prove to be specifically gainful in the **isolation of certain unsaturated fatty acids** subjects to *perfect/normal autoxidation at room temperature* (Brown, 1942). Meaningful success and progress has been duly accomplished by carrying out the *chromatography of methyl esters of fatty acids* upon a **silica column under an inert atmosphere of N_2 gas**.

Saturated Fatty Acids: Considerable research and meticulous organized studies revealed an array of **saturated fatty acids** ranging between C_4 to C_{26} (*identified as the constituents of fats*) duly recorded in Table 3.1.

S.No.	Acid	No. of C atoms	Formula	MP(°C)	BP(°C)
	Table 3.1: Saturated Fatty Acids				
1.	**Butyric**	4	$CH_3(CH_2)_2COOH$	−4.7	163
2.	**Isovaleric**	5	$(CH_3)_2CHCH_2COOH$	−51	174
3.	**Caproic**	6	$CH_3(CH_2)_4COOH$	−1.5	205
4.	**Caprylic**	8	$CH_3(CH_2)_6COOH$	16.5	237
5.	**Capric**	10	$CH_3(CH_2)_8COOH$	31.3	269
6.	**Lauric**	12	$CH_3(CH_2)_{10}COOH$	43.6	102/1mm.
7.	**Myristic**	14	$CH_3(CH_2)_{12}COOH$	58.0	122/1 mm
8.	**Palmitic**	16	$CH_3(CH_2)_{14}COOH$	62.9	139/1mm
9.	**Stearic**	18	$CH_3(CH_2)_{16}COOH$	69.9	160/1 mm
10.	**Arachidic**	20	$CH_3(CH_2)_{18}COOH$	75.2	205/1 mm
11.	**Behenic**	22	$CH_3(CH_2)_{20}COOH$	80.2	
12.	**Lignoceric**	24	$CH_3(CH_2)_{22}COOH$	84.2	
13.	**Cerotic**	26	$CH_3(CH_2)_{24}COOH$	87.8	

---

* Since the *methyl esters of fatty acids* are more prone to separation by **fractional vacuum distillation**.

Remarks: These essentially include:

1. **Palmitic Acid** – It is having the widest occurrence and also a component of almost all fats. Besides, it is the chief component (**15–45%**) of the fatty acids of **Palm Oil** (growth abundantly in the entire West Africa *viz.,* **Nigeria**), fruit fat of *Elaeis guineensis*.

2. **Lauric Acid** – It is named after the *laurel family*, from which it was first isolated (1842); and forms the most frequently encountered **saturated acids**, such as:
   - **palm-pernel oil (52%),**
   - **coconut oil**, and
   - **babassu oil**.

3. **Myristic Acid** – It may be isolated readily from the seed fat of '**Nutmeg**'.

4. **Stearic Acid** – It usually occurs in large quantum (**10–30%**) in the *animal fats* e.g., Beef, Lard, Lamb, Goat-meats, but is mostly found in *traces* in **vegetable fats**.

Unsaturated Fatty Acids (or Olefinic Fatty Acids) : The unsaturated fatty acids having *less than 10 C-atoms* have not yet been found in nature; whereas, the $C_{10}-$, $C_{12}-$, and $C_{14}-$ acids do occur only in traces in *a few selected fats only*. Table 3.2 records some **olefinic acids** with known and proven chemical structures.

Table 3.2. Unsaturated Fatty Acids*

S.No.	Acid	Carbon Atoms	Formula	MP (°C)
1.	D⁹-Decylenic	10	$CH_2 == CH (CH_2)_7 COOH$	–
2.	Stillingic	10	$CH_3 (CH_2)_4 CH==CHCH==CHCO_2H$ (*cis, trans*)	–
3.	D⁹-Dodecylenic	12	$CH_3CH_2CH==CH (CH_2)_7 COOH$	–
4.	Palmitoleic	16	$CH_3(CH_2)_5CH==CH(CH_2)_7COOH$ (*cis*)	–
5.	Oleic	18	$CH_3(CH_2)_7CH==CH(CH_2)_7COOH$ (*cis*)	13, 16
6.	Ricinoleic	18	$CH_3(CH_2)_5CH(OH)CH_2CH==CH(CH_2)_7COOH$ (*cis*)	50
7.	Petroselinic	18	$CH_3(CH_2)_{10}CH==CH(CH_2)_4COOH$ (*cis*)	30
8.	Vaccenic	18	$CH_3(CH_2)_5CH==CH(CH_2)_9COOH$ (*cis and trans*)	–
9.	Linoleic	18	$CH_3(CH_2)_4CH==CHCH_2CH==CH(CH_2)_7COOH$	–5
10.	Linolenic	18	$CH_3CH_2CH==CHCH_2CH==CHCH_2CH==CH(CH_2)_7COOH$	–11
11.	Eleostearic	18	$CH_3(CH_2)_3(CH==CH)_3(CH_2)_7COOH$ (cis, trans, trans)	49
12.	Licanic	18	$CH_3(CH_2)_3(CH==CH)_3(CH_2)_4CO(CH_2)_2COOH$	75
13.	Parinaric	18	$CH_3CH_2(CH==CH)_4(CH_2)_7COOH$	86
14.	Gadoleic	20	$CH_3(CH_2)_9CH==CH(CH_2)_7COOH$	–
15.	Arachidonic	20	$CH_3(CH_2)_4(CH==CHCH_2)_4(CH_2)_2COOH$	–
16.	Cetoleic	22	$CH_{3C}(CH_2)_9CH==CH(CH_2)_9COOH$	–
17.	Eruic	22	$CH_3(CH_2)_7CH==CH(CH_2)_{11}COOH$ (*cis*)	33.5
18.	Selachloeic (or) Hervonic	24	$CH_3(CH_2)_7CH==CH(CH_2)_{13}COOH$ (*cis*)	39

* The symbol is used as an abbreviation for a double bond; the superscript is gives the position of the first carbon of the double bond.

1.2. Physiological Functionalities of Lipids

A critical survey of literature would ultimately reveal that there are in all *seven* altogether different and important **physiologic functionalities of lipids** as detailed under:

1. **As a Vital Source of Energy –** The most widely recognized and accepted pivotal role of 'lipids' is as a '**source of energy**'. It has been determined that **1g of lipid** (or *fat*) provides **9 k cal**; whereas, *1g of carbohydrate* or *1g of protein* gives **4.5 k cal**. Hence, **edible fat (or lipid)** is an outstanding source of energy especially for those who are young and active *e.g.,* children, students, atheletes, workers and the like. In the same vein, for elderly people one should preferably make use of:

   - **skimmed milk,**
   - **double-toned milk,**
   - **fat-free 'yoghurt',** and
   - the latest **'cholesterol-free milk'.**

Various Functionalities of Lipids: These essentially include:

(a) **Lipid (fat)** may be duly stored in *almost anhydrous conditions* and also in *unlimited quantum practically.*

NOTE   **Perhaps 'lipid' could be regarded as the most concentrated form wherein the so-called '*potential energy*' is stored with utmost liberty.**

(b) **Lipid yields more heat (energy) per gramme –** In reality, **lipid** yields more heat per gramme when burnt. Besides, *human bodies* do store certain '*carbohydrates*' in the form of '*glycogen*' (in the **liver**) for delivering instant energy as and when required, but the energy stored as the **fats (adipose tissues)** is also equally important (*i.e.,* energy produced by 1g of fat is first the double of carbohydrates).

(c) **Availability of Fatty Acids in a Modified Compact Form –** The **fatty acids**, in general, along with their so-called '*flexible backbones*' may be easily stored in a modified **compact form** in comparison to the respective:
   'highly spatially oriented and rigid glycogen structure'. Therefore, the phenomena of '**storage of fat**' in the humans (as the *adipose tissues*) does provide critically the *perfect economy* in terms of both:

   - **Weight** and
   - **Space**

Besides, the aforesaid *three* reasons [(a) through (c)] there exist *two* other prevalent good reasons for the **storage of fat** to serve as an '*excellent form of energy*', such as:

   ❑ Since '*fat*' *is water insoluble* and furthermore once it has been carried up to the '**fat depots**' in the body by the so-called highly *specialized proteins right into the plasma – it is unlikely to* '*break loose*' and creep into the **watery body-fluids** that virtually bath the **adipose tissues ultimately**.

   ❑ Thus, '**fat**' remains both as a '**stable**' and '**fixed reservoir of energy**'– unless and until it gets duly metabolized by the enzymes that hydrolyze it to **glycerol** and **fatty acids**. In fact, the *enzymes* are perfectly under the command and control of an array of hormones; and hence, are adequately activated under the such parameters where the body is duly engaged in an *enhanced energy expenditure mode*.

NOTE **Ever though the countless meritorious plus points the 'carbohydrates' and Not fats** – remains the most preferred source of energy (fuel) for the body; and hence, any attempt to oxidise appreciable amount of fat without the concomitant degradation of sufficient quantum of *carbohydrate* may finally lead to some serious consequences.

2. **Fat Provides Excellent Insulation** – Based on the underlying fact that fat serves as a **bad-conductor of heat**, it categorically provides an **excellent insulation**. Therefore, in a *cold environment*, in which *heat is prominently lost* to its immediate surrounding, it caters for *two* **most vital solutions to combat cold:**
   - as an *insulating blanket*; and
   - an *extra source of energy*.

3. **Fat as Padding for Internal Organs** – Various vital and important **internal organs,** such as:
   - **Brain**
   - **Heart**
   - **Kidneys** and
   - **Nervous Tissue,**

   are found to be rich in certain **lipids** (*fats*) – a crucial aspect that indicates the importance of such chemical entities to the '*life-support system*' in humans.

4. **Lipids as Important Building Blocks of Physiologically Active Materials** – There are quite a few compounds that have been duly derived from the **lipids** which are proved to be the so-called '**building blocks of physiologically active materials**' *viz.*, **acetic acid (CH_3COOH)** is utilized in the body to synthesize:
   - **cholesterol**, and
   - **allied compounds (hormones);**

5. **Dietary Lipids as a Source of Essential Fatty Acids** – Though there are a host of *important functions* of the **essential fatty acids (EFAs)**, yet hardly any of them is well-defined in the literature(s):
   (i) **EFAs** are established to be the *integral constituents* of the **structural lipids** of both:
      - **cell membrane,** and
      - **mitochondrial membrane.**
   (ii) **EFAs,** are also found abundantly in the '**reproductive organs**' in high concentrations.
   (iii) **EFAs** are also present in the **phospholipids** (in **brain**) particularly at the *2-position*;
   (iv) **Arachidonic Acid (plus allied C-2 fatty acids)** do invariably give rise to cluster of **physiologically active** chemical entities termed as the *Postage Landins*;
   (v) **EFAs** are involved prominently in the following *two critical aspects*:
      - **genesis of fatty livers**; and
      - **metabolism of cholesterol.**
   (vi) **Cure of Certain Types of Eczema in Infants** – Feeding of an admixture of **EFAs** do help to cure certain types of eczema in infants perceptively.

6. **Lipoproteins: Integral Constituents of Cell Walls** – The **lipids** that are duly present in the **lipoproteins** constituting the respective '**cell wall**' are recognized to be the '**phospholipids**'. Because, the **lipids** are *water insolubles*, they serve as an '**ideal barrier**' for complete *check* and *prevention* of the so-called **water-soluble substances** from crossing over freely between:

- **intra-cellular fluids**, and
- **extra-cellular fluids**.

7. **Role of Dietary Fats** – For a *good and healthy heart* the physicians all over the world recommend strongly such **refined edible oils** for daily use at home:
    - **Rice-Bran oil,**
    - **Sun-Flower oil,**
    - **Olive oil,**
    - **Corn oil,**
    - **Mustard oil,**
    - **Sesame oil,**
    - **Soybean oil**, and the like

The '**dietary fat**' is also found to be absolutely necessary to for the absorption of sufficient quantum **EFAs** plus **fat-soluble vitamins** (Vit. A, D, E, K) from the GI-tract.

2. CLASSIFICATION OF LIPIDS

The '**Lipids**' are broadly classified into the following *three* **major categories**, namely:

(a) **Simple Lipids,**
(b) **Compound Lipids,**
(c) **Derived Lipids,**

which shall now be discussed individually in the sections that follows:

2.1. Simple Lipids [Fats and Oils]

The **simple lipids** usually upon *hydrolysis* give the following products of hydrolysis:

- **one or more fatty acids**, and
- **an alcohol**.

Alternatively, the **simple lipids** are nothing but the designated esters of *fatty acids* and *alcohols*, such as:

- **Neutral Fats** and   • **Waxes**

2.1.1. Neutral Fats

1. **Preamble** – In reality, the '**neutral fats**' are the esters of various **fatty acids** and **glycerides** (*tri-hydric alcohols*). These are also termed as '**triglycerides**' – since *three* **moles of fatty acids** usually get condensed with one mole of glycerol to form fat.

Examples: Formation of **Tributyrin** $[(C_3H_7COO)_3C_2H_5]$ by the condensation of *3-moles of butyric acid* with *one mole of glycerol*, as given under:

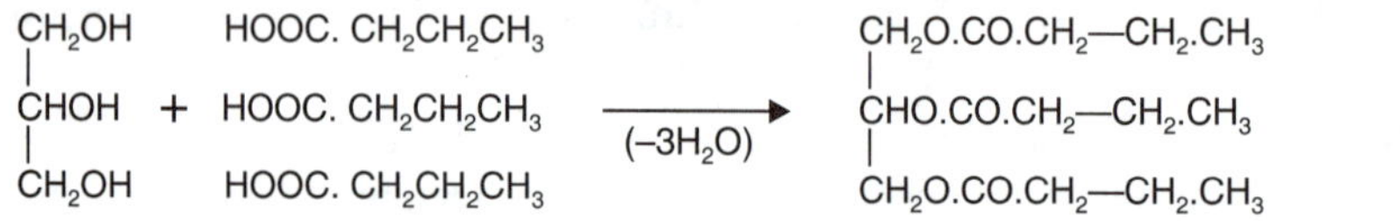

$$
\begin{array}{lll}
CH_2OH & HOOC.\,CH_2CH_2CH_3 & \\
| & & \\
CHOH & +\ HOOC.\,CH_2CH_2CH_3 & \xrightarrow{\ (-3H_2O)\ } \\
| & & \\
CH_2OH & HOOC.\,CH_2CH_2CH_3 &
\end{array}
\qquad
\begin{array}{l}
CH_2O.CO.CH_2—CH_2.CH_3 \\
| \\
CHO.CO.CH_2—CH_2.CH_3 \\
| \\
CH_2O.CO.CH_2—CH_2.CH_3
\end{array}
$$

Glycerol   Butyric Acid   **Tributyrin [Glyceryl tributyrate]**

Likewise, **palmitic acid** gets condensed with glycerol to form **Tripalmitin**, as shown below:

$$
\begin{array}{l}
CH_2OH \quad\quad HOOC.(CH_2)_{14}.CH_3 \\
| \\
CHOH \quad + \quad HOOC.(CH_2)_{14}.CH_3 \quad \xrightarrow[(-3H_2O)]{} \\
| \\
CH_2OH \quad\quad HOOC.(CH_2)_{14}.CH_3
\end{array}
\quad
\begin{array}{l}
CH_2O.CO.(CH_2)_{14}.CH_3 \\
| \\
CHO.CO.(CH_2)_{14}.CH_3 \\
| \\
CH_2O.CO.(CH_2)_{14}.CH_3
\end{array}
$$

Glycerol   **Palmitic Acid**   **Tripalmitin [Glyceryl tripalmitate]**

Hydrolysis of Neutral Fats – Nevertheless, the *hydrolysis* of **Neutral fats** give: **Fatty Acids** and **Glycerol**. Besides, such *hydrolysis* is invariably carried out by the help of specific *enzymes* – 'Lipases', as show below:

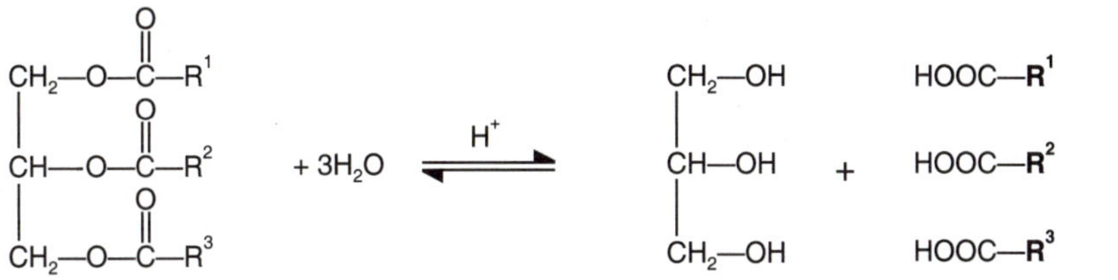

$$
\begin{array}{l}
CH_2{-}O{-}\overset{\overset{O}{\|}}{C}{-}R^1 \\[4pt]
CH{-}O{-}\overset{\overset{O}{\|}}{C}{-}R^2 \quad + 3H_2O \\[4pt]
CH_2{-}O{-}\overset{\overset{O}{\|}}{C}{-}R^3
\end{array}
\quad \underset{}{\overset{H^+}{\rightleftharpoons}} \quad
\begin{array}{l}
CH_2{-}OH \\[4pt]
CH{-}OH \quad + \\[4pt]
CH_2{-}OH
\end{array}
\quad
\begin{array}{l}
HOOC{-}R^1 \\[4pt]
HOOC{-}R^2 \\[4pt]
HOOC{-}R^3
\end{array}
$$

Triglyceride   **Glycerol**   **Three Different Fatty Acids**

Observations: Following are certain vital observations:

- **Simple Triglyceride** – When all the *three* resulting **fatty acids** are found to be *absolutely identical* in nature.

- **Mixed Triglyceride** – When the resulting **fatty acids** are not identical in nature – the **triglyceride** is said to be a **mixed triglyceride**.

- **Neutral Fats** – are largely comprised of the so-called **mixed glycerides**, since these **glycerides** do possess:
    - **completely devoid of 'free acid'**, or
    - **total absence of '*basic moieties*'**.

Characteristic Features of Fats: These essentially include:

1. **Fats** are usually found to be as **solids** or **liquids** at the *normal* **room temperature**.
2. **Fats** are usually lighter than **water (sp. gr. = 0.86)**.
3. **Melting Point of Fats** – It solely depends upon the following *two* **cardinal factors:**
    - **chain-length of fatty acids**; and
    - **degree of saturation of fatty acids**.
4. The **mp of 'fats'** have always been found to be higher than their respective '**solidification points**', such as:

    Tristearin – *Melts* at **72°C**, but *solidifies* on cooling at **52°C**.

    Differences Between Fats and Oils – The major points of **difference between fats and oils** are as given under:

S.No.	Fats	Oils
1.	Solids or semisolids at room temperature.	Liquids at room temperature.
2.	Chemically – the fats have larger quantum of *saturated fatty acids e.g.*, **stearic acid, palmitic acids**; and melts at *higher temperatures*	Oils do possess a larger proportion of **unsaturated fatty acids** *e.g.*, oleic acid, linolenic *temperatures.* acid; and melts at *lower temperatures*.
3.	**Fats** usually occur in **animals** in a layer beneath the skin and around certain internal organs *viz.*, **heart, kidneys**; and also in between the *muscular tissues*.	**Oils** mostly occur in **plants**, particularly in: seeds, kernels, fruits (*coconut oil* and *olive oil*)

Fats and Oil – The **fats and oils** are invariably derived from **animals** and **plant sources,** such as:

- **Lard** (Pig fat) • **Tallow** (Beef and Mutton fats) • **Olive oil** (Plant Fruits) • **Coconut oil** (Fruit fat) • **Linseed oil** • **Cotton seed oil** • **Sunflower seed oil** • **Soyabean oil** • **Groundnut oil** • **Corn oil** • **Sesame oil** • **Mustard oil,** • **Rice Bran oil** and the like.

2. **Extraction of Fats and Oils** – A survey of literature would reveal that the *four* most important procedures for the extraction of fats and oils are:

- **Rendering,**
- **Pressing,**
- **Solvent Extraction,** and
- **Refining,**

which shall now be treated individually as under:

(a) **Rendering** – The *animal tissues* are cut into small pieces along with the **liver, kidney, stomach, spleen, heart** etc., and are **boiled with clean** water preferably in *SS-vessels* or **steamed adequately** for several hours at a stretch. Thus, the **fat melts down** and being lighter than water floats at the surface of the water as a **transparent layer of liquid**. The molten fat floating at the surface is **removed** carefully, **filtered,** and **stored** properly.

(b) **Pressing** – Various selected *plant portions e.g.,* **seeds, kernels, fruits containing** *vegetable fats and oils* are duly crushed by :

- **heavy mechanized rollers,** or
- **pressed under hydraulic press,** or
- **continuous motorized expellers,**

whereby the oozed out '**oil**' gets separated leaving behind the *oil-cake* (which is being used as '**cattle feed**').

 **NOTE**   The squeezed out '*oil-cake*' obtained above may be further *heated* and *pressed* to recover the 'oil' – known as the '*Second Grade Oil*' in trade that invariably comprises with:

- **undesirable ingredients** and • **strong odour.**

(c) **Solvent Extraction** – In this case, the pressed cake, after adequate hot pressing, has been properly crushed and extracted with *pure* **organic solvents**, for instance:

- **carbon tetrachloride** • **n-Hexane** • **Petroleum ether,** so as to obtain the residual vegetable oil.

(d) **Refining** – The **refining** is an important step in the extraction of edible **oil and fat** from natural sources. Thus, the *crude* **oil or fat** is carefully treated with a little *alkali* (NaOH or KOH) to :
- **neutralize the 'free' fatty acid**, and
- **coagulate any colloidal impurities**.

Bleaching – Bleaching refers to a process whereby –'**a compound or a mixture of compounds are capable of removing undesirable colour from a substance either:**
- **by adsorption (a physical phenomenon)**, or
- **by an oxidative process**.

In actual practice, the product to be purified is duly bleached at **70–80°C** by making use of *activated animal charcoal*, or *Fuller's Earth* (a natural deposit), or *Plaster of Paris* (a. deposit). After thorough *agitation* or *stirring* the **decolourized oil** is filtered through a *Filter Press* (under positive pressure); and subsequently, *deodourized* by passing **super-heated steam** through it for **30–60 minutes**. The **decolourized** and **deodourized** edible oil is brought to the room temperature rapidly and stored in air-tight dry glass bottles or jars so as to preserve its natural freshness and goodness.

Nomenclature of Fats and Oil – In a broader perspective, it has been observed that when '*Glycerol*' – a *trihydric alcohol,* essentially its esters, comprises *three* **acid residues**; and thus, two distinct situations may arise:
- *First* : **Formation of Simple Glyceride** – *i.e.*, if the **3-acid residues** are exactly the same; and
- *Second* : **Formation of Mixed Glyceride** – *i.e.*, if the **3-acid resides** are altogether different from one another.

Simple Glycerides – These are invariably named according to the respective *inherent* **acid residues** attached to them duly by replacement of the terminal '*i.e.* acid' by '*in*', such as:
- **Glycerol tributyric Acid :**   **Tributyrin;**
- **Glycerol tristearic Acid :**   **Tristearin.**

Examples: Following are a few typical examples:
(i) **Stearin** – is the *glyceride* of **stearic acid** – a saturated fatty acid (**solid**);
(ii) **Palmitin** – is the *glyceride* of **palmitic acid** – a saturated fatty acid (**solid**);
(iii) **Olein** – is the glyceride of **oleic acid** – an unsaturated fatty acid (**liquid**).

Mixed Glycerides – These are usually comprised with *more than one acid residue*; and hence, are mostly present in naturally occurring *oils and fats*, for instance:
- **mixed glyceride,**
- **oleo-palmito-stearin,**
- **stearo-dipalmitin,** and
- **palmito-distearin,**

that are found to be specifically present in '**lard**' (*i.e.*, the *fat* derived from '**pigs**').

Besides, for the **mixed glycerides** – the *exact positions* and *their respective names* (of the '**acid moieties**') are duly specified as : **1, 2, 3 or α–, β– or α'– and the like.**

Examples: Following are two classical examples:

$$CH_2OCOC_3H_7$$
$$|$$
$$CHOCOC_{15}H_{31}$$
$$|$$
$$CH_2OCOC_{17}H_{33}$$

Glyceryl oleobutyropalmitate
[Present-in Butter Fat (Ghee)]

$$CH_2OCOC_{17}H_{35}$$
$$|$$
$$CHOCOC_{15}H_{31}$$
$$|$$
$$CH_2OCOC_{17}H_{35}$$

β-Palmito-α,α'-distearin

3. Physical and Chemical Properties of Fats

(a) **Physical Properties of Fats** – In a broader perspective, most 'fats' (with a few '*exceptions*') which are outright derived from '*animal sources*' are invariably found to be in the form of **solids** at ordinary room temperature (20 ± 2°C); and the *ones* obtained particular from:

- *plant sources* viz., soybean, olive, mustard, corn, cottonseed, sesame, sunflower, groundnut (peanut), rice bran, palm, linseed, coconut; and
- *fish sources* viz., codliver, mackerel, salmon* etc.

 NOTE In true sense, the so-called '*Liquid Fats*' are often termed as *oils* – even though they are *esters of glycerol i.e.,* very much akin to '*solid fats*'.

Interestingly, the structural difference prevailing between the:

- **Solid Fats** and
- **Liquid Oils**,

actually depend solely upon the '**degree of unsaturation**' of the *inherent* **fatty acids**.

Points to Ponder – There are *two* **important points**:

1. The so-called **physical properties** pertaining to the '*fatty acids*' are usually carried over to the **physical properties** of the corresponding '*triglycerides*'.
2. The '**solid animal fats**' mostly consist of the '*saturated fatty acids*'; whereas, the respective '*vegetable oils*' do contain *higher quantum* of the **unsaturated fatty acids**.

Table 3.3 records the **average percentage of fatty acid** contents of certain common '**Fats and Oils**'.

Table 3.3 The Average Percentage of Fatty Acids and Iodine Number of Certain Commonly Available Fats and Oils

S. No.	Type of Fats/Oils	Fats/Oils	Lauric Acid (%)	Myristic Acid (%)	Palmitic Acid (%)	Stearic Acid (%)	Oleic Acid (%)	Linoleic Acid (%)	Linolenic Acid (%)	Other Acid (%)	Iodine Number
I	Vegetable oils	• Coconut oil	45.4	18.0	10.5	2.3	7.5	–	–	16.3	10
		• Corn oil	–	1.4	10.2	3.0	49.6	14.3	–	1.5	123
		• Cottonseed oil	–	1.4	23.4	1.1	22.0	47.8	–	3.4	106
		• Linseed oil	–	–	6.3	2.5	19.0	24.1	47.4	0.7	179
		• Olive oil	–	–	6.9	2.3	84.4	4.6	–	1.8	81
		• Palm oil	–	1.4	40.1	5.5	42.7	10.3	–	–	54
		• Peanut oil	–	–	8.3	3.1	56.0	36.0	–	6.6	93

(Contd.....)

* **Fish Oil:**They usually serve as a rich source of **Omega-3 triglycerides** known to be useful as *lipid lowering drug substances* and are often '**liquids**'.

	• Sunflower oil	–	6.8	–	–	18.6	70.1	3.4	1.1	145
	• Soyabean oil	0.2	0.1	9.8	2.4	28.9	52.3	3.6	2.7	130
II Animal fats	• Beef tallow	–	6.3	27.4	14.1	49.6	2.5	–	0.1	50
	• Butter (Ghee)	2.5	11.1	29.0	9.2	26.7	3.6	–	17.9	36
	• Human fat	–	2.7	24.0	8.4	46.9	10.2	–	7.8	68
	• Lard	–	1.3	28.3	11.9	47.3	60	–	5.0	59

NOTE : **(1) It may be observed explicitly that even the 'solid fats' do contain certain unsaturated fatty acids; and also the 'liquid fats' contain some saturated fatty acids.**

(2) Essential fatty acids viz., linoleic acid and linolenic acid (unsaturated fatty acids) should be included essentially in a 'normal diet', because the human body is not able to synthesize them from the precursors.

Salient Features: These essentially comprise:

1. **Coconut oil** contains only a *small quantum of unsaturated fatty acids*; and hence, it normally occurs as a '*liquid*' not due to the presence of several **olefinic (*double*) bonds,** but on account of its having a rich content of *low molecular weight fatty acids* (*i.e.,* mostly **lauric acid**).

2. The so-called '**polyunsaturated fatty acids**' (**PUFAs**), usually present in **oils** having an average of **more than one double (olefinic) bond per fatty acid chain** *e.g.,* *olive oil, sunflower oil* (used in regular/routine diet) – are believed to prevent heart attacks significantly.

(b) **Chemical Properties of Fats –** Following are the *six* vital and important **chemical properties of fats**, namely:

- **Hydrolysis,**
- **Oxidation,**
- **Hydrogenation,**
- **Drying,**
- **Hydrogenolysis,** and
- **Formation of Acrolein,**

which will be treated individually in the sections that follows:

❏ **Hydrolysis** – It refers to the phenomenon whereby the molecules are duly cleared into their constituents by the interaction of water.

Thus, when '**fats**' are subjected to *hydrolysis* with either:

- **Enzyme Lipase or** • **An Alkali (NaOH/KOH),**

they usually yield **fatty acids** plus **glycerol**. Besides, it has been duly taken cognizance of the fact that 'lipases' do bring about the *critical splitting of fats* through several sequential steps as indicated under:

Triglycerides → Diglycerides → Monoglycerides → Fatty Acids + Glycerol

NOTE **The 'Lipases' do exert their optimized activity within the temperature ranging from 0° – 40°C.**

Thus, we may have the following expressions:

$$CH_2.\overset{\overset{\displaystyle O}{\|}}{C}.OR_1 \quad CH.\overset{\overset{\displaystyle O}{\|}}{C}-OR_2 \quad CH_2.\overset{\overset{\displaystyle O}{\|}}{C}-OR_3$$

A Triglyceride   A Diglyceride   A Monoglyceride   Glycerol [A trihydric alcohol]

With reactions: $\xrightarrow[{[-R_3COOH]}]{H_2O/Lipase;}$, $\xrightarrow[{[-R_2COOH]}]{H_2O/Lipase;}$, $\xrightarrow[{[-R_1COOH]}]{H_2O/Lipase;}$

Remark – The 'fats' are invariably hydrolyzed with **alkali** (**NaOH/KOH**) thereby giving rise to the formation of *'free fatty acids'* that eventually interact with alkali to yield their respective salts. The *'resulting salts'* are known as ' **soaps**'; and hence, the underlying phenomenon is termed as '**saponification**'.

Example: The so-called *alkaline hydrolysis* of the respective **oils and fats** usually forms *two* distinct products of reaction:

- **Glycerol** and • **Soaps** (Salts of 'fatty acids')

Thus, we may express as follows:

$$\begin{array}{l} CH_2.O.COR \\ CH_2.O.COR' \\ CH_2.O.COR'' \end{array} + 3\ NaOH \longrightarrow \begin{array}{l} CH_2OH \\ CHOH \\ CH_2OH \end{array} + \begin{array}{l} R.COONa \\ R'.COONa \\ R''.COONa \end{array}$$

Oil or Fat   Glycerol   Soaps

Saponification: **Saponification** relates to the alkaline hydrolysis of an ester, such as: hydrolysis of *glyceryl tristearate* with NaOH yielding **Na-stearate** (*i.e.*, the 'soap') plus **glycerol** (*i.e.*, the **by product**).

In other words, by analogy any *interaction* of an '*ester*' with an '*alkali*' to yield:

- **an alcohol** and • **Salt of an acid,**

is invariably referred to as the '**saponification reaction**' obviously, the large number of common soaps are, in fact, duly designated as the:

'**mixtures of sodium salts of C$_{12}$ to higher fatty acids**'

Formation of Soaps: In general, '**soap**' refers to any chemical entity (compound) of:

- **one or more fatty acids**, or
- **their equivalents,**

with an *alkali*. It also relates to a '**detergent**' that is much used in an array of such products as:

- **Liniments** • **Enemas** • **Making of Pills** • **Antacids**
- **Antiseptic** and • **Mild Aperient** (*laxative*).

In a broader perspective, the '**soaps**' are composed of several molecules that usually consist of **large hydrocarbon moieties**, namely:

- *lipophilic* (**fat-loving**) **moieties,**
- *hydrophilic* (**water-loving**) **moieties,** and
- **polar moiety.**

However, the inherent '*lipophobic segment*', which is invariably referred to as the respective '*hydrophobic*' (**water-hating**) **moiety** that eventually:

- **dissolves in fats/oils** or
- **greasy substance**;

whereas, the *polar segments* are found to be water-soluble.

Micelle – Interestingly, the **soap molecules**, after being subjected to dissolution in water, they do aggregate to form a '*less ordered structure*' – termed as **micelle**. In other words, a **micelle** refers to an agglomeration of amphiphillic (*surfactant*) molecules in a dispersion medium (*solvent*) having a diameter on the order of **50 angstroms (or 5 nm)**.

In true sense, a '**micelle**' represents a *roughly spherical assembly of large molecules* (see Fig. 3.1), that essentially depict both **polar and nonpolar regions**. It has been duly established that the respective region of the '**micelle**' may be responsible for absorbing other *adjuvant non polar entities*; and hence, hide them strategically under the **cover of the water solution**.

> **NOTE** Therefore, the cleaning characteristic features of '*soaps*' solely depend upon their ability to form emulsions with fat-soluble substances; and thus, the solid particulate matters of *diet* become dispersed in the emulsion.

Hydropkilic

Lipophobic
(Hydrophobic)

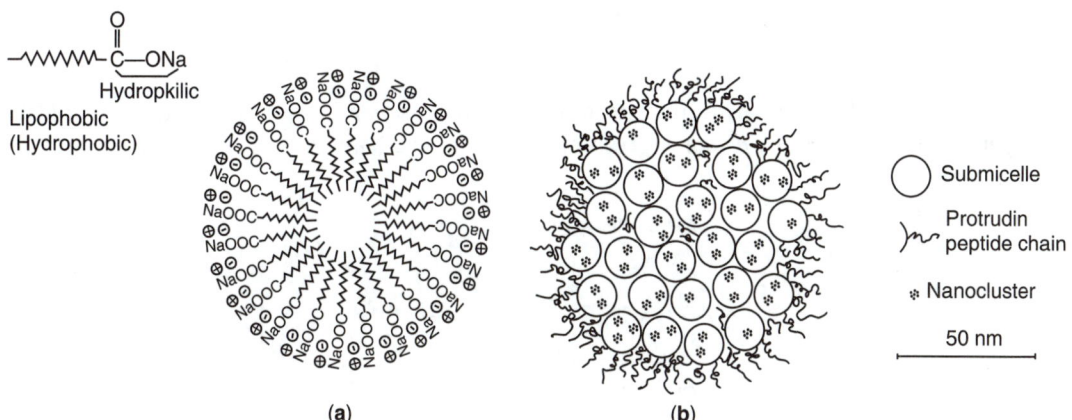

○ Submicelle

⌒ Protrudin
peptide chain

✳ Nanocluster

⊢———⊣ 50 nm

(a) (b)

Fig. 3.1 (a) The Critical Formation of a '*Micellel in Water*', **(b)** Cross-section via a Tentative Model of a '*Casein Micelle*'.

It is, however, pertinent to state at this point in time that the '**soaps**' are never proved to be an effective agent in water containing *salts of the heavier metals* (*viz.*, Pb^{++}-salts; Hg^{++}-salts; Fe^{++}-salts; Ca^{++}-salts) – since they get duly precipitated in their respective **insoluble salts**. Amazingly in such instances, the '**detergents**'* prove to be an excellent and *effective* **cleaning agent**.

- **Oxidation** – It is, however, pertinent to state here that the **oxidation** of *fats* in the air often takes place together with **hydrolysis**. In fact, **oxidation** occurs invariably at the *double bonds* (**olefinic**

---

* *Detergents:* These refer to the '**surfactants**'–employed as a **cleansing agent**. *Examples*-include: **soaps** with elongated alkyl chains attached to polar moieties *viz.*, sulphonic acids. Sodium Lauryl sulphate $[CH_3(CH_2)_{10}CH_2O.SO_3^- Na^+]$ designates a typical detergent.

bonds) of the respective **'fatty acids'** *viz.*, **oleic acid, linoleic acid, linolenic acid** etc., thereby giving such *chemical entities* as:

- **Short-chain fatty acids   and   • Aldehydes**

Thus, the resulting fatty acids generated in the course of *hydrolysis phenomenon* critically imparts:

> ➤ **less palatable taste**, and
> ➤ **perceptively changed 'odour' of fat (known as rancidity*)**

Importantly, one may reasonably prevent the process of **'oxidation'** by carefully incorporating a small quantum of some *organic substances* invariably termed as **'antioxidants'**.

Examples: A few typical commonly employed **antioxidants** are:

- **Natural Antioxidants** *viz.*, Tocopherol (Vitamin E)**;
- **Commercial Antioxidants** *viz.*, *tert*-Butyl p-cresol (A), and Nordihydroguiaretic Acid (B)

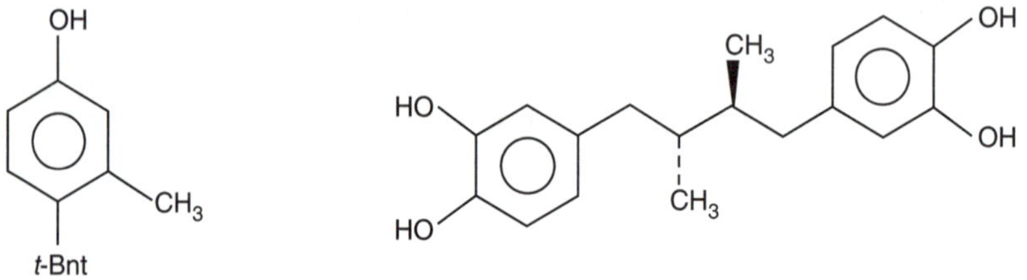

t-Bnt
(a)   **(b)**

Explanation: In *biochemical terms*, the phenomenon of *oxidation* relates to a so-called **direct attack by oxygen (O₂)**. However, the *primary* product, thus obtained, happens to be a *'peroxide'* [$-O°_2$] as shown below:

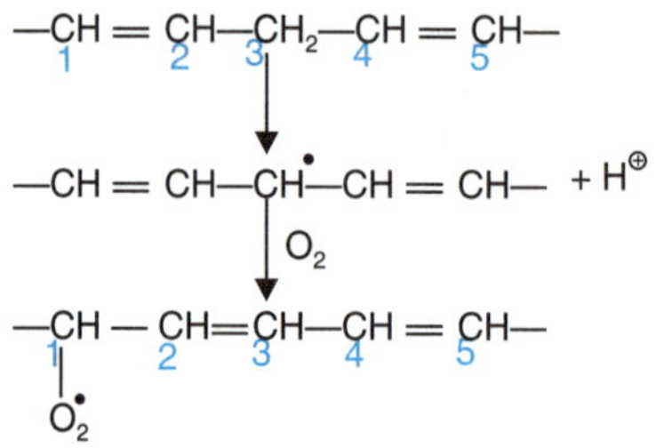

Comments: These essentially include:

(i) The **'initial radical'** may subsequently **'activate'** another *acylchain*; and hence, **propagate the reaction in turn**.

(ii) The *shift of the double bond* from the **initial state** to the **final state** has taken place from **C–1/C–2 to C–2/C–3**

❑ **Hydrogenation** – It refers to a reaction in which a reactant gets reduced by the **catalytic addition of molecular hydrogen (H₂)** to the easily reducible functional moieties *viz.*, across:

---

* **Rancid:** It refers to an offensive (unpleasant) smell or taste *e.g.*, a *vegetable oil* which has undergone an **oxidative degradation**.

** **Tocopherol:** It occurs in natural edible oil *viz.*, **Rice Bran Oil**.

- *carbon-carbon double bonds*;
- *carbon-oxygen double bonds* (*ketones*: $>C=O$); and
- *carbon-nitrogen double bonds* (*imines*: $> C = NH$).

In fact, the **catalytic hydrogenation** of the respective *unsaturated plant fats* that get duly converted into rather more saturated and solid fats. In usual practice, the **catalytic hydrogenation** is accomplished using:

- **finally divided Ni**, or
- **Palladium (Pd) metal**

Commercial Utilization of Hydrogenation – In a broader perspective, **hydrogenation** is usually carried out over finely divided Ni, in the particular **commercial production** of such commonly used *edible* **improvized fats** all over the world as:

- **Margarine** *i.e.*, butter substitute waster from *animal* or *vegetable* fats; and
- **Vegetable shortening** *i.e.*, hydrogenated vegetable fats used in *Bakery Products* mostly and *Confectionaries*.

Examples:

❏ **A general example:**

$$\boxed{\text{Vegetable Oil}} \xrightarrow[\substack{\text{Ni–Powder;} \\ \text{(Pressurized} \\ \text{Vessel)}}]{\text{H}_2\text{–gas;}} \boxed{\text{Vegetable Fat}}$$

[*Liquid Form*]   [*Solid Form*]

❏ **A typical example**

$$
\begin{array}{l}
\text{CH}_2\text{O.CO.C}_{17}\text{H}_{33} \\
| \\
\text{CHO.CO.C}_{17}\text{H}_{33} \\
| \\
\text{CH}_2\text{O.CO.C}_{17}\text{H}_{33}
\end{array}
\; + 3\text{H}_2 \;
\xrightarrow[\text{Hydrogenation}]{\text{Catalytic}}
\begin{array}{l}
\text{CH}_2\text{O.CO.C}_{17}\text{H}_{35} \\
| \\
\text{CHO.CO.C}_{17}\text{H}_{35} \\
| \\
\text{CH}_2\text{O.CO.C}_{17}\text{H}_{35}
\end{array}
$$

Triolein [*Liquid Oil*]   **Tristearin [*Solid Fat*]**

Points to Ponder

1. Normally the **hydrogenation** is done on a commercial scale to produce the *solid-shortening* available in the stores under an array of **Branded Products** *viz.*, **Dalda®**; **Rath®**; **Gagan®** etc.

NOTE   **Particular precautionary measure must be taken by the manufacturers to avoid *complete hydrogenation* of all the double bonds since a fat with almost no double bonds at all shall be too brittle.**

2. Interestingly, the so-called '**partial hydrogenation**' distinctly gives rise to a product having the *perfect and proper* consistency for use in cooking in the household.
3. **Margaine** (or **Artificial Butter**): During its large-scale production, the **hydrogenation** process may be *stopped immediately at a semisolid fat stage*.
4. **Totally Hydrogenated Fats**– These are mostly solid, brittle, and unpaltable; and hence, could be used for making *candles* only.

5. Housewives, all over the world, generally give an overall preference to the available '**solid fats**' rather than the '**liquid oils**' for routine cooking purposes. Therefore, tonnes of refined/purified such vegetable oils:

- **Cotton seed oil**
- **Groundnut oil** and **Palm oil**

are adequately converted to *solid edible fat* (or **vegetable ghee**) by means of the **hydrogenation process** as discussed earlier.

6. **Hydrogenation**–It has been observed that **hydrogenation** does convert only a *certain* **portion of the glycerides of unsaturated fatty acids into the corresponding saturated fatty acids**.

- **Drying**– There are a host of vegetable oils that essentially have *glycerides of unsaturated fatty acids viz.*,
- *Two* **Double Bonds** *e.g.*, **linoleic acid**; and
- *Three* **Double Bonds** *e.g.*, **linolenic, α-Oleostearic**, and **licanic acid.**

Amazingly, these foretold fatty acids do have the inherent property of *absorbing oxygen (O_2) gradually* from the environment (air); and subsequently, undergoing **natural polymerization** to yield a *hard transparent coating*; and hence, are used in **making paints** and oil cloth. In fact the underlying process is termal as 'drying'; and the resulting oils as the '**drying oils**'.

Salient Features: These essentially include:

1. In case, the acid comprises critically the – '**conjugate-system of double bonds**', such as:
   - **α-Eleostearic Acid** and
   - **Licanic Acids**,

   that are present in **Tung oil** and **Oiticiea oil**.

2. Importantly, the '**drying phenomenon**' occurs rather more readily and efficiently *vis-a-vis* the **glycerids** consisting of the **non-conjugated double bonds** *viz.*,
   - **Linseed Oil** and
   - **Perilla Oil**,

   that eventually contain *Linoleic Acid* and *Linolenic Acids*.

 NOTE   **It is, however, pertinent to state here that both *Tung Oil* and *Oiticica oil* are found to be superior to *Linseed Oil*, but their actual usage as the 'drying oils' have been restricted grossly due to their exhorbitant high-cost.**

3. **Mechanism of Exerting the Drying Effect** – The most probable mechanism of exerting the drying effect critically is caused by virtue of the particular *addition of oxygen (O_2)*, to the respective *unsaturated linkages** present strategically in the *esters of the oil (soap)* plus the effect of '**polymerization**' to a certain extent.

In addition, the ensuring '**drying phenomenon**' gets duly *catalyzed* (**energised**) by the crucial incorporation of *siccatives* (or *driers*), such as: **salts of Pb^{2+} and Mn^{2+}**. Nevertheless as per the *behavioural profile i.e.* when exposed to air the '**oils**' could be categorized into *three* distinct types:

- **drying oils,**
- **semi-drying oils**, and
- **non-drying oils**,

which shall now be treated individually as under:

---

*   That is, the **double bonds** (or **olefinic bonds**).

(a) **Drying oils** – Following are the two glaring examples of the '**drying oils**', namely:

- **Linseed oil** : *Linum usitatissimum* Linn, (*Fam:* **Liliaceae**); and
- **Hempseed oil** : *Cannabis sativa* Linn. (***Fam: Moraccae***),

which primarily do possess the *glycerides* of **linoleic** and **linolenic acids. Moreover**, on being exposed to atmospheric O_2 (air) – these get dried up and form a '**solid-elastic film**'.

Important Points

1. The major criteria of a '**good drying oil**' being that it usually gets dried up within a short span of **4-5 hours.**

2. **Linoxyn** – refers to a good quality *oxidized* and *polymerized* **linseed oil**, that critically has a *thick, brownish* and *elastic substance* which finds its enormous use in the commercial production of:

- **Linoleum** • **Lincrust** and • **Oil-Cloth.**

(b) **Semi-Drying Oils** – Following are the *two* important examples of the '**semi-drying oils**', such as

- **Sunflower Oil** : *Helianthus annus* Linné (*Fam:* **Compositae**); and
- **Cotton seed Oil** : *Gossypium hirustum* Linné (*Fam:* ***Malvaceae***),

which possess a **low-content** of *linolenic acid* in comparison to the *drying oils*; whereas, they do contain a **high content** or *linoleic acid* than that of the *non-drying oils*.

(c) **Non-Drying Oils** – The *two* most prominent examples of the '**non-drying oils**' are as given under:

- **Olive Oil** : *Olea europaea* Linne., (*Fam:* **Oleaceae**); and
- **Almond Oil** : *Prunus amygdalus* Batsch., (*Fam:* **Rosaceae**),

which largely contain the '**trioleins**'. However, on exposure to light and prolonged storage the aforesaid *two* **non-drying oils** become '**rancid**'.

Why do the 'oils' render rancid on storage? This could be due to the **influence of microorganisms** whereby the oil gets duly decomposed into:

- **Glycerol** and • **Polyunsaturated Fatty Acids (PUFAs)**

Thus, **PUFAs** get critically oxidized into the respective *aldehydes* and *carboxylic acids* having *much lesser C-atoms* in the residual molecules; whereas, the so-called *saturated fatty acids* duly get decomposed under the influence of the ensuing microorganisms to result into the formation of *ketones* respectively.

Examples: **Methyl heptyl ketones** do have a *perceptive unpleasant (disagreeable) rancid odour*. The actual decomposition takes place *via* the **β-oxidation modality** as given below

$$CH_3.(CH_2)_4.\ CH_2.CH_2.COOH \xrightarrow[\text{Oxidation}]{+\ O_2;} CH_3.(CH_2)_4.\overset{\overset{\displaystyle O}{\|}}{C}-CH_2.COOH \xrightarrow[\text{Decarboxylation}]{-\ CO_2;} CH_3.(CH_2)_4.\overset{\overset{\displaystyle O}{\|}}{C}.CH_3$$

Heptanoic acid

Methyl amyl ketone

Another major factor responsible for the so-called '**unpleasant taste**', particularly in the '*animal fats*', is perhaps due to the *hydrolysis* that is critically brought about by ***certain typical enzymes*** *viz.,* '*lipase*', present inherently in them. The ultimate hydrolysis of the *triglycerides* may distinctly yield **short-chain** **fatty acids** – that are responsible for attributing the overall '**bad odours**'. Therefore, in order to check and prevent the **rancidity**, it is very desired that the **fats and oils** must be stored under complete *refrigerated conditions* (since at *low temperatures* the foretold reactions are rather *slower on l* *sluggish*). Besides, the **fats and oils** should be stored in **amber – coloured glass bottles** or **opaque plastic bottles/jars**, which may retard the phenomenon of 'oxidation' generally caused by **UV/sun-light** to a great extent.

Caution: In order to prevent the *oxidation of fats and oils*, one may add certain approved and recommended '*antioxidants*' *viz.*, **Carotenoids, Vitamin E (Tocopherols), Vitamin C (Ascorbic Acid), Phenols, Tannins,** and **Hydroquinones**.

> **NOTE** In general, the '*vegetable fats*', namely; **Corn Oil, Groundnut Oil, Soyabean Oil, Olive Oil, Rice-Bran Oil, Sunflower Oil, Coconut Oil, Sesame Oil, Mustard Oil, Palm Oil,** may be preserved for a much longer duration than the corresponding '*animal fats*' (*viz.*, **Cod Liver Oil, Lard, Tallow etc.,**).

However, **biochemically** one may explain the '**oxidation process**' as the effective **direct attack by oxygen (O_2). Interestingly, the *initial product* is mostly a 'peroxide'**–as shown under:

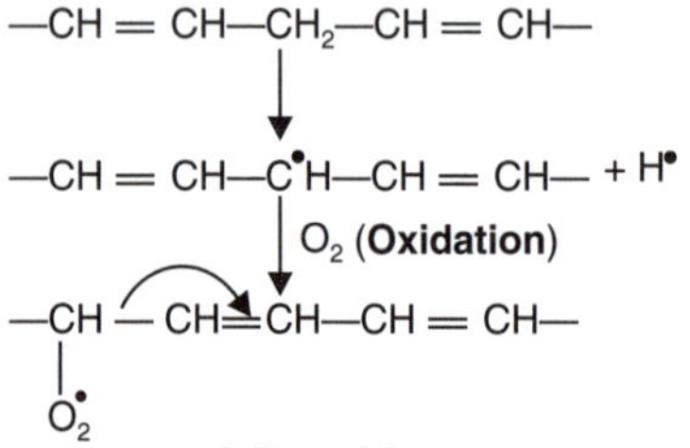

$$-CH=CH-CH_2-CH=CH-$$

$$\downarrow$$

$$-CH=CH-\overset{\bullet}{C}H-CH=CH- + H^\bullet$$

$$\downarrow O_2 \text{ (Oxidation)}$$

$$-CH\overset{\frown}{-}CH\overset{\equiv}{=}CH-CH=CH-$$
$$\underset{\overset{|}{\overset{\bullet}{O}_2}}{}$$

A Peroxide

Comments: These essentially include:

➢ The *initial* **peroxide radical** may eventually acitivate another *acyl chain*; and

➢ Consequently, help to propogate the reaction promptly.

> **NOTE** The critical shift of the '*double-bond*' (as marked above) takes place ultimately by the ensuring '*oxidation process*'.

The Fatty Acids– The *naturally occurring 'fatty acids' are of two* **types**, namely:

❑ *Plant Origin*: These invariably comprise **even-number of C-atoms**. The major possible explanation for this being that only *even-numbered fatty acids* are usually found in such '*fats*' since the human body builds these acids exclusively from the **acetic-acid units**; and, therefore, put the **C-atoms** in *two-at-a time* predominantly.

❑ **Bacterial Origin:** fatty acids are mostly found to be **highly unsaturated** in nature.

Points to Ponder– These essentially comprise:

1. First and foremost, it is worthwhile to mention that since almost all the fats essentially contain '**glycerol**' (*i.e.*, *trihydric alcohol*),–their individual characteristic features have been found to differ as per the prevailing **nature of fatty acids present in them**.

Examples: Following are some typical examples:

(i) **Higher Animals** – The **fatty acids** found in the *higher animals* do have a tendency to retain critical *chain lengths with:*, **16, 18,** or **20 C-atoms** (except in the '**milk fat**'–which essentially contains the **saturated fatty acids;** and the so-called C_{10}-C_{14} **fatty acids** do make up 10% of the molar total).

(ii) **Humans** – Amazingly, **oleic acid** (an **unsaturated fatty acid**) designates the most abundant fatty acid in the humans, – which could be seen in both:

- **adipose tissues** (*depot fat*), and
- **milk fat.**

 NOTE — In *human fat*, the presence of '*Butyric Acid*' (*i.e.*, the 4 C-chain shortest fatty acid) is found to be in abundance.

General Formula of Fatty Acids – It is represented as given under:

$$CH_3 - (CH_2)_n - COOH \text{ [or R - COOH]}$$

where, **'R'–designates** the *hydrocarbon chain*. The '**acyl radical**' is: $CH_3\text{-}(CH_2)_n\text{-}CO\text{-chain}$; whereas the value of '**n**' varies from : '*zero*' in **acetic acid** to '*86*' in **mycolic acid.** Carboxyl (–COOH) moiety in a Fatty Acid – is indeed a *fairly strong polar moiety*, which eventually gets ionized in *water* at the **intracellular pH** with the loss of a **proton (H^+).** Thus, we may express it as under:

$$R\ COOH \rightleftharpoons R\ COO^- + H^+$$

Status of 'R'–the Hydrocarbon Chain: It has been duly ascertained that the **hydrocarbon chain** (**R**) is *insoluble in water;* and hence, is **nonpolar** in nature. Based on the aforesaid facts it is obvious that the '**fatty acids**' do possess both these characteristic features, namely:

- **hydrophilic**, and
- **hydrophobic.**

Importantly, at the crucial **physiological pH** the so-called '**fatty acids**' invariably exist in solution as the typical **ionized alkyl carbohydrates**, such as:

Generalized Form	:	**R – COO⁻**
Stearate	:	$CH_3(CH_2)_{16} - COO^-$;
Palmitate	:	$CH_3(CH_2)_{14} - COO^-$;
Arachidate	:	$CH_3(CH_2)_{18} - COO^-$;
Myristate	:	$CH_3(CH_2)_{12} - COO^-$;
Laurate	:	$CH_3 (CH_2)_{10} - COO^-$;
Caprylate	:	$CH_3(CH_2)_6 - COO^-$;
Caproate	:	$CH_3(CH_2)_4 - COO^-$;
Butyrate	:	$CH_3(CH_2)_2 - COO^-$;

Saturated and Unsaturated Fatty Acids: The, saturated-fatty acids do have exclusively the '*single bonds*' in the **hydrocarbon chain (R)**. The **unsaturated–fatty acids** do have at least one '*olefinic or double bond [C=C]*' in the *hydrocarbon chain* (**R**).

The various **saturated-fatty acids** along with their *formula* and respective *common sources(s)* are recorded in Table 3.4. These are usually found to be as **solids** at the room temperature since they perceptively possess a *regular nature of their aliphatic chain* that **permits the molecules to be packed in a close, compact, and parallel alignment**, as shown below:

Remarks: It is pertinent to state here that in the **saturated fatty acids** the ensuring **interactions**, caused due to the **van der Waals forces**, between the neighbouring chains are rather *weak in nature*, but the *regular consistent packing* permits these forces to operate over a large extended segment of the *carbon-chain* so as to accomplished a substantial quantum of energy actually required to melt them eventually.

Table 3.4: Important Saturated Fatty Acids usually Found in Plants and Animals with their Formulae and Common Sources

S.No.	Name of Fatty Acid	General Formula	Common Sources
1	**Butyric Acid**	$CH_3-(CH_2)_2-COOH$	Dairy Butter
2	**Caproic Acid**	$CH_3-(CH_2)_4-COOH$	Dairy Butter
3	**Caprylic Acid**	$CH_3-(CH_2)_6-COOH$	Palm Oil; Coconut Oil
4	**Lauric Acid**	$CH_3-(CH_2)_{10}-COOH$	Laurel (Kernel) Oil
5	**Myristic Acid**	$CH_3-(CH_2)_{12}-COOH$	Oil of Nutmeg
6	**Palmitic Acid**	$CH_3-(CH_2)_{14}-COOH$	Animal Fat; Palm Oil
7	**Stearic Acid**	$CH_3-(CH_2)_{16}-COOH$	Cocoa Butter
8	**Arachidic Acid**	$CH_3-(CH_2)_{18}-COOH$	Groundnut Oil

The '**unsaturated fatty acids**', in contrast, are invariably found in the form of **liquid acids** at the ordinary room temperature, which is predominantly based upon the underlying fact that the respective *cis*-**olefinic bond (or double bond)** *interrupt the so-called regular packing of the chains*.

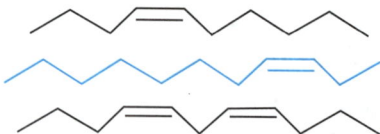

Obviously, in such typical instances one may require much less energy to melt them. In other words, '**the greater is the degree of unsaturation, – the lesser would be the mp**', which could be further expatiated by the fact that – *each double –bond introduces more disorder with regard to the packing of the molecules.*

Besides, the **unsaturated fatty acids** do have essentially either **one or more olefinic (double) bonds**; and the *general formula* is expressed as under:

$$R - CH = CH - (CH_2)_n - COOH$$

wherein, the number of **olefinic double bonds** range between **1 to 6**; and could be **positioned** after the **9th, 12th, 15th, and 18th C-atoms**. Table 3.5 records the major unsaturated fatty acids commonly available in naturally occurring **plant species** and some *animal species* as well.

Table 3.5. Some Important Unsaturated Fatty Acids Available in Plants and Animals

S.No.	Name of Fatty Acid	General Formula	Common Sources
1	Erucic Acid	$CH_3(CH_2)_2 CH = CH(CH_2)_{11} COOH$	Mustard Oil
2	Linoleic Acid	$CH_3(CH_2)_4 CH = CH CH_2 CH = CH (CH_2)_7$ $-COOH$	Soyabean Oil, Linseed Oil
3	Linolenic Acid	$CH_3CH_2CH = CH_2CH = CH CH_2CH = CH(CH_2)_7 - COOH$	Hempseed Oil, Linseed Oil
4	Oleic Acid	$CH_3(CH_2)_7 CH = CH (CH_2)_7 -COOH$	Olive Oil, Pork Fat
5	Palmitoleic Acid	$CH_3(CH_2)_5CH = CH(CH_2)_7 - COOH$	Sardine Oil; Milk Fat

SOME CHARACTERISTIC FEATURES OF UNSATURATED FATTY ACIDS

Following are some of the critically observed characteristic features of the *Unsaturated Fatty Acids*, namely;

(i) **Presence of *cis*–Olefinic Bond (Double Bond) in Monounsaturated Fatty Acids –** They invariably comprise a *cis*-**olefinic bond** in a restricted number of preferred positions in the carbon-chain.

Examples: These essentially include:

- **Polyunsaturated fatty acids (PUFAs)** usually possess 2 to 6 *cis*-**olefinic bonds** which are not *conjugated in nature.*

- A few **unsaturated fatty acids** *viz.*, **Linoleic Acid Linolenic Acid**, and **Arachidonic Acid** are essentially required for the *normal health of animals*; and hence, have got to be duly supplemented to the '*organism*' in the **normal diet.**

> **NOTE** In fact, such fatty acids are usually termed as: *Essential Fatty Acids*.

(ii) **Polyunsaturated Fatty Acids [PUFAs] –** Such **fatty acids** that essentially comprise *more than one olefinic bond* are known as **polyunsaturated fatty acids (PUFAs).** Thus, as per the actual *number of olefinic bonds* – the ensuring **polyunsaturated fatty acids (PUFAs)** are termed as follows:

- **Monoenoic** (with *one* **olefinic bond**);
- **Dienoic** (with *two* **olefinic bonds**),
- **Trienoic** (with *three* **olefinic bonds**),
- **Tetraenoic** (with *four* **olefinic bonds**).

(iii) **Microbes: Devoid of PUFAs –** Importantly, the microbes are practically devoid of **PUFAs.** However, **vaccenic acid** designates as the **principal PUFA** in *microbes*.

$$CH_3 - (CH_2)_5 CH = CH - (CH_2)_9 - COOH$$
Vaccenic Acid

(iv) **Abundant Presence of Unsaturated Fatty Acids in the *Living Organisms*:** The *living organisms* show an abundant presence of **unsaturated fatty acids** in them *vis-a-vis* the saturated fatty acids. Importantly, the so-called **unsaturated fatty acids** may possess either:

- an *odd number* of C-atoms; or
- an *even number* of C-atoms.

Odd-Numbered Unsaturated Fatty Acids – They mostly occur in '*all cells*' – as the prominent **minor components** of the *total fatty acids*. Besides, they are indeed more prevalent in *plant products*, but to a lesser extent in *animal products*.

Even-Numbered Unsaturated Fatty Acids – They are most commonly found in *animal species*, which could be due to the underlying fact that the **animal cells**, in particular, are able to *synthesize the fatty acids by the critical* polymerization of '*two C-units*' present in them.

(v) **Decrease in mp (°C) with Increase in Number of Olefinic Bond** – It has been observed, in general, that the **mp (°C)** of incuent *unsaturated fatty acids* are definitely **lower in number** *vis-a-vis* the respective *saturated fatty acids*; also they usually occur as '**liquid oils**' rather than '**solid fats**' at the prevailing **room temperature (20 ± 2°C)**.

Noticeably, the so-called '*plant triglycerides*' do have a relatively large proportion of the **unsaturated fatty acids** *e.g.*,

- **Oleic Acid**   • **Linoleic Acid**   and   • **Linolenic Acid**

Therefore, obviously these fatty acids essentially exhibit *lower mp (°C)* and are obtained as *liquids* at *room temperatures*. The '*animal triglycerides*', on the other hand; do have a relatively *higher proportion* of the **saturated fatty acids** *e.g.*,

- **Palmitic Acid**   and
- **Stearic Acid**.

In addition, the **animal triglycerides** have a distinct higher **melting points (°C)** and are mostly available as:

- '**Solid fats**' – at room temperature, or
- '**Semisolid fats**' – at room temperature.

(vi) **Zigzag Configuration in Hydrocarbon Chain of Saturated Fatty Acids** – It has been proved beyond any reasonable doubt that the respective **hydrocarbon chain of a saturated fatty acid** essentially inherits a **zigzag configuration** having the **bond angle** between the *carbon-carbon being 109°*. Thus, the *18C fatty acid*: **stearic acid** may be illustrated explicitly as follows:

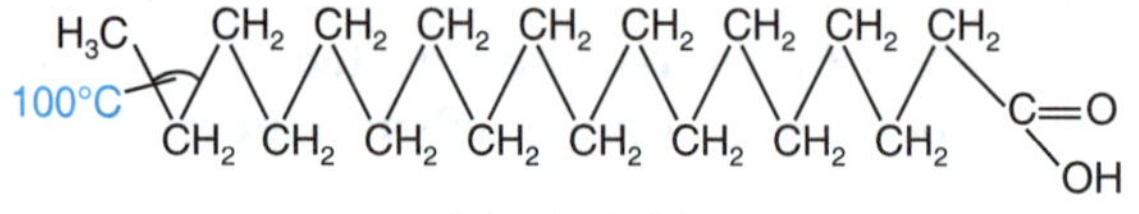

Stearic Acid

Introduction of '*One*' Olefinic Bond – The introduction of *one* **olefinic bond** (or *double bond*) between **C–9** and **C–10** in the unsaturated fatty acid '*oleic acid*' molecule, in fact, *caused a bend in the molecule*.

Introduction of '*Two*' Olefinic Bonds – The introduction of *two* **olefinic bonds** (or *double bonds*) between **C–6** and **C–7**, and **C–9** and **C–10** in the unsaturated fatty acid '*linoleic acid*' helps to cause **further bending** of the ensuring **hydrocarbon chain**.

Occurrence of Geometric Isomerism in Fatty Acids – One may critically observe the unique occurrence of **geometric isomerism** in the '**fatty acids**' particularly in such **fatty acids** that critically possess the **olefinic bonds**. Nevertheless, a good number of the **unsaturated fatty acids** invariably occur in the relatively *less stable cis-isomeric forms* rather than the *more stable trans-isomeric forms*.

Thus, we may have the following structure of '**Oleic Acid**' along with **geometric configurations** *viz.*,

- *cis*-Oleic Acid **(A)**; and
- *trans*-Oleic Acid **(B)**.

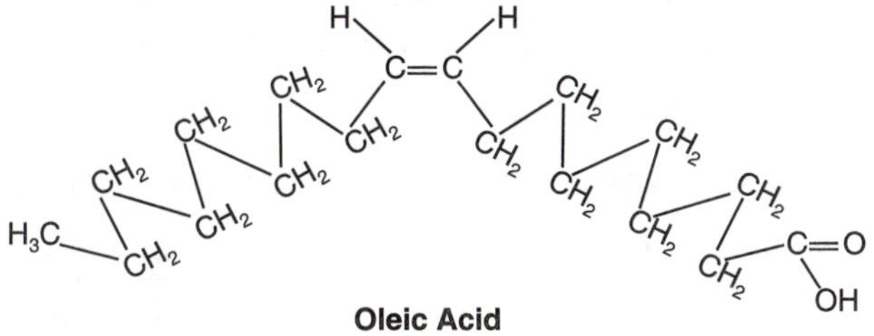

Oleic Acid

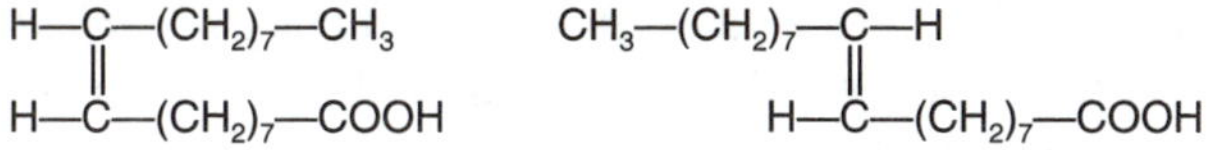

H—C—(CH$_2$)$_7$—CH$_3$
H—C—(CH$_2$)$_7$—COOH
(a)

cis-Oleic Acid

CH$_3$—(CH$_2$)$_7$—C—H
H—C—(CH$_2$)$_7$—COOH
(b)

trans-Oleic Acid

Salient Features of Fatty Acid – These essentially comprise :

(i) The **polyunsaturated fatty acids (PUFAs),** in general, may possess
- *conjugated* double – bond system ; or
- *nonconjugated* double – bond system

(ii) α – **Elaeostearic Acid:** It essentially possess a **conjugated triene double-bond system** as given under:

$$\text{CH}_3\text{—(CH}_2)_3\text{—CH}\overset{\textit{trans}}{=}\text{CH.CH}\overset{\textit{cis}}{=}\text{CH.CH}\overset{\textit{trans}}{=}\text{CH (CH}_2)_7\text{—COOH}$$

> **NOTE**
> The presence of the *conjugated double bond system* has been observed to be much more reactive thereby, ultimately leading to the '*polymerization* of fatty acids.'

(iii) **Linoleic Acid:** It possesses a **nonconjugated double bond system** as given below:

—CH$_2$—CH=CH—CH$_2$—CH=CH—CH$_2$—

Linoleic Acid

> **NOTE**
> In the *nonconjugated system* the methylene (–CH$_2$–) moiety critically possess double bonds on its either sides.

(iv) **Tuberculostearic Acid:** Anderson (1929–36)* isolated from several strains of **tubercle bacilli** and also from *B. leprae*, an *optically active*, *odd-numbered fatty acid*, **tuberculostearic acid**, which was duly proven to be **10-methylstearic acid** by its due **oxidation and synthesis** procedures, as shown under:

$$\text{CH}_3\text{—(CH}_2)_7\text{—}\overset{*10}{\underset{\underset{\text{CH}_3}{|}}{\text{CH}}}\text{—(CH}_2)_8\text{—COOH} \xrightarrow[\text{(Oxidation)}]{\text{[O]}} \text{CH}_3(\text{CH}_2)_7\text{—}\overset{\overset{\text{O}}{||}}{\text{C}}\text{—CH}_3 + \text{HOOC (CH}_2)_7\text{COOH}$$

Tuberculotearic Acid
[C$_{19}$H$_{28}$O$_2$;mp : 10-11°C]

Methyl-*n*-octyl ketone

Azelaic Acid

---

* **Anderson R.J :** 1870 Harna, Sweden, Ph.D., Cornell; Yele Univ.

Furthermore, using the *same source* Anderson isolated altogether another substance known as **phthio acid** which eventually gives rise to the formation of **typical tubercular lesions** upon injection in animals. The material was subsequently found to be a mixture. However, one of its components known as **mycolipenic acid** ($[\alpha]_D + 8°$; λ_{max} 2200Å) – shown to have the following structure (Polgar, 1954):

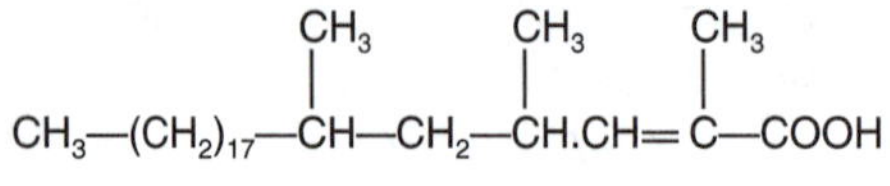

$$
\begin{array}{ccc}
& CH_3 \quad CH_3 \quad CH_3 \\
& | \qquad | \qquad | \\
CH_3—(CH_2)_{17}—CH—CH_2—CH.CH= & C—COOH
\end{array}
$$

Mycolipenic Acid

$$
\begin{array}{ccc}
CH_3 \quad CH_3 \quad CH_3 \\
| \qquad | \qquad | \\
CH_3 (CH_2)_n.CH.CH_2.CH—CH_2.CH—COOH
\end{array}
$$

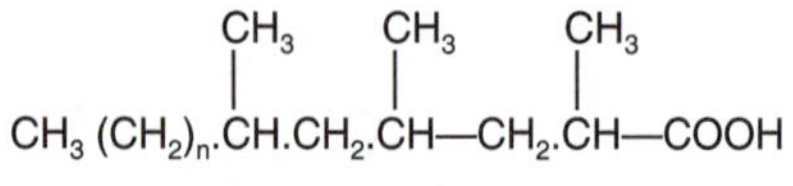

Mycoeranic Acid
[A levorotatory component]
where n=21

(v) **Chaulmoogric Acid:** There are a few naturally occurring *branched-chain acids* exhibiting specifically a definite **physiological activity profile**. In fact, there are *three* such acids which comprise the **cyclopentenyl ring system**, namely:

- **Chaulmoogric Acid** • **Hydnocapric Acid** and • **Gorlic Acid**

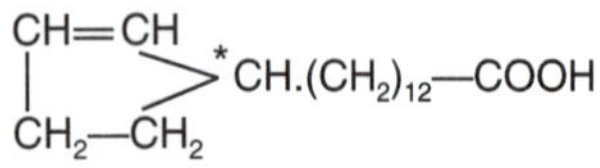

$$
\begin{array}{l}
CH=CH \\
| \qquad \quad \!\!\! >^*CH.(CH_2)_{12}—COOH \\
CH_2—CH_2
\end{array}
$$

Chaulmoogric Acid
($C_{16}H_{28}O_2$)
$[\alpha]_D + 56°$; mp=71°C

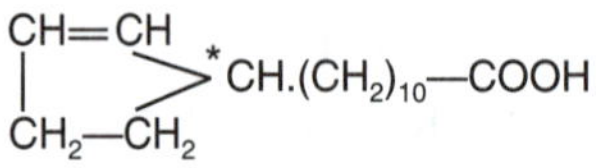

$$
\begin{array}{l}
CH=CH \\
| \qquad \quad \!\!\! >^*CH.(CH_2)_{10}—COOH \\
CH_2—CH_2
\end{array}
$$

Hydnocarpic Acid
($C_{16}H_{28}O_2$)
($[\alpha]_D + 68°$; mp=59-60°C)

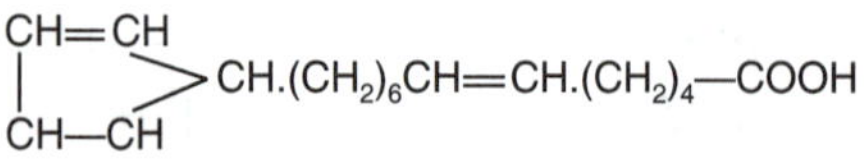

$$
\begin{array}{l}
CH=CH \\
| \qquad \quad \!\!\! >CH.(CH_2)_6CH=CH.(CH_2)_4—COOH \\
CH—CH
\end{array}
$$

Gorlic Acid
($C_{16}H_{28}O_2$)
($[\alpha]_D + 61°C$; mp=6°C)

Chaulmoogric Acid – is obtained from *Chaulmoogra Oil*, a native (**East Indian**) name for the fat of *Hydrocarpus Kurzii*. In fact, the **hydnocarpus oils** have been used extensively since centuries by **Hindus** and **Chinese** populace for the critical treatment of *'leprosy'* (*i.e.*, a disease produced by an **acid-fast bacillus**)

 NOTE | The *three* aforesaid acids do occur as major components of the seed fats of the family *flacourtiaceae* exclusively.

Isomerism in Unsaturated Fatty Acids – Interestingly, the causation of *distinct and different* **configuration(s)** very much around the *olefinic (double) bond* – the so-called **unsaturated fatty acids** do exhibit the *cis-trans* isomerism (also known as the **geometrical isomerism**). In other words, it relates to the type of *stereoisomer* resulting from differing arrangement of the functional moieties either:

- **on the same sides of a double bond** (*cis*-isomer), or
- **on the opposite sides of a double bond** (*trans*-isomer).

Examples: Following are *two* classical **examples of isomerism** in the **unsaturated fatty acids**, such as:

- **Oleic Acid** (*cis* – form) and • **Elaidic Acid** (*trans* – form)

$$
\begin{array}{l}
CH_3—(CH_2)_7—C—\boxed{H} \\
\qquad \qquad \quad \| \\
HOOC—(CH_2)_7—C—\boxed{H}
\end{array}
$$
Located on same Side of double bond

Oleic Acid
[*cis*–Form]

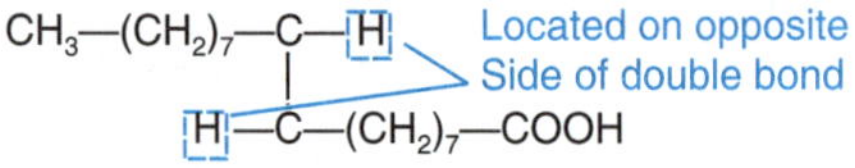

$$
\begin{array}{l}
CH_3—(CH_2)_7—C—\boxed{H} \\
\qquad \qquad \quad | \\
\boxed{H}—C—(CH_2)_7—COOH
\end{array}
$$
Located on opposite Side of double bond

Elaidic Acid
[*trans*–Form]

Comment: Of the *two* **different isomeric forms** – it has been observed duly that:

> ➤ *cis*-**Isomers** – represent usually the so-called naturally occurring fatty acids; and

> ➤ *trans*-**Isomers** – are **not** usually utilized by the ensuing organism.

U-Shaped Arachidonic Acid: There are quite many 'geometric isomers' pertaining to the '**acids**' that essentially exhibit much *greater degree of unsaturation*. Besides, the **unsaturated long chain fatty acids** that specifically occur in nature mostly belong to the *cis*-**form** only; and as such the molecules are invariably bent strategically at the **olefinic (double) bond** *e.g.*, **Arachidonic Acid** (which is U-shaped) as shown under:

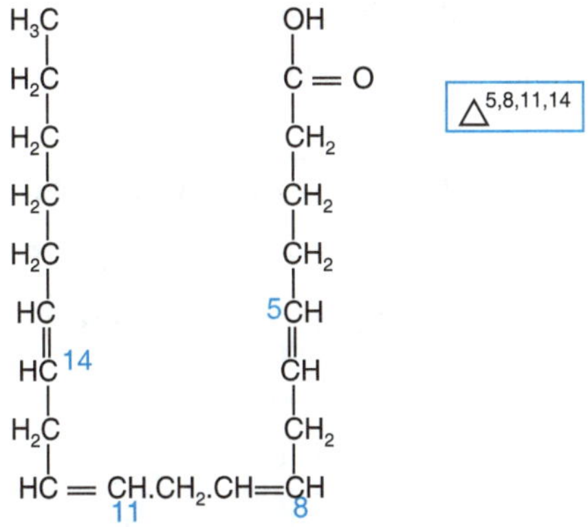

Arachidonic Acid
[U-Shaped]

Essential Fatty Acids [EFAs] – The **EFAs** refer to the group of *polyunsaturated fatty acids* (*PUFAs*) derived exclusively from the '*plant sources*' that must be supplied in the *human normal 'diet'* – since they cannot be synthesized in the body.

Based on the scientific evidences from the literature it has been duly established that the '**human body**' is capable of converting the **stearic acid** (*i.e.*, a *saturated fatty acid*) into the respective **oleic acid** (*i.e.*,) an *unsaturated fatty acid*) by the critical *insertion of a* **double bond** into the former. Nevertheless, it is absolute **not possible** to insert any further **olefinic (double) bonds**; and, therefore, the resulting unsaturated fatty acid, **oleic acid** cannot be converted to:

- **Linoleic Acid** - **Linolenic Acid** and - **Arachidonic Acid**

   possessing 2, 3, and 4 double (olefinic) bonds respectively.

Points to Ponder –

1. The aforesaid *three* typical unsaturated fatty acids are found to be quite essential for both:
   - **normal growth of skin tissues**, and
   - **healthy rejuvination of skin manifestations**.

2. As these **PUFAs** are incapable of being synthesized in the human body, they by all means should be duly provided in the form of a **regular healthy diet**; and hence, these are collectively termed as the:

   Essential Fatty Acids (EFAs).

3. **EFAs** invariably occur in an array of known **vegetable oils**, such as:
   - **Corn Oil** - **Cottonseed Oil** - **Groundnut Oil** - **Sunflower Oil** - **Olive Oil**
   - **Rice-Bran Oil** - **Soyabean Oil** - **Sesame Oil and** - **Mustard Oil**

> **NOTE**
> The so-called 'animal fats' *viz.*, Butter Oil (Ghee), Lard, Tallow etc., are the poor sources of EFAs.

4. **EFAs** are found to be not only of *special interest* but also of *most advantageous significance* in the overall and effective control–management of the '*hypercholesterolaemia*', – which is intimately associated with:

   • **atherosclerosis** (*i.e.*, hardening of arteries in the heart thereby causing partial blockade and the normal flow of blood away from the heart); and

   • to fight against the '*battle of bulge*' *i.e.*, to reduce the *body's accumulated fat* (or *adipose tissues) to* make life better and prolonged.

Table 3.6 records the composition of fatty acid profile in various important fats derived from '**plant**' and '**animal**' origin.

S.No.	Oil/Fat	Saturated	Monounsaturated	Polyunsaturated
Table 3.6: The Composition of Fatty Acids in Various Plant Oils and Animal Fats				
			Percentage of Total Fatty Acids	
I	**Plant Oils**			
	• Olive Oil	20	26	54
	• Corn Oil	15	31	53
	• Soyabean Oil	14	24	53
	• Saft Margarine	23	22	52
II	**Animal Fats**			
	• Butter Fat	60	56	04
	• Pork Fat	59	39	02
	• Beef Fat	53	44	02
	• Chicken Fat	39	44	21

Large-Scale Production of Higher Fatty Acids – The *large-scale (or commercial) production of higher fatty acids* is also sometimes referred to as the '**hydrolysis of fats'.** Thus, the actual hydrolysis of so-called '*fats*' to the respective **fatty acids** (*i.e.*, *fat-splitting*) may be accomplished effectively by either:

   • **Batch Process**, or
   • **Continuous Process**.

Nevertheless, one may take cognizance of the fact that the comparison to the specific '**Batch Saponification Process**', the actual production of soap by the continuous '**Fat-Splitting Process**, followed immediately by its *critical neutralization*' (**with NaOH**) has eventually turned the **entire process** much *shorter and convenient*.

Since the ensuing '**fatty acids**' are invariably, subjected to further *refining processes by means of* '*distillation*', – they are invariably produced from the '*low-grade fats*', namely:

   • **low-grade tallow**,
   • **bone and skin fat (grease)**,
   • **low-grade vegetable oils**, and
   • '*acid-oils*'–**not suitable for manufacture of soap.**

In usual practice, before one subjects the *fat* to the '**actual *fat-splitting phenomena***', it is mostly refined by agitating the *diluted sulphuric acid* (H_2SO_4) (that helps in the critical *coagulation of proteins*); and ultimately washed repeatedly with *soft-water*.

Modern Fat-Splitting Procedure – It is normally carried out under the following *two* specific parameters:

- **total devoid of any '*catalyst*'**, and
- **under reasonable high-pressure**.

Alternatively, in a typical '**non-catalytic procedure**', the respective '***autoclave***' is duly filled up to **2/3rd with the purified fat**, and the incorporation of *air-free condensate* obtained duly from a previously *hydrolyzed product*. The respective *valves* of the autoclave are duly closed and *steam* is injected progressively. However, the entire on-going process is performed usually in **2-stages** :

- **First-stage** – production of *glycerol-water*; and
- **Second-stage** – *water* being replaced appropriately.

Continuous Fat-Splitting Procedure – The latest procedure being the **continuous fat-splitting technique** which are undertaken at **slightly higher** *temperatures* and *pressures* *vis-a-vis* those employed in the '**batch processes**'. The **hydrolysis procedure** is being carried out in relatively **tall and slender SS-tanks** duly provided with several *SS-perforated plates*.

Modus Operandi – The various sequential steps involved are as follows:

1. **Pre-heated** fat is duly pumped right into the bottom of the **SS-tank** *via* a strategically placed '**distributor**' device.
2. **Pre-heated water** is subsequently pumped *via* the top of the **SS-tank**.
3. Now, **super-heated steam** (through *coiled-coil*) is to be injected into the SS-tank to maintain:
   - **pressure upto 600–700 lb.. in^{-2}**; and
   - **temperature to ~ 280°C**.
4. The **fatty acids**, thus liberated, being *lighter than water floats up* in the **SS-tank** – which could be removed in a *separate SS-Vessel*; whereas, the *liberated* glycerol, being *heavier than water* settles down along with the water in the **SS-tank** which may be removed periodically through a valve.
5. Thus, the **fatty acids** recovered by either of the *two* aforesaid procedures (*viz.*, **Batch Process** and **Continuous Process**) may be subjected *perfect* **drying process** accomplished by:
   - **controlled heating under reduced pressure**, and
   - **refining by distillation under high vacuum***.

 **NOTE** **The routine careful hydrolysis of such vital sources as: bone, *skin-grease*, and *tallow* yields an admixture of *three* most vital and important fatty acids *viz.*,**

- Palmitic and stearic Acids (Solids),
- Oleic Acid (liquid)

Physical and Chemical Tests of Fats and Oils – The ultimate industrial and marketed potential (value) of a specific **fat** or **oil** has been entirely based upon its **composition and purity**. Hence, a plethora of **physical and chemical tests** have been put forward to ascertain the ensuring potential of fat or oil.

---

* That is, **by** making use of '**Dowtherm**' – (a *heat-transfer agent*) as the **heating medium**.

Physical Tests – These essentially comprise:

- **mp(°C)** • **Specific Gravity (sp. gr.)** • **Viscosity** • **Weight per mL and**
- **Refractive Index (RI),**

and are duly provided in latest ***Merck Index*** as well as various *Official Compendia viz.*, IP, BP, USP, Eur. P., etc.

Chemical Tests – These predominately include:

- **Acid Value** • **Saponification Value** • **Iodine Value** • **Acetyl Value**
- **Polenske Value** and • **Reichert Meissl (RM–Value),**

which shall now be treated individually as under:

(i) **Acid Value** – The 'acid value' of a **fat** or an **oil** may be defined as – '**the number of milligrammes of potassium hydroxide required to neutralize the free organic acid present in 1g of fat or oil'**.

It is usually determined by dissolving an aliquot (*i.e.*, *weighed quantity*) of fat or oil in ethanol and titrating against a *standard alkali (0.1 M KOH)* using **Phenolphthalein** as an indicator.

>
> **1. *Acid Value* categorically indicates the precise and exact quantum of *free-fatty acids* present in Fat or Oil.**
> **2. A relatively *high acid-value* is indicative critically of a *stale fat or oil* (stored under quite improper parameters).**

(ii) **Saponification Value:** It may be defined as – '**the number of milligrammes of potassium hydroxide required to saponify completely 1g of fat or oil *i.e.*, to neutralize completely the fatty acids resulting from complete hydrolysis of 1g of fat or oil'**.

- ❑ **Saponification value** may be duly calculated by refluxing a weighed quantity (**1 to 2g**) of the **fat** or **oil** with a known excess of *standard alcoholic caustic potash (KOH) solution*; and **back-titrating** the excess of *alkali (KOH)* with a standardized acid.

- ❑ **Saponification value** invariably provides a vivid idea with respect to the *m.w. of fat or oil*. Thus, the *smaller the saponification value* – the *higher would be the respective molecular weight (m.w.)*.

- ❑ **Saponification value** also finds its enormous utility in calculating the precise and exact amount of **alkali required** for the conversion of a definite amount of **fat** or **oil** into the '**soap'**; and also in detecting the degree of **adulteration of fat or oil** either by:

  - **lower** or • **higher – saponification value.**

- ❑ The actual difference between the '**saponification value**' and '**acid value**' is commonly called the **ester value of the fat or oil.**

(iii) **Iodine Value:** It may be defined as – '**the number of grammes of iodine taken up by 100 g of fat or oil'**. Since *iodine* fails to react rapidly; hence, **iodine monochloride (ICl)** is being employed in actual practice.

Importance of Iodine Value – These essentially comprise:

  ➢ It may be regarded as a critical measure of its extent of unsaturation; and thus, provides explicitly an idea of its **drying character profile**.

  ➢ Besides, it also aids vehemently in finding the *probable adulteration in a fat or oil*; and also pre-judging its ensuing **stability profile** for the **preparation of soap**.

Determination of Iodine Value – Following are the various sequential steps involved for the determination of iodine value;

- An aliquot of **fat or oil** is carefully transferred to an *Iodine Flask* (250 mL), dissolved in 15mL of $CHCl_3$ or CCl_4, and treated with 25 mL of **Wij's Reagent*** (IP, 2007).
- The glass stopper of the flask is duly moistened with *KI–solution*, and the *Iodine Flask* is kept in *dark* for 30–6 **minutes**.
- The resulting solution is now mixed with 15 mL of 10% (w/v) aqueous solution of KI and litrated carefully with **standard *thiosulphate solution* (0.1M)** using freshly prepared *starch as an* indicator (end point: colourless to dark blue).
- A '**blank**' determination is also carried out (*i.e.*, **devoid of oil or fat**). Thus, the actual observed difference between the *volume of thiosulphate solution (mL) 0.1M* in the above *two* **titrations** will clearly indicate the **equivalent quantum of I_2 – absorbed** (by the *analyte*)

Thus,
$$\text{Iodine value} = \frac{(a-b) \times 1.27}{W}$$

where,

a = Reading of Blank Titration,

b = Reading of Actual Titration,

W = Weight of Fat or Oil Taken

(**The milliequivalent of iodine per mL = 1.27**)

Non-Drying Oils *viz.*, **Mustard Oil, Peanut Oil** – shows the I_2–value ranging between **85 to 105**;

Semi-Drying Oil *viz.*, **Sesame Oil, Cottonseed Oil**–shows the I_2-value ranging between **105 to 120**; and

Drying Oils *viz.*, Soyabean Oil, Linseed Oil–shows the I_2–value ranges **above 120**.

\multicolumn{4}{c}{**Table 3.7 Saponification Value and Iodine Value of Some Selected Fats and Oils**}			
S.No.	**Fats or Oils**	**Saponification Value**	**Iodine Value**
I	**Animal Fats**		
	• Butter	216–235	26–45
	• Lard	193–200	40–66
II	**Vegetable Fats**		
	• Coconut	246–250	8–10
	• Palm	196–210	48–58
III	**Animal oils**		
	• Lard oil	190–195	46–70
	• Whale oil	188–194	110–150
	• Fish oil (Sardines)	105–195	120–192

(Contd.....)

---

* **Wij's Reagent:** It is prepared by (a) dissolving 8g of ICl_3 in 250 mL glacial active acid; and (b) dissolving 9g of **KI** in 500 mL of glacial active acid; and mixing the two solutions.

IV	Vegetable oils		
	• Castor	176–187	81–91
	• Olive	185–200	74–94
	• Cotton Seed	191–196	108–115
	• Linseed	190–196	175–200

(iv) **Acetyl Value:** The '**acetyl value**' relates to a measure of the **alcoholic (–OH) moiety** present in **fat** or **oil**. It may be defined as '**the number of milligrammes of potassium' hydroxide (KOH) required to neutralize acetic acid (CH_3COOH) liberated duly by the saponification of 1g of completely acetylated oil or fat**'.

(v) **Polenske Value:** The **Polensake Value** may be defined as '**the number of mL of 0.1M potassium hydroxide (KOH) solution needed to neutralize the steam-volatile, but water insoluble fatty acids obtained duly from the distillate of 5g of hydrolyzed fat or oil**'.

 NOTE   *Water-insoluble fatty acids* are obtained meticulously by the careful filtration of the distillate in *Reichert-Meissl Value*. These fatty acids are dissolved in ethanol and titrated against standardized 0.1M KOH solution.

(vi) **Reichert-Meissl Value:** The **Reichert-Messl Value** actually relates to the precise measure of *volatile fatty acids* present duly as the respective '**glycerides**' present in **fats** and **oil**.

It may be defined as – '**the number of mL of 0.1 M potassium hydroxide (KOH) solution required to neutralize the distillate of 5g of hydrolyzed fat or oil**'.

The **RM-value** is invariably employed for the precise determination of '*purity*' of **butter ghee**.

RM-values of some marketed **fats and oils** are as given under:

- **Pure Butter, Ghee**   : **RM–value** ranges between 20–30;
- **Coconut Oil**   : **RM–value** ranges between 6–8.5; and
- **Palm Kernel Oil**   : **RM–value** ranges between 4–7.

2.1.2. *Waxes*

The waxes are substances which contain both saturated and unsaturated hydrocarbons together with certain substitution by oxygen.

Another school of thought relate the '**waxes**' as–'**the esters of high molecular weight fatty acids with alcohols other than glycerol (*true waxes*) or with sterols *e.g.*, cholesterol (sterylesters).**'

In general, the *constituent fatty acids* and *alcohols* do have C-atoms ranging between 24–36.

Variants of Waxes: Following are some known variants of waxes:

❑ **Carnauba Wax [or Brazil Wax]:** It is obtained as an exudate from the pores of the leaves of the *Brazilian wax palm tree, Copernicia* prunifera (Muell) HE Moore, [*Copernica cerifera* (Arruda da Camara) Mart.], *Palmae*. The inherent hardness and high-polish capability of this important wax can be ascribed to the critical presence of esters of the so-called **hydroxylated unsaturated fatty acid** having about **12 C-atoms** in the *acid chain*. Besides, it is composed of:

- **Myricyl Alcohol (30 C-atoms)**, and
- **Cerotate (26 C-atoms)**.

❑ **Bees Wax [or Yellow Beeswax]:** It is a substance obtained from bee honeycombs. **Bees wax** essentially consists of esters of **straight-chain monohydric alocohols** with even-numbered carbon-chains from C_{24} to C_{36} esterified with straight-chain acids also having even-number of C-atoms up to C_{36} (some C_{18}-hydroxy acids), such as:

- **Triacontanol hexadecanoate**, and
- **Hexacosanol hexacosanoate**.

In **bees wax**, the fatty acid constituent is a rather *smaller chain acid viz*, **Palmitic Acid (16°C)**; and alcohol is **myricyl (30 C-atoms)**.

Ambretolide – It is a hydroxy acid that occurs in the seeds of *Abelmosehus esculentus*; which is responsible for its inherent characteristic odour of the seed.

$$H_2C \underset{(CH).(CH_2)_5-C}{\overset{(CH_2)_7-CH_2}{\diagdown}} O$$

Ambretolide

Table 3.8 enlists some of the abundantly available vital **alcohols** and **fatty acids** in waxes together with their **molecular formulae.**

			Table 3.8 Certain Abundantly Available Important Alcohols and Fatty Acids in Waxes		
S.No.	**Alcohols**	**Molecular Formula**	**S.No.**	**Fatty Acids**	**Molecular Formula**
1	Lauryl	$C_{12}H_{25}OH$	1	Myristic	$C_{13}H_{27}COOH$
2	Cetyl	$C_{16}H_{33}OH$	2	Palmitic	$C_{15}H_{31}COOH$
3	Octadecyl	$C_{18}H_{37}OH$	3	Cerotic	$C_{25}H_{51}COOH$
4	Carnaubyl	$C_{24}H_{49}OH$	4	Melissic	$C_{29}H_{59}COOH$
5	Ceryl	$C_{26}H_{53}OH$			
6	Myricyl	$C_{30}H_{61}OH$			
7	Cocceryl	$C_{30}H_{60}(OH)_2$			

NOTE Interestingly, in certain specific waxes the *alcohol component* may be '*cholesterol*'; and such waxes are generally found in '*blood*'.

Salient Features of Waxes: Following are some of the salient features of waxes:

1. Since the waxes are established to be the **fully reduced hydrocarbon chains** that are found to be *water insoluble* and *quite resistant* to the usual *atmospheric oxidation*. They do have a very high mp. Due to these typical characteristic features they are mostly used for polishing they wooden **furnitures.** Besides, these waxes are found to be absolutely **inert chemically**; and hence, are not digested by the so-called **fat-splitting enzymes.** However, the presence of *hot ethanolic (alcoholic)* **KOH solution** may expedite the **phenomenon of fat-splitting**.

2. The **waxes**, in general, do possess a relatively **higher mp** *vis-a-vis* the **fats** (60–100°C); and hence, are **harder in texture.** In fact, the **animals** and **plants** make use of them usually as the **protective coatings.** Besides, the *leaves* **of most plant species** are duly coated with '**wax**'*.

---

* That is, cause protection from **microbial attacks** and to conserve moisture.

Interestingly, one may also observe the presence of waxes in:

- **feathers of birds,***
- **furs of animals, and**
- **ears of humans protected by ear-wax**.

3. Importantly, the **waxes** are not saponified readily *vis-a-vis* the **fats**; and hence, not attacked by **lipase** (an *enzyme*) easily. *Saponification of waxes*;

- may be **saponified** by prolonged boiling with alcoholic KOH,
- more conveniently **saponified** by treatment of a solution of **wax** in: *Petroleum Ether, Absolute Ethanol*, and *metallic* Na (*i.e.*, with **CH_3–ONa sodium methoxide**)
- *Saponification products of waxes* designate the **water-soluble soaps** (*i.e.*, the respective **sodium salts of higher fatty acids**).
- **Water-insoluble long-chain alcohols** invariably-occur in the 'unsaponifiable fraction'.
- **waxes comprise of 31–55% pf 'unsaponifiable matter'**; whereas, the **fats** and **oils** do contain only **1–2%** of **unsaponifiable matter.**
- **waxes largely serve as *water-barriers*** particularly in **inserts, birds**, and **furred animals**.
- **waxes** also serve as the *protective coating* on the *fruits* and *leaves*.
- **Lanolin** (a *wool-fat*) is usually liberated in the skin of a large segment of **fur-bearing animals**.
- Some important waxes have been duly derived from various **plant sources** *viz.*,
  - ➤ **Carnauba wax** (a Brazilian Palm Tree);
  - ➤ **Lanolin** (from *Sheep's wool*);
  - ➤ **Beeswax** (from *bee's honey-comb*); and
  - ➤ **Spermaceti** (from *whales*).
- **waxes** do find their enormous usage in the preparation of a variety of such useful products as:
  - ➤ **Ointments,**
  - ➤ **Cosmetics,**
  - ➤ **Polishes, and**
  - ➤ **Perfumed Christmas Candles.**

NOTE However, the '*Paraffin waxes*' are not '*esters*', but are mixtures of *high-molecular weight alkanes*.

Some Common Commercially Available Waxes: Following are some of the important common commercially available waxes, namely:

- **Spermaceti** • **Sperm Oil**   • **Jojoba oil**   • **Lanolin (Wool Wax)** and
- **Cholesteryl Palmitate**

which shall now be discussed individually in the sections that follows:

❑ **Spermaceti** – The **permaceti** is obtained abundantly from the **head of the '*sperm whals*'**. It is found to be a rich source of **cetyl alcohol ($C_{16}H_{38}OH$)** and **palmitic acid (CH_3–$(CH_2)_{16}$–COOH)**]:

---

* That enables the '**ducks**' to swim in water freely.

Palmetic Acid ⟍   ⟋ Cetylalohol Acid

$$CH_3-(CH_2)_{14}.\overset{\overset{\displaystyle O}{\|}}{C}-OC_{16}H_{33}$$

Cetyl Palmitate

NOTE Spermacetic is mostly employed in making *'candles' physeter macrocephalus –* is a source of pure *cetyl palmitate.*

❑ **Sperm Oil** – The **sperm oil** refers to a *liquid wax*, which occurs mostly with *spermaceti* in the *sperm whale* as the **esters** of *fatty-acids*.

Sperm oil is a thin yellow-liquid having slightly fishy odour (if not of a good quality): d = 0.875–0.844; saponification value 123–147; Iodine value 80–84; soluble in ether and $CHCl_3$, but insoluble in water, cold ethanol, and Petroleum ether.

Uses: Sperm oil finds its abundant usage in:

- **Lamps** - **Hardening Steel** - **Soap Production** and - **Lubricator.**

NOTE *Sperm Oil* hardly turns 'gummy' in nature – as many oils usually do.

❑ **Jojoba Oil** – The **Jojoba oil** is a *liquid wax* derived from the seeds of *Simmondsia chinensis* (Link) Scheider [*Fam: Buxaceae*]. The natural plant is a **bushy shrub** native to the arid regions of **Northern Mexico** and to the **Southwestern United States**.

A survey of literature reveals that the **jojoba seeds** invariably comprise **45 to 55%** of an *'Ester Blend'* (*not the glycerides*) that is a liquid at *ambient temperatures*. On being subjected to *hydrolysis* one may obtain the following *major components*, namely:

- **Eicosanoic Acid (35%)** – a C_{20} – unsaturated acid;
- **Eicosenol (22%)** – a C_{20}– unsaturated alcohol; and
- **Docosanol (21%)** – a C_{22}– unsaturated alcohol.

However, the careful *hydrogenation* of **'jojoba oil'** gives rise to the formation of a *'crystalline wax'* which critically has the **appearance** and **characteristic features** of **spermaceti** (described above).

NOTE The 'Jojoba Oil' and its respective hydrogenated structural analogues are found to be extremely useful *emolients* and *excellent pharmaceutical adjuncts.*

Eicosanoic Acid
(Arachidic Acid)

$H_3C-(CH_2)_{21}-OH$

Docosanol

❑ **Lanolin [Wool Wax]** – **Lanolin** refers to the *purified fat like substance obtained from the wool of sheep*, which is being used extensively in **hydrophilic ointment bases and salves.*** Amazingly, it readily forms an **emulsion with water**; and, therefore, based on this wonderful phenomenon it is now a lot easier and convenient for **drug substances** that are soluble in water to be incorporated into **salves*.**

---

* **Salves:** It refers to the **stiff ointment** or **cerate** (*viz.*, **Galen's cold cream**) usually applied to *wounds* or *sores*.

❑ **Cholesteryl Palmitate [Ester of Cholesterol and Palmitic Acid] – cholesteryl palmitate** is found in *blood plasma* and designates '**wax**' having the following structure:

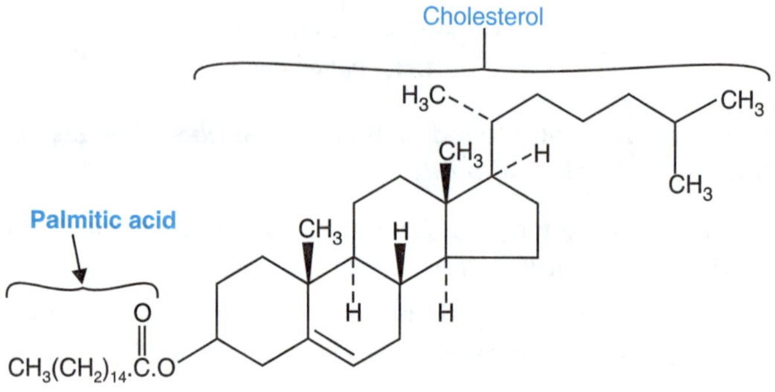

Cholesteryl Palnitate

Physiological Importance of Waxes – There are *six* most important physiological functionalities of the '*waxes*' as stated under:

1. **Waxes** predominantly serve as an excellent *protective agent* particularly on the out surfaces of **plants** and **animals**.

2. Generally, the waxes are abundantly found on the surface of: **Feathers (in *Birds*)** and **Hairs (in *Animals'*)**, which provide adequate *softness* and sufficient *pliability*.

3. **Waxes** do help a long way particularly to the so-called *aquatic animals viz.*, duck, penguin etc., from becoming wet. As and when they come out of water, these *aquatic animals* just shake herself once vigorously and soon they become quite dry.

4. *Waxes* (**natural gifted protectives**) found as a *thin-coating* on the *surfaces of plants*: **fruits**, **leaves**, **beans**, actually protect them from **excessive dissipation of moisture content** – thereby rendering them to *survive in hot summer season*. Besides, such plants, namely:

   • **Palm Trees** • **Cactus** and • **Date Palm**,

   may live for much longer duration in deserts *viz.*, **Sahara Desert**, **African Desert**, **Arabian Desert**, and **Indian Desert** even without rains.

5. Importantly, the **waxy coating** predominantly protects the plant from rendering profusely infected due to *microbial flora* and *fungal attacks*, which may prove to be **detrimental** for their ultimate healthy growth and survival as well.

6. **Waxy coating** on several fruits, such as:

   • **Apples** • **Plums** • **Appricot**, • **Citrus Fruits** • **Kiwi Fruit** • **Pears**
   • **Avocado** and • **Custard Apple**,

   do prevent them from undergoing '*dehydration*'; and thus, they may be stored for much longer periods of time. Moreover, the **waxy coating** also provides appreciable protection of such fruits from certain **hazardous organisms** that are responsible for causing spoilage due to decay.

2.2. Compound Lipids

The compound lipids usually refer to such typical lipids that essentially possess certain additional elements or moieties in addition to the respective:

- **fatty acids**, and
- **alcohols**

Importantly, the so-called '**additional group**' may invariably comprises such elements are:

- **Nitrogen (N)** • **Phosphorus (P)** • **Sulphur(s)** and • **Protein**

Variants in Compound Lipids – There are *two* different types of **Compound Lipids**, such as:

    (a) **Phospholipids**, and

    (b) **Glycolipids**,

which shall now be discussed briefly in the sections that follow.

2.2.1. Phospholipids

In a broader perspective, **phospholipid** relates to a chemical entity (compound) consisting of either:

- **an amide (–CO–NH–)**; or
- **an ester (–CO–OR)**,

of a *fatty acid*; and an '**ester**' or **phosphoric acid** with such substances as:

- **Glycerol** • **Sphingol (Sphingosine)** • **Choline** and • **Ethanolamine**.

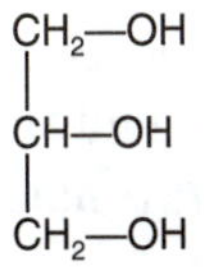

| Glycerol | Sphingol [Sphingosine] | Choline | Ethanolamine |

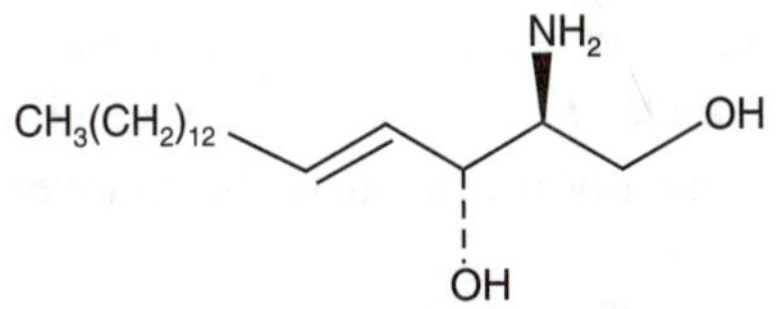

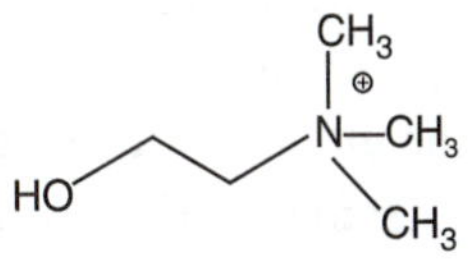

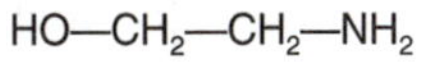

Alternatively, the **Phospholipids** are referred to as the lipids having:

- **Glycerol** • **Phosphoric Acid** and • **Fatty Acids**.

Obviously, these may be regarded as '**Fats**' wherein one of the **fatty acid components** has been duly replaced by a phosphoric acid (H_3PO_4).

There are *five* different types of '**Phospholipids**', namely:

   ➤ **Lecithins (Phosphatidyl Choline)**,

   ➤ **Cephalins (Phosphatidyl Enthanolamines)**,

   ➤ **Plasmalogens**,

   ➤ **Phosphoinositides**, and

   ➤ **Phosphosphingosides**,

which will be duly described in the following sections:

2.2.1.1. Lecithins [Phosphotidyl Choline, Lecithol, Vitellin, Kelecin, Granulestin]

Lecithin represents a mixture of '**phospholipids**', which is critically composed of **fatty acids, glycerol, phosphorus**, and **choline** (or *inositol*), and are found in all *living-cell membranes* (**plants and animals**). It is indeed a *significant constituent* of the **nervous tissue** and **brain substance**. It designates a mixture of the so-called '**diglycerides**' of fatty acids *viz.*,

- **Stearic Acid** • **Palmitic Acid** and • **Oleic Acid**,.

that are eventually linked to the **choline ester of phosphoric acid** (*i.e.*, the *commercial grade of Lecithin**) contains up to **2.2% P.**

Structure of Lecithins

Lecithins

Characteristic Features – These essentially include:

1. **Lecithins** are usually obtained as *yellowish grey solid.*
2. **Lecithins** on being exposed to air soon get darkened rapidly in colour and also absorb moisture from the atmosphere thereby resulting into the formation of a **dark greasy mass**.
3. **Solubility Profile – Lecithins** are soluble in *solvent ether* and *ethanol* but insoluble in *acetone*.
4. **Action of Enzymes on Lecithins –** It has been duly observed that the **lecithins** are subjected to *drastic cleavage* under the influence of the **enzyme** *lecithinase* to form '**lysolecithin**', ultimately.
5. The occurrence of the **enzyme** *lecithinase* is critically prevalent in the following *two* **natural sources**, namely:
    - **Bee Venom**   and   • **Cobra Venom**

NOTE   Thus, when a human being receives a stung, by the 'bee' or a bite from the snake 'cobra' (*i.e.*, these venoms are injected into the *blood stream*, it immediately brings forth a *rapid halmolysis of the RBCs* due to the presence of '*lysolecithins*'.

Thus, we may have the following expressions:

Lecithin   →[**Lecithinase** [An Enzyme]]→   **Lysolecithin** $+ R_2\text{—COOH}$

Occurrence of Lecithins – Lecithin relates to an extremely important constituent of the 'lipoproteins' – specifically the **chylominerons***. Interestingly, the '**egg yolk**' is known to be the

---

*   **Chylomicron:** It relates to the '**lipoprotein**' synthesized in the *intestinal epithelial cells* and composed of the *triglycerides, fats, cholesterol, phospholipids,* and *proteins.*

richest source of lecithin; and hence, also has a vital and important role in the respective **metabolism of fat in the liver**. It has been observed that at the critical physiological pH in the body – the *positive* and *negative* moieties located strategically in the 'lecithin' are almost – '*ionized*' to the same extent; and there is **no net charge**. Hence, **lecithin** is regarded to be a '*neutral* **phospholipid**'.

NOTE The '*lecithins*' present in the human membranes are found to be prevalent in more than twenty different forms.

CONSTITUTION OF LECITHIN

In order to determine the **constitution of 'lecithin'**, one may have to take into consideration the following **cardinal aspects** sequentially:

1. **Hydrolysis of Lecithin** – The careful **hydrolysis of lecithin** by heating with either *dilute alkalines* or *dilute acids* usually give: **choline, fatty acids**, and *optically active* **glycero-α-phosphoric acid**. Based on these scientific revelations one may infer that the **lecithins** are duly established to be the **esters of glycerol** whereby:

   ➢ **2-hydroxyl (OH) moities get adequately esterifeid by fatty acids**; and

   ➢ **3rd has been esterified with phosphoric acid,**

   and thus, the latter one gets transformed into an **ester with choline**.

   Importantly, the precise and exact nature of fatty acids has been duly found to depend upon the *exact source of the lecithins*. Thus, one may express the general *structure for the* **lecithins** as given under:

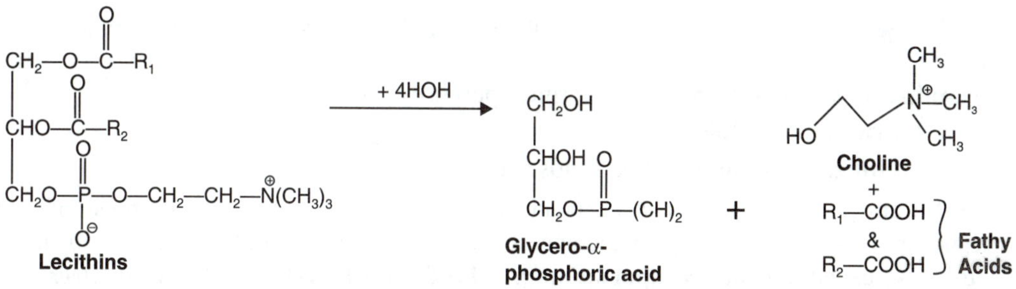

2. **Rare Presence of Glycerol-β-Phosphoric Acid** – In the **lecithia** molecules the **β-hydroxy moiety of glycerol** is found to be rarely esterified by *phosphoric acid (H_3PO_4)*; and, therefore, such highly specific **lecithins** do give rise to the formation of the so called '**optically** *inactive* **glycerol-β-phosphoric acid**' upon hydrolysis:

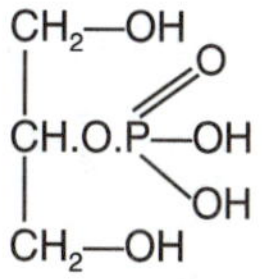

Glycerol-β-phosphoric acid

---

* That is, the product of commerce is predominantly **soyabean Lecithin** duly obtained as a by-product in the production of *soybean oil*.

3. **Fatty Acids Present in Lecithins from Animal Tissues** – The **lecithins** derived from the *animal tissues* are usually found to be

- **Stearic Acid** • **Palmitic Acid** • **Oleic Acid** • **Linoleic Acid**
- **Arachidonic Acid** and • **Clupanodonic Acid**.

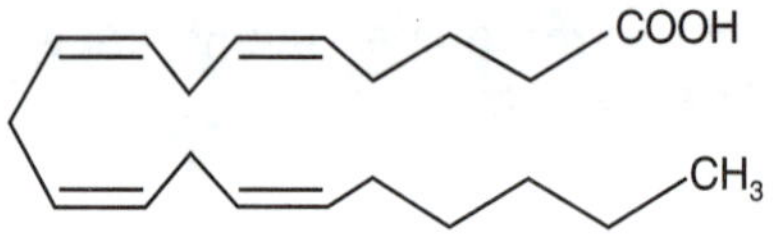

Arachidonic Acid

Important Point – It has been established with ample evidences that there could exist **only** *two* **fatty acid molecules** in a respective '**lecithin**' entity, which fact may be further substantiated based on the following classical instance:

❑ **Tetrabromide of a '*dioleolecithin*' carefully isolated from an 'egg-yolk' that has a close resemblance to the lecithin obtained from '*brain*' (of aquatic/terrestrial animal species) comprising exclusively *palmitic acid along with synthetic dictearyl lipid*; and

❑ **Fatty acid radicals of '*lecithins*' could be present either saturated form or unsaturated form (though mostly they are available as *mixed fatty acids*).**

3. **Confimation of the proposed constitution of 'Lecithins'** – It has been duly ascertained by its *actual synthesis* particularly in the form of a '*double salt*' with cadmium chloride ($CdCl_2$). Some after the removal of **cadmium (Cd)** by means of *ammonium carbonate [$(NH_4)_2CO_3$]*, – one may eventually isolate *pure* **lecithin** wherein all the **N-content** is solely due to the '*choline*'.

Comments – These essentially include:

1. **Lecithin** designates as one of the most common and prevalent form of the **phospholipid** in *animal species.*

2. **Lecithins** are largely required for *two* critical purposes:
   - **transport of lipids**, and
   - **utilization of other allied lipids (in the 'liver').**

3. Besides, any particular factor(s) that appreciably interferes with the **on-going synthesis of choline*** in *vivo*, may also cause a definite hinderance to the **overall synthesis of lecithins**, thereby indulging into the so-called **normal transportation profile of lipids both to and from the '*liver*'**.

> **NOTE** It would ultimately result into the accumulation of the '*lipid content*'; and hence, could be responsible for developing a condition termed as '*fatty liver*'.

2.2.1.2. Cephalins [Kephalins Phosphatidyl ethanolamine;]

Cephalins, in general, represent a **phospholipid** which could be either:

- **a phosphatidyl serine**; or
- **a phosphatidylethanolamine**

In a broader perspective, the '**cephalins**' represent a group of **phospholipids** usually found in *all living organisms*. It is indeed a significant constituent of:

---

* That is, an integral constituent of the '**lecithins**'.

- **nervous tissue**, and
- **brain substance**.

However, the 'cephalins' invariably consist of these vital and important constituents:

➢ **Glycerophosphoric Acid** – wherein the *two* **free hydroxy (OH) moieties** are duly extended with the so-called *long-chain fatty acid residues*; and

➢ **Ethanolamine**,

that eventually forms an ester-linkage with the **phosphate (PO_4^{3-}) moiety**.

It has been duly established that –

- **α-Isomers*** – designate as the derivatives of **α-glycerophosphoric acid** *q.v.*; and
- **β-Isomers** – designate as the structural analogues of **β-glycerophosphoric acid** *q.v.*

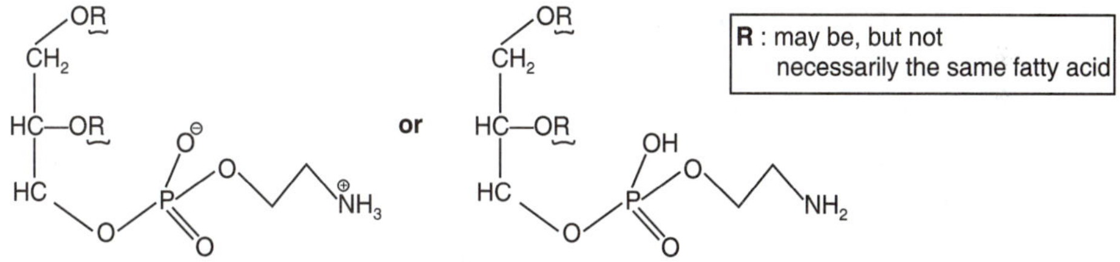

α-Cephalin
[Phosphatidyl Ethanolamine]

Occurrence of Cephalins – The cephalins invariably occur in the *animal tissue* in close association with the **lecithins** (see **Section 2.2.1.1**). Besides, they also occur in the '**soyabean oil**'.

However, the major glaring difference between **cephalins** and **lecithins** is the prevalent nature of the '*nitrogen base*.' Thus, the **cephalins**, in general, comprise of:

- **Ethanolamine (colamine)**, and
- **Serine (instead of 'choline')**

Ethanolamine
(Colamine)

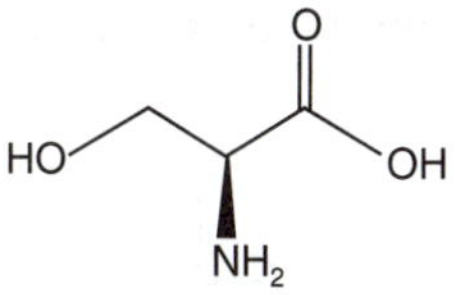

and

Serine

Importantly, the **fatty acid** components of the 'cephalins' do essentially consist of:

- **Stearic acid** • **Oleic acid** • **Linoleic acid and** • **Arachidonic acid**

Interestingly, **phosphatidyl serine** usually comes into being in relatively lesser quantum in practically *all tissues*. Moreover, it designates an **acid phospholipid** since it critically bears an overall '*net negative charge*' at the so-called prevailing *physiological pH;* and it also has the '*serine*' as the requisite **polar chemical entity** – as given below:

---

* The '**natural products**' normally occur in the **α-form** only; whereas, the **β-form** is now known to be an **artifact**.

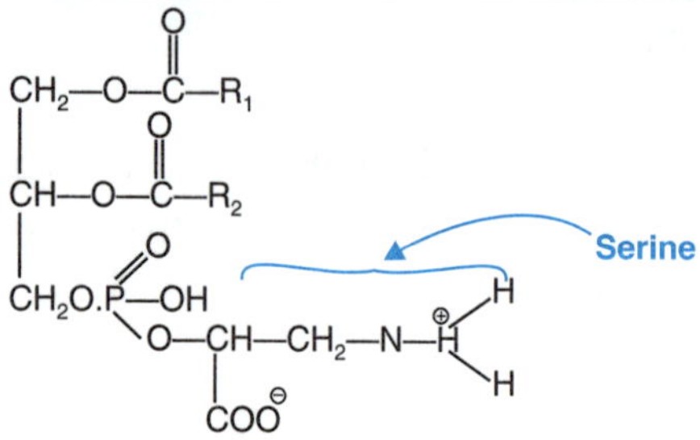

Phosphatidyl Serine

2.2.1.3. Plasmalogens

The 'plasmalogen' refers to any of the various phospholipids in which the group at of one *C–1 of glycerol* happens to be an **ether-linked alcohol** in place of an **ester-linked fatty acid**. It is mostly found in the following three body constituents, namely:

- **myelin sheaths of nerve fibres,**
- **cell membranes of muscle**, and
- **platelets.**

Plasmogens are also knows as: **Profibrinolysin**, and **Plasma Trypsinogen**. It relates to the *circulating plasma precursor* (*zymogen*) or **inactive form of plasmin**.

It is found to be soluble *below* pH 5 and *above* pH 9; and sparingly soluble at **intermediate pH values**. However, it shows maximum stability at **pH ranging between 2 and 3**; and exhibits resistant to heat <**pH 5**. It is virtually soluble in *all* **lipid solvents**.

A survey of literature reveals that the 'plaminogens' do occur abundantly in *brain* and *muscles*. They also occur in the *seeds* of higher plants. Importantly, in this particular **phospholipids**, one may critically observe the presence of a *complex aldehyde* attached strategically to the respective α-**carbon atom** (or at **C–1) of the glycerol**. The remaining molecule has been identified duly to be *exactly the same* as in **Lecithin** or **Cephalin** (see *sections 2.2.1.1 and 2.2.1.2*). **Plasminogen** has the following chemical structure.

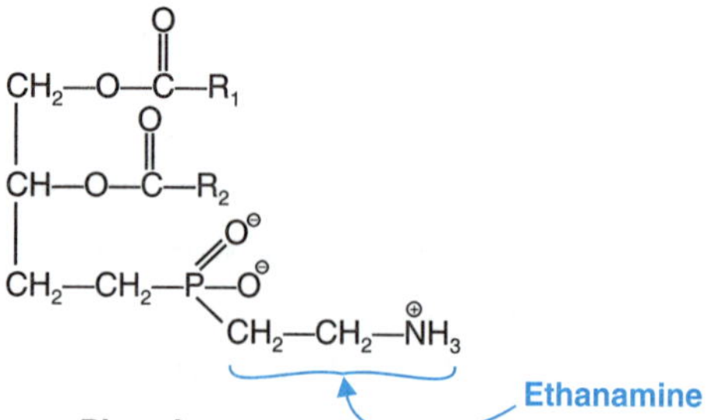

Plasminogen

2.2.1.4. Phosphoinositides

– The **phosphoinosides** represent such typical *phospholipids* that essentially have **inositol** (a *hexahydric alcohol*):

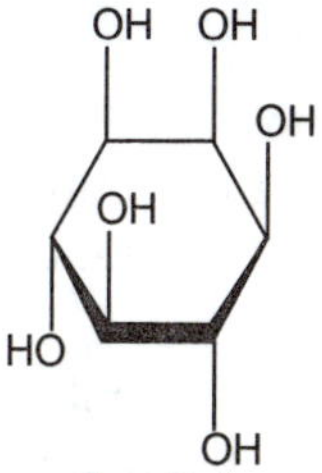

Inositol

These could be in existence in *two* forms, namely:

- **Monophosphoinositides** and
- **Diphosphoinositides**

and their respective chemical structures as given under:

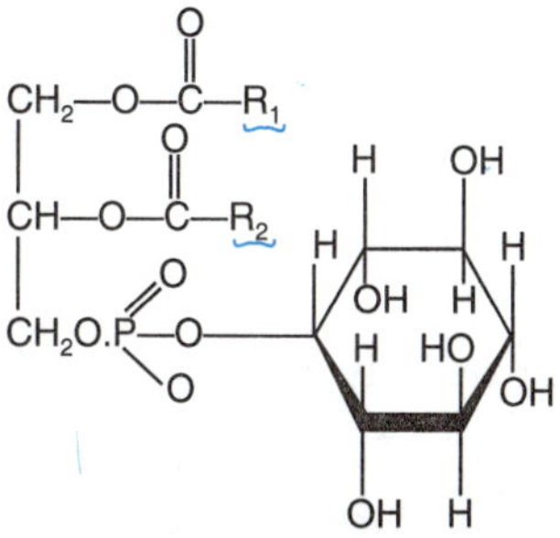

Monophosphoinositides

[It is widely dictributed in
both *animal* and *plant* species]

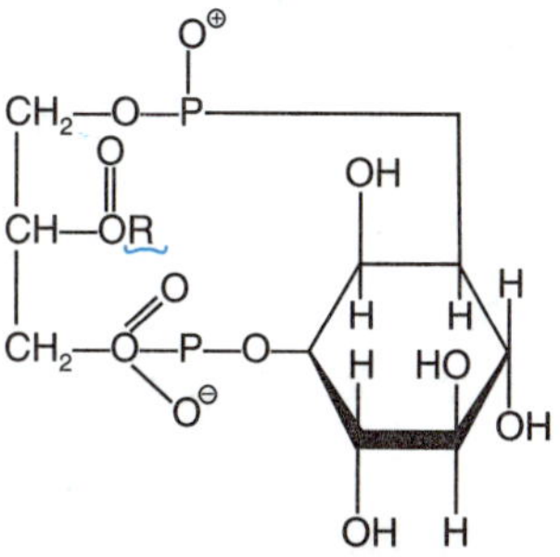

Diphosphoinositides

[It is commonly to be seen
in the *brain tissues* solely.]

2.2.1.5. Phosphosphingosides

The **phosphosphingosides** do represent the *phospholipids* in which the **glycerol** has been duly replaced by either:

⊙ **Sphingosine**

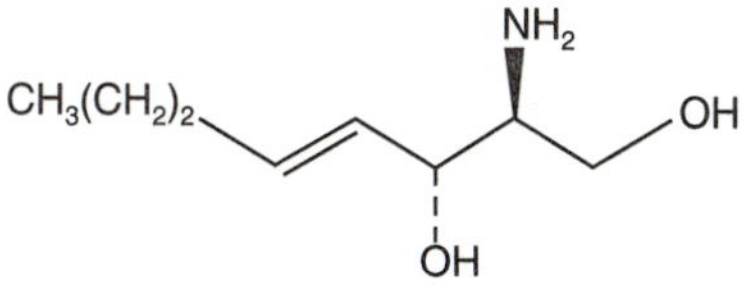

to yield

Phytosphingosine

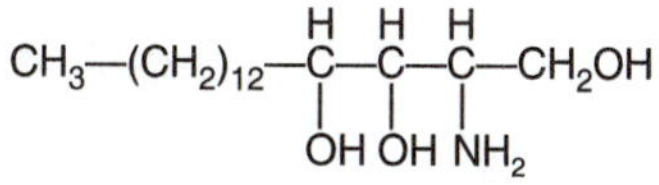

OR

⊙ **Phytosphingosine***

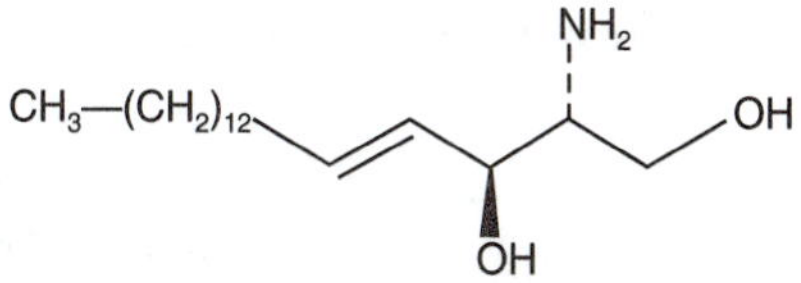

to yield

Sphingotine

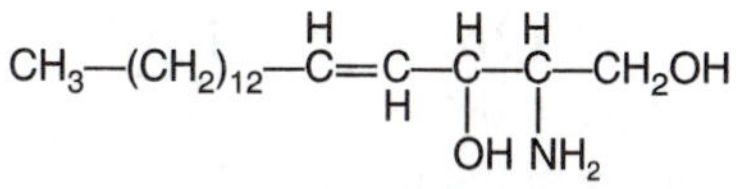

Interestingly, the aforesaid lipids usually occur in **nervous tissues**; and do lack profusely in **microorganisms** and **plant species**.

---

* A NUX–Vomica Alkaloid.

2.2.2. Glycolipids

A 'glycolipid' refers to a lipid containing the so-called 'carbohydrate moieties', usually 'galactose' but also *glucose*, *inositol*, or *others*. Thus, it can describe those lipids derived from either **glycerol** or **sphingosine** – *either with or without phosphates*. Besides, the specific term is invariably used to signify the particular:

'**sphingosine structural analogues lacking the phosphate groups (glycosphingolipid).**'

Importantly, in '**cerebrosides**'[*]–the *fatty acid* pertaining to the *ceramide component* may either comprise:

- **18-carbon chain**, or
- **24-carbon chain**,

whereas, the *latter chains* – are found exclusively in '**glycolipids**' *i.e.*, the so-called *complex-lipids*.

It has been duly reported that either a '**glucose**' or a '**galactose**' carbohydrate unit gainfully gives rise to the formation of a *β-glycoside linkage* having the respective *ceramide segment* of the molecule. However, the **cerebrosides** usually exist in:

- **brain abundantly (up to 7% of the dry weight); and**
- **nerve synapses.**

In true sense, the cerebrosides (*or glycosphingosides*) are duly designated as the **complex** lipids comprising essentially:

- ➢ **sphingosine,**
- ➢ **a fatty acid**, and
- ➢ **a sugar** (*glucose or galactose*),

and they actually differ from a good number of the so-called '**complex lipids**' in having complete absence of the **phosphoric acid** (H_3PO_4). However, one may write the most probable arrangement of the component units present in the *molecular formula* as shown under:

CH_3—$(CH_2)_{12}$—CH
‖
CH—CH—CH—CH_2O—CH—$CHOH.CHOH.CHOH$—CH—CH_2OH
| | |_____|
HO NH O
|
$O{=}C{-}R$

Cerebroside

Important Points– These essentially include:

1. The **cerebrosides** do have such known acids as:
   - **Lignoceric acid** • α-**Hydroxylignoceric (cerebronic) acid** • **Nervonic acid** and
   - α-**Hydroxynervonic acid.**
2. The '**gangliosides**' (*i.e.*, the respective lipids of the '**ganglia**') are found to be even more complex in nature (Klenk, 1942)[**]. Besides, these (*gangliosides*) also contain a **fatty acid**, **sphingosine, neuraminic acid** ($C_{10}H_{19}NO_2$) [whose structure not yet established], and **hexose** (both *glucose* and *galactose*).

---

[*]   **That is, a glycoside composed of a fatty acyl sphingolanide and a sugar (mainly galactose).**

[**] **Klenk B: 1896 Pfalzgrafenweiler, Germany, Ph.D., Tijbingen, University, Cologne.**

The 'gangliosides'* are the *tetraglycosyl ceramides* to which are attached strategically the so-called: **1, 2 or 3 residues of sialic acid**.

Molecular Formula – of 'Ganglioside' as given under:

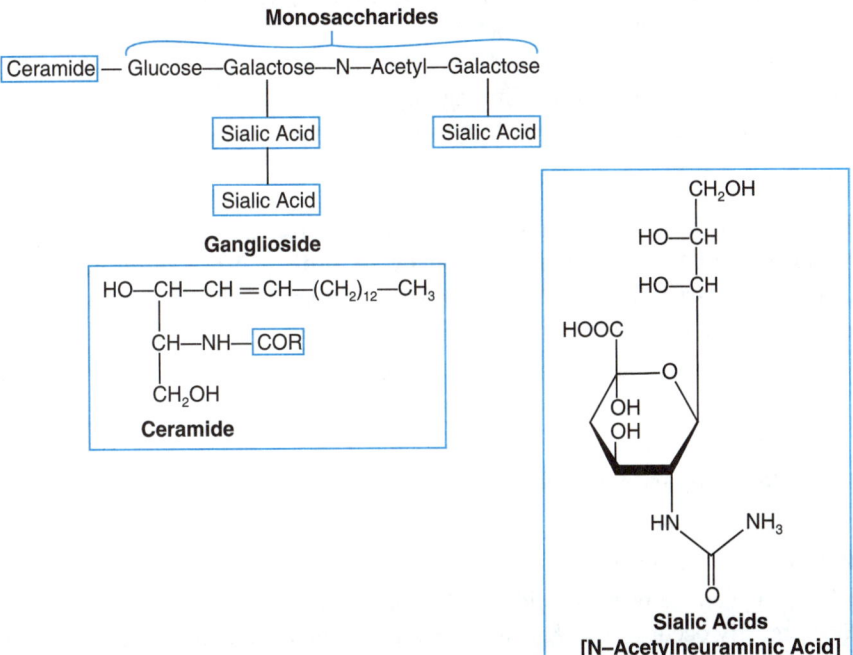

(1) *Ceramides* joined by typical glycosidic bonds to 2, 3, or 4 monosaccharide units are usually known as the '*ceramideoligosaccharides*'.
(2) The so-called tetraglycosyl ceramides, having four monosaccharide units, do some as the *precursors of gangliosides*.

2.3. Derived Lipids

In reality, the '**derived lipids**' essentially comprise specifically the **hydrolytic products** of lipids together with an array of *lipid-like* **chemical entities**, such as:

- **Alcohols** • **Aldehydes** • **Carotenoids** • **Adrenocorticoids** • **Oils** • **Ketones**
- **Hydrocarbons** • **Sterols** • **Bile Acids** • **Proctaglandins** and • **Sphingolipids**.

However, in a broader perspective the classical and major category of the derived lipids is the '*steroids*' that are duly represented by the presence of a **cyclopentanophenanthrene** nucleus as given below:

Ring A,B,C = Phenanthrene
Ring D = Cyclopentano

Cyclopentanophenanthrene Nucleus

---

* They usually act as the '**antogens**'.

Classification Derived Lipids: The '**derived lipids**' may be broadly classified under the following *five* major sub-heads,

(a) **Sterols** [*viz.*, Phytosterols (**Ergosterol, α-Spinasterol**); Zoosterols (Cholesterol);]
(b) **Steroidal Hormones** [*viz.*, Adrenocorticoid Hormones; Sex Hormones];
(c) **Bile Salts** [*viz.*, Glycocholate; Taurocholate];
(d) **Prostaglandins** [*viz.*, Prostaglandin E (PGE)]; and
(e) **Sphingolipids** [*viz.*, Sphingosine; Sphingomyclin],

which shall now be discussed briefly in the sections that follows:

2.3.1. *Sterols*

Sterols may be defined as – '**terpenoids, solid, unsaturated steroid alcohols with a hydroxy (–OH) moiety located at the C–3 position that occur both in** '*free-state*' **or** '*as esters*' **or as** '**glycosides**', **and are classified as per the inherent organism in which they invariably exist as: Phytosterols, Zoosterols, and Mycosterols.**'

Occurrence – The **sterols** occur abundantly in *plants*, animals, and *microorganism*. Besides, these are also found in several other *typical sites*, for instance:

- **cell membranes,**
- **cellular compartments**, and
- **adipose tissues.**

Characteristic Features – These essentially include:

1. **Sterols** – in general appear as solid wax-like substances.
2. **Sterols** – mostly occur as *alcohols* and are quite abundant either *as such* or *as esters of fatty acids*.
3. They are freely soluble in fat solvents.
4. **Sterols** do not undergo saponification (**unlike other lipids**); and, therefore, by means of thus process they may be easily separated from other **lipids**.

❑ ERGOSTEROL

Chemical Structure

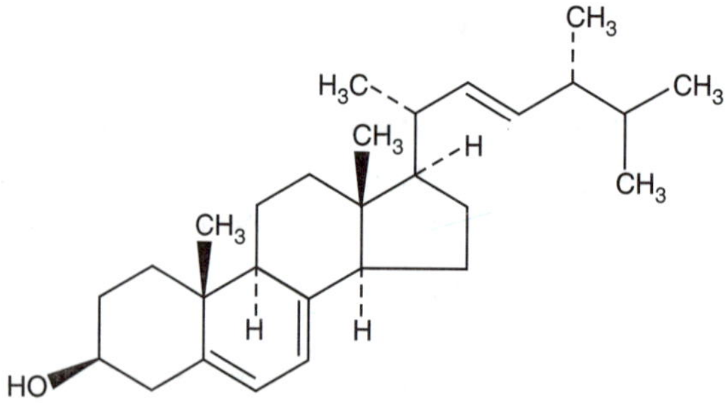

Ergocterol

Ergosterol is regarded to be the most important of the **provitamin D.** It is usually obtained from the *yeast* cells which eventually synthesizes it right from the **simple sugars** *viz.*, **glucose.** The **damp yeast** yields **2.5% ergosterol**. It occurs in food only in small quantum. **Ergosterol** is duly isolated from the so-called **parasitic fungus** *Claviceps purpurea* (*Ergot*)

❑ **α-Spinasterol** *Syn*: **Bessisterol; Hitodesterol; α-Spinasterin**

Chemical Structure

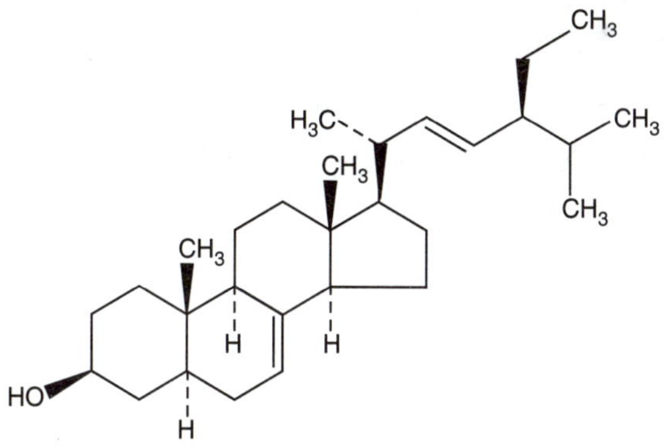

α-Spinasterol

α-Spinasterol is usually extracted from the *spinach leaves. It is found to be stereoisomeric* with **chondrillasterol**. In addition, it is also obtained from several other known sources *viz.*, **cabbage stigmasterol, soybeans, and coconut**.

❑ **Cholesterol [*Syn*: Cholesterin]**

Chemical Structure

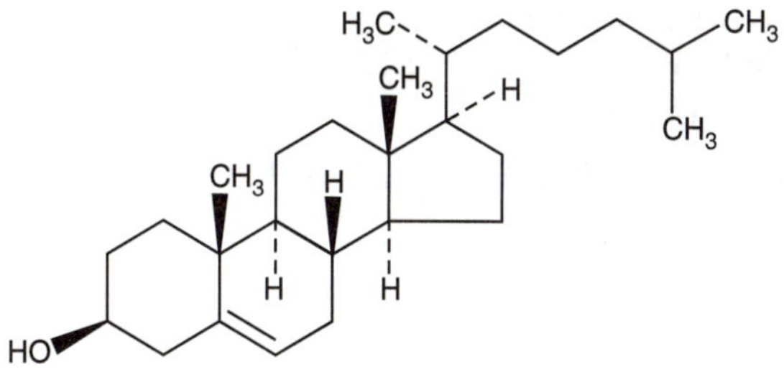

Cholesterol

Cholesterol is the principal **sterol** of the *higher animals*. It is mostly found in **all body tissues:** especially in:

- **Brain**
- **Spinal Cord and**
- **Animal Fats or Oils**.

Amazingly, **cholesterol** is the main constituent of the human **gallstones**. It is prepared on a large-scale from the **spinal cord** of cattle by particular petroleum ether (60 – 80°C) extraction of the *unsaponifiable fraction*.

2.3.2. Steroidal Hormones

The **steroidal hormones** are of *two* types, namely:

- **Adrenocorticoid Hormones**, and   • **Sex Hormones**.

2.3.2.1. Adrenocorticoid Hormones

The **adrenocorticoid hormones** also known as: **Corticosteroids** or **Adrenocorticoid Steroids** – actually belong to the adrenal cortex. Following are some of the cardinal **adrenocortical hormones**, namely:

(a) Short-to-Medium Acting Glucocorticoids

❑ **Hydrocortisone [*Syn:* Compound F; Reichstein's Substance M]**

Chemical Structure

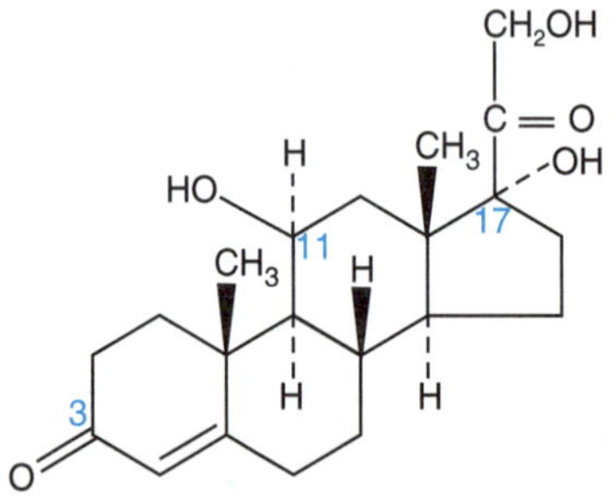

Hydrocortisone

Characteristic Features– These essentially include:

1. It has been duly established that the pharmacological activity of **hydrocortisone** is mostly mediated by the so-called **cytoplasmic glucocorticoid** receptors
2. It possesses the following features:
   - **Plasma Half-life**   : 1.5–3 Hrs.
   - **Biological Half-life**   : 8–12 Hrs; and
   - **Volume of Distribution (Vd)**   : 0.3–0.5L. kg^{-1}

(b) Intermediate–Acting Glucocorticoids

❑ **Triamcinolone**

Chemical Structure

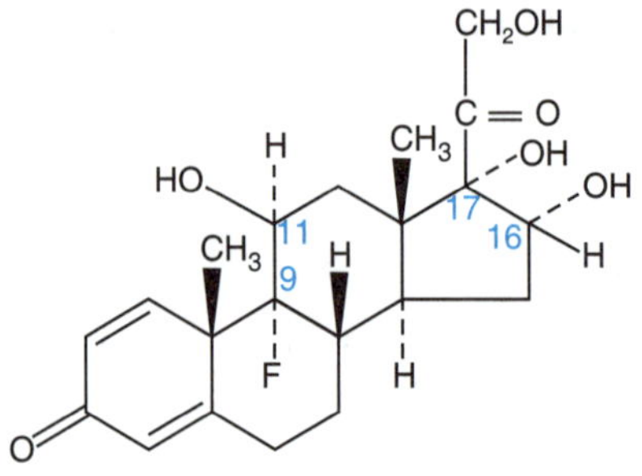

Triamcinolone

Salient Features

1. It is found to be 7–13 times more potent **hydrocortisone**.
2. **Triamcinolone** is completely devoid of *mineralocorticoid* activity and other observed side-effects of *hydrocortisone*.
3. The *oral administration* of this drug usually helps to retain the survival of the *first*-**pass** *via* the liver than the **hydrocortisone** (*i.e.*, it shows more quantum of drug in the blood level).
4. **Triamcinolone** exhibits the following critical aspects:
   - **Plasma Half-life**: ~ 5 Hrs;
   - **Biological Half-life** : varies between 18–36 Hrs; and
   - **Volume of Distribution (Vd)** : ranges between 1.4 –2.1 L. kg^{-1} (which being dose dependent).

(c) Long-Acting Glucocorticoids

❑ Betamethasone

Chemical Structure

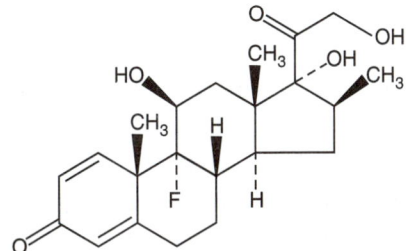

Betamethasone

Characteristic Features – These essentially comprise:

1. **Betamethasone** is found to be an extremely potent **long-acting glucocorticoid**.
2. If has a 20–30 folds increased therapeutic activity *vis-a-vis* **cortisol**.
3. This drug causes rarely any sort of indication of either *sodium* or *water* **retention** and **potassium depletion** – on being subjected to treatment in conjunction with either:
   - **cortisone**, or
   - **other adrenal corticoids**.
4. **Betamethasone** explicitly shows the following *three* critical aspects, namely:
   - **Plasma Half-life** : ~6.5 Hrs.;
   - **Biological Half-life** : ranges between 36–54 Hrs., and
   - **Volume of Distribution (Vd)** : 1.8 L. kg^{-1}

2.3.2.2. Sex Hormones

The 'Sex Hormones' do control and regulate the development of both *main* and *secondary* **sexual features**. In true sense, they also modulate the so-called **sex-related functionalities**, such as:

   - **menstrual cycle** (*in women*), and
   - **production of** *sperm* and *eggs*.,

There are *three* major types of **sex hormones**, namely:

> ➢ **Androgens (or Male Sex Hormones);**
> ➢ **Oestrogens (or Female Sex Hormones);** and
> ➢ **Gestogens (comprised in pregnancy only).**

2.3.2.2.1. *Androgens*

❑ **Testosterone [Androderm; *trans* – Testosterone; Androderm]**

Chemical Structure

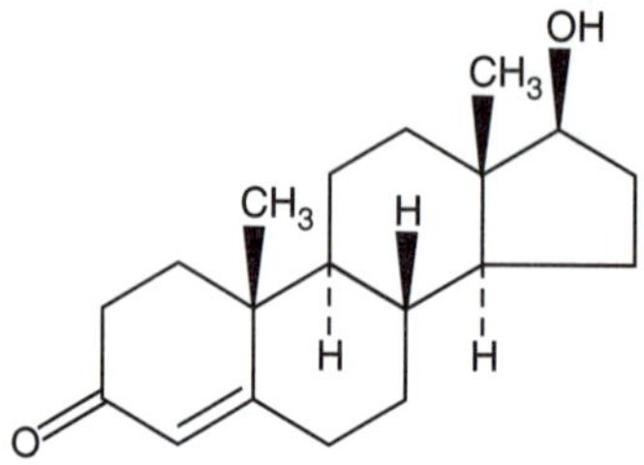

Testosterone

Testosterone refers to the principal hormone of the *testes* and produced duly by the **interstitial cells**.

Characteristic Features – These essentially comprise:

1. It is primarily responsible for the control, management, and development of the **male-sex organs**.
2. It also helps in the so-called maintenance as well as function of the **make** *secondary* **sex-characteristics**.
3. It does help to increase the size of the **serotum phallus**, **seminal vesicles**, **prostrate glands**, and largely enhances the sexual activity profile in **adolescent males**.
4. **Testosterone** finds its abundant utility in the male for such critical and vital functions:
   - **hypogonadism;**   •   **eunachoidism,**   and   •   **male celimacteric**

2.3.2.2.2. *Oestrogens*

❑ **Oestrone [Estrone; Follicular Hormone; Folliculin,]**

Chemical Structure

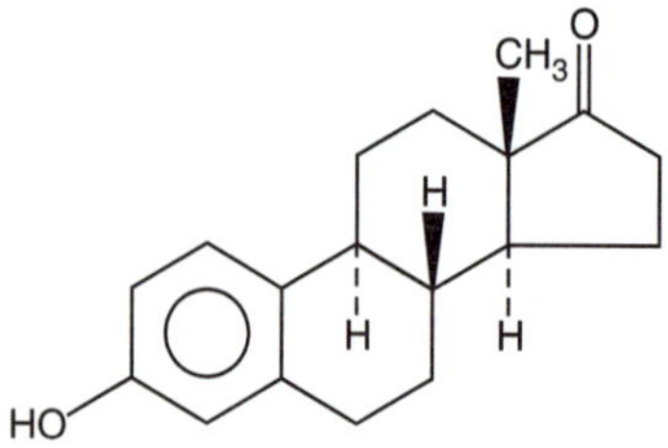

Oestrone

Estrone is a metabolite of 17β-estradiol and possesses significantly less biological activity. It is usually isolated as the *d*-form (*i.e.*, *d*-estrone).

Estrone occurs in the pregnancy urine of women and males. It also occurs in the *follicular liquor* of a variety of *animal species*, in *human placenta*, in *urine of bulls and stallions*, and in *Palm kernel oil*.

2.3.2.2.3. Gestogens

❑ **Progesterone [Corpus Luteum Hormones; Gestone; Utrogestan]**

Chemical Structure

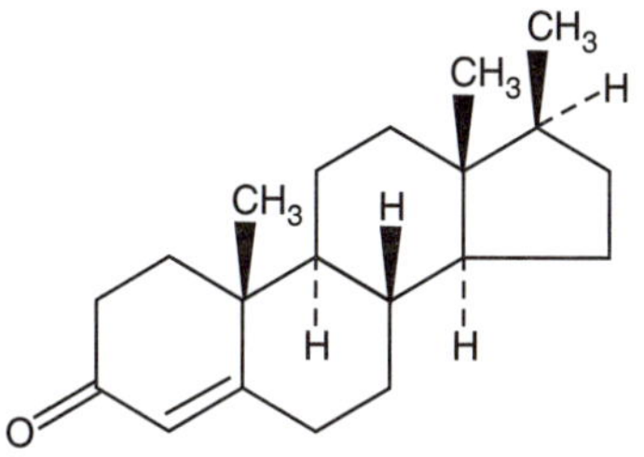

Progesterone

Progesterone relates to the active principle of the corpus *leuteum*, duly secreted during the **latter half of the menstrual cycle**; and if the pregnancy ensues, secretion continues. Besides, it exerts an *anti-ovulatory effect* when administered during days **5 to 25** of the normal menstrual cycle. It is usually isolated from the *corpus leuteum of pregnant cows*.

It is used to **control habitual abortion** and to **suppress or synchronize estrns**.

2.3.3. Bile Salts

The '**bile salts**' are known to be the *oxidative products* of **cholesterol**. In fact, the conversion of **cholesterol** takes place at *two* **sequential stages** as given below:

Cholesterol   Stage–1   Stage–2

Stage–1: Conversion of cholesterol to the respective **trihydroxy derivative**; and

Stage–2: Terminal of the aliphatic *side-chain* in **cholesterol** gets duly oxidised to the respective **carboxylic acid**.

Remark – The *terminal* **carboxylic acid (–COOH) moiety** in turn yields a corresponding **amide (–CO–NH₂) linkage** with either of the *two* following **amino acids**, namely:

H₂N—CH—COOH   OR   H₂N—CH₂—SO₃H

Glycine   **Taurine**
[α-**Amino acetic acid**]

Thus, we may have the following *two* recognized '**Bile Salts**', namely:

Glycocholate	**Taurocholate**
[Glycocholic Acid]	[Taurocholic Acid]

Characteristic Features of Bile Salts: These essentially comprise:

1. **Bile Salts** are excellent detergents.
2. Importantly, *one end of the molecules* (see above) (*i.e.*, the *aliphatic side-chain*) behaves as the **strongly hydrophilic** segment since it predominantly holds the *negative charge*; whereas, the rest of the molecule bears the typical *hydrophobic character*.
3. Based on the above facts, the '**bile salts**' may be able to disperse adequately the so-called **dietary lipids**, right inside the *small intestinal canal* in the form of a '**finely dispersed emulsion**' which ultimately go a long-way to help the on-going **digestion phenomenon**.
4. As the '**bile salts**' are virtually eliminated *via* the faeces, they also help in the substantial removal of the '**excess of cholesterol**' in *two* known ways, namely:
   - **they undergo cleavage automatically to yield such products containing 'cholesterol***; and'**
   - **they predominantly solubilize the deposited cholesterol in the form of bile-salt-cholesterol products**.

2.3.4. Prostaglandins

The '**prostaglandins**' designate the class of *fatty acids* specifically derived from **arachidonic acid** by the *cyclization* phenomenon to form a **5-membered ring** in the close vicinity of the '*middle of the fatty acid chain*'. Besides, they represent a *class of hormnones* that essentially possesses an array of physiological effects, such as:

   - **vasodilation**, and
   - **smooth muscle contraction**.

Examples: **PGF$_1$, PGE$_1$, and PGA$_2$.**

The '**prostaglandins**' were first and foremost discovered, when it was duly demonstrated that the '*seminal fluid*' particularly caused a *hysterectomized uterus* (*i.e.*, **excision of the uterns**) to *undergo contraction*. Interestingly, the very *name* suggests that these **substances** (or *chemical entities*) are invariably a product given by the '**prostate gland**'. It has been reported that in **adult males** the seminal fluid does secrete up to 0.1 mg of **prostaglandin per day**.

---

* That is, **cholesterol** gets eliminated *via* the **bile salts**.

NOTE **Nevertheless, a small quantum of *prostaglandins* are duly present throughout the entire human body in either sex.**

Synthesis of Prostaglandins – take place *in vivo* from the **arachidonic acid** due to the critical **ring closure** at : C-8 and C-12 as given under:

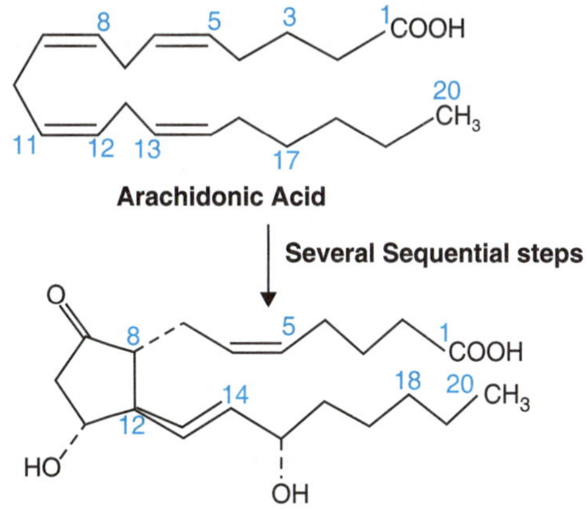

Arachidonic Acid

Several Sequential steps

Prostaglandin E₂
[PGE₂]

It has been observed that the **prostaglandin E group** critically possesses a **carbonyl (>CO) function** located strategically at **C-9**, with the *subscript* thereby indicating the exact number of **olefinic (double) bonds** present in it.

Prostaglandin F group (PGF₂α) has *two* **hydroxyl (–OH) groups** located strategically at positions **C-9** and **C-11** as shown under:

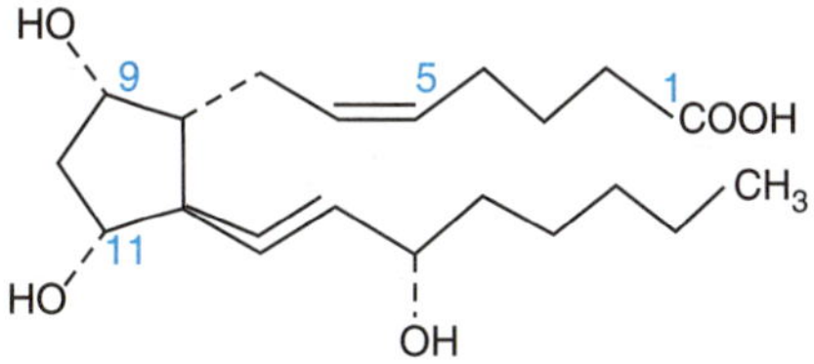

Prostaglandin F₂α
[PGF₂α]

The '**prostaglands**' as a group do exhibit a plethora of effects upon the *body chemistry*, and invariably act as the gainful **mediators of hormones** in the human body.

Examples: Both **PGE₂** and **PGF₂α** particularly *induce labour*; and hence, are used to afford a *therapeutic abortion* – in an early pregnancy. Obviously, one may critically take cognizance of the fact that

- **PGE₂** – help to lower the **BP**; and
- **PGF₂α** – causes an **hypertensive activity profile**.

However, **PGE₂** also finds its usage for the cure and relief from **asthma** as an '*aerosol inhaler*'– by opening up the congestion in the **bronchial tubes** by relaxing its adjoining muscles. Besides, **PGE₂** also aids as a **decongestant** by dilating the *nasal passages* by causing definite constriction of the respective **blood vessels**.

2.3.5. Sphingolipids

These are a specific type of a **lipid** solely derived from the *amino alcohol* termed as **sphingol** (or **sphingosine**).

The 'sphingolipids' refer to the **specific coating of *nerve axons*** (*i.e.*, **myclin**) which essentially contains an altogether different kind of a *complex lipid* invariably termed as the '**sphingolipids**'. However, it has been duly established that in the **sphingolipids** the *alcohol segment* is nothing but *sphingosine* as given under:

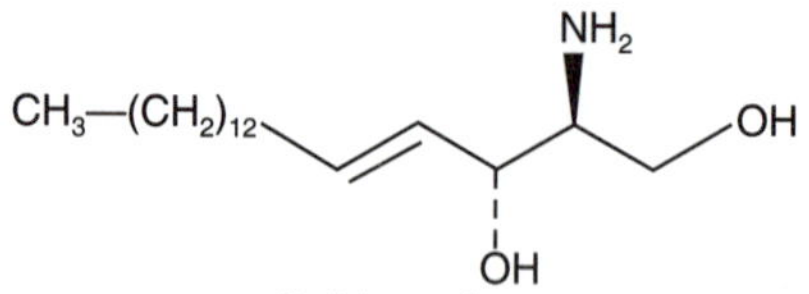

Sphingosine

It has been observed that a **long-chain fatty acid** is duly attached to the **amino(–N) moiety** by the critical formation of an amide (–CONH–) linkage. Thus, the **hydroxyl (–OH) moiety** located strategically at the terminal of the chain gets duly **esterified by 'phosphorylcholine'**.

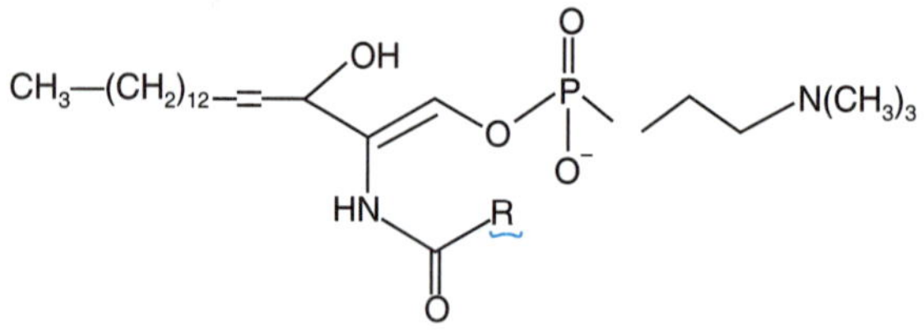

R = Different fatty acids
   viz., stearic palmitic, lignoceric, nervonic acids

Sphingomyelins

The '*schematic* **chemical structure**' of *sphingomyelin* is given below :

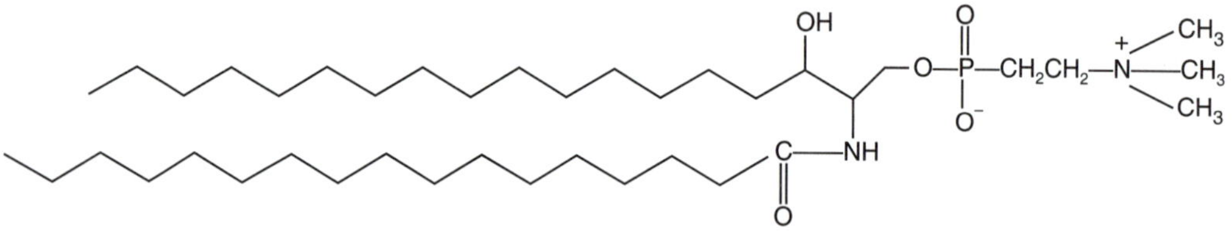

Sphingomyelin (Schematic Structure)

Ceramide Portion of Molecule – Importantly, the specific combination of **fatty acids** and **sphingosine** is quite often referred to as the '**ceramide portion of the molecule**' – since several of these **chemical entities (compounds)** are usually contained in the '**cerebrosides**'.

> **NOTE**   The '*ceramide portion*' of the so-called *complex lipids* may comprise various fatty acids; however, the *stearic acid* occurs chiefly in the '*sphingomyelins*'.

FURTHER READING REFERENCES

Bennett H	:	**Industrial Waxes, Vols. I–II,** 2nd edn., Chemical Publishing Co., Inc., New York, 1975.
Brown JH	:	**Jojoba Liquid Wax – Substitute for Spermaceti**, Manuf. Chem. Aerosol News, **50** (6), 47, 1979.
Gunstone FD *et al.*	:	**The Lipid Handbook,** Chapman and Hall, London (UK), 1986.
Kolattukudy PE (Ed.)	:	**Chemistry and Biochemistry of Natural Waxes**, Elsevier Scientific Publishing co., New York, 1976.
Pace-Asclak C and Granstorm E (Eds.)	:	**Prostaglandins and Related Substances**, Elsevier Scientific Publishing Co., New York, 1983.
Wills Al (Ed.)	:	**CRC-Handbook of Eicosanoids : Prostaglandins and Related Lipids, Vols. I–II,** CRC-Press, Boca Raton, Florida (USA), 1987.

REVIEW QUESTIONS

1. What are **Lipids**? How would you expatide the presence of **Hydrophilic Nature of Hydrocarbon Stractures in Lipids**? Explain

2. Discuss the following :
    (a) **Oceurrence of Lipids**
    (b) **Physiolegical Functioualities of Lipids**

3. Write short notes on any **Two** of the following:
    (a) **U-Shaped Arachidonic Acid**
    (b) **Isomerism in Unsaturated Fatty Acids**
    (c) **Iodine Value**
    (d) **Ambretolide**

4. Give a comprehensive account on **Waxe**s with special reference to the following:
    • **Spermaceti**
    • **Sperm Oil**
    • **Lanolin (Wool Wax)** and
    • **Cholesteryl Palmitate**

5. What are **Phospholipids**? Discuss brifly any **two** of the following:
    (a) **Lecithins**
    (b) **Cephalins**
    (c) **Plasmalogens**
    (d) **Phosphophigosides**

6. What are **Derived Lipids**? Discues with at least **Two** typical examples.

7. Deseribe the **Androgens, Oestrogens,** and **Gestogeus** with the help of suitable examples.

Contents at a Glance

Polynuclear Aromatic Hydrocarbons

4

1. INTRODUCTION

In general, all such *organic* (or *aromatic*) *hydrocarbons* having essentially ***more than*** one **aromatic nucleus** are invariably termed as '**polynuclear aromatic hydrocarbons**'.

In actual practice, these ***hydrocarbons*** may be categorized into the following *three* sub-heads, namely:

(a) **Polyphenyl Compounds** (or **Isolated Systems**),

(b) **Condensed Ring Systems,** and

(c) **Anthracene**,

which shall now be discussed separately in the sections that follows.

2. POLYPHENYL COMPOUNDS [or ISOLATED SYSTEMS]

There are quite a few important members belonging to the **Polyphenyl Compounds (or Isolated Systems)** as given under-based upon the various evidences from an array of evidences from the literature, such as:

- **Diphenyl** (or **Biphenyl**)
- **Diphenic Acid** (or **Diphenyl –2, 2′ – dicarboxylic acid**)
- **Benzidine** (or **4, 4′–Diaminobiphenyl**),
- **Diphenylmethane**,
- **Triphenylmethane** (or **Tritane**),
- **Triphenyl carbinol**
- **Triphenylmethyl Chloride** (or **Trityl Chloride**),
- **Tetraphenylmethane**, and
- **Hexaphenylmethane**,

which shall now be treated individually in the following sections

2.1. Diphenyl [Syn: Biphenyl; Bibenzene; Phenylbenzene]

Diphenyl usually occurs in small quantum in naturally occurring '*coal tar*'.

Chemical Structure

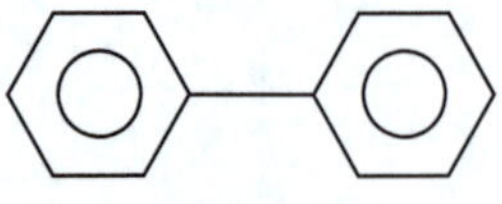

Diphenyl

Methods of Preparation – Diphenyl may be prepared by the following *six* different methods as elaborated under:

(a) **Fittig's Reaction (1863)*** – The **Fittig Reaction** critically involves the treatment of *bromobenzene* with *Na-metal* in *ether solution*. Thus, we may have the following reaction:

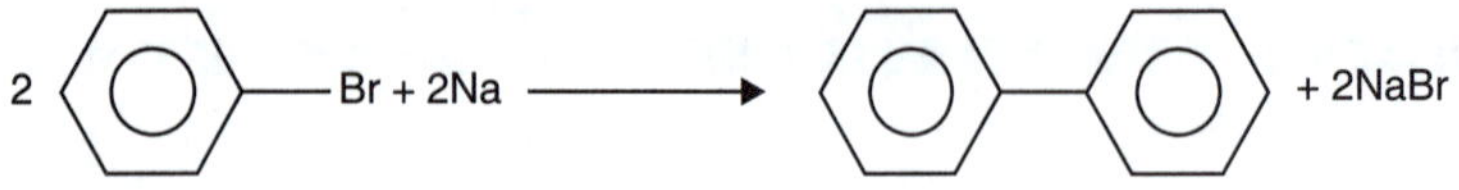

Bromobenzene   **Diphenyl**

(b) **Ullmann's Biaryl Synthesis (1904, 1905)**** – **Diphenyl** may also be prepared by heating together *iodobenzene* with finely divided *Cu-powder* in a *sealed tube* – that eventually yields ~ 80% of the product as shown below:

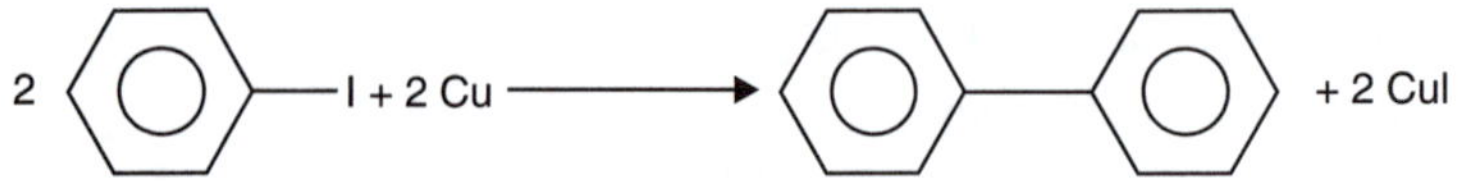

Iodobenzene   **Diphenyl**

Comments: Importantly, the **aryl-chlorides** and **aryl-bromides** may react only when there exists a specific **negative moiety** located strategically either at *ortho*-or **para**-position to the respective **halogen-atom**.

Example: When *O-chlorobenzene* is subjected to heating with finely divided **Cu-powder**, it gives rise to the formation of **2-dinitrodiphenyl** as shown under:

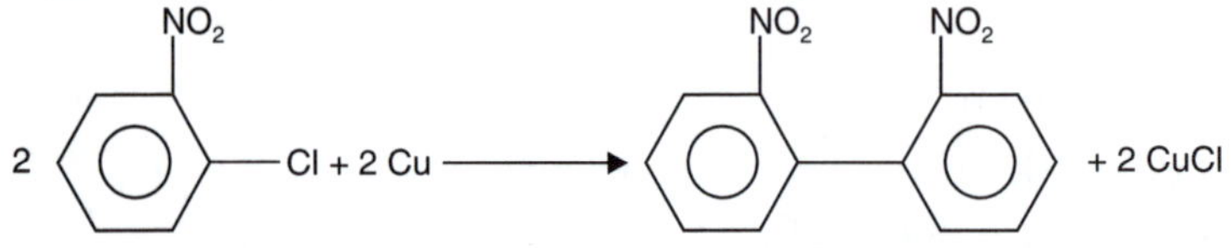

O-**Chlorobenzene**   **2 Dinitrodiphenyl**

(c) **Busch's Method (1936)** – The **diphenyl** is duly prepared by refluxing an alkaline solution of bromobenzene in ethanol with hydrazine ($H_2N – NH_2$) in the presence of **Pd-catalyst** upon the calcium carbonate [$Ca(CO_3)_2$] support – as given below:

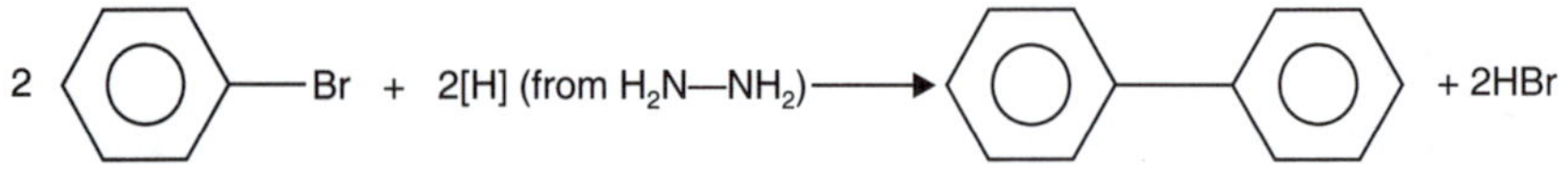

Bromobenzene   **Hydrazine**   **Diphenyl**

---

*   Fittig R (1835–1910): Hamberg, Germany;

** Ulmann F: *Ann*. 332: 38, 1904; Ulmann F and Sponagel P: *Ber*, **38**: 2211, 1905

(d) **From Benzenediazonium Chloride [or Sulphate]: Diphenyl** (or **Biphenyl**) may be prepared from **benzenediazonium chloride/benzene diazonium sulphate** by carefully heating it in redistilled ethanol with *finely divided Cu-powder* – as shown below:

Method-1

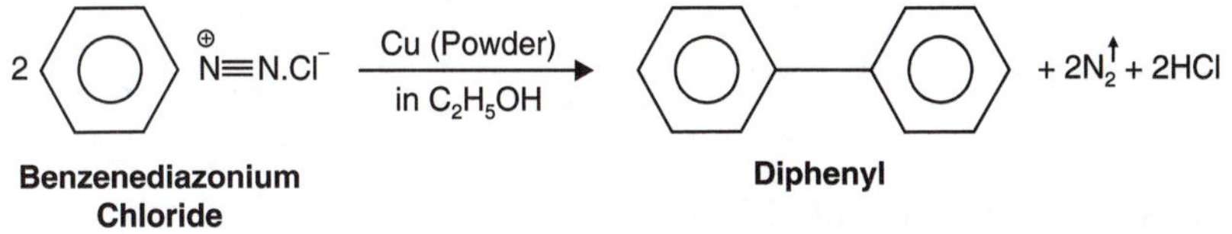

Benzenediazonium
Chloride

Diphenyl

Remark: The yield of this reaction is almost **25%**, that could be enhanced upto **50%** by warming heating together an *admixture* of: **Aniline, Phenylnitrite,** and **Benzene** followed by **refluxing**.

Method-2: The *diazotization* of **benzidine** with sodium nitrite and hydrochloric acid (<5°C); and allowing the **resulting diazonium chloride solution** to remain in contact with **hypophosphorus acid** (H_3PO_2) to yield **diphenyl** – as stated under:

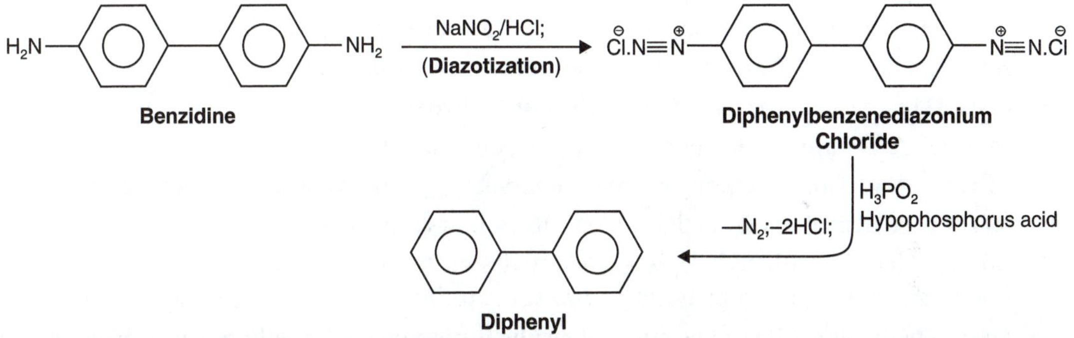

Benzidine

Diphenylbenzenediazonium
Chloride

H_3PO_2
$-N_2; -2HCl;$ Hypophosphorus acid

Diphenyl

(e) **Commercial Methods of Preparation** – There are, infact, *two* known *commercial methods of preparation* of **diphenyl** :

Method–1: Diphenyl may be prepared commercially by allowing the *vapours of pure* **benzene** (C_6H_6) *via* a *pre-heated* (600–800°C) *'iron tube'* duly packed with *Pumice Powder**.

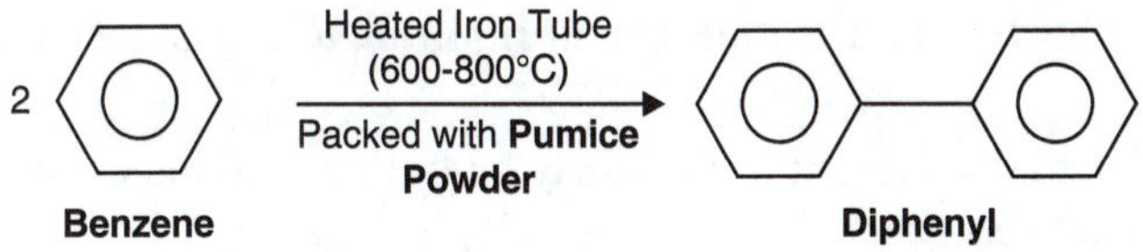

Benzene

Diphenyl

Method-2: It may also be produced on a large-scale by mixing together the '**Benzene Vapours**' (*preheated to 650°C*) along with **superheated steam** at **1000–1100°C**; and passing the admixture of vapours *via* a **steel vessel coated inside with a thin film of ferric oxide** (Fe_2O_3).

(f) **Diphenyl from Grignard's Reagent** – It is prepared by the action of a **Grignard reagent** *phenylmagnesium bromide* upon *bromobenzene* in the critical pressence of a small quantum of **cobaltous chloride** ($COCl_2$) – as shown under:

---

* **Pumice Powder:** A very finely divided lightweight glass used for the smothering or polishing surfaces.

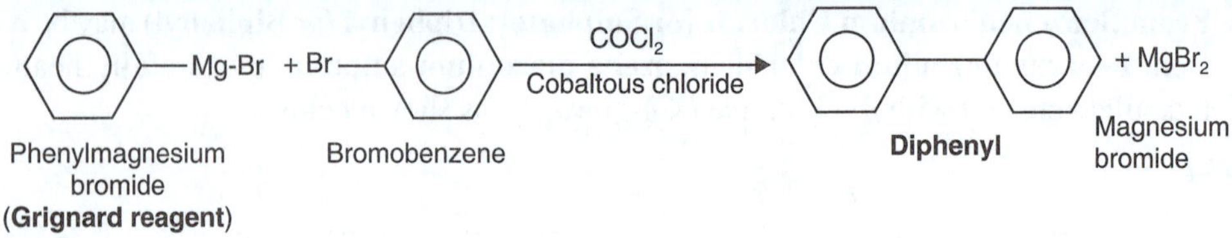

Phenylmagnesium bromide (**Grignard reagent**) + Bromobenzene → **Diphenyl** + Magnesium bromide

Characteristic Features

These essentially include:

1. **Diphenyl** is obtained as colourless leaflets.
2. It has a pleasant odour.
3. Its **density** is 1.041.
4. It has **mp: 69–71°C**; and **bp: 254–255°C***.
5. It has n_D^{77} 1.588.
6. **Solubility Profile** – It is fairly soluble in alcohol and ether; but insoluble in water.

Caution: Its major potential symptoms** of overexposure are confined to:

- **Headache**   • **Nausea**   • **Fatigue**   • **Numb Limbs**
- **Liver Damage** and   • **Irritation of Eyes and Throat**.

Substitution Reactions: Following are the *five* cardinal **substitution reactions**, namely;

- **Nitration** • **Chlorination** • **Sulphonation** • **Oxidation** and • **Reduction**

which shall now be treated briefly in the sections that follows separately.

Preamble: Since, the **phenyl (C_6H_5–) moiety** designates a typical *ortho-* and *para-* **directing functional moiety**, it distinctly entails *two* **substitution reactions**, for instance:

➤ *First* **Substitution** – it is an **electrophilic substitution** that essentially provides the *para-***substituted product**; and certainly to a relatively lesser degree the so-called *ortho-***substituted product.**

➤ *Second* **Substitution** – irrespective of the underlying fact that whether the *aforesaid ensuing moiety* is either *electron releasing* or *electron attracting* – there prevails the critical introduction of a **second substitution** in the respective *unsubstituted ring* in the corresponding *ortho* and *para*-positions.

Nevertheless, the foretold statement of facts may be further ascertained by the help of the **above said** *five* **specific chemical reactions**.

❑ **NITRATION: Diphenyl** on being subjected to heating with a mixture of concentrated *nitric acid* (HNO_3) and *sulphuric acid* (H_2SO_4) to obtain *para-***nitrophenyl** (or *4-nitrophenyl*) as the major product plus a small quantum of the corresponding *isomer ortho-***nitrophenyl** (or *2-nitrophenyl*). Furthermore, when the **4-nitrophenyl** is nitrated progressively it gives rise to the formation of **4, 4′–dinitrophenyl** plus a certain small proportion of:

- **2, 4′–dinitrophenyl**; and   • **2, 2′– dinitrophenyl**.

---

*   It is found to be stable at a temperature higher than its BP.

**   **NIOSCH: Pocket Guide to Chemical Hazards, (DHHS/NIOSH**, 97–140, 1997) p: 120.

Thus, we may have the following sequence of reactions:

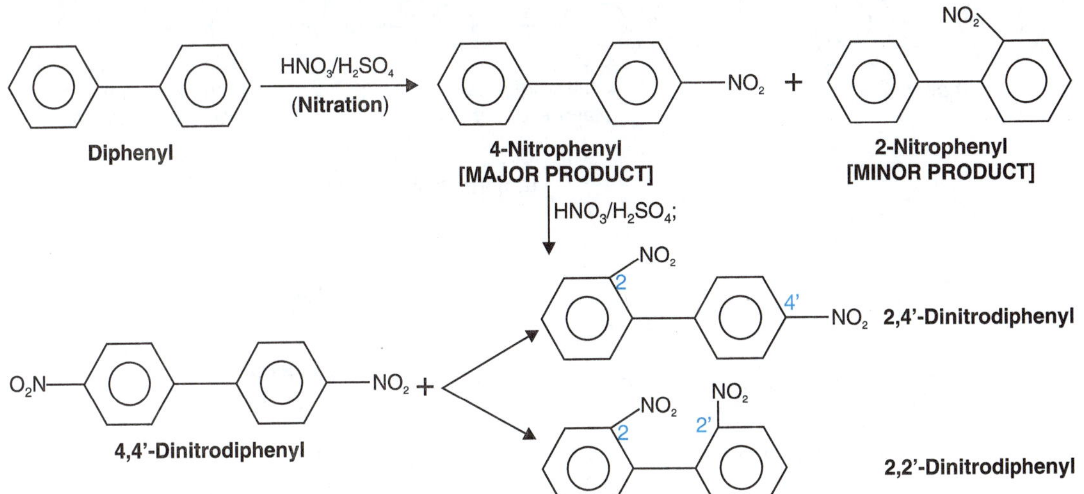

Diphenyl — HNO₃/H₂SO₄ (Nitration) → **4-Nitrophenyl [MAJOR PRODUCT]** + **2-Nitrophenyl [MINOR PRODUCT]**

HNO₃/H₂SO₄;

4,4'-Dinitrodiphenyl + → **2,4'-Dinitrodiphenyl** / **2,2'-Dinitrodiphenyl**

Impact of Charge Distribution Concept: Importantly, as per the impact of the ensuring *charge distribution concept*, one may observe critically that an inherent *p*-**nitro moiety** is duly expected to direct the **incoming group** to the corresponding *m*-**position** in the *other adjoining phenyl ring*, but it eventually exerts direction to the actual *p*-**position** – as depicted below explicitly:

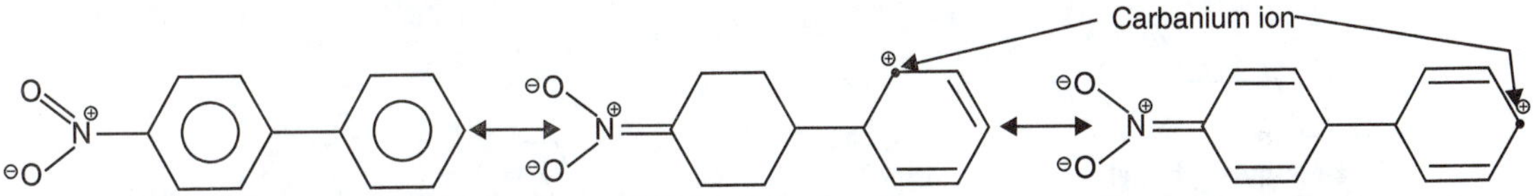

Carbanium ion

Explanation – A most probable and logical explanation for the aforesaid **transformations** would be that taken into consideration the various observed stabilities of the *carbanium ion* intermediates, there prevails much *larger dispersal of charge* in the respective *p*-(or *O*–) **ion** *vis-a-vis* the *m*-position. As a result, the *former* is certainly found to be rather '**more stable**'; and hence, the **orientation** to the *p*-**position** is favoured more generously. Thus, we may have the following sequential reactions:

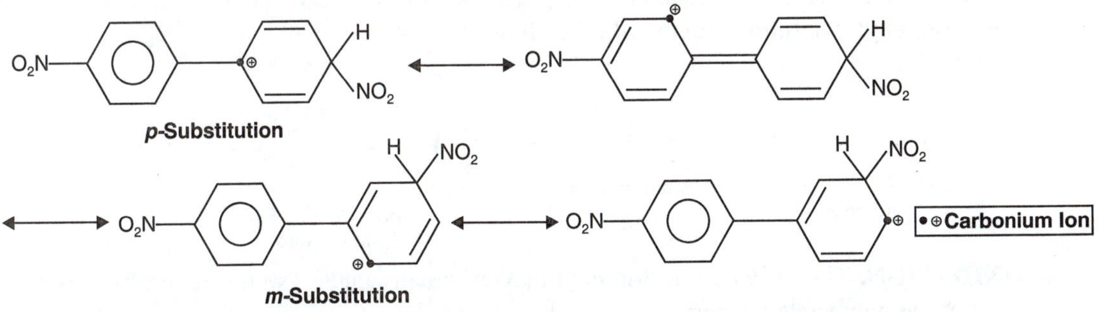

p-Substitution

m-Substitution

Carbonium Ion

❑ **Chlorination:** Importantly, **diphenyl** undergoes **chlorination** to produce 4-chlorophenyl, as the major product, which upon further *chlorination* yields **4, 4'-dichlorodiphenyl** as elaborated under:

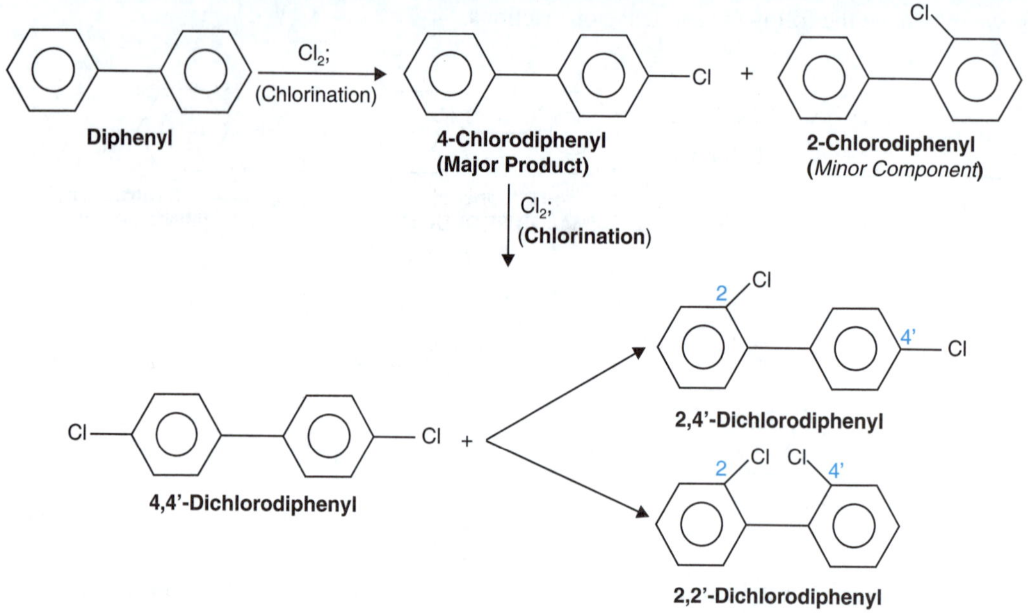

Diphenyl → **Cl₂; (Chlorination)** → **4-Chlorodiphenyl (Major Product)** + **2-Chlorodiphenyl (*Minor Component*)**

Cl₂; (Chlorination)

4,4'-Dichlorodiphenyl + → **2,4'-Dichlorodiphenyl** / **2,2'-Dichlorodiphenyl**

Important Observations – In another critical situation when either *4-methyl-*or *4-hydroxy-diphenyl* is duly *chlorinated* gives rise to the formation of **3-chloro-4-methyldiphenyl** as the major product as shown under:

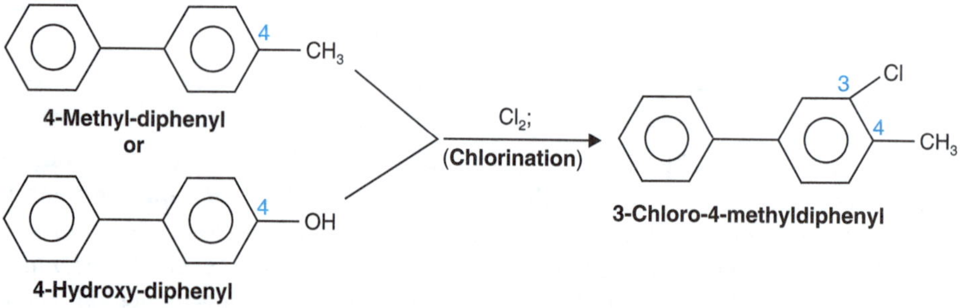

4-Methyl-diphenyl or / **4-Hydroxy-diphenyl** → **Cl₂; (Chlorination)** → **3-Chloro-4-methyldiphenyl**

❑ **SULPHONATION:** The *sulphonation* of **diphenyl** with hot concentrated *sulphuric acid* (H_2SO_4) yield **diphenyl-4-sulphonic acid** with the elimination of a mole of **water (H_2O)** as depicted below:

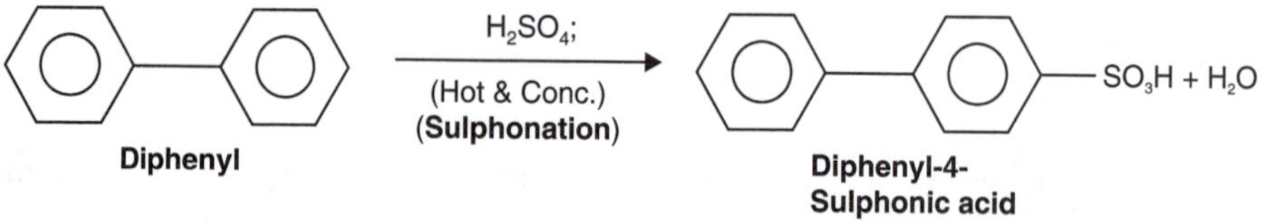

Diphenyl → **H_2SO_4; (Hot & Conc.) (Sulphonation)** → **Diphenyl-4-Sulphonic acid** — $SO_3H + H_2O$

❑ **OXIDATION:** The *oxidation* of **diphenyl** may eventually undertake *two* divergent course of reactions as enumerated under:

• **Oxidation by Chronic Acid** – **Diphenyl** yields *two sets of products of oxidation reaction,* namely:

➤ **Carbon dioxide (CO$_2$) + Water (H$_2$O)** – as **major product**; and

➤ **Benzoic Acid (C$_6$H$_5$–COOH)** – as *minor product*.

• **Ozonolysation (with O$_3$) at – 20°C** – diphenyl gives only *one major product* **Benzoic Acid** (~ **86%**).

❑ **Reduction: Diphenyl** on being subjected to *reduction* (or **hydrogenation**) in the presence of *Raney Ni* yields **bicyclohexyl**; whereas, the *intermediate product* (*viz.*, *phenylcyclohexyl*) could not be isolated at all. Thus, we may have the following reactions:

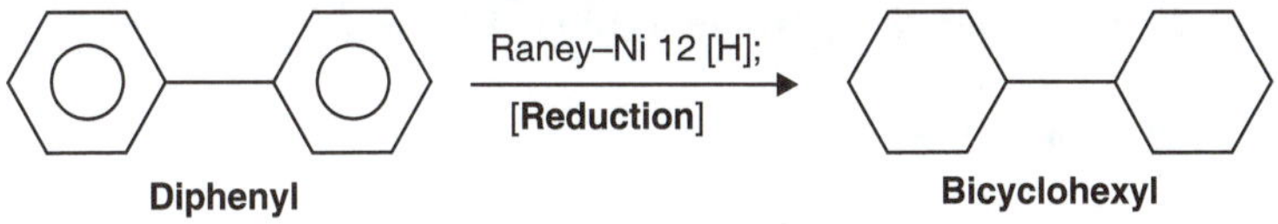

Raney–Ni 12 [H];
[Reduction]

Diphenyl   **Bicyclohexyl**

Uses

1. **Diphenyl** finds its use as a **heat-transfer agent**.
2. It is being employed as an important and effective '**fungistat**' for *oranges* (applied usually to *inside of shipping containers or wrappers*).
3. The **chlorinated diphenyls** are used as '**plasticizers**'.

2.2. Diphenic Acid [Syn: a, ac–Bibenzoic Acid, Diphenyl 2, 2c–dicarboxylic Acid]

Diphenic acid is a prepared from *diazotized anthranilic acid* by treatment with a **cuproammonia–sulphite reducing agent**.

Chemical Structure

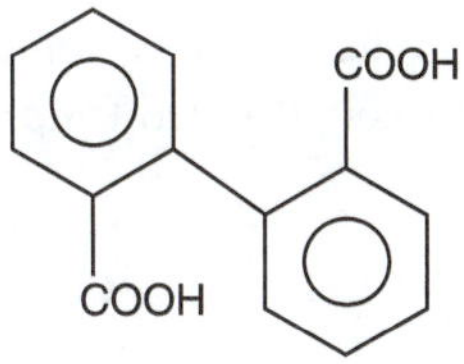

COOH

COOH

Diphenic Acid
[1:1'–Biphenyl]–2,2'–dicarboxylic acid;

Preparation of Diphenic Acid : It is prepared by means of any of the *three* established methods of preparation, namely:

• **From Diazotized Anthranilic Acid**,
• **From Phenanthraquinone**, and
• **From Methyl–O–Iodobenzoate**,

which shall now be discussed individually in the sections that follows:

2.2.1. From Diazotized Anthranilic Acid

The **diazotized anthranilic acid** when subjected to treatment with *ammoniacal cuprous oxide* it forms **diphenic acid** ranging between **72–84%**, as shown under:

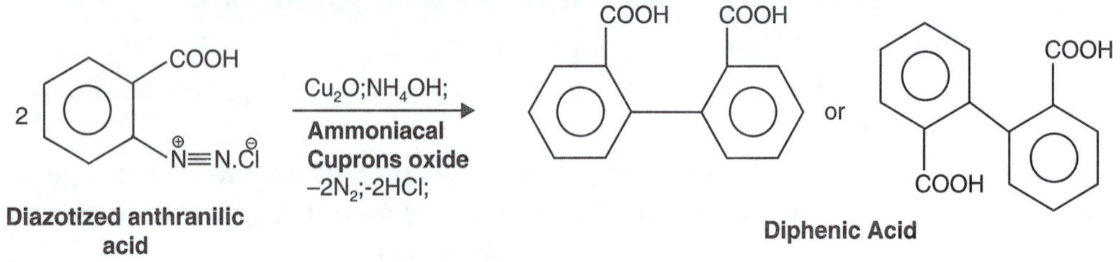

Diazotized anthranilic acid → Diphenic Acid

Cu₂O;NH₄OH;
Ammoniacal Cuprons oxide
–2N₂;-2HCl;

NOTE   *O*-Aminobenzoic acid is called the anthranilic acid.

2.2.2. From Phenanthraquinone

Oxidation of *phenanthraquinone* by either of the following *two* **methods** namely:

- **Potassium dichromate ($K_2Cr_2O_7$) and sulphuric acid (H_2SO_4); or**
- **Hydrogen peroxide (50% H_2O_2) in glacial acetic acid (CH_3COOH),**

yield **diphenic acid** (~ **68%**), as given under:

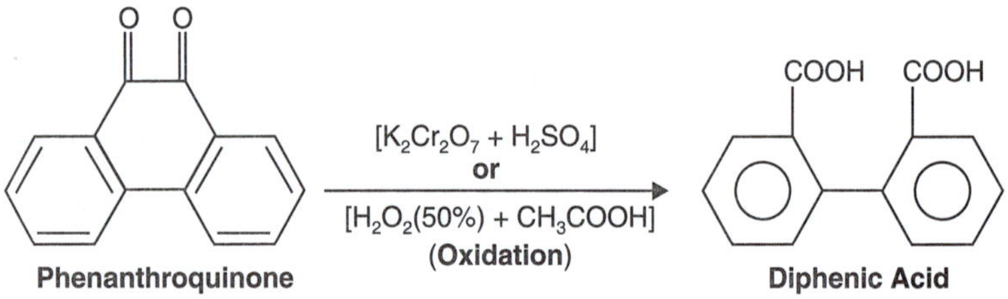

Phenanthroquinone → Diphenic Acid

$[K_2Cr_2O_7 + H_2SO_4]$
or
$[H_2O_2(50\%) + CH_3COOH]$
(Oxidation)

2.2.3. From Methyl–O–iodobenzoate

Ulmann reaction (1904, 1905*) yield the respective **methyl ester of diphenic acid** when **methyl** *O*-**iodobenzoate** is heated carefully with finely divided *Cu-powder*, as stated below:

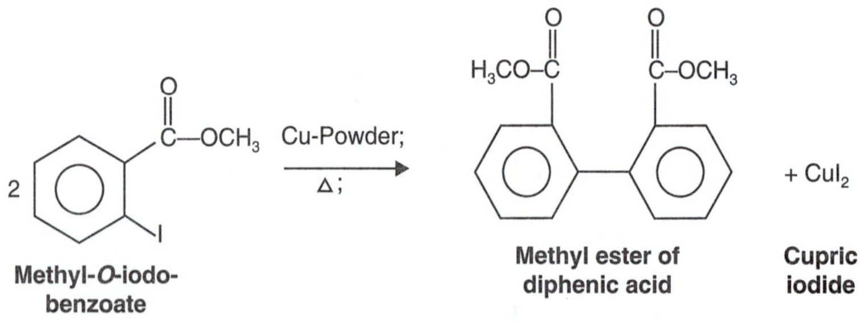

Methyl-*O*-iodo-benzoate → Methyl ester of diphenic acid + Cupric iodide

Cu-Powder;
Δ;

+ CuI₂

Characteristic Features

Following are the *six* **cardinal characteristic features** of '*diphenic acid*':

1. **Formation of Diphenic Anhydride – Diphenic acid** on being heated with *acetic anhydride* (or *acetyl chloride*) yields **diphenic anhydride (mp: 217°C).**

---

* Ulmann F., *Ann.*, **332**:38 (1904); Ulmann F. and Sponge P., *Ber.*, **38**:221 (1905).

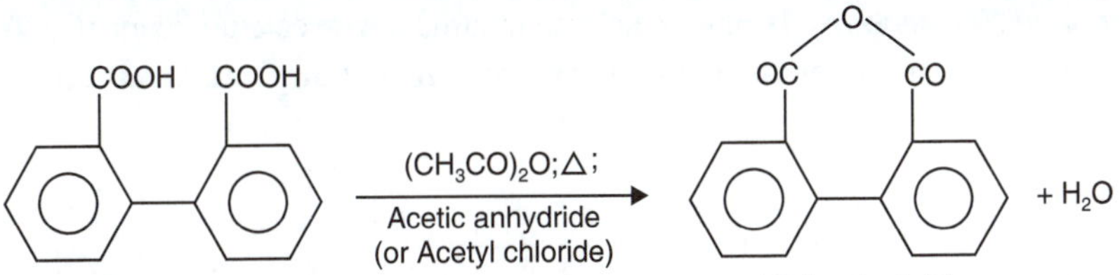
COOH
COOH
(CH3CO)2O;Δ;
Acetic anhydride
(or Acetyl chloride)
O
OC
CO
+ H2O

Diphenic Acid **Diphenic Acid**

2. ***Formation of Fluorenone-4-carboxylic acid*** – **Diphenic acid** when heated with **concentrated sulphuric acid (36 m)** gives rise to the formation of **fluorenone-4-carboxylic acid (mp 227°C)** as stated below:

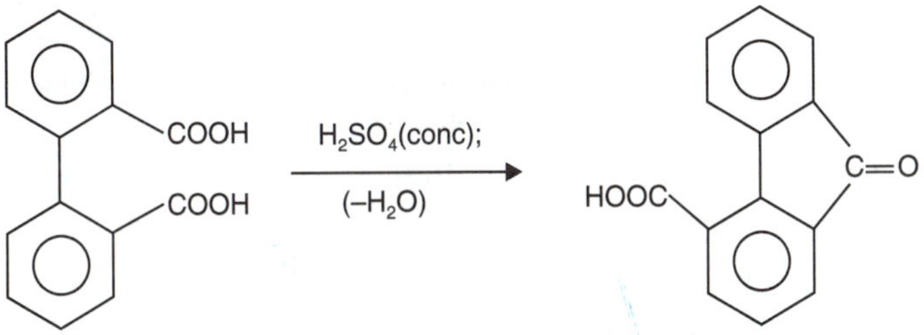
COOH
COOH
H2SO4(conc);
(–H2O)
HOOC
C=O

Diphenic Acid **Fluorenone-4-carboxylic Acid (mp; 227°C)**

Comment – Thus, one may critically observe a '*ring closure*' involving elimination of **water** between **one of the carboxyl (–COOH) moieties** and an ***ortho* position** in the adjoining ring is brought about by treatment of **diphenic acid** with *concentrated sulphuric acid (36 m)* to produce **fluorenone-4-carboxylic acid.**

3. ***Formation of Phthalic Acid*** – Oxidation of diphenic acid with *potassium permanganate ($KMnO_4$)* yields **phthalic acid** (mp~23) as shown under:

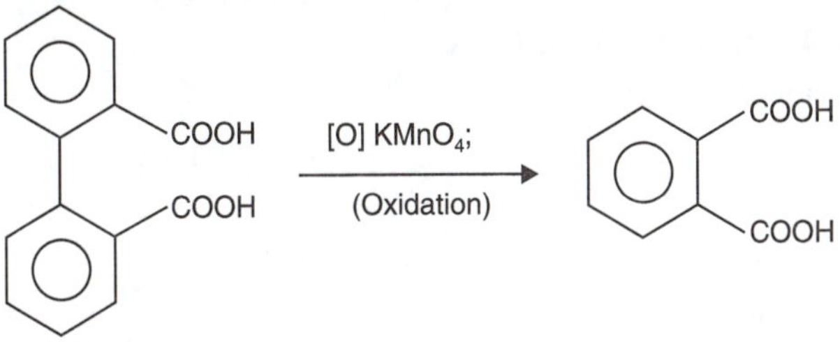
COOH
COOH
[O] KMnO4;
(Oxidation)
COOH
COOH

Diphenic Acid **Phthalic Acid**

4. ***Formation of Diphenyl*** – **Diphenic acid** on being subjected to distillation with **sodalime** (*i.e.*, an admixture of *calcium oxide, CaO* with sodium hydroxide, NaOH, 5–20% and containing 6–18% H_2O) gives **diphenyl** (or *biphenyl*) as stated below:

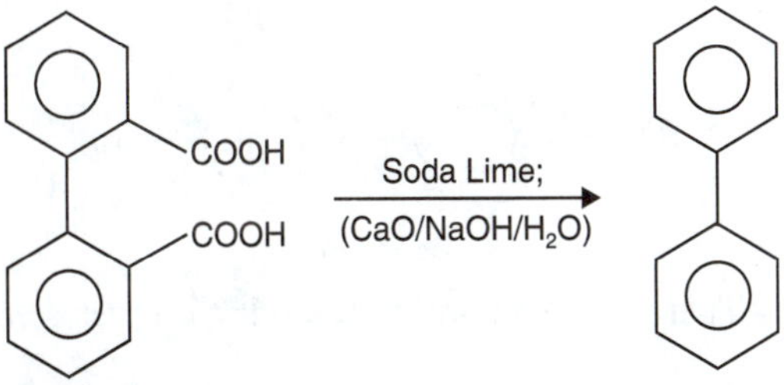
COOH
COOH
Soda Lime;
(CaO/NaOH/H2O)

Diphenic Acid **Diphenyl**

5. *Formation of Fluorenone* – **Diphenic acid** on treatment with calcium oxide (CaO) yields *calcium diphenate*, which on further distillation forms the **fluorenone** plus *calcium carbonate* – as given below:

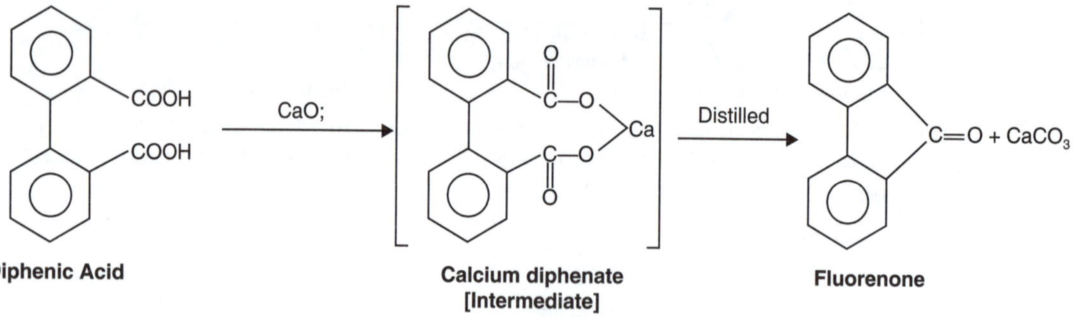

Diphenic Acid   **Calcium diphenate**   **Fluorenone**
                                          [Intermediate]

Comments – The respective transformation of **diphenic acid** is duly accomplished by *distillation with lime*, that eventually results in the closure of a **5-membered ketone ring** (see **calcium diphenate** – an '**intermediate**') by loss of a mole of *calcium carbonate* and production of **fluorenone**.

6. *Physical Properties of Diphenic Acid* – These essentially include:
   - Crystals are obtained from hot water.
   - Hisher melting modifications: mp 171–172°C.
   - Substantially soluble in hot-water.
   - **Solubility Profile** – soluble in ethanol, methyl ethyl ketone, acetone, acetic acid, isopropanol.

Uses

1. Diphenic acid finds its use as an '*intermediate*' for surface coatings.
2. It is used as the **lubricating oil additives**.
3. It also finds its abundant usages in:
   - **cosmetics**,
   - **surfactants**,
   - **plasticizers**, and
   - **textile chemicals**.

2.3. Benzidine [or 4, 4'-Diaminophenyl]

Zinin (1845) first discovered **benzidine**.

Chemical Structure

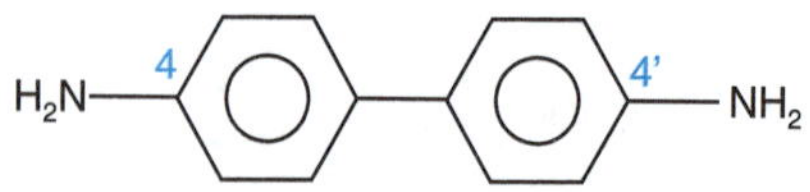

Benzidine

Preparation of Benzidine – There are *two* known methods for the *preparation of benzidine*:

❑ **Method–1 Benzidine** may be produced by the careful **reduction** of **nitrobenzene** with **Zn/NaOH**; and the resultant hydrazobenzene (C_6H_5–NH, NH–C_6H_5) is heated with an acid:

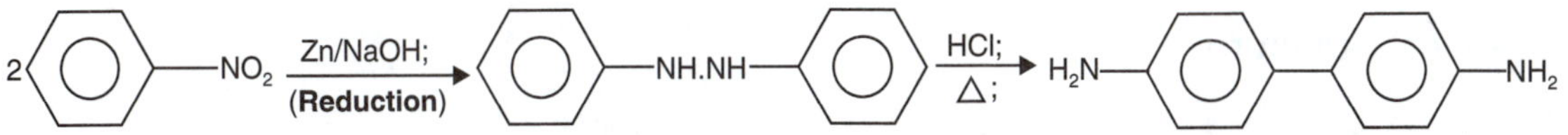

Nitrobenzene   Hydrazobenzene   Benzidine

❑ **Method–2** *Hydroxyamino* benzene on acidification yields *p-amino-phenol*. *Hydrazoazobenzene* on treatment with an acid gives benzidine (or **4, 4′–diminophenyl**)–as given under:

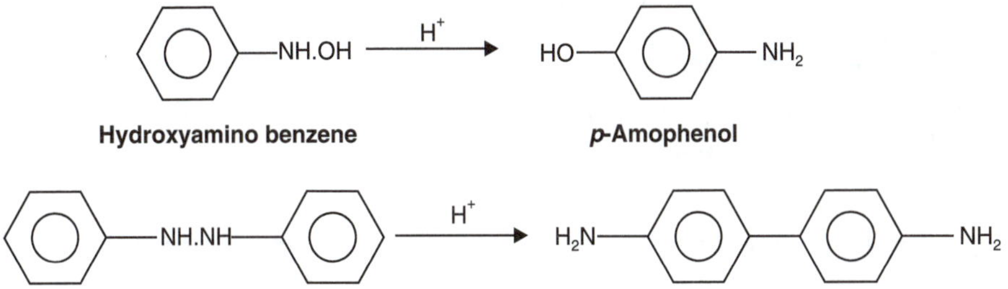

Hydroxyamino benzene   *p*-Amophenol

Hydrazobenzene   Benzidine
[4,4'–Diamophenyl]

Characteristic Features

1. **Benzidine** occurs as a white or slightly reddish crystalline powder, which usually darkens on exposure to air and light.

   Caution: An array of typical *potential systems of overxposure* do include: *Haematuria*; *Secondary Anemia from Haemolysis*; *Acute cystisis*; *Acute liver disorder*; *Dermatitis*; and *Irregular and painful urination**.

2. It has **mp: 115–120°C;** (when heated slowly); anhydrous product has **mp: 128°C**; and rapidly heated shows **bp ~ 400°C**.

3. **Solubility Profile** – 1g dissolves in *2.5L cold water*; *107 mL boiling water; 5 mL boiling ethanol*; and solvent ether 50 mL.

   Storage Conditions – Keep well closed and protected from light.

Uses

1. Benzidine finds its abundant usage in the direct commercial production of the '**Azo-Dyes**' (*viz.*, **Congo Red**).

2. It is used as a *sensitive reagent* for the detection of **hydrogen peroxide (H_2O_2)** in *fresh milk*.

3. *Benzidine* also helps in the detection of *blood*.

4. *Benzidine Dihydrochloride* salt is invariably used for:

   • **Quantitative assay of sulphates (SO_4^{2-})**; and

   • As a '*reagent*' for the presence of metals.

---

* **NIOSCH –Pocket Guide to Chemical Hazards** [DHHS/NIOSCH: **97**:140, p.26. 1997]

2.4. Diphenylmethane

Diphenylmethane designates the most frequently encountered *benzenoid hydrocarbons*.

Chemical Structure

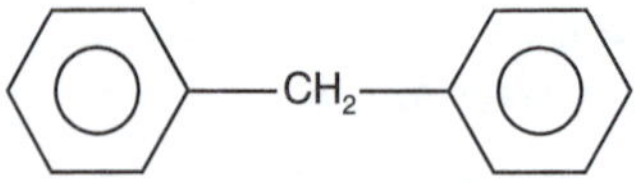

Diphenylmethane

Preparation of Diphenylmethane The preparation of **diphenylmethane** may be achieved by any one of the following *four* **recognized methods:**

2.4.1. From Grignard's Reagent

Benzyl chloride on being warmed with **phenylmagnesium bromide**, a *Grignard's Reagent* between 60 to 70°C, it yields **diphenylmethane**, as given below:

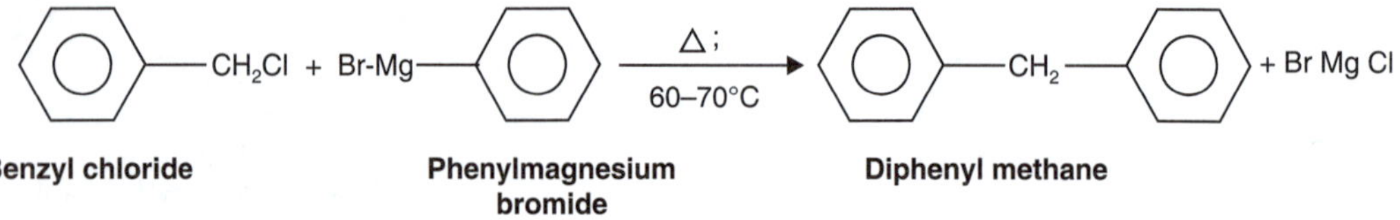

Benzyl chloride   **Phenylmagnesium**   **Diphenyl methane**
                             bromide

Thus, *magnesium bromochloride* gets eliminated as a by product.

2.4.2. From Friedel Craft's Reaction*

The *condensation* of **benzene** with *benzyl chloride* in the presence of anhydrous aluminium chloride $(AlCl_3)$ gives rise to the formation of **diphenylmethane** with the elimination of a mole of hydrochloric acid (HCl) as given below:

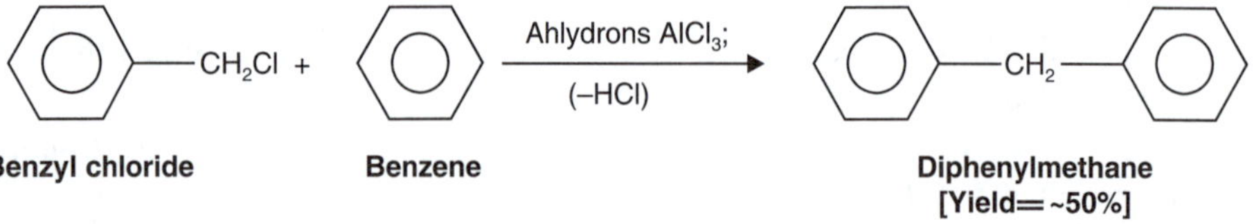

Benzyl chloride   **Benzene**   **Diphenylmethane**
                                                         [Yield = ~50%]

2.4.3. From Benzophenone

The careful reduction of *benzophenone* with hydroiodic acid (HI) and red phosphorus under pressure at 160°C undergoes the **Wolff–Kishner reduction**** with the elimination of a mole of water (H_2O).

However, the *reduction* may also be carried out by using a mixture of **LiACH$_4$–ACCl$_3$** on *benzophenone*.

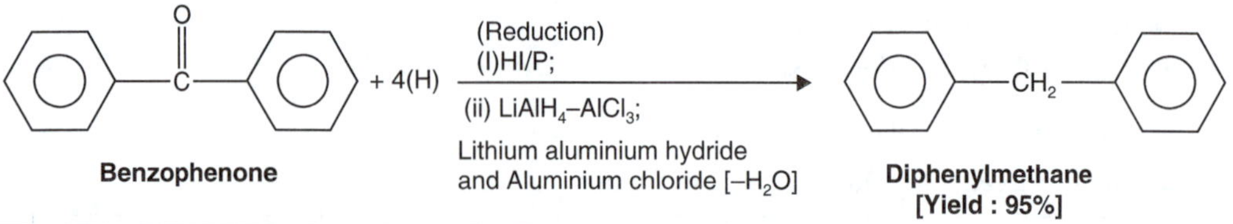

Benzophenone   Lithium aluminium hydride   **Diphenylmethane**
                             and Aluminium chloride [–H$_2$O]   **[Yield : 95%]**

---

*   **Friedel C and Crafts JM:** *Compt Rend,* **84**: 1392, 1450 (1877).

** Wolff L: *Ann,* **394:**86, 1912; Kishner N: *J Rus Phys Chem Soc.,* **43**:582, 1911.

2.4.4. From Benzene

The meticulous *condensation* of *two* **moles of benzene** with *one mole* of **formaldehyde** in the critical presence of a few drops of concentrated sulphuric acid (36 M) yields **diphenylmethane** with the loss of a mole of water-as depicted under:

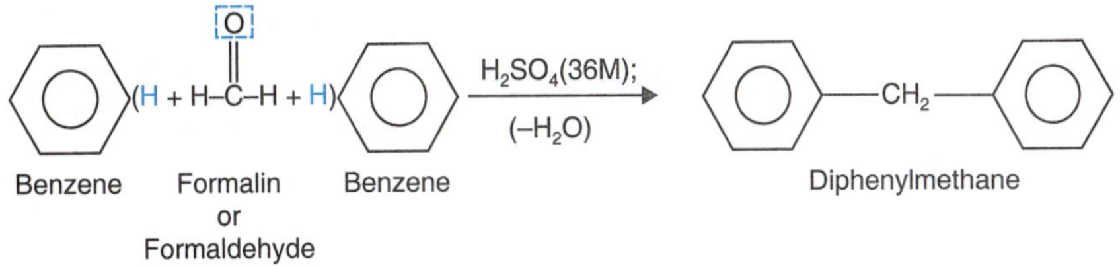

| Benzene | Formalin or Formaldehyde | Benzene | | Diphenylmethane |

Chemical Reactions of Diphenylmethane – Following are the *five* important and vital **chemical reactions** of **diphenylmethane**.

1. **Nitration Reaction** – In fact, the *nitration* of *diphenyl-methane* takes place at *two* **different stages:**
   - **First** – the formation of *p*-**nitrophenylmethane** occurs; and
   - **Secondly** – the further *nitration* of the resulting product yields **p, p′–dinitrophenylmethane**.

Thus, we may have the following sequence of reactions:

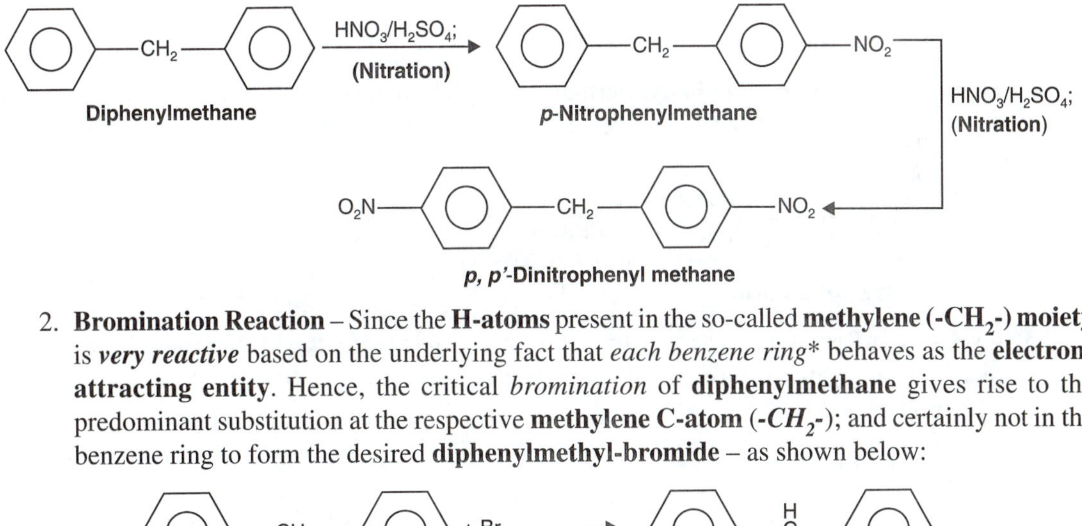

2. **Bromination Reaction** – Since the **H-atoms** present in the so-called **methylene (-CH$_2$-) moiety** is *very reactive* based on the underlying fact that *each benzene ring** behaves as the **electron-attracting entity**. Hence, the critical *bromination* of **diphenylmethane** gives rise to the predominant substitution at the respective **methylene C-atom (-CH$_2$-)**; and certainly not in the benzene ring to form the desired **diphenylmethyl-bromide** – as shown below:

Thus, a mole of **hydrobromic acid (HBr)** gets eliminated.

Hydrolysis of Diphenylmethyl bromide – The *hydrolysis* of the resulting product (**diphenylmethyl bromide**) by means of **alkali** rapidly gives rise to the formation of **benzhydrol** (or **diphenylcarbinol**) as stated under:

---

* Perhaps due to the presence of **π-electron clouds** present on either sides of the **benzene ring**.

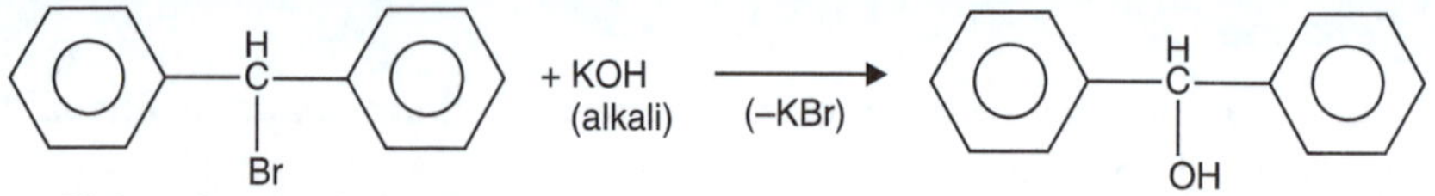

Diphenylmethyl bromide + KOH (alkali) (−KBr) → **Benzhydrol**
[or Diphenylcarbinol]
[mp : 69°C; bp$_{748}$: 298°C]

3. **Chlorination Reaction – Diphenylmethane** on being subjected to **chlorination reaction** in the presence of **UV-light** (**sunlight**) yields **diphenyldichloromethane** with the loss of **two** moles of HCl – as shown below:

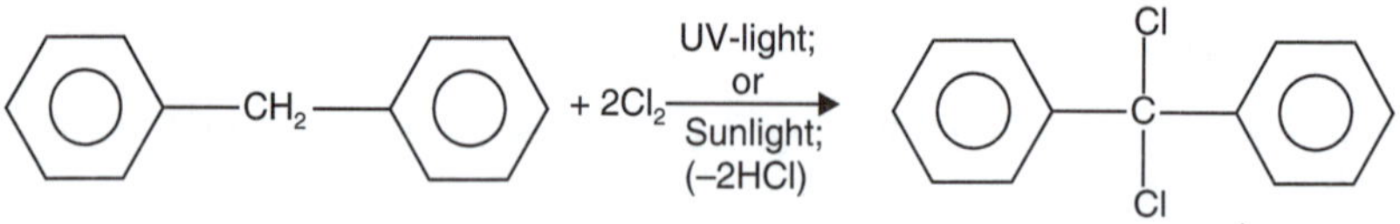

Diphenylmethane + 2Cl$_2$ → (UV-light; or Sunlight; (−2HCl)) **Diphenyldichloromethane**

Alternatively, the *'chlorination'* accomplished in the presence of *Fe/I$_2$* as a *catalyst*; *p,p'*–**dichlorodiphenylmethane** is duly produced with the elimination of *two* moles of HCl.

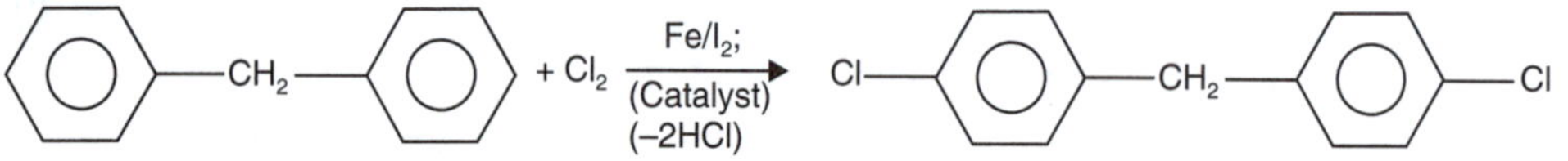

Diphenylmethane + Cl$_2$ → Fe/I$_2$; (Catalyst) (−2HCl) **p,p'–Dichlorodiphenylmethane**

4. **Oxidation Reaction – Diphenylmethane** when duly oxidized with *chromic acid* (*CrO$_3$*) and *glacial acetic acid* (*CH$_3$COOH*) gives rise to the formation of benzophenone – as given under:

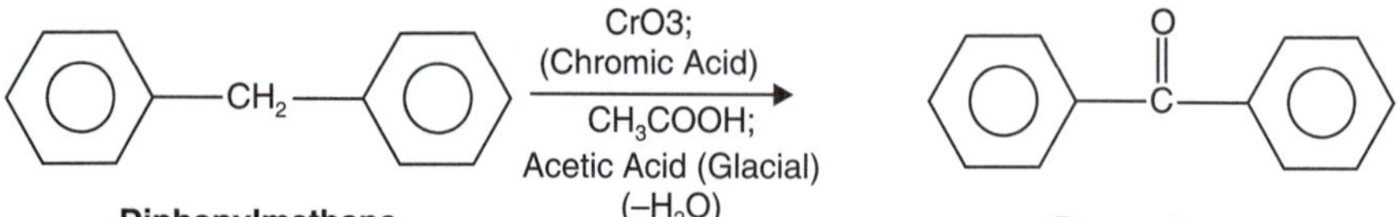

Diphenylmethane → CrO3; (Chromic Acid) CH$_3$COOH; Acetic Acid (Glacial) (−H$_2$O) **Benzophenone**

5. **Cyclization Reaction** – The vapourized form of diphenylmethane when passed *via* a red-hot metalic tube – yields *fluorene* (due to *cyclization*) and a mole of **hydrogen** (*H$_2$*) is obtained as a by product:

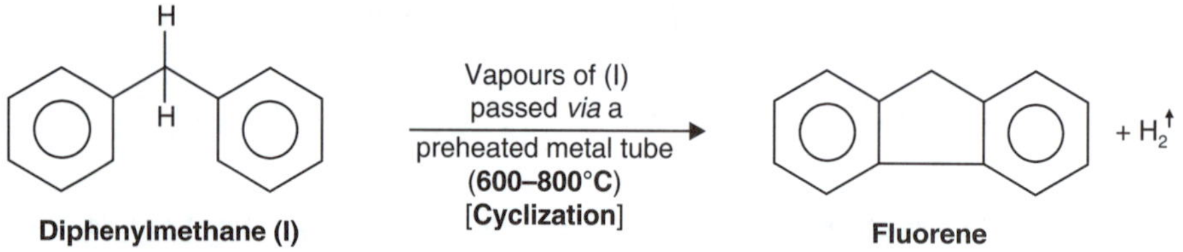

Diphenylmethane (I) → Vapours of (I) passed *via* a preheated metal tube (600–800°C) [Cyclization] **Fluorene** + H$_2$↑

Physical Parameters of Diphenylmethane – These essentially include:

1. It is obtained as **orthorhombic needles**.
2. **Diphenylmethane** has the inherent **odour of oranges** (citrus flavour)
3. It has mp: 25.9°C and bp$_{100}$: 186.3°C.

4. It has d_4^{10} 1.3421 (solid) and n_D^{120} 1.57683.

5. **Solubility Profile** – It is found to be freely soluble in ethanol, solvent ether, chloroform, hexane, and benzene; insoluble in **liquefied ammonia (NH$_3$)**.

Uses

1. Since **diphenylmethane** possesses a typical characteristic **germanium–like odour** – it is abundantly used as a perfume in toilet soaps (*i.e.*, bathing soaps).

2.5. Triphenylmethane (or Tritan)

Chemical Structure

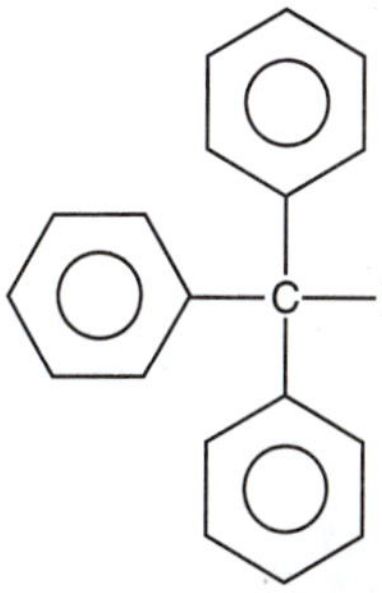

Triphenylmethane

Preparation of Triphenylmethane : The desired compound may be prepared by any one of the following *three* procedures:

(a) **From Benzaldehyde–Benzaldehyde** on being subjected to *condensation* with benzene at **250–270 °C**, in the presence of zinc chloride (ZnCl$_2$) gives rise to the formation of **triphenylmethane**–as stated below:

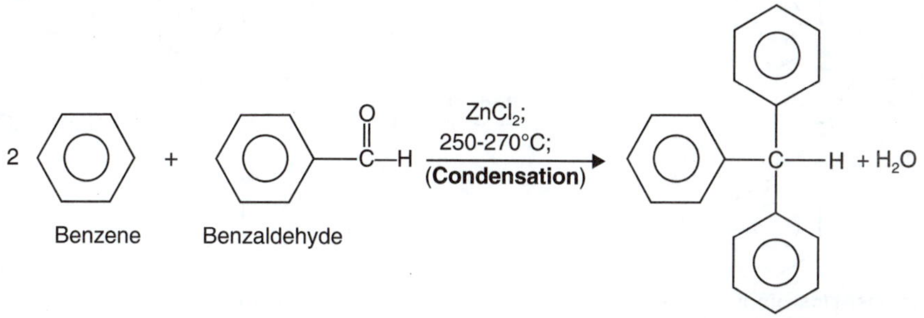

Benzene   Benzaldehyde   Triphenylmethane

Thus, a mole of water gets eliminated.

(b) **Friedel-Craft's Reaction*–**Benzene when condensed with either *chloroform* (CHCl$_3$) or *benzal chloride* (C$_6$H$_5$–CHCl$_2$) in the critical presence of anhydrous aluminium chloride (AlCl$_3$)–it yields **triphenylmethane**–as given under:

---

* Friedel C and Crafts JM: *Compt Rend*, **84:** 1392, 1450, (1877).

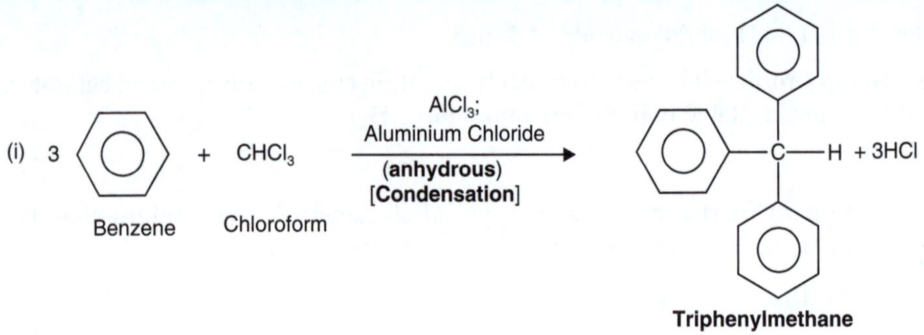

(i) 3 Benzene + $CHCl_3$ Chloroform $\xrightarrow[\text{[Condensation]}]{\substack{AlCl_3; \\ \text{Aluminium Chloride} \\ \textbf{(anhydrous)}}}$ Triphenylmethane $+ 3HCl$

Thus, *three* moles of HCl get eliminated.

(ii) 2 Benzene + Benzal Chloride $CHCl_2$ $\xrightarrow[\textbf{(Condensation)}]{\substack{ZnCl_3; \\ \text{Zinc Chloride} \\ \text{(anhydrous)}}}$ Triphenylmethane $+ 2HCl$

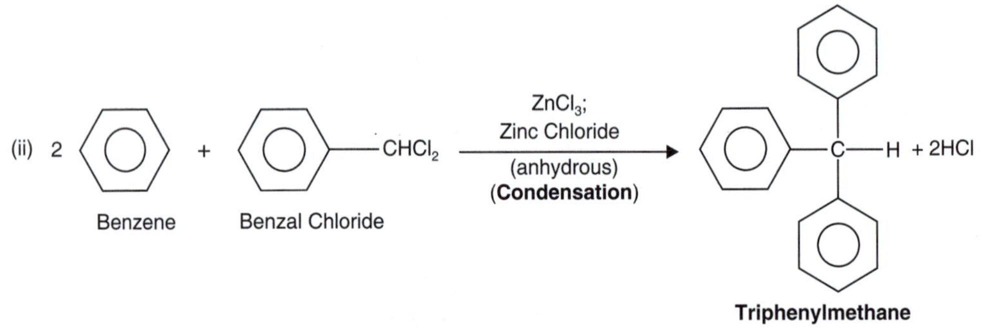

Thus, *two* moles of HCl get eliminated.

(iii) **From Triphenylcarbinol**–Reduction of **triphenylcarbinol** using *formic acid (HCOOH)* gives triphenylmethane plus a mole each of *carbon dioxide (CO_2)* and *water (H_2O)* as the by products– as shown under:

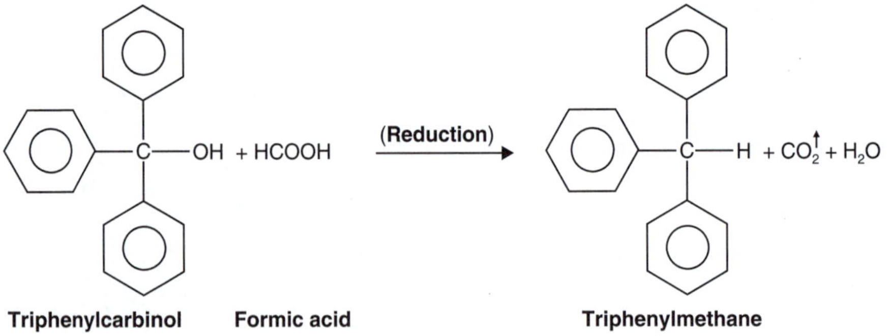

Triphenylcarbinol C—OH $+ HCOOH$ Formic acid $\xrightarrow{\textbf{(Reduction)}}$ C—H $+ CO_2\uparrow + H_2O$ Triphenylmethane

Characteristic Features of Triphenylmethane–These essentially include:

1. **Triphenylmethane** occurs as an *orthorhombic pyramidal*, solvated crystals containing one mole of benzene.
2. It has **mp: 78.2°C**.
3. It dries on exposure to **air**.
4. When it is dry its mp: **93.4°C**.
5. It mostly occurs in a **stable form:** there are *two metastable forms* (*i.e.*, slight margin of stability of a substance that alters into another substance as conditions change).

6. It has d_4^{100} 1.0134; pb_{760}: 360°C; and n_D^{100}. 1.59546.

7. **Solubility Pofile:** Triphenylmethane is very soluble in ether, hot ethanol, and chloroform; quite soluble in petroleum ether, benzene, and carbon disulphide (CS_2); and slightly soluble in glacial acetic acid.

8. **Reactivity of the *tertiary* H-atom of methane in Triphenylmethane:** The critical presence of *three* **phenyl moieties in the triphenylmethane** molecule, – the *tertiary* **H-atoms of methane** present in the said chemical entity gets eventually *quite reactive in nature*. Perhaps, this could be the reason it actually undergoes *oxidation* very rapidly with such reagents as:
   - **Lead dioxide (PbO_2) or** - **Chromic acid (Cr O_3)**

   to result into the formation of **triphenylcarbinol [$(C_6H_5)_3C–OH$].**

9. Following are *four* typical and classical interactions of triphenylmethane, namely:
   - treatment with sodium metal in dry ether or liquid ammonia gives **triphenylmethane sodium (I)**;
   - **nitration of triphenylmethane** with HNO_3/H_2SO_4–affords the introduction of **one nitro(NO_2) moiety** entering each phenylring at the *para-position specifically* **(II),**
   - **bromination** of the parent compound yields **triphenylmethylbromide (III)**, and
   - **interaction** with **PbO_2 and CrO_3** gives **triphenylcarbinol (IV),**

which may be summarized as under:

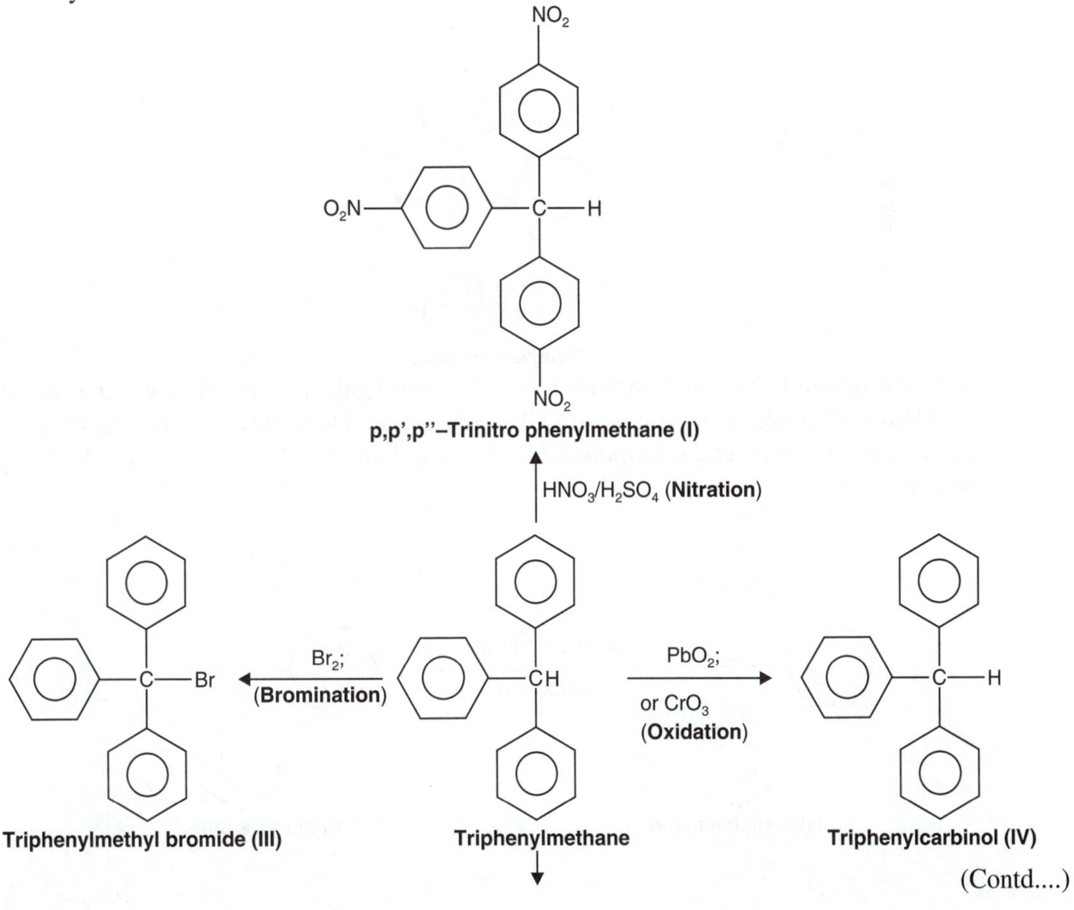

p,p',p''–Trinitro phenylmethane (I)

HNO_3/H_2SO_4 **(Nitration)**

Triphenylmethyl bromide (III)   Triphenylmethane   Triphenylcarbinol (IV)

Br_2; **(Bromination)** PbO_2; or CrO_3 **(Oxidation)**

(Contd....)

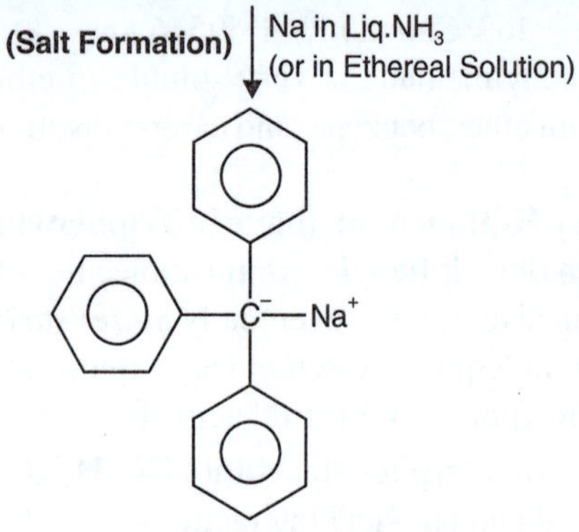

(Salt Formation) | Na in Liq.NH$_3$
(or in Ethereal Solution)

Triphenylmethyl sodium (I)

Uses

Triphenylmethane is the parent substance in the production of the so-called '**triphenylmethane dyes**'.

2.6. Triphenylcarbinol [*Syn:* Triphenylmethanol; Tritanol;]

Chemical Structure

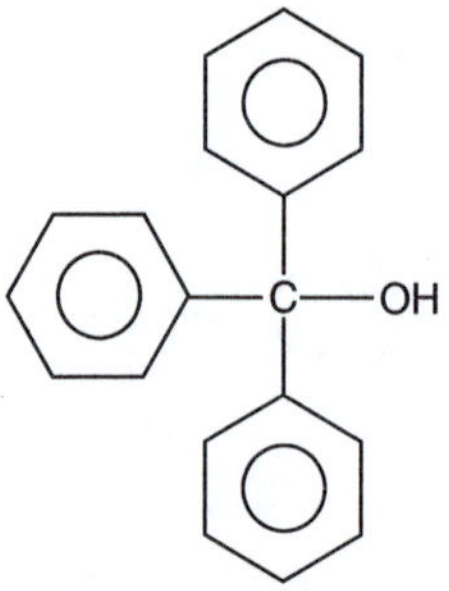

Triphenylcarbinol

Preparation of Triphenylcarbinol–It may be prepared by *two* distinct chemical procedures, namely:

(a) **Oxidation of Triphenylmethane with Chromium Trioxide [CrO$_3$] Triphenylcarbinol** may be prepared by the careful *oxidation* of **triphenylmethane** using *chromium trioxide* (CrO$_3$) as shown below:

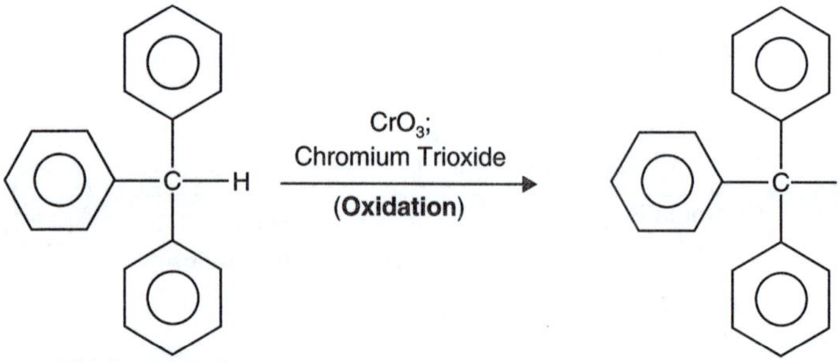

CrO$_3$;
Chromium Trioxide
(Oxidation)

Triphenylmethane   **Triphenylcarbinol**

It is produced upto a yield of 85%.

(b) **Preparation of Phenylmagnesium Bromide (*Grignard reagent*) followed by its Hydrolysis–**
When phenylmagnesium bromide is treated with either *benzophenone* or *ethyl benzoate* it gives the desired **Grignard reagent** phenylmagnesium bromide, which now on *hydrolysis* gives **triphenylcarbinol** with the elimination of a mole of *magnesium hydroxy bromide*, as given under:

(i) From Benzophenone

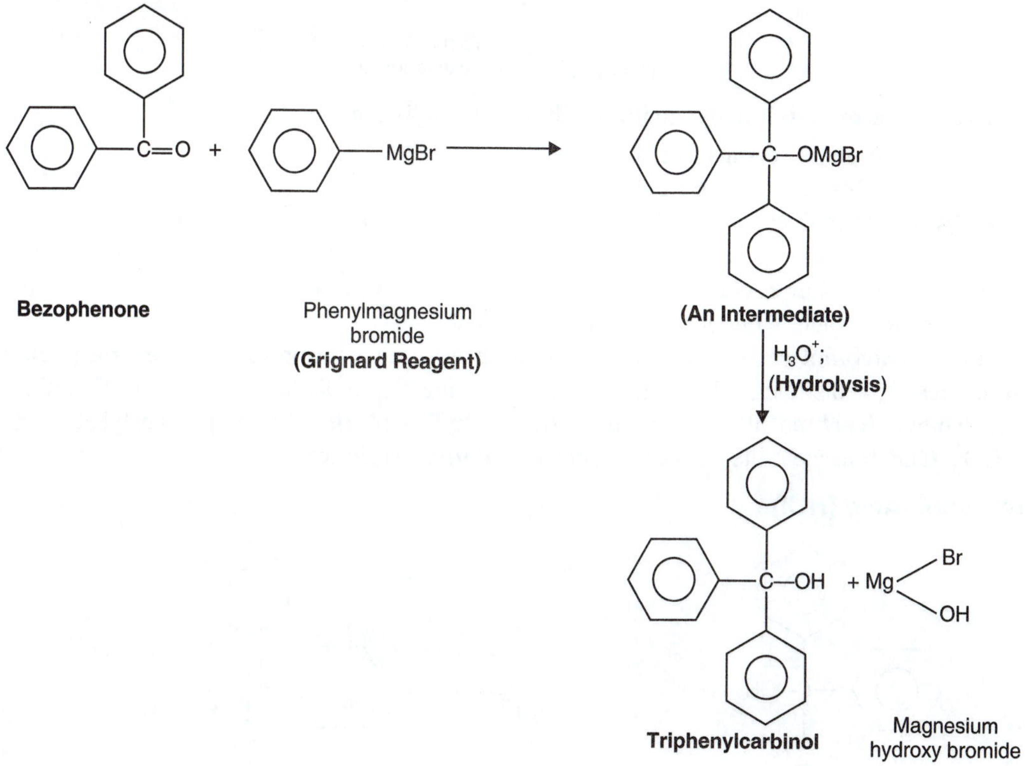

Bezophenone   Phenylmagnesium
bromide
(Grignard Reagent)   **(An Intermediate)**

H_3O^+;
(Hydrolysis)

Triphenylcarbinol   Magnesium
hydroxy bromide

(ii) From Ethylbenzoate

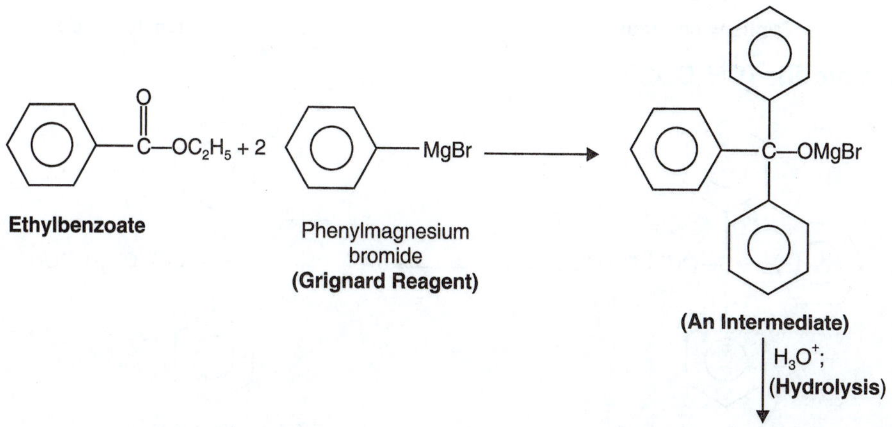

Ethylbenzoate   Phenylmagnesium
bromide
(Grignard Reagent)   **(An Intermediate)**

H_3O^+;
(Hydrolysis)

(Contd....)

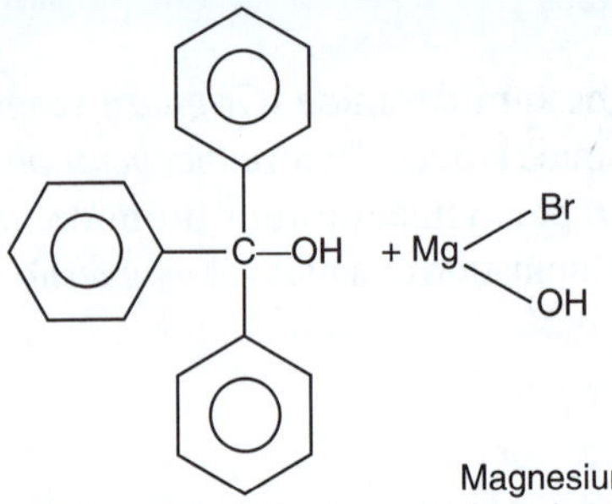

Triphenylcarbinol   Magnesium
hydroxy bromide

Characteristic Features of Triphenylcarbinol–These essentially include:

1. It occurs as trigonal crystals from benzene.
2. It has d_4^{20} 1.1999; and mp: 164.2°C.
3. **Triphenylcarbinol** gets distilled between 160–80°C without any decomposition whatsoever.
4. **Solubility Profile**–It is easily soluble in ethanol, ether, and benzene. It gets solubilized in concentrated sulphuric acid (H_2SO_4) with a distinct intense yellow colouration. It is also soluble in glacial acetic acid without producing any colour.
5. **Triphenylcarbinol** is rendered extremely reactive by virtue of the presence of the *three* **phenyl moieties** located strategically on the **C-atom** *carrying the alcoholic (OH) moiety*. Therefore, the **triphenylcarbinol** almost reacts instantly with **hydrochloric acid (HCl)** or **acetyl chloride** (**CH₃–COCl**) to yield the respective *triphenylchloride*–as shown below:

(i) Hydrochloric Acid (HCl)

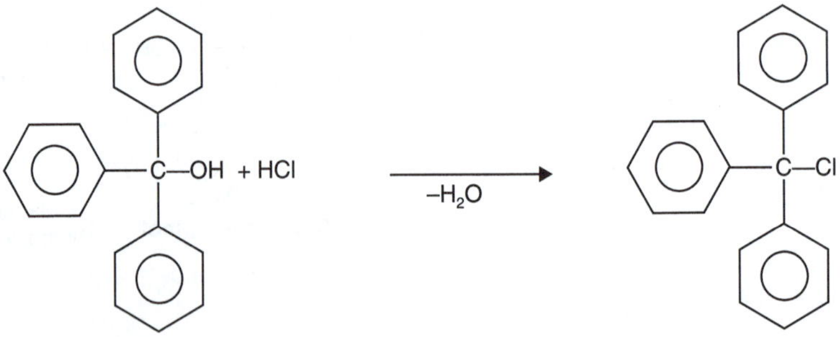

Triphenylcarbinol   Triphenylchloride

(ii) Acetyl Chloride (CH₃COCl)

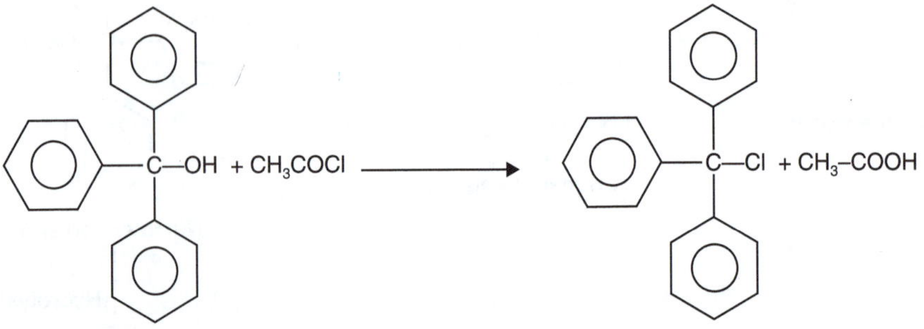

Triphenylcarbinol   Triphenylchloride

6. **Reduction with Formic Acid (HCOOH) or Lithium Aluminium Hydride (LiAlH$_4$)**–The critical reduction of carbinol with either *formic acid* or *lithium aluminium hydride* to yield **triphenylmethane** – as given under:

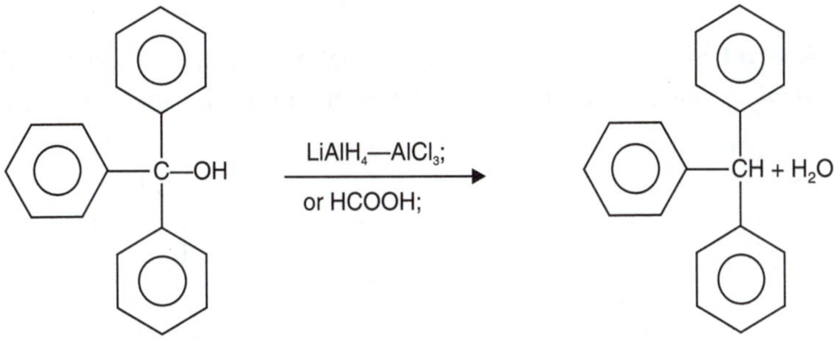

| Triphenylcarbinol | Triphenylmethane |

Thus, a mole of water gets eliminated.

7. **Condensation with Aniline Hydrochloride (C$_6$H$_5$–NH$_2$.HCl) gives *para*-Amino tetraphenylmethane**

The condensation of **triphenylcarbinol** with **aniline HCl*** yields *p*-aminophenylmethane as given below:

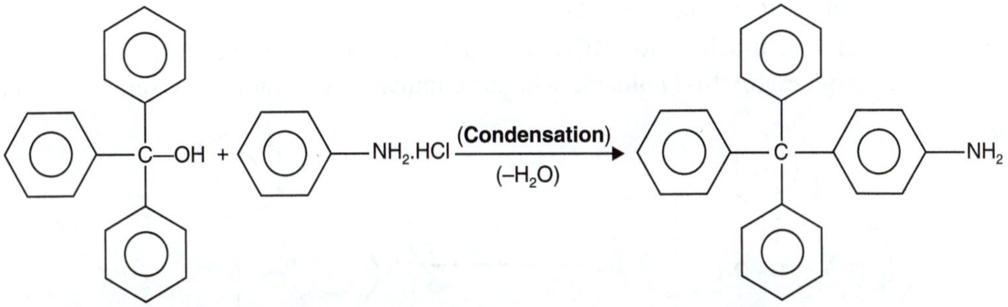

Triphenylcarbinol   Aniline hydrochloride   *p*-Aminophenylmethane

Thus, a mole of water gets eliminated from the reaction mixture.

2.7. Triphenylmethyl Chloride [or Trityl Chloride]

Chemical Structure

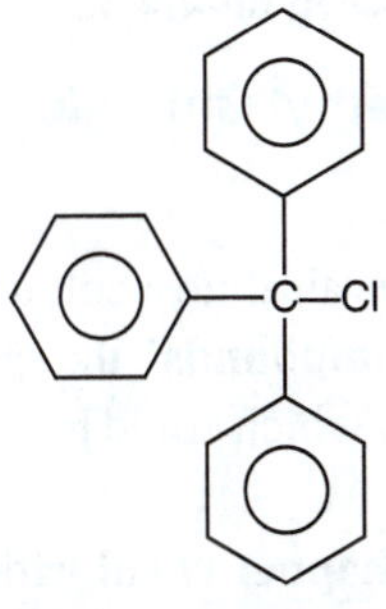

Triphenylmethyl chloride

---

* **Aniline** being insoluble in water, its hydrochloride salt is always used in such reactions, for its solubility in water.

Preparation of Triphenylmethyl Chloride–The **triphenylmethyl chloride** may be prepared from any one of the following *two* methods:

(a)   **Reaction of Triphenylcarbinol with Acetyl Chloride (CH_3COCl) or Hydrochloric Acid (HCl)**

(i)   **From Acetyl Chloride (CH_3–CO–Cl)**–Interaction of *triphenylcarbinol* with *acetyl chloride* yields *triphenylmethyl chloride* with the elimination of a mole of acetic acid,–as shown under:

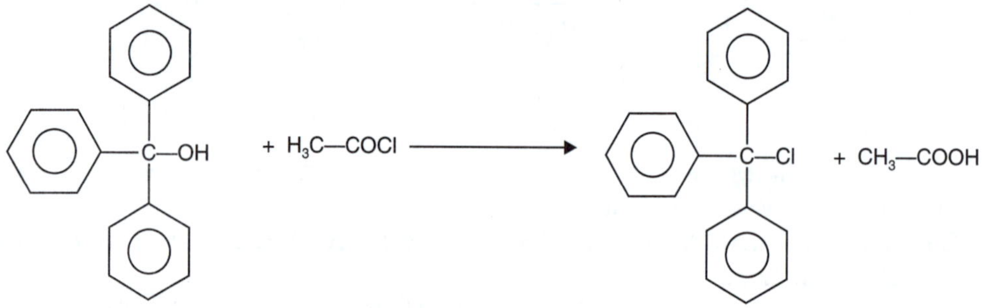

| Triphenylcarbinol | Acetyl chloride | Triphenylmethyl chloride | Acetic acid |

The yield of the product is 93.95%.

(ii)   **From Hydrochloric Acid (HCl)**–Reaction of *triphenylcarbinol* with *hydrochloric acid* gives **triphenylmethyl chloride** with the elimination of a mole of water,–as given under:

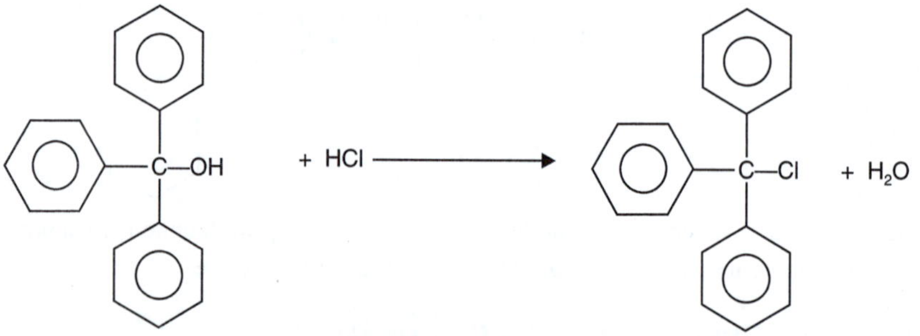

| Triphenylcarbinol | Hydrochloric Acid | Triphenylmethyl chloride |

The yield of the resulting product ranges between **66–84%.**

Characteristic Features of Triphenylmethyl Chloride

These essentially comprise:

1.   **Triphenylmethyl chloride** designates an almost colourless crystalline solid having **mp:112°C.**
2.   Very much akin to other '**triphenyl compounds**', the specific *chlorine atom* in **triphenylmethyl chloride** appears to be quite reactive, which could be exemplified by the following *two* **typical examples**.

(i)   **Treatment with water (H_2O)**–Triphenyl chloride when boiled with water it duly forms **triphenyl carbinol** *i.e.*, the correspondings *alcohol* – as stated under:

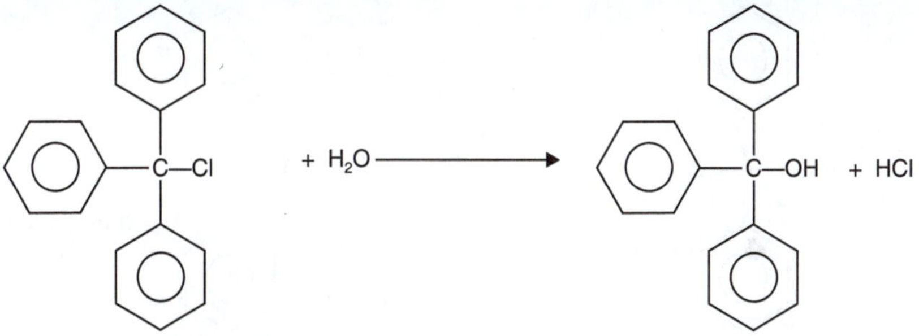

Triphenyl Chloride   Triphenylcarbinol

(ii) **Treatment with ethanol (C$_2$H$_5$–OH)–Triphenyl chloride** on being treated with ethanol gives rise to the formation of ethyltriphenyl–**methyl ether** with the elimination of a mole of HCl,–as given under:

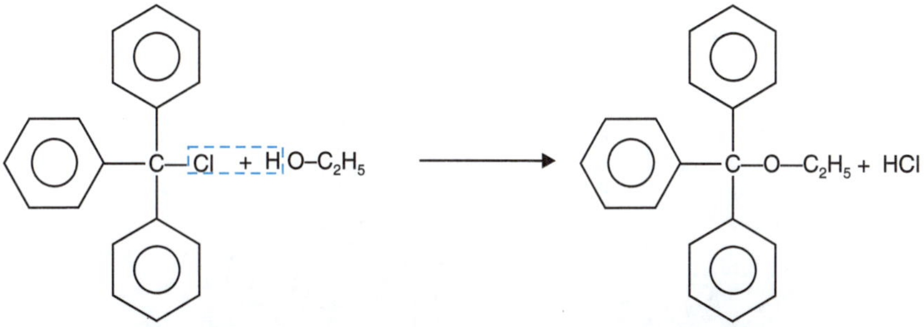

   Triphenyl chloride   Ethanol   Ethyltriphenylmethyl
                                                          ether

3. **Reaction with Magnesium and subsequently Hydrolysis**-When **triphenyl chloride** is first reacted with magnesium and then subjected to *hydrolysis*-it yields **triphenylmethane**,–as given below:

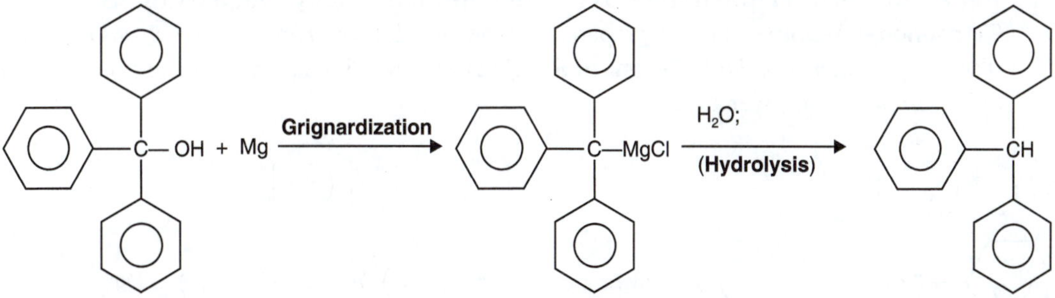

Triphenylcarbinol   Triphenylmethyl-   Triphenylmethane
                               magnesium chloride
                               [Grignard Reagent]

4. **Ionization due to dissolution in liquefied sulphur dioxide (SO$_2$)–Triphenylmethyl chloride** on dissolution in *liquefied sulphur dioxide* undergoes critical **'ionization'** to yield the respective **triphenylcarbonium ion** plus the **chloride ion,**–as shown under:

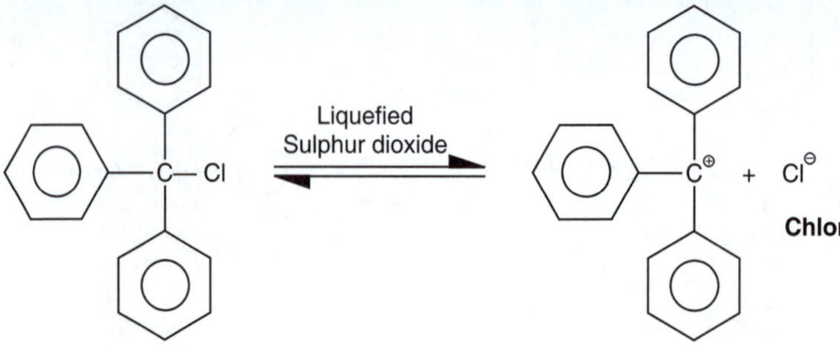

Triphenylmethyl chloride   **Triphenylcarbonium Ion**

2.8. Tetraphenylmethane

Chemical Structure

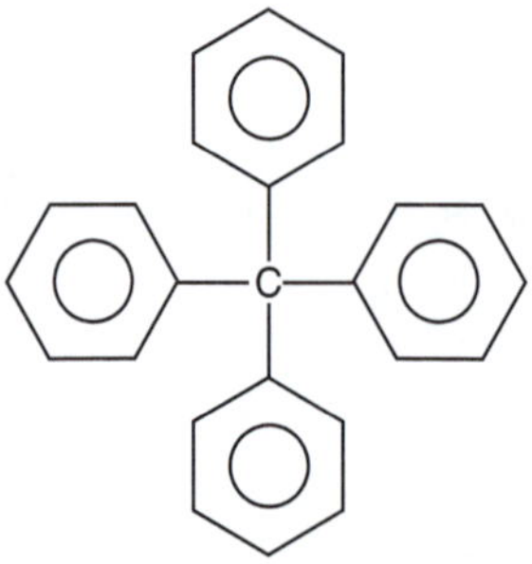

Tetraphenylmethane
[mp : 285°C; bp : 431°C]

Preparation of Tetraphenylmethane–Following are the *two* known methods for the preparation of **tetraphenylmethane**:

(a) **Reaction with Triphenylmethyl Chloride** and **Phenylmagnesium Bromide**– **Tetraphenylmethane** may be prepared by the reaction of *triphenylmethyl chloride* and *phenylmagnesium bromide* (a **Grignard Reagent**) with the elimination of a mole of *magnesium chloro bromide*,–as given below:

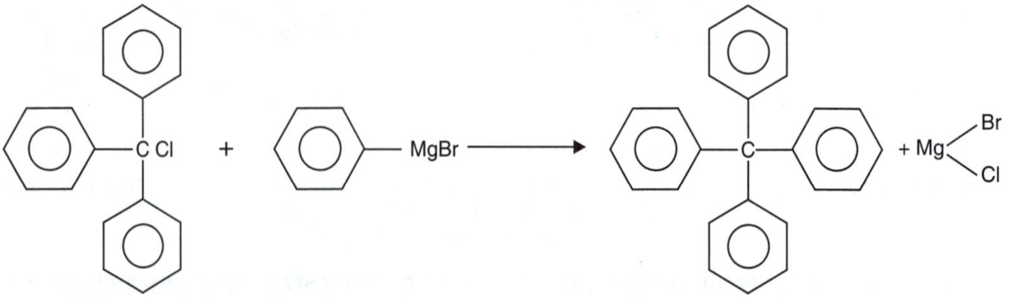

Triphenylmethyl   Phenylmagnesium   **Tetraphenylmethane**   **Magnesium**
chloride   bromide   **bromochloride**
                    (Grignard Reagent)

(b) **Diazotisation of *para*–Aminotetraphenylmethane followed by boiling with Ethanol–**
It essentially involves *two* sequential steps, namely:
- **Preparation of *para*-aminotetraphenylmethane by the condensation of triphenylcarbinol with aniline hydrochloride;** and
- **Diazotization of the resulting product with NaNO$_2$/H$_2$SO$_4$(at 0–5°C) followed by boiling with ethanol.**

(I)

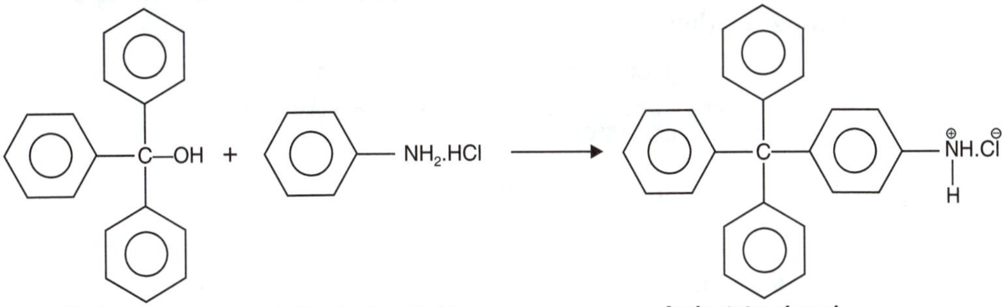

| Triphenylcarbinol | Aniline hydrochloride | *p*-Aminotetraphenyl-methane chloride |

(ii)

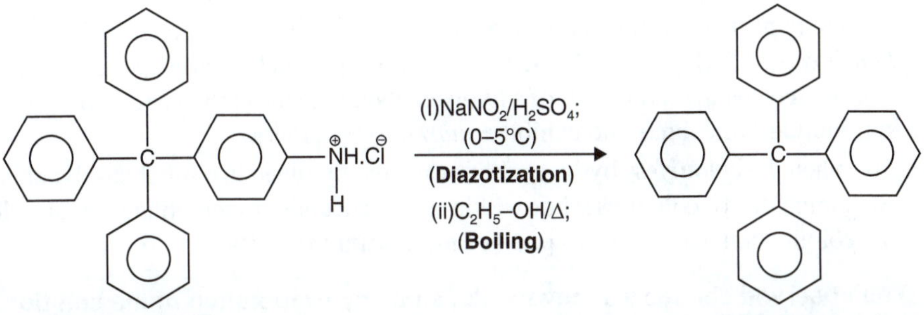

(I)NaNO$_2$/H$_2$SO$_4$;
(0–5°C)

(Diazotization)

(ii)C$_2$H$_5$–OH/Δ;
(Boiling)

p-Aminotetraphenyl-
methane chloride

Tetraphenylmethane*

2.9. Hexaphenylethane

Chemical Structure

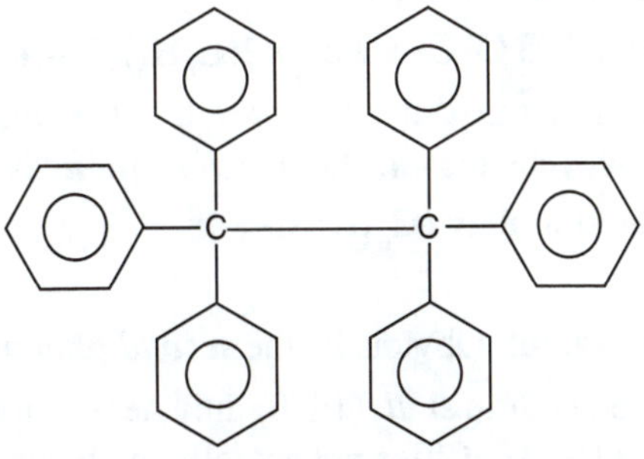

Hexaphenylethane

---

* It is a fairly stable crystalline solid substance.

Gomberg (1900)* was pioneer in the preparation of an authentic '**free radical**'. Evidence from the literature reveals that Gomberg had synthesized successfully **tetraphenylmethane** (mp: 285°C); and hence, was interested to observe that the **tetranitro derivative** of this *hydrocarbon*, that critically possess '**no H-atom**' on the *methane C-atom*, did not give any *colour reaction* with **ethanolic KOH** and thereby duly contrasted with **trinitrotriphenyl methane**.

Therefore, in an attempt to determine whether the ensuing difference was '*general in nature*', Gomberg sought to synthesize the so-called *completely* **phenylated ethane**, '**hexaphenylethane**' by the following **coupling reaction**:

$$2(C_6H_5)_3\,CCl\ +\ 2Ag\ \longrightarrow\ (C_6H_5)_3\,C\!\!-\!\!C(C_2H_5)_3 + 2AgCl\!\downarrow$$

Triphenylchloromethane   **Hexaphenylethane**
(**mp : 113°C**)

Characteristic Features of Hexaphenylethane

These essentially include:

1. **Hexaphenylethane** mostly occurred as a high-melting, sparingly soluble, white solid resembling *tetraphenylmethane* (its respective structural analogue) with respect to the *physical* properties as well as its *inert character*.

2. A solution of *triphenylchloromethane* in benzene with finely divided *Ag* or *Zn* in an atmosphere of *carbon dioxide* (CO_2) gave a **yellow-coloured solution**, which upon evaporation in the absence of air, –deposited *colourless crystals* of a **hydrocarbon** with the particular composition expected for **hexaphenylethane**, but it did exhibit *remarkable reactivity*.

Interestingly, when the **colourless hydrocarbon** was first dissolved in a suitable solvent, the resulting solution was momentarily colourless, but within a few seconds became tinged with **yellow** (and subsequently the colour soon deepened to a point of **maximum intensity**).

NOTE	The observed change was reversible, since by evaporation of the solution, the so-called *colourless hydrocarbon* could be recovered conveniently.

*Gomberg's Interpretation–*According to Gomberg the actual results are as stated under:

"The experimental evidence…forces one to the conclusion that one may have to deal at this point in time with a '*free radical triphenylmethyl* [$(C_6H_5)_3C–$]'. Thus, the ultimate action of Zn results in a first possible abstraction of the halogen, as shown under:

$$\textbf{2(C}_6\textbf{H}_5)_2\ \textbf{C–Cl + Zn} \rightarrow \textbf{2(C}_6\textbf{H}_5)_3\textbf{C – + ZnCl}_2$$

The above reaction could be further expatiated as a result of the critical removal of the *halogen atom* from **triphenylchloromethane**, whereby the *4th valence of methane* is found to be duly bound either:

- **to take up the complicated moiety ($C_6H_5)_3C–$; or**
- **to remain as such**,

i.e., having the **C-atom as trivalent**. Obviously, the *second phenomenon* holds good."

3. Lankamp *et al.* (1968) and Guhrie *et al.* (1970) duly demonstrated that the respective *dimer* of **triphenylmethyl radical [$(C_6H_5)C°$]** is not actually the **hexaphenylethane** but is:
   1-diphenylmethylene-4-tritylcyclohexa–2, 5-diene

---

* **Gomberg M(1866–1947): Ph.D, Michigan (Prescott); Michigan,** *J Am Chem. Soc.,* **69:** 1921 (1947).

Unexpected Stability Profile of Triphenylmethyl Radical Relative to Hexaphenylethane–It has been duly established that the unexpected stability of the **triphenylmethyl radical** relative to 'hexaphenylmethane' is solely attributed to the phenomenon of *resonance*; and hence, a plethora of such '**resonating structures**' do contribute vehemently to the so called '**resonance hybrid**', –as depicted under:

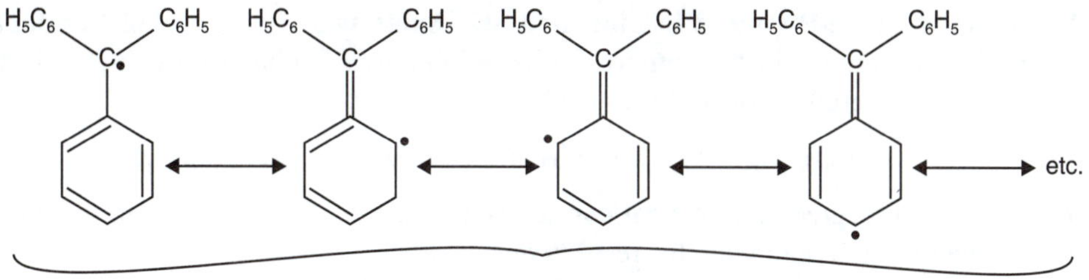

Different *resonating structures* of *triphenylmethyl radical*

Since the ensuing '*triphenylmethyl radical*' is proved to be an extremely reactive entity, it does react with a number of reagents to result into the formation of the corresponding *triphenylmethyl structural analogues*.

Examples: Following are a few classical examples:

(i) **Reaction with Oxygen (O_2) yields a colourless 'Peroxide'–**

$$(C_6H_5)_3 \, C_\bullet \xrightarrow{O_2;} (C_6H_5)_3 \, C—O—O_\bullet \xrightarrow{(C_6H_5)_6 \, C_2}$$

Peroxide radical

$$(C_6H_5)_3 \, C—O—O—C \, (C_6H_5)_3 + (C_6H_5)_3 \, C_\bullet$$

A Colourless Peroxide

(ii) **Reaction with Iodine (I_2) yields Triphenylmethyl Iodide–**

$$2(C_6H_5)_3 \, C^\bullet + I_2 \rightleftharpoons 2(C_6H_5)_3 \, I$$

Triphenylmethyl Iodide

(iii) **Interaction with Nitric Oxide (–NO) yields Nitrosotriphenyl Methane–**

$$(C_6H_5)_3 \, C^\bullet + NO \longrightarrow 2(C_6H_5)_3 \, C—N{=}O$$

Nitrosotriphenyl methane

NOTE The actual formation of '*Nitrosotriphenyl methane*' still remains to be ascertained.

(iv) **Reaction with Na-metal yields *Triphenylmethylsodium*–**

$$(C_6H_5)_3 \, C^\bullet + Na \longrightarrow (C_6H_5)_3 \, C^-_\bullet Na^+$$

Triphenylmethylsodium

The resulting product is a **brick-red solid** exhibiting an excellent '*electrical conductivity*'.

(v) **Reduction of Salts of Ag, Au, and Hg to the Respective Metals–**In fact, the *triphenylmethyl free radical* serves as a vital and powerful *reducing agent*–that helps to cause reduction of the salts of Ag, Au, and Hg into the respective **metals**, as shown under:

$$(C_6H_5)_3 \, C^\bullet + AgCl \longrightarrow (C_6H_5)_3 \, CCl + Ag$$

**Triphenylmethyl-
chloride**

$$2(C_6H_5)_3 \, C^\bullet + HgCl_2 \longrightarrow 2(C_6H_5)_3 \, CCl + Hg$$

**Triphenylmethyl-
chloride**

4.   **Hexaphenylethane Dissolved in Liquefied SO_2 Exhibits a High Level of Conductivity*–**
The critical dissolution of **hexaphenylmethane** in *liquefied sulphur dioxide (SO_2)* exhibits a high degree of overall conductivity, as depicted under:

$$(C_6H_5)_3{-}C^\bullet \, C{-}(C_6H_5)_3 \text{ or } [(C_6H_5)_6C_2] \rightleftharpoons (C_6H_5)_3 \, C^\oplus + (C_6H_5)_3C^\ominus$$

Remarks–The high degree of **electrical conductivity** is certainly caused due to the above cited dissociation of **hexaphenylethane** into the respective:

- **triphenylmethyl carbocations $[(C_6H_5)_3C^\oplus]$**; and
- **triphenylmethyl carbanions $[(C_6H_5)_3C^\oplus]$**.

Dauben *et al.* (1960) proved the very existence of the **triphenylmethyl carbonium ion** by actually carrying out the preparation of the following *two* **distinct chemical entities (compounds)**, namely:

➤   **$(C_6H_5)_3C^\oplus.ClO_4^-$: Triphenylmethyl perchlorate**–obtained as a *crystalline substance*; and

➤   **$(C_6H_5)_3C^\oplus.BF_4^-$: Triphenylmethyl fluoroborate**–obtained as a *crystalline product*.

Besides, the resulting *carbonium anions* do have a tendency to abstract the typical '**hydride ions**' from the respective hydrocarbons having critical *low* **nucleophilic power**.

3. THE CONDENSED RING SYSTEMS

The **condensed ring systems** relate to the reaction in which either *two or more* organic chemical entities (molecules) are duly connected to form a *layer molecule*, such as: **naphthalene, anthracene, anthraquinone**, and, **phenanthrene**. Sometimes, it is also referred to as '**polymerization**'.

Alternatively, these invariably include such compounds wherein *two or more* **benzene rings** are meticulously fused together in the *ortho*-**position** *i.e.*, the respective **benzene ring** has *two* C-atoms in common.

Following are the *four* cardinal members belonging to the category of **condensed ring systems**, namely:

- **Naphthalene**,   • **Anthracene**,   • **Anthraquinone**, and   • **Phenanthrene**,

which will now be discussed individually in the sections that follows:

3.1. Naphthalene [*Syn*: Naphthalin; Naphthene; Tar Camphor]

Preamble–Garden (1819) first and foremost isolated the organic chemical entity (compound) from the '**coal tar**' (where it occurs as the major constituent of '**coal tar**'). The '*dry coal tar*' contains nearly **11% of naphthalene**. It crystalizes from the '**middle**' or '**carbolic oil**' fraction of the duly distilled tar.

---

*   **Conductivity:** The quantitative expression of the flow of an *electrical current across a substance* is referred to as **conductivity**.

*Purification–***Naphthalene** is purified by *hot pressing*, which may be followed by washing with H_2SO_4, **NaOH**, and **water**; and finally by:

- **fractional distillation**, or
- **sublimation**

Isolation from Petroleum–Naphthalene is obtained from petroleum product containing naphthalene.

*Modus Operandi–***Naphthalene** mostly contained in the crude petroleum is duly converted into an *aromatic oil* usually found to be rich in:

- **Naphthalene** and
- **Dimethylnaphthalene**,

by allowing to pass *petroleum fractions* right through a **heated catalyst** (*viz.*, **Cu**) at 700°C under the atmospheric pressure. Importantly, the **dimethylnaphthalene** thus obtained is converted subsequently into **naphthalene** by means of the so-called *hydrodealkylation technique* [*i.e.*, by heating with **hydrogen** under pressure ranging between *150–1000 psi*]–at nearly 750°C in the presence of a *metal-oxide catalyst*.

CONSTITUTION OF NAPHTHALENE

The constitution of '**naphthalene**' may be adequately elucidated based upon the following elaborated facts, namely:

1. **Molecular Formula–**Based on the various scientific and analytical data, the *molecular formula* of **naphthalene** has been found to be $C_{10}H_8$.

2. **Presence of Benzene (Phenyl) Ring–**Since **naphthalene** critically exhibits most of the '**aromatic characteristic features of benzene**', such as:
   - **Halogenation**
   - **Nitration** and
   - **Sulphonation**,

that eventually reveals and ascertains the presence of at least **one benzene nucleus** in the **naphthalene nucleus** the *two* **adjacent positions**.

In the particular instance when **naphthalene** is being subjected to *oxidation*, it specifically gives rise to the formation of a known dicarboxy aromatic acid called '**phthalic acid**', Hence, it reveals predominantly that the '**naphthalene molecule**' does comprise a '**benzene nucleus**',–which is duly substituted at the *two* **adjacent positions.** Hence, based upon the aforesaid scientific evidences one may now draw the so-called '***partial structure* of naphthalene**' as given below:

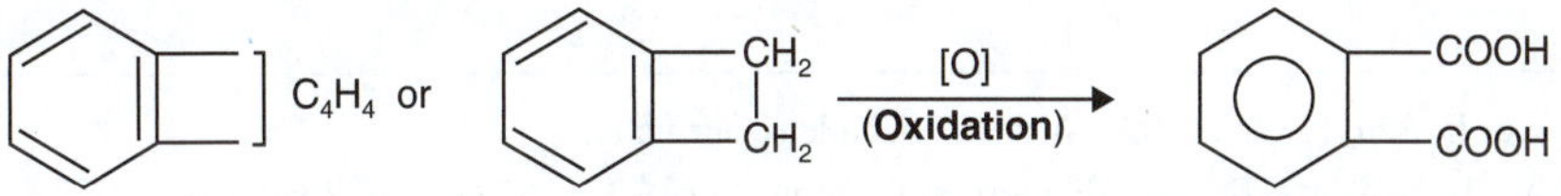

Phthalic Acid

3. **Specific Presence of Two-Benzene Rings Fused at the α-position**–Erlenmayer (1866)* duly proposed the following **structure (A)** in analogy to the *structure of benzene* as suggested by Kekule (1865)**:

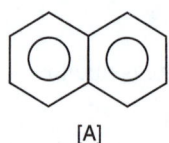

[A]

❑ Nevertheless, Graebe (1869)*** put forward a rather more direct, acceptable, and conclusive proof with regard to the presence of 2-benzene rings duly fused together at the respective α-position exclusively. The actual method followed by Graebe is stated explicitly as under:

❑ **Naphthalene**–upon *nitration* gives **nitronaphthalene**, which upon subsequent *oxidation* yields *O*-**nitrophthalic acid**.

Comment–The above reaction clearly reveals that the *inherent nitro* (–NO₂) *moiety* is located strategically in the **'benzene ring'**; besides, the **side-chains do precisely undergo oxidation**.

❑ **Nitronaphthalene**–upon *reduction* gives rise to the formation of **aminonaphthalene**, which on further *oxidation* yields **phthalic acid** (an *atomatic dicarboxylate*).

Comment–Thus, it reveals vehemently that the so-called **'benzene-ring'** present in *phthalic acid* is duly derived by the *oxidation of aminonaphthalene*. However, it is pertinent to state here that this is not the *same ring* as that contained originally in the **nitro moiety in nitronaphthalene;** *i.e.*, **naphthalene** does comprise essentially of *two* **benzene rings**.

Importantly, the foretold conclusive inference is solely based upon the fact that an *amino* (–NH₂) *functional moiety attached to the benzene-ring* predominantly renders the latter **extremely sensitive to oxidation**.

In fact, the above *two* **glaring facts** may be duly summarized in the following *scheme of reactions:*

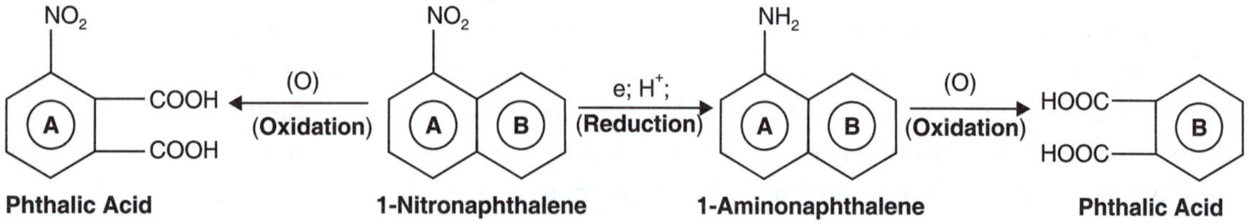

Naphthalene

HNO₃/H₂SO₄;
(Nitration)

NO₂	NO₂	NH₂

Phthalic Acid ← (O) (Oxidation) ← 1-Nitronaphthalene → e; H⁺; (Reduction) → 1-Aminonaphthalene → (O) (Oxidation) → Phthalic Acid

---

* Erlenmeyer E, *Sen*, (1825–1909). Wehen/Wisbaden, Germany.

** Kekule A: (1829–96); b.Darmstadt, Germany; Bonn; *Ber.*, **23**, 1265 (1890); **29**:197 (1896).

*** Graebe C; (1841–1927); b. Frankfurt, Germany.

Remarks: From the above **scheme of reactions** one may draw *two* **cardinal** conclusions:

- The *oxidation* of 1-nitronaphthalene reveals critically that the '**ring-A**' carrying the '**nitro** (–NO$_2$) **moiety**' almost remain intact; whereas, the '**ring-B**' gets eliminated completely; and

- The *oxidation* of 1-aminonaphthalene shows explicitly that the '**ring-A**' bearing the '**amino** (–NH$_2$) **moiety**' gets destroyed completely; whereas, the '**ring-B**' almost remains intact.

Therefore, the said *two* benzene rings '**A**' and '**B**' are duly fused at the *α-position* in the **naphthalene molecule**.

SYNTHESES OF NAPHTHALENE

Naphthalene may be synthesized by any one of the following *four* methods, namely:

(i) **Fittig Synthesis (1885):** Fitting R (1835–1910); b. Hamburg, Germany; Ph.D., Gottingen; Straibourg

(ii) **Haworth Synthesis (1932):** Haworth R D (b. 1898) Cheadle (Cheshire); Ph.D., and D.Sc, Manchester (U.K).

(iii) **From 4-Phenylbut-1-ene; and**

(iv) **From *O*-Xylene bromide**

which shall now be discussed individually in the sections that follows:

(a) Fittig Synthesis

First of all **4-phenylbut-3-enoic acid** is warmed carefully with sulphuric acid to obtain 1-naphthol (or α-naphthol), which is subsequently distilled with Zn-dust to produce **naphthalene:**

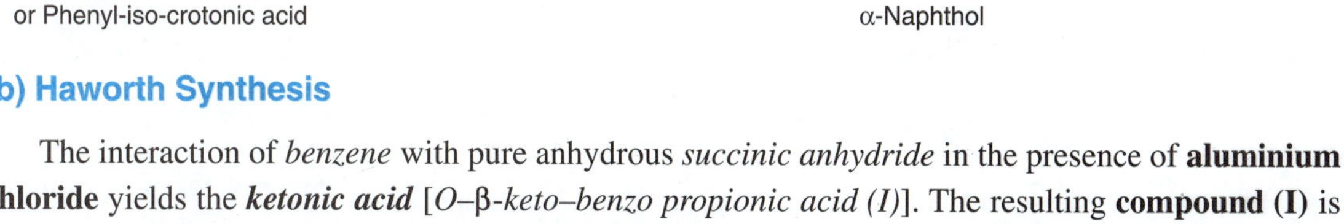

4-Phenylbut-3-enoic acid	An Intermediate	1-Naphthol	**Naphthalene**
or β-Benzylidene-propionic acid	(A *keto*-derivative)	or	
or Phenyl-iso-crotonic acid		α-Naphthol	

(b) Haworth Synthesis

The interaction of *benzene* with pure anhydrous *succinic anhydride* in the presence of **aluminium chloride** yields the *ketonic acid* [*O*–β-keto–benzo propionic acid (*I*)]. The resulting **compound (I)** is subjected to reduction by the **Clemmensen method** to produce the corresponding *O*-phenylbutanoic acid (II). The ring-closure phenomenon is duly affected by heating **compound (II)** with *concentrated sulphuric acid (H$_2$SO$_4$)* to give *O*-tetralone (III), which is further reduced by the **Clemmenson method** to yield **tetrahydronaphthalene (IV)**. Finally, the **compound (IV)** on being *dehydrogenated* by heating it with *palladized carbon (Pd-C)* or *selenium powder* gives the desired compound **naphthalene**.

All the aforesaid **sequential reactions** could be summarized as stated under:

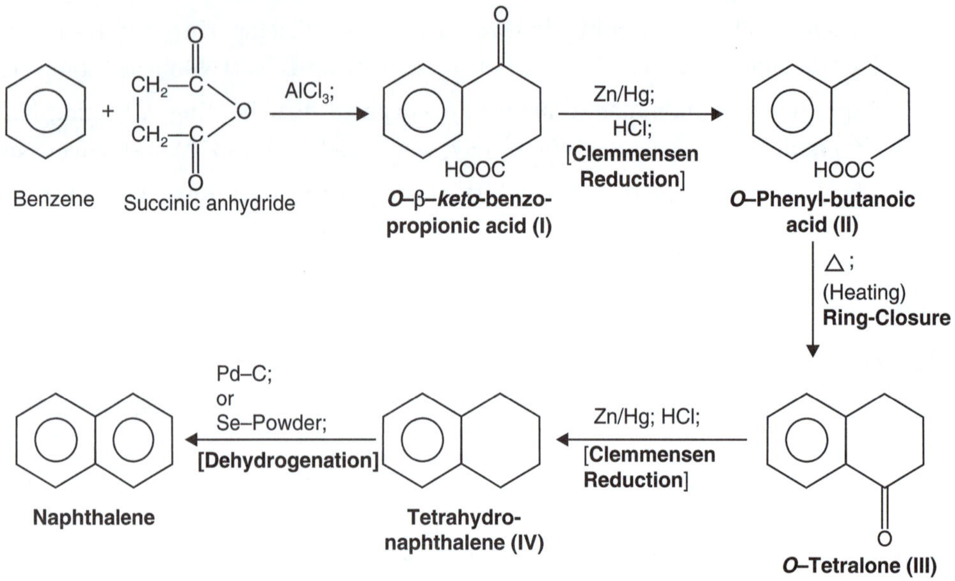

Amazingly, the ring closure of **4-phenylbutanoic acid (II) to form α-tetralone** may also be accomplished by the help of **Friedel-Craft's reaction** upon the *acid-chloride analogue* [using aluminium chloride (AlCl$_3$)] as shown below:

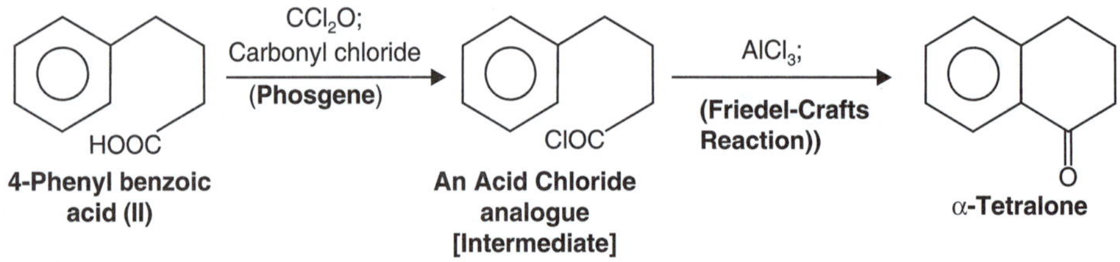

(c) From 4-Phenylbut-1-ene

Naphthalene can also be prepared by passing **4-phenylbut-1-ene** over red-hot calcium oxide (CaO)– as given under:

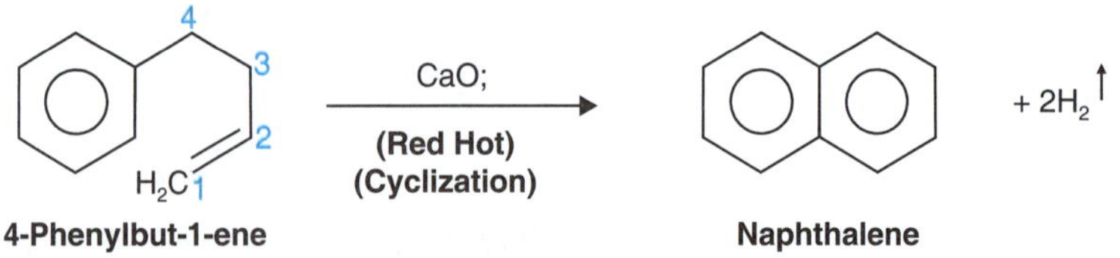

The above reaction occurs due to the phenomenon of *cyclization* whereby *two* moles of hydrogen are eliminated.

(d) From O-Xylene bromide

The **O-Xylene bromide** is treated with *disodio-ethanetetracarboxylic ester* to give the respective **tetrahydronaphthalene tetracarboxylic ester**, which on refluxing with an **acid (H⁺)** yields the corresponding *dicarboxylic acid*. Finally, the **silver-salt of the resulting product** gives **naphthalene** upon heating, as stated below:

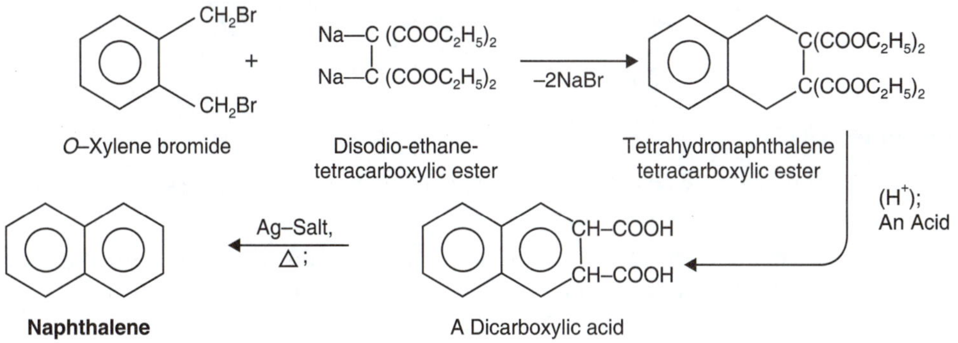

O–Xylene bromide	Disodio-ethane-tetracarboxylic ester	Tetrahydronaphthalene tetracarboxylic ester

Naphthalene	A Dicarboxylic acid

Exact Position of the Double Bonds (or Olefinic Bonds) in Naphthalene

Based solely upon the *physiochemical* evidences, such as: **heat of combustion** etc., it revealed beyond any reasonable doubt that **naphthalene** designates critically a *resonance hybrid* of the following *three* **major resonating structures A, B, and C:**

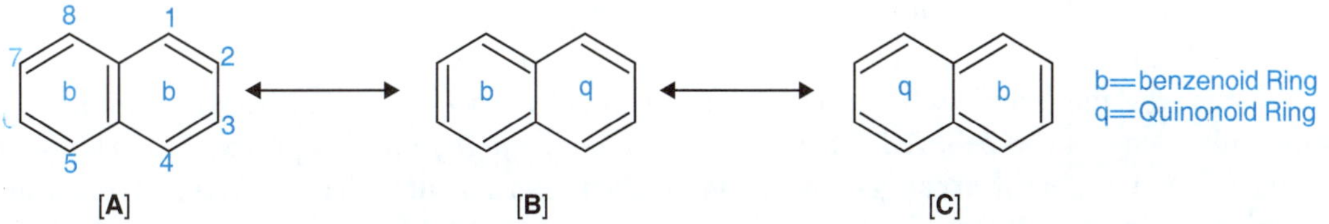

b=benzenoid Ring
q=Quinonoid Ring

[A]   [B]   [C]

NOTE The prevailing resonance energy of '*naphthalene*' has been found to be *255.2 kj. mol⁻¹*.

> **NOTE** The prevailing resonance energy of '*naphthalene*' has been found to be *255.2 kj. mol^{-1}*.

(i). Explanation for the Phenomenon of Resonance in Naphthalene

Interestingly, one may broadly explain the phenomenon resonance in **naphthalene** as expatiated below:

1. In the aforesaid structures one may observe that the bond between C-1 and C-2 is an *olefinic bond* (or *double bond*) in the structures **A** and **B** and a *single bond* in strucutre C. Therefore, it may be assumed appropriately that:

   - an *equal contribution* of each of the above 3-resonating structures (*viz.*, A, B, and C); and
   - it overall represents a *2/3rd double bond character and 1/3rd single bond character* [see structure (D)].

Comment: Obviously, the **C-1 and C-2 bond** would be presumed to be *shorter* in length *vis-a-vis* the **C-2 and C-3 bond**, which has been duly verified by **X-Ray diffraction*** studies of the *crystalline* **naphthalene**. Besides, the determined bond length between **C-1 and C-2** is **1.361Å**; whereas, the bond length between **C-2 and C-3** is **1.421Å**, as shown in **structure (E):**

---

* Kar A: **Pharmaceutical Drug Analysis**, 3rd edition, New Age International, New Delhi, 2013.

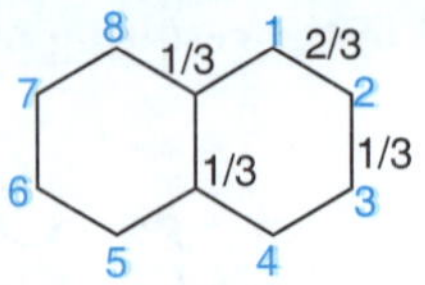

[D] 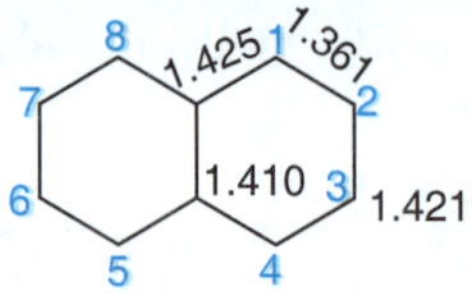   [E]

(ii). Fries Rule (1935)*

Fries critically compared the possible rearrangement of double bonds in the so-called '**polynuclear chemical entities**' having *benzoquinone nucleus*.

Therefore, the respective rearrangement in the aforesaid structures (**B**) and (**C**) virtually corresponds to the already established structures pertaining to **α-benzoquinone** [see structure (F)]:

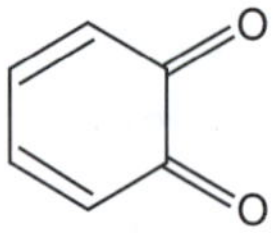

[F]

Remarks: The resonance energy of '**naphthalene**' stands at **255.2 kJ. mole^{-1}** and is definitely larger than *benzene* (**163.2 kJ. mole^{-1}**) *i.e.*, more the number of resonating structures more will be the resonance energy.

Nevertheless, the **quinones** are proved to be much more reactive in comparison to an '**aromatic compound**'. Hence, Fries believed vehemently that a *fairly stable form* of a **polynuclear compound** fails to consist of a **specific arrangement** resembling very much akin to the one present in the respective '**quinone structure**'. Thus, based solely upon this overall conclusion, **Fries** formulated a '*rule*' – usually termed as the **Fries Rule**.

According to the **Fries Rule:**

"**The most stable rearrangement of a polynuclear compound designates actually that '*form*' which essentially possesses the maximum number of '*rings in the so-called benzenoid profiles*' i.e., three olefinic (double) bonds in each individually ring**".

Comment: It is, however, pertinent to state here that out of the *three* aforesaid *canonical forms of the resonance* [*viz.*, (**A**), (**B**), and (**C**)] *hybrid* of '**naphthalene**', the particular **form (A)** having *two* **typical inherent benzenoid rings** is predominantly found to be **more stable** *vis-a-vis* the **forms (B) and (C)** with one **benzenoid ring**.

(iii). Explanation Based on *Atomic Orbitals*

Now, based upon the particular explanation pertaining to the consideration of the '**atomic orbitals**'—one may identify the crucial strategical presence of *ten* **C-atoms** located in the available corners of the *two* **fused haxagonal rings (in 'naphthalene')**. Importantly, *each C-atom* is duly linked to the other **3 C-atoms** by means of the **σ-bonds** that are eventually formed from the critical **overlapping of triogonal sp^2 orbitals**. Besides, it may be observed explicitly that all **C-and H-atoms** are lying exclusively in a '*single plane*'.

---

* Fries K: b.1875, Kisdrich/Rhine, Germany (Ph.D)

Interestingly, there exists a *cloud of π-electrons* duly formed by the overlapping of the *p-orbitals* and almost shaped like the *numerical figure* '8'–both above and below the plane of the C-atoms. Furthermore, the said **cloud of π-electrons** is justifiably assumed to be made of *two* **partially over-lapping sextets π-clouds** embracing legltimately the *six-electrons*–that prominently do have a *pair of π-electrons in common*, as illustrated under:

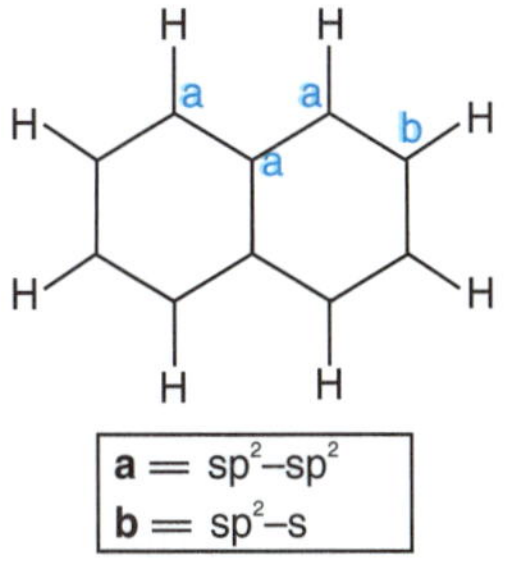

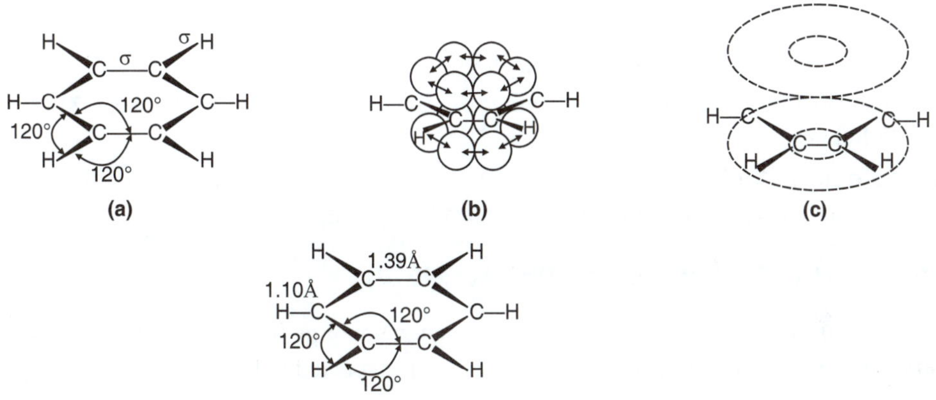

a = sp²–sp²
b = sp²–s

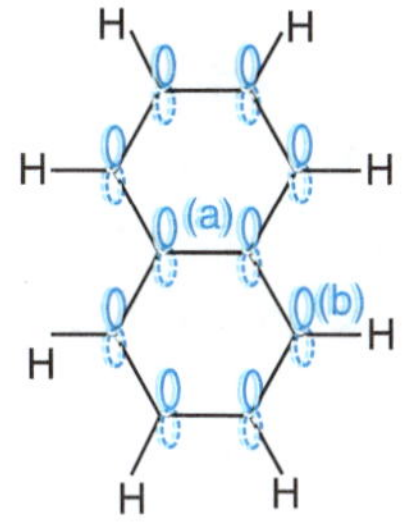

(a)

(b)

(c)

(d)

BENZENE MOLECULE

(a) Only σ-bonds are shown; (b) Only p-orbitals overlap to form σ-bonds, (c) π-Clouds above and below the plane of the ring; (d) Shape and size

(iv). Correctness of Naphthalene (A) Formula

Amazingly, one may even prove the correctness of **naphthalene (A)** formula by the aid of **X-ray diffraction analysis** (Robertson, 1951), that showed vividly the following specific aspects in **naphthalene:**

❑ the 1, 2-(α: β)-bond distance = 1.36Å; and

❑ the 2, 3-(β: β)-bond distance = 1.42Å.

Thus, the difference from *benzene* where all the **C–C bond** lengths are **1.48Å** (*i.e.,* a value that ranges between:

• **C-C single bond = 1.54Å,** and

• **C=C double bond = 1.33Å (see figure 'E').**

Remark: The formula of **naphthalene (A)** mostly fulfils and satisfies the critical/specifc requirements of an **organic chemist** that there prevails a distinct difference between the **α:β–and the β:β–bonds** because the former retains the typical **larger double-bond character**.

PHYSICAL PROPERTIES OF NAPHTHALENE

Following are the important physical properties of naphthalene, namely:

(i) **Naphthalene** occurs mostly as monoclinic prismatic plates from ether or by sublimation. It may also be sold as: white scales, powder, balls, or cakes.

(ii) It has mp = 80.2°C.

(iii) It has the characteristic odour of **'moth balls'**.

(iv) It even gets volatalized at room temperature.

(v) **Naphthalene has d^{20}_4 1.162 and d^{100}_4 0.9628.**

(vi) It sublimes significantly at temperatures above mp.

(vii) It is volatile with steam.

(viii) It has bp_{760} = 217.9°C; bp_{100} = 145.5°C.

(ix) It exhibits flash points: **Open-Cup Method** = 79°C; and **Closed-Cup Method** = 88°C.

(x) Its autoignition stands at 567°C.

(xi) It exhibits puple fluorescence in Hg-light.

(xii) It shows UV-absorption bonds having several characteristic bonds between **217.5 and 320 nm in hexane**.

(xiii) **Solubility Profile:** It is insoluble in water; but 1 g dissolves in 13 mL ethanol methanol; in 3.5 mL benzene of toluene; in 8mL of olive oil or turpentene oil.

CHEMICAL PROPERTIES OF NAPHTALENE

The **'naphthalene'** molecule comprises *two* **benzene rings** fused together at the **α-position spe-cifically**. Hence, the actual overall resonance energy is expected to be:

$$2 \times 150.6 = 301.2 \text{ kJ. mole}^{-1}$$

However, the *observed resonance energy* of **naphthalene** is found to be **255.2 kJ. mole^{-1}** (*i.e.,* nearly 15% less than the aforesaid value. It obviously implies that **naphthalene** is *less aromatic* in character *vis-a-vis* **benzene molecule** (*i.e.,* each ring in **naphthalene molecule** essentially bears an inherent *resonance energy* of **127.6 kJ mole^{-1}**). As a result, **naphthalene** will have marked and pro-nounced **lower aromatic characteristic features** than *benzene*; and, therefore, shall be definitely more reactive in nature.

The following various proven and established scientific evidences based upon the *reactions* of **naphthalene** will be able to substantiate its chemical properties squarely:

Addition Reactions of Naphthalene : These essentially include:

□ Addition of H-atom (Reduction)

Naphthalene on being subjected to *reduction* yield a good number of **products of napthalene;** however, the precise and exact nature of the **resulting product(s)** solely depends upon the critical nature of the *reducing agent* being employed, which could be seen from the under-mentioned array of typical example, namely:

(a) **Reduction of Naphthalene with Na-metal/Ethanol (absolute)–Naphthalene** upon reduction with freshly cut pieces of *Na-metal and absolute ethanol* yields, **1, 4-dihydronaphthalene** (or 1,4–dialin)–as given below:

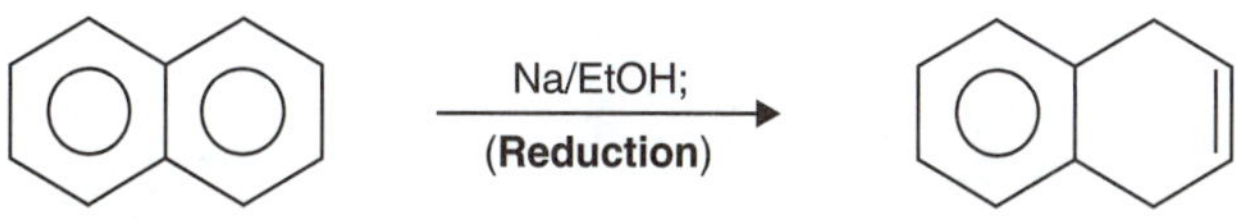

Naphthalene   1,4-Dialin or
[1,4-Dihydronaphthalene]

The resulting product **1, 4-dialin** is found to be quite unstable and gets isomerized readily to the respective **1, 2-dialin** when subjected to heat with *ethanolic sodium ethoxide*.

(b) **Reduction of Naphthalene with Na-metal/Isopentanol–Naphthalene** on being subjected to reduction with freshly cut pieces of *Na-metal and isopentanol (anhydrous)* gives rise to the formation of **1, 2, 3, 4-tetrahydronaphthalene** (or **tetraline**)–as shown under:

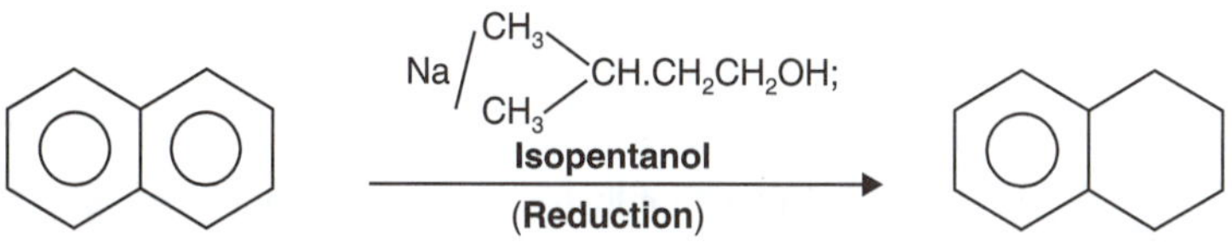

Naphthalene   Tetralin or
[1,2,3,4-Tetrahydronaphthalene]

(c) **Catalytic Reduction of Naphthalene (under pressure)**–When **naphthalene** undergoes *catalytic reduction* under pressure to yield **tetralin** at the *first-stage*, which on further *reduction* gives **decalin** (or **decahydronaphthalene**) at the *second-stage* –as given under:

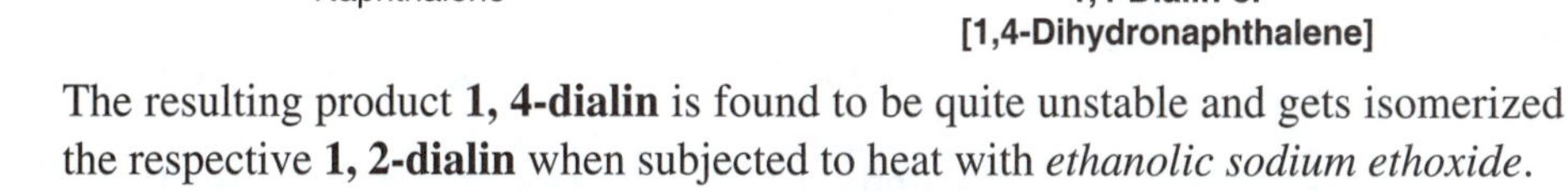

Naphthalene   Tetralin or   Decalin or
[1,2,3,4-Tetrahydronaphthalene]   [Decahydronaphthalene]

Comment–Thus one may eventually obtain *two* **geometrical forms of decalin** *viz.*,

- *cis*–**Decalin (bp = 193°C)**; and
- *trans*–Decalin (bp = 185°C).

Importantly, with **Ni (as a catalyst)**, we would lay bounds on the ***trans*-decalin** exclusively; and when Pt (as a catalysed) we would obtain *cis*-decalin.

| NOTE | **Nevertheless, the commercial product happens to be an admixture of both *cis*- and *trans*-decalin, and finds its enormous usage as a favoured solvent for *varnishes* and *laquers* etc.** |

❑ Addition of Chlorine

Naphthalene interacts with *dry chlorine* (Cl$_2$) *to yield the following two* **specific addition products**.

- **naphthalene dichloride [C$_{10}$H$_8$Cl$_2$]**; and
- **naphthalene tetrachloride [C$_{10}$H Cl$_4$]**.

Importantly, in both these chemical entities (compounds) the *chlorine atoms* are duly present in the **'same ring'**: that may be confirmed by the underlying fact that upon *oxidation* both the aforesaid compounds give **phthalic acid**. The various reactions may be expressed as under:

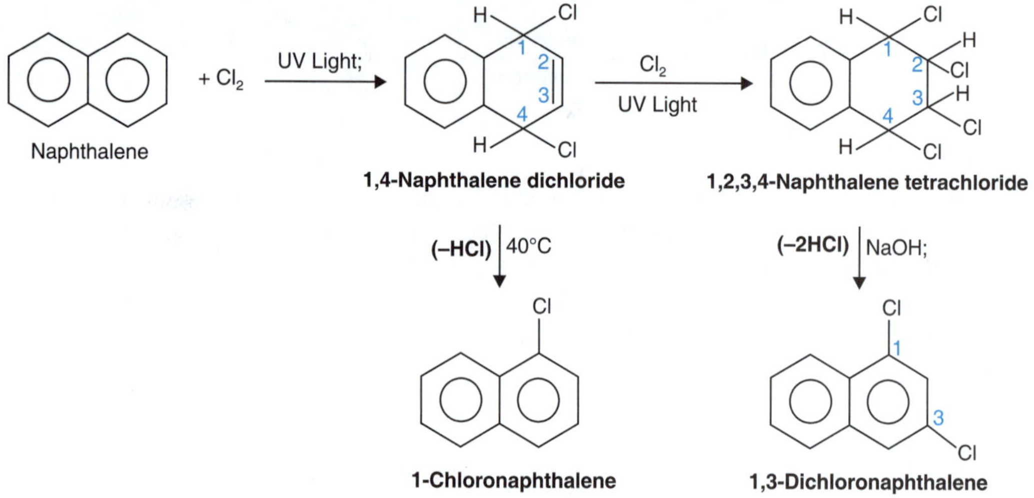

Naphthalene   + Cl₂   UV Light;

1,4-Naphthalene dichloride   **1,2,3,4-Naphthalene tetrachloride**

(–HCl) | 40°C   (–2HCl) | NaOH;

1-Chloronaphthalene   **1,3-Dichloronaphthalene**

Thus, when 1, 4-naphthalene dichloride is heated at 40°C, it loses a mole of *hydrogen chloride* (HCl); and heated with alkali (NaOH) loses *two* moles of hydrogen chloride (HCl) to yield an *admixture of dichloronaphthalenes* and the respective **1, 3–isomer predominating**.

❑ Addition of Ozone (O₃)

Ozone (**O₃**) aptly adds on to the **napththalene** molecule to yield the corresponding **naphtalene diazomide** which on subsequent hydrolysis gives rise to the formation of **phthalaldehyde**, as given below:

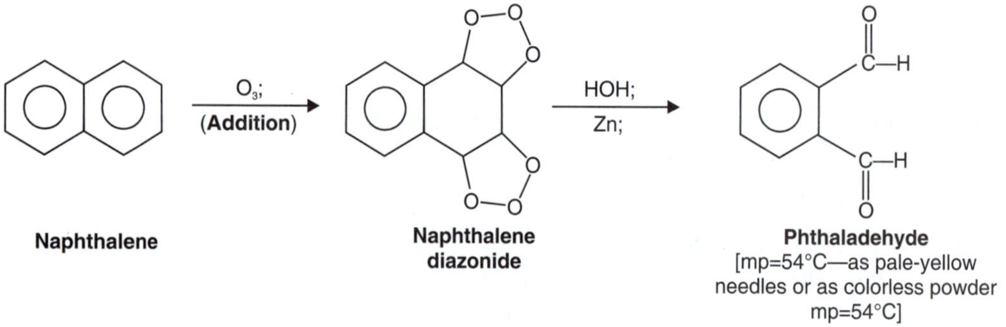

Naphthalene   **Naphthalene diazonide**   **Phthaladehyde**
[mp=54°C—as pale-yellow needles or as colorless powder mp=54°C]

❑ Addition of Sodium

Naphthalene in *dioxan* when treated with freshly-cut pieces of *sodium-metal*, it produces a **green coloured solid substance,** as shown under:

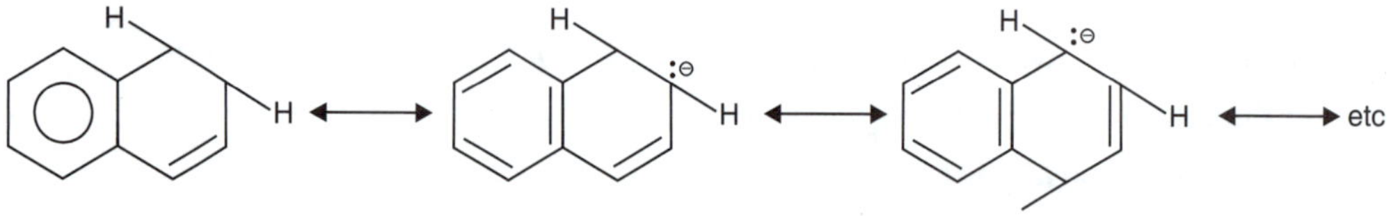

Explanation–It could be explained as a '**free-radical mechanism**' that has been produced due to the critical *transfer of an electron pair* (**:**) from the Na-metal to the corresponding **lowest vacant molecular orbital in the naphthalene molecule** (extreme structure on right-hand side), which gets duely stabilized by **resonance** as shown above.

Substitution Reactions of Naphthalene

Since **naphthalene** essentially comprises *two* **benzene rings** (*in a fused form*) its actual orientiation seems to be more complicated *vis-a-vis benzene*. It has been duly proven that the very *first in-coming moiety is invariably entering the 1-position* in the **naphthalene** molecule, but having the following *two* **glaring exceptions**, namely:

❑ *Sulphuration* **carried out at high temperature**; and

❑ **Friedel-Craft's reaction yields the *two* derivatives as the major products.**

Following are a few '**substitution reactions**' of **naphthalene**:

(a) **Nitration–Naphthalene** on being treated with HNO_3/H_2SO_4 reacts at an ambient temperature to produce the respective **1-nitronaphthalene**, as shown under:

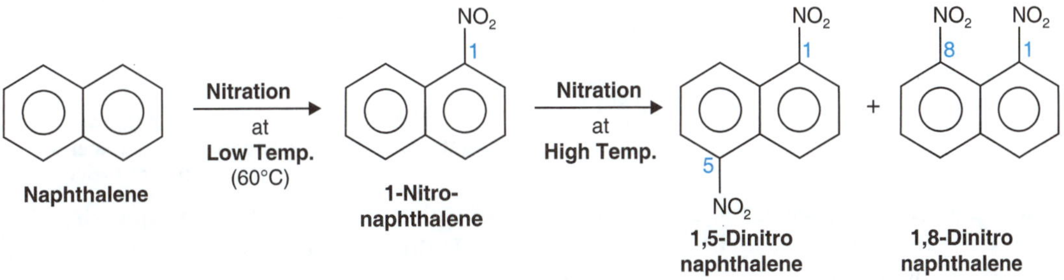

Naphthalene — Nitration at Low Temp. (60°C) → 1-Nitro-naphthalene — Nitration at High Temp. → 1,5-Dinitro naphthalene + 1,8-Dinitro naphthalene

❑ **Important Observations**

- **1-Nitronaphthalene**–a yellow solid (mp = 60°C)
- Nitration of **naphthalene** at *high temperature* (*>150°C*) one may obtain an admixture of **1, 5-dinitro**–and **1, 8-dinitronaphthalenes**.

(b) **Sulphonation–Naphthalene** on being treated with concentrated sulphuric acid (H_2SO_4) at 40°C gives the *1-derivative* as the **major product**; whereas, at **160°C** it yield the respective *2-derivative* as the **major product**–as stated under:

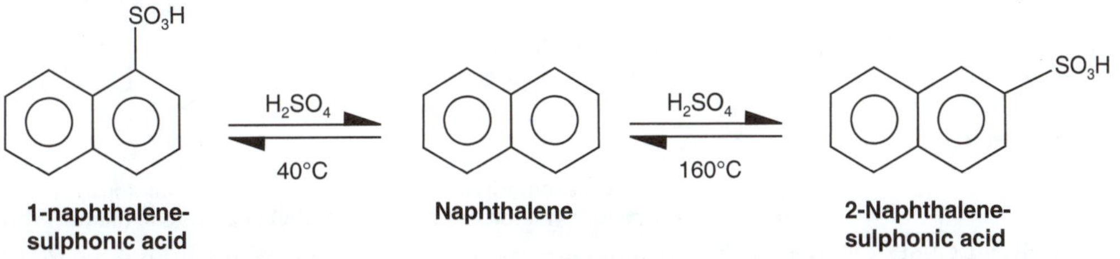

1-naphthalene-sulphonic acid ⇌ (H_2SO_4, 40°C) Naphthalene ⇌ (H_2SO_4, 160°C) 2-Naphthalene-sulphonic acid

Explanation: 1-Naphthalene sulphonic acid when heated with concentrated sulphuric acid *first* gets converted to **naphthalene** (**at 40°C**), but when heated further to **160°C** yields the corresponding **2-naphthalene sulphonic acid**.

4

The above sequence of reaction reveals that the *former* is the **kinetically controlled product**; whereas, the **latter** being the **thermodynamically controlled product**. Obviously, the particular reason for the *greater stability profile* of the respective '**2-acid**' is not yet fully understood. Neverthless, one could suggest a probable reason that there exists a **greater repulsion** in the said **1-acid** by virtue of the presence of **H-atom** located strategically at the **C-8 position**.

(c) **Halogenation–** It has been observed that **bromine (Br_2)** reacts with **naphthalene** even without the presence of an usual **Lewis acid catalyst**.

Examples: **Naphthalene** when *brominated with boiling carbon tetrachloride (CCl_4) solution*, it gives **1-bromonaphthalene**. Nevertheless, on being *further bromination* it critically gives rise to the formation of *two* products, namely:

- **1, 4–dibromonaphthalene (as *major product*)**; and
- **1, 2–dibromonaphthalene (as *minor product*)**

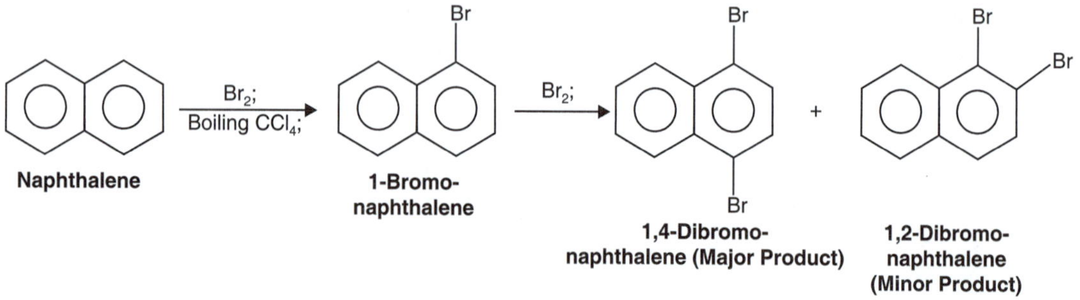

Naphthalene → 1-Bromo-naphthalene → 1,4-Dibromo-naphthalene (Major Product) + 1,2-Dibromo-naphthalene (Minor Product)

Remark: **Naphthalene** when subjected to *chlorination* with sulphuryl chloride (**1 equivalent**) in the presence of **aluminium chloride ($AlCl_3$)**–it yields **1-chloronaphthalene**. Nevertheless, with **2 equivalents** of *sulphuryl chloride* (at **100–140°C**) it gives **1, 4-dichloronapthalene**.

(d) **Chloromethylation–**The *chloromethylation* of **naphthalene** is usually carried out using an admixture of *paraformaldehyde, HCl, glacial acetic acid,* and *phosphoric acid* to obtain the following *two* products:

- **1-chloromethyl naphthalene (*major product*)**; and
- **1, 5–dichloromethyl naphthalene (*minor product*)**

Thus, we may have the following expression:

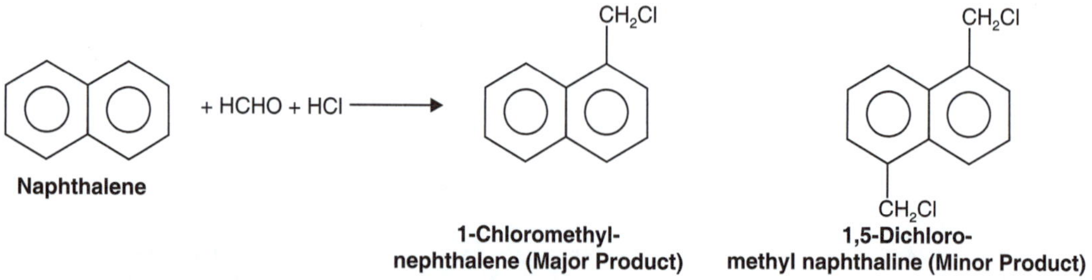

Naphthalene + HCHO + HCl → 1-Chloromethyl-nephthalene (Major Product) + 1,5-Dichloro-methyl naphthaline (Minor Product)

(e) **Friedel-Craft's Reaction–**Naphthalene on being subjected to treatment with *methyl iodide (CH_3I)* gives an admixture of **1-and 2-methylnaphthalene**. However, with **ethyl bromide (CH_5Br)**–it only yields **2-ethylnaphthalene**. But with *aluminium chloride ($AlCl_3$)* and *alcohols–* it largely produces **2, 6–dialkylnaphthalenes**. Interestingly, with *boron trifluorides (BF_3)* and *alcohols–*it gives **1, 4–dialkyl naphthalene** as the *major product* exclusively.

Besides, the '*acylation*' of **naphthalene** is carried out with *acetyl chloride* (CH_3COCl) in the critical presence of *aluminium chloride* ($AlCl_3$)–it gives an admixture of **1- and 2-ketones**. However, the exact percentage yield solely depends upon the nature of the **solvent** being employed.

Exaqmple: **Naphthalene** on being subjected to **Friedel–Craft's reaction** either in the presence of:

- **acetylchloride (CH_3COCl) in carbon disulphide (CS_2)**, or
- **aluminium chloride ($AlCl_3$),**

to obtain an admixture of **1- and 2-methyl naphthalene ketones**–as given under:

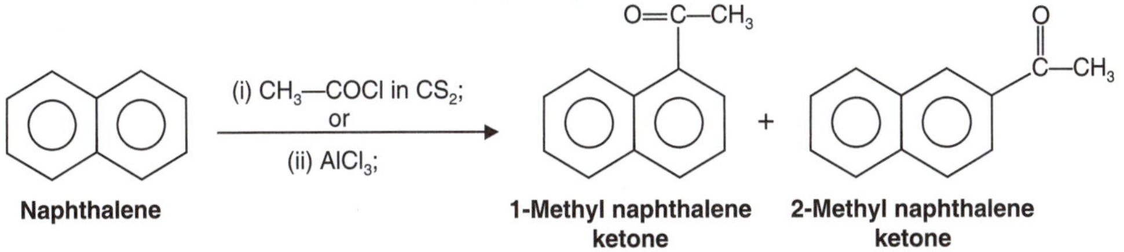

| | **1-Methyl naphthalene ketone** | **2-Methyl naphthalene ketone** |

Remark–Nevertheless, the said *two* **products** *i.e.,***1- and 2-methylnaphthalene ketones** are duly formed in a *ratio* of **3:1 in CS$_2$ and 1 : 9 in C$_6$H$_5$–NO$_2$ (nitrobenzene).**

Oxidation Reactions of Naphthalene

Naphthalene upon careful *oxidation* in the presence of an admixture of **mercuric sulphate (Hg_2SO_4)** or by air in the presence of **vanadium pentoxide (V_2O_5)** to yield **phthalic anhydride** as the end product, as given below:

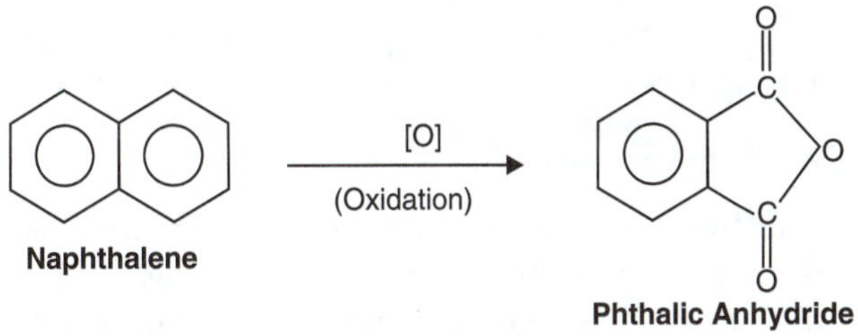

However, **naphthalene** specifically gives rise to the formation of a variety of *end-products*–which is solely based on the various types of **oxidising agents being utilized**, such as:

(i) **With KMnO$_4$ (potassium permanganate) in acidic solution(H_2SO_4)**–It yields **phthalic acid** as the **end-product:**

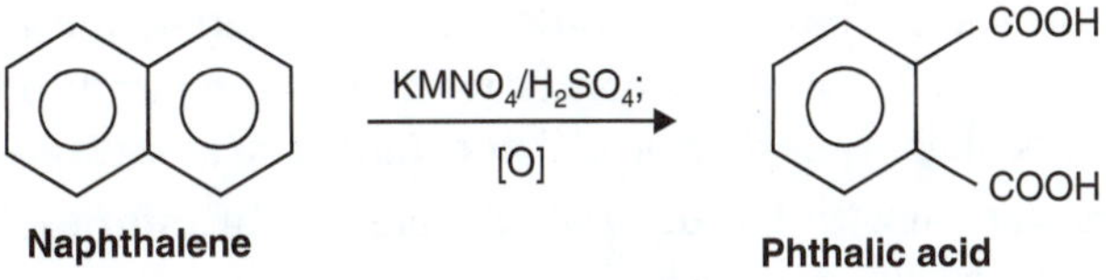

(ii) **With alkaline potassium permanganate ($KMnO_4$)**–It gives rise to the formation of **phthalonic acid**–as shown under:

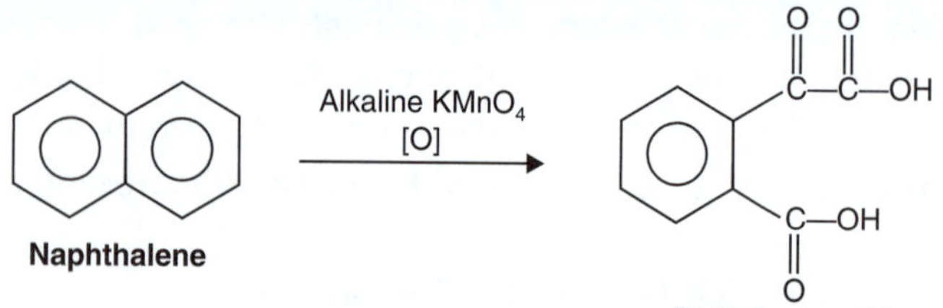

Naphthalene → Alkaline KMnO₄ [O] → **Phthalonic acid [a triketo acid]**

(iii) **With Chromic Acid (H₂CrO₄)**–In this particular instance, naphthalene gets duly oxidized to produce **1, 4-dinaphthaquinone**–as given below:

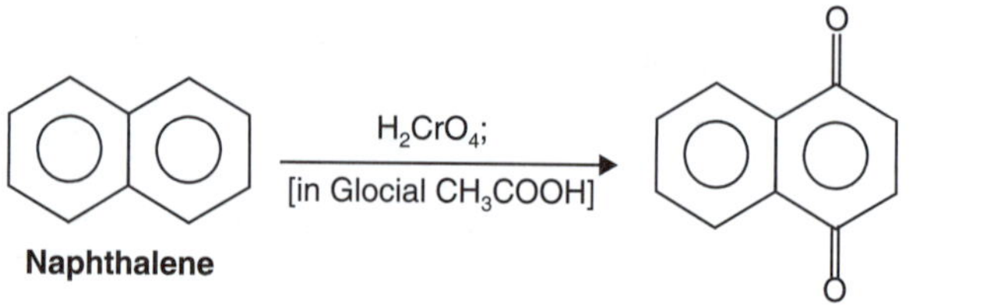

Naphthalene → H_2CrO_4; [in Glocial CH_3COOH] → **1,4-Dinaphthaquinone**

Uses of Naphthalene: These essentially include:

1. **Naphthalene** finds its enormous use as a **moth repellent** and insecticide.
2. It has also been used as a *dusting powder*.
3. The reduced derivatives of **naphthalene** *viz.*, **Decalin** and **Tetralin** are mostly used as **motor fuels**, **solvents**, and **lubricants**.
4. **Naphthalene** is being employed internally as an *intestinal antiseptic* and *vermicide*.
5. **Naphthalene** is largely used to *manufacture phthalic anhydride* which is a vital and important chemical in the dye industry.

3.2. Anthracene [C₁₄H₁₀] (Syn: Green Oil)

Preamble–They term **anthracene** finds its generic nomenclature from the *Greek* word–"Anthrax" meaning *"coal"*. It is a **tricyclic hydrocarbon** found within the range ~**0.5 to 1.1 per cent** in the so-called '**anthracene oil**' fraction of the **coal-tar distillation product** invariably collected between 270° to 370°C.

Major Impurities in Anthracene–Following are a few *major known impurities* in **anthracene**, namely:

• **Carbazole** • **Chrysene** • **Paraffins** and   • **Phenanthrene**

Preparation of Anthracene–The derived '**anthracene oil fraction**' is cooled vigorously when the solid mass of anthracene separates out, which may be removed subsequently using a '**filter press**' so as to obtain a crude product containing **carbazole** and **phenanthrene** as *major contaminants*.

Recovery of Anthracene from the Crude Product–Pure form of **anthracene** is invariably recovered by any one of the following specified procedures:

(a) **Washing with 'Solvent Naphtha' followed by 'Pyridine'**– The *crude* cake of anthracene is powdered and washed:

- *First*–with '**solvent naphtha**'–to dissolve the **phenanthrene** and filtered, and
- *Secondly*–the resulting product is washed with **pyridine** to dissolve the **carbazole**.

However, the **anthracene** is duly purified by the process of '**sublimation**'.

(b) The resulting product, obtained duly after the removal of **phenanthracene** [*as in* (*a*) *above*], is carefully fused with potassium hydroxide (**KOH**) to produce the respective salt **potassium carbazole**; whereas, the so-called '**unreacted anthracene**' could be easily sublimed out of the melt and recovered.

(c) **Pureset form of Anthracene**–It may be duly prepared by the meticulous *reduction* of **anthraquinone** with *Zn-dust* and *ammonia* (*NH₃*), as given below:

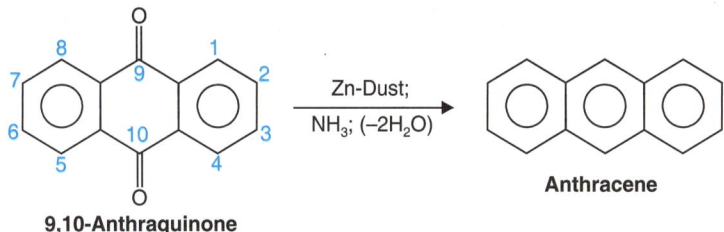

9,10-Anthraquinone → Zn-Dust; NH₃; (−2H₂O) → Anthracene

CONSTITUTION OF ANTHRACENE

The **constitution of anthracene** has been elucidated properly based on the various scientific and experimental results as described under in a sequential manner:

1. **Molecular Formula**–Based upon the various analytical data and findings the *molecular formula of anthracene* is found to be $C_{14}H_{10}$.

2. **Presence of Benzene Rings**–It has been established that very much similar to both *benzene* and *naphthalene*–the **anthracene** undergoes an array of usual **substitution reactions**. However, such glaring facts do support equivocally the view that **anthracene** more or less shows a close resemblance to: **Benzene** and **Naphthalene**.

Bromination of Anthracene–**Anthracene** on being subjected to *bromination* gives **bromoanthracene ($C_{14}H_9Br$)** which on fusion with KOH (potassium hydroxide) yields **hydroxyanthracene ($C_{14}H_9OH$)**. Thus, the resulting product ($C_{14}H_9OH$) upon vigorous oxidation yields **phthalic acid** together with a relatively smaller quantum of *O*-**benzoyl benzoic acid**.

These sequence of reactions may be expressed as under:

$$C_{14}H_{10} \xrightarrow[\text{(Bromination)}]{Br_2;} C_{14}H_9Br \xrightarrow[\text{(Fusion)}]{KOH;} C_{14}H_9OH \longrightarrow$$

Anthracene   Bromoanthracene   Hydroxyanthracene

Vigorous Oxidation →

Phthalic acid   +   *O*-Benzoylbenzoic acid

Remarks–Based upon the aforesaid reactions one may conclude obviously that **anthracene** does contain at least *two* **benzene rings** having the following *skeleton structure* (*I*):

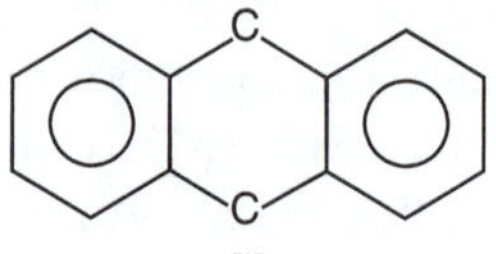

[I]

Importantly, the presence of *two* **benzene rings** has been further ascertained by the underlying fact that **anthracene** upon *oxidation* with *dichromate* ($K_2Cr_2O_7$) *and sulphuric acid* (H_2SO_4) gives rise to the production of **9, 10–anthraquinone**–that eventually on fusion with *KOH at 280°C* yields *two* **moles of benzoic acid**–as depicted below:

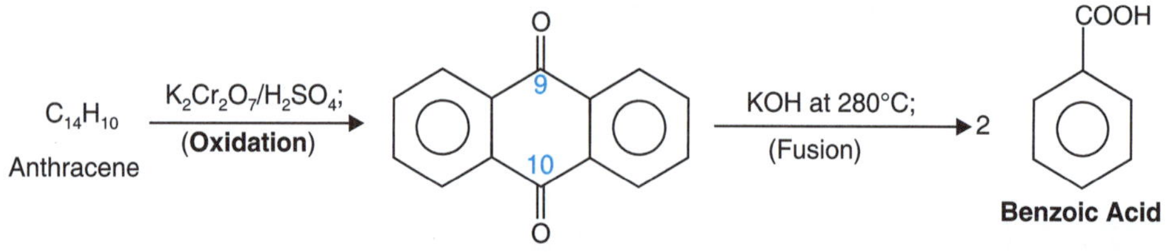

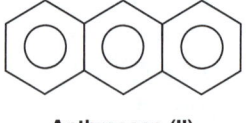

$$C_{14}H_{10}$$
Anthracene

$\xrightarrow[\textbf{(Oxidation)}]{K_2Cr_2O_7/H_2SO_4;}$

$\xrightarrow[\textbf{(Fusion)}]{\text{KOH at 280°C;}}$ 2

COOH

Benzoic Acid

9,10-Anthraquinone

| **NOTE** | It confirms the presence of *two* benzene rings in the *anthracene molecule*. |

3. **Presence of Three Benzene Rings**–Interestingly, the *skeleton* of **anthracene** (**I**) comprises *fourteen* **C-atoms**. In order to fit in squarely the **10 H-atoms** and also retain appropriately the so-called **quadrivalency of each C-atom**–it becomes absolutely essential that the specific C-atom located strategically in between the *2 benzene rings* should be present in the form of a **closed ring system** *i.e.*, the **anthracene** molecule is made up of **3 benzene rings** duly fused in a linear fashion as given under in **structure** (**II**):

Anthracene (II)

4. **Confirmation of Anthracene Structure (II)**– Importantly, the probable **structure (II)** assigned to **anthracene** has been duly ascertained by means of the usual *rational synthesis of certain anthracene derivatives*.

Examples–When **2 moles of** *O*-**bromo benzyl bromide** is made to react with *metallic Na* (freshly cut pieces) it yields distinctly **9, 10-dihydroanthracene** which is found to be quite identical with the one formed by the *reduction of anthracene*.

The various sequential reactions may be expressed as under:

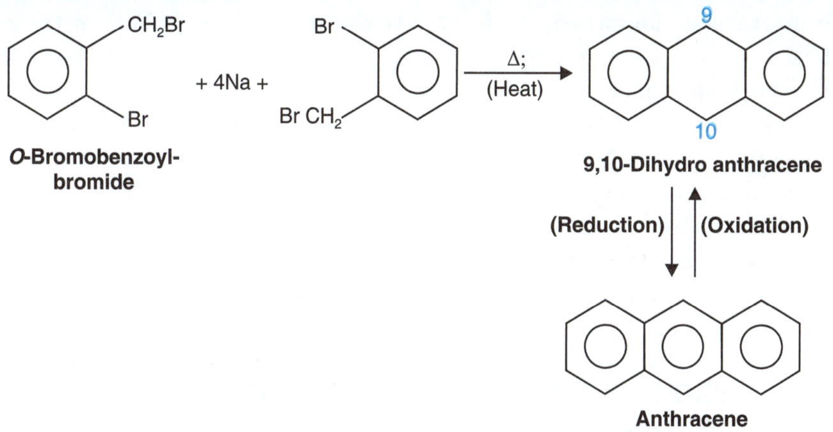

O-Bromobenzoyl-bromide

9,10-Dihydro anthracene

(Reduction) | (Oxidation)

Anthracene

5. **Position of Olefinic Bonds (or Double Bonds) in Anthracene**–The synthesis of **anthracene** by the *Friedel-Craft's condensation* method between *benzene* and *acetylene tetrabromide* suggests vehemently that the following **structure (III)** for **anthracene** having a *parabond*–as shown under:

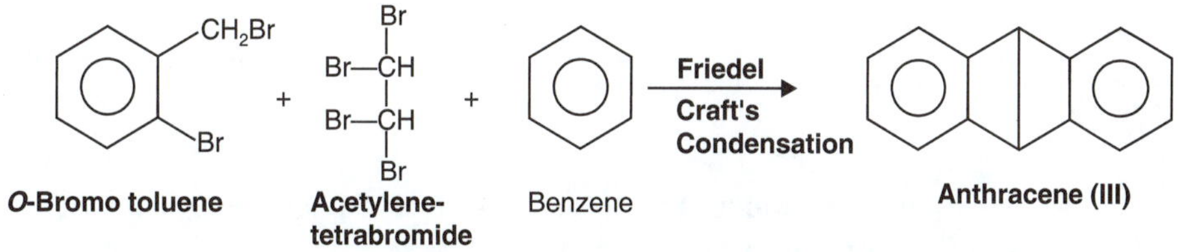

O-Bromo toluene   Acetylene-tetrabromide   Benzene   Anthracene (III)

Nevertheless, the above **structure (III)** for **anthracene** has been pronounced to be an *unsuitable structure*–based upon the following *three* **cardinal reasons**, such as:

(a) *Synthesis of Anthracene by Diel's–Alder Reaction*–It essentially involves **1, 4-naphthaquinone** and **butadiene** to obtain **9, 10–anthraquinone** by subjecting to *oxidation with chromium trioxide* (CrO_3) *in glacial acetic acid*; and ultimately distilling the resulting product with *Zn-dust* reduction helps to eliminate the *para-bond* completely.

The various reactions may be expressed as under:

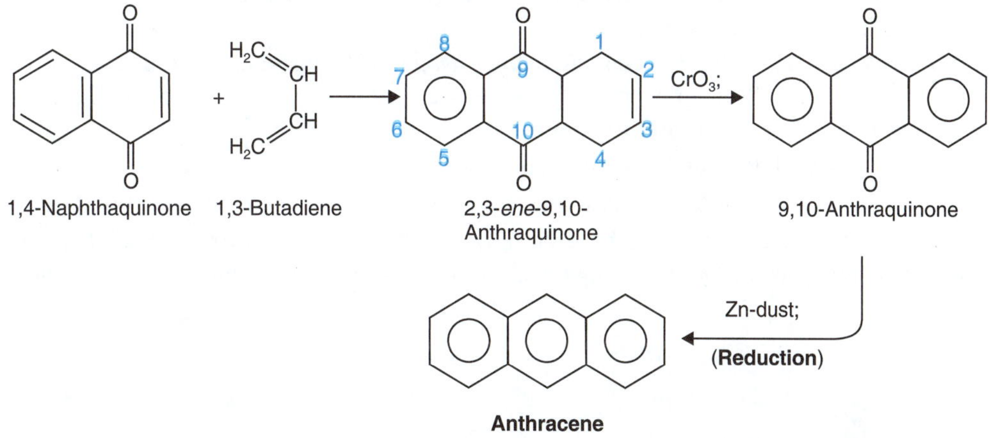

1,4-Naphthaquinone   1,3-Butadiene   2,3-*ene*-9,10-Anthraquinone   9,10-Anthraquinone

Zn-dust;
(Reduction)

Anthracene

(b) Since **anthracene** adds on to the **maleic anhydride molecule** particularly in the *'middle ring'*– it critically serves as a **'conjugated diene'** (*i.e.*, *Diel's–Alder reaction*); and hence, the formation of the *'para-bond'* in the *middle ring is totally absent*.

(c) Importantly, the *X-ray diffraction analysis* of the **anthracene** *molecule reveals explicitly that* **all the C-atoms** duly present in the **anthracene** molecule do lie mostly in the *same plane*. In addition, the prevailing distance existing between the *para* **C-atoms** in each ring has been vividly shown to be the same as in *benzene*.

Remark–Based on the aforesaid *sequence of reactions* one may rightly confirm the **structure (II)** assigned to *'anthracene'* wherein the *three* benzene rings are fused strategically at the *ortho*-**positions** predominantly.

ANTHRACENE: DESIGNATED BEST AS A 'RESONANCE HYBRID OF FOUR CHEMICAL STRUCTURES'

Interestingly, based on the aforesaid *analogy* and *scientific reasonings related to 'benzene'*–one may draw the conclusion that **'anthracene'** may be best regarded as a **'resonance hybrid'** of the following *four* **chemical structures [A, B, C, and D]**: [*"b"* : **Benzenoid Ring; "q"** Quinonoid Ring]

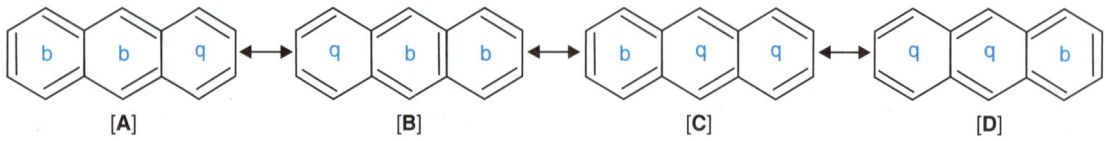

[A]   [B]   [C]   [D]

Amazingly, the above *four* resonance structures [*A through D*] have been supported grossly by the following *three* **vital and important facts,** namely:

❑ The **anthracene** molecule essentially comprises *three* **benzene rings**; and; therefore, it critically possesses **(3 + 1) or 4 principal contributing structures;**

❑ **The anthracene** molecule fits appropriately in **(4n + 2) π-electron rule** (having **14π electrons**) *i.e.*, n = 3; (4n + 2) = (4 × 3 + 2) or **14**; and

❑ The actual *resonance energy* of **anthracene** is found to be **351.1 kg. mol.$^{-1}$**

6. **Position and Nature of Olefinic Bonds (or Double Bonds)**– An elaborate study pertaining to the **X-Ray Diffraction (XRD)** pattern of **anthracene** critically brings forth the following *three* **glaring facts,** namely:

(i) Practically all of the **C-C covalent linkages** in **anthracene** have been found to be little shorter in length in comparison to the respective *C-C single bonds*, but definitely longer than the *alkene C= C olefinic bond*.

(ii) Obviously, all the bonds are not exhibiting the same length.

(iii) Explicitly the **C$_1$–C$_2$ bond** is shorter *vis-a-vis* the corresponding **C$_2$–C$_3$ bond**.

Furthermore, the prevailing **'bond lengths'** as well as the **'double bond characteristic feature'**–that may be apprehended by crucially examining the aforesaid contributing structures as stated below:

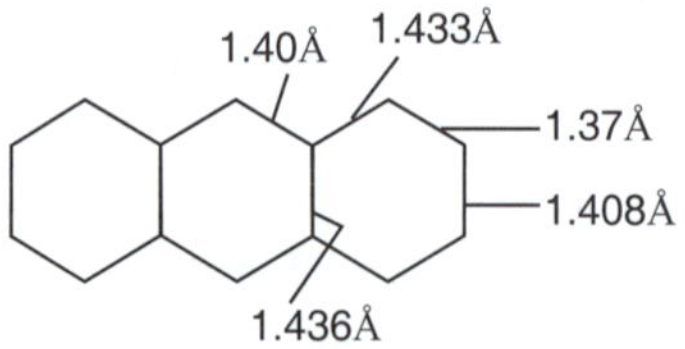

1.40Å   1.433Å

1.37Å

1.408Å

1.436Å

Thus, based on the foregoing discussion one may safely infer that–

"there exists a significant amount of bond fixation in anthracene".

Further Important Points– These essentially include:

(a) The C_1–C_2 bond is observed to be *double in* 3 out of 4 resonance structures; and, therefore, possesses **3/4th double bond character**.

(b) Besides, the C_2–C_3 bond is found to be *double* in **one**; and *single* in **3 resonance structures**. Hence, it exhibits the **1/4th double bond characteristic feature**. Thus, one could also calculate the so-called **double-bond character** of other bonds in a similar manner as shown under:

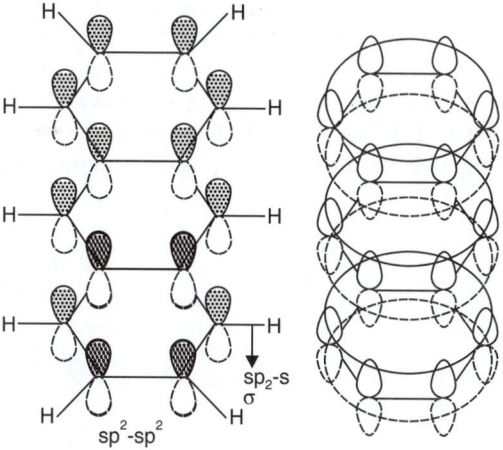

MOLECULAR ORBITAL STRUCTURE OF ANTHRACENE

In a broader perspective the structure of **anthracene** resembles very close to both '*benzene*' and '*naphthalene*'–that may also be explained explicitly on the basis of '*Molecular Orbital Theory*'. Obviously, the *fourteen* **C-atoms** duly forming the **3D-benzene ring systems** are found meticulously in a state of sp^2-*hybridisation*; and hence, the molecule is observed to be *planar in characteristics*.

The Fig. 4.1 clearly depicts the involvement of the so-called **unhybridized p-orbital** up each C-atom specifically in its critical lateral overlap along with its neighbours; and thus, the overall net result would appear as a result in the form of a **continuous π-electron cloud** hovering both above and *below the plane of the C-atoms*. Importantly, the entire cloud may be regarded to be made up of *3-partially overlapping sexlets* (*i.e.*, the *π-clouds* comprising *six* clouds) that essentially do possess **2-pairs of common π-electrons**. Nevertheless, the overall 3 C-rings are observed to be adequately sandwitched between the *two* **lobes of the cloud** (see Fig. 4.1), which eventually renders in lowering of overall energy to a certain extent.

Fig. 4.1. Naphthalene Orbital Structure: Cloud made up of *three* partially overlapping sexlets having 2-pairs of common p-electrons.

Interestingly, the usual representation of the **anthracene molecule** along the same analogy of **benzene** and **naphthalene** may be duly represented as under:

Anthracene

Comment–Obviously, the circles inside the *benzene rings* are solely indicative of the prevailing **π-electron clouds**.

SYNTHESIS OF ANTHRACENE

There are *four* known methods usually adopted for the *synthesis* of **anthracene, namely:**

- ❏ **By Friedel–Crafts Reaction (using benzyl chloride, or phthalic anhydride)**
- ❏ **By Elbs Reaction (*via* polynuclear hydrocarbon),**
- ❏ **By Fittig Reaction (using *O*-bromobenzyl bromide), and**
- ❏ **From Naphthalene (*via* Diel's-Alder reaction, or Michael Condensation),**

which shall now be discussed separately in the sections that follows:

1. **By Friedel-Crafts Reaction**–**Anthracene** may be prepared by the use of the **Friedel-Crafts reaction** in *two* ways, such as:
   - • **from benzyl chloride, and**   • **from phthalic anhydride.**

   (a) **From Benzyl Chloride:** When benzyl chloride is treated carefully in the presence of *anhydrous aluminium chloride (AlCl₃)* to produce **9,10-dihydroanthracene** that eventually loses *two* **H-atoms** to form **anthracene** as shown under:

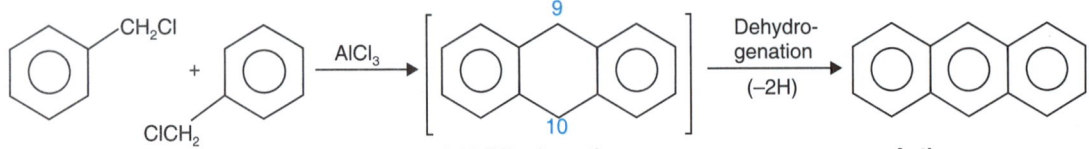

| Benzylchloride [2Moles] | 9,10-Dihydroanthracene (Intermediate) | Anthracene |

   (b) **From Phthalic Anhydride:** The treatment of an admixture of **phthalic anhydride** and **benzene** with anhydrous aluminium chloride (AlCl₃) yields *O*-benzoyl benzoic acid, which on subsequent heating with **concentrated H₂SO₄ at 100°C** gives **anthraquinone**. Finally, the resulting product with distillation along with **Zn-dust** gives the desired product **anthracene**-as shown below:

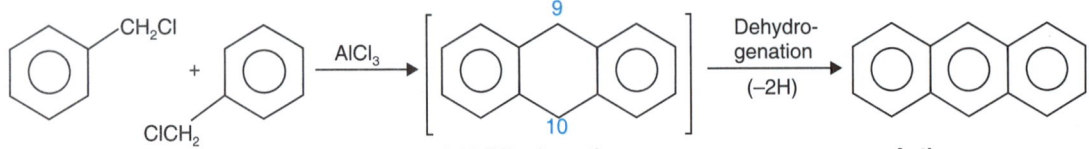

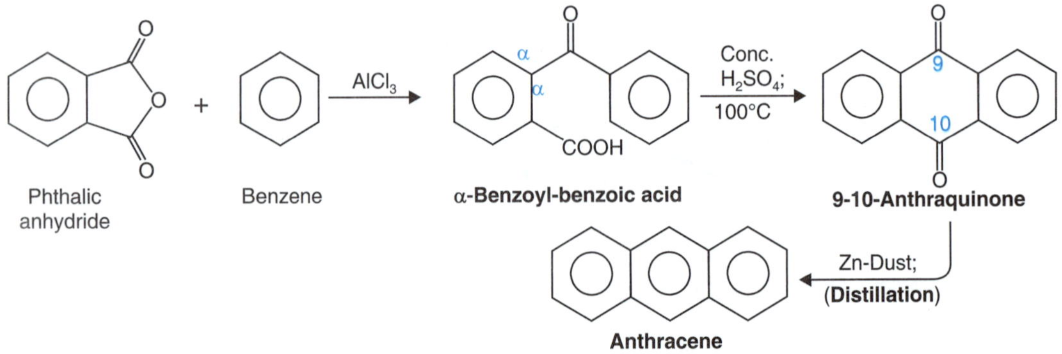

Phthalic anhydride   Benzene   α-Benzoyl-benzoic acid   9-10-Anthraquinone

Zn-Dust; (Distillation)

Anthracene

2. **By Elbs Reaction***–In this particular instance, a **polynuclear hydrocarbon** (*viz.*, *O*-methylbenzophenone) is prepared initially by the *pyrolysis of* a **diaryl ketone** possessing either a *methylene* or *methyl* moiety present *ortho* to the respective **carbonyl (>C=O) moiety.**

Example: O-**Methyl benzophenone** (*i.e.*, a *polynuclear hydrocarbon*) on heating loses a mole of water to give **anthracene:**

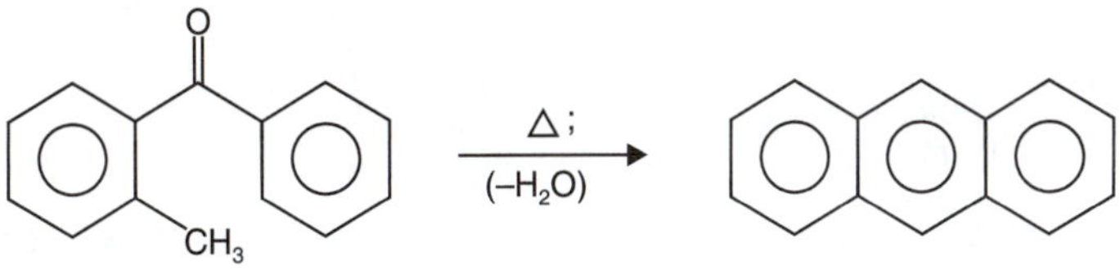

O-Methyl benzophenone   Anthracene

Mechanism of Elbs Reaction

The various steps involved in the **Elbs Reaction** are as depicted under in a sequential manner:

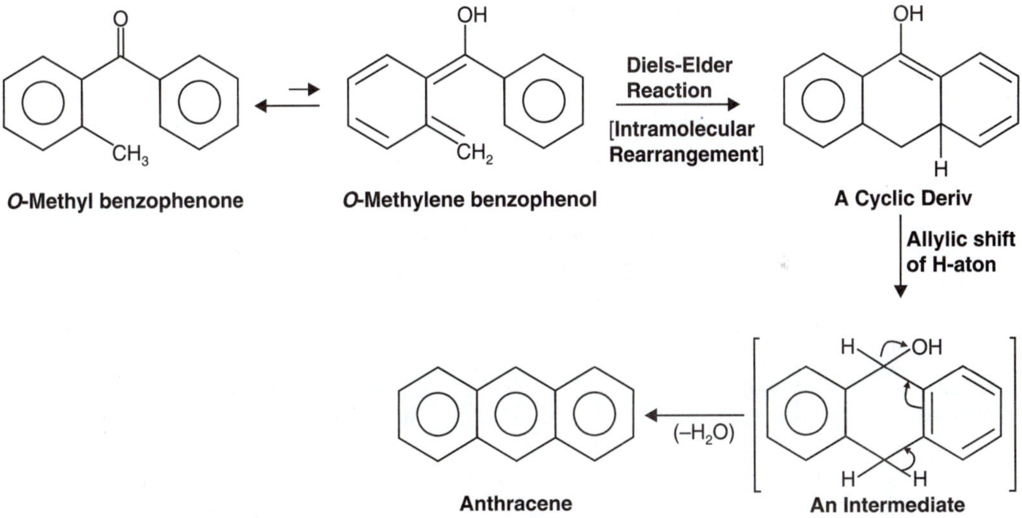

O-Methyl benzophenone   O-Methylene benzophenol   A Cyclic Deriv

Anthracene   An Intermediate

All the aforesaid steps are self-explanatory.

3. **By Fittig Reaction****– In this specific instance, *O*-**bromobenzyl bromide** (which is duly prepared by the careful *bromination* of *O*-**bromotoluene** with 1 mole of *bromine* at 150°C) is heated with *Na*-in the presence of solvent *ether* to yield **9,10-dihydroanthracene** that eventually on *mild oxidation* gives rise to the formation of **anthracene**–as shown under:

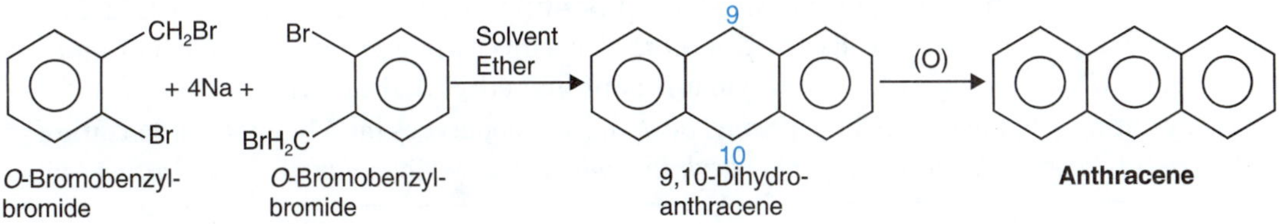

O-Bromobenzyl-   O-Bromobenzyl-   9,10-Dihydro-   Anthracene
bromide   bromide   anthracene

---

*   **Elbs K: J Prakt Chem., 48: 179(1893);**

** **Fittig R: 1835-1910; b. Hamburg (Germany); Ph.D; Gattingen; Strasbourg**

Special Observation–It has been amply proven that the above cited method, certain *phenanthrene* is also produced simultaneously as a by product as given under:

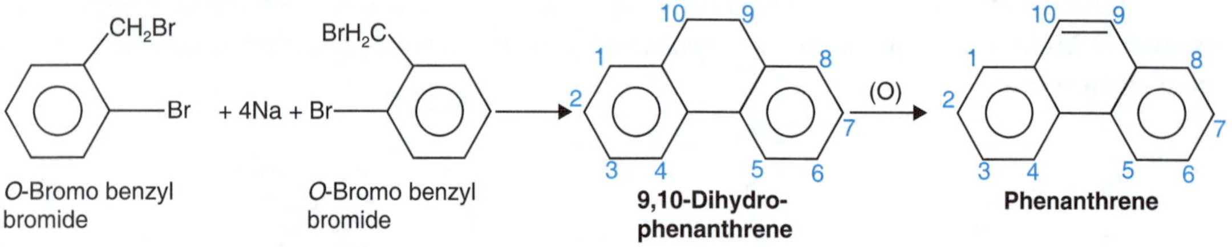

| O-Bromo benzyl bromide | O-Bromo benzyl bromide | 9,10-Dihydro-phenanthrene | Phenanthrene |

4. **From Naphthalene–Naphthalene** may also be converted into **anthracene** almost quantitatively by means of the following *two* methods, namely:

- **Diels-Alder Reaction**, and
- **Michael condensation,***

which shall now be treated separately as under:

(i) **Diels–Alder Reaction–** The interaction between *1,4-naphthaquinone* and *1,3-butadine* leads to the **Diels-Alder reactions** to yield dihydro **9,10-anthraquinone,** which upon *oxidation* with chromic acid (CrO_3) in glacial acetic acid (CH_3COOH) to give **9,10-anthraquinone.** Finally, the resulting product on being subjected to reduction with **Zn-dust** gives the desired product **anthracene**–as given under:

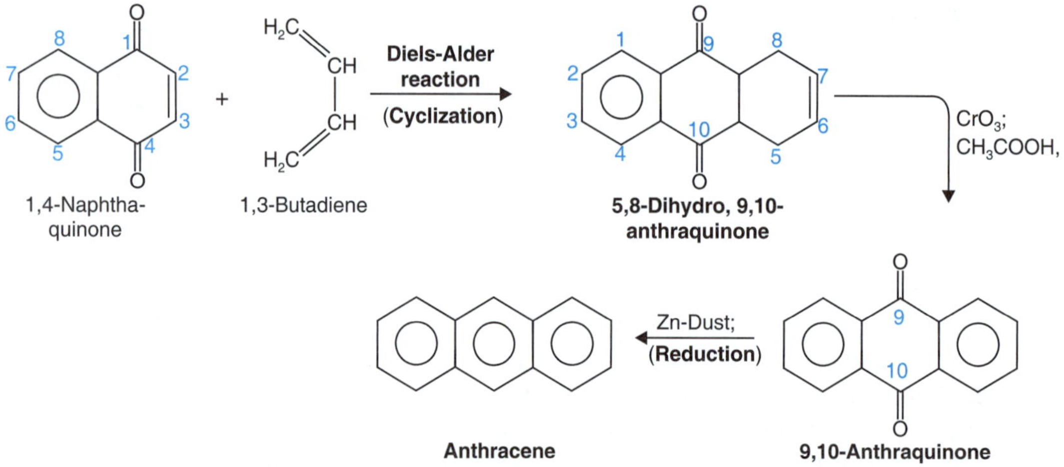

| 1,4-Naphtha-quinone | 1,3-Butadiene | 5,8-Dihydro, 9,10-anthraquinone |

Anthracene   9,10-Anthraquinone

(ii) **Michael Condensation–Naphthalene** when treated with H_2SO_4 at 180°C, followed by **NaCN,** and finally subjected to *hydrolysis* yields **β-naphthyl carboxylate,** which on reduction with Na-Hg/H_2O gives **5,8-dihydro-β-naphthyl carboxylate** and this with ethanol undergoes *esterification* to produce **5, 8-dihydro-β-naphthyl ethyl carboxylate.**

The resulting ester on treatment with *ethyl acetoacetate* undergoes the **Michael condensation**** to yield *6-(α-ethylaceto-acetate)-7-ethyl carboxylate naphthalene,* which undergoes *cyclization* (with the

---

*   Michael A: *J Prakt Chem.*, [2]**35**, 349 (1887).

**   That is, it involves the condensation of an α,β-uasaturated carbonyl compounds with an **active methylene moiety** *e.g.,* **acetoacetic ester.**

loss of a mole of ethanol) to give **6,8-diketo-5-ethylcarboxylate–9,10-dihydro anthracene**, which ultimately on reduction with **Zn-dust distillation** yield **anthracene**. These reactions are summarized as under:

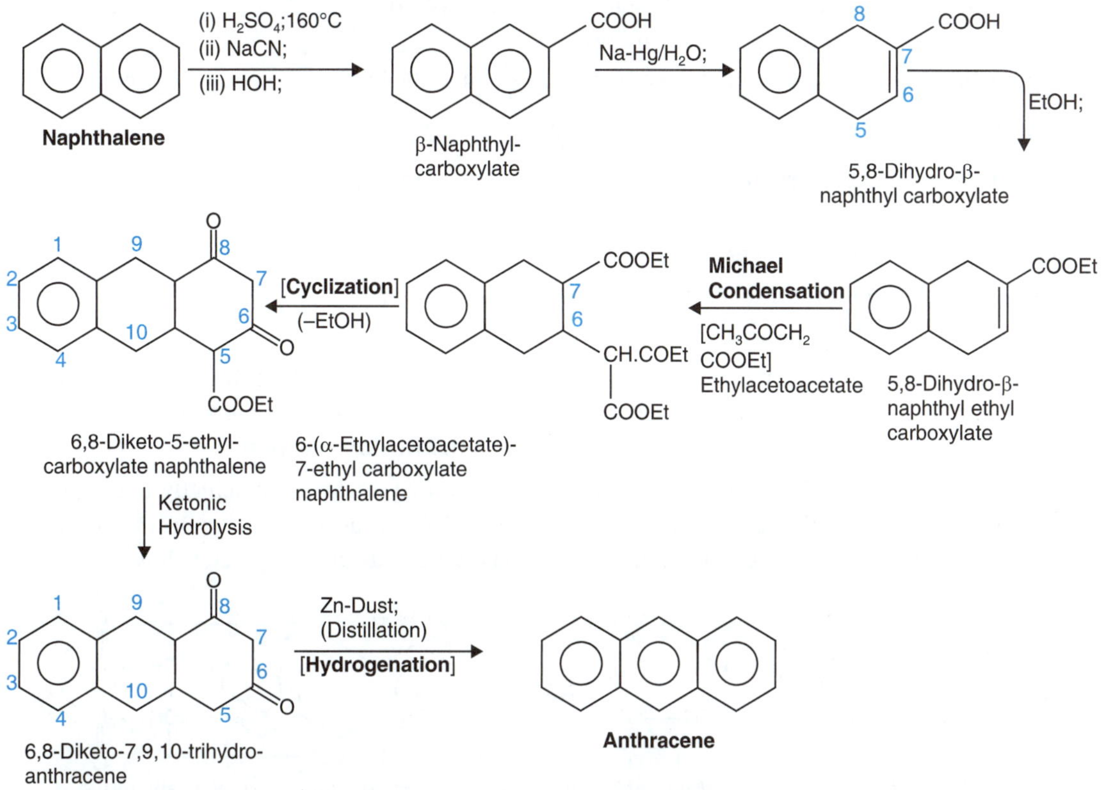

Naphthalene

(i) H$_2$SO$_4$;160°C
(ii) NaCN;
(iii) HOH;

β-Naphthyl-carboxylate

COOH

Na-Hg/H$_2$O;

COOH

8
7
6
5

EtOH;

5,8-Dihydro-β-naphthyl carboxylate

COOEt

Michael Condensation

[CH$_3$COCH$_2$ COOEt]
Ethylacetoacetate

5,8-Dihydro-β-naphthyl ethyl carboxylate

COOEt
7
6
CH.COEt
COOEt

6-(α-Ethylacetoacetate)-7-ethyl carboxylate naphthalene

[Cyclization]
(–EtOH)

O
8
7
6
5
O
1 9
2
3
10
4
COOEt

6,8-Diketo-5-ethyl-carboxylate naphthalene

Ketonic Hydrolysis

O
1 9 8
2 7
3 6
10
4 5
O

6,8-Diketo-7,9,10-trihydro-anthracene

Zn-Dust;
(Distillation)
[Hydrogenation]

Anthracene

Remark–Thus, **anthracene** may be synthesized by the help of *four* aforesaid routes of synthesis.

Characteristic Features of Anthracene

We may deal with the specific characteristic features of **anthracene** separately under the following *two* heads:

- **Physical characteristics**, and
- **Chemical characteristics**.

❑ Physical Characteristics

These essentially include:

1. Anthracene usually occurs as *monoclinic plates* from ethanol.
2. It sublimes on heating.
3. In **purest form anthracene** appears as colourless having **violet fluorescence.**
4. In *impure state* (due to the presence of: **Tetracene** and **Naphthacene**) it appears mostly *yellow* with **green fluorescence.**

5. **Anthracene** is found to be strongly **triboluminescent** and **triboelectric**.
6. It shows: d_4^{27} 1.25; mp: 218°C; bp_{760}: 342°C.
7. **Solubility Profile:** It is insoluble in an aqueous phase. However, 1 g-dissolves in 67 mL absolute ehtnol; 70 mL methanol; 62 mL benzene; 85 mL chloroform; 200 mL ether; 31 mL carbon disulphide; 86 mL CCl_4; and 125 mL toluene.
8. **Anthracene** gets darkened on exposure to UV (sunlight) light.

❑ **Chemical Characteristics**

The **chemical characteristics** of **anthracene** centres around the following *specific zones* in the molecule, such as:

*   **9, 10-positions: very reactive;**
*   **Olefinic (double) bonds: unsaturated features**; and
*   **Aromatic properties: due to 3 fused benzene rings**.

In this manner, **anthracene** shows a marked and pronounced tendency towards the so-called *'addition reactions'*. However, it has been extablished to be **less aromatic *vis-a-vis*** naphthalene. Besides, **anthracene** accounts for critical **electrophilic substitution reaction**.

1. **Addition of Maleic Anhydride**–It has been shown that **anthracene** actually undergoes the **Diels-Alder reaction** strategically at the **C-9 and C-10** positions to yield the respective **endoanthracene maleic anhydride** (*i.e.*, a **Diels-Alder-adduct**)–as given below:

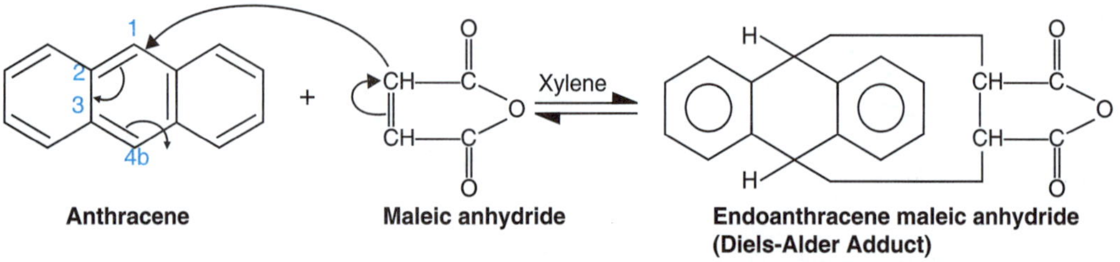

| Anthracene | Maleic anhydride | Endoanthracene maleic anhydride (Diels-Alder Adduct) |

2. **Reduction**–Anthracene when subjected to reduction with **Na** and **isopropanol** yields **9,10-dihydroanthracene**. The resulting compound is *not fluorescent* which upon:
    *   **heating**, or
    *   **treatment with H_2SO_4(conc.)**,
    loses *two*-**H-atoms** to reform **'anthracene'**–as shown below:

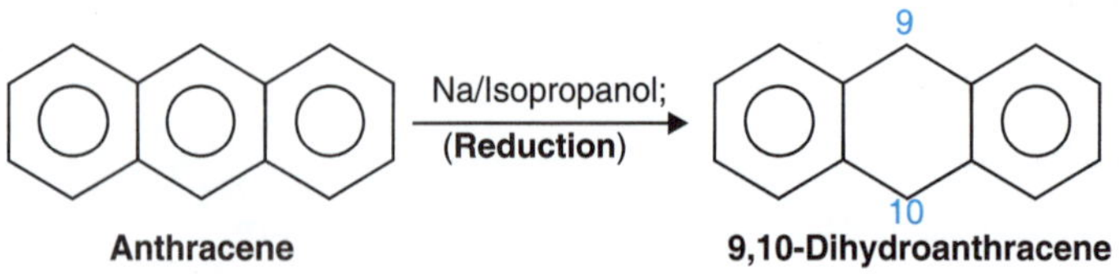

| Anthracene | | 9,10-Dihydroanthracene |

Remark: Nevertheless, when **anthracene** gets *reduced catalytically* in the presence of **Ni** between **200–300°C** gives: **tetra, hexa-and octa-hydroanthracene** and ultimately the product **perhydroanthracene** $(C_{14}H_{24})$:

Thus, we may have the following expressions–

$$C_{14}H_{10} \xrightarrow[2H_2;]{Ni;} C_{14}H_{14} \xrightarrow[2H_2;]{Ni;} C_{14}H_{16} \xrightarrow{H_2} C_{14}H_{18} \xrightarrow{3H_2;} C_{14}H_{24}$$

| Anthracene | Tetra-hydro-anthracene | Hexa-hydro-anthracene | Octa-hydro-anthracene | Perhydro-anthracene |

3. **Halogenation**– A chilled solution of **anthracene** in *carbon disulphide* (CS_2) on being subjected to the passage of chlorine (Cl_2) it yields **anthracene dichloride**, which loses a mole of *HCl* either:

   • **on heating,   or**   • **treatment with alkali (NaOH/KOH),**

to produce **9-chloroanthracene** finally –as given below:

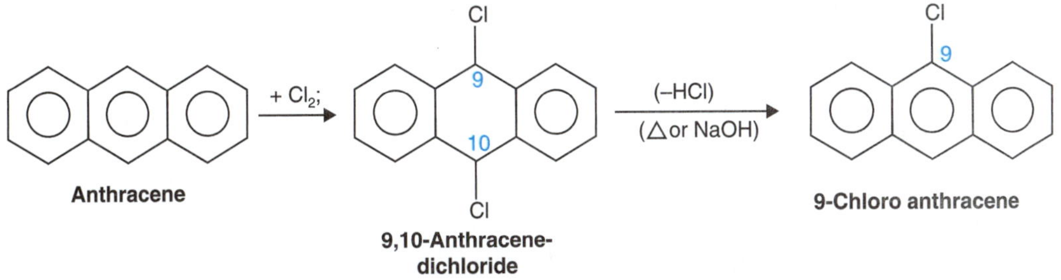

Anthracene +Cl₂; 9,10-Anthracene-dichloride (–HCl) (△or NaOH) 9-Chloro anthracene

Comment–Likewise, the bromination of anthracene in boiling carbon tetrachloride (CCl_4) yields **9, 10-dibromoanthracene**, which eventually loses, a *halogen acid* by virtue of the ensuing '*1,4-elimination*' rather *slowly* at the room temperature ($20 \pm 2°C$) and *rapidly on heating gently to produce the respective* **9-bromoanthracene.**

4. **Formylation**–**Formylation** relates to the introduction of an '*aldehyde*' group in a molecule. In this particular instance **anthracene** may be *formylated at the 9-position* in the presence of $POCl_3$ and *heat*–as shown under:

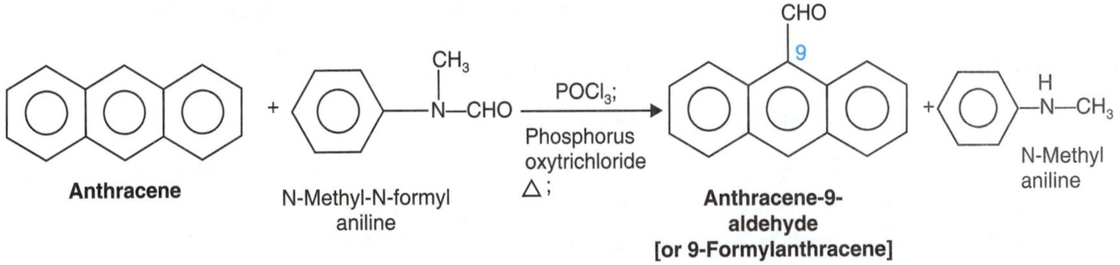

Anthracene + N-Methyl-N-formyl aniline $\xrightarrow[\Delta;]{POCl_3;\ \text{Phosphorus oxytrichloride}}$ Anthracene-9-aldehyde [or 9-Formylanthracene] + N-Methyl aniline

5. **Nitration**–In usual practice, the **nitration** takes place in *two* altogether different situations as stated under:

(a) **With aqueous HNO₃**–It oxidizes **anthracene** to **anthraquinone** as shown below:

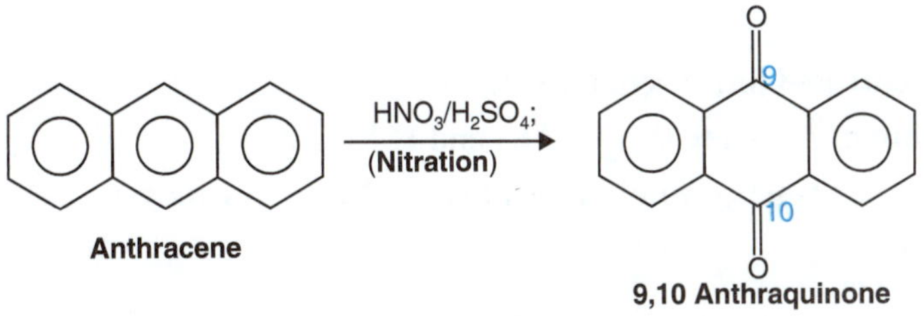

Anthracene $\xrightarrow[\text{(Nitration)}]{HNO_3/H_2SO_4;}$ 9,10 Anthraquinone

(b) **With HNO$_3$ in Acetic Anhydride–Anthracene** undergoes *nitration* with *HNO$_3$ in acetic anhydride* to yield *two* distinct products:

- **9-anthracene (mp: 145°C)**; and
- **9,10-dianitro anthracene (mp: 294°C)**.

Thus, we may have:

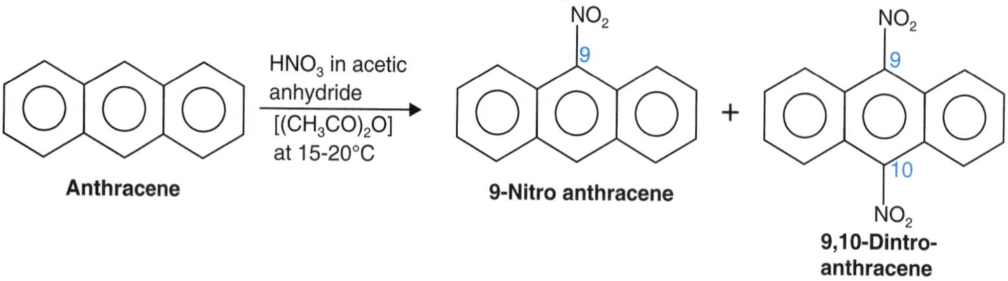

HNO$_3$ in acetic anhydride [(CH$_3$CO)$_2$O] at 15-20°C

Anthracene 9-Nitro anthracene 9,10-Dintro-anthracene

6. **Sulphonation**–It has been duly established that **anthracene** undergoes **sulphonation** readily to form an admixture of:

- **1-anthracene sulphonic acid**, and
- **2-anthracene sulphonic acid**.

Anthracene H$_2$SO$_3$; (Sulphonation) (–H$_2$O) 1-Anthracene-sulphonic acid 2-Anthracene-sulphonic acid

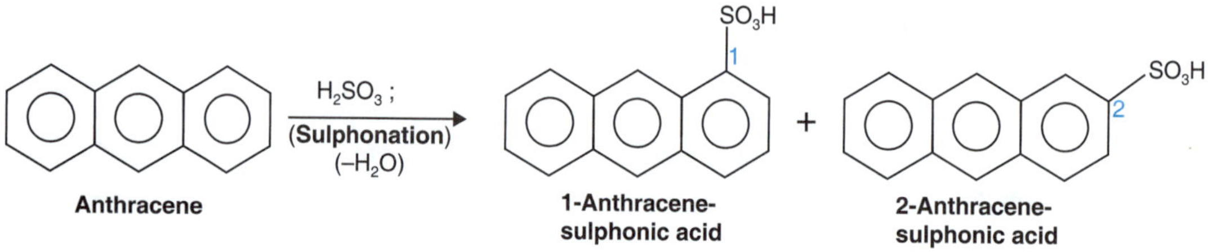

Remarks–Following are a few critical remarks and observations:

1. A small quantum of **disulphonic acid** is also formed simultaneously.
2. However, the **2nd position (C-2)** is found to be rather **more favourable at high temperatures**.
3. Equal proportion of **1- and 2-anthracene sulphonic acids** are formed when *sulphonation is being carried out with* **H$_2$SO$_4$ in glacial acetic acid (CH$_3$COOH)**.
4. It has also been observed quite evidently that there prevails an absolutely little tendency to undergo rearrangement to the corresponding **2-anthracene sulphonic acid**.

7. **Oxidation to form 1,8-Disulphonic Acid and 2,7-Disulphonic Acid; Anthracene** when subjected to oxidation with either:

- **Concentrated nitric acid (HNO$_3$)**, or
- **Sodium dichromate (Na$_2$Cr$_2$O$_7$) and H$_2$SO$_4$**, or
- **air in the presence vandium pentoxide (V$_2$O$_5$) as catalyst**.

it yields anthraquinone:

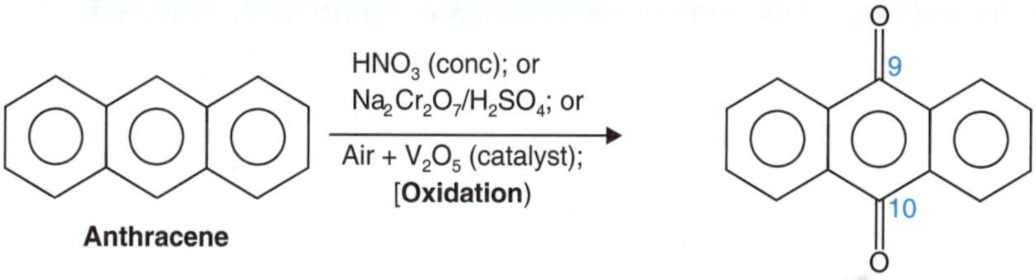

Anthracene

HNO$_3$ (conc); or
Na$_2$Cr$_2$O$_7$/H$_2$SO$_4$; or
Air + V$_2$O$_5$ (catalyst);
[**Oxidation**]

9,10-Anthraquinone

Uses

1. **Anthracene** is primarily used in the large-scale production of **anthraquinone.**
2. **Anthraquinone** find its abundant usage in the production of **dyes** *viz.*, **Alizarin**.

ANTHRAQUINONE

[*Syn*; **9,10-Anthracenedione; 9,10-Anthraquinone; 9,10-Dioxoanthracene**].

   Preamble: Anthracene on being subjected to heating in the presence of concentrated **nitric acid** (**HNO$_3$; 12 M**) it does not, as expected normally, give a *nitro*-**derivative**, but it gets oxidized to anthraquinone *i.e.*, a **diketone**:

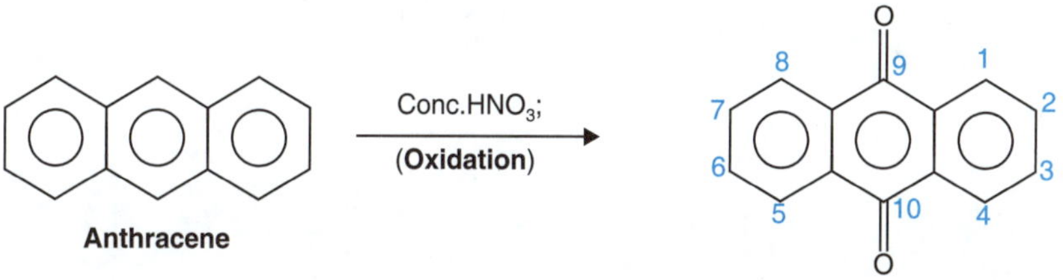

Anthracene

Conc.HNO$_3$;

(**Oxidation**)

9,10-Anthraquinone
(A diketone)

   Graded Reduction of Anthraquinone–It has been observed that the **graded reduction** of **anthraquinone** yields:

*   *First stage*–**Oxanthrone** or its *tautomer, 'anthraquinol'* (**9,10-anthracenediol**):

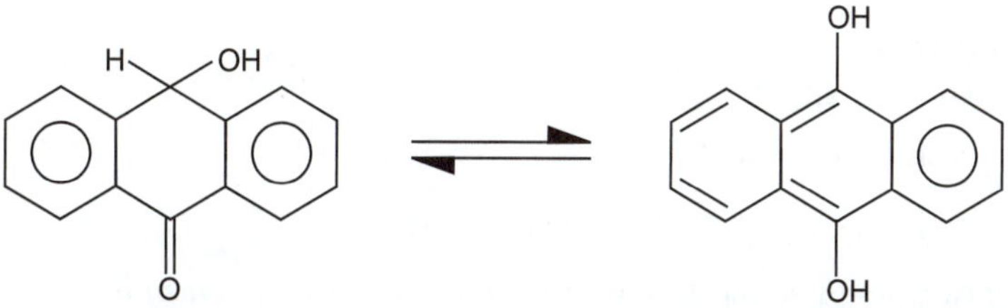

Oxathrone

Anthraquinol
[9,10-Anthracenediol]

- *Second stage*–it yields anthrone or its *tautomer*, 'anthrol' (**anthranol**).

Anthrone   Anthrol

Isomeric Quinones of Anthracene–In fact, there are *nine* possible **isomeric quinones of anthracene**; of which only *three* are recognized, namely:

⊙   1,2–Anthraquinone

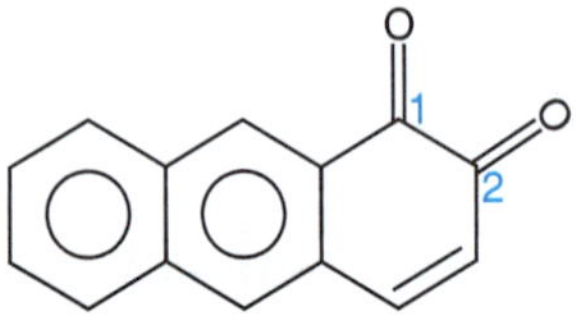

⊙   1,4–Anthraquinone

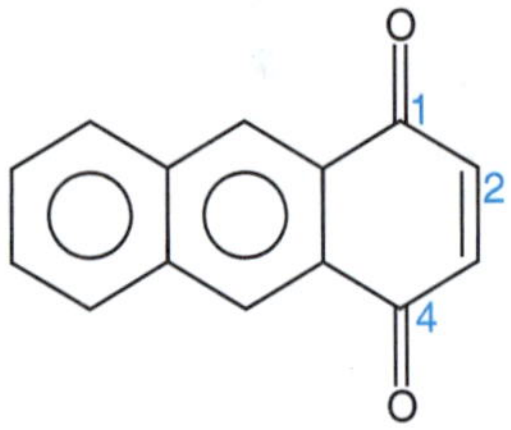

⊙   9,10–Anthraquinone

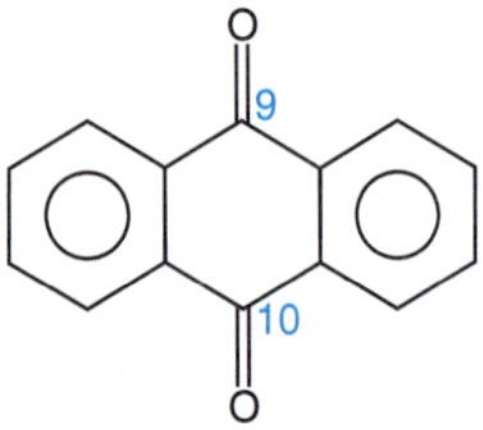

Preparation of Anthraquinone–There are *three* known methods for the preparation of **anthraquinone**, such as:

- **From, 'Anthracene'**
- **From 'Phthalic Anhydride'**, and
- **From '1,4-Naphthaquinone'**,

which will be discussed briefly in the section that follows:

(a) **From Anthracene**–First and foremost (*during the early nineteenth century*) **anthraquinone** was largely produced on a commercial scale by the *oxidation* of **anthracene** using an admixture of **sodium dichromate ($Na_2Cr_2O_7$)** and **concentrated sulphuric acid (H_2SO_4)** upto a yield of **90%**—as given under:

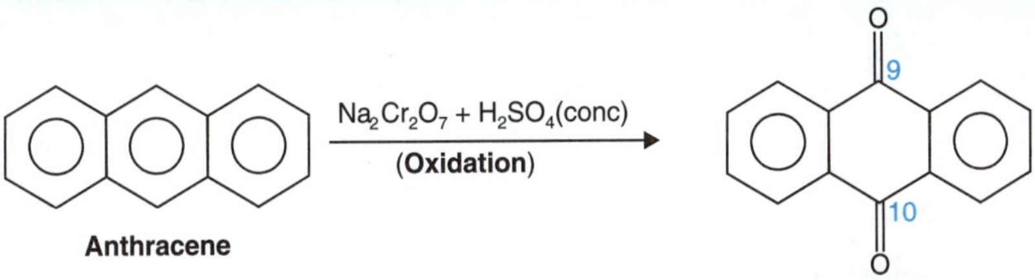

Anthracene → **Anthraquinone [9,10-Anthraquinone]**

$Na_2Cr_2O_7 + H_2SO_4$(conc) (Oxidation)

Remarks–Soonafter it was considered to be a lot convenient and cost-effective to oxidize the *crude* **anthracene** comprising **carbazole** as an inherent impurity (*when isolated from coal tar*) in the presence of *air* and **vanadium pentoxide (V_2O_5)** at a temperature ranging between 300–500°C. However, under such experimental conditionalities, **anthracene** undergoes *oxidation* to the corresponding 'anthraquinone'; whereas, the **carbazole** (*impurity*) gets duly *oxidized* to **carbon dioxide (CO_2)**.

Thus, we may have the following expression:

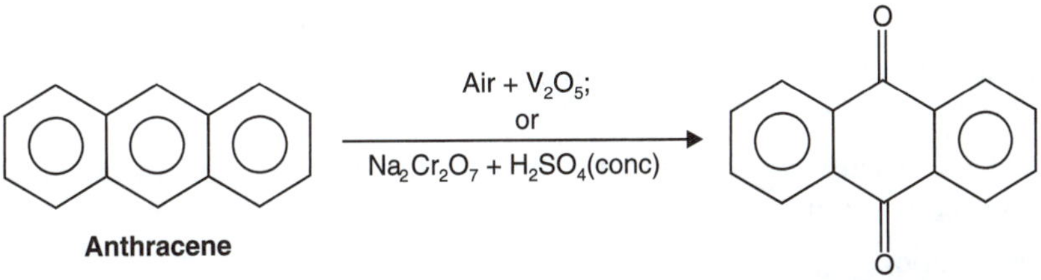

Anthracene → **Anthraquinone [9,10-Anthraquinone]**

Air + V_2O_5; or $Na_2Cr_2O_7 + H_2SO_4$(conc)

(b) From Phthalic Anhydride–The large-scale production of **anthraquinone** from **phthalic anhydride** is carried out in its **benzene solution** by treatment with *aluminium chloride ($AlCl_3$)* when **α-benzoylbenzoic acid** is produced. The resulting product on further heating with concentrated *sulphuric acid (H_2SO_4)* at 100°C yields **anthraquinone**, as shown under:

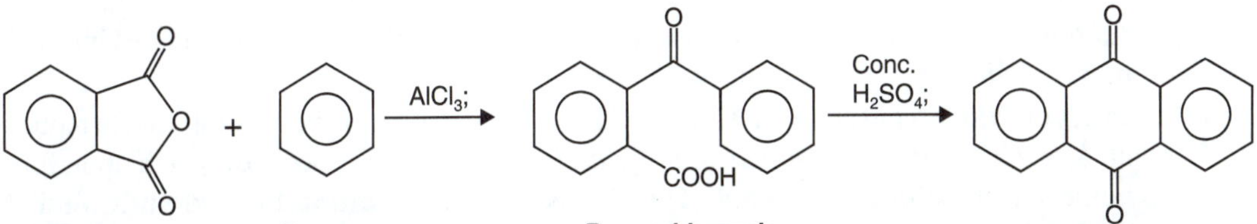

Phthalic anhydride + Benzene $\xrightarrow{AlCl_3;}$ α-Benzoyl-benzoic acid $\xrightarrow{\text{Conc.} H_2SO_4;}$ Anthraquinone

NOTE In case, we replace *benzene* with either toluene or chlorobenzine, we may subsequently obtain methylanthraquinone or chloroanthraquinone respectively– that is being used solely in the *manufacturer of dyes*.

(c) From Naphthaquinone–It critically involves the **Diels-Alder reaction*** taking place between *anthraquinone* and *butadiene* duly followed by *oxidation* of the resulting products with chromium trioxide (CrO_3) in glacial acetic acid–as given under:

---

* Diels O and Alder K: *Ann*, **460**: 98(1928); 470: 62(1929); *Ber*; **62**: 2081, 2087 (1929).

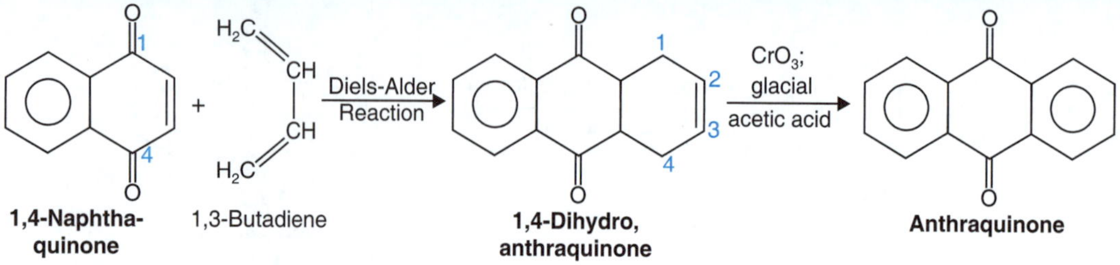

| 1,4-Naphtha-quinone | 1,3-Butadiene | 1,4-Dihydro, anthraquinone | Anthraquinone |

Characteristic Features of Anthraquinone

❑ **Physical Characteristics**–These essentially comprise:

1. **Anthraquinone** occurs as a light yellow, slender monoclinic prisms by sublimation *in vacuo*.
2. It is obtained as almost colorless, orthorhombic, bipyramidal crystals from $H_2SO_4 + H_2O$.
3. It has d_4^{20}1.42–1.44, mp: 286°C, bp$_{760}$: 377°C.
4. **Solubility Profile**–It is insoluble in **water**. Solubility (g/100g) in **ethanol** at 18°C 0.05; at 25°C 0.44; in boiling ethanol 2.25; in **solvent ether** at 25° 0.11; in **CHCl$_3$** at 20 0.61; at 40° 1.00; at 60° 1.60; in **benzene** at 20° 0.26; at 40°C 0.50; at 60° 1.00; at 80° 1.80; in **toluene** at 25°C 0.30.

❑ **Chemical Characteristics**–Following are the *five* important **chemical characteristics** of **anthraquinone**, namely:

- **Reduction,**
- **Distillation with Zn-dust,**
- **Nitration,**
- **Sulphonation,** and
- **Halogenation,**

which shall now be treated individually in the sections that follows:

1. **Reduction:** The effective and gainful *reduction* of **anthraquinone** may be affected by any one of the *four* different methods described below:

(a) **Reduction with Sn and HCl (in CH$_3$COOH)**–*Anthraquinone* on being subjected to reduction with S_n + *HCl in glacial acetic acid* gives *anthrone* (as an intermediate) which upon further treatment with boiling alkali (NaOH) undergoes **tautomerization** to yield **anthranol**. The various sequence of reactions are as given under:

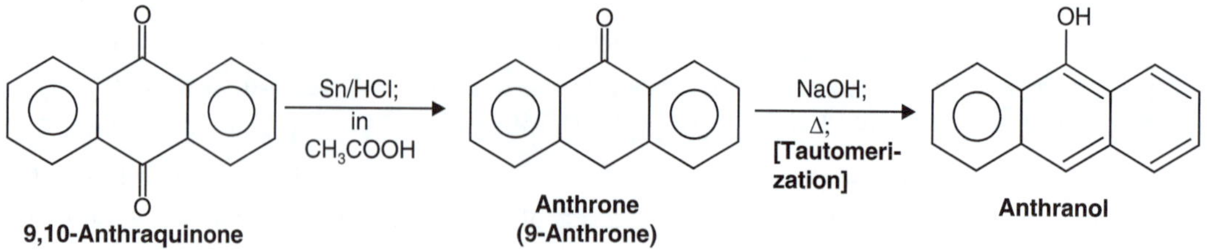

| 9,10-Anthraquinone | | Anthrone (9-Anthrone) | | Anthranol |

(b) **Reduction with Zn-dust and HCl (in CH$_3$COOH)**–*Anthraquinone* (**2-moles**) when reduced with *Zn-dust and HCl in glacial acetic acid* it yields a **bianthryl**–as given below:

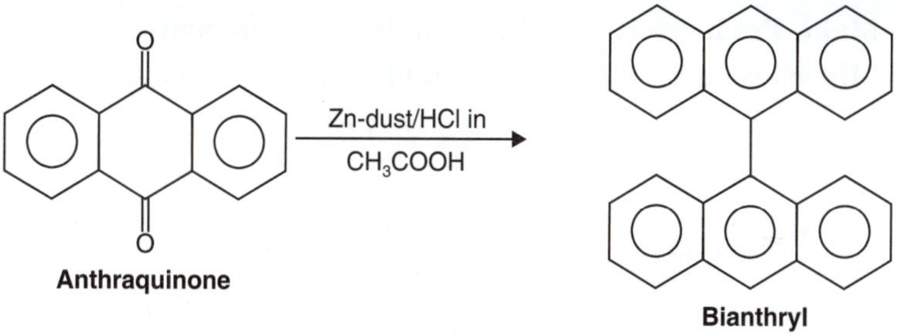

Anthraquinone

Zn-dust/HCl in CH₃COOH

Bianthryl

(c) **Reduction with Zn-dust and aqueous NaOH**–*Anthraquinone* on being reduction with *Zn-dust and aqueous NaOH* gives rise to the formation of **anthraquinol**. However, the resulting compound (**anthraquinol**) undergoes *tautomerization* in an acidic environment to yield **oxanthrol**–as shown below:

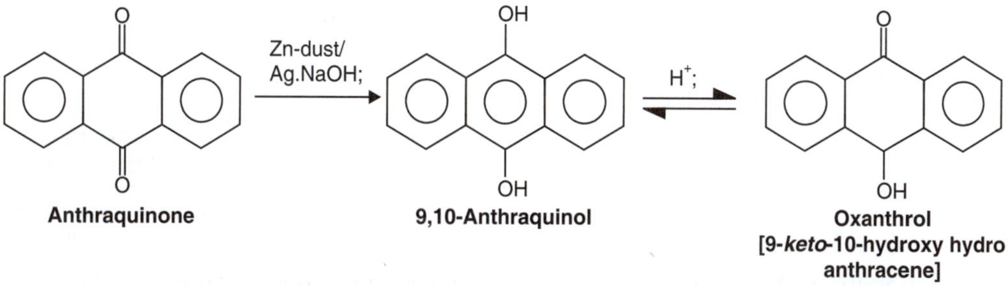

Anthraquinone

Zn-dust/ Ag.NaOH;

9,10-Anthraquinol

H⁺;

Oxanthrol
[9-*keto*-10-hydroxy hydro anthracene]

(d) **Reduction with Zn-dust and aqueous NH₄OH**–*Anthraquinone* on being subjected to treatment with *Zn-dust and aqueous NH₄OH* gives **9,10-dihydroanthrol**–as shown under:

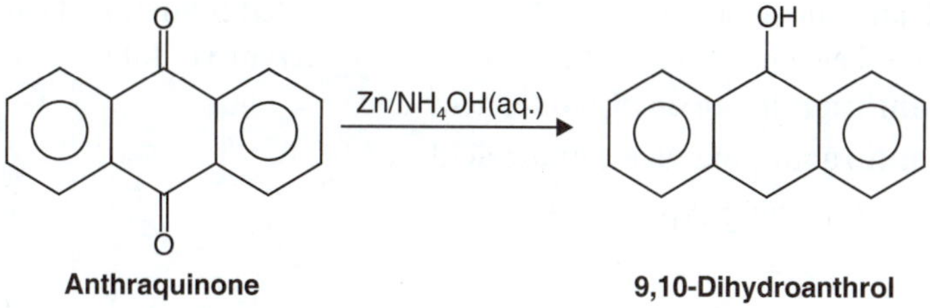

Anthraquinone

Zn/NH₄OH(aq.)

9,10-Dihydroanthrol

2. **Distillation with Zn-Dust**–Amazingly, when **anthraquinone** is distilled with *Zn-dust*, it yields **anthracene**– as given below:

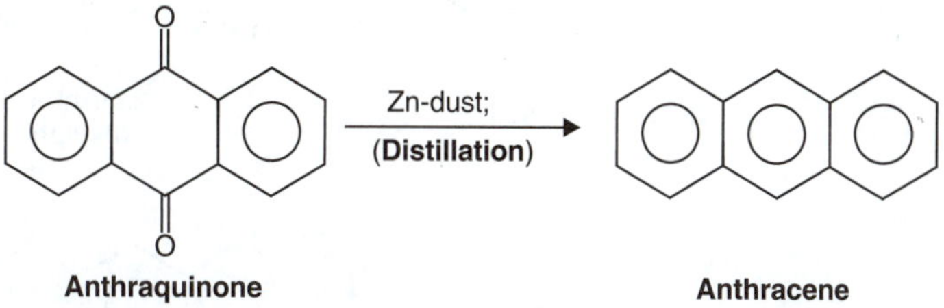

Anthraquinone

Zn-dust; **(Distillation)**

Anthracene

3. **Nitration**–*Anthraquinone* when undergoes *nitration* with an admixture of *nitric acid* (HNO_3) *and sulphuric acid* (H_2SO_4) it gives *1-nitroanthraquinone*. However, the resulting compound on further nitration yields:

- **As Major Products–1, 5- and 1, 8-dinitroanthraquinones**, and
- **As Minor Products–1, 4- and 1, 7-dinitroanthraquinones.**

Anthraquinone → **1-Nitroanthraquinone** (HNO₃/H₂SO₄; Nitration) → Further Nitration → **1,6-Dinitroanthraquinone** + **1,8-Dinitroanthraquinone**

Major Products of Nitration of Anthraquinone

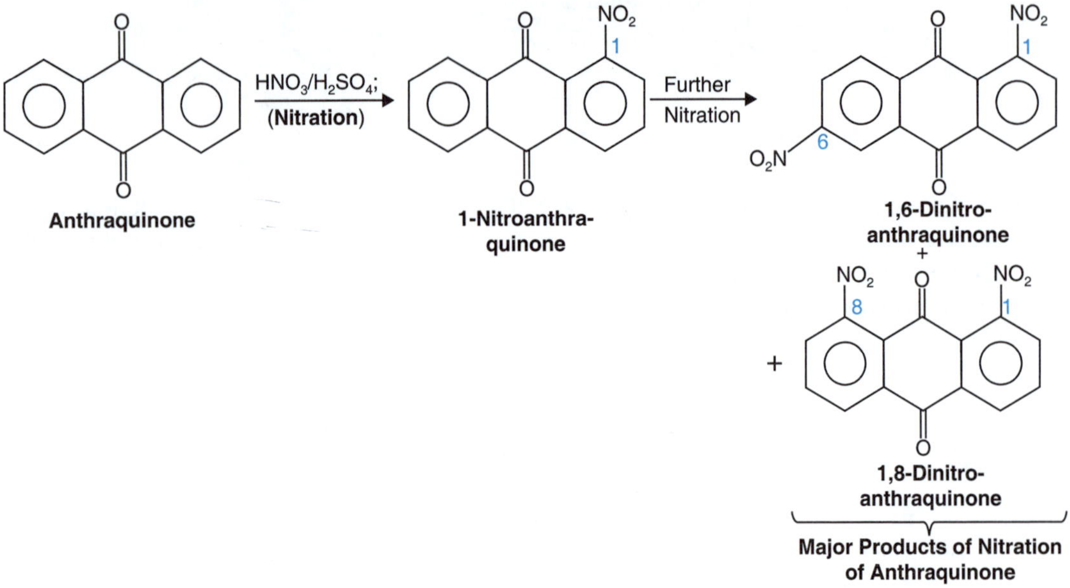

4. **Sulphonation–** In general, to afford the *sulphonation* of **anthraquinone** with the help of *concentrated H₂SO₄* seems to be a rather difficult task. Nevertheless, its effective *sulphonation* with oleum at 160°C to produce **2-sulphonic acid derivative** as the **primary product** together with a small amount of the *l-isomer*. Besides, the so-called continuous heating with oleum may ultimately yield an admixture of the following *two* **derivatives** (almost to an equal amount):

- **2, 6-anthraquinone disulphonic acid**; and
- **2, 7-anthraquinone disulphonic acid**.

Thus, we may have the following expressions:

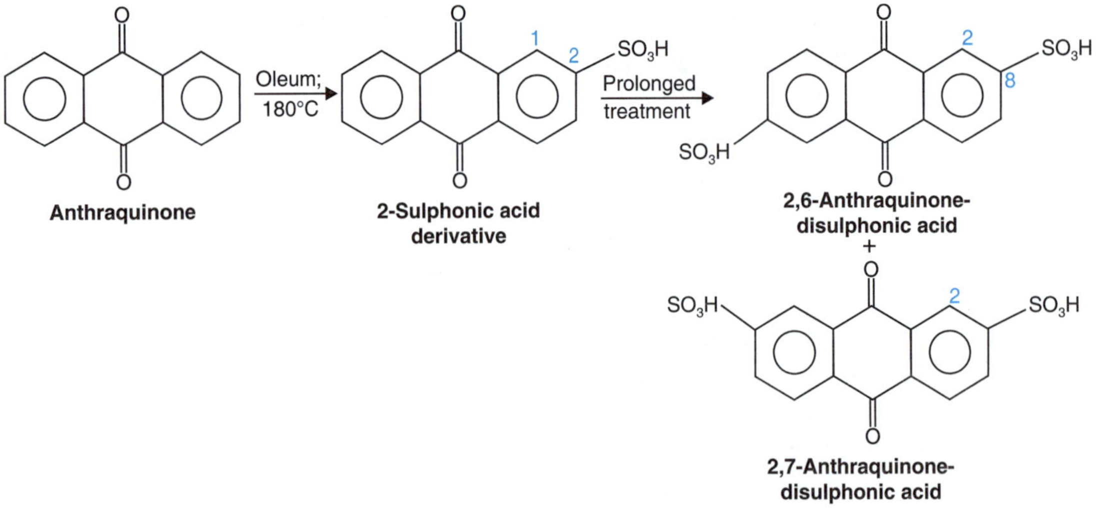

Anthraquinone → Oleum; 180°C → **2-Sulphonic acid derivative** → Prolonged treatment → **2,6-Anthraquinonedisulphonic acid** + **2,7-Anthraquinonedisulphonic acid**

Sulphonation of Anthraquinone with HgSO₄–In a particular instance when the sulphonation of anthraquinone is carried out with *mercuric sulphate (HgSO₄)*–the **sulphonation** adopts an altogether different route thereby producing **1-anthraquinone sulphoric acid,** which on a prolonged treatment yield a mixture of:

- **1, 5-anthraquinone disulphonic acid**; and
- **1, 8-anthraquinone disulphonic acid,**

as illustrated under:

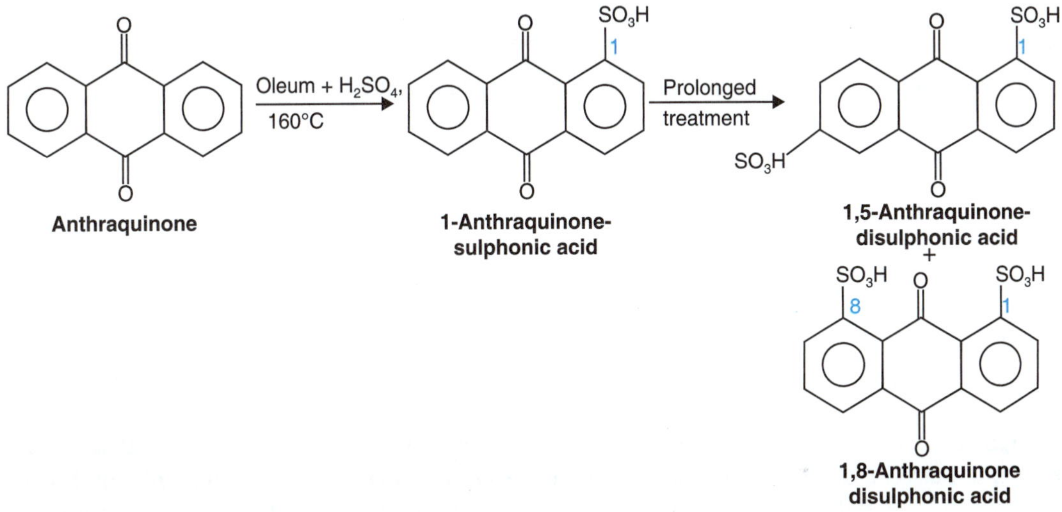

Anthraquinone → (Oleum + H₂SO₄, 160°C) → 1-Anthraquinone-sulphonic acid → (Prolonged treatment) → 1,5-Anthraquinone-disulphonic acid + 1,8-Anthraquinone disulphonic acid

Remark–Rapid displacement of the sulphonic acid moiety located strategically at **1- or 2- position** by *Cl-atom* to produce the respective **1-chloroanthraquinone**-as shown under:

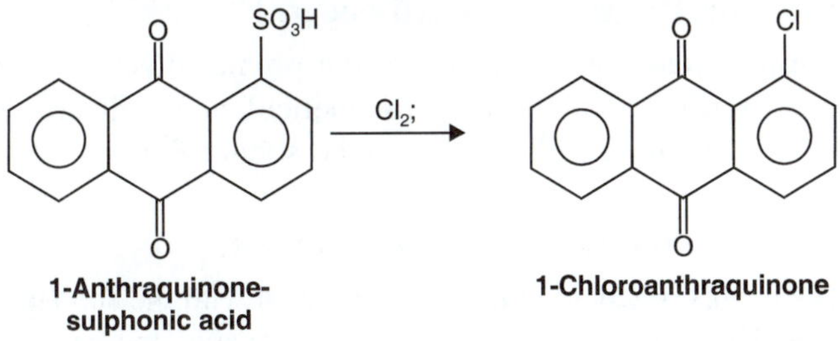

1-Anthraquinone-sulphonic acid → (Cl₂;) → 1-Chloroanthraquinone

5. **Halogenation**–It has been duly proven that **anthraquinone** may be *halogenated* with utmost difficulty. Thus, one may safely conclude that the *direct halogenation* of *anthraquinone* to obtain *mono*-**halogenoanthraquinone** is not possible at all.

3.3. Phenanthrene

Preamble–**Phenanthrene** is isomeric with *anthracene* and it usually occurs in *coal tar* (~4%), q. v., and also in *products of incomplete combustion.*

Importantly, **phenanthrene** vividly cites an example of an *angular polynuclear hydrocarbon*. Besides, it is found to be structurally related to several naturally occurring **plant alkaloids**, such as:

- **Morphine** (an *opium alkaloid*):

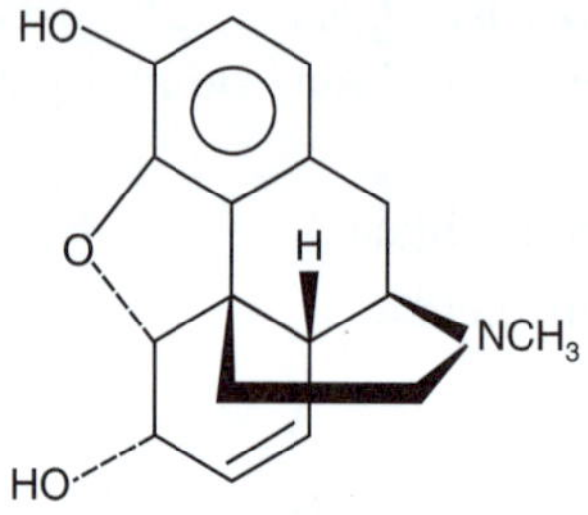

Morphine

- **Cholesterol** (a *steroidal alcohol*)-occurs in animal and milk fat.

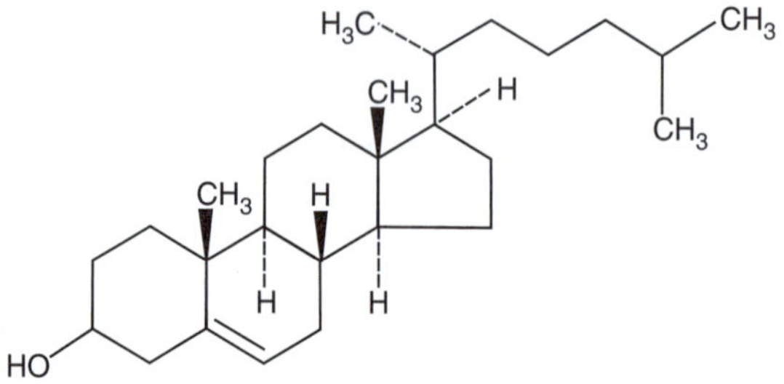

Cholesterol

Comment–In true sense, **phenanthrene** enjoys a far more importance *vis-a-vis* the **anthracene**; and hence, could also be obtained from *coal tar* in the **green-oil fraction**. Interestingly, **phenanthrene** is ascertained to be a definitely more important chemical entity than the *anthracene* from a **biological aspect** because its '*nucleus*' is prevalent in a variety of **vital natural products**, namely:

➢ **Alkaloids: Codeine, Thebaine, Morphine**; and
➢ **Steroids: Testosterone, Progesterone, Cortisone** etc.

Isolation of Phenanthrene–In actual practice, the **phenanthrene** is isolated right from the '*anthracene oil*' i.e., a specific fraction of **coal-tar** meticulously collected between the range of temperature **270–370°C**. Thus, the **solid-cake** obtained soonafter upon cooling *anthracene* does contain:

- **Phenanthrene**   - **Anthracene** and   - **Carbazole.**

Now, the **solid-cake** is pulverized and thoroughly washed with the **solvent naphtha** so as to dissolve the **phenanthrene**. The resulting clear solution when evaporated under vacuo yields **phenanthrene.**

CONSTITUTION OF PHENANTHRENE

The **constitution of phenanthrene** has been duly elucidated based on the various analytical data and scientific observations as enumerated under:

1. **Molecular Formula–**The *molecular formula* based on the analytical data is found to be $C_{14}H_{10}$.

2. **Presence of Diphenyl Ring System in Phenanthrene–**Phenanthrene on being subjected to *oxidation* with chromic acid (H_2CrO_4) [obtained from $K_2Cr_2O_7$ and $H_2SO_4(98\%)$], it gives **9,10-**

Phenanthraquinone, which is obtained as an *orange color product* (mp: 206°C). Now, further *oxidation* with **hydrogen peroxide (H_2O_2)** in acetic acid solution results in the *critical cleavage* of the '*quinone ring system*' to produce **diphenic acid (diphenyl, α, α'-dicarboxylic acid)**. The resulting compound on distillation with *sodaline* gives **diphenyl**.

Thus, we may have the following expression:

$$\text{Phenanthrene} \xrightarrow[H_2CrO_4]{[O]} \text{9,10-Phenanthraquinone} \xrightarrow[[O]]{H_2O_2 + CH_3COOH;} \text{Diphenic Acid} \xrightarrow{\text{Soda-lime}} \text{Diphenyl}$$

Now, based on the above sequel of reactions, it may be inferred that **phenanthrene** is related structurally to **diphenyl** (which has an *established chemical strcture*); and, therefore, the former (*i.e.,* **phenanthrene**) should essentially possess a '**diphenyl skeleton**' whose *two ortho-positions* found to be on the *same side* and are duly substituted as given in the following **structure (A):**

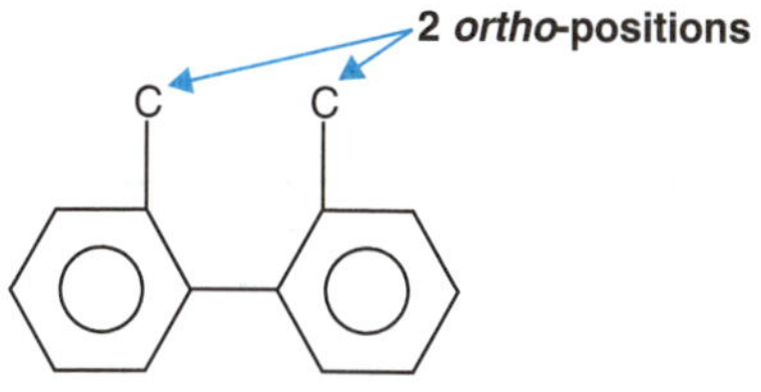

[A]

The above *C and H skeleton* is equivalent to $C_{14}H_8$; whereas, the **molecular formula of phenan-threne** is found to be $C_{14}H_{10}$. Thus, we may have to account for the remaining **2 H-atoms** only. Amazingly, the so-called only **2 H-atoms** are to be duly fitted by means of the proper maintenance of the '*tetravalent state*' of the aforesaid **2 C-atoms** by affording the meticulours closure of the '**middle ring**' so as to form the new structure of **phenanthrene (B):**

Phenanthrene
[B]

1,10-Phenanthraquinone

Diphenyl

Diphenic Acid

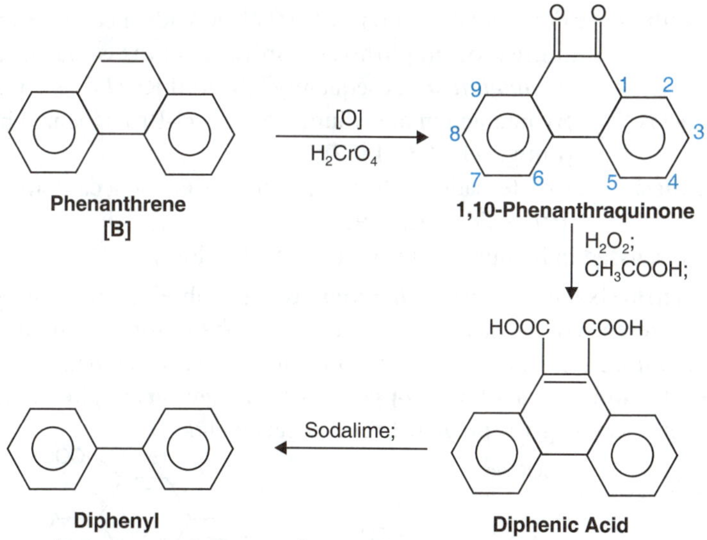

Remarks–The proposed structure of **phenanthrene (B)** is quite capable of an appropriate and logical explanation for the ultimate formation of **diphenyl** *via* the generation of *1,10-phenanthraquinone* and *diphenic acid.*

3. **Exact Position of the Olefinic (Double) Bonds**– In true sense, the **phenanthrene** is adequately and justifiably designated as a '*resonance hybrid*' of the following *five* **structural variants:**

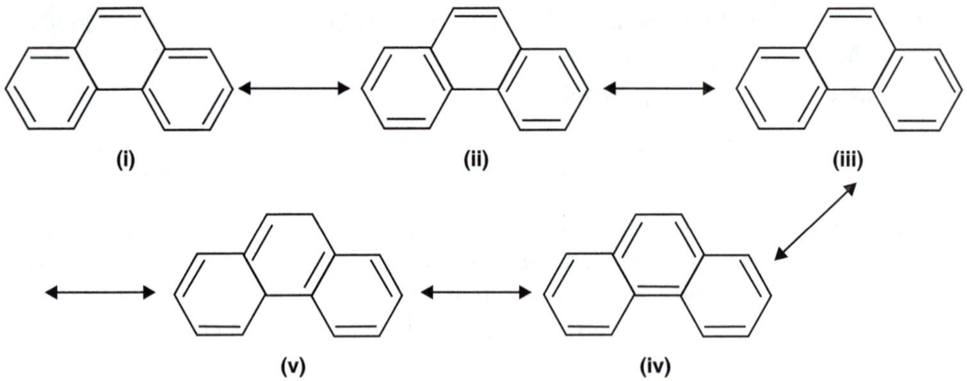

(i)   (ii)   (iii)

(v)   (iv)

However, the *olefinic bond character* of the various bonds is duly depicted in the following structure:

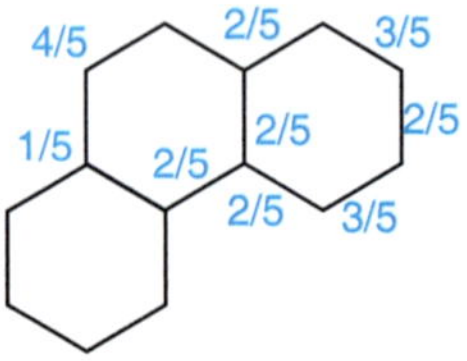

4/5   2/5   3/5
2/5   2/5
1/5   2/5
2/5   3/5

Resonance Energy = 38.7 kcal.mol^{-1}

Olefinic Bond Character in Phenanthrene.

4. **Synthesis of Phenanthrene**: The ultimate **structure (ii)** for **phenanthrene** has been duly confirmed by **synthesis** *via four* different routes, namely:

(a) Pschorr Synthesis : From *O*-Nitrobenzaldelyde and Phenyl acetic acid, *Pschorr*: *Ber,29*:500,1989;

(b) Haworth Synthesis : From condensation of Naphalene with snccinic anhydride to yield *two* moles of naphthoylpropionic acids-separated-converted into *phenanthrene* subsequently [HaworthRD:*J Chem Soe,1932*, 1125,2717;

(c) Fittig Synthesis : By interaction of sodium metal on *O*-Bromobenzyl bromide [Fittig:*Ber,5*:933,1872];

(d) Stobbe Synthesis : Condensation between succinic ester and a carbonyl compound [Stobbe H:*Ber, 26*:2312,1893].

which shall now be discussed individualy in the sections that follows:

4.1. Pschorr Synthesis-The *O*-**nitrobenzaldehyde** is subjected to heating with **sodium β-phenylacetate** in the presence of *acetic anhydride*, it yields –**phenyl-nitro cinnamic acid,** which upon subsequent *reduction-diazotization* gives rise to the formation of respective **diazonium salt**. The resulting salt when reacted with **sulphuric acid** and **copper (Cu) powder** forms **phenanthrene-9-carboxylic acid**, which on vigorous heating gives **phenanthrene**-as stated under:

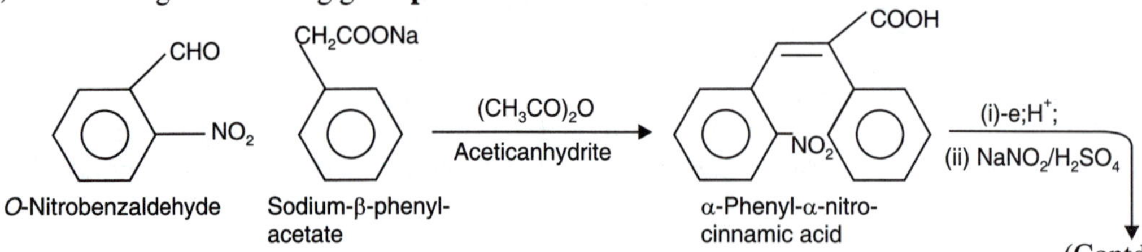

CHO   CH$_2$COONa   COOH

NO$_2$

$\xrightarrow[\text{Aceticanhydrite}]{(CH_3CO)_2O}$ NO$_2$ $\xrightarrow[\text{(ii) NaNO}_2\text{/H}_2\text{SO}_4]{\text{(i)-e;H}^+;}$

O-Nitrobenzaldehyde   Sodium-β-phenyl-acetate   α-Phenyl-α-nitro-cinnamic acid

(Contd.....)

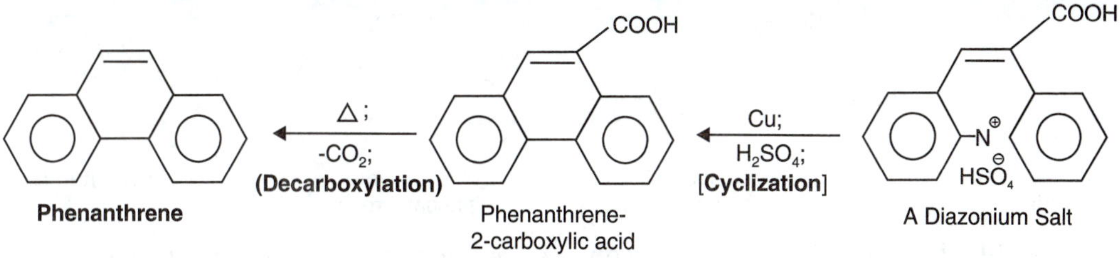

Phenanthrene   **Phenanthrene-2-carboxylic acid**   **A Diazonium Salt**

The various sequential steps are self-explanatory,

4.2. Haworth Synthesis–The condensation of **naphthalene** with **succinic anhydride** in the critical presence of *aluminium chloride (AlCl₃)* in a solution of *nitrobenzene* gives two moles of **naphthoylpropionie acid.** The resulting product may be separated duly and subjected to conversion into **phenathrene** as given under:

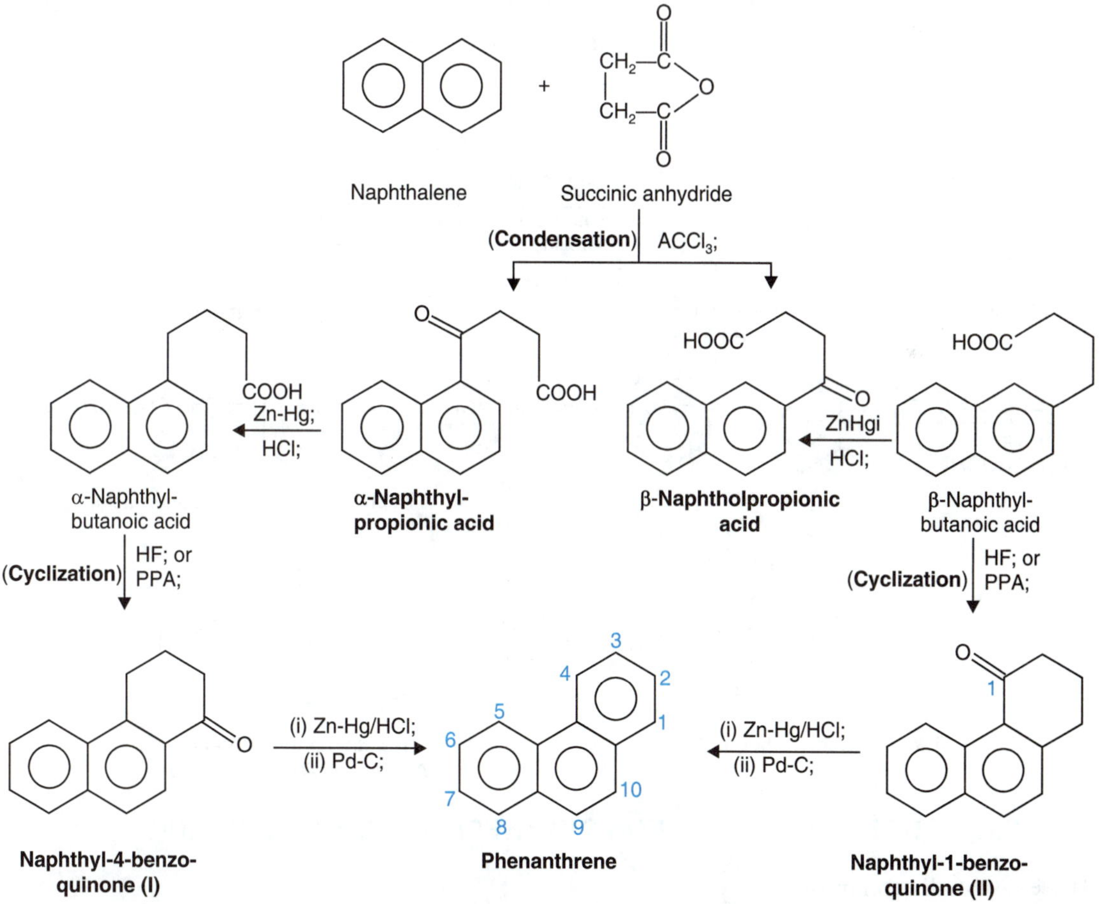

Naphthalene   **Succinic anhydride**

(Condensation) ACCl₃;

α-Naphthyl-butanoic acid   **α-Naphthyl-propionic acid**   **β-Naphtholpropionic acid**   **β-Naphthyl-butanoic acid**

(Cyclization) HF; or PPA;

Naphthyl-4-benzo-quinone (I)   **Phenanthrene**   **Naphthyl-1-benzo-quinone (II)**

4.3. Fittig Synthesis-The interaction of *two* moles of –*bromo-benzyl bromide* with freshly cut and dried pieces of *Na-metal* yields **9,10-dihydrophenanthrene** (as an *intermediate*) which on mild-oxidation yields **phenanthrene**-as shown below:

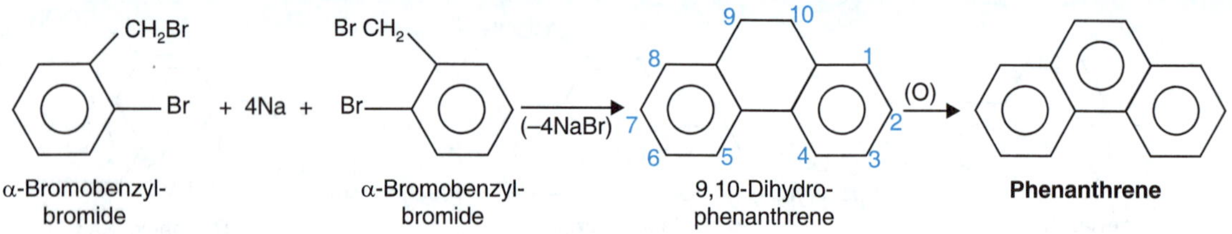

CH2Br
Br
+ 4Na +
Br CH2
Br
(–4NaBr)
(O)
α-Bromobenzyl-bromide
α-Bromobenzyl-bromide
9,10-Dihydro-phenanthrene
Phenanthrene

4.4. Stobbe Synthesis-It essentially involves a *condensation phenomena;* and hence, is chiefly employed for the typical **synthesis of phenanthrene structural analogues**. Importantly, it critically affects the condensation between:

- **A succinic ester, and**
- **A carbonyl (>C=O) chemical entity,**

in the crucial presence of **potassium tert-butoxide [$(CH_3)_3COK$]**. Thus, the *propionic acid residue* is duly introduced at the very site of the prevailing carbonyl (>C=O) moiety of the ***aromatic ketone.***

Thus, we may have the following expressions:

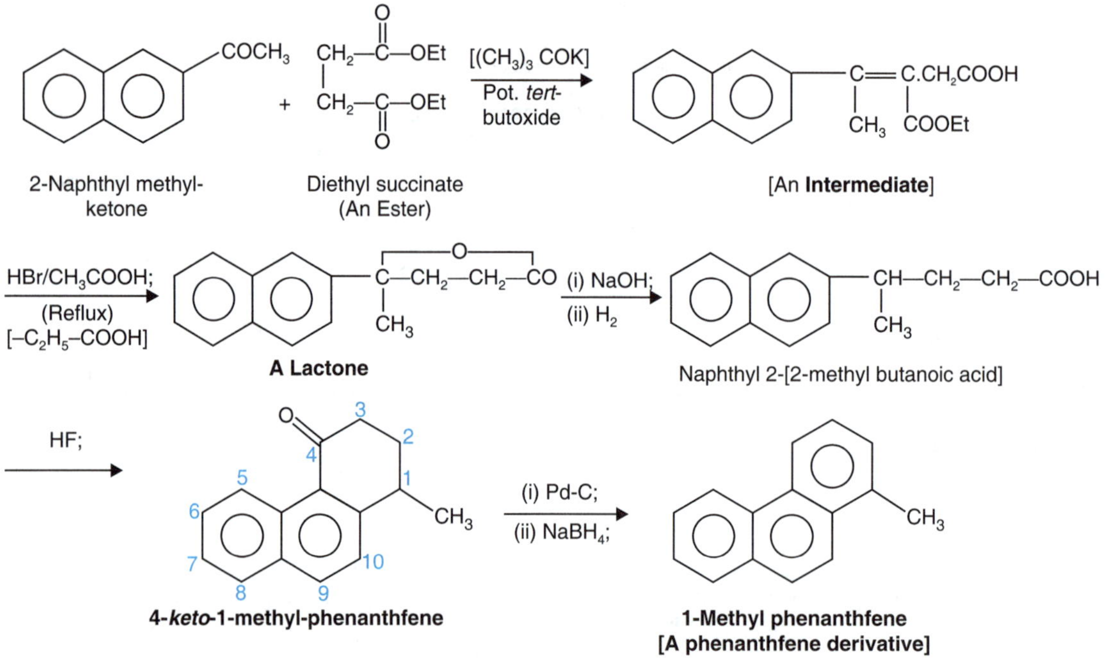

COCH3
[(CH3)3 COK]
Pot. tert-butoxide
2-Naphthyl methyl-ketone
Diethyl succinate
(An Ester)
[An Intermediate]
HBr/CH3COOH;
(Reflux)
[–C2H5–COOH]
A Lactone
(i) NaOH;
(ii) H2
Naphthyl 2-[2-methyl butanoic acid]
HF;
(i) Pd-C;
(ii) NaBH4;
4-keto-1-methyl-phenanthfene
1-Methyl phenanthfene
[A phenanthfene derivative]

TWO EXCEPTIONAL METHODS FOR SYNTHESIS OF PHENANTHRENE

These essentially comprise:

- **Irradiation Method, and**
- **Dehydrogenation Method.**

Irradiation Method- When either of the *two* **geometrical isomers of stilbene**, namely:

- *cis–stilbene,* and
- *trans–stilbene,*

is subjected to the *irradiation,* it yields **phenanthrene**-as shown under:

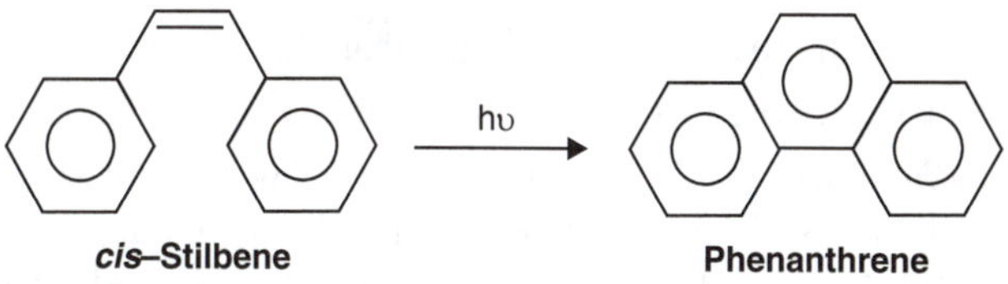

cis-Stilbene   Phenanthrene

Dehydrogenation Method-The tratement of *ortho*-dimethyl benzene with pure S-powder yields **Phenanthrene** – as stated below:

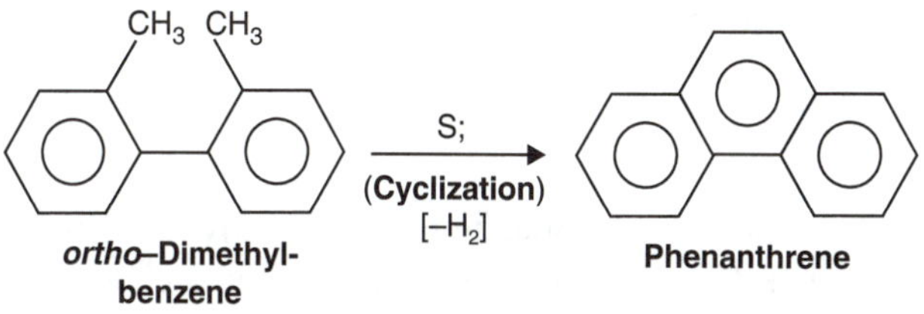

ortho–**Dimethyl-benzene**   Phenanthrene

ISOMERISM IN PHENANTHRENE

The structural formula of **phenanthrene** may be duly written in any of the following *three* ways, such as:

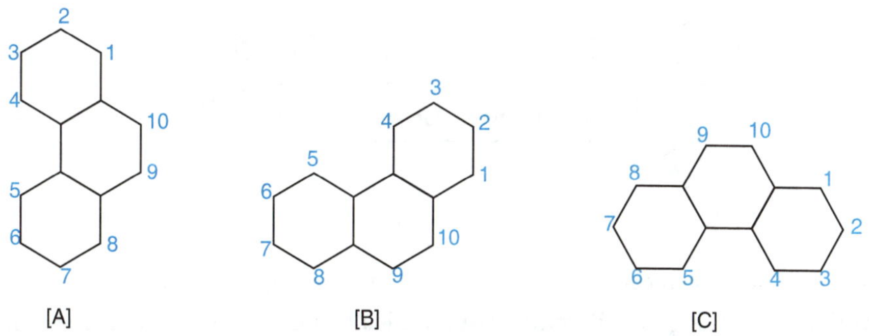

[A]   [B]   [C]

Monosubstitution Derivatives of Phenanthrene- do exhibit *five* distinct isomers at positions:**1,2,3,4,and 9.**

Disubstitution for two Identical Substituents-invariably give rise to *twenty five* possible isomers.

NOTE Since, there exists a wide possibility of an avalanche of isomers, the so-called structural analogues of phenanthrene are prepared usually *via* the synthesis route preferentially *vis-à-vis* the adaptation of any direct substitution in the so-called *phenanthrene nucleus.*

Characteristic Features of Phenanthrene

The characteristic features of **phenanthrene** may be featured in *two* different aspects:

❑ Physical Characteristics

Threee essentially include:

1. **Phenanthrene** occurs as monoclinic plates from ethanol.

2. It has d^{25} 1.179, mp:100°C, bb:340°C, and sublimes in a *high vaccum.*

3. *Solubility Profile: Phenanthrene* is almost insoluble in water, but soluble in organic solvent-especially in aromatic hydrocarbons. Besides 1g dissolves in 60ml cold ethanol, 10mL boiling 95%(v/v) ethanol, 25mL absolute, 2.4mL toluene or CCL_4, 2ml benzene, 1ml CS_2, and 3.3mL ahydrons ether. It is also soluble in glacial acetic acid. Amazingly, its respective solutions do exhibit a *blue fluorescence.*

4. *Molecular Compounds*-**Phenanthrene** forms molecular compounds with **picric acid, picryl chloride, dinitrobenzene, and similar nitro compounds.**

❑ Chemical Characteristics

They essentially comprise:

1. Very much akin to *anthracene,* it is found to be very reactive specifically at the **C-9 and C-10 positions.** Hence, **phenanthracene** possesses almost similar chemical characteristics features to *anthracene.*

2. **Addition Reactions**-following are the *three* known **addition reactions,** namely:

   (a) **Addition of Hydrogen (Reduction): Phenanthreine** undergoes rapid *eatalytic reduction* (by making use of CuO *(copper oxide)* amd *chromic oxide (Cr_2O_3)*] to yield **9,10-dihydrophenanthrene**, as given under:

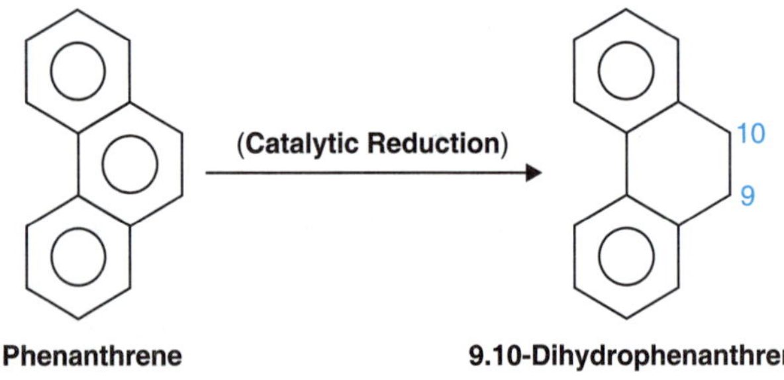

| Phenanthrene | (Catalytic Reduction) | 9.10-Dihydrophenanthrene |

   (b) **Addition of Halogens:** Bromine (**Br_2**) critically adds on to phenanthrene rapidly to form **9,10-dibromophenanthracene (or 9,10-Phenanthracene dibromide),** as given below:

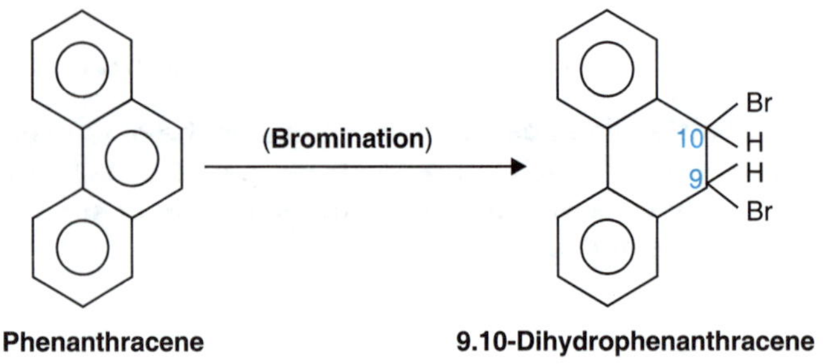

| Phenanthracene | (Bromination) | 9.10-Dihydrophenanthracene |

(c) **Addition of Ozone (O₃):Phenanthrene** undergoes *ozonolysis* to give *dialdehyde*, which upon subsequent oxidation yields **diphenic acid** as given under:

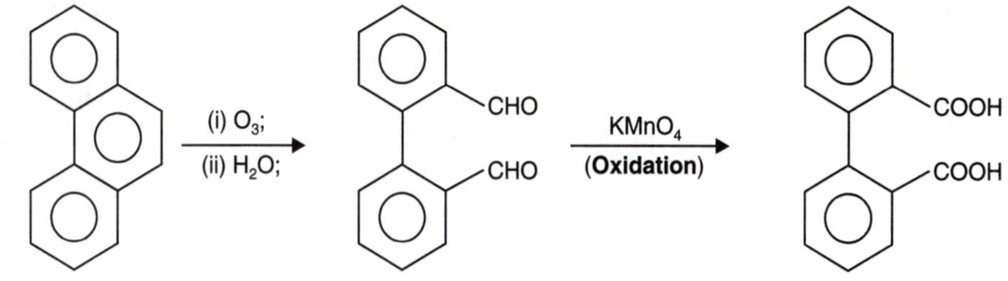

| Phenanthrene | A Dialdehyde | Diphenic Acid |

3. Oxidation Reaction: Phenanthrene on being subjected to **oxidation reaction** with *peracetic acid [CH₃-COO-OH]* gives rise to the formation of **diphenic acid** – as shown below:

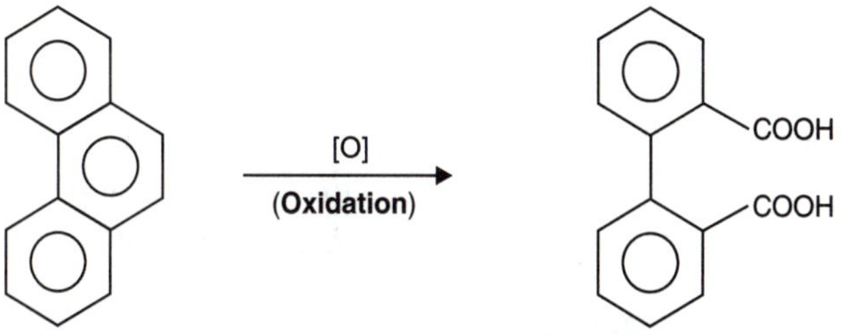

| Phenanthrene | Diphenic Acid |

4. Substitution Reaction: Importantly, **Phenanthrene** invariably behaves almost similar to:

• **'Aromatic Hydrocarbons'**

and hence, favourably undergoes the so-called **substitution reactions**. Nevertheless, one may not be able to produce solely the **'mono-substitution products'**, since the **'di-substitutions products** are also produced inadvertently. Obviously, there is an exception to the aforesaid episode whereby the **9-bromophenanthrene** that is normally generated by the careful incorporation of *bromine* to the corresponding refluing solutions of **phenanthrene** in CCl₄ (carbon tetrachloride). Thus, when **9-bromophenanthrene** is heated along with *cuprous cyanide [Cu (CN)₂]* at 260°C, -it results into the production of **9-cyanophenanthrene,** which is duly *hydrolyzed* to yield **Phenanthrene-9-carboxylic acid**-as given under:

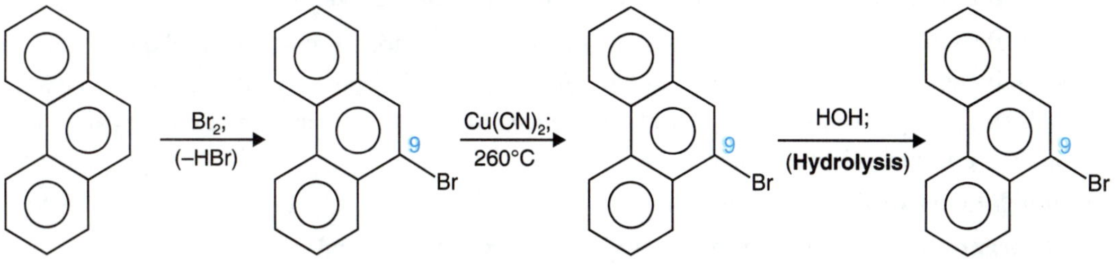

| **Phenanthrene** | 9-Bromo-phenanthrene | 9-Cyano-phenanthrene | Phenanthrene-9-carboxylic acid |

Comments:

1. **Phenanthrene** also gets *chloromethylated* at the 9-position.
2. **Phenanthrene** on being *nitrated* gives an admixture of **3-mononitro derivatives**; and the **3-isomer** is mostly predominating.
3. **Phenanthrene** when treated with *N-methyl-N-aldehyde pyridine* yields **phenanthrene-9-aldehyde** and **N-methyl pyridine** as shown below:

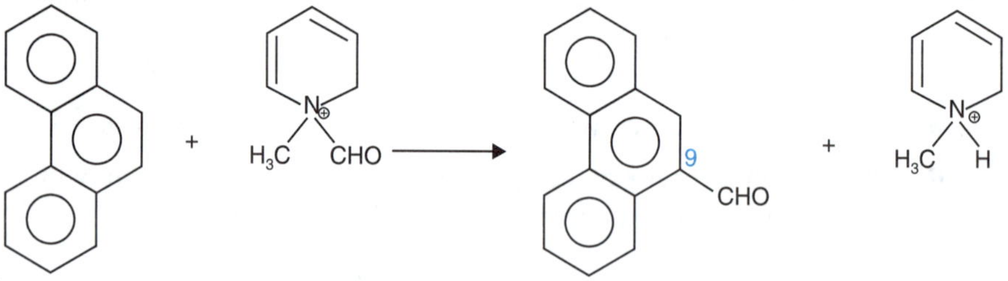

| Phenanthrene | N-Methyl-N-aldehyde pyridine | Phenanthrene-9-aldehyde | N-Methyl-pyridium cation |

4. Likewise, **phenanthrene** when *sulphonated* gives a mixture of **1-,2-,3-,4-, and 9-phenanthrene sulphonates** respectively; and the ultimate ratio of these isomers solely depend upon the prevailzing temperature at which the *sulphonation* was carried out.

4. THE CARCINOGENIC HYDROCARBONS

In a broader sense, the so-called **carcinogenic hydrocarbons** invariably refer to such agents *viz.*, **tobacco smoke, certain specific chemicals** (*benzene, saccharine, phenacetin etc.,*) **asbestors fibres, cement dusts, cool dusts, and high-dose radiations** that do have the critical property of cansing **cancer (neoplasm)**.

What is cancer (neoplasm)? The common terminology utilized to relate to a *malignant* tumour, irrespective of the tissue of origin.

'Malignancy' indicates that:

- the *tumour* is capable of progressive growth, unrectrained by the capsule of the parent organ, and/or
- that is capable of distant spread *via* lymphatics or the blood-stream, ensuring in the critical development of secondary deposits of tumour termed as *'metastases'*.

Nevertheless, microscopically, the neoplasm (cancer) cells invariably appear to be altogether different from the equivalent normal cells in the so-called *'affected tissues'*. Especially they may exhibit explicity a *lesser degree of differentiation* (*i.e.*, **they are more 'primitive'**), characteristics indicative of the following two prominent features, namely:

- much faster proliferative rate; and
- disorganized alignment in relationship to other cells or blood vessels.

Diagnosis- of 'cancer' solely rests upon the observation of these typical **microscopic characteristic features observed in biopsies**, *i.e.*, tissue removed surgically for such specific examinations).

4.1. Genesis of Carcinogenic Hydrocarbons

The actual overwhelming interest in the specialized domain of **'carcinogenic hydrocarbons'** attained tremendous critical attention and concern when it was duly established that such personnels working in such areas which are *directly exposed to coal-tar* do have a tendency to acquire *skin cancer* sooner or latter. Subsequently, several **British Chemists*** suggested categorically that the **cancer** is caused most perspectively by the presence of certain **'active high-boiling aromatic hydrocarbons'** in *coal tar.* Heiger (1930) reported that the **carcinogenic tars** (*i.e.,* responsible for causing *cancerons growth*) do exhibit a **fluroscence spectrum** having specific bands at **400,428, and 440 nm** *e.g.,* **1,2-benzanthracene,** thereby *ascertaining* the underlying fact, either this *parent chemical entity* or its *structural analogues,* may cause cancer ultimately.

Follow-up Revelations-A host of *polynuclear hydrocarbons* were identified that proved to be *a potent source to cancer (neoplasm),* for instance:

- **20-Methylcholanthrene [or 3-Methylcholanthrene]**

- **1,2:5,6-Dibenzanthracene**

- **9,10-Dimethl-1,2 benzanthracene**

---

* Kennaway and Cook (1930)–at the **Royal Cancer Hospital,** London (UK) by producing *cancer* successfully in the *mice* by exposure to **coaltar.**

4.2. Structural of Polynuclear Compounds *vis-a-vis* Carcinogenic Activity Profile

In fact, the exhaustive structures of the **polynuclear compounds** *vis-à-*vis **the carcinogenic activity profile** *i.e.*, the cancerous growth upon the *laboratory experimental animals*, such as: **mice or rabbits-** one may adopt either of the following two well-defined procedures:

Procedure-I: The solutions of *polynuclear hydrocarbon* is duly prepared either in **acetone or benzene**, which is then applied on the shaved *skin* of *mice* or *rabbit* in a predetermined time-schedule (periodically). Thus, if the **'test compound'** possesses **carcinogenic potency**, the *cancerous growth of the skin* (also called'**epitheliomas**') or *appearance of benign tumours* (also known as **'papilomas'**) will occur prominently.

❑ *For more potent test compounds*- the observed growth is quite rapid and fast; and

❑ *For less potent test compounds*- sufficient time-lag is usually observed for the development of tumours.

Procedure-II: First of all the solutions of **hydrocarbons** under investigation is duly prepared in **tricaprylin** solvent. Now, the resulting solutions is carefully injected by a *single* **subcutaneous injection** into the body of an *experimental laboratory animal.*

Comments- In case, the **hydrocarbon** exhibits *carcinogenic activity* a tumour of the connective tissues **(sarcoma)** must appear on the injected area often a span of several *weeks or months.*

Overall Observations- Bearing in mind the net results of the aforesaid *two* **procedures I & II**, one may explicitly conclude the so-called *overall effectiveness* of the **'test hydrocarbons'** causing **neoplasm (cancer)** in the following order:

> **'9,10-dimethyl>5,9,10-trimethyl-1,2-borzan-thracene>20-methyl chloranthrene>3,4- benzpyrene>1,2:5,6-dibenzoanthracene'.**

4.3. Critical Effect of Methyl(-CH₃) Moiety on the Carcinogenic Activity Profile of Poly- nuclear Hydrocarbons

Importantly, it has been well established and proven that the **1,2-dibenzanthracene** is regarded to be the *parent hydrocarbon* of almost all the known vital and **important carcinogenic hydrocarbons.** Amazingly, the *parent hydrocarbons* as such does not possess any *cancer-producing activity* (whatsoever); nevertheless, it is definitely **rendered active carcinogenically**-as and when a methyl moiety is being introduced strategically in the said compound at **C-10, C-9, or C-5,** such as:

• **5,10-dimethyl-1,2-benzanthracene;**

• **10-methyl-1,2-benzanthracene cholanthrene; and**

• **3,4-benzpyrene,**

whose chemical structures are as given under:

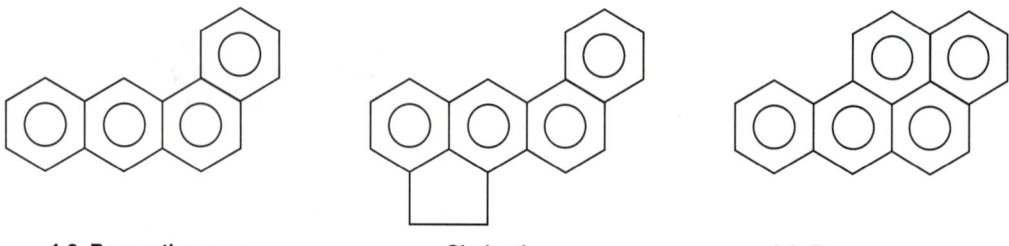

| 1,2–Benzanthracene | Cholanthrene | 1,2–Benzpyrene |

Remarks- These essentially comprise:

1. Besides, the exact position of the methyl (–CH$_3$) group, the total number of *methyl groups* are duly observed to influence the **overall activity profile** of a **hydrocarbon.**

Example- **Phenanthrene** is found to be **'inactive'**; whereas, its *methyl derivatives* are **'active'** in the following order:

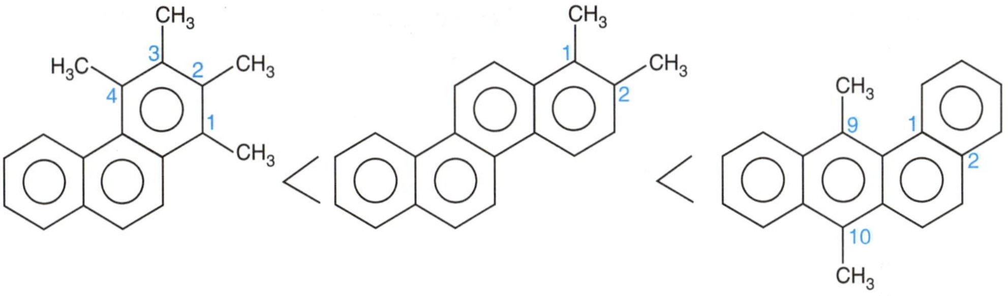

| 1,2.3,4-Tetra-methyl phenanthrene | 1,2.-Dimethyl-chrysene | 9,10-Dimethyl-1,2-benzanthracene |

2. Exactly, in the same manner **anthracene** is **'inactive'**; however, its respective **methyl (-CH$_3$)** **derivatives** *e.g.,* **6-methyl,** and **2,6-dimethyl analogues-** are found to be quite potent in the *increasing order*-as shown below:

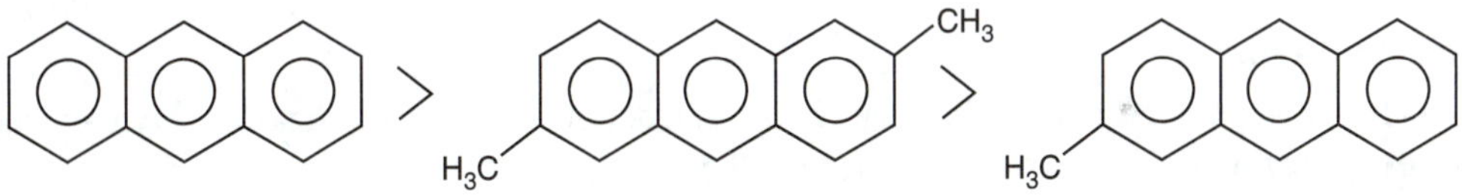

| Anthracene [*Inactive*] | 2,6-Dimethylanthracene [*Active*] | 6-Methyl anthracene [*Active*] |

*Remarks-*The aforesaid examples explicity makes it evident that the inherent *'carcinogenic activity* profile critically gets enhanced with an increase in the prevailing number of the *methyl moieties.* Nevertheless, the overall experimental observations ultimately lead to the results that the presence of the **methyl (–CH$_3$) moiety** may also largely minimize the *actual potency of the chemical entity (compound)* in causing the **cancerous growth.**

*Example-*The *two* **compounds**, namely: **3,4,8,9-dibenzpyrene** and **3,4,9,10-dibenzpyrene** are found to be *'active'* **carcinogenically;** however, their corresponding **methyl (–CH$_3$)** derivatives are invariably observed to be either:

• **Less** *'active'*, **or**
• **Absolutely** *'inactive'*.

4.4. Specific Examples of Certain Carcinogenic Hydrocarbons

A survey of literature would reveal the presence of the following *four* typical examples of the so-called **'carcinogenic hydrocarbons'**, such as:

(i) **20-Methylcholanthrene [or 3-Methylcholanthrene],**
(ii) **1,2-Benzpyrene [or 3,4-Benzpyrene],**

(iii)   **1,2,5,6-Dibenzanthracene [or Dibenj (a.j) anthracene}, and**
(iv)   **9,10-Dimethyl-1,2-benzanthracene,**

which shall now be discussed individually in the sections that follows:

4.4.1. *20-Methylcholanthrene* [or *3-Methylcholanthrene*]

Wieland and Dane (1933)* and Cook and Hallewood (1934)** discovered successfully that **20-methylcholanthrene** possesses exceptionally **carcinogenic** characteristics profile.

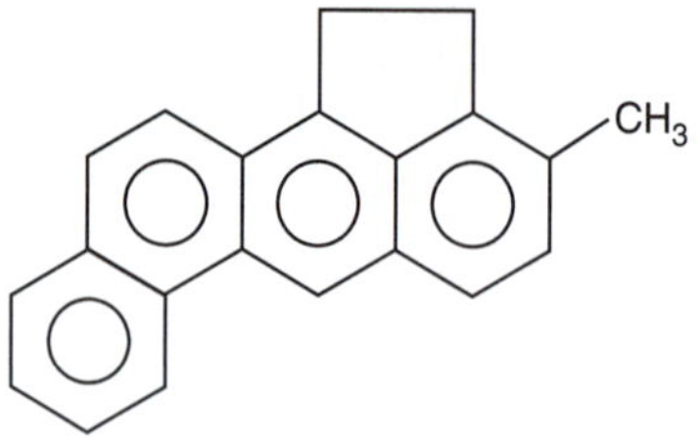

20-Methylcholanthrene

*Comments-*It was proposed that **20-methylcholanthrene** is generated duly in the body by the respective metabolism of:

- **Deoxycholic Acid**   •   **Cholic Acid**   or   •   **Cholesterol.**

Nevertheless, its actual metabolism has never been proved *in vivo.* Interestingly, therte are certain reactions that are duly involved in the so-called *metabolic conversion of some steroids into the desired* **methylcholanthrene.**

Examples- In actual practice, a host of known **chemical** reactions, such as:**Cyclization**, **Hydrogenation,** and **Dehydrogenation** do get involved in such synthetic routes of these chemical entities:

(a) 20-Methylcholanthrene from Deoxycholic Acid

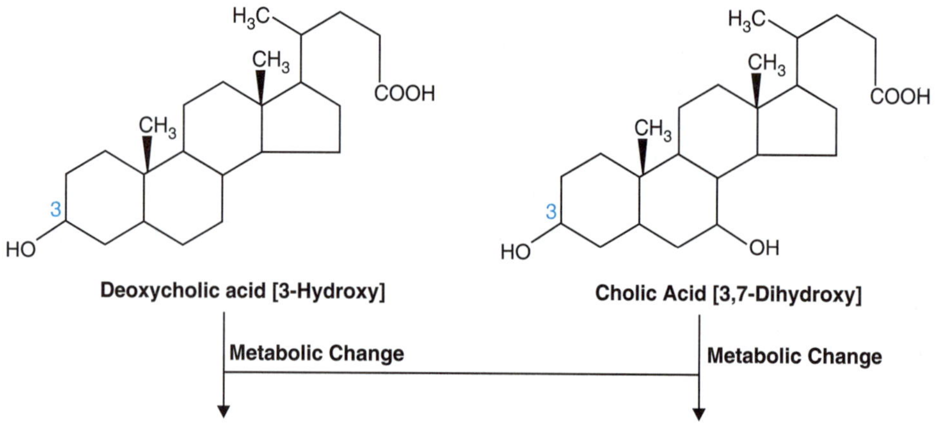

Deoxycholic acid [3-Hydroxy]   **Cholic Acid [3,7-Dihydroxy]**

Metabolic Change   **Metabolic Change**

(Contd.....)

---

*   Wieland and Dane : *Physiol phem,* 219:240,1933.
**   Cook and Haslewood:*J chem..sore,*428, 1934.

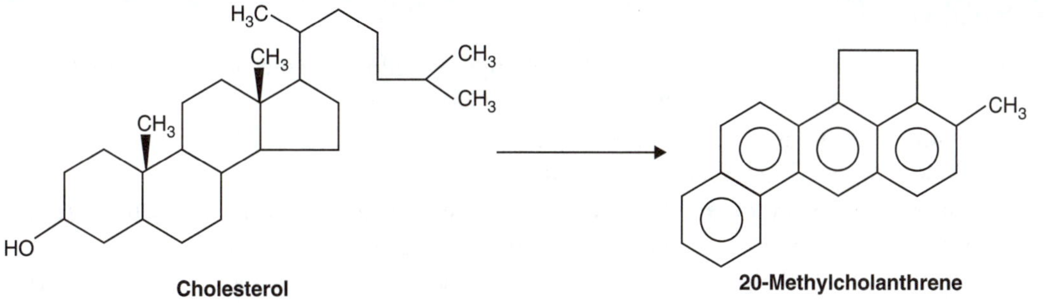
H3C
CH3
CH3
CH3
CH3
HO
CH3
Cholesterol
20-Methylcholanthrene

Thus, one may obtain **cholesterol** *via* the **metabolic change** either from **deoxycholic acid** *[3-Hydroxy]* or **cholic acid** *[3,7-dihydrgen]* which upon further transformation yields in **20 -methylcholanthrene**.

Characteristics Features- These essentially include:

1. It is obtained as a pale yellow, slender prisms from an admixture of *benzene* and *ether.*
2. It has mp:179-180°C and bp:280C.
3. It has d^{20}:1.28.

Synthesis of 20-Methylcholanthrene-Fieser and Seligman (1936)* put forward the synthesis of **20-methylcholanthrene** starting from ***p*-chlorotoluene** involving the following sequential steps:

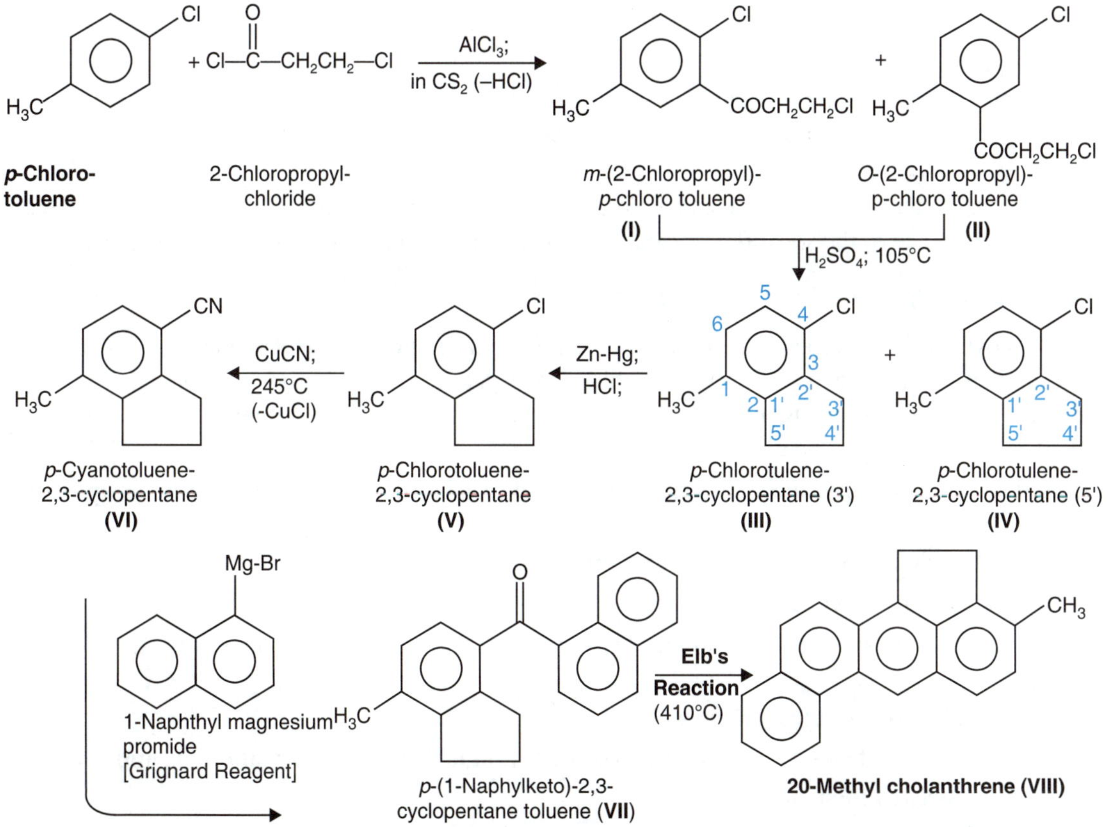
Cl
O
+ Cl—C—CH2CH2—Cl
AlCl3;
in CS2 (–HCl)
H3C
Cl
COCH2CH2Cl
H3C
Cl
H3C
COCH2CH2Cl
p-Chloro-
toluene
2-Chloropropyl-
chloride
m-(2-Chloropropyl)-
p-chloro toluene
(I)
O-(2-Chloropropyl)-
p-chloro toluene
(II)
H2SO4; 105°C
CN
CuCN;
245°C
(-CuCl)
H3C
Cl
Zn-Hg;
HCl;
H3C
5
6
4
Cl
3
1
2
1'
2'
3'
5'
4'
H3C
+
Cl
H3C
p-Cyanotoluene-
2,3-cyclopentane
(VI)
p-Chlorotoluene-
2,3-cyclopentane
(V)
p-Chlorotulene-
2,3-cyclopentane (3')
(III)
p-Chlorotulene-
2,3-cyclopentane (5')
(IV)
Mg-Br
1-Naphthyl magnesium
promide
[Grignard Reagent]
O
H3C
p-(1-Naphylketo)-2,3-
cyclopentane toluene (VII)
Elb's
Reaction
(410°C)
CH3
20-Methyl cholanthrene (VIII)

* Fieser and Seligman: *J Am Chem Soc*, **58:**3482,1936.

Explanation

1. Interaction of *p*-chlorotoluene and β2-chloropropyl chloride in the presence of AlCl$_3$ in *carbon disulphide (CS$_2$)* yields **compounds (I)** and **Compound (II).**

2. Both compounds **(I)** and **(II)** on treatment with H$_2$SO$_4$ at 105°C gives *two* respective **cyclopentone (3') (III)** and **cyclopentone (5') (IV).**

3. The resulting **compounds (III) and (IV)** on being subjected to reduction with *Zn-Hg* in HCL yields *p*-**chlorotoluene -2,3-cyclopentane (V),** which upon tratement with *cuprous cyanide (CuCN)* at 245°C loses a mole of *cuprons chloride (CuCl)* to give *p*-**cyanotoluene-2,3-cyclopentane(VI).**

4. **Compounds (VI)** when treated witrh a **Grignard reagnent** (1-naphthyl magnesium bromide) produces **compound (VII),** which finally undergoes the **Elb's Reaction** at 410C to give the desired product **20-methylcholanthrene (VIII).**

Another Chemical Structure for 20-Methylcholanthrene.

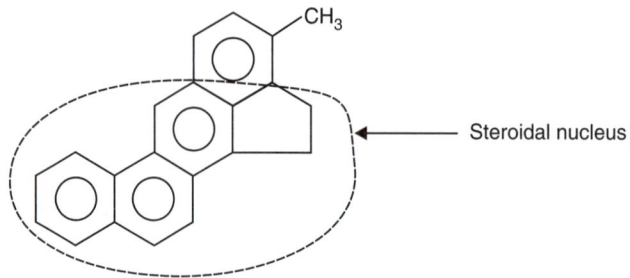

Obviously, the above alternative chemical structures assigned to **20-methylcholanthrene** vividly shows the presence of the **steroidal nucleus** *i.e.,* **cyclopentanophenanthrene nucleus** as *marked.*

4.4.2. 1,2-Benzpyrene [or 3,4-Benzpyrene]

It enjoys the glory of being the first and foremost **'carcinogenic hydrocarbon** ever isolated from *coal-tar* that critically exhibits an appreciably strong pharmacologic activity. Later on, it was duly established to be the **'main agent'** for causing cancer in such individuals engaged exclusively in the so-called *coal-mines* across the globe.

Occurrence- **1,2-Benzpyrene** is invariably produced in the course of *pyrolysis* of an array of such substances as:

- **Cellulosic hydrocarbon materials, and**
- **Complex hydrocarbon substances.**

Nevertheless, in small amount it indeed shows its prevalent existence in the following substances, namely:

- ❑ **Exhausts of petrol engines,**
- ❑ **Burning of wood (*viz.,* domestic cooking, jungle fire-as seen in California and Australia,**
- ❑ **Scorched food, and**
- ❑ **Stone dust.**

Synthesis of 1,2-Benzpyrene

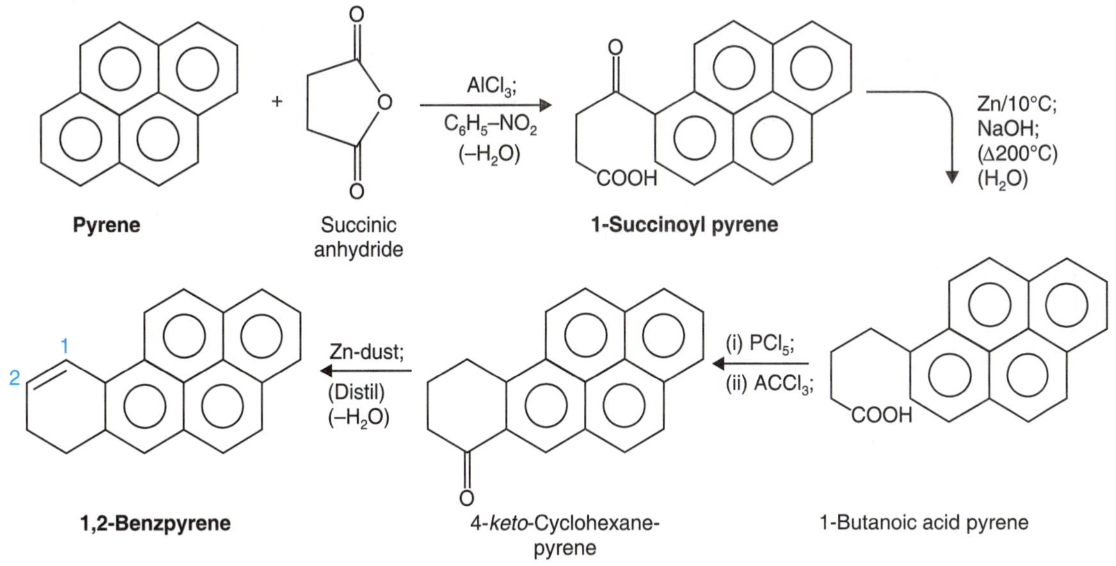

Pyrene + **Succinic anhydride** → **1-Succinoyl pyrene**

AlCl₃; C₆H₅–NO₂ (–H₂O)

Zn/10°C; NaOH; (Δ200°C) (H₂O)

1,2-Benzpyrene ← Zn-dust; (Distil) (–H₂O) ← **4-keto-Cyclohexane-pyrene** ← (i) PCl₅; (ii) ACCl₃; ← **1-Butanoic acid pyrene**

Explanation- The various steps involved may be explained as under:

1. Interaction of **pyrene** and **succinic anhydride** in the presence of aluminium trichloride in nitrobenzene loses a mole of water to yield **1-succinoyl pyrene,** which upon treatement with Zn (at10°C) followed by heating with NaOH at 200°C gives a mole of **1-butamoic acid pyrene.**

2. The resulting compound on reaction with PCl₅ followed by AlCl₃ undergoes cyclization to produce 4-**keto-cyclohexane pyrene,** which on distillation with *Zn-dust* gives **1-2-benzpyrene.**

> **NOTE** **The position e-1 1,2 benzpyrene is found to be the most *'reactive position'.***

4.4.3. *1,2,5,6-Dibenzanthracene [Dibenz (a.j) anthracene]*

Incidently, this specific aromatic hydrocarbon is also present legitimately in **coal tar products.** Clar(1929)* and Benchman(1937)** proposed the synthesis of **1,2,5,6-dibenzanthracene** starting from the condensation of **β-naphthylcarbonyl chloride** and **β-methylnaphthalene** as given under. However, one may obtain **2-napthoyl chloride** (or β-napthylecarbon chloride) by treating *2-naphthoic acid* with a chlorinating agent **PCl₅** (phosphorous pentachloride):

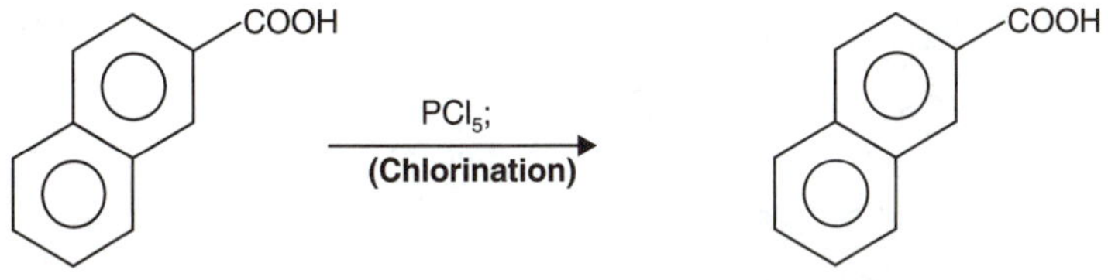

2-Naphthoic acid (A) → PCl₅; (Chlorination) → **2-Naphthoic Chloride (B)**

(Contd.....)

---

* Clar:*Ber,* **62**:350,1378 (1929).

** Bechmann,*J Org chem.,* **1:**347 (1937).

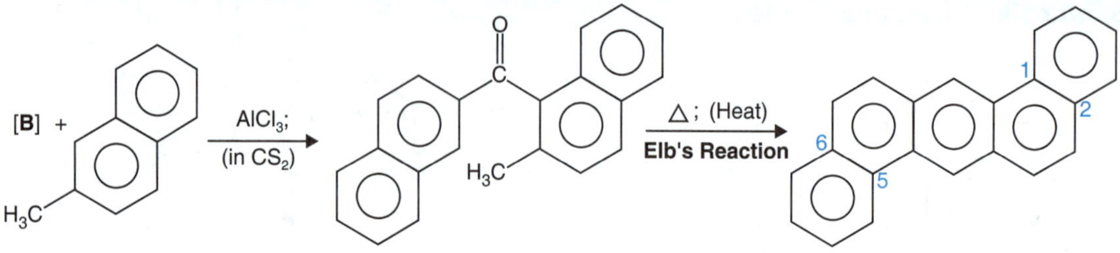

[B] + 2-Methyl-naphthalene → (AlCl₃; in CS₂) → Methyl-dinaphthyl ketone (C) → (Δ; (Heat) Elb's Reaction) → 1,2,5,6-Dibenzanthracene

Thus, **2-naphthoyl chloride (B)** reacts with **2-methylnapthalene in the presence of AlCl₃ in CS₂ to yield methyl-dinaphthylketone (C).**

The product (C) on heating undergoes **Elb's reaction** to give **1,2,5,6-dibenzanthracene.**

Characteristic Features

1. It is obtained as *plates* or *leaflets* from acetic acid.
2. The crystals may be either *monoclinic* or *orthorhombic* in nature.
3. It undergoes sublimation and melts at 266°C.
4. It is soluble in petroleum ether, benzene, toluene, xylene, oils, and other organic solvents; slightly soluble in alcohol and ether, and insoluble in water.

Caution: It is proclaimed to be a *carcinogenic agent*.

4.4.4. 9,10-Dimethyl-1,2-Benzanthracene

Newman (1938)* proposed the synthesis of **9,10-dimethyl-1,2-benzanthracene** from the starting material *9,10-dimethyl-9,10-dihydroxy-9,10-dihydro-1,2-benzanthracene* as shown below:

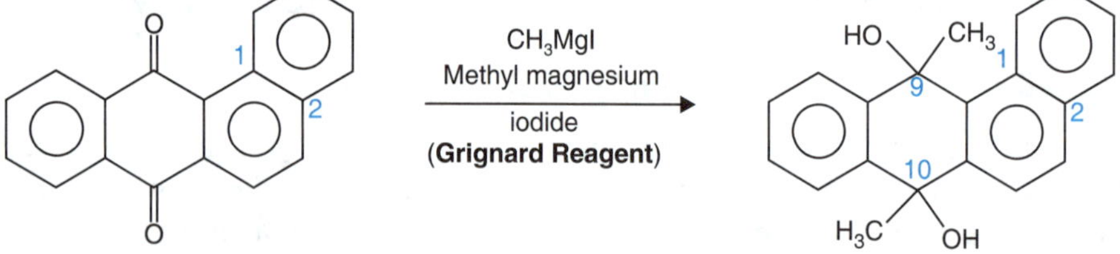

1,2-Benzanthra-quinone → (CH₃MgI Methyl magnesium iodide (**Grignard Reagent**)) → 9,10–Dimethyl-9,10-dihydroxy-9,10-dihydro-1,2-benzanthracene (**A**)

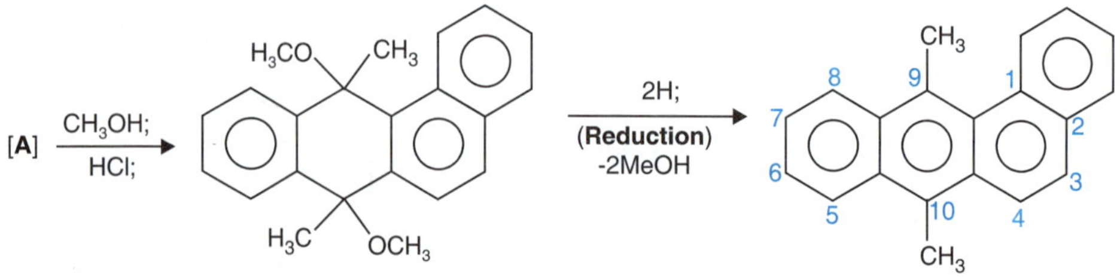

[A] → (CH₃OH; HCl;) → 9,10-Dimethyl-9,10-dimethoxy 9,10-dihydro-1,2-benzanthracene (**B**) → (2H; (**Reduction**) -2MeOH) → 9,10-Dimethyl-1,3-benzanthracene (**C**)

---

* Newman, *J Am Chem Soe.*, **60**:1023(1938).

Explanation

1. 1,2-Benzanthracene undergoes *Grignardization* with CH_3MgI to yield the **product (A)**, which on further treatment with methanol in an acidic environment yields the respective **9,10-dimethoxy derivative (B).**

2. The resulting **product (B)** on being subjected to reduction loses *two* moles of methanol to give the desired **product (C)** *i.e.*, **9,10-dimethyl-1,3-benzanthracene.**

Characteristics Features

1. It occurs as plates, leaflets from acetone-alcohol having a faint greenish-yellow tinge.
2. It has mp:122-123°C.
3. It shows maximum fluorescence at **440nm**.
4. **Solubility Profile** – it is freely soluble in **benzene;** moderately soluble in **acetone;** slightly soluble in **alcohol;** and insoluble in **water.**

However, it may be solublized in water by **purines** *eg.*, **caffeine, tetramethyluric acid.** Besides, the **nucleorides, adenosine**, and **guanosine** do exhibit a **solvent action** predominantly.

FURTHER READING REFERENCES	
House H and House O	: **Modern Synthetic Reactions,** 2nd edn., Benjamin, Menlo Park, 1972.
Simon J	: **The Total Synthesis of Natural Products, Vols-1-5,** Wiley Interscience, New York, 1973-1983.
Tedder JM, Neehvatal A, and Jubb AH	: **Basic Organic Chemistry. Part-5**, Industrial Products, Wiley London (UK),1975.
Vogel BS *et al.*	: **Vogel's Textbook of Practical Organic Chemistry,** 4th edn., Longman, London (UK), 1978.
Warren S	: **Designing Organic Synthesis**, Wiley, Chichester, 1978.

REVIEW QUESTIONS

1. What do you mean by **Polynuclear Aromatic Hydrocarbons?** How can we classify them? Give a brif account of the **Polyphenyl Compounds** and support your answer with a few typical examples.

2. How do we prepare the following compounds:
   - (a) **Diphenic Acid**
   - (b) **Benzidine**
   - (c) **Triphenylmethane**
   - (d) **Triphenylcarbinol**

3. Give a comprehensive account on the 'Condensed Ring Systems. Explain your answer with suitable examples

4. Discuss the chemical properties of **Naphthalene**.

5. Describe the **Constitution of Anthracene.**

6. What are the various methods for the **Preparation of Anthraquinone?** Give examples wherever necessary.

7. Elaborate on the "**Coustitution of Phenanthrene**".

8. Write a detailed account on the 'Carcinogenic Hydrocarbons'

Contents at a Glance

5 Enzymes, Coenzymes, and Fermentation

[A] ENZYMES

1. INTRODUCTION

Enzymes may be defined as the *catalysts* duly elaborated by the living organisms that eventually control and modulate the several processes intimately associated with life.

Amazingly, many of them do possess high degree of *specificity* with regard to the *substances (substrates)* whose reactions they catalyze predominantly; and, therefore, these are invariably termed by the addition of the suffix-*ase* to the *root of the name of the substrate.*

Sunner* succeeded in isolating a pure enzyme for the first time in 1926 and named it *'urease'*- since it is specific to the substrate urea and catalyzes *hydrolysis* to CO_2 and NH_3 ultimately. Subsequently, to Sunner's leading research and findings broadly paved the way for **'more enzymes'** that were isolated substantially in a much more **pure form.** Importantly, quite a few of them are *all proteins;* however, some do comprise a typical *non-proteinoid prosthetic moiety* found to be absolutely essential for inheriting the **activity profile**,

2. ENZYME VARIANTS

In usual practice, the **enzymes** fall into *two* **broad categories,** namely:

- **Hydrolytic Enzymes,** and
- **Oxidative Enzymes.**

2.1. Hydrolytic Enzymes

. The **hydrolytic enzymes** refer to such enzymes that control specifically the phenomenon of *hydrolysis* (and *resynthesis)* of such substances as:

- **Esters** • **Carbohydrates** • **Proteins** and • **Amides.**

Obviously, these appear to be **simple proteins**:

2.2. Oxidative Enzymes

The **oxidative enzymes** relate to such enzymes that critically control various **oxidation-reduction reactions.** Besides, these enzymes do contain the so-called *prosthetic moieties* that are essential for the various **oxidation-reduction processes** duly controlled and modulated by the **enzymes.**

---

* Sunner JH:(1887-1955) b.Canton, Mass; Ph.D., Harvard, N.Y.State, Agr.coll, Eornell (USA); Nobel Prize 1946.

Table: 5.1 records the **enzymes** of either variants (as stated above) which have been meticulously isolated in sufficiently pure form. Interestingly, it does not register such instances wherein the *prosthetic moiety* is duly known but the **'enzyme'** has not get been characterized as a **complete conjugated protein state.**

TABLE:5.1: THE HYDROLYTIC ENZYMES

S. No	Name	Substrates	Products
I	**ESTERASES**		
	• Acetylesterase	Esters of acetic acid	CH_3COOH + EtOH
	• Cholinesterase	Acetylcholine	CH_3COOH + Choline
	• Lipase	Glycerides	Fatty Acids + glycerol
	• Phosphatases	Phosphate Esters	H_3PO_4 + Alcohols
II	**CARBOHDRASES**		
	• Maltase	Maltose	Glucose
	• Amylase	Starch	Maltose
	• Lactase	Lactose	Galactose + Glucose
III	**PROTEASES & PEPTIDASES**		
	• Renin	Casein	Paracasein
	• Pepsin	Proteins	Proteoses, peptones
	• Trypoin	Proteins	Polypeptides + Amino acids
	• Carboxypeptidase	Carboxypolypeptides	Amino acids
IV	**PHOSPHORYLASES**	Polysaccharides	Hexose phosphate
V	**AMIDASES**		
	• Urease	Urea	$CO_2 + NH_3$
	• Arginase	Arginine	Urea + Ormittine

Table:5.2 enlists the various enzymes belonging to the class of **oxidative enzymes:**

TABLE: 5.2: THE OXIDATIVE ENZYMES

S.no	Name	Substrate	Products
I	**DEHYDROGENASES**		
	Lactic dehydrogenase	Lactic acid	Pyruvic acid
	Alcohol dehydrogenase	Ethanol	Acetaldehyde
	Succinic dehydrogenase	Succinic acid	Fumaric acid
II	**OXIDASES**		
	Tyrosinase	Tyrosine	Melanin
	Ascorbic and oxidase	Ascorbic acid	Dehydroascorbic acid
III	**CATALASE**	$2H_2O_2$	$2H_2O+O_2$

3. SALIENT FEATURES OF ENZYMES

These essentially include such salient features as:

(i) Practically **all enzymes** do exhibit **stereochemical specificity;** and hence, the enzyme-**'lactic acid dehydrogenase catalyzes'** the oxidation of only **L-lactic acid** but not of **D-lactic acid.**

(ii) *Certain enzymes* particularly show *absolute specificity.* whereby they do control the reaction of *only* **one substrate** *viz.,* **urease**.

(iii) A few **enzymes** explicity exhibit only the so-called **'linkage-specificity'**, such as:

- 1some **'esterases'** are capable of promoting the hydrolysis of *any ester* irrespective **of the basic structures of the respective acid** and **alcohol components.**

- usually the more prevalent requirements do fall in between the **absolute type** and the *linkage type i.e.,* the said **'enzyme'** not only needs a **specific linkage** but also **certain functional moieties as well** that are located strategically in the *vicinity of this linkage.*

- Bergmann (1937) carried out the pioneering work on **'group specificity'** using some *synthetic peptides*- that ultimately led to the fact that various **proteinases** do have *definite needs* so as to catalyze the **hydrolysis of peptide bonds**.

4. CLASSIFICATION OF ENZYMES

In a broader perspective, the **enzymes** may be classified under the *six* major heads, namely:

- **Oxidoreductases,**
- **Transferases,**
- **Hydrolases (Hydrolytic Enzymes),**
- **Ligases,**
- **Isomerases, and**
- **Ligases (Synthetases),**

which shall now be treated separately with *main classes* and *sub-classes* supported by suitable examples:

4.1. Oxidoreductases

They actually designate the type of **enzymes** which particularly catalyzes **oxidation-reduction reactions** *viz.,* Dehydrogenase, Oxidase, and Oxygenase.

In other words, the **oxido reductases** refers to such **enzymes** that are involved intimately in the following *two* **chemical reactions,** namely:

- **Biological oxidations*, and**
- **Biological reductions**.**

---

* **Oxidation:** it means the addition of O_2, or removal of H_2, or removal of electrons, or increase in bonds to more electronegative atoms, or decrease of bonds to less-electronegative atoms. However, in most of the **oxidation reactions** involve either the enhancement in bonds to O_2 or reducing in bonds to H_2.

** **Reduction:** It means removal of O_2, or addition of H_2, or addition of electrons, or gain-in-bonds to less electronegative atoms, or decrease of bonds to more electronegative atoms, However, in most of the reduction reaction involve either the gain in bonds to H_2, or loss of bonds to O_2.

In general, the **oxidoreductases** essentially comprises an array of *sub-classes,* for instance:

- **Dehydrogenases**
- **Oxidases**
- **Oxygenases**
- **Hydroxylases** and
- **Hydroperoxidases.**

❑ **DEHYDROGENASES-** The **enzymes** the critically catalyze the *removal of H_2* from **one substrate** and eventually pass it onto a **second substrate.** Alternatively, they are *incapable* of passing the H_2 almost directly to O_2 even though the different **dehydrogenases** do actually differ in the **structure of the apoenzyme**. In fact, they do have the **same *co-enzyme,*** such as :

- **Nieotinamide adenine dinucleotide (NAD+), or**
- **Nicotinamide adenine dinucleotide phosphate (NADP+).**

Thus, we may express the **reaction of the dehydrogenases** schematically as depicted under:

Sub AH_2 ←→ E-NAD$^+$ ←→ Sub BH_2

| Sub = Substrate |
| E = Enzyme |

Sub A ←→ E-NADH+H$^+$ ←→ Sub B

Remarks- The above schematic **eqation** vividly depicts that the **substrate A** transfers a **H-atom** with its *bonding pair of electrons* and a *proton (H^+)-* to the respective **enzyme (E)** having. NAD+- as the co-enzyme.

Thus, the reduced form of **enzyme (E-NADH$^+$H$^+$)** in turn helps successfully to transfer its H_2 to another **substrate B.**

Example – **Nicotinamide** *i.e.,* it serves as the *oxidizing agent* or *hydrogen carrier* in either **NAD$^+$** or **NADP$^+$** as shown under:

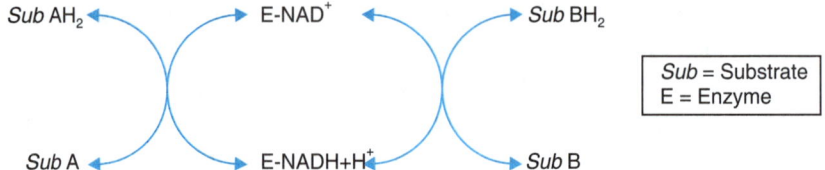

[R : Designates the '*rest portion*' of either the NAD$^+$ or NADP$^+$ molecule]

❑ **OXIDASES-** These represent the **enzymes** that solely catalyze the critical removal of:

- **H-atom from a substrate, and**
- **Ultimately pass it directly to* O_2.**

Further more, very much akin to the dehydrogenases- there exist several *classes of oxidases* that eventually differ from one another in the specific structure of the **apoenzyme**, however, do possess the **same co-enzyme** (*i.e.,* in this instance, it is found to be **Flavin-adenine dinucleotide (FAD).** Thus, one may represent explicit the *oxidation of a* **substrate** schematically as under:

---

* *i.e.,* one may take cognizance of the fact that the **dehydrogenases** are unable to pass **H-atom** *directly to O_2.*

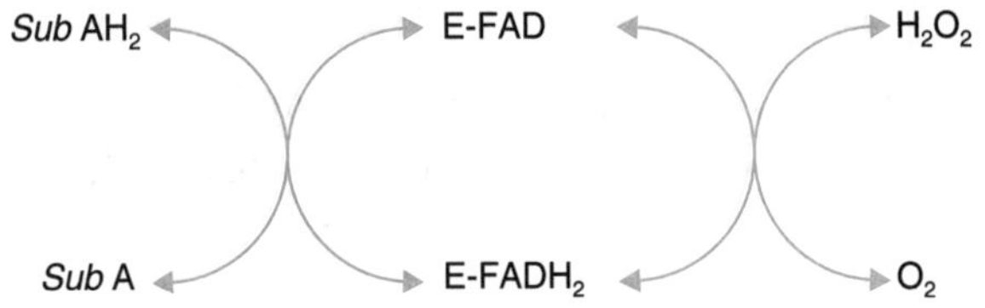

$$Sub\ AH_2 \rightleftharpoons E\text{-}FAD \rightleftharpoons H_2O_2$$

$$Sub\ A \rightleftharpoons E\text{-}FADH_2 \rightleftharpoons O_2$$

Comments- Obviously, in the **oxidases**, the specific *'flavin group'* present in the **coenzyme FAD** eaters as the **hydrogen carrier** as depicted in the following:

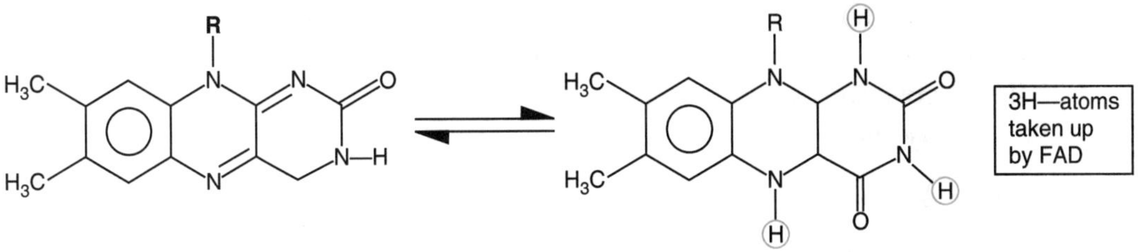

FAD (*Coenzyme*)   Hydrogenated FAD
(as H-carrier)

3H—atoms taken up by FAD

NOTE The 'oxygenases' **represent the *subclass of enzymes* that catalyzes the incorporation of O₂, directly into the *substrate*. Besides, the 'oxygenases' do require the crucial presence of *metal ions viz.*, Fe, Cu.**

Main Classes and Sub-Classes	Examples
1(a) Acting on the CH-OH moiety of 'donors':	
• with NAD or NADP as acceptor	Alcohol/Lactate dehydrogenases
• with O₂ as acceptor	Glucose oxidase
(b) Acting on the CHO or CO moiety of 'donors':	
• with NAD or NADP as acceptor	Glyceraldelhyde-3-phosphate dehydrogenase
• with O₂ as acceptor	Xanthine oxidase
(c) Acting on the CH-CH moiety of donors:	
• with NAD or NADP as acceptor	Dihydrouracil dehydrogenases
• with cytochrome	Acyl-CoA dehydrogenase
(d) Acting on CH-NH₂ moiety of donors:	
• with NAD or NADP as acceptor	Glutamate dehydrogenase
• with O₂ as acceptor	Amino acid oxidases
(e) Acting on the heme-groups of donors:	
• with O₂ as acceptor	Cytochrome oxidase
(f) Acting on H₂O₂ as electron acceptor:	Catalyse

4.2. Transferases

Enzymes that catalyze the transfer of a '*group*' from one molecule to another*.

In fact, these enzymes specifically catalyze the transfer of a **group (X)** right from *one substrate to another substrate.*

Thus, we may have:

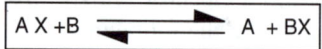

$$A X + B \rightleftharpoons A + BX$$

Besides, the **transferases** may be further classified into various **sub-classes** based exclusively upon the nature of *the transferring moiety (X),* such as:

- **NH$_2$-moiety** *(transaminase), and*
- **PO$_4$-moiety** *(transphosphorylation).*

Main Classes and Sub-Classes	Examples:
1 (a) Transferring C$_1$-Moieties:	
• Methyltransferases	Guanidoacetate methyltransferace.
• Hydroxymethyltransferases	Serinehydroxymethyltransferase
(b) Acyl transferases	Choline acyltransferase
(c) Glycosyl transferases	Phosphorylases
(d) Transferring amino moiety (Amino transferases)	Transaminases
(e) Transfer of P-containing groups	Hexopinases

4.3. Hydrolases [Hydrolytic Enzymes]

The **enzyme** that catalyzes the removal of a *moiety* by making use of **water** *e.g.,* **Esterase, Protease, and Carbohydrase.**

In other words, the **hydrolases** do refer to such **enzymes** that particularly catalyze the *hydrolysis process i.e.,* involves in the **direct addition of water molecule(s)**-which takes place *across the bond* that eventually gets cleaved. Importantly, the **hydrolases** are further categorized into several *sub-classes* based upon the nature of either:

- **The group undergoing hydrolysis, or**
- **The specific bond being hydrolyzed,**   • **Acid anhydride**
- **Ester**   • **Ether**   • **Glycosyl (C–C)**
- **C-Halide and**   • **Peptide bonds.**

Therefore, the various **sub-classes** of this particular class are duly represented by:

- *Esterases*   • *Etherases*   • *Peptidases*
- *Glycosidases*   • *Phosphotidases and*   • *Thiolecterases.*

---

* That is, one of *six classes* of enzymes recognized by the **Enzyme Commission of the International Union of Biochemistry (IUB)** *e.g.,* Aminotransferase, Acetyltransferase.

Remarks- Besides, the precise nomenclature of these **enzymes** they are invariably the name of the *substrate* duly followed by **'ase'**, *viz*:

- **Dipeptidase** • **Penicillinase** and • **Urease.**

 **NOTE** In fact, the nomenclatures of these *enzymes* are normally indicative of the exact type of reaction being catalyzed and do have the *unusual endings* (i.e., the recommended name of *Lysozyme is 'muramidase')*. Besides, quite many *peptidases* that end with *'in' (viz.,* chromotrypsin, papain, renin).

Main Classes and Sub-clsses:	Examples
1(a) Acting on Ester Bonds:	
• Carboxylic ester hydrolases	Esterases; Lipases
• Phosphoric Monoester Hydrolases	Phosphatases
(b) Cleaving Glycosides:	
• Glycosidases	Amylase; β-Glycosidases.
• N-Glycosidases	Nucleosidases
(c) Cleaving Peptide Linkages:	
• α-Aminopeptide amino acid hydrolases	Leucine aminopeptidase
• α-Carboxypeptide amino acid hydrolases	Carboxypeptidases
• Peptidyl peptide hydrolases (Endopeptidases)	Pepsin; Trypsin; Chymotrypsin.

4.4. Lyases

The **'Lyases'** refers to an enzyme catalyzing the removal of a group from a molecule by nonhydrolytic means. Furthermore, the **Lysases** is regarded to be a *smaller class of enzymes* which catalyze critically the precise removal of a rather **small molecule** right from a **substrate molecule**. As the on-going reactions are **reversible in nature;** hence, the **lysases** can also be regarded to catalyze the specific addition of *smaller molecules to the substrate molecule.*

Thus, we may have:

$$\underset{\displaystyle >\!\!\!\underset{|}{C}\!-\!\!\underset{|}{C}\!\!\!<}{\overset{X \quad Y}{}} \rightleftharpoons \; >\!\!C\!=\!\!C\!\!< \; + X\!-\!Y$$

Remarks- The **'Lysases'** are further classified based on the inherent linkages being attacked, such as the following **bonds:**

- **C—C** • **C—O** • **C—N** • **C—S** and • **C—Halide**

Examples-This specific group essentially includes:

- **Aldolase**
- **Fumarate Hydratase**
- **Pyruvate decarboxylase** and
- **Enclase.**

Main Classes and Sub-classes	Example
1(a) C-C Lysases:	
• Carboxyl-lysases	Pyruvate decarboxylase
• Aldehyde-lysases	Aldolases
(b) C-O Lysases	
• Aldehyde-lysases	Fumarase (fumarate hydratase)
(c) C-N Lysases	
• Ammonia lysases	Histidase (Histidine ammonialyase)

4.5. Isomerases

The **isomerase** refers to an enzyme that critically catalyzes the change of *one isomer* into *another isomer*. A few typical examples are as stated below:

- *cis-trans* **isomerase,**
- *D*-or *L*-**amino acid racenase,**
- **mutase, and**
- **epimerase.**

In true sense, this particular class essentially includes all enzymes that *catalyze* isomerization *i.e.,* 'interconversion of *Optical, Geometrical, or Positional Isomers.*

 **NOTE** **The great variety of enzymes belonging to this specific category absolutely necessiates a logical *sub-classification* that are based mostly on the different types of reactions involved, such as:**

- *Intramolecular transferases and*
- *Racemases.*

Main Classes and Sub-classes	Example:
1 (a) Racemases and Epimerases:	
• Acting on amino acids	Alamnine racemerase Ribulose-5P-epimerase
• Acting on carbohydrates	Matheylacetoacetate isomerase.
(b) *cis-trans*-**Isomerases:**	
(c) Intramolecular oxidoreductases:	
• Interconversion of aldoses and ketoses.	Triose phosphate isomerase, glucose phosphate isomerase.

4.6. Ligases (Synthetases)

The **ligase** relates to an enzyme catalyzing the joining of two chemical entities (compounds) wherein an *energy source (viz., ATP)* is essentially required.

In other words, these enzymes do catalyze the *synthesis reactions* by the help of joining *two molecules* being coupled with the breakdown of:

- **A pyrophosphate linkage of adenosine triphosphate (ATP); and**
- **An adenosine diphosphate (ADP).**

In this manner, the formation of the **malonyl-CoA** from **acetyl-CoA** in the crucial presence of *acetyl-CoA carboxylases a*s the **'catalyst'** serves as an important example. However, the **ligases** also needs *'biotin'* as a **co-enzyme.**

Main Classes and Sub-Classes	Examples
1 (a) Forming C-O Bonds:	
• Amino acids-RNA Ligases	Amino acid activating enzymes
(b) Forming C-S Bonds:	
• CoA-Ligases	Succinic thiokinase
(c) Forming C-N Bonds:	
• Acid-ammonia ligases	Peptide synthetase,
• Acid-amino acid ligases	Glucathione synthetase.
(d) Forming C-C Bonds:	
• Carboxylases	Acetyl Co-A carboxylases.

5. CHARACTERISTICS FEATURES OF ENZYMES

A survey of literature would reveal that the so-called **characteristic features of enzymes** centres around *two* aspects, namely:

- **Enzyme Efficiency, and**
- **Enzyme Specificity,**

which shall now be discussed separately in the sections that follows:

5.1. Enzyme Efficiency

It has been duly ascertained that a host of vital and important biochemical characteristics of **'living cells'** is its inherent ability to accomplish various rapid **chemical conversions** at *temperatures* invariably *below 40°C*; whereas, the simultaneous *chemical* reactions in a laboratory (*i.e. in vitro*) do proceed at relatively **high temperatures**. Interstingly, this specific characteristic feature of the **'living cell'** is accomplished perceptively due to its inherent *'catalytic nature of enzymes'*. Nevertheless, the various experiments carried out with an array of enzymes, isolated duly from *cells* and *tissues,* have revealed explicitly that-

'efficiency of enzymes is far greater *i.e.* 10^8 times (one hundred million times*) vis-à-vis* the *chemical catalysts'*.

Thus, we may have the following schematic expression:

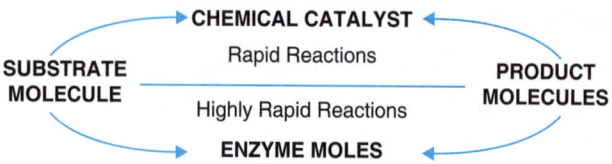

Comment- Amazingly, the rate of *enzyme-controlled reactions* is intimately dependent upon such critical conditionalities as:

- **Temperature**
- **Concentration and**
- **pH**

It has been observed that usually a *negligible effect* could be visible in the region of 0°C, **enzyme activity** gets enhanced rapidly with *rise-in-temperature* to a point (the so-called **'optimum temperature'**, usually between 35 to 50°C)-and beyond which the **enzyme efficiency** begins seriously to be affected adversely.

NOTE **Most *'enzymes'* either gets destroyed or inactivated by being heated at about 75°C.**

Observations- Obviously, the *velocity of the chemical reaction* duly catalyzed by the **'enzymes'** is explained adequately due to *the underlying fact that the 'enzymes' very much akin other catalysts* do help in the reduction of the **energy of activation*** **(Ea)** essentially needed for the said reaction to proceed. However, in a **laboratory chemical reaction**, the *energy of activation* gets decreased by increaseing the prevailing temperature of the reactions. As the living, *organisms (enzymes)* do function at temperatures normally **below 40°C-** perhaps owing to the sensitivity of *several cellular materials* to **high temperatures** (or the *method of heating for increasing the reaction velocity*) is not suitable. Therefore, whenever **the 'energy of activation of a reaction'** is reduced, it definitely becomes a lot easier for the *reactant molecules* to gain the **minimal quantum of energy** that is required critically for enabling them to react; and hence, the overall observed **'reaction velocity'** gets increased.

The above revelation of facts may be quite evident from the Table:5.2.

TABLE:5.2: Important Inherent Relation Between Energy of Activation and Reaction Velocity

S.no	Chemical Reaction	catalyst		Ea(cal.mole^{-1})	Relative Reaction Velocity
I	Decomposition of H_2O_2	•	NO	18,000	1
		•	pt	12,000	26,000
			ctalase (enzyme)	600	>10^{10}
II	Hydrolysis of Sucrose	•	Acid	26,000	1
		•	Invertase(enzyme)	11,000	>10^{11}

---

* That is, before a **chemical reaction** comes into play, the reacting molecules should gain a certain mini-mum of energy- usually known as the **'Activation Energy'** Thus, greater the **'activation energy'**-the slower would be the *rate of reaction* at a given temperature.

Fig:5.1 illustrated the graphic representation for the vital inherent relationship existing between the **energy of activation** and **reaction velocity.**

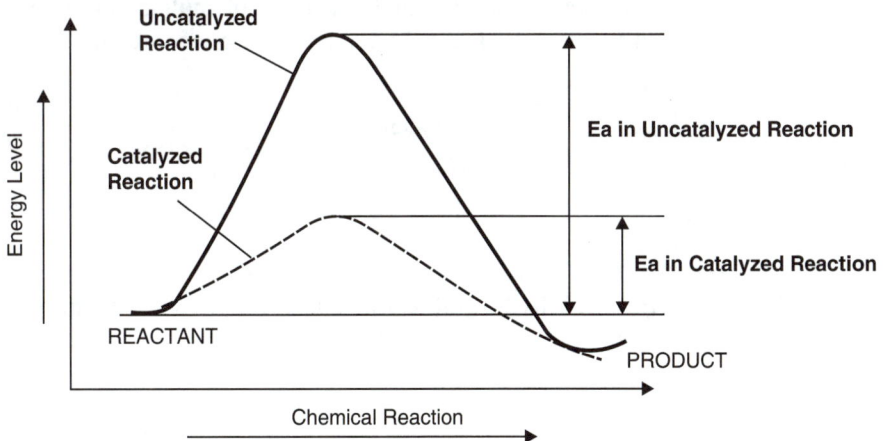

Fig : 5.1: Graphic Representation for the Vital Relationship existing between Energy of Activation and Reaction Velocity.

Explanation

1. Conversion of reactant to product (*uncatalyzed reaction*) the **energy of activation (Ea)** is much higher (*see Fig:5.1*).
2. Conversion of reactant to product (*catalyzed reaction*) the **energy of activation (Ea)** is almost 1/3rd (see **Fig:5.1**).

5.2. Enzymes Specificity

Importantly, the so-called **'enzyme specificity'** does play even a much bigger role *vis-à-vis* the relatively very **high-catalytic activity** of the **enzymes** generally.

Based on the scientific evidences, there are *three* **distinct types** of the **enzymatic specificity**, namely:

- **Stereochemical specificity,**
- **Reaction specificity, and**
- **Substrate specificity.**

Obviously, this particular feature *(criteria)* of enzymes may give rise to a splendid explanation for the following *two* glaring aspects, such as:

- ➢ **Integration, and**
- ➢ **Organization**,

of the large extent of altogether divergent **chemical reactions** which normally take place in the **living cells.**

5.2.1. *Stereochemical Specificity (Stereospecificity)*

In general, a plethora of the **biological reactions** are mostly specific in nature as far as the critical **production of the stereoisomers**.

Examples: following are *two* typical examples:

(a) **Conversion of Succinic Acid to Fumaric Acid-** *i.e., succinic acid* gets *dehydrogenated* to give exclusively **fumaric acid;** and absolutely *no trace* of **maleic acid** is produced (that usually takes place during the **chemical dehydrogenation phenomenon**)-as shown under:

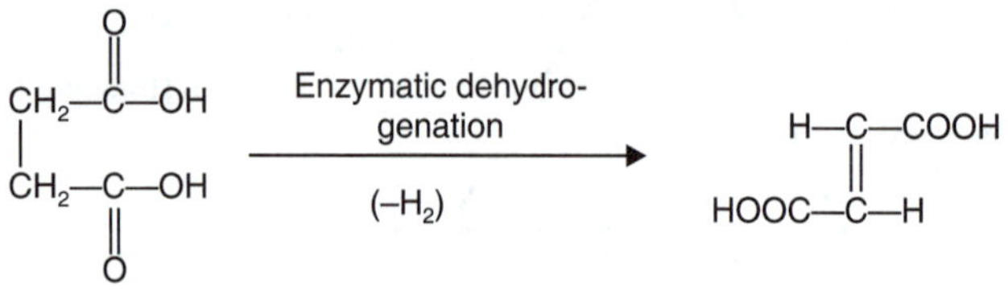

Succinic Acid   Fumaric Acid

(b) **Conversion of D-glyceraldehyde-3-phosphate and dihydroxyacetone phosphate by enzyme 'aldolase' to fructose-1,6-diphosphate.**

Thus, we may have the following expressions:

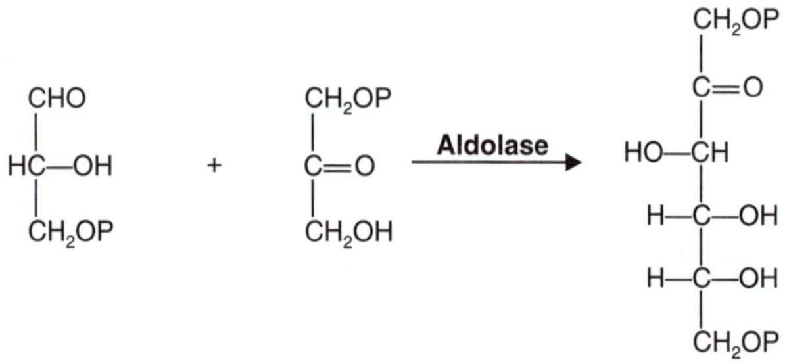

D–Glyceraldehyde-   Dihydroxyacetone   **D–Fructose-1,6-diphosphate**
3-phosphate   phosphate

Explanation

1. D-Glyceraldehyde-3phosphate and dihydroxyacetone phosphate are duly combined (in the presence of *'aldolase'*) to yield **D-fructose-1,6-diphosphate**- that could be predominantly *one of the 4-possible optical isomers* which could have been generated in the course of the **chemical reaction.**
2. It is pertinent to state here that the **enzymes** are highly specific and critical in the oxidation of both **D-and L-amino acids.**

> **NOTE**   **A particular enzyme will oxidize only one of the *two* optical isomers.**

5.2.2. Reaction Specificity

The **enzymes** are found to be *highly specific i.e.,* **one enzyme** invariably catalyzes **only one reaction** which the substrate can undergo.

Example: **Oxaloacetic acid*** may undergo several reactions *e.g.,*

- **reduction – to yield malic acid,**
- **decarboxylation – to give pyruvic acid,**

---

* That is, an important intermediate in **'metabolism'**.

- **amination – to produce aspartic acid, and**
- **acetic acid – to generate citric acid.**

The aforesaid reactions may be expatiated as under:

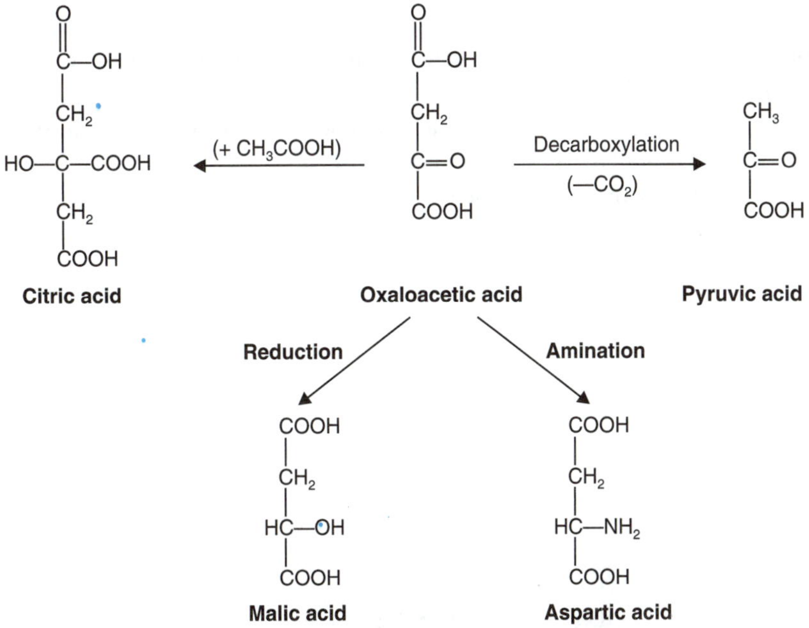

$$
\underset{\textbf{Citric acid}}{
\begin{array}{c}
\overset{O}{\overset{\|}{C}}-OH \\
| \\
CH_2 \\
| \\
HO-C-COOH \\
| \\
CH_2 \\
| \\
COOH
\end{array}}
\xleftarrow{(+\ CH_3COOH)}
\underset{\textbf{Oxaloacetic acid}}{
\begin{array}{c}
\overset{O}{\overset{\|}{C}}-OH \\
| \\
CH_2 \\
| \\
C=O \\
| \\
COOH
\end{array}}
\xrightarrow[(-CO_2)]{\text{Decarboxylation}}
\underset{\textbf{Pyruvic acid}}{
\begin{array}{c}
CH_3 \\
| \\
C=O \\
| \\
COOH
\end{array}}
$$

Reduction ↙   ↘ Amination

$$
\underset{\textbf{Malic acid}}{
\begin{array}{c}
COOH \\
| \\
CH_2 \\
| \\
HC-OH \\
| \\
COOH
\end{array}}
\qquad\qquad
\underset{\textbf{Aspartic acid}}{
\begin{array}{c}
COOH \\
| \\
CH_2 \\
| \\
HC-NH_2 \\
| \\
COOH
\end{array}}
$$

Enzymatic Reactions Involving the Oxaloacetic Acid

Comment: In the above set of reactions, each of the reaction pertaining to **oxaloacetic acid** gets catalyzed critically by its own separate/specific enzyme- that prevalently catalyzes *only that reaction and none of the other reactions*. It has been observed beyond any reasonable doubt that the so-called *enzyme-catalyzed reactions* are more or less of a **'quantitative nature'** *vis-à-vis* the corresponding analogous **chemical reactions** usually carried out in the *laboratory* by other agents, this could be due to the following *two* **possible reasons:**

➤ **Partly on account of the *reaction specificity;* and**
➤ **Partly by virtue of the fact they often come into play at *low temperature only.***

5.2.3. Substrate Specificity

Importantly, the degree of observed **substrate specificity** does vary from *enzyme to enzymes.*

However, certain **hydrolases** (see *section 4.3*) are comparatively found to be:

- **non-specific in character,**
- **a few of them may require substrate* containing some groups, and**
- **certain other enzymes do react very particularly with one-substrate** solely.**

---

* That is, **Group Specificity** viz., **β-galactosidase** and **α-glucosidase** that eventually cleave all **β-galactosidases** and glucosidases respectively.

** Such as: **Urease** that catalyzes the hydrolysis of **urea** only.

Thus, we may have the following expression:

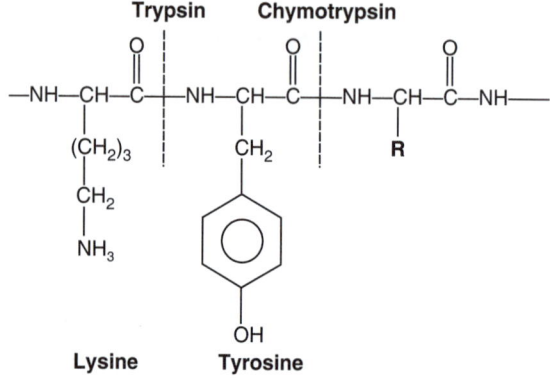

$$H_2N-\overset{\overset{\displaystyle O}{\|}}{C}-NH_2 \xrightarrow[\text{(Hydrolysis)}]{H_2O;} 2\,NH_3\uparrow + CO_2\uparrow$$

Urea

Important Examples- There are *three* important examples of *group specificity e.g.*, **Trypsin and Chymotripsin.**

    (a) Trypsin Hydrolases: These residues are obtained from **lysine** and **arginine**, which could be shown as under-'**specificity of proteolytic enzymes'.**

    (b) Chymotrypsin Hydrolases: These residues are only of the **aromatic amino acids**, such as: **Phe; Tyr; and Try.**

The Specificity of Proteolytic Enzymes

NOTE   A meticulous investigative study of the effects of these *highly specific enzymes* may provide useful information(s) with respect to the precise arrangement of amino acids residues in a *protein.*

    (c) Invertase Hydrolyses Several Fructosides – Likewise, '**invertase**' happens to be a *hydrolyzing enzyme* that is solely responsible for the **hydrolysis of many fructosides** *e.g.*,

- **sucrose, and**
- **methyl fructoside**.

The aforesaid reactions may be stated as under:

    ■ Glucose–O–fructose $\xrightarrow[\text{(H}_2\text{O)}]{\text{invertase}}$ Glucose + Fructose

        Sucrose

    ■ H_3C–O–fructose $\xrightarrow[\text{(H}_2\text{O)}]{\text{invertase}}$ CH_3OH + Fructose

        Methyl fructoside

[B] COENZYMES

1. INTRODUCTION

The **coenzymes** refers to a *small, organic, non-protein molecule* that eventually functions as a **reactant** or a **factor** which should by all means be present for an enzyme to function in its **catalystic role.**

Sometimes, it is regarded to be a *'loosely bound'* **prosthetic moiety*** of an **enzyme**. Importantly, the critical combination of a **coenzyme** *(nonprotein)* and its respective **apoenzyme'** *i.e.*, the **complete enzyme.**

In a broader perspective, one may rightly proclaim that *'all enzymes'* are nothing but the **protein molecules**. Obviously, certain *enzymes* designate the **'simple proteins'**; whereas, the others represent the **'conjugated proteins'.** Thus, we may have:

Holoenzyme $\rightleftharpoons$ Apoenzyme + Coenzyme

Alternatively, the **'coenzyme'** may also be defined as: **"a non-proteinous substance absolutely necessary and critical for the activity of the enzyme'.**

> **NOTE** The critical presence of both an *enzyme* and a *coenzyme* are an absolute necessity to eater as a *'catalyst'* in a biochemical reaction *in vivo.*

It has been amply proven and demonstrated that the **'coenzymes'** do represent the **organic molecules** having the dimension almost *intermediate* lying between the following *two* **entities**, namely:

- **Small-molecule intermediary metabolites**- that eventually serve as the *substrates of various enzymatic reactions*, and
- **Macromolecular proteins** (*i.e.*, pertaining to large molecules or polymers *viz.*, nucleic acids and proteins).

Alternatively, The **'coenzyme'** is invariably considered to be an *easily* **dissociable segment** (sometimes termed as the *prosthetic portion or moiety)* attached intimately to the **protein component** *(apoenzyme)* to represent into the formation of the *complete, enzymatically active*, and *conjugated protein* (known as the *'holoenzyme').*

Nevertheless, each **'coenzyme'** (or *cofactor*) serves precisely both as an:

- **Acceptor** and/or
- **Donor**

of certain particular kind of an **atom** or **cluster of atoms** to be removed either from or added on to a *small-molecule substrate* involved solely in a reaction catalyzed by the **holoenzyme.**

Example: Folic acid **coenzymes** either *donate* or *accept* the so-called **single-carbon units** (generally at different stages of oxidation) usually in a good number of **enzymatic reactions** combined together as a *single-carbon unit metabolism.*

Newer concept of *'coenzyme* – Importantly, the **'coenzymes'** are nothing but a *second substrate i.e.*, a **cosubstrate.** Obviously, it could be attributed perhaps due to the following *two* reasons, namely:

(a) Most prevalently, the so-called **'chemical changes'** taking place in the *coenzyme* precisely do counterbalance those occurring in the **substrate.**

Example: **Transphosphorylation reactions*** involved critically in the *metabolism of sugars (carbohydrates).* Thus, for each and every molecule of sugar moiety undergoing **transphosphorylation-** *one mole of ATP* gets duly **dephosphorylated** thereby converted to **ADP-** as shown under:

---

* That is, the *non-protein moiety* of a conjugated protein or enzyme.

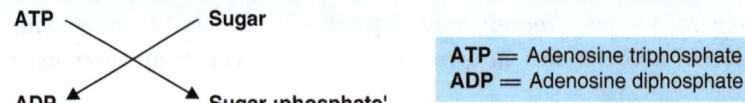

ATP
Sugar
ADP
Sugar ;phosphate'

ATP = Adenosine triphosphate
ADP = Adenosine diphosphate

(b) The underlying superb significance of the **'coenzymes'** being that their inherent reactions could certainly enlighter the so-called ***physiological importance***.

Example: The critical ability of ***'muscle'*** to afford the **anaerobic conversion** of *pyruvic acid* to *lactic acid* suggests vehemently that the *incumbment phenomenon* neither:

- **resides in *pyruvic acid* nor in lactic acid;** and
- **said reaction serves squarely for the particular conversion of *NADH to NAD$^+$*.**

NOTE **it is a worthwhile to state here that without the indulgence of *NAD$^+$* the glycolysis process fails to take place-thereby ceasing the *anaerobic ATP synthesis* predominantly; and hence, the muscular work almost stops apparently.**

2. COENZYME VARIANTS

Following are the vital and important **coenzyme variants**, namely:

- **Thiamine pyrophosphate (Caboxylase),** • **Coenzyme I,**
- **Coenzyme II,** • **Coenzyme A,**
- **Flavin Mononucleotide (FMN) [Riboflavin Phosphate],**
- **Flavin Adenine Dinucleotide (FAD),**
- **Codecarboxylase (Pyridoxal Phosphate),** • **Coenzymes Q,**
- **Adenosine Monophosphate (AMP), Adenosine Diphosphate (ADP), Adenosine Triphosphate (ATP),**
- **Glucose-1,6-diphosphate,** • **Ascorbic Acid Oxidase, and**
- **Uridine Diphosphate Glucose (UDPG),**

which shall now be discussed briefly in the sections that follows:

2.1. Thiamine Pyrophosphate (Carboxylase)

Chemical Structure

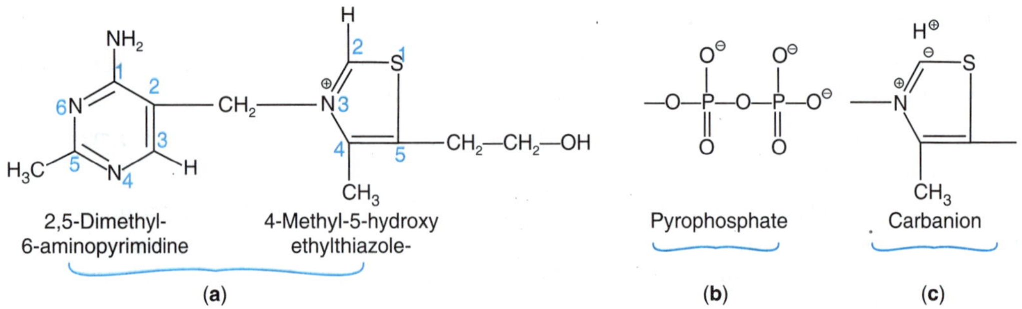

NH2
6N
2
1
3
H3C
5
N4
H
CH2
N3
2
S
1
H
4
5
CH3
CH2—CH2—OH
2,5-Dimethyl-
6-aminopyrimidine
4-Methyl-5-hydroxy
ethylthiazole-
O⊖
O⊖
—O—P—O—P—O⊖
O
O
Pyrophosphate
H⊕
⊖
S
⊕
—N
CH3
Carbanion
(a)
(b)
(c)

* That is, a specific phenomenon that essentially involves the transfer of a **phosphate (PO_4^{3-}) moiety** (being catalyzed by the **enzymes)** from a **donor** to a suitable **acceptor.**

Thiamine (A): The vitamin (B); In **Thiamine Diphosphate,** the OH group is duly replaced by 'pyrophoshate; C:Carbanion form.

Synthesis of Thiamine Pyrophosphate

The most common and prevalent synthesis of **thiamine pyrophosphate** essentially involves the treatment of **5-β-bromomethyl) thiamine bromide (A)** with **silver pyrophosphate (B)** in Phosphoric acid (H_3PO_4) solution at 100°C (Weil-Malherbe,1940)*. Now from the resulting solutions the *coenzyme carboxylase* is carefully precipitated as its **Ag-Salt** and then as the **Phosphotingstate.**

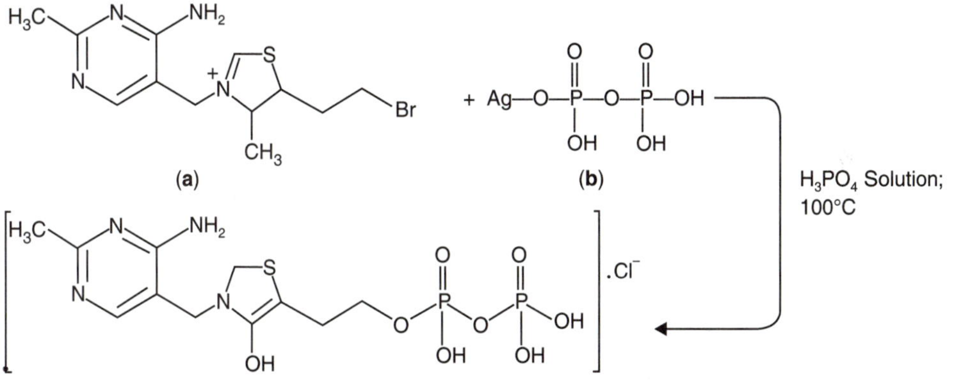

Thiamine Pyrophosphate

How does it act? In fact, the *enzyme* **carboxylase** helps to break down the **pyruvate** into *acetaldehyde* in the *fermentation* phase and also in the respective **carbohydrate metabolism.** Interestingly, the above reaction comes into play in the critical presence of the *coenzyme carboxycase (thiamine pyrophosphate).*

Thus, we may have the following expression:

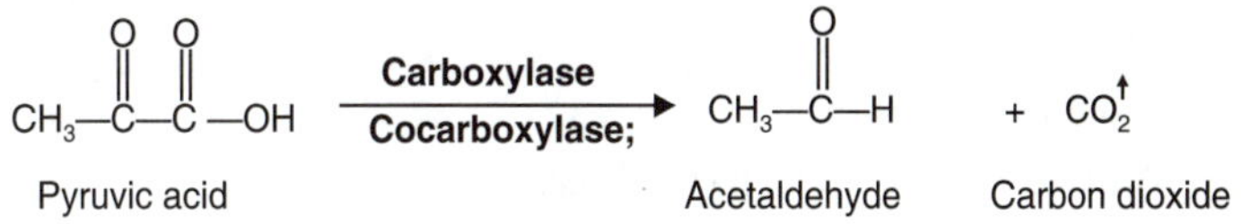

Pyruvic acid   Acetaldehyde   Carbon dioxide

Mechanism-The underlying mechanism of the **carboxylase** solely rests upon the ability of ionization of the proton at **position C-2** in the *thiazolium ring system.* Nevertheless, the crucial **'liability'** of the respective **H-atom** is by virtue of its *rapid replacement* by **denterium (D)-**as and when the **thiazolium salts** are dissolved suitably in an *acidified* **denterium oxide (D_2O)-**as shown below:

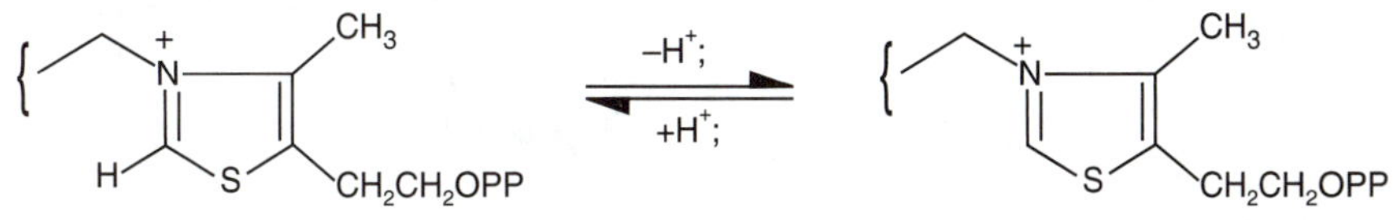

Carboxylase   **Deprotonated carboxylase**

---

* Wail-Malherbe: Biochem J., **54:**1980,1940.

2.2. Coenzyme I [NAD; B-NAD; NAD⁺]

It is one of the biologically active forms of **nicotinic acid** that occurs abundantly in the '*living cells*' primarily in the so-called *oxidized state*. Besides, it largely caters as a coenzyme of the *dehydrogenases* especially in these *two* biological transformation, such as:

- *Primary* alcohols, and
- *Secondary* alcohols.

NAD invariably acts as a *hydrogen acceptor*-forming **NADH** which then serves as a *hydrogen donor* in the specific **respiratory chain**.

Chemical Structures

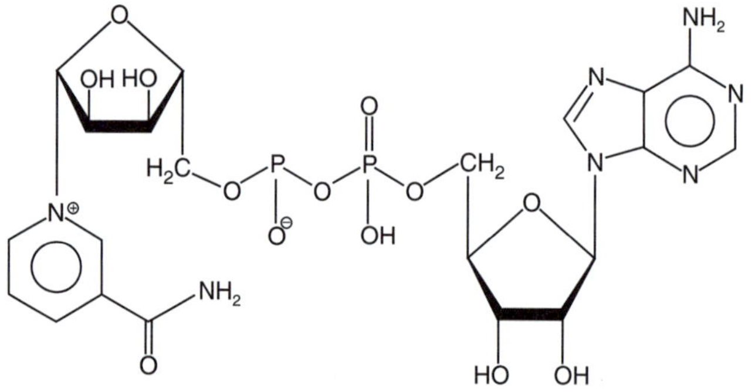

Coenzyme I

CONSTITUTION OF COENZYME I

The probable chemical structure of **coenzyme I** has been duly elucidated by actually performing its '*hydrolysis*' under *different experimental parameters:*

❏ **Hydrolysis in the Acidic Environment** –Under this condition the **coenzyme I** gives rise to the formation of:

- one mole of adenine,
- two moles of D-ribose-5-pyrophosphate, and
- one mole of nicotinamide,

as given under:

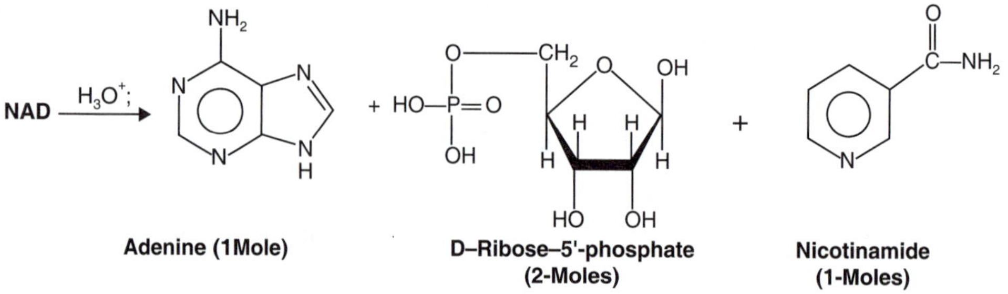

| Adenine (1Mole) | D–Ribose–5'-phosphate (2-Moles) | Nicotinamide (1-Moles) |

Comment-The aforesaid reaction reveals explicity that **NAD** has **one adenine, two ribose, two phosphate,** and **one nicotinamide**.

❑ **Mild Alkaline Hydrolysis of NAD-**In this particular instance **coenzyme I** yields:
- **one mole of adenosine diphosphate-ribose;** and
- **one mole of nicotinamide,**

as shown below:

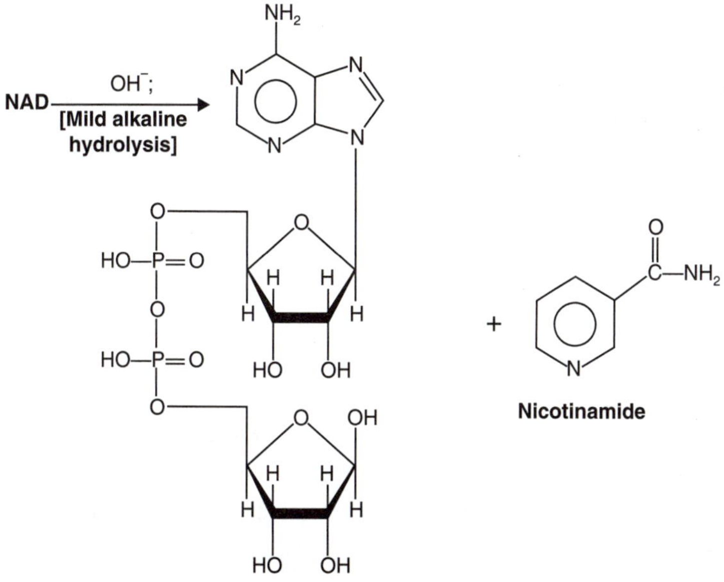

Adenine diphosphate–Ribose

Nicotinamide

However, under **vigorous alkaline parameters-**the **NAD** produces one mole of **adenosine diphosphate (ADP)** as depicted under:

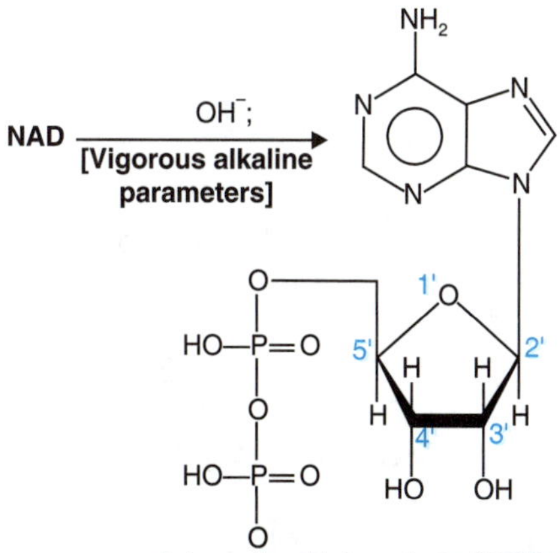

Adenasine diphosphate (ADP)

Remark-Based on the aforesaid **hydrolytic reactions** it may be observed vividly that:
- **adenosine molecule,** and
- **ribose molecule,**

are joined together at the respective **5'-position** by a *pyrophosphate moiety.*

❑ **Enzymatic Hydrolysis of Coenzyme I**-The **enzymatic hydrolysis of coenzyme I** yields the following *two* **distinct products.**

- **Adenosine,** and
- **Nicotinamide riboside,**

as given below:

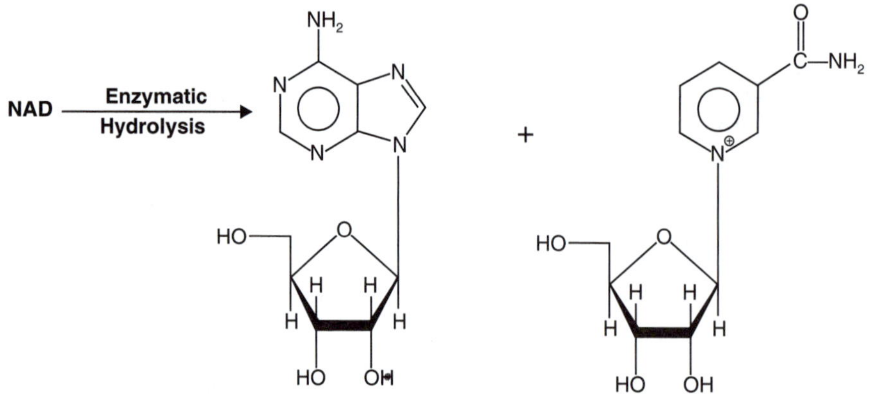

Adenosine   Nicotinamide riboside

Probable Chemical Structure of Coenzyme I-Thus, based upon the aforesaid *degradation reactions* it could be possible to assign a probable chemical structure of **coenzyme I** molecule, which eventually consists of:

- **one mole of adenosine, and**
- **one mole of nicotinamide riboside,**

obviously *via* their **5'-positions** by a **pyrophosphate moiety**.

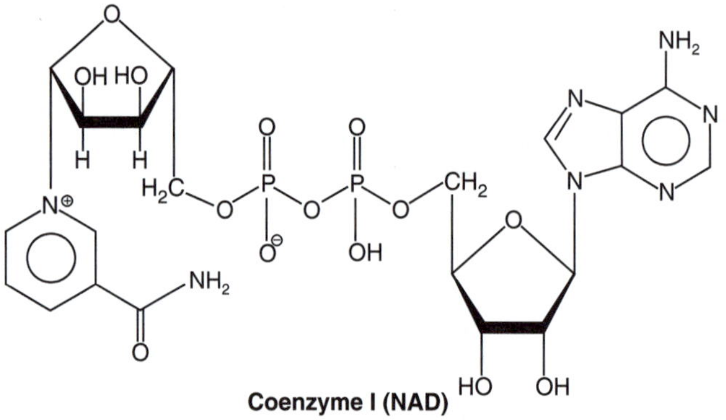

Coenzyme I (NAD)

SYNTHESIS OF COENZYME I

Hughes *et al.* (1957)* synthesized successfully **coenzyme I** from *three* **chemical entites:**

***

* Hughes *et al:JChem Soe.,* 3733,1957.

- **D-Ribose**
- **Nicotinamide and**
- **Adenosine-5'-phosphate.**

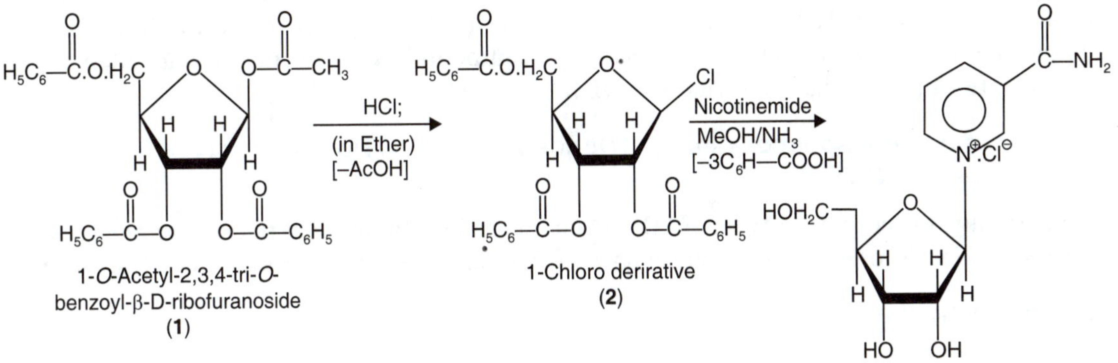

1-*O*-Acetyl-2,3,4-tri-*O*-benzoyl-β-D-ribofuranoside
(1)

1-Chloro derirative
(2)

N–Chloro-nicotinamide
riboside (3)

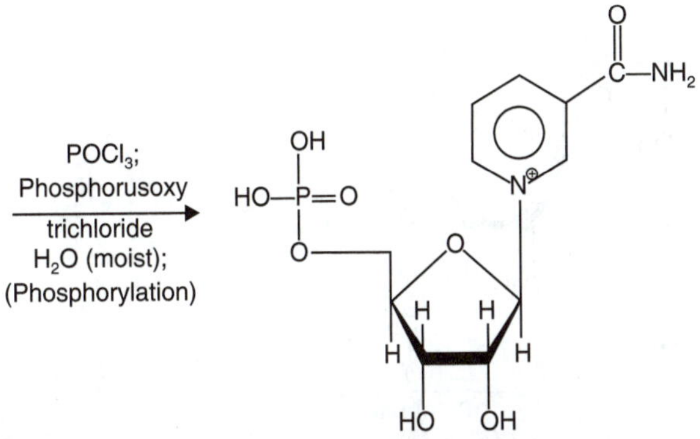

Nicotinamide ribocide
–5'-phosphate (X)

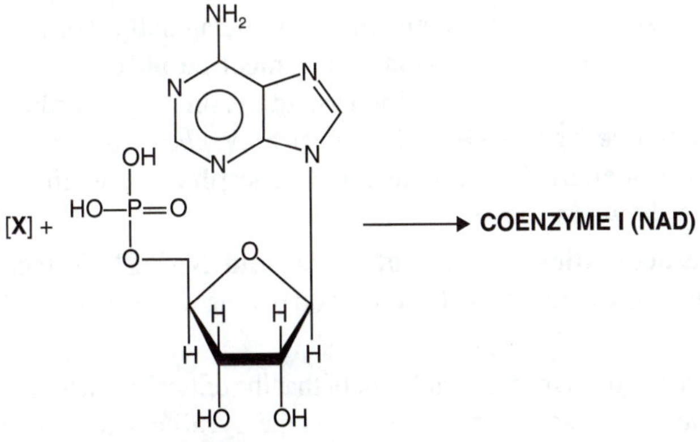

Adenosine-5'-phosphate (4)

[X] + ⟶ COENZYME I (NAD)

Explanation–1-*O*-Acetyl-2,3,4-tri-*O*-benzoyl-β-D-ribofuranoside (**1**) on treatment with - HCl in ether eliminates a mole of acetic acid to give the corresponding **chloro derivatives (2)**. The **compound(2)** on reaction with *nicotinamide* in ammoniacal methanol loses 3 moles of benzoic acid to yield **N-chloronicotinamide riboside (3)**. The resulting **compound (3)** upon phosphorylation with **POCL₃** in the presence of moisture (H₂O) gives rise to the formation of nicotinamide riboside-5'-phosphate (**X**). Finally, the **compound (X)** on interaction with **adenosine-5'-phosphate (4)** yields the **coenzyme I (NAD)**.

2.3. Coenzyme II [Phosphocozy-mase; Triphosphopyridine Nucleotide (TPN); Codehy drogenase II; NADP⁺]

Importantly, the **coenzyme II (or NADP⁺)*** does contain an **additional phosphate moiety** than *coenzyme I molecule* located strategically at *position 2'* of the inherent **ribose molecule of adenosine**.

Chemical Structure

Coenzyme II

CONSTITUTION OF COENZYME II

It has been duly established that **coenzyme II** on being subjected to *hydrolysis* by a **nucleotide phosphatase,** it gives rise to the formation of the **nicotinamide nucleotide** together with *another product of reaction* that has been duly identified as: **adenosine-2',5'-diphosphate.** However, the *latter product* which on **selective** *hydrolysis* with respect to *5'-phosphate moiety* gives an altogether **new product** that has been identified as: **adenosine-2'-phosphate.** Therefore, the probable structure of **coenzyme II** may be shown as above.

Biological Characteristics Features of NAD⁺ and NADP⁺: In fact, both these *enzymes viz.,* **Coenzyme I (NAD⁺)** and **Coenzyme II (NADP⁺)** do belong to the specific class of '**hydrogen transfer enzymes'**.

It has been proven beyond any reasonable doubt that the critical reduction of either **NAD⁺** *(Coenzyme I)* or **NADP⁺** *(coenzyme I)* invariably comes into play in the **pyridine ring system** *(i.e., The basic nucleus of the said enzymes)* present in the **niacinamide molecule** itself.

---

* **NADP⁺**- Nicotinamide adenine dinucleotide phosphate.

Thus, we may have the following expressions:

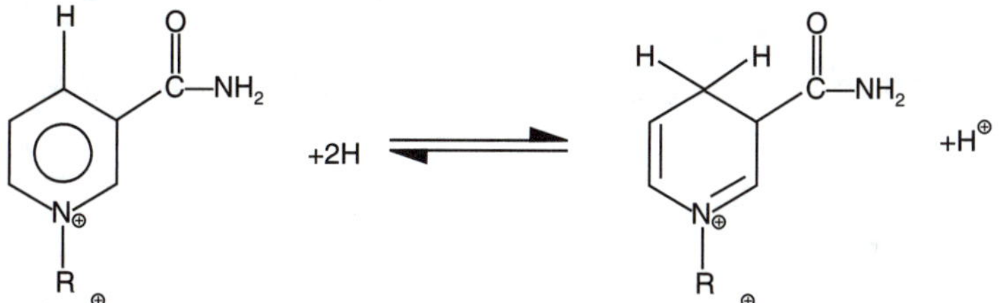

$$\text{NAD}^{\oplus} \quad +2H \rightleftharpoons \quad \text{NADH}^{\ominus} \quad +H^{\oplus}$$

Comment- **NAD⁺ (coenzyme I)** does possess an *inherent* and *typical pair of* **enantiotrophic (*prochiral*) faces**, Thus, in a particular instance when **NAD⁺** accepts a **hydride ion (H⁻)-** it readily forms the **NADH;** and hence, it predominantly comprises the so-called **enantiotropic (prochiral) hydrogens** located strategically at the **C-4 poisition:**

Thus, we may have the following expressions:

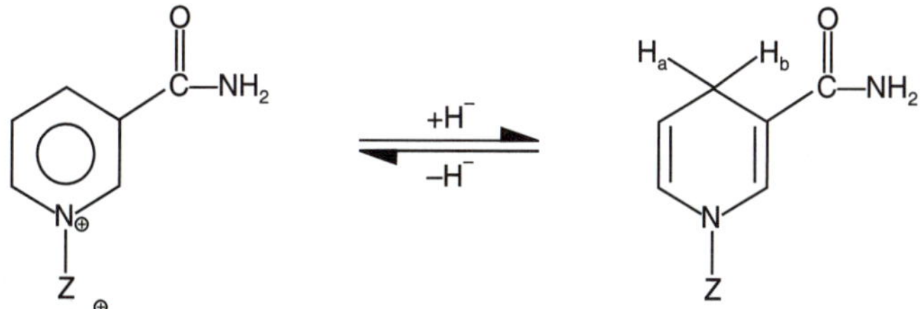

$$\text{NAD}^{\oplus} \quad \underset{-H^-}{\overset{+H^-}{\rightleftharpoons}} \quad \text{NADH}$$

Observation-Based on the experimental data one may safely conclude that the respective **NAD⁺** enzyme complex is invariuably a typical **stereospecific entity** – wherein, only one *hydrogen atom (Ha or Hb)* is being engaged in the reaction exclusively. Besides, the **specific face of NAD⁺** - that gets eventually attacked; and the **particular hydride ion (H⁻)** from **NADH⁺** gets duly transferred solely depends on the **nature of enzyme.**

The following Table:5.2 records some of the known *substrates* that are specifically attacked by **NAD⁺** and **NADP⁺**

TABLE:5.2 Some Known Substrates Attached by NAD⁺ and NADP⁺

S.No.	Name of Substrate	Chemical Formula	Nomenclature of Coenzyme
1	Ethanol	CH_3-CH_2-OH	NAD⁺
2	Lactic acid	$CH_3-\overset{OH}{\underset{}{C}}-COOH$	NAD⁺
3	Malic acid	$HOOC-\overset{H}{\underset{OH}{C}}-CH_2-COOH$	NAD⁺
4	Isocitric Acid	$HOOC-\overset{OH}{\underset{H}{C}}-\overset{H}{\underset{COOH}{C}}-CH_2-COOH$	NAD⁺(or NADP⁺)

(Contd....)

* Baddiley *et al.*, *Nature*, **171**,76,1953

| 5 | Glycerol | $CH_2OH—CHOH—CH_2OH$ | NAD^+ |
| 6 | Glutamic acid | $HOOC—CH_2—CH_2—\overset{\displaystyle NH_2}{\underset{\displaystyle H}{C}}—COOH$ | NAD^+ (or $NADP^+$) |

2.4. Coenzyme A [CoA-SH]

The **coenzyme A** relates to an acyl carrier molecule which essentially comprises a **3'-phosphate derivatives of ADP** linked to *panlotheric acid via* a **phosphate ester bondage.** Besides, the incumbent **pantothenic acid** is duly attached to **β-mercaptomethylamine** by an **amide bondage.**

Importantly, the **coenzyme A** critically forms a part of an *enzyme* that eventually catalyzes the so-called **biological acetylation reactions** *i.e.,* its grossly aids in the actual transference of the *acetyl (-C-CH₃) moiety* from one chemical entity to another.

Baddily *et al.* (1953)* were pioneer in establishing the structure of **coenzyme A** through a series of *selective enzymatic degradations*.

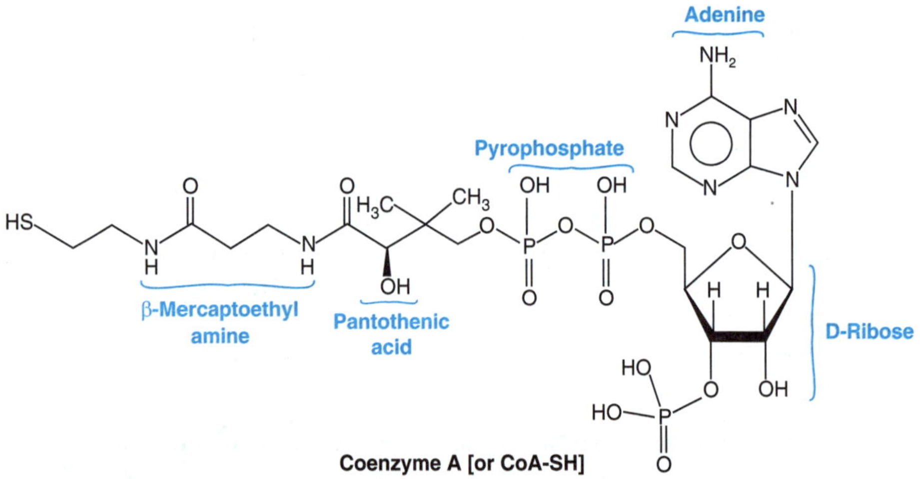

Coenzyme A [or CoA-SH]

Based on the survey of literature and scientific evidences it has been duly proven that **coenzyme A** does participate practically in most of the so-called *biosynthetic processes*, *invivo-* that usually come into play by way of **2 C-units**. In this manner, the **coenzyme A** gets involved perceptively in:

- **Carbohydrate metabolism; and**
- **Biosynthesis of fatty acids, terpenoids and steroids,**

spread out exhaustively in the **plant and animal kingdoms.**

Coenzyme A: Biosynthesis of Fatty Acids- The **acetyl coenzyme A[CoA-S.COCH₃]** predominantly undergoes the phenomenon of **'self-condensation'** to result into the formation of **acetoacetyl-coenzyme A-**that gets converted respectively into the **butyryl coenzyme A** by means of various sequential reactions-as shown under:

---

* Baddiley *et al., Nature,***171**,76, 1953

$$CoA—S.CO.CH_2 H+ CoA—S.CO.CH_3 \rightleftharpoons CoA—S.CO.CH_2.COCH_3 + CoA—SH$$

Acetyl coenzyme A   (Hydration) $\Big\downarrow$ (H)   **Coenzyme A**

$$\underset{\text{(Hydration)}}{\overset{\text{(H)}}{\diagup}} CoA—S.CO.CH=CH.CH_3 \xleftarrow{—H_2O} \overset{OH}{\overset{|}{CoA—S.CO.CH_2.CH.CH_3}}$$

A Hydroxy derivative

CoA—S.CO.CH$_2$.CH.CH$_3$
Butyryl coenzyme A

Comment: From the aforesaid sequence of reactions it may be observed that *two* **C-atoms** are being specifically added onto the *acetyl-CoA* to produce **'butyryl-CoA'**. Thus, the resulting **'butyryl-CoA'** again gets condensed with another **acetyl-CoA** and subsequently forms the desired **butyroacetyl-CoA**. In this way, the successive repetition of such sequences would ultimately lead to the formation of a **long-chain fatty acid.**

Thus, we may have the following expressions:

$$CoA—S.CO.CH_2 CH_2CH_2H + CoA—S.COCH_3 \longrightarrow CoA—S.CO.CH_2 CH_2CH_2COCH_3$$

Butyryl coenzyme A   **Acetyl coenzyme A**

(i) **Reduction;**
(ii) **Dehydration (–H$_2$O)$_2$**
(iii) **Reduction;**

$$CoA—SH + HOOC.(CH_2)_4CH_3 \xleftarrow[\text{(Hydrolysis)}]{H_2O;} CoA—S.CO(CH_2)_4CH_3$$

Coenzyme A   A Fatty Acid   **Butyroacetyl—CoA**

STRUCTURES OF COENZYME A-The *enzymatic hydrolysis* of **coenzyme A** gives to the formation of:

- **pantothenic acid,**
- **adenosine phosphate (ADP), and**
- **a S-containing unidentified product.**

However, it was further observed that *one mole of* **adenosine** plus *three moles of* **phosphate** were duly produced from each mole of **'pantothemic acid'.**

Important Observations- These essentially include:

1. Out of the said *three* **moles of phosphates** one may categorically observes:
   - *one* **mole was found in a 'monophosphate'**; and
   - *two* **moles as an integral part of the** *pyrophosphate ester.*

2. Based on the detailed **'enzymatic investigative studies'** it revealed explicity that the **'monophosphate ester moiety'** was found to be located strategically at the *C-3' position of* **ribose segment of adenosine moiety of the coenzyme A.**

3. The *alkaline hydrolysis* of **coenzyme A** specifically yielded **pantothenic acid-4'-phosphate** which indicates critically that the inherent *pyrophosphate moiety* is duly present as a **'bridge'** located between the following *two* **moieties**:
   - **5'-position of adenosine; and**
   - **4'-position of pantothenic acid.**

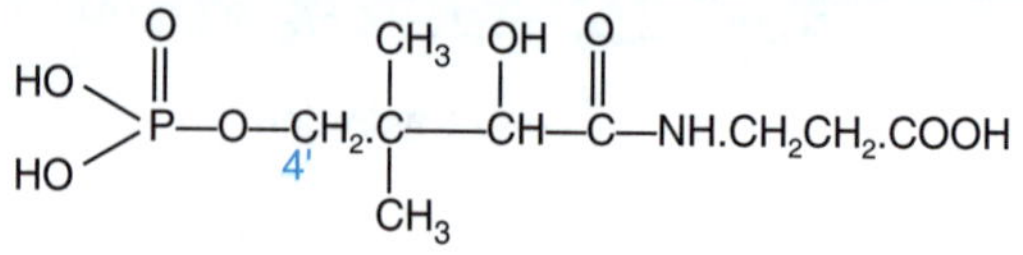

Pantothenic acid—4'—phosphate

4. However, the structure of the **sulphur ('S')** bearing chemical entities was duly established as **the 2-aminoethanethol** since it was established to be the so-called *degradation* **product of pantetheine** (i.e., *a compound obtained from coenzyme A).*

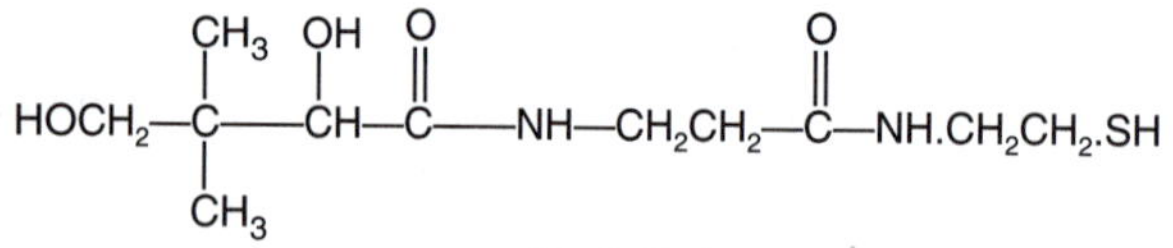

Pantetheine

Remark: Interestingly, the synthesis and characterization of **pantetheine** revealed the presence of an *amide (-CONH-) linkage* (duly formed by the *amino moiety of aminoethanettiol* and the *carboxyl moiety of pantothehic acid)* in the **coenzyme A molecule.** (see section 2.4)

SYNTHESIS OF COENZYME A

Moffatt and Khorana (1959)* by the condensation of *adenosine-2'(3'),5'-diphosphate* with *morpholine* to give the ***activated nucleotide 5'-phosphomorpholidate (I)-*** which on further condensation with **pantetheine-4'-phosphate (II)** yields the diphosphate **condensed product (III)** with the elimination of a mole of *morpholine.* The resulting **product (III)** on being subjected to *hydrolysis* in an acidic environment forms an admixture of **coenzyme A** and **iso-coenzyme A.**

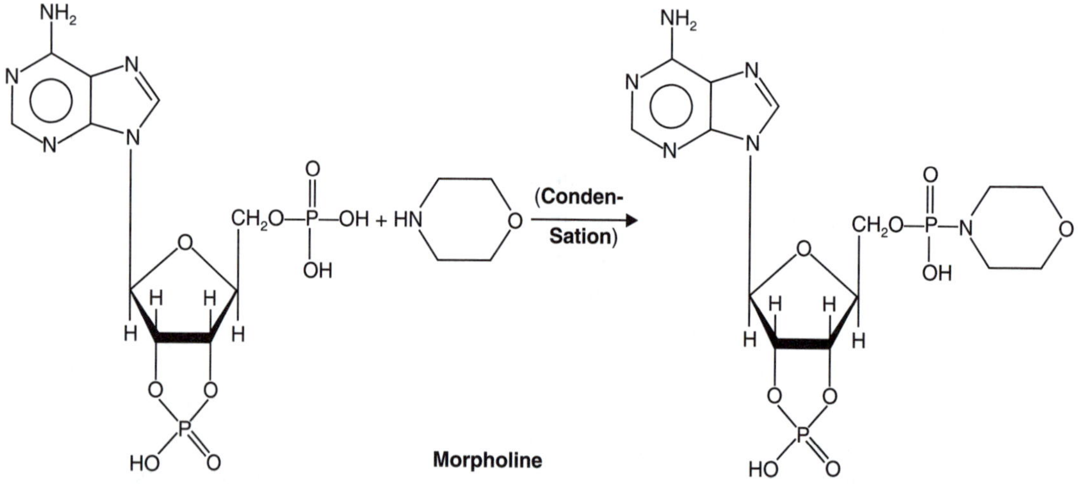

Nucleotide-5'phosphoridate

Morpholine

(I)

(Conden-
Sation)

(Contd....)

---

* Moffatt and Khorana, *J Am Chem Soe.,* **81** : 65,1959.

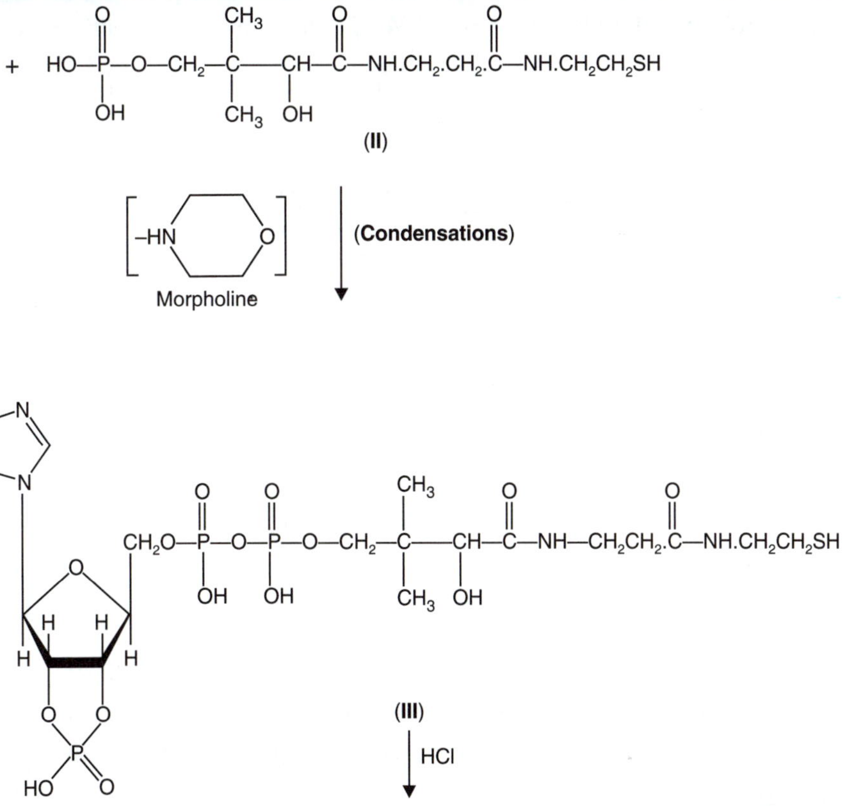

(II)

Morpholine

(Condensations)

(III)

HCl

COENZYME A (3'–phosphate) + ISO-COENZYME A (2'–phosphate)

2.5. Flavin Mononucleotide [FMN] (Riboflavin Phosphate)

Flavin mononucleotide (FMN) designates one of the *bioactive* forms of riboflavin. It occurs as the prosthetic group for several *oxidoreductases* that are involved intimately in **biological energy metabolic phenomenon**. It is found to be the *cofactor* for:

- **Blue-light photoreceptor** and
- **Phototropin.**

Following is the chemical structures of **FMN**:

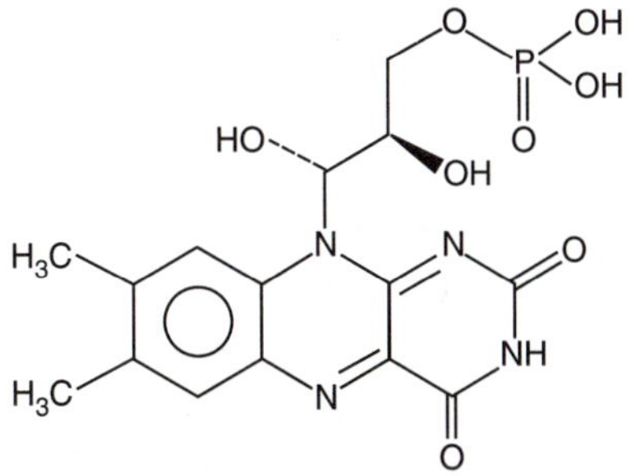

Flavin Mononucleotide (FMN)
[Riboflovin-5'-phosphoric acid]

Kuhn (1936)* successfully synthesized **FMN** in the following sequential manner:

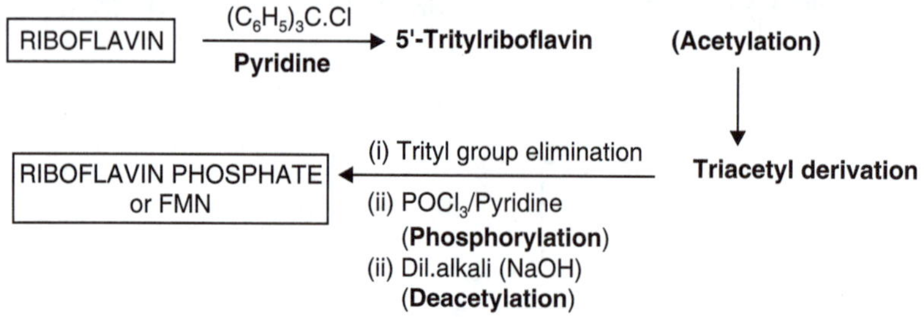

FMN serves as an excellent *vitamin* (**enzyme cofactor**).

FMN serves as an excellent *vitamin* (**enzyme cofactor**).

2.6. Flavin Adenine Dinucleotide (FAD) [*Sym.*Alloxan Adenine Dinucleotide]

Flavin adenine dinucleotide (FAD) relates to one of the most prevalent and biological active forms of *riboflavin*. **FAD** essentially possess the following *cardinal aspects:*

➢ Redox cofactor involved in biological energy metabolism;

➢ Gets reduced to *FADH$_2$* so as to carry the high energy electron as part of the so-called *electron-transport chain;* and

➢ Serves as the *prosthetic moiety* of certain *flavoproteins* including such vital entities as:

- **D-amino acid oxidase**
- **Glucose oxidase**
- **Glycine oxidase**
- **Fumaric hydrogenase**
- **Histaminases and**
- **Xanthine oxidase**

Chemical Structure

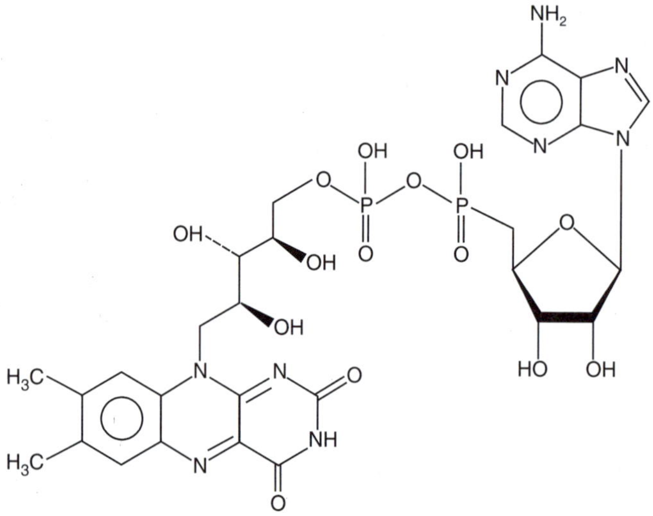

Flavin adenine dinucleotide (FAD)

---

*   Kuhn R: b 1900; Vienna Ph.D., mMunich (Willstatter); Zurich, Hcidelbarg, KWI, Med.Res., Nobel Prize 1938.

SYNTHESIS OF FAD- Moffatt and Khorana (1958)* put forward the synthesis of **FAD** by the interaction of *monothallous* salt of *riboflavin phosphate (I) and 2',3'-O- propylidene adenosine 5'-benzylphosphorochloridate (II)*- as given under:

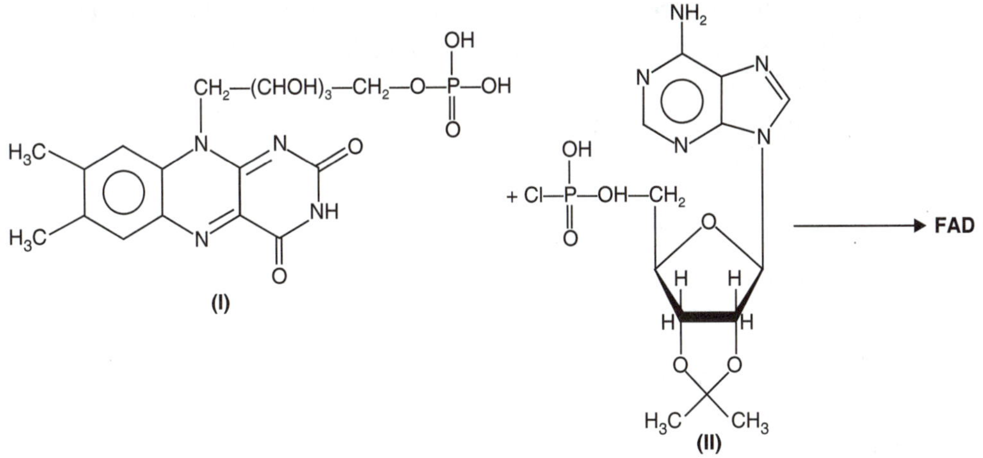

(I)

(II)

FAD finds its use as the **vitamin (enzyme cofactor).**

2.7. Coenzymes Q [Ubiquinones]

The **coenzyme Q (Ubiquinones)** represent a group of *lipid-soluble* **benzoquinones** involved intimately in the *electron transport phenomenon* **in mitochondria** *i.e.*, in the oxidation of succinate or reduced **inactive adenine dinucleotide (NADH)** *via* the so-called **cytochrome system.** However, it invariably occurs mostly in the **aerobic organisms** from *bacteria* to *higher plants and animals.* The **survey** of literature reveals that there exist *ten* **different coenzymes Q** (*viz.*, Q_1 to Q_{10}) which essentially differ in the exact number of **isoprenoid** $\left[\begin{array}{c} CH_3 \\ -H_2C=C-CH=CH_2- \end{array} \right]$ **units** located strategically in the side-chain at **C-6 position** of the **quinine molecule**:

Chemical Structure

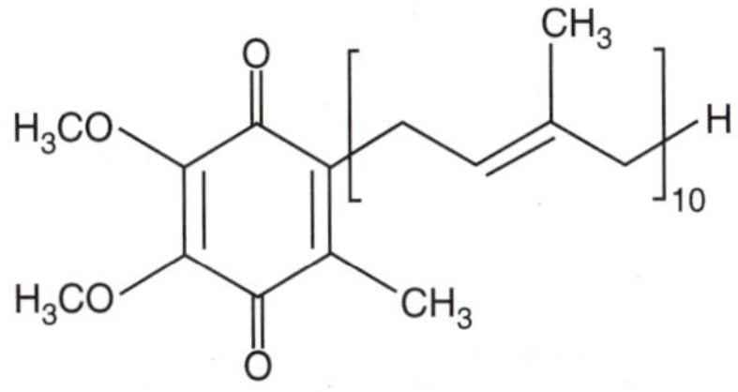

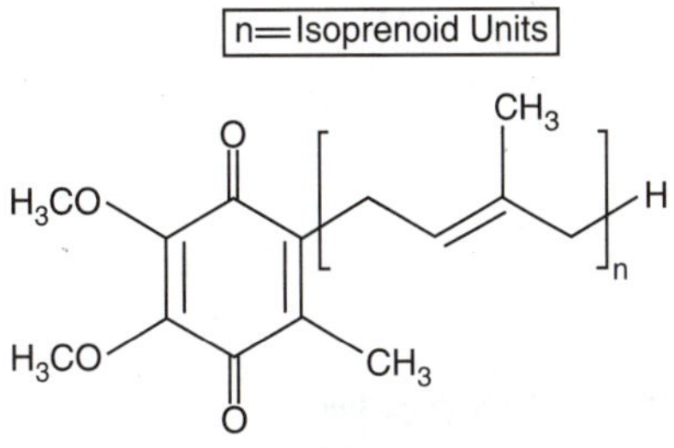

n=Isoprenoid Units

Ubiquinone 50   or   **Ubiquinone Q**

Coenzyme Q has been reported to occupy a key position in the *respiratory phenomenon of the mitochondria*. Besides, it is also actively engaged in the critical **biosynthesis of starch.**

---

*Moffatt and Khorana: *Jam Chem Soe.*, **80**:3756,1958.

SYNTHESIS OF COENZYME Q

Interstingly, **2,3-dimethoxy-5-methyl-1,4-benzoquinone (I) [or Coenzyme Q$_0$]** happens to be the recognized *key intermediate* in the specific synthesis of *coenzyme Q group.*

Hence, there are *two* **distinct** methods proposed for their actual synthesis, namely:

□   **Coenzyme Q$_0$** may be synthesized may be synthesized either from:

  •   **Cresol or**

  •   **Methyltrimethylgallate*,**

that is expressed as under:

Method-1: *From Cresol*

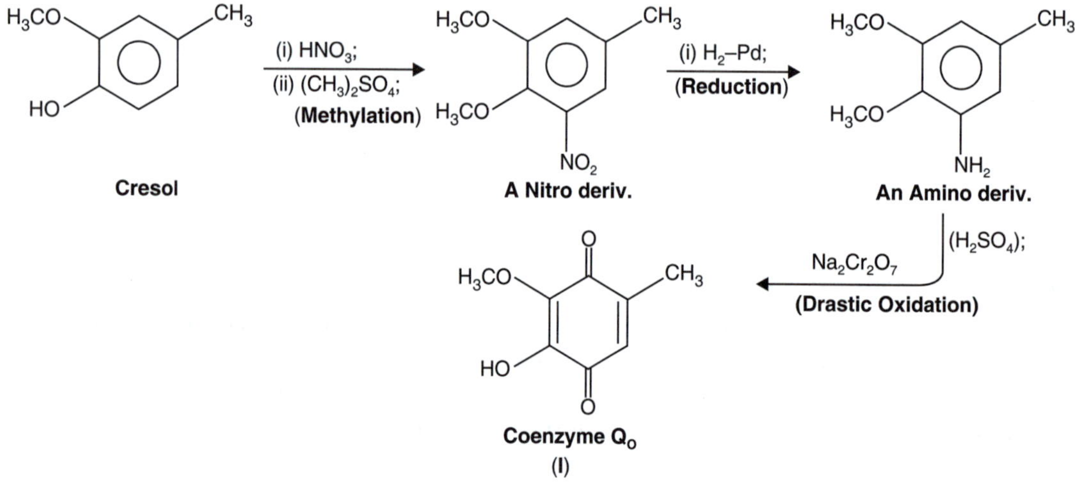

Cresol   (i) HNO$_3$; (ii) (CH$_3$)$_2$SO$_4$; (Methylation) → **A Nitro deriv.**   (i) H$_2$–Pd; (Reduction) → **An Amino deriv.**

Na$_2$Cr$_2$O$_7$ (H$_2$SO$_4$); (Drastic Oxidation) → **Coenzyme Q$_0$** (I)

Explanation: - **Cresol** on being subjected to *nitration* followed by *methylation* yield a **nitro derivative**, which on *reduction* gives the corresponding **amino derivatives**. Finally the resulting amino derivative on being oxidized vigorously yields **coenzyme Qo.**

Method-2: From Methyltrimethylgallate

Methyltrimethyl gallete   Cu-Chromate; H$_2$;(Reduction) → **3,4,5-Trimethaxy totuene**   *p*–NO$_2$ *para*-Nitro-diazine

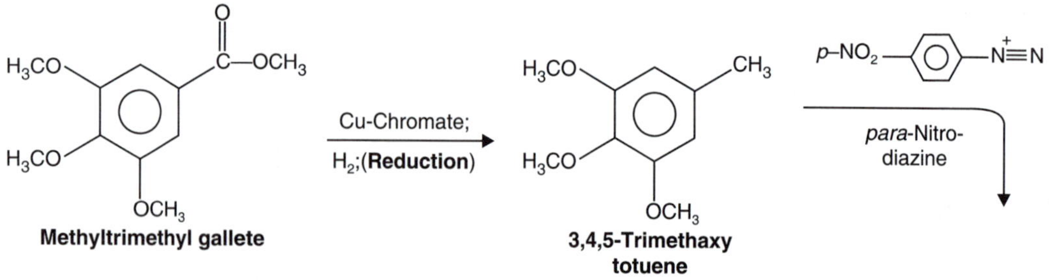

(Contd....)

---

*   **Isler (1961)**-a chemist of the Hoffmann La Roche Laboratories at Switzerland.

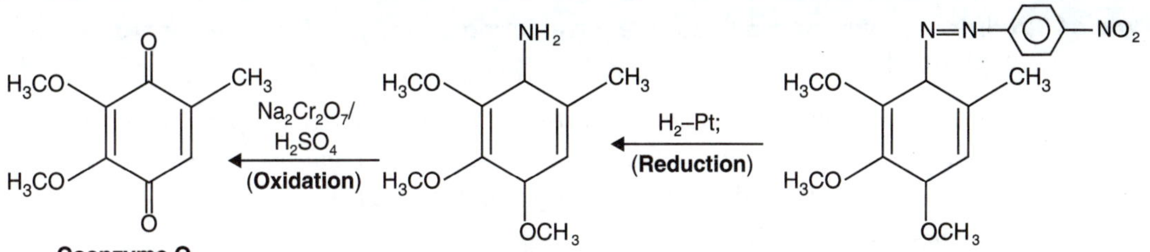

Coenzyme Q$_o$ (I) — 2-Amino deriv. — 2-p-Nitrophenyl diazene derivative.

Explanation- Methyltrimethyl gallate on reaction with **Cu-chromate** followed by *reduction* yields **3,4,5-trimethoxy toluene,** which on being subjected to the treatment with *p*-**nitrophenyldiazine** gives **2-*p*-nitrophenyl diazene derivative.** This product on further *reduction* with H$_2$-Pd yields the **2-amino derivatives** which on oxidation (vigorous) with Na$_2$Cr$_2$O$_7$/H$_2$SO$_4$ produces the desired product **Coenzyme Qo (I).**

2.8. Adenosine Monophosphate (AMP), Adenosine Diphosphate (ADP), Adenosine Triph-osphate (ATP)

In a broader sense, these **coenzymes** are invariably involved in the process of ***transphosphorylation.***

ATP is a *coenzyme* valuable in the transfer of *phosphate bond energy*. However, after stimulation of the *mamalian skeletal muscle* the **ATP** gets duly hydrolyzed to **ADP** by the **myosin-actin complex** unless the hydrolysis is prevented articulately. by injection of **magnesium sulphate.**

Structure of ATP

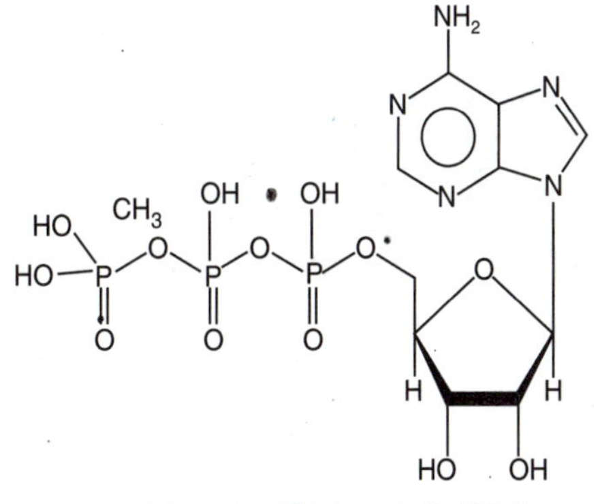

Adenosine Triphosphate (ATP)

Salient features of ATP- These. essentially comprise:

1. It is a **polyelectrolyte molecule** bearing in all four negative (-ve) charges. As it critically possesses a **low molecular weight-**it has the inherent ability to move around almost *freely inside the cells.*
2. **ATP** crucially acts a vital **'mediator'** to attain the specific energy derived from *one reaction*; and subsequently transfers this acquired energy to drive **another reaction.**
3. **ATP** is synthesized in an array of *metabolic reaction episodes* pertaining to the cell, namely:
   - **Cyclic phosphorylation,**
   - **Noncyclic phosphorylation, and**
   - **Oxidative phosphorylation.**

| NOTE | In order to overcome a host of *'energy barriers'*- the much-needed energy should be supplied. |

4. **ATP** provides the energy in most *biosynthetic processes* particularly in the so-called **transphosophorylation reactions** in the critical presence of an *appropriate enzyme;* and hence, gets converted ultimately into the **adenosine diphosphate (ADP).**

Thus, we may have the following expression:

$$\text{ROH} + \textbf{ATP} \longrightarrow \text{R-O-PO(OH)}_2 + \textbf{ADP}$$

Importantly, **ADP** also behaves as a *phosphorylating agent;* and it eventually gets converted into the so-called **adenosine monophosphate (AMP)-** as shown below:

$$\text{ROH} + \textbf{ADP} \qquad \text{R—OPO(OH)}_2 + \textbf{AMP}$$

Following is a rather not so-common reaction of **ATP** that is related to **pyrophosphorylation phenomenon** *viz.,*

$$\text{ROH} + \textbf{ATP} \longrightarrow \text{R—OPO(OH)—O—PO(OH)}_2 + \textbf{AMP}$$

Important Observations- In the particular instance, when the *structural formulae* of **ATP, ADP, and AMP** are Observed meticulously, one may explicity take cognizance of the fact that-

'**phosphate bondage in AMP is duly linked by a *normal ester bondage;* whereas, the so-called *terminal phosphate moieties* in ADP and ATP are attached to a *phosphate moiety* by means of an 'acid anhydride bond'.**

It has also been observed that the **'net free energy'** is usually employed to drive forward the *'couped reactions'.*

Besides, the **'acid anhydride bonds'** are largely known as the *energy-rich bonds,* which are mostly designated by the **symbol (~).**

Examples: Both **ATP** and **ADP** may be represented as given under:

ATP: Adenine-Ribose-O—PO(OH)—O—PO(OH)—O—PO(OH)$_2$

ADP: Adenine-Ribose-O—PO(OH)—O—PO(OH)$_2$

2.9. Glucose-1,6-diphosphate

Glucose-1,6-diphosphate normally acts as a **coenzyme** of *phosphoglucomutase* in the actual conversion of **glucose-6-phosphate to glucose-1-phosphate.**

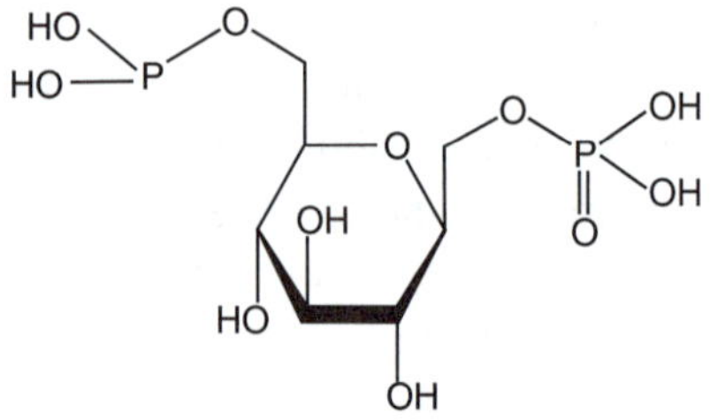

Glucose-1,6-diphosphate

2.10. Ascorbic Acid Oxidase

The *ascorbic acid oxidase* usually acts as a specific **coenzyme** in the *oxidation process of ascorbic acid* to its ultimate conversion to the **dehydroascorbic acid**-as given under:

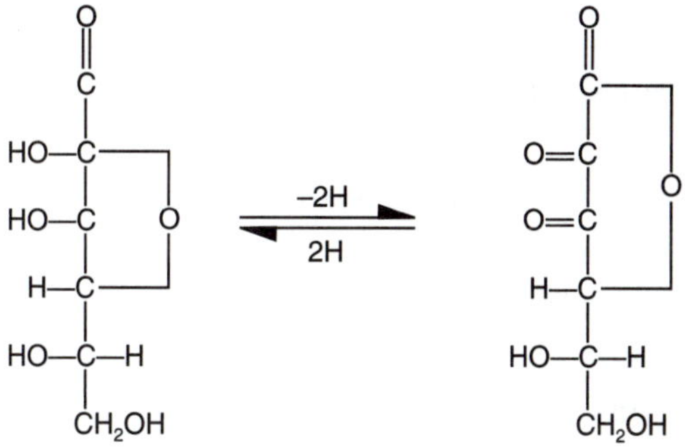

L–Ascorbic Acid   **L–Dehydroascorbic Acid**

2.11. Uridine Diphosphate Glucose (UDPG)

The **uridine diphosphate glucose (UDPG)** has been duly isolated from the *two* prevalent sources, namely:

- **yeast cells, and**
- **animal tissues.**

Interstingly, the said *coenzyme* has the following established and recognized **chemical structures:**

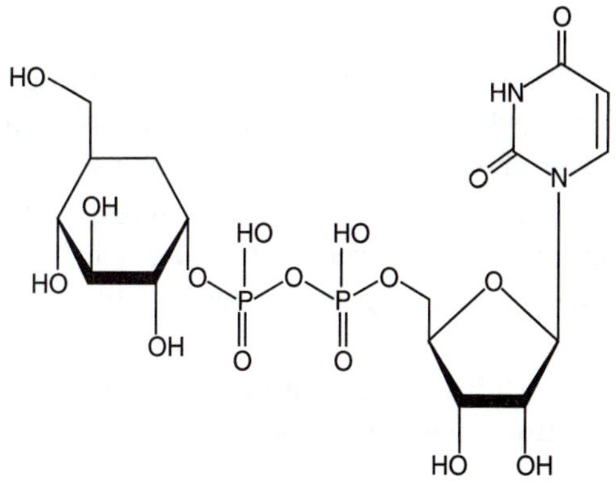

Uridine Diphosphate Glucose (UDPG)

UDPG designates the *coenzyme* of the **galactowaldenase system** which eventually catalyzes the conversion of :

Galactose–1–phosphate ⟶ Glucose–1–phosphate

FERMENTATION

Fermentation designates the *anaerobic metabolism* or *degradation process* of the carbohydrates (sugars). It is an *energy-yielding phenomenon* wherein the organic molecules do serve as:

- **electron donors, and**
- **electron acceptors.**

Alternatively, the terminology **'fermentation'** may also be applicable on a rather broader sense to a spectrum of oxidation processes-that do refer to the *various aspects metabolism* particulary confined to different species of:

- **Bacteria (microbes) and**
- **Fungi**

As stated the majority of *oxidation processes* actively associated with **'fermentation'** do go a long way in carrying out the decomposition of *organic compounds* (chemical substances) into simpler ones owing to the activity of microorganisms/yeast cells.

Following are a few widely known **'fermentations'**, namely:

- **Alcoholic Fermentation,**
- **Butyric Acid Fermentation,**
- **Oxalic Acid Fermentation, and**
- **Citric Acid Fermentation.**

ALCOHOL FERMENTATION-It is perhaps one of the oldest and best-known typical example of **'fermentation'**- that is usually carried out from a variety of such *raw materials* as:

- ❑ **Molasses-** a by product obtained from the *sugar industry* containing nearly 8-10% of not-so-easily recoverable sugar.
- ❑ **Potato starch-** a good source of natural good quality starch, being used in Russia since ages, to perhaps the famous alcoholic beverages **'VODKA'** by means of alcohol fermentation using **yeast cells.**
- ❑ **Red Sugar Starch-** It serves as another naturally occurring good quality starch obtained from vegetable source, which is being used extensively since ages in **CUBA,** to prepare **alcoholic drinks** by alcohol fermentation by making use of specific yeast cells.
- ❑ **Malt sugars-** the starch present in the staple food (cereal) **'Barley'** is first converted into the **malt sugar** by a specialized technique of **'malting'** whereby the natural starch present in **'whole barley grains'** is duly subjected to **alcohol fermentation** using a specific **yeast strain**.

Comments: Obviously, the so-called **'alcohol fermentation'** usually makes use of *'sugars'* viz, **Glucose, Fructose, Mannose, and Galactose-** that are decomposed by certain **yeasts** to *ethanol* as per the equation:

Besides, in the **alcohol fermentation** yielding *'ethanol'* as the **main product;** wheras, a host of other substances, such as:

- **Acetaldehyde**
- **Glycerol**
- **Pyruvic Acid**
- **Higher Alcohols (viz., aryl alcohol)**
- **Succinic Acid**

- **Tyrosol and**
- **Fused oil (*viz.*, admixture of isoamyl alcohol plus *d*-amyl alcohol).**

Interestingly, a large number of the **fermentation reactions** are specifically **'anaerobic'** in nature; however, a few are also found to be **'aerobic'** in nature.

MECHANISM OF ALCOHOL FERMENTATION

It is however, worthwhile to mention at the very outset that the phenomenon of **'alcohol fermentation'** is understood to be an extremely *complicated process*. Nevertheless, it may be thoughtfully categorized into *two* **major segments,** namely:

- ❏ **Glycolysis, and**
- ❏ **Conversion of Pyruvic Acid into 'Ethanol'.**

(a) **Glycolysis:** The reactions directly associated with the phenomenon of **'glycolysis'** may be further sub-divided into **two categories:**

First category- It essentially deals with the critical conversion of *most of the sugars present eg.,*

- **Starch**
- **Glucose**
- **Fructose**
- **Mannose and**
- **Galactose**

[Fructose-1,6-diphosphate due to the involvement of enzymes]

Following are the *five* **typical examples:**

- ◼ **Starch** + Pi **Phosphorylase** $\xrightarrow{}$ Glucose–1–phosphate $\xrightarrow{\text{Phospho gluco-mutase}}$ Glucose–6–Phosphates

- ◼ **Glucose** + ATP **Glucokinase** $\xrightarrow{}$ Glucose–6–phosphate + ADP

- ◼ **Fructose** + ATP **Fructokinase** $\xrightarrow{}$ Fructose–6–phosphate + ADP

- ◼ **Glucose**–6–phosphate **Phosphohexose** $\xrightarrow{}$ Fructose–6–phosphate

- ◼ **Fructose**–6–phosphate + ATP **Phosphofructokinase** $\xrightarrow{}$ Fructose_1,6–diphosphate + ADP

Second Category-The reactions pertaining to the *second* category invariably starts with the critical splitting of:

'Fructose-1,6-diphosphate'

into the three respective **carbon compounds** together with:

- **3-phosphoglyceraldehyde (A), and**
- **Dihydroxyacetone-3-phosphate (B),**

In the presence of the enzyme **aldolase**. Consequently, these *two* **sugar entities** are duly interconverted *via* the *enzyme* **phosphotriose-isomerase**-as given under:

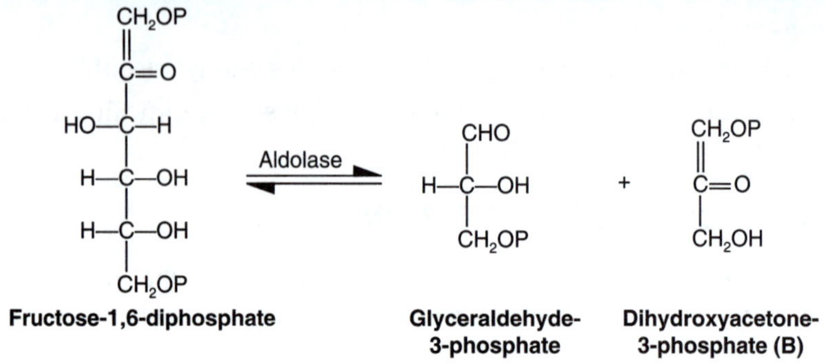

Fructose-1,6-diphosphate → (Aldolase) → Glyceraldehyde-3-phosphate + Dihydroxyacetone-3-phosphate (B)

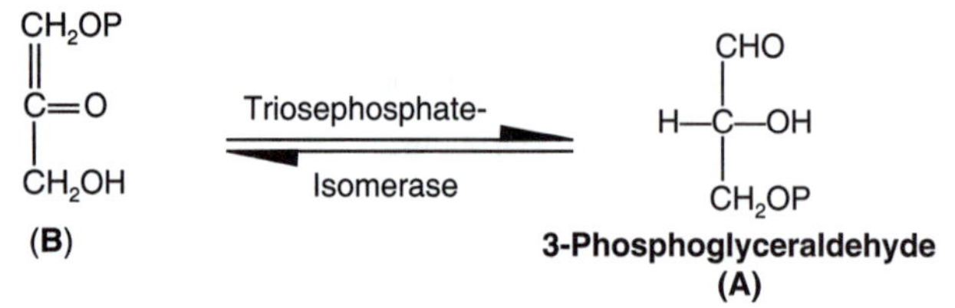

(B) → (Triosephosphate-Isomerase) → 3-Phosphoglyceraldehyde (A)

Furthermore, the **compound (A)** produced in the above step, in the presence of **NAD+ (Coenzyme I)**-that usually serves as a **H-acceptor** and **inorganic phosphate** plus **phosphoglyceraldehyde dehydrogenase**, thereby finally converting into:

- **1,3-diphosphoglyceric acid, and**   • **NAD+ into NADH+H+.**

Consequently, **1,3-diphosphoglyceric** acid loses *one* **phosphate moiety** to **ADP** to yield **3-phosphoglyceric acid and ATP.**

Importantly, the *enzyme* **phosphoglyceral kinase** solely helps to mediate this *transformation* predominantly. In this way, one **ATP** is produced. Hence, the resulting **3-phosphoglyceric acid (PGA)** gets duly transformed into **2-phosphoglyceric acid** in the presence of the *enzyme enolase*.

Thus, ultimately **2-phosphophenol pyruvic acid** loses a *phosphate moiety* to **ADP**-thereby giving rise to the critical formation of **pyruvic acid** and **ATP** in the presence of the *enzyme* **pyruvate kinase.**

Thus, we may have the following expressions:

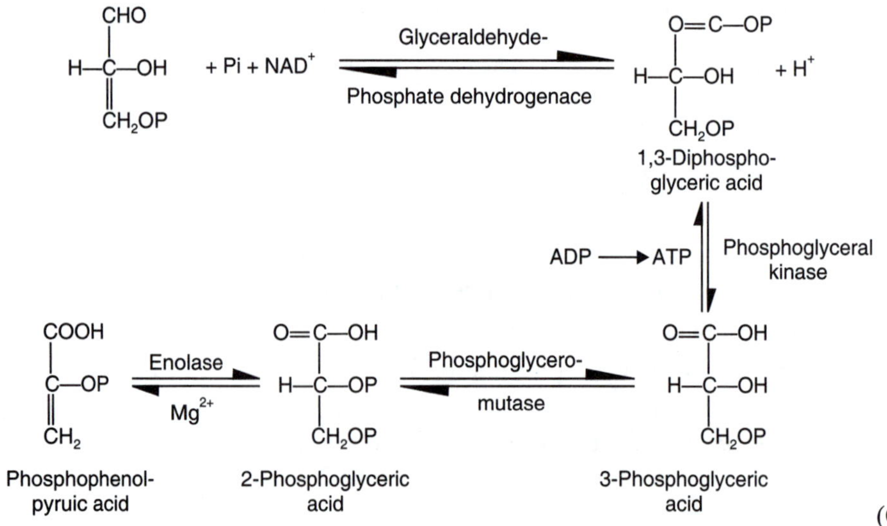

(Contd....)

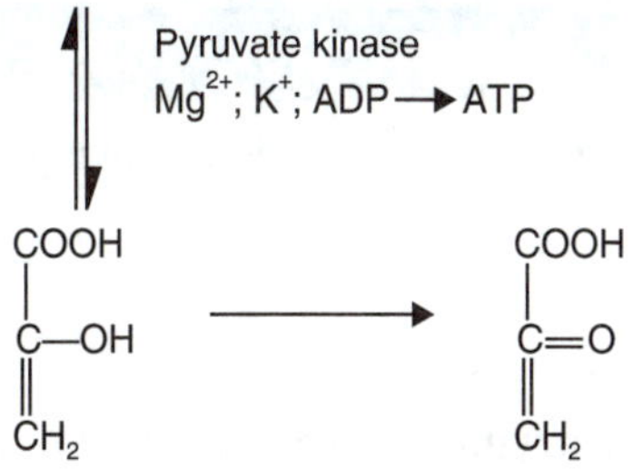

$$\text{Pyruvate kinase}$$
$$Mg^{2+}; K^+; ADP \longrightarrow ATP$$

COOH	COOH
|	|
C—OH	C=O
||	||
CH$_2$	CH$_2$
Enolpyrurate	**Pyruric Acid**

(b) Conversion of Pyruvic Acid to Ethanol

It has been duly observed that in **yeasts** and **higher plants** the *pyruvic acid* gets duly converted to acetaldehyde (CH$_3$CHO) in the presence of *pyruvic carboxylase* thereby eliminating a **carbon dioxide (CO$_2$) mole. The acetaldehyde (CH$_3$CHO)** so generated then accepts gracefully the H-atom from **NADH+H+** to yield **CH$_3$CH$_2$OH (Ethanol) and NAD+ (coenzyme I)** respectively.

The *alcohol dehydrogenase* thus mediates this particular reaction- as illustrated below:

■ $CH_3CO.COOH \xrightarrow[\text{Thiamine pyrophosphate}]{\text{Pyruvic dehydrogenase}} CH_3CHO + CO_2\uparrow$

■ $CH_3CHO + NADH + H^+ \xrightarrow{\text{Alcohol dehydrogenase}} CH_3CH_2OH + NAD^+$
$\quad\quad\quad\quad\quad\quad\quad\quad\quad\quad\quad\quad\quad\quad\quad\quad\quad\quad$ **Ethanol   Coenzyme I**

■ $CH_3COCOOH + NADH + H^+ \xrightarrow[\text{Tpp}]{\text{Decarboxylase DH}} CH_3CH_2OH + NAD+CO_2\uparrow$
$\quad\quad\quad\quad\quad\quad\quad\quad\quad\quad\quad\quad\quad\quad\quad\quad\quad\quad\quad\quad$ **Ethanol   Coenzyme I**

FURTHER READING REFERENCES

1. Avison AWD and Hawkins JD : **The role of Phosphoric Esters in Biological Reactions,** *Quart Rev.,* **5**:171,1951.

2. Bell DJ : **An Introduction to Carbohydrate Biochemistry**, University Tutorial Press, London (UK), 1948.

3. Cox DR : **Planning of Experiments,** John Wiley & Sons, New York, ;221,1965.

4. Davies O : **Design and Analysis of Industrial Experiments,** Hefner, New York, 1963.

5. Ochoa S : **Enzymatic Mechanisms in the Citric Acid Cycle, Advances in Enzymology, 15**:183, 1954.

REVIEW QUESTIONS

1. What are **Enzymes**? Give a brief account on the following topies :
   - (a) **Hydrolytic Enzymes**
   - (b) **Oxidative Enzymes**
   - (c) **Salient Features of Enzymes**

2. How would you classify the **Enzymes**? Give at least one example from each category you have mentioned.

3. Discuse the characteristic features of enzymes with particular reference to the following aspects:
   - (i) **Enzyme Efficiency**
   - (ii) **Enzyme Specificity**

4. Give a comprehensive description of the **Coenzymes** including :
   - (a) **Coenzyme Variants**
   - (b) **Constitution of Coenzyme I**

5. Write the chemical strurcture of **Coenzyme II**. Discues in details the **Constitution of Coenzyme II.**

6. Attempt **short notes** on any *two* of the following :
   - (i) **Coenzyme A**
   - (ii) **Flavin Mononucleotide [FMN]**
   - (iii) **Synthesis of Coenzyme Q**

7. Explain briefly the **Mechanism of Alcohol Fermentation** by citing suitable examples.

Contents at a Glance

Prostaglandins (PGE₂ and PGF₂ₐ)

6

1. INTRODUCTION

The **prostaglandins** duly represent a class of fatty acids that are derived meticulously from *arachidonic* acid by the cyclization phenomenon to form a **5-membered ring system** in the vicinity of the fatty acid chain, a class of hormones that essentially possesses an array of *physiological effects* including:

- **Vasodilation,** and
- **Smooth-muscle contraction.**

Examples: PGE_1 , PGE_2, and $PGF_{2\alpha}$.

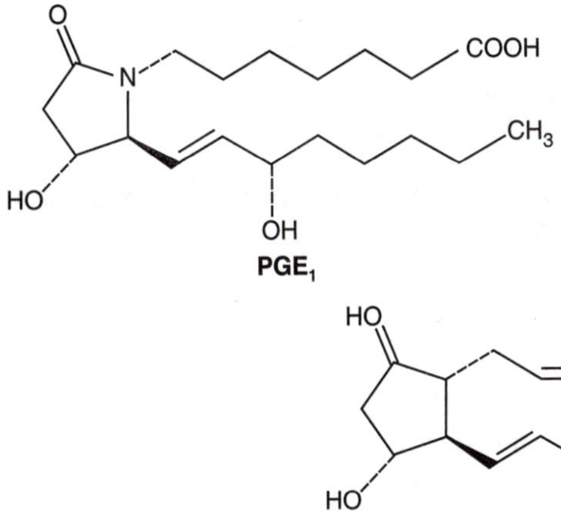

PGE₁

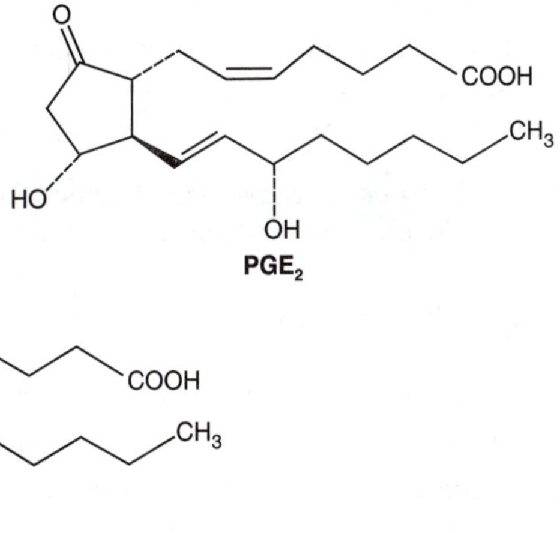

PGE₂

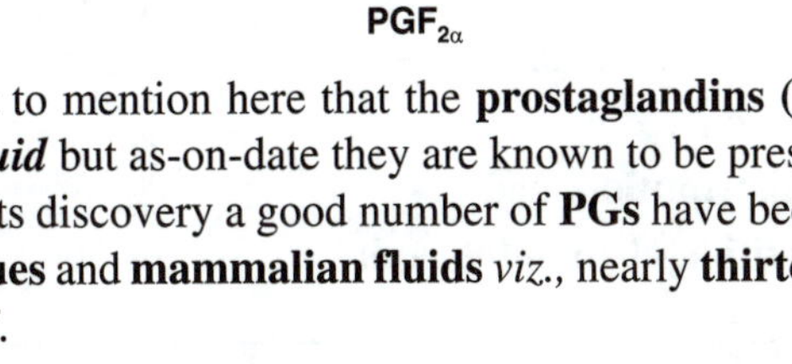

PGF₂ₐ

It is, however, pertinent to mention here that the **prostaglandins (PGs)** were first and foremost discovered in the *seminal fluid* but as-on-date they are known to be present in almost all the so-called **mammalian tissues**. Since its discovery a good number of **PGs** have been duly isolated and identified in both the **mammalian tissues** and **mammalian fluids** *viz.,* nearly **thirteen** different **PGs**- are isolated from the *seminal fluid* itself.

PG Variants- The **PGs** have been categorized into *eight* altogether different groups, namely: **'PGA, PGB, PGC, PGD, PGE, PGF, PGG, and PGH'.**

Nevertheless, there are certain **PGs** that have been studied exhaustively, such as:

'**PGE$_1$, PGE$_{2\alpha}$,PGF$_{1\alpha}$,PGF$_{2\alpha}$,PGA$_1$,PGA$_2$,PGB$_1$, PGB$_2$**'.

Importantly, the aforesaid **PGs** do differ from one another in their *chemical structures*; however, they do have certain *common structural features*, for instance:

- **All** *PGs* do belong to the class of *'eicosa fatty acids'* **with a cyclopentane ring;** and
- **They do possess a** *hydroxyl (–OH)* **group at C-15 and a** *trans*-**double bond at C-13.**

Following are the *two* typical examples of **eicosa fatty acids**:

- **Arachidonic Acid (I)** and   •   **Prostanoic Acid (II)**

(I)   (II)

Occurrence of PGs Kurzrok and Lieb (1930) first reported the presence of **prostaglandins (PGs)** in the necessary *genital glands* of humans (*viz.*, **testes**). Later on, Goldblatt and Euler (1935) pronounced gracefully that the extracts derived from either:

- *Sheep* **vesicular glands**, and
- *Human* **seminal fluids**,

ultimately showed almost identical therapeutic effects upon the **uterine smooth muscles**. It was, however, further ascertained that the extended investigative studies carried out on a variety of *animals* exhibited the **maximum quantum of PGs** in the **sheep vesicular glands** only.

> **NOTE**   The human seminal fluid (plasma) contains the maximum quantum of prostaglandins
> (PGs), which ranges between 50-60 mg.mL^{-1}.

2. ISOLATION OF PGs

Prostaglandin E$_1$-isolated from sheep seminal vescicle tissue (Bergstrom *et al.*, 1962)*.

Prostaglandin E$_2$- isolated from sheep prostrate gland (Bergstrom;1962)**.

Prostaglandin Factor (PGF)-Bergstrom and Sjovall (1957) reported the critical elaborated procedure for the preparation of pure crystalline **prostaglandin factor (PGF)** by carrying out the dried and homogenized glands obtained from sheep with **95% (v/v) ethanol**. Subsequently, the ethanol extract was carefully concentrated under vaccuo and treated further as follows:

- **subsequent extractions,**
- **counter-current distribution,** and
- **reverse-phase partition chromatography.**

---

  * Bergstrom *et al*: *Acta.Chem Scand,* **16**:501, 1962.

** Bergstrom *et al: Biochem Biophysics Acta.,* 207, 1962.

Ultimately, the gainful usage of the **'Column Chromatography'** upon the *activated* silica gel of the respective:

'bioactive fractions in an acidic environment'

using the *requisite solvent* as the eluant gave ~2mg of pure **PGF** obtained in the form of:

- **colourless small needle-like crystals and**   • **mp:102-103°C**

NOTE   Thus, one may also isolate *PGE* very much in the same manner as that of *PGF*.

3. CLASSIFICATION AND NOMENCLATURE OF PROSTAGLANDINS [PGs]

First and foremost one must have the crystal-clear concept and understanding of the fact that the so-called *parent skeleton* in the **prostaglandins (PGs)** do possess the following *two* **well-recognized segments,** namely:

❏ **A completely saturated cyclopentane comprising a C-20 hydrocarbon skeleton invariably termed as the 'prostane'; and**

❏ **A related carboxylic acid skeleton usually known as the prostanoic acid.**

Hence, the above *two* recognized and established structural segments may be given as shown below:

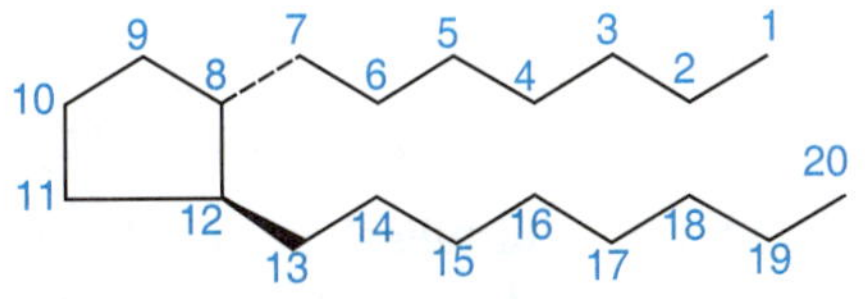

Prostane

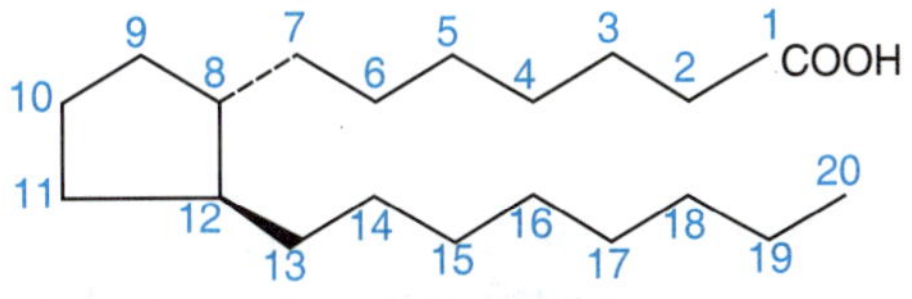

Prostanoic Acid

NOMENCLATURE Interestingly, the nomenclature of the **prostaglandins (PGs)** is based exclusively upon the *inherent* **nature of the functionalities** in the so-called **5-membered cycloheptane ring system** (as shown under). Therefore, based upon the aforesaid factual delebrations the **prostaglandins (PGs)** are meticulously classified into *six* **well-established and broadly known families**–as illustrated under:

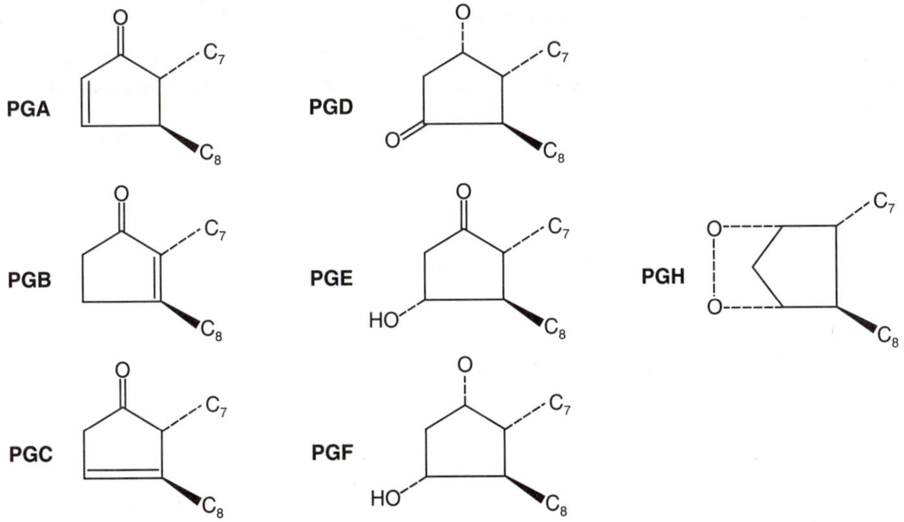

PGA

PGD

PGB

PGE

PGH

PGC

PGF

Obviously, the **chemical entities** (*compounds*) having greater solubility in *solvent ether (Et-O-Et)* are usually called as **PGEs;** whereas, the greater solubility bearing compounds dissolved in **phosphate buffers** are termed as **PGFs** (since '**F**' stands for '*fosfate*' in Swedish).

It is, however, pertinent to state here that in the *two* **aforesaid families PGA and PGB**–the letters '**A**' and '**B**' are normally employed for compounds that are yielded from the reaction of **PGE** with the *acid* and *base* respectively. However, the various on-going reactions occurring amongst the '*various families*', such as:

- **PGA**   • **PGB**   • **PGE** and   • **PGF**

are depicted explicitly as given under:

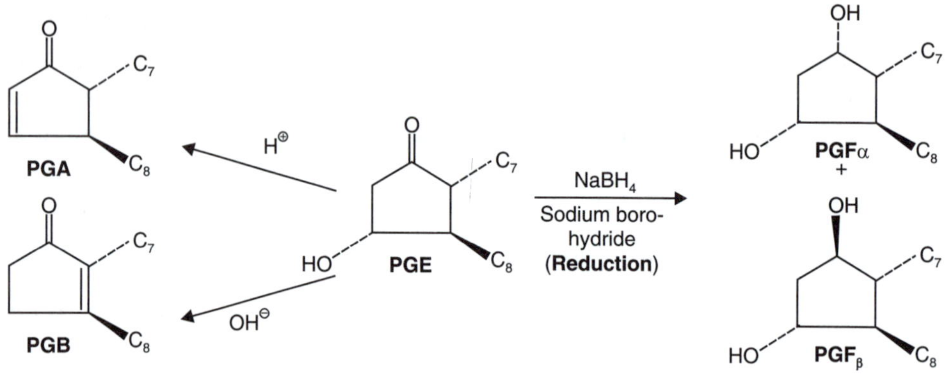

Remarks: The particular *subscript numerals* (appearing in the '**various families**') are usually employed so as to denote precisely **the actual number of double bonds** existing in the *side-chains* of the **molecule.**

Example: Amazingly, **PGE₁** does possess only one double bond existing between **C-13 and C-14** in the form of the *trans*-double bond; **PGE₂** inherits an *additional* double bond appearing between **C-5 and C-6;** and finally **PGE₃** is having a *cis*-double bond between **C-17 and C-18.**

> **NOTE :** (1) Since the α-hydroxyl (OH) group is duly located at C-15 in most of the naturally occurring '*bioactive prostaglandins (PGEs);* hence, no particular mention is even made of this feature in the nomenclature.
>
> (2) To show clearly the nature of the modified substituents of functionalities, we may add certain prefixes and suffixes (prefix follows the modifications).

Structures of Six Important Prostaglandins (PGEs)-These include essentially the following PGEs, namely:

(1) 13,14-Dihydroprostaglandin (PGE₂)

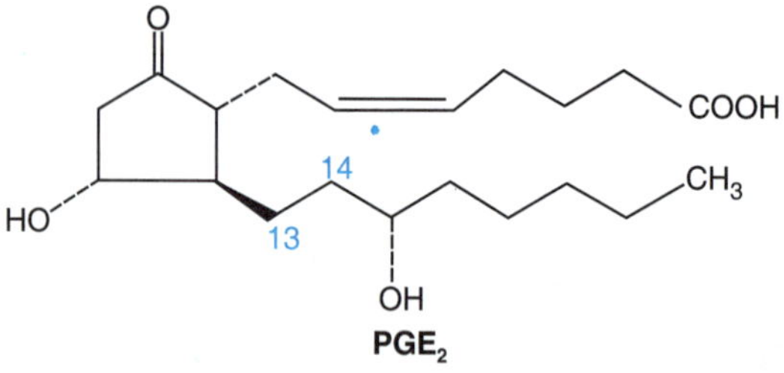

In **PGE₂,** one may vividly observe:

> **Prefix:** dihydro; and

Modified Moiety: Saturation of C-double bond between **C-13 and C-14**.

(2) 8-Azaprostaglandin (PGE₁)

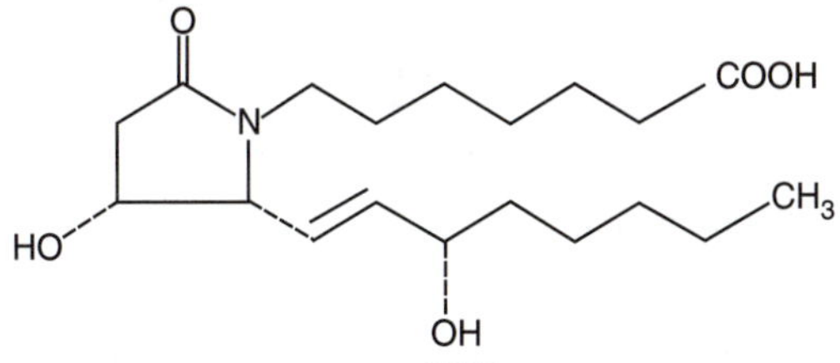

PGE₁

In **PGE₁**, one may visualize clearly:

> **Prefix**: Aza; and

Modified Moiety: Replacement of methylene group with a N-atom.

(3) 11-Deoxy-11-oxaprostaglandin

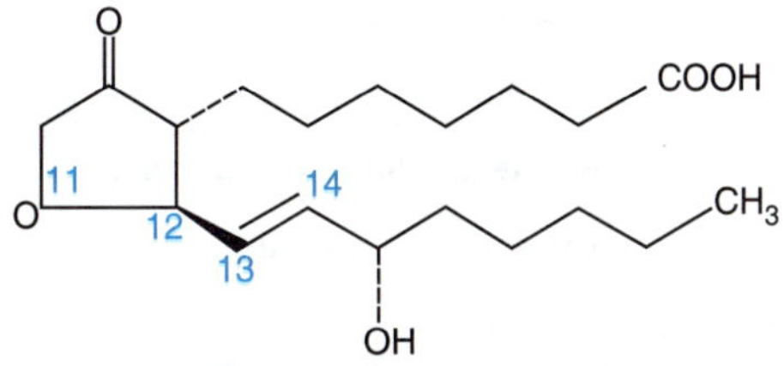

11-Deoxy-11-oxaprostaglandin

It essentially has the following:

Prefix : Oxa; and

Modified Moiety: Replacement of methylene group with a O-atom.

(4) 19-Oxoprostaglandins

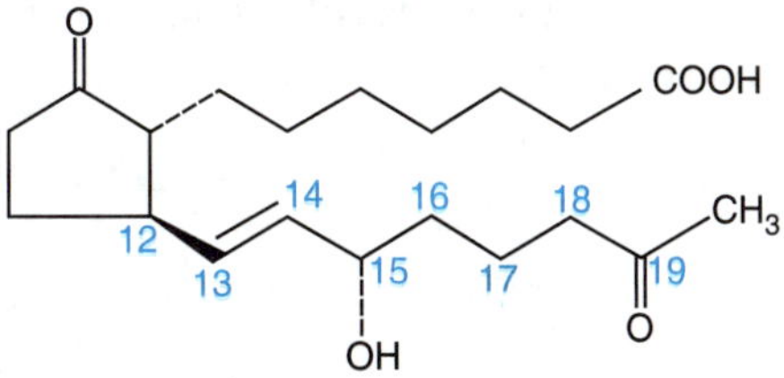

19-Oxoprostaglandins

It has the following essential attachements:

Prefix: Oxo; and

Modified Moiety: Replacement of methylene (–CH₂–) group with a carbonyl (>C=O) group.

(5) *Ent*-Prostaglandin

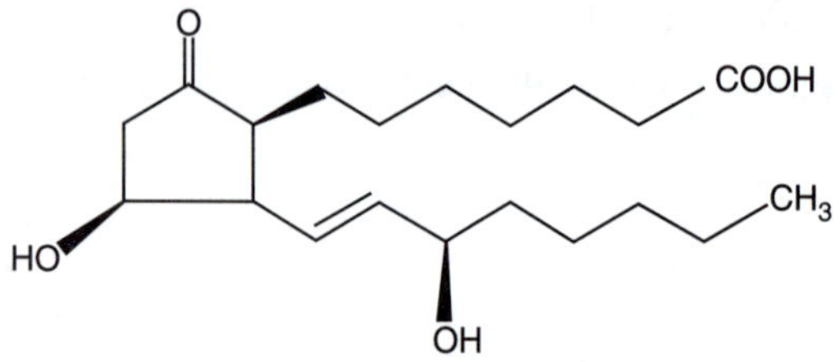

Ent-Prostaglandin

it has the following essential attachements:

Suffix: Ent; and

Modified Moiety: Reversal of all the **chiral (asymmetric) centres** of a **natural prostaglandin**.

(6) Prostaglandin E$_1$ Methyl Ester

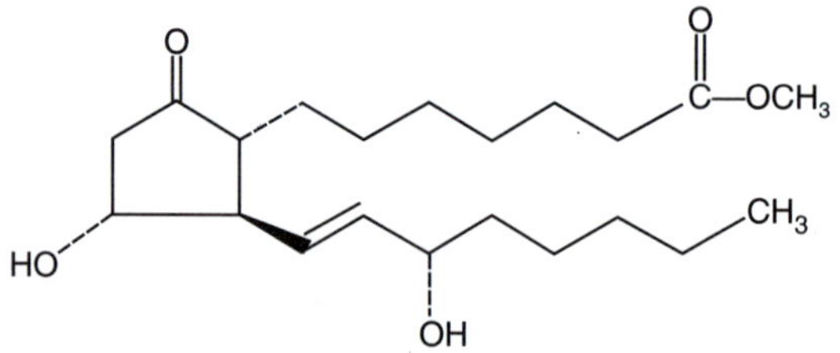

Prostaglandin E$_1$ Methyl Ester

It prominently has the following cardinal attachements:

Suffix: Carboxylic ester; and

Modified Moiety: Esterification of the carboxylic (COOH) moiety.

4. CONFIGURATIONS IN RING SYSTEM AND SIDE-CHAIN OF PGEs

Anderson was pioneer in putting forward the *four* **fundamental stereoisomeric hydrocarbon skeletons**, namely:

- **Prostane** • **Ent-prostane**   •   **Isoprostane** and   • **Ent-isoprostane**

that are found abundantly in the **prostaglandins (PGEs)** the structures/configurations of the above *four* **stereoisomeric hydrocarbon skeletons** are as shown below:

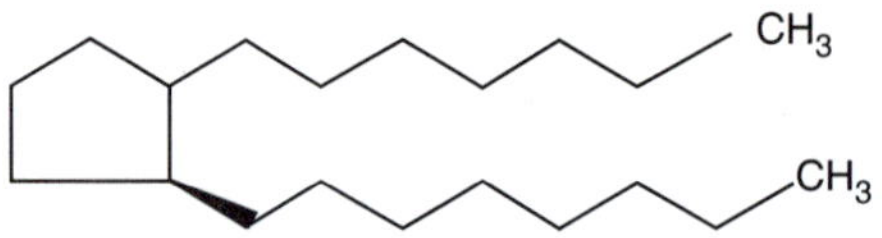

Prostane

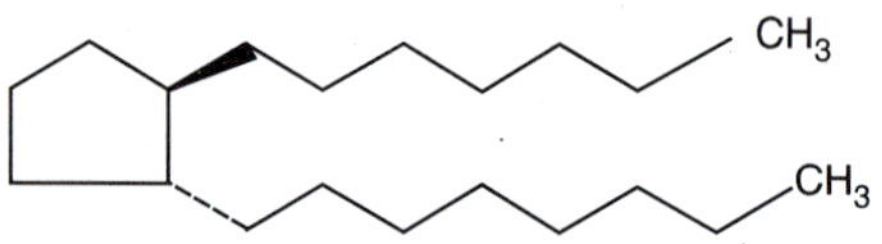

Ent-Prostane

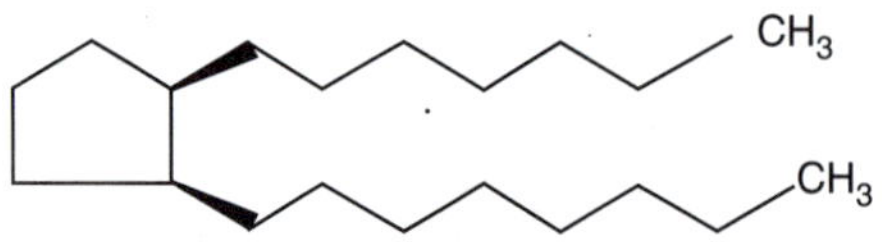

Isoprostane

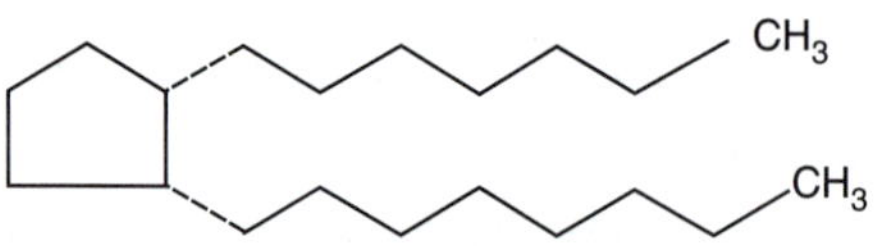

Ent-isoprostane

Thus, the **stereochemistry** of the **various substituents** located strategically on the **cyclopentane ring system** are designed intelligently as stated under:

➤ *First* – when there exists a particular substituents oriented above the plane of the **5-membered cyelopentane ring system:** which is designated to exhibit a *β-configuration;* and is duly shown by a **solid-line bond**;

➤ *Second* – when there occurs a specific substituent oriented below the plane of the **5-membered cyclopentane ring system**: which is duly designated to have *α-configuration*; and is shown by a **dotted-line bond;** and

➤ *Third* – when the prevailing configuration of the substituent is not known; and hence, is normally shown with a **wavy-line bond** (~~).

Absolute Configuration-In usual practice, the so-called conventionally drawn *chemical structures* of the **prostaglandins (PGs)** are invariably considered to be the **absolute configuration** of the 'prostanoids'- duly derived from the **mammalian sources**.

BIOGENESIS OF PGE$_2$ and PGF$_{2\alpha}$- The structures of **PGE$_2$** and **PGF$_{2\alpha}$** are already provided in the introduction of this chapter. In a broader perspective, the so-called **microsomal-enzyme system** invariably termed as the *'prostagladin synthetase'* is indeed distributed abundantly in the *mammalian tissues*. Obviously, the above enzyme initiates the desired *biosynthesis* of the **prostaglandins (PGs)** right from the **essential fatty acids** *e.g.,*

'the arachidonic acid cascade'.

However, it is pertinent to state at this point in time that the **prostaglandins (PGs)** are not usually stored in tissues but are grossly *biosynthesized;* and subsequently, get duly *released* **on demand** by typical **physiological stimuli** (*in vivo*).

The Role of Arachidonic Acid–It has been established adequately that **arachidonic acid** serves as the *precursor* of **PGE$_2$** and **PGF$_{2\alpha}$**.

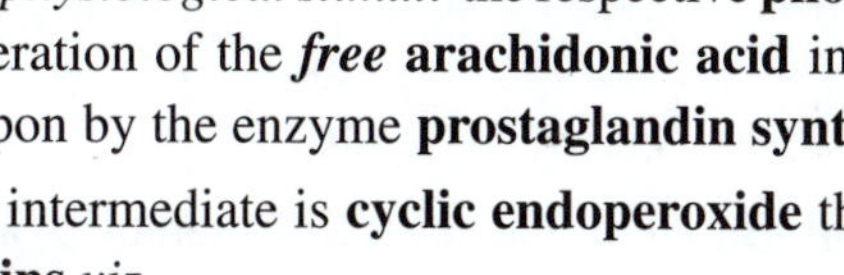

Arachidonic Acid

Interestingly, the **arachidonic acid** is duly stored as a **phospholipid** in the *animal tissues*. Subsequently, on demand by *the physiological stimuli*-the respective **phospholipase A** gets duly activated that eventually helps in the liberation of the *free* **arachidonic acid** in the biological system. Finally, this *acid* is specifically acted upon by the enzyme **prostaglandin synthetase**.

Remarks- Neverthless, the key intermediate is **cyclic endoperoxide** that eventually gets converted to the corresponding **prostaglandins** *viz.,*

• **PGE$_2$**
• **PGF$_{2\alpha}$** and
• **PGD$_2$**.

BIOSYNTHESIS OF PGE$_2$, PGF$_{2\alpha}$, and PGD$_2$.

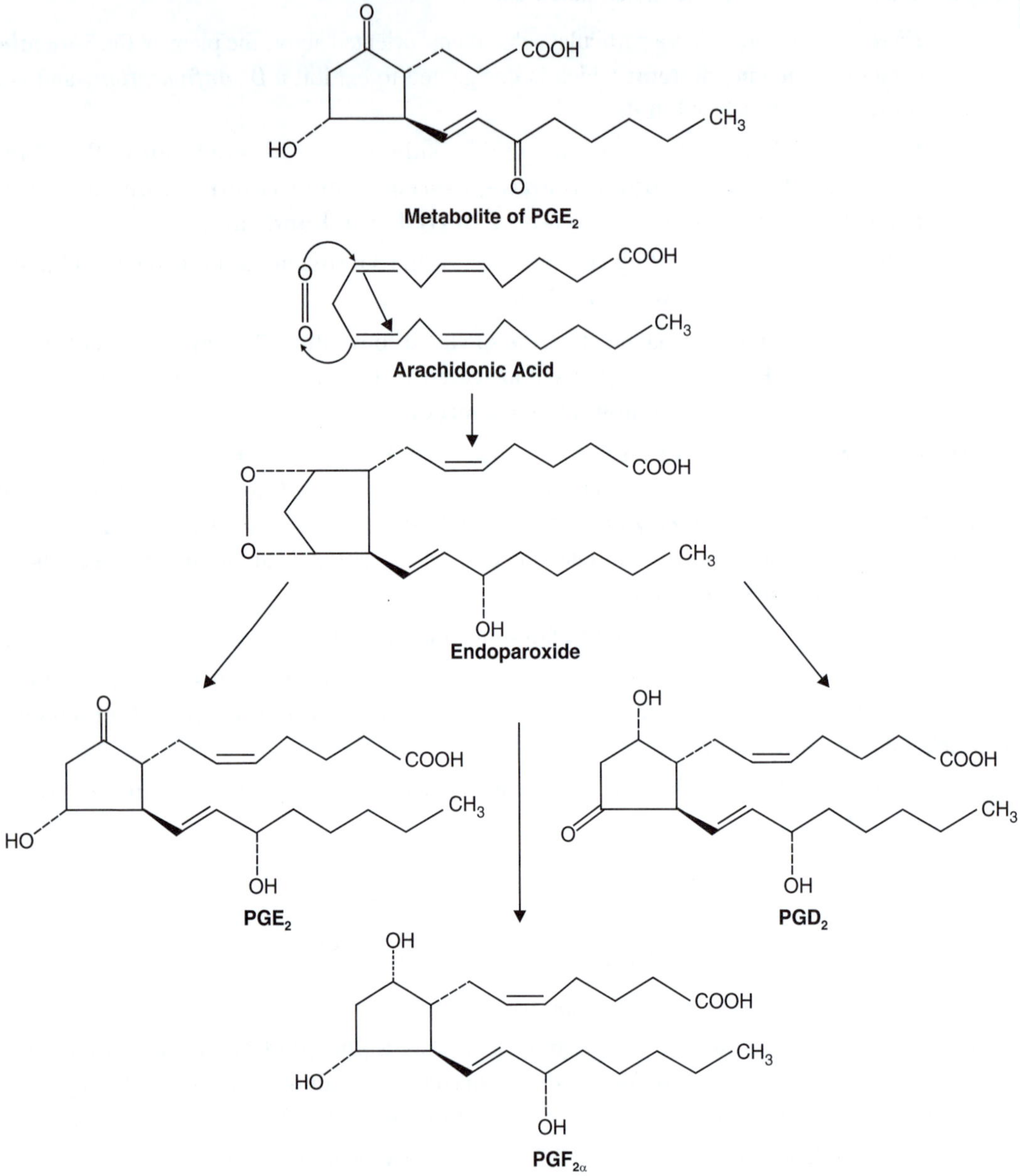

Metabolite of PGE$_2$

Arachidonic Acid

Endoparoxide

PGE$_2$

PGD$_2$

PGF$_{2\alpha}$

Prostaglandin Metabolism–Based on the survey of literatures and scientific evidences one may take cognizance of the underlying revelations that –

'the prostaglandins (PGs) get inactivated due to an array of enzymatic reactions'

which ultimately limit their activity profile perceptively.

Thus, there are *two* appreciative prominent **enzymatic reaction**, such as:

1. those intimately involving the *oxidation phenomenon* pertaining to the C-15 allylic hydroxyl (OH) moity by the specific enzyme *15- hydroxygenases* and
2. the critical follow up reduction of the *double* bond between C-13 and C-14 by the enzyme **Δ^{13} reductase.**

   *Comment :*In this manner, the so-called **β(beta)** and **ω(omega) oxidation** do come into play predominantly.

PHYSIOLOGICAL FUNCTIONALITY OF PROSTAGLANDINS (PGs)

Following are the *five* most vital and prominent physiological functionalities of **PGs**, namely:

1. The prostaglandins (**PGs**) do serve as the local modulators of the cellular functionalities.
2. PGs–in fact, play a major role in the critical regulation of:
   - **Endocrine System**   • **Nervous System**   • **Digestive System**   • **Haemostatic Mechanism**   • **Respiratory System**   • **Cardiovascular System (CVS)** and
   - **Renal functions.**
3. **PGs-**also control meticulously both:
   - **Lipid Metabolism,** and   • **Carbohydrate Metabolism**
4. In a particular instance, when there occurs an obvious change either in the **prostaglandin production** or **prostaglandin metabolism**-it may finally cause certain *typical physiological disorders*, for instance:
   - **Bronchial Asthma**   • **Fever**   • **Hypertension**   • **Pain**
   - **Ulceration** and   • **Inflammation Symptoms.**
5. Importantly, the therapeutic effects related to :
   - **Anti-inflammatory**   • **Antipyretic and Analgesic**

are prominently caused and ascribed to the exclusive reduction and inhibition of the **prostaglandin biosynthesis** by such **drug substances.**

Examples: Aspirin (Acetylsalicylic Acid) prevents solely the functionality of the so-called *'arachidonic acid cascade'.*

Table: 6.1 records some of the physiological effects of the **'prostaglandins (PGs);**

Table:6.1:Certain Typical Physiological Effects of Prostaglandins (PGs)

S. No.	Prostaglandin Variants	Physiological Systems	Physiological Effects	Therapeutic Uses
1.	**PGE$_1$**	Cardiovascular	• Dilation of blood Vessels	• Congestive heart failure (CHF) •· Anti-hypertensive
2.	**PGE$_1$,PGE$_2$,PGA**	Gastrointestinal (GI) treatment	• Inhibits acid secretion	• In gastric and peptic ulcers.
3.	**PGE$_1$,PGD$_2$, Prostacyclin**	Platelets	• Inhibits platelets aggregation	• Reduces and prevents blood-clot formation.
4.	**PGE$_2$**	Renal Function	• Modulates renal blood flow	• In impaired renal function.

(Contd....)

| 5. | PGE$_2$ and PGF$_2$ | Reproductive system | • Stimulation of uterine muscles | • For induction of labour pain
• Medical termination of pregnancy. |
| 6. | PGE$_1$, PGE$_2$ | Respiratory system | • Relaxation of bronchial tissues | • In asthma
• In bronchoconstriction. |

5. SYNTHESIS OF PGE$_2$ AND PGF$_{2\alpha}$

It has been reported that in the particular presence of the well-known *microsomal enzyme PGH$_2$ synthase+*, and **arachidonic acid**- that eventually undergoes the so-called *oxidative-cyclization* to give rise to the formation of *parent* **prostaglandin (PGH$_2$)**. In fact, it predominantly caters as the precursor of various other **prostaglandins(PGs)**, **prostacyclins**, and **thromboxanes**–as depicted in the following **synthesis** of the said *three* **products** from *arachidonic acid:*

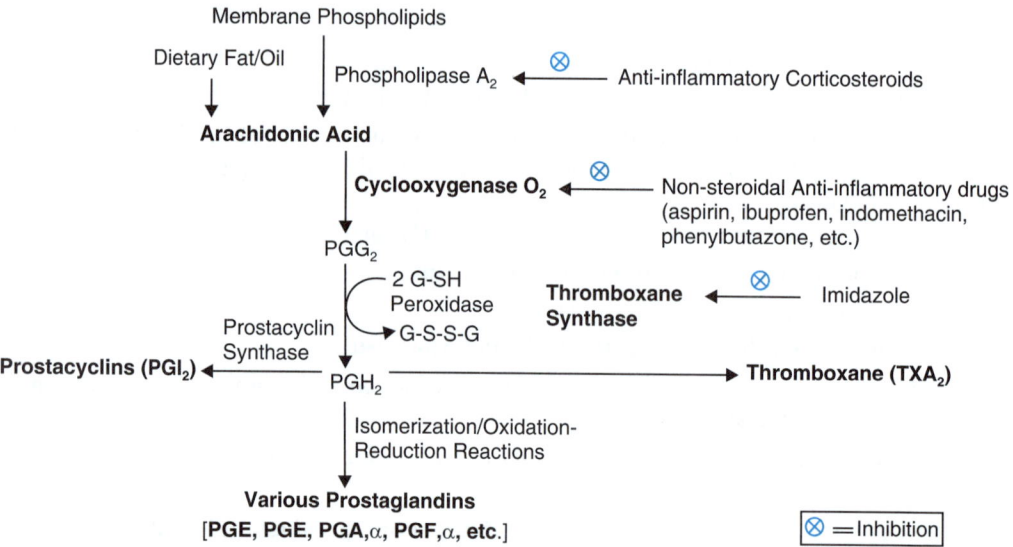

Explanations-Thus, we may put forward the explanations for the formation of such *products of reactions* as:

- ❑ **Prostacyclins**-These are duly synthesized from the **parent prostaglandin (PGH$_2$)** specifically in the *endothelium of blood vessels, kidneys,* and *heart*. The **prostacyclic synthase** (a *microsomal enzyme*) helps in the actual conversion of **PGH$_2$** into **PGH** (as shown in the above syntheses). Besides, **PGI$_2$** and **thromboxanes** do exhibit the explicit *antagonistic activity profiles*; whereas, **PGI$_2$** shows the following *two* **actions:**
  - **Inhibits platelet aggregation**, and
  - **Brings forth vasodilation.**

---

* It is also known as the **PG-endoperoxide synthase complex** or **COX.**

Remarks: These essentially includes:

- **PGI_2** – **caused the inhibition of platelet aggregation and minimizes the bp by affording vasodilation; and**
- **Prostacyclins** – **also causes an appreciable stimulation of the gastric secretion.**

Thromboxanes-In true sense, the **thromboxanes** are duly synthesized from **PGH_2** (*i.e.,* **parent prostaglandin**) in the ***blood platelets*** (**thrombocytes).** Interestingly, the enzyme **Thromboxane synthase** (an enzyme found abundantly bin the blood platelets *i.e.*, thrombocytes)-obviously helps in the conversion of **PGH_2** into **TxA_2** (*i.e.*, **thromboxane A_2**)-as shown in the above syntheses). In addition, the **thromboxanes** are found to be totally responsible for the following ***three*** **activities**, such as:

- **Constriction of blood vessels,**
- **Platelet aggregation,** and
- **Mobilization of the intracellular Ca^{2+} ions.**

Besides, the **thromboxanes** do causes an effective **vasoconstriction** and the **platelet aggregation.**

Remarks: These essentially comprise:

- By inhibiting the synthesis of **thromboxanes**–the ***aspirin*** does inhibit the platelet aggregation; and hence, *checks and prevents the blood coagulation*; and
- Hence, ***aspirin*** in relatively low doses (75-150 $mg.day^{-1}$) is recommended invariably to *reduce the risk of heart attacks* (MI) in eoronary heart diseases (CHD)–since it minimizes the risk of possible blood-clot formation in the coronary arteries.

Corey *et al.* (1969) first and foremost reported the **sterospecific synthesis of PGE_2** and **$PGF_{2\alpha}$** involving the ***bicyclo*** **[2,2,1] heptene intermediate**- as shown under:

COREY'S SYNTHESIS OF PGE_2 AND $PGF_{2\alpha}$

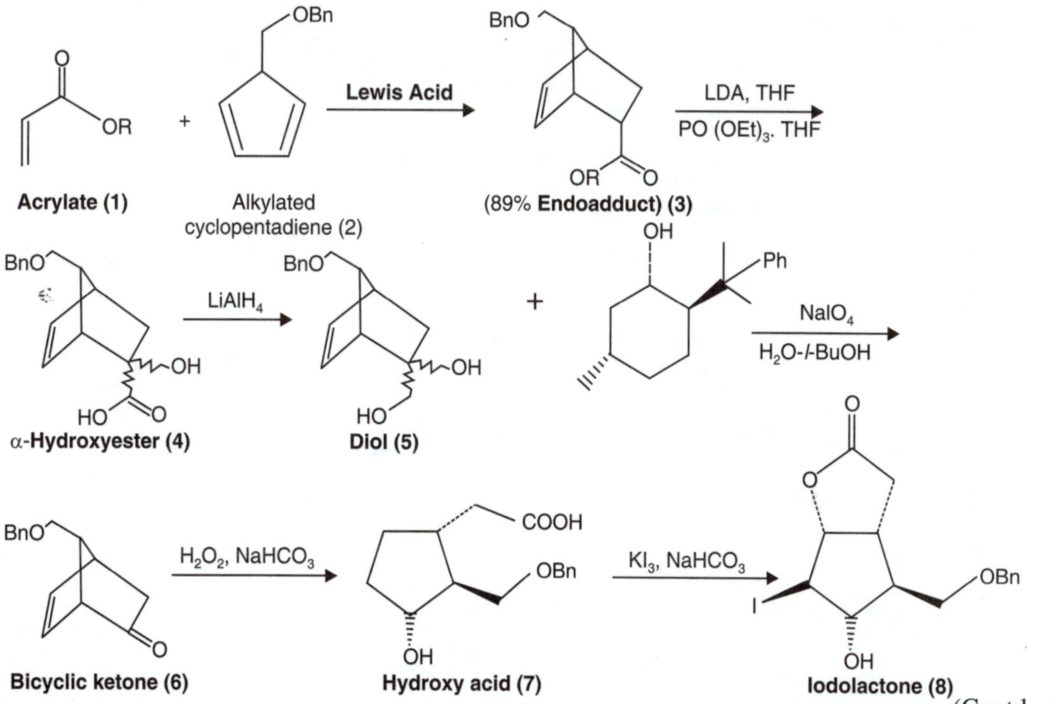
OBn
BnO
O
OR
+
Lewis Acid
LDA, THF
PO (OEt)3. THF
OR
O
Acrylate (1)
Alkylated cyclopentadiene (2)
(89% Endoadduct) (3)
BnO
LiAlH4
BnO
OH
Ph
+
NaIO4
H2O-t-BuOH
OH
HO
O
α-Hydroxyester (4)
OH
HO
Diol (5)
BnO
H2O2, NaHCO3
COOH
OBn
KI3, NaHCO3
O
O
OBn
I
O
OH
OH
Bicyclic ketone (6)
Hydroxy acid (7)
Iodolactone (8)

(Contd....)

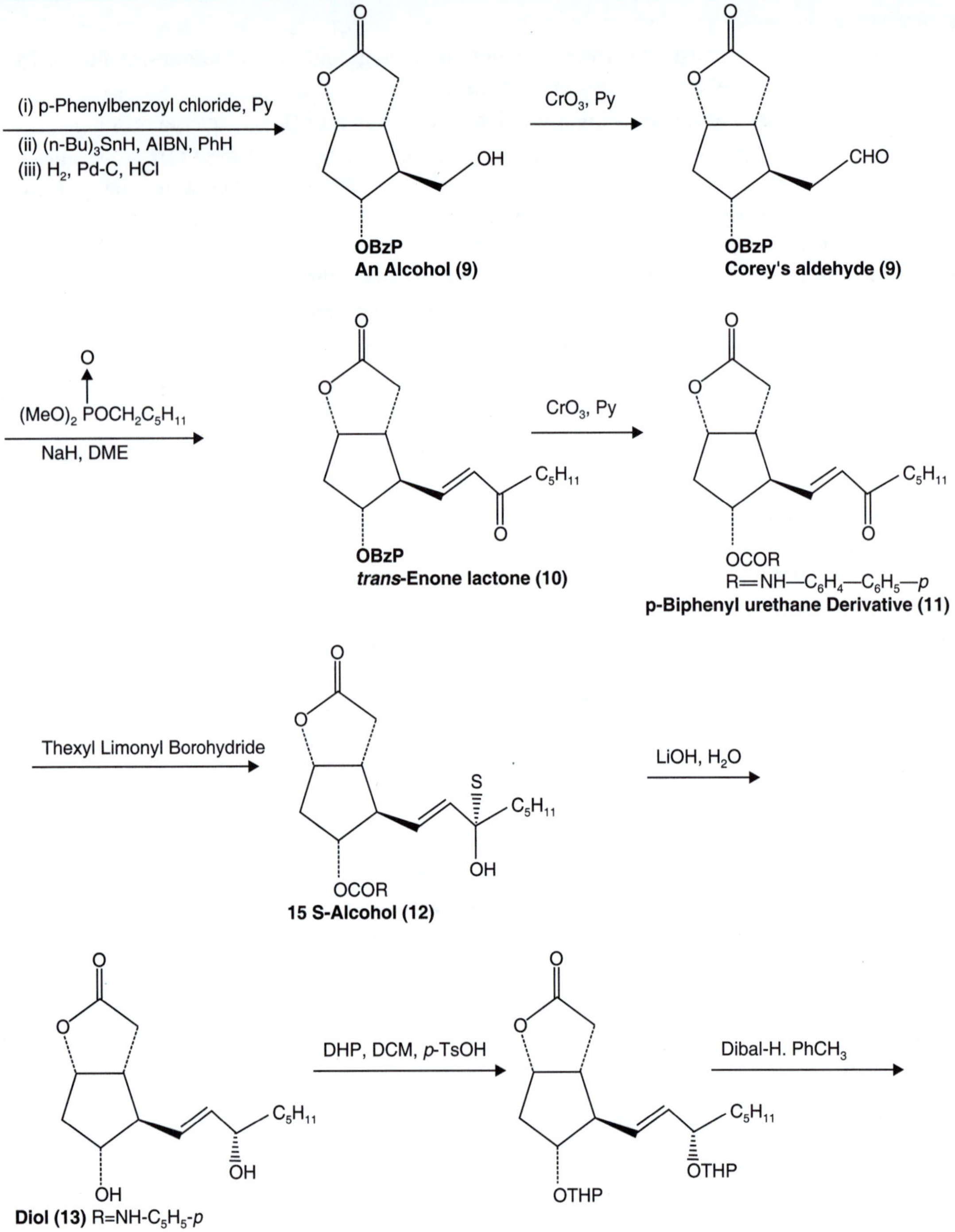
(i) p-Phenylbenzoyl chloride, Py
(ii) (n-Bu)$_3$SnH, AIBN, PhH
(iii) H_2, Pd-C, HCl
OH
OBzP
An Alcohol (9)
CrO_3, Py
CHO
OBzP
Corey's aldehyde (9)
(MeO)$_2$ POCH$_2$C$_5$H$_{11}$
NaH, DME
C_5H_{11}
OBzP
trans-Enone lactone (10)
CrO_3, Py
C_5H_{11}
OCOR
R=NH—C$_6$H$_4$—C$_6$H$_5$—p
p-Biphenyl urethane Derivative (11)
Thexyl Limonyl Borohydride
S
C_5H_{11}
OH
OCOR
15 S-Alcohol (12)
LiOH, H_2O
C_5H_{11}
OH
OH
Diol (13) R=NH-C$_5$H$_5$-p
DHP, DCM, p-TsOH
C_5H_{11}
OTHP
OTHP
Dibal-H. PhCH$_3$

(Contd....)

Ph₂P CO⁻₂

DMSO

AcOH

A Hydroxy acid (15)

(i) CrO₃, H₂SO₄-acetone
(ii) AcOH

R=NH-C₆H₅-*p*
Lactol (14)

C₅H₁₁

OTHP
OTHP

PGE₂α (16)

PGF₂ (17)

Explanation-The various steps involved sequentially in Corey's synthesis of **PGE₂** and **PGF₂** are duly explained as under:

1. First step essentially involves the **Lewis acid catalyzed asymmetric Diel's-Alder reaction** between the **acerylate (1)**-*optically active* and the **alkylated cyclopentadiene (2)** to give the *endo-***adduct (89%0 (3).**

2. The resulting *enolate* of the *endo-***adduct (3)** is carefully added to an oxygenated solution of *THF (tetrahydrofuran)* along with *triettyl phosphate* to yield α-**hydroxyl ester (4)**, which upon reduction with *lithium aluminium hydride* (**LiAlH₄**) gives rise to the formation of a **diol (5)** plus the regeneration of the **chiral catalyst** in the reaction mixture.

3. **Diol (5)** on being subjected to reaction with *sodium periodate* (**NaIO₄**) (an oxidizing agent) gives the **optically active bicyclic ketone (6),** which on further oxidation followed by *saponification* gives the corresponding **hydroxyl acid (7).**

4. The resulting **product (7)** on treatement with aqueous *potassium triodide* (**KI₃**) yields the *optically active* **iodolactone (8)**which critically possesses the *same optical rotation* as that of the materials by the resolution of the *racemic* **hydroxyl acid** with **amphetamine.**

5. The *optically active* **iodolactene (8)** is duly esterified with *p*-phenylbenzoyl chloride followed by *deiodination* with *tri-n-butyl tin hydride*; and also the eleavage of the ensuring **benzyl-ether** with H2/pd-C gives the respective **alcohol** (9), which is *oxidized* with **Collin's reagent** (CrO₃;Py) to yield the *optically active* **Corey's aldehyde (9).**

6. **Corey's aldehyde (9)** with interaction of **dimethyl-2-oxo-heptyl phosphonate** [in the presence of sodium hydride (NaH) and dimethoxy ethane (DME)] yields the *trans*-**enone lactone (10),** which upon meticulous conversion to *p*-**biphenyl urethane derivative (11)**- followed by careful *asymmetric reduction* with **hexyl limonyl borohydride** gives rise to the formation of **15 $-alcohol (12)** as the *major product.*

7. The removal of the *p*-biphenylurethane from the **15 $-alcohol (12)** by means of *hydrolysis* with aqueous **lithium hydroxide (LiOH)** gives the **diol (13),** which is duly protected as the respective **tetrahydroperanyl ethers (THP).**

8. The resulting **lactone (diol)13** is now subjected to reduction with **Dibal-H** to yield **Lactol (14).**

9. **Lactol (14)** when subjected to **Wittig condensation** with *5-triphenylphosphonium pentanoic acid (15)* and **sodium methyl sulpfinyl methride** gives the corresponding **hydroxyl acid (15).**

10. Finally, the crucial *deprotection* of the **hydroxyl acid (15)** with acetic acid gives rise to the formation of **PGF$_2$ (16);** whereas, on the other hand the careful *oxidation* followed by *hydrolysis* yields **PGE$_2$ (17).**

FURTHER READING REFERENCES

Brickner SJ *et al*	:	**J Med Chem., 39**:673, 1996.
Chen. Z *et al*.	:	**Prac wat Acad Sci (USA), 99**; 8306,2002.
Clere J *et al*	:	**Cell Biol Interenet., 27**: 619,1997.
Fretland A J and Omiecinski CJ	:	**Chem Biol Interact, 129**: 41, 2000.
Parli CJ and Schmidt B	:	**Res Commun Chem Pathol Pharmacol., 10**:601, 19757.
Purke DV	:	**Biochem J, 77**:493,1960.
Saljoughian M and William PG	:	**Curr Pharm Design., 6**:1029,2000.
Schill G *et al*	:	**Progress in Drug Metabolism, Vol.2, Wiley, New York**, 1977.
Silber B *et al*	:	**J Pharm Sci, 71**:699, 1982.
Smith DA *et al*	:	**Drug Metab Rev., 16**: 365, 1985.

REVIEW QUESTIONS

1. What are **Prostaglandins**? Give the chemical structures of :
   PGE$_1$, PGE$_2$ and **PGE$_{2\alpha}$**

2. Discuss the **classification** and **nomenclature of Prostaglandins** [PGEs] and support your answer with suitable examples

3. Write **short notes** on the following :
   (a) **Configuration in Ring Systems and Side-Chain of PGEs**
   (b) **Biogenesis of PGE$_2$ and PGF$_{2\alpha}$**

4. Describe the **Prostaglandin Metabolism** based on certain typical examples

5. Give a detailed account on the **Corey's Synthesis of PGE$_2$** and **PGF$_{2\alpha}$** along with proper explanations of the course of reactions undertaken briefly.

Contents at a Glance

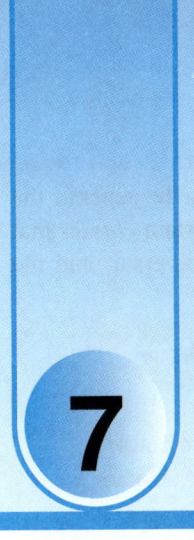

Biosynthesis of Flavonoids

1. INTRODUCTION

The term **flavonoids** refers to a **C-15 skeleton** having a **chroman ring system** essentially bearing an aromatic ring in **position 2,3, or 4.** Besides, there are **two benzene rings** that are duly present in the so-called **flavonoid nucleus,** namely:

- *First-* **derived from the condensation of** *acetate units (i.e., three* **malonyl units); and**
- *Second-* **derived from the well-known** *Shikimic acid pathway (i.e.,* **cinnamic acid).**

Amazingly, these **two aforesaid units** are joined duly by **a C3 structures**–which may be either **open** or **closed.**

It is, however, important to state here that the **flavonoid nucleus** is invariably attached to a *sugar moiety*–thereby resulting into the formation of a **water-soluble glycoside.**

Another school of thought would like to visualize the entire episode emerging from the *biogenetic point of view* whereby it could be quite feasible and possible to divide the **C-15 carbon framework of the flavonoids** into *two* **precise segments,** such as:

- ❑ *One segment*-having *6 C-atom* that forms the *ring A;* and
- ❑ *Second segment*- having *9 C-atoms* (C6-C3) usually termed as the so-called **phenylpropanoid moiety.**

Interestingly, the **phenylpropanoid moiety** usually serves as the established *precursor* of an array of naturally occurring chemical entities (compounds), namely:

- **Amino acids;** and
- **Phenolic compounds.**

> **NOTE** Importantly, the *biogenesis of Ring 'A'* comes into play *via* the polyketide route (or acetate pathway); whereas, that of *Ring 'B'* occurs *via* the *Shikimic acid pathway.*

1.1. Chemistry and Distribution of Flavonoids

In a broader perspective the **flavonoids** are structurally derived from the *parent compound* **flavones,** that invariably occurs as:

- **A white mealy farinaceous upon the** *primula* **plant species; and**
- **Most of them do share intimately a plethora of common characteristics features.**

Table:7.1 records the characteristic features of the *different* flavonoid classes. In general, the **flavonoids** are predominantly present in almost all *vascular plants*, but there are certain *classes* that occurs more abundantly than others; whereas, the **flavones** and **flavonols** are universal, and the **isoflavones** and **biflavonyls** are reported in only a **few plant families.**

Table:7.1: Characteristics Features of Various Flavonoid Classes.

S. No.	Flavonoid Class	Characteristics Features	Distribution
1	Anthocyonins	Water soluble; visible λ_{max} 515-545nm; Mobile in BAW on paper*.	Scarlet,red,blue,maure flower pigments; also in leaf plus other tissues.
2	Biflavonyls	On BAW chromatograms-dull** absorbing spots of very high RF.	Mostly colourless, and mainly confined to the **gymnosperms.**
3	Chaloones and Aurones	Instantly give red colouration with ammonia (colour change may be seen *in situ*). Visible λ_{max} ranges between **370-410 nm.**	Yellow flower pigments but occasionally present in other tissues.
4	Flavonols	Acid hydrolysis shows bright yellow spots in UV-light on the **forestral Chromatograms**; Spectral λ_{max}: 350-386 nm.	Occurs chiefly as the colourless copigments in both *cyanic* and *acyanic* flowers.
5	Flavones	Acid hydrolyzed product shows a *dull absorbing brown spots* on the **forestral chromatograms**; spectral λ_{max}: **330-350 nm.**	Same as the **flavonols.**
6	Flavanones	Give intense red colouration with mg/HCl; An exceptional *bitter taste.*	Colourless; found in *leaf* and *fruit* (mostly in the *citrus* fruits).
7	Glycoflavones	Contain C-C sugar; Mobile in water -unlike the **normal flavones.**	Same as the **Flavonols**
8	Isoflavones	Mobile on paper in water; No particular **colour tests** are yet known.	Colourless; often found in *root*; only common in *one family* i.e., the *Leguminoseae.*
9	Proanthocyanidins	Usually yields **anthocyanidins** (colour extrastable easily in **amyl alcohol**) when the respective tissue is heated for 30 minutes in 2 M HCl.	Mostly colourless, found in **heartwoods** and in the leaves of the **woody plants.**

Flavonoids on Water-Soluble Compounds–Since the **flavonoids** are invariably found as the *water-soluble compounds*–these may be extracted conveniently using **70% (v/v) ethanol;** and thus remain in the so-called **aqueous layer**-following the usual partitioning of this extract with **petroleum ether.** Importantly, the **flavonoids** are *phenolic* in nature; and hence, change in colour on being subjected to treatment with a *base* or with *ammonia i.e.*, they may be detected promptly either on the **chromatograms** or in the **solution form.**

---

* Betalains do replace anthocyanins as purple pigments in *centrospermae*; and may be distinguished by very low mobility in BAW at spectral range 530-554 nm.

** Certain extremely methylated flavones do behave in a similar manner.

Table:7.2 Records the typical colour properties of the 'flavonoids' both in **visible and UV-light:**

TABLE:7.2: Typical Colour Properties of Flavonoids In Visible and UV-Light

Colour in Visible Range	All Alone	Colour in UV-Light Treated with Ammonia	Inference(s)
Orange Red	• Dull orange • Red or Mauve	• Blue	➢ Anthocyanidin-3-glycosides
Mauve	• Fluroscent • Yellow or Pink	• Blue	➢ Most anthocyanidin-3,5-diglycosides
Bright Yellow	• Dark brown or blacl • Bright yellow or yellow-green	• Dark brc ; black • Dark red or Bright orange • Bright orange or red	➢ 6-Hydroxylated flavonoids and flavones; ➢ Certain chalcone glycosides ➢ Most chalcones ➢ Aurones
Very Pale Yellow	• Dark brown	• Bright yellow or yellow brown • Clear yellow green or dark brown	➢ Most flavonol glycosides. ➢ Most flavones glycosides Biflavonoids and rare substituted flavones.
None	• Dark Mauve • Faint Blue • Dark Mauve	• Faint brown • Intense blue • Pale yellow or yellow-green	➢ Most isoflavones and flavonols. ➢ 5-Desoxyisoflavones and 7,8-dihydroxy flavonones. ➢ Flavonones and flavonoid-7-glycosides.

Remarks- Following are some of the so-called **'general chromatographic sprays',** such as:

- 5% (v/v) aqueous ethanol;
- $AlCl_3$-Response: Yellowish-green fluroesence in UV-light;
- Aqueous $FeCl_3/K_3Fe(CN)_6$ (1:1)-Response: Blue on an yellow background;
- **Folin Reagent**-Response $(+NH_3)$: Blue on a white-background.

Flavonoids Comprise Conjugated Aromatic Systems–Based on these findings and observations the **flavonoids** predominantly show the *intense absorption bands* both in the **visible and UV-regions** of the *electromagnetic spectrum*-as given in Table:7.3.

TABLE:7.3 The Spectral Characteristic Features of Major Flavonoid Classes.

Sr. No.	Principal Maxima $[\lambda_{max}(nm)]$	Subsidiary Maxima (with Relative % age Intensities)	Inference (s)
1.	475-560	*ca.* 275(55%)	• Anthocyanins
2.	390-430	240-270 (32%)	• Aurones
3.	365-390	240-260 (30%)	• Chalcones
4.	350-390 250-270	*ca* 300(40%)	• Flavonols

(Contd....)

5.	330-350 250-270	Absent	•	Flavones and Biflavonyls
6.	275-290 ca. 225	310-330 (30%)	•	Flavones and flavononls
7.	255-265	310-330 (25%)	•	Isoflavones

Salient Features of Flavonoids–These essentially include:

   (i) **Flavonoids** are invariably present in *plants* and bound to **sugar moieties** as the *'glycosides'* (or **plant glycosides**).

   (ii) Any one **'flavonoid aglycone'** may normally exist in a *single plant species* in the form of **many glycosidic combinations**.

   (iii) Perhaps due to this specific logistic reason,–when the **flavonoids** are being analyzed it becomes almost preferable as well as advisable to examine thoroughly the so-called **'aglycones'** (*i.e.*, the *non-sugar moiety*) duly present in the *hydrolyzed* plant extracts before actually assessing the:

'complexity of glycosides which could be present duly in the *original* plant extract'.

2. (A) BIOGENESIS OF RING 'A' IN FLAVONOIDS [ACETATE PATHWAY;POLYKETIDE PATHWAY]

Malonate is known to be a *precursor* of **fatty acids**; and hence, serves as the *precursor* of **C-6 polyketides.** Fig:7.1 depicts the actual route followed by the **acetate pathway** thereby expatiating the **biogenesis of ring 'A'.**

(B) BIOGENESIS OF RING 'B' IN FLAVONOIDS [SHIKIMATE PATHWAY; SHIKIMIC ACID PATHWAY]

The condensation of **D-erythrose-4-phosphate (1) with phosphophenol pyruvate (PEP) (2)** yields **3-deoxy-D-arabinoheptulosonate-7-phosphate (DHAP) (3)** after due catalyzation with an *inorganic phosphate (Pi)*-as shown in Fig:7.1. thus, the *cyclization* of **DHAP (3)** catalyzed by *3-dehydroquinate synthase* (loses a mole of HOH) to form an *alicyclic* **intermediate (4).**

Subsequently, the enzyme **3-dehydroquinate dehydratase** is capable of catalyzing the *dehydration* to introduce a *double-bond* between **C-4** and **C-5** to afford **3-dehydroshikimate (5).**

The stereoselective and reversible reduction of **C-3 carbon moiety** duly catalyzed by the enzyme **shikimate dehydrogenase** takes place to give rise to the formation of **shikimate (6).**

NOTE   **In this specific instances the *enzyme* dehydrogenase is NADP+; and the reaction critically involves the transfer of a *'hydride'* right from the *nicotinamide ring* into NADPH.**

The **shikimate (6)** gets duly converted to **shikimate-3-monophosphate (7)** by involving the action of the enzyme *shikimate kinase* (thereby **ATP** gets converted to **ADP**). Furthermore, the **product (7)** on condensation with **phosphophenol pyruvate (PEP)** and **an inorganic phosphate (Pi)** plus the enzyme **5EPSP-synthase** yields **5-enolpyruvyl-3-phospho shikimate (8).**

Finally, it involves *dehydration* to introduce a *second* **double bond** in the **6-membered ring** to give **chlorismate (9).** (The *dephosphorylation* is duly catalyzed by **chlorismate synthase enzyme).**

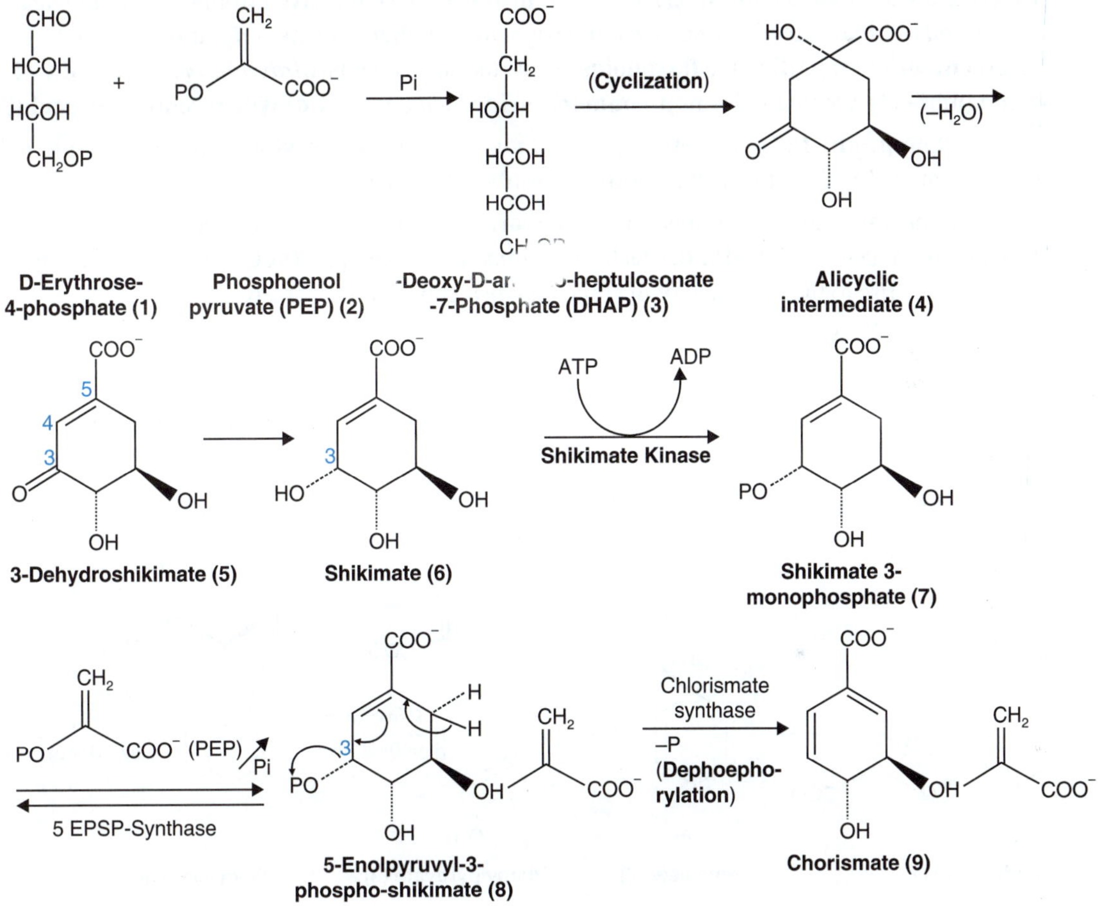

Fig. 7.1: Biosynthesis of Chlorismate

3. BIOSYNTHESIS OF FLAVONOLS FROM *TRANS*-CINNAMATE

The *enzyme* **chlorismate mutase** helps to catalyze the conversion of **Chlorismate (1)** to **perphenate (2)** by the aid of a *pericyclic reaction mode* usually termed as the **unimolecular –intramolecular rearrangement.**

The resulting **perphenate (2)** gets further transformed into **phenylpyruvate (3)** due to the catalytic action of the enzyme *perphenate dehydratase* yields the **phenylalanine (4).**

> **NOTE** In the aforesaid transamination reaction the phenylpyruvate (3) gets specifically catalyzed by the *enzyme* aromatic amino acid amino transferase which yields *phenylalanine* (4). Besides, in the said *trans*-amination reaction the glutamate does behave as the '*donor*' of an amino (NH2) moiety.

The formation of *trans*-einnamate (5) takes place by the *enzymatic deamination* of **phenylalanine (4)** duly catalyzed by the *enzyme* **L-phenylalanine ammonia lyase (PAL)-** that occurs stereospecifically with the critical loss of an *amino (NH_2) moiety* and **pro-S-hydrogen** derived from the *L-amino acid* to give rise to *trans*-cinnamate (5). Amazingly, it designates the **phenyl propanoid, moiety [C-6 to C-3]** i.e., the **precursor of ring 'B'** in the **flavonoids.** Thus, the subsequent *hydroxylation* at **C-2 and C-4** of *trans*-cinnamate (5) eventually forms *p*-coumaric acid (6) and the **2,4-dihydroxycinnamic acid (7).**

The resulting **product (7)** is carefully *glycosylated* to give **2-glycosyl-4-hydroxycinnamic acid (8),** which upon *cyclization* affords the **coumarin umbelliferone (9).**

Other important Transformations-The *p*-coumaric acid (6) on being subjected to *demethylation* yields **3,4-dihydroxycoumaric acid (10),** which on *cyclization* gives **saffrole (11).** Consequently, *P*-coumaric acid (6) also gives rise to the formation of the following *two* compounds, namely:

- **Eugenol (12),** and
- **Vanillin (13).**

Fig:7.2 illustrates vividly the biosynthesis of the **falvonols** (*viz.,* **eugenol, vanillin**) from *trans*-einnamate as shown under:

From Chlorismate to Coumarin Umbelliferone

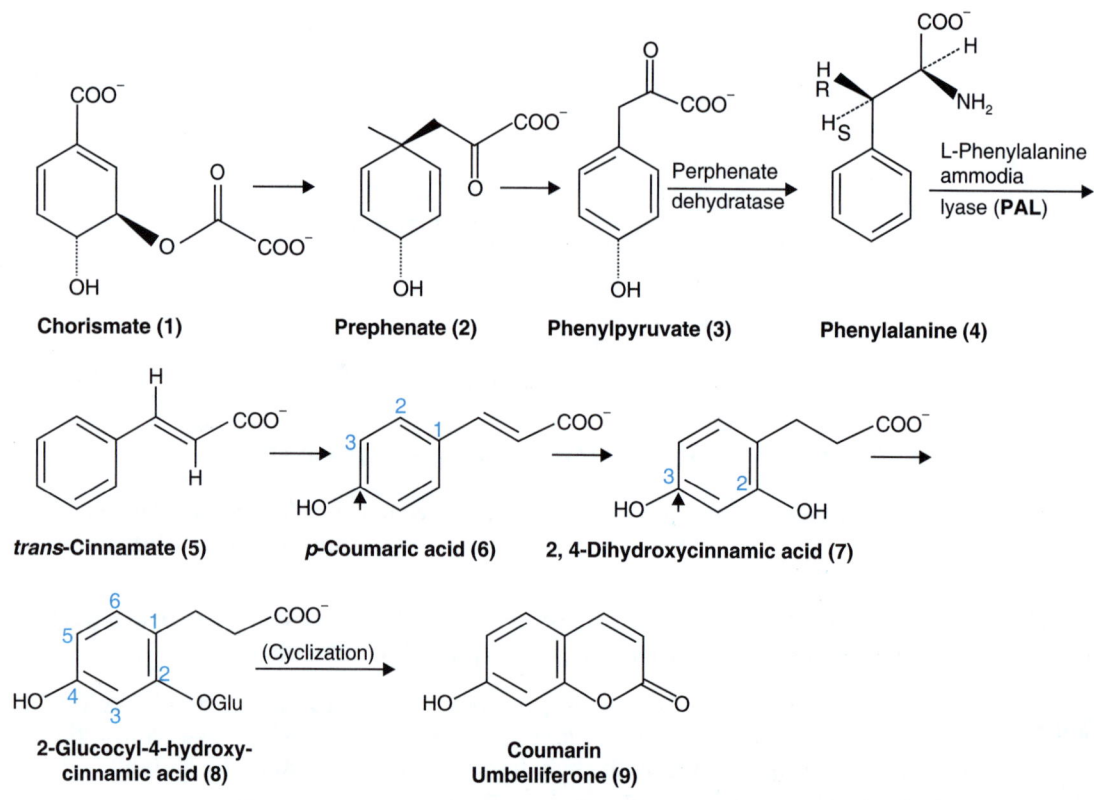

Chorismate (1)   Prephenate (2)   Phenylpyruvate (3)   Phenylalanine (4)

trans-Cinnamate (5)   *p*-Coumaric acid (6)   2, 4-Dihydroxycinnamic acid (7)

2-Glucocyl-4-hydroxy-cinnamic acid (8)   Coumarin Umbelliferone (9)

From P-Coumaric Acid to Phenolic Compounds *viz.*, Saffrole, Eugenol, and Vanillin

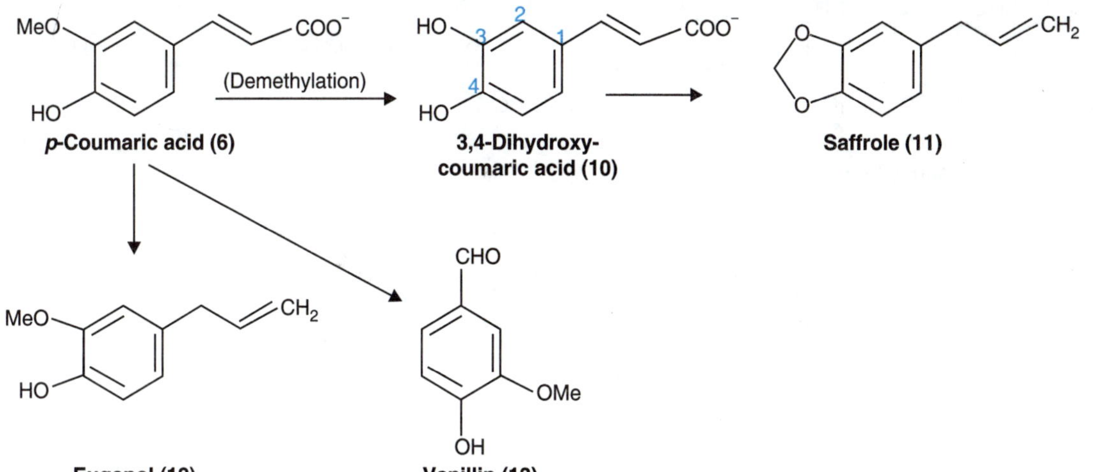

Fig. 7.2: Biosynthesis of the Flavonols from *trans*-Cinnamate.

4. BIOSYNTHESIS OF THE FLAVONOIDS FROM *TRANS*-CINNAMATE VIA CHAL-CONE SYNTHASE (CHS) ENZYME

First of all *trans*-cinnamate **(1)** gets converted to *p*-coumaric acid **(2)** by **hydroxylation.** *Secondly,* the careful *condensation* of **3-moles of malonyl coenzyme A** (acetate units) **(3)** –**coumaroyl coenzymes A (4)** duly catalyzed by **chalcone synthase (CHS)** yields an **adduct (5).** Thirdly, the resulting **adduct (5)** gets duly converted to *chalcone,* **Narigenin(6),** as shown in Fig: 7.3.

The resulting product **narigenin (6)** upon *decarboxylation* yields **stilbene (7);** whereas, *cyclization* of the same gives **flavones (8).** Treatement of **flavonone (8)** with the enzyme *chalone isomerase* **(CHI)** yields **dihydroflavonol (9).** Amazingly, further *hydroxylation* of **product (9)** gives rise to the formation of the following *two* **flavonols,** namely:

- **3,4-Dihydroxy Flavonol (10)** and
- **3,4,5-Trihydroxy Flavonol (11).**

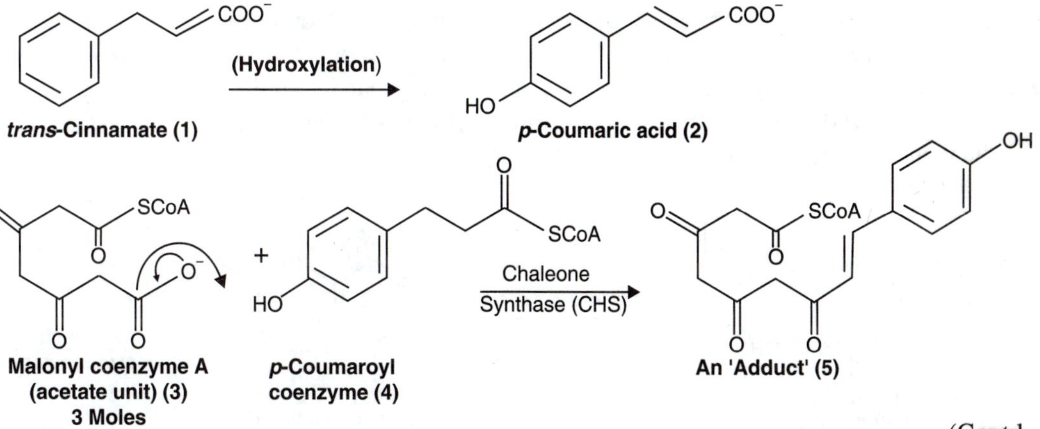

(Contd....).

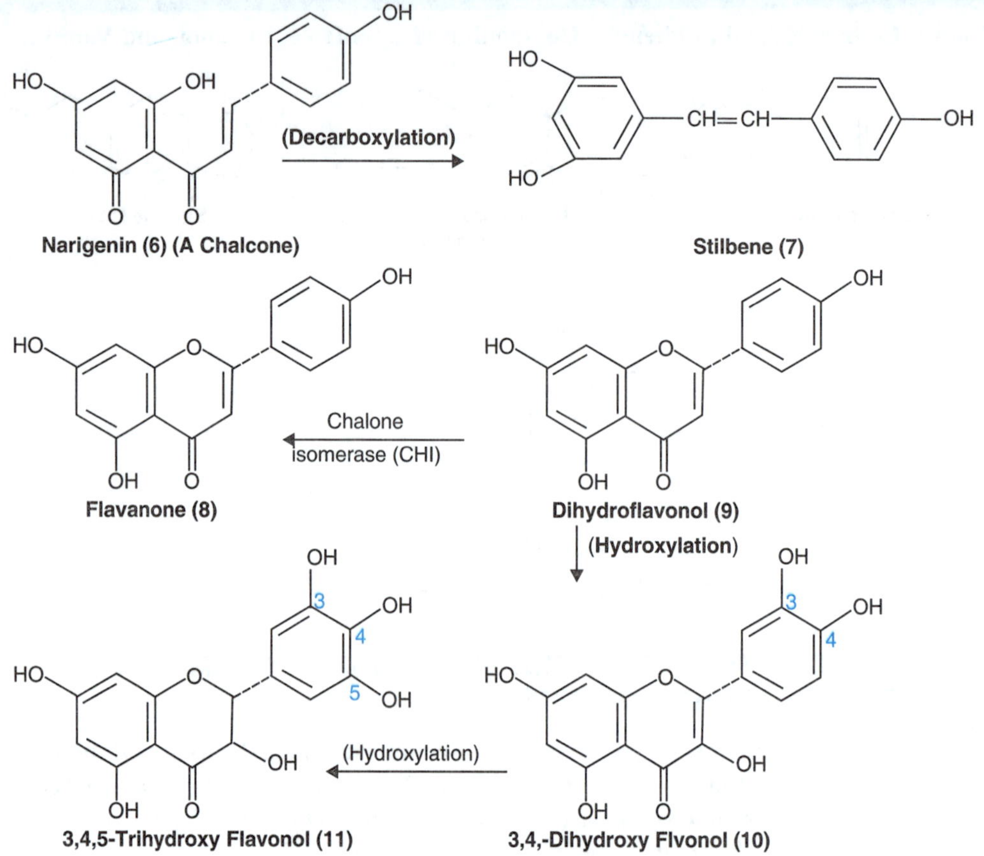

Fig : 7.2 : Biosynthesis of the Flavonols from *trans–*Cinnamate.

BIOSYNTHESIS OF ISOFLAVONES

It has been duly established that the **ehalcone** more or less behaves as an *effective intermediate* for the critical and specific **biosynthesis of isoflavones**- as shown in Fig.7.4.

The very *first step* shows the actual conversion of **chalcone (1)** into the respective **chalcone oxide (2)** *i.e.,* an *oxidative product*. Subsequently, **chalcone oxide (2)** undergoes *intra*-molecular **rearrangements** according to the proven mechanisms duly put forward in order to explain the so-called **acid-catalyzed rearrangement** of the ensuing α,β-**epoxy ketones** to give an **aldehyde derivative (3),** which upon *cyclization* yields the **isoflavone (4).**

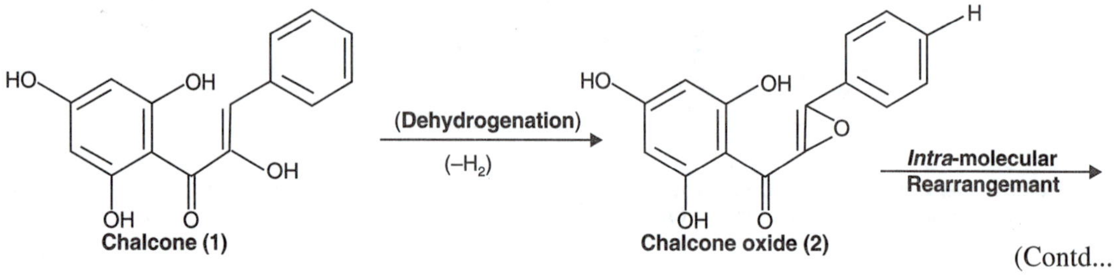

(Contd....)

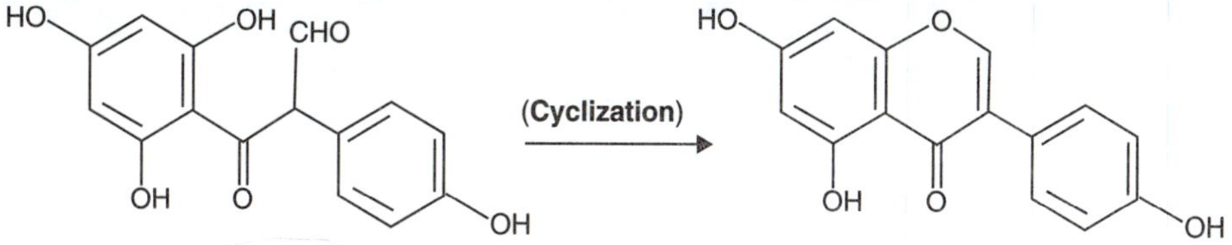

Fig. 7.4: Biosynthesis of the Isoflavones.

FURTHER READING REFERENCES

Forkmann G and Heller W : **Biosynthesis of Flavonoids: Comprehensive Natural Product Chemistry, Vol.1,** Elsevier, Amsterdam,pp:713-748,1999.

Harborne JB and Wilkams CA : **Advances in Flavonoid Research Since 1982,**

Harborne JB and Williams CA : **Advances in Flavonoid Research Since 1992, Phytochemistry,55:**481-504,2000.

Hemingway RW and Lake PE (Eds) : **Plants Polyphenols: Synthesis Properties, Significance,** Plenium Press, New York,1992.

Matern U et al: **Biosynthesis of Coumarine** : **Comprehensive Natural Products Chemistry: Vol.1.,** Elsevier, Amsterdam, PP: 623-637,1999.

Stafford HA and Ibrahim RK (Eds) : **Recent Advances in Phytochemistry Vol:26, Plenium Press,New York,1992.**

REVIEW QUESTIONS

1. What are **Flavonoids?** Enumerate briefly the *two* **biogentic point of view** of the **flavonoid nucleus**.
2. Discuss the **Chemistry and Distribution of Flavonoids** with perticular reference to its **Class, Characteristic Features,** and **Distribution.**
3. Enumerate the following aspects explicitly:
    (a) **Typical Colour Properties of Flavonoids in Visible and UV-Light.**
    (b) **Spotral Characteustic Features of Major Flavonoid Classes**
4. Give a comprehensive account on the **'Biogenesis of Ring 'B' in Flavonoids**.
5. Describe the biosynthesis of **'Flavonols'** from *trans*-**Cinnanate** supported by necessary explanatory notes.
6. Discuss thr **Biogenesis of Flavonoids from *trans*-Cinnamate *via* Chalcone Syntase (CHS) Enzyme.**
7. How would you explain the **Biosynthesis of Isoflavones?**

Index

469

A

G

H

U

V

W

X

Y

Z